AF836293

HANDBOOK OF BIOPOLYMERS AND BIODEGRADABLE PLASTICS

Plastics Design Library

PLASTICS DESIGN LIBRARY (PDL)
PDL HANDBOOK SERIES

Series Editor: Sina Ebnesajjad, PhD
President, FluoroConsultants Group, LLC
Chadds Ford, PA, USA
www.FluoroConsultants.com

The **PDL Handbook Series** is aimed at a wide range of engineers and other professionals working in the plastics industry, and related sectors using plastics and adhesives.
PDL is a series of data books, reference works and practical guides covering plastics engineering, applications, processing, and manufacturing, and applied aspects of polymer science, elastomers and adhesives.

Recent titles in the series
Brandau, Stretch Blow Molding, Second Edition
ISBN: 9781437735277

Chandrasekaran, Rubber Seals for Fluid and Hydraulic Systems
ISBN: 9780815520757

Ebnesajjad, Handbook of Adhesives and Surface Preparation
ISBN: 9781437744613

Grot, Fluorinated Ionomers, Second Edition
ISBN: 9781437744576

Kutz, Applied Plastics Engineering Handbook
ISBN: 9781437735147

Kutz, PEEK Biomaterials Handbook
ISBN: 9781437744637

McKeen, Fatigue and Tribological Properties of Plastics and Elastomers, Second Edition
ISBN: 9780080964508

McKeen, Film Properties of Plastics and Elastomers, Third Edition
ISBN: 9781455725519

McKeen, Permeability Properties of plastics and Elastomers, Third edition
ISBN: 9781437734690

McKeen, The Effect of Creep and Other Time Related Factors on Plastics and Elastomers, Second Edition
ISBN: 9780815515852

Sastri, Plastics in Medical Devices
ISBN: 9780815520276

Tolinski, Additives for Polyolefins
ISBN: 9780815520511

Wagner, Multilayer Flexible Packaging
ISBN: 9780815520214

Woishnis & Ebnesajjad, Chemical Resistance, Volumes 1 & 2 – Chemical Resistance of Thermoplastics
ISBN: 9781455778966

Woishnis & Ebnesajjad, Chemical Resistance, Volume 3 – Chemical Resistance of Specialty Thermoplastics
ISBN: 9781455731107

To submit a new book proposal for the series, please contact
Sina Ebnesajjad, Series Editor
sina@FluoroConsultants.com
or
Matthew Deans, Senior Publisher
m.deans@elsevier.com

HANDBOOK OF BIOPOLYMERS AND BIODEGRADABLE PLASTICS

PROPERTIES, PROCESSING, AND APPLICATIONS

Edited By

Sina Ebnesajjad

ELSEVIER

Amsterdam • Boston • Heidelberg • London • New York • Oxford
Paris • San Diego • San Francisco • Singapore • Sydney • Tokyo
William Andrew is an imprint of Elsevier

William Andrew is an imprint of Elsevier
The Boulevard, Langford Lane, Kidlington, Oxford, OX5 1GB, UK
225 Wyman Street, Waltham, MA 02451, USA

First published 2013

Copyright © 2013 Elsevier Inc. All rights reserved.

No part of this publication may be reproduced or transmitted in any form or by any means, electronic or mechanical, including photocopying, recording, or any information storage and retrieval system, without permission in writing from the publisher. Details on how to seek permission, further information about the Publisher's permissions policies and our arrangement with organizations such as the Copyright Clearance Center and the Copyright Licensing Agency, can be found at our website: www.elsevier.com/permissions

This book and the individual contributions contained in it are protected under copyright by the Publisher (other than as may be noted herein).

Notices

Knowledge and best practice in this field are constantly changing. As new research and experience broaden our understanding, changes in research methods, professional practices, or medical treatment may become necessary.

Practitioners and researchers must always rely on their own experience and knowledge in evaluating and using any information, methods, compounds, or experiments described herein. In using such information or methods they should be mindful of their own safety and the safety of others, including parties for whom they have a professional responsibility.

To the fullest extent of the law, neither the Publisher nor the authors, contributors, or editors, assume any liability for any injury and/or damage to persons or property as a matter of products liability, negligence or otherwise, or from any use or operation of any methods, products, instructions, or ideas contained in the material herein.

British Library Cataloguing in Publication Data
A catalogue record for this book is available from the British Library

Library of Congress Cataloguing in Publication Data
A catalog record for this book is available from the Library of Congress

ISBN: 978-1-4557-2834-3

For information on all William Andrew publications
visit our website at **store.elsevier.com**

Printed and bound in the United States

12 13 14 15 10 9 8 7 6 5 4 3 2 1

Working together to grow
libraries in developing countries

www.elsevier.com | www.bookaid.org | www.sabre.org

ELSEVIER BOOK AID International Sabre Foundation

Contents

Preface

This book is about biobased and biodegradable polymers and plastics. It covers a fairly broad range of biopolymers with a strong focus on plastics, simply because of the large global consumption and impact of the latter on the environment and different life forms on the Earth. No matter where in the world plastics objects are thrown out, eventually they find their ways to the oceans and continents around the globe.

There are many sustainability issues which have been driving the development of monomers and biodegradable polymers from renewable plant resources. Some of those issues better known by the public include the cost of the traditional raw material source petroleum, global warming and environmental damage. A less understood problem is the extent of post-use pollution caused by plastic objects. Of all the plastics, metals and papers collected for recycling only an estimated 25% actually makes its way to reuse. The rest are disposed of because contamination renders them unusable.

A poorly addressed issue is the containers, bags, bottles, toys and other plastic objects that litter the roadsides and have been finding their way into the oceans. There are now five massive "Garbage Patches" in the Pacific, Atlantic and Indian Oceans in which immense quantities of plastic objects have gathered in the swirling vortex of these oceans' currents. Typically, plastic objects appear submerged just under the water surface. The Great Pacific Garbage Patch is one of the largest trash gyres, which takes up a large area of the Pacific Ocean estimated twice the size of the continental United States. The marine animals and birds ingest the plastics, mistaken for food, resulting in a build up of toxins, starvation and premature death. The ocean currents have been depositing massive quantities of intact and ground plastics on beaches of Pacific Islands such as Hawaii.

The estimated weight of the plastics in the Pacific Garbage Patch is 3.5 million tons. A cleanup of garbage in the oceans is impractical at present, yet a typical piece of plastic requires 600 years of exposure to the atmospheric elements before it completely degrades. The most practical action has been to focus on reduction in the disposal of plastics and the development of commercial biodegradable plastics with relatively short lives. Ideally, a wholly plant-driven biodegradable plastic would decompose to carbon dioxide and water after a *short* exposure to the weather elements. It would be carbon dioxide neutral by being plant derived.

There are numerous books about biopolymers, covering the scientific research that is enabling the new generation of plastics. The goal in this handbook is to bring together some of the core knowledge in the field to provide a practical and wide-ranging guide for engineers, product designers and scientists involved in the commercial development of biopolymers and bioplastics, and their use in applications as varied as drinks bottles, medical devices and automotive manufacturing. The handbook includes a broad selection of material previously published in a number of Elsevier books; some of this material has been updated specially for this book. In addition, a section on polylactic acid (PLA), its synthesis, properties and applications, appears in print for the first time—material that will be included in a forthcoming book on PLA.

This book provides information about polymeric biomaterials: plant-derived polymers, methods of manufacture, applications and disposal. Whole chapters describe biodegradable and biobased polymers and plant polymer resources, demands, and sustainability. Separate chapters cover PLA, starch, cellulose and polymers based on plant oils, and their applications. The use of natural polymers in medicinal chemistry and tissue engineering has been covered in some detail.

Disposal methods covered here include composting, direct biodegradation and measurement tools for the biodegradability of polymers and plastics. One chapter has been devoted to compostable polymer materials definitions, structures and methods of preparation. Another chapter describes biodegradability testing of compostable polymer materials.

The editor wishes to thank the authors who have generously contributed material to this book. They are experts in their fields and provide valuable information and insights into the polymers of the future. The contributors include: X. S. Sun, A. R. Rahmat, L. T. Sin, W. A. W. A. Rahman, A. Gandini, M. N. Belgacem, W. He, R. Benson, L. Jiang, J. Zhang, A. J. F. Carvalho, A. Dufresne, L. Avérous, E. Rudnik, R. P. Wool, A. Nussinovitch, K. Pal, A. T. Paulson, D. Rousseau, L. M. Grover, A. M. Smith, M. Gomes, H. Azevedo, P. Malafaya, S. Silva, J. Oliveira, G. Silva, R. Sousa, J. Mano, R. Reis, A. Kramschuster and L. S. Turng.

Sina Ebnesajjad
May 2012
Chadds Ford, Pennsylvania, USA

1 Overview of Plant Polymers: Resources, Demands, and Sustainability

Xiuzhi Susan Sun

OUTLINE

Advances in petroleum-based fuels and polymers have benefited mankind in numerous ways. Petroleum-based plastics can be disposable and highly durable, depending on their composition and specific application. However, petroleum resources are finite, and prices are likely to continue to rise in the future. In addition, global climate change, caused in part by carbon dioxide released by the process of fossil fuel combustion, has become an increasingly important problem, and the disposal of items made of petroleum-based plastics, such as fast-food utensils, packaging containers, and trash bags, also creates an environmental problem. Petroleum-based or synthetic solvents and chemicals are also contributing to poor air quality. It is necessary to find new ways to secure sustainable world development. Renewable biomaterials that can be used for both bioenergy and bioproducts are possible alternatives to petroleum-based and synthetic products.

Agriculture offers a broad range of commodities, including forest, plant/crop, farm, and marine animals, that have many uses. Plant-based materials have been used traditionally for food and feed and are increasingly being used in pharmaceuticals and nutraceuticals. Industrial use of agricultural commodities for fuels and consumer products began in the 1920s, but they were soon replaced by petroleum-based chemicals after World War II because of low cost and durability of petrochemicals. This chapter focuses on bio-based polymers derived from plant-based renewable resources, their market potential, and the sustainability of the agriculture industry of the future.

The three major plant-based polymers are protein, oil, and carbohydrates. Starch and cellulose, also called polysaccharides, are the main naturally occurring polymers in the large carbohydrate family. Agricultural fiber is also a member of the carbohydrate family. Natural fiber such as flax, hemp, straw, kenaf, jute, and cellulose consists mainly of cellulose, hemicellulose, and lignin, but is usually listed as a material when used as a fiber in composites.

Corn, soybean, wheat, and sorghum are the four major crops grown in the United States (Table 1.1), with total annual production of about 400 million metric tons (800 billion pounds) in the year 2000. Annually, 10–15% of these grains are used for food, 40–50% for feeds, and the rest could be for various industrial uses. Based on U.S. Department of Agriculture statistics, the total land used for crops is about 455 million acres, which is about 20% of the total usable land (Fig. 1.1) [1]. Including other crops, such as rice, barley, peanuts, and canola, the United States has the potential to produce about 550 million metric tons of grains and legumes. At least 150 million metric tons of grains and legumes are available for nonfood industrial uses. In general, seeds make up

Ebnesajjad: Handbook of Biopolymers and Biodegradable Plastics. http://dx.doi.org/10.1016/B978-1-4557-2834-3.00001-X
© 2013 Elsevier Inc. All rights reserved. Adapted from a chapter in: Wool, *Bio-based Polymers and Composites* (2005).

Table 1.1 Production of Selected Grains and Legumes (Million Metric Tons)

	Wheat	Soybean	Corn	Sorghum
World production	578	172	585	55
United States	60 (2nd)	75 (1st)	253 (1st)	12 (1st)
Other countries	99.6 (1st)	37 (2nd)	106 (2nd)	9 (2nd)
	China	Brazil	China	India
	37 (3rd)	15.4 (4th)	40 (3rd)	2.8 (6th)
	France	China	India	China

Sources: From Ref. [31] and USDA World Agriculture Production, July 27, 2001.

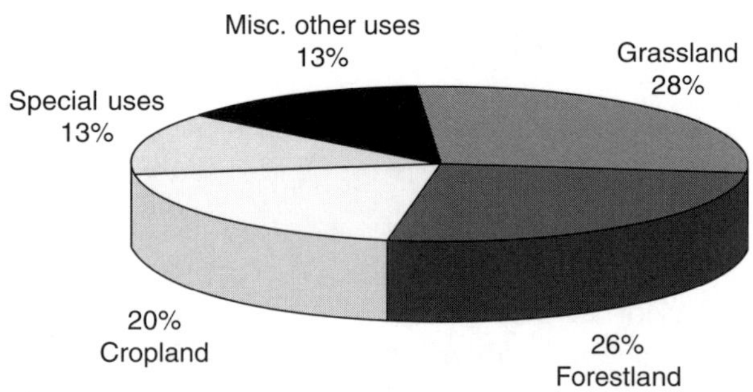

Figure 1.1 Land use and distribution. Total useful land in the United States is about 2.3 billion acres.

about 45–52% of the dry mass of a plant. This means that there is the potential to produce about 400 million metric dry tons of cellulosic sugar-based biomass (agriculture fiber residues) annually in the United States alone based on the total production of corn, soybean, wheat, and sorghum. Including other crops, plants, and forest products, the total annual US production of cellulosic sugar-based biomass could be about 800 million dry tons.

1.1 Plant Proteins

Plant proteins are amino acid polymers derived mainly from oilseeds (i.e., soybeans) and grains (i.e., wheat and corn), and are usually produced as by-products of processing oils and starches (Table 1.2). The potential US protein production is about 120 billion pounds of soybean meal containing about 50% protein, about 20 billion pounds of wheat gluten containing about 70% protein, and about 40 billion pounds of corn gluten containing about 65% protein. Of the corn protein, about 30% is a functional protein called *corn zein protein* [2]. Plant proteins are widely used as major ingredients for food, feed, pharmaceuticals, nutraceuticals, paper coating, textile sizing, and, increasingly, adhesives. Plant proteins are complex macromolecules that contain a number of chemically linked amino acid monomers, which together form polypeptide chains, constituting the primary structure. The helix and sheet patterns of the polypeptide chains are called secondary structures. A number of side chains are connected to the amino acid monomers. These side chains and attached groups interact with each other, mainly through hydrogen and disulfide bonds, to form tertiary or quaternary structures. These proteins often have large molecular weights, in the range of 100,000–600,000 Dalton (Da) (Dalton = grams per mole), which makes them suitable for polymers and adhesives.

Proteins can be modified by physical, chemical, and enzymatic methods. Modification results in structural or conformational changes from the native structure without alteration of the amino acid sequence. Modifications that change the secondary, tertiary, or quaternary structure of a protein molecule are referred to as *denaturation* modifications [3]. The compact protein structure becomes unfolded during denaturation, which is accompanied by the breaking and reforming of the intermolecular and intramolecular interactions [4].

Physical modification methods mainly involve heat [5] and pressure [6] treatments. Heat provides the protein with sufficient thermal energy to break hydrophobic interactions and disassociate the subunits [5]. The disassociation and unfolding expose the hydrophobic groups previously enclosed within the contact area between subunits or on the interior of the folded molecules. For example, soybean protein disassociates and coagulates at high pressure and exhibits large hydrophobic regions and high viscosity [6].

Table 1.2 Average Composition of Cereal Grains and Oilseeds (% Dry Weight Basis)

Cereal Grains	Protein	Fat	Starch	Fiber	Ash	Source
Wheat	12.2	1.9	71.9	1.9	1.7	[45]
Rye	11.6	1.7	71.9	1.9	2.0	[45]
Barley	10.9	2.3	73.5	4.3	2.4	[45]
Oats	11.3	5.8	55.5	10.9	3.2	[45]
Maize	10.2	4.6	79.5	2.3	1.3	[45]
Millet	10.3	4.5	58.9	8.7	4.7	[45]
Sorghum	11.0	3.5	65.0	4.9	2.6	[45]
Rice	8.1	1.2	75.8	0.5	1.4	[45]
Oilseeds						
Soybean	51–70[1]	18–26	—	6.5	3.7–7.4	[47]
Rapseed	36–44[1]	38–50	—	12–18	7.4–8.8	[47]
Sunflower	20.8	54.8	18.4	2.1	3.4	[47]
Peanut	30	50	14	2.9	3.1	[47]
Canola	22.0	41.0	22	10.0	5.0	[46]
Caster bean	12–16	45–50	3–7	23–27	2	[47]
Cottonseed	22	19.5	35	19.0	4.5	[46]
Copra	4.6–8.0	68–79	17.4–21	4.6–7.7	2.4–3.7	[47]
Safflower	21	41.0	14.5	19.0	4.5	[46]
Linseed	22–26	41.5–45.5	27–31	5.5–9.7	4.3–2.7	[47]
Sesame	20	52	23	—	5.6	[47]

[1] Oil-free basis.
Sources: From Refs. [45], [46], and [47].

Chemical modification methods may cause alteration of the functional properties, which are related closely to protein size, structure conformation, and the level and distribution of ionic charges. Furthermore, chemical treatments could cause reactions between functional groups, resulting in either adding a new functional group or removing a component from the protein. Chemical modification methods include acetylation, succinylation, phosphorylation, limited hydrolysis, and specific amide bond hydrolysis. Acetylation is the reaction between a protein amino, or a hydroxyl group, and the carboxyl group of an acetylating agent. The acetylation reaction can modify the surface hydrophobicity of a protein [7]. Succinylation converts the cationic amino groups in the protein to an anionic residue, which increases the net negative charge, resulting in an increase in hydrophobicity under specific succinylating conditions [8]. This treatment also increases the viscosity [9]. Phosphorylation is another effective method to increase negative charges, thereby affecting gel-forming ability and cross-linking [10]. Gel-forming ability can also be increased by alkylation treatment [8]. Chemical hydrolysis is one of the most popular methods for protein modifications by acid-based agents. For example, peptide bonds on either side of aspartic acid can be cleaved at a higher rate than other peptide bonds during mild acid hydrolysis [11]. The hydrophobicity of a protein greatly increases under specific conditions of mild acid hydrolysis [12, 13].

1.2 Plant Oils

Plant oils, such as soy oil, corn oil, and flax oil, can be derived from many crops (Table 1.2). The United States has the potential to produce about 30 billion pounds of soy oil, 25 billion pounds of corn oil, and many billion pounds of oils from other oilseeds as listed in Table 1.2. Plant oils are triglycerides and contain various fatty acids. Soybean, a major oil plant, contains about 20% oil. Soy oil is inexpensive in the United States, sold at about $0.20/lb. Refined soy oil contains more than 99% triglycerides and about eight major fatty acids, including linoleic, oleic, linolenic, palmitic, and stearic acids [14]. These fatty acids differ in chain length, composition, distribution, and location. Some are saturated, and some are unsaturated, which results in differences in the physical and chemical properties of the oil. Control of the fatty acid distribution function is essential to optimize polymer properties. Such plant oil can be physically treated and chemically modified to meet specific industrial applications [15].

Adhesives and resins can be derived from bio-based oils using similar synthetic techniques to those used with petroleum polymers. Many active sites from the triglycerides, such as double bonds, allylic carbons, and ester groups, can be used to introduce polymerizable groups. Wool and coworkers [16] prepared soy oil-based resins by functionalizing the triglycerides. This was accomplished by attaching polymerizable chemical groups, such as maleinates and acrylic acid, or by converting the unsaturated sites to epoxies or hydroxyl functionalities, making the triglycerides capable of polymerizing via ring-opening, free-radical, or polycondensation reactions.

The second method of producing resins from oil is to reduce the triglycerides into monoglycerides. Polymerizable groups, such as maleate half esters, can be attached to the monoglycerides, allowing them to polymerize through free-radical polymerization [17].

The third method is to functionalize the unsaturated sites and reduce the triglycerides to monoglycerides, which can form monomers by reacting with maleic anhydride, allowing polymerization by free-radical polymerization [18,19]. Such reactions produce bio-based polymers that have properties and costs comparable to those of petrochemical-based adhesives and composite resins.

1.3 Plant Starches

Starch is a carbohydrate polymer that can be purified from various sources with environmentally sound processes and green engineering. Corn, wheat, sorghum, and potato are all major resources, containing about 70–80% starches (Table 1.2). The potential US starch production is about 455 billion pounds each year, obtained from wheat, corn, and sorghum. However, only 5 billion pounds of starch are produced annually in the United States, mainly from corn. These starches have been used in the food industries, as well as in the paper and other nonfood industries. This number is expected to increase to about 10 billion pounds in the near future with the development of biopolymers, such as poly(lactic acid) (PLA), as substitutes for petroleum-based plastics [20]. Ethanol production from starch as a liquid fuel substitute will also increase until new hydrogen- and methanol-based fuels become viable in the next 10–20 years.

Starch is a polysaccharide of repeating glucose monomers and is a mixture of two polymers: linear amylose linked by α-1,4-bonds and branched amylopectin linked by α-1,6-bonds. At a given molecular weight, amylose swells to a much larger volume in solution than amylopectin [21], but the more amorphous amylopectin absorbs the more water than amylose at elevated temperatures [22]. Linear amylose polymers can also align their chains faster than branched amylopectin polymers. The branched amylopectin can have an infinite variety of structures, depending on the frequency of branching and the length of the branched chains. Different physical properties are associated with these various structures. These molecules can be cross-linked by themselves, or with other multifunctional reagents. As the cross-linking increases, the cross-linked polymer becomes less water soluble.

Many modified starches are produced by chemical substitution of the hydroxyl groups attached to the starch molecules. The type of modification, degree of substitution, and modification conditions will greatly affect the characteristics of the final modified starch and, consequently, product quality. Four major starch modification methods have been used [23]: (1) pregelatinization, such as disintegration of the crystalline starch granules by heat, high pH, or shear force, to obtain water-soluble amorphous products; (2) degradation of starch by acids or enzymes to reduce the molecular weight, resulting in a lower viscosity; (3) chemical substitution by either esterification with acid anhydrides or by etherification with epoxide compounds; and (4) cross-bonding with bifunctional esterifying or etherifying reagents to increase the starch molecular weight, resulting in a higher viscosity.

1.4 Agricultural Fibers and Cellulose

Agricultural fibers include crop residuals, such as straw, stems, hulls, and milling by-products (e.g., brans) from wheat, corn, soybean, sorghum, oat, barley, rice, and other crops. The major chemical composition of these fibers is similar to wood fibers and includes cellulose, lignin, and pentosan. Wood fiber contains about 40–45% cellulose, 26–34% lignin, and 7–14% pentosan. In comparison, wheat straw contains about 29–35% cellulose, 16–21% lignin, and 26–32% pentosan [24]. Wheat straw also contains about 0.6–3.6% protein [25]. Other cereal straws, such as rice, barley, oat, and rye, have chemical compositions similar to that of wheat straw [26]. Large quantities of agricultural fibers are available for biofuels and bioproducts. For example, about 400 million metric tons (800 billion pounds) of dry crop residues are available, based on the grain production of corn, soybean, wheat, and sorghum at a straw-to-seed ratio of 45–52% [27–31]. Among these residues, about 150 billion pounds are wheat straw [32]. Wheat straw is usually used for fuel, manure, cattle feed, mulch, and bedding materials for animals [33]. Particleboard can be prepared using wheat straw [34–36], sunflower stalks [37], rice straw, cotton stalks, sugarcane bagasse, flax [38], maize husks, and maize cobs [33].

Natural fibers can be used for composites as harvested, or they can be used as raw materials for cellulose production. Cellulose can be modified into cellulose esters, such as cellulose acetate, cellulose acetate propionate, and butyrate, which are currently used as major components of thermoplastics. Cellulose, a major component of natural fibers, occurs in nature largely in crystalline forms made up of partially aligned or oriented linear polymer chains, consisting of up to 10,000 β-1, 4-linked anhydroglucose units. Cellulose chains are compacted aggregates, held together by hydrogen bonds forming a three-dimensional structure, which imparts mechanical strength to cellulose and contributes to its biodegradation and acid hydrolysis [39]. Hemicellulose is mainly composed of β-1, 4-linked D-xylopyranoyl units with side chains of various lengths containing L-arabinose, D-glucuronic acid, or its 4-*O*-methyl ether, D-galactose, and possibly D-glucose [40]. Lignin is mainly made up of phenylpropane units. Lignin is encrusted in the cell wall and partly covalently bonded with hemicellulose polysaccharides. Lignin is often a by-product of cellulose or paper pulping manufacture. It is inexpensive and mainly used for fuel and reformed composite materials [41]. Lignin may also have a potential use in adhesives. It can be functionalized to make it soluble in composite resins and be used as a comonomer and interfacial agent for natural fibers and soy-based resin composites.

1.5 Market Potential for Plant Polymers

Materials and composites used for construction, automobile parts, furniture, packaging, utensils, printing, coatings, and textile sizing represent a large market (about $100 billion) that includes a broad variety of products, such as adhesives, resins, plastics, binders, fibers, paints, inks, additives, and solvents. For example, about 20 billion pounds of adhesives are used annually in the United States. Among those adhesives, about 8 billion pounds are formaldehyde-based adhesives, 3.5 billion pounds are thermoset- and thermoplastic-based adhesives, 7.5 billion pounds are latex based, 0.5 billion pounds are isocyanate based, and the rest are various adhesives with unique applications. The latex-based adhesives are mainly used for packaging, coatings, labeling, inks, paints, office glues, furniture, furnishings, and similar uses. The formaldehyde-based adhesives primarily include urea formaldehyde and phenol formaldehyde adhesives, which are mainly used for plywood, particleboard, medium-density fiberboard, and oriented strand board for construction and furniture. Generally, the adhesive is about 5–20 wt% of a wood-based composite material used in construction, with the rest of the composite comprised mainly of fiber materials. With an average of 10% adhesive used in such composites, the total annual fiber demand is about 150 billion pounds.

The demands for thermoplastic resin are another indicator of market potential. Narayan [20] did a search in 1994 and found that about 54.2 billion pounds of thermoplastic resins and 11.1 billion pounds of styrene-based latex resins were produced in 1992 in the United States. These resins are used

Table 1.3 Thermoplastic Resin Uses and Distributions

Thermoplastic Resins	Amount (Billions of Pounds)	Styrene-Based Latex	Amount (Millions of Pounds)
Packing	18.2	Adhesives, inks, and coatings	461
Building and construction	7.6	Furniture and furnishings	369
Electrical and electronic	2.6	All other	313
Furniture and furnishings	2.4		
Consumer and Institutional	5.9		
Industrial	0.6		
Adhesives, inks, and coatings	1.2		
Transportation	2.5		
Exports	6.8		
All other	6.6		

Source: Facts and Figures of the U.S. Plastics Industry, Society of the Plastic Industry, 1993.

mainly in packaging, construction, furniture, and adhesives (Table 1.3). About 22 billion pounds of plastic waste was discharged in 1992 with an annual rate of increase of 5.9% [42]. This figure is expected to reach 42 billion pounds by 2007. Based on U.S. Environmental Protection Agency (EPA) statistics, about 10 million pounds of plastic wastes are produced aboard government ships [20]. These wastes can be used as an indicator for market potential for both bio-based and biodegradable materials.

An example of disposable materials produced from thermoplastics is given in Fig. 1.2. These thermoplastics are commonly used for packaging containers, films, closures, foams, cutlery, utensils, loose fill, and other applications. Many other single-use or short-term-use items, such as diapers, personal and feminine hygiene products, masks, gowns, gloves, and even computer hardware and television frames, all have market potential for bio-based materials.

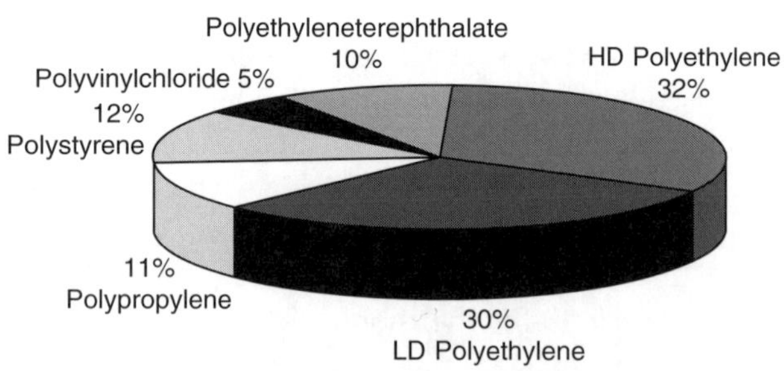

Figure 1.2 Uses and distributions of disposable plastic materials. Total disposable plastics are about 13,655 million pounds. *(Source: Facts and figures of the U.S. Plastics Industry, Society of the Plastic Industry, 1993.)*

1.6 Sustainable Agriculture Industry of the Future

Durability, compatibility, affordability, and sustainability are the challenges of converting renewable resources into industrial materials. Sustainable development provides growth of both ecological integrity and social equity to meet basic human needs through viable economic development over time. When a new material is designed and manufactured, one consideration should be sustainability, including resource availability, land use, biodiversity, environmental impact, energy efficiency, soil conservation, and the impact on the social community. Besides a favorable life cycle analysis, research and development of bio-based products should consider the limits that will maintain sustainable development. The design of bio-based materials should favor increased materials supplements, optimized land use, improved plant biodiversity, minimized environmental pollution, and improved energy efficiency, while at the same time meeting consumer demands.

About 467 million dry tons of biomass are available for energy use, including energy crops (switch grass, hybrid poplar, and willow), forest residues, mill residues, sludge, biogas, and other wastes [43]. In addition, about 550 million dry tons of crop residues are produced annually in the United States, based on the total grain and legume production [31]. Some of these residues need to be returned to fields to maintain soil quality (such as soil carbon balance), and some are used for manure or animal bedding, but approximately 70% of these crop residues may be available for energy uses. Burning of residual natural fibers is now forbidden in most Western countries, and their utilization in materials as proposed herein has a double environmental benefit.

The total amount of energy consumption in the United States is about 100 quadrillion Btu annually [44]. About 40% of the Btu comes from petroleum oils. The total estimated cellulosic sugar-based biomass available for biofuel is about 467 million dry tons in addition to 385 million dry tons of crop residues. Based on current technology, biomass materials would contribute about 10–15% of the total energy annually used in the United States [43]. To make sugar-based cellulosic biomass economically viable for energy, advanced technology is being developed to convert these biomasses into biofuels at higher efficiency. In addition, plant production needs to be increased at least three- to fourfold during the next 40 years to meet national biofuel needs. It makes an excellent environmental sense to utilize grains and waste agriculture fibers for materials and fuels that otherwise would be derived from petroleum. However, such energy and material conversions should be done in a sustainable green engineering manner such that a gallon of ethanol fuel does not require a gallon of petroleum to produce.

The total estimated market potential for bioproducts could be about 160 billion pounds (about 80 million metric tons). There are about 250 million metric tons of grains and legumes potentially available annually in the United States for industrial products. Polymers from grains and legumes require much less energy to convert into useful materials for some, but not all, bioproducts. Protein, oil, carbohydrates, and/or their monomers, including amino acids, fatty acids, sugars, and phenolics, are all important platforms as coproducts of a feedstock system and meet the large demands for bioproducts, including adhesives, resins, composites, plastics, lubricants, coatings, solvents, inks, paints, and many other chemicals (Fig. 1.3). Plant materials can rarely

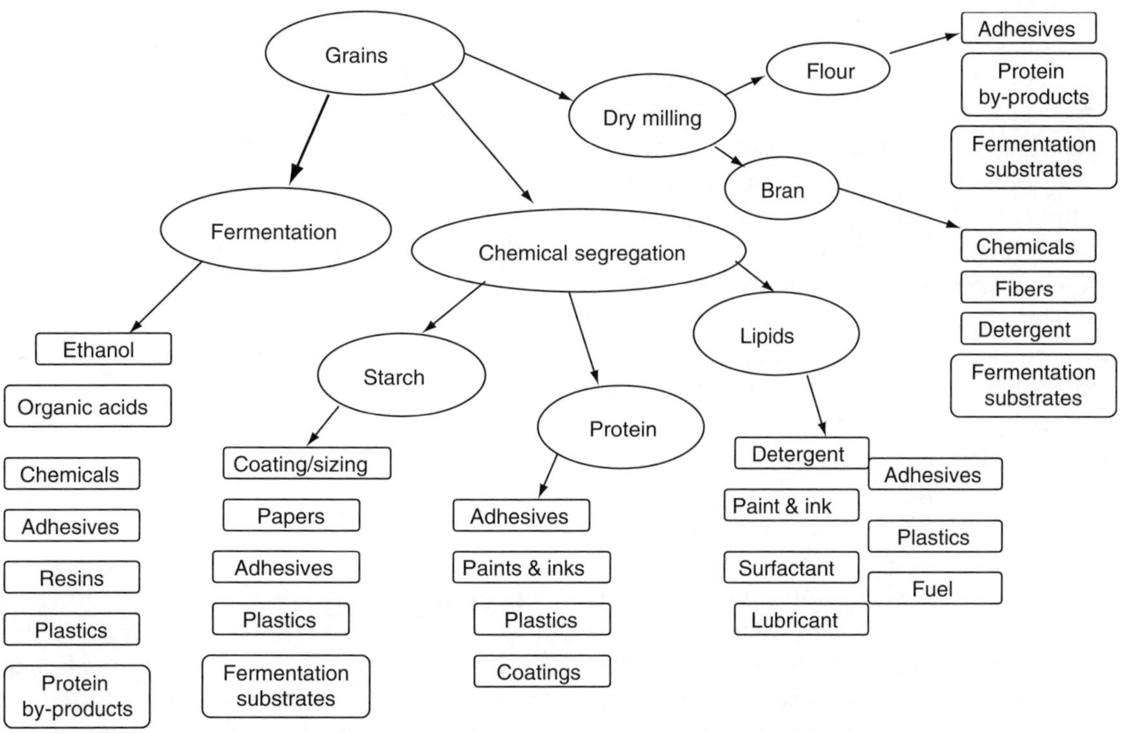

Figure 1.3 Diagram of possible industrial products from biorefinery process of grains and legumes. Application potentials are beyond those listed in the diagram.

be used in their natural form, but they can be converted into functional polymers for consumer products after bioconversions, reactions, and modifications with physical, chemical, enzymatic, and genetic approaches.

Plant material structures are genetically controlled, which consequently affects product performance. Plant materials are studied in this book in relation to their product performance. Proteins are complex macromolecules that contain a number of amino acid monomers linked by amide bonds. The sequences of these amino acids and composition determine protein structure, functional groups, and conformational structures that affect both processing and quality of the end product. The triglyceride oil molecular structure is essentially that of a three-arm star, where the length of the arms, the degree of unsaturation, and the fatty acids' content and distribution are the important structural variables for product quality. Advanced technologies, such as biopolymer simulation and modeling, surface structure analysis, chemical structure analysis, synthesis, thermal phase transitions, and rheological behavior analysis, should be used to obtain the information required to better understand bio-based polymers.

Research and development for a sustainable agricultural industry for plant-based materials and composites include five major units: plant science, production, bioprocessing, utilization, and products designed to meet society's demands. Based on several road maps developed by federal funding agencies for bio-based materials research, we summarize the critical research needs and directions as follows: Research efforts in plant science should focus on genomics, enzymes, metabolism, and bioinformatics. This allows for a better understanding of gene regulation, plant metabolic pathways, carbon flow, functional genomics, molecular evolution, and protein/oil/carbohydrate formation, which helps in developing tools and technologies for functional gene markers, gene switching, gene screening and sequencing, and gene manipulation. Research efforts in production focus on plant and grain quality consistency, unit costs, yield, infrastructure, and designed plants. It is important to produce components with favorable traits, improve yields, understand interactions of genotypes with environment, control bio-based polymer and compound quality, develop harvesting technologies, and use land economically. For bio-based polymer and materials science and engineering, attention should be given to bio-based polymer chemistry, reactions and modification pathways, processing technologies, enzyme metabolism for bioconversion, bioseparation, molecular structure and functional performance, scale-up, economics, and infrastructure. Understanding these areas will allow for the development of new technologies for cost-effective conversion of plant materials into functional industrial materials. Plant materials utilization focuses on market/function identification, bioproduct designs, new bio-based materials development, performance definition, life cycle analysis and cost−value analysis, material standards improvement, new analytical method development, infrastructure and distribution system, and the main driver, economics.

1.7 Conclusion

Plant protein, oil, starch, lignin, and cellulosic materials are all important platforms for bioproduct applications. Agricultural commodities typically cannot be used as they appear in nature. They need to be converted into functional polymers and materials by various technologies including chemical reactions, fermentation, and modifications. Research efforts need to focus on interdisciplinary approaches that integrate plant science, production, processing, and utilization. Integrated research teams in the areas of materials science and engineering, plant science, biochemistry/chemistry, and economics should be assembled in collaboration with universities, institutions, national laboratories, and industries to achieve what we need in this and coming centuries.

Acknowledgment

The authors thank Dr. Forrest Chumley for his thorough review of this chapter.

References

[1] M. Vesterby, K.S. Krupa, Major Uses of Land in the United States. Statistical Bulletin No. 973, U.S. Department of Agriculture, Washington, DC,, 2001.

[2] R. Shukla, M. Cheryan, Ind. Crops Prod. 13 (2001) 171–192.
[3] N.S. Hettiarachchy, U. Kalapathy, D.J. Myers, J. Am. Oil Chem. Soc. 72 (12) (1995) 1461–1464.
[4] G. Careri, A. Giansanti, E. Gratten, Biopolymers 18 (1979) 1187–1203.
[5] E. Niwae, T. Wang, S. Kanoh, et al., Bull. Japanese Soc. Sci. Fisheries 54 (10) (1988) 1851.
[6] N. Kajiyama, S. Isobe, K. Uemura, et al., Int. J. Food Sci. Technol. 30 (2) (1995) 147.
[7] S.H. Kim, J.S. Rhee, J. Food Biochem. 13 (3) (1989) 187.
[8] S.H. Kim, J.E. Kinsella, Cereal Chem. 63 (1986) 342.
[9] S.H. Kim, J.E. Kinsella, J. Food Sci. 52 (5) (1987) 1341.
[10] F.S. Frederick, in: B.J. Hudson (Ed.), Biochem. Food Proteins, Elsevier Science Publishers, Ireland, Ltd., 1992.
[11] K.K.R. Han, G. Biserte, Int. J. Biochem. 15 (1983) 875.
[12] N. Matsudomi, T. Sasaki, A. Kato, et al., Agric. Biol. Chem. 49 (5) (1985) 1251.
[13] J.R. Wagner, J. Gueguen, J. Agr. and Food. Chem. 43 (8) (1995) 1993–2000.
[14] K. Liu, Soybeans: Chemistry, Technology, and Utilization, International Thomson Publishing, New York, 1997.
[15] R.P. Wool, S.H. Kusefoglu, G.R. Palmese, et al., U.S. Patent 6,121,398; 2001.
[16] R.P. Wool, S.H. Kusefoglu, G.R. Palmese, et al., U.S. Patent 6,121,398; 2000.
[17] D.J. Tracker, G.W. Borden, O. Smith, W. U.S. Patent 3,979,270; 1976.
[18] D.J. Tracker, G.W. Borden, O. Smith, W. U.S. Patent 3,931,075; 1976.
[19] R.P. Wool, S.N. Khot, J.J. LaScala, et al., Affordable Composites and Plastics from Renewable Resources; Part I: Synthesis of Monomers and Polymers, 177–204; Part 2: Manufacture of Composites, 205–224, Advancing Sustainability through Green Chemistry and Engineering, R. L. Lankey and P. T. Anastas, Eds., ACS Symposium Series 823, Washington, DC, 2002.
[20] R. Narayan, Polymeric materials from agriculture feedstocks, in: Fishman, et al. (Eds.), Polymers from Agricultural Coproducts. ACS Symposium Series 575, Washington, DC, 1994.
[21] D.J. Thomas, W.A. Atwell, Starches: Practical Guides for the Food Industry, American Association of Cereal Chemists, St. Paul, MN, 1997.
[22] R.F. Tester, W.R. Morrison, Cereal Chem. 67 (6) (1990) 551.
[23] R.L. Whistler, J.N. BeMiller, Carbohydrate Chemistry for Food Scientists, American Association of Cereal Chemists, St. Paul, MN, 1999.
[24] G.S. Karr, Acetylation of Ground Wheat Straw for the Production of Strawboard. M.S. Thesis, Kansas State University, Manhattan, 1998.
[25] J.E. Atchison, J.N. McGovern, History of paper and the importance of agro-based plant fibers, in: F. Hamilton, B. Leopold (Eds.), Pulp and Paper Manufacture, Secondary Fibers and Agro-Based Pulping, vol. 3, TAPPI Press, Atlanta, 1983.
[26] R.M. Rowell, Opportunities for Lignocellulosic Materials and Composites. ACS Symposium Series 476, American Chemical Society, Washington, DC, 1992. 12–27.
[27] R.C. Sharma, E.L. Smith, Crop Sci. 26 (1986) 1147–1150.
[28] K.M. King, D.H. Greer, Agron. J. 78 (1986) 515–521.
[29] I. Rajcan, M. Tollenaar, Field Crops Res. 60 (1999) 245–253.
[30] W.T. Schapaugh, J.R. Wilcox, Crop Sci. 20 (1980) 529–533.
[31] World Agricultural Supply and Demand Estimates, World Agricultural Outlook Board, U.S. Department of Agriculture, Washington, DC; most recent edition: WASDE–399, June 11, 2003.
[32] J. Zucaro, R. Reen, The second forest: filling the wood source gap while creating the environmental performance board of the 21st century. developing composites from wheat straw,, in: Proceeding of 29th International SymposiumParticleboard/Composite Materials, Washington State University, Pullman, WA, 1995, pp. 225–231.
[33] A. Sampathrajan, N.C. Vijayaraghavan, K.R. Swaminathan, Bioresour. Technol. 40 (3) (1992) 249–251.
[34] G. Han, C. Zhang, D. Zhang, et al., J. Wood Sci. 44 (4) (1998) 282–286.
[35] G. Karr, X. Sun, Indust. Crops Products 12 (2000) 19–24.
[36] X. Mo, J. Hu, X. Sun, et al., Indust. Crops Products 14 (2001) 1–9.

[37] P. Khristova, N. Yossifov, S. Gabir, et al., Cellul. Chem. Technol. 32 (3–4) (1998) 327–337.
[38] G. Heslop, in: Proceeding of 31st International Particleboard/Composite Materials Symposium, 8–10 April 1997, Washington State University, 1997, pp. 109–113.
[39] D. Theander, Review of straw carbohydrate research, in: R.D. Hill, L. Munck (Eds.), New Approaches to Research on Cereal Carbohydrates, Elsevier Science Publishers, Amsterdam, 1985, p. 217.
[40] J.M. Lawther, R.C. Sun, W.B. Banks, J. Agric. Food Chem. 43 (1995) 667.
[41] J.L. McCarthy, A. Islam, in: W.G. Glasser, R.A. Northey, R.A. Schultz (Eds.), Historical, Biological, and Materials Perspectives, American Chemical Society, Washington, DC, 2000, pp. 2–99. Chapter 1.
[42] Anonymous, Plastics Eng. 50 (1994) 34.
[43] Biobased Products and Bioenergy Road Map, Oct 2002, United States, Department of Energy.
[44] Annual Energy Review, Energy Information Administration. Washington, DC (2001), http://www.eia.doe.gov/emeu/aer/pdf/03842001.pdf, 2001.
[45] R. Lasztity, The Chemistry of Cereal Proteins, second ed. CRC Press, Boca Raton, FL, 1995.
[46] E.W. Lusas, in: K. Kulp, G. PonteJr. (Eds.), Oilseeds and Oil-Bearing Materials. In Handbook of Cereal Science and Technology, second ed. Marcel Dekker, New York, 2000, pp. 297–362.
[47] D.K. Salunkhe, J.K. Chavan, R.N. Adsule, et al., World Oilseeds Chemistry, Technology, and Utilization, Van Nostrand Reinhold, New York, 1991.

2 Overview of Poly(lactic Acid)

Lee Tin Sin, Abdul R. Rahmat and Wan A.W.A. Rahman

OUTLINE

2.1 Background to Biodegradable Polymers

People have been using polymers for thousands of years. In ancient times, natural plant gum was used to adhere pieces of wood in house building. When the ancients started to explore the oceans, natural plant gum was applied as a waterproof coating to boats. At that time people did not know the extent to which polymers could be put to use, so their use was limited to very specific applications. Of course, the ancients depended on plant-derived polymers. No modifications were made to their formulation, nor were polymers synthesized to improve applications.

Natural rubber has been known about since 1495, when Christopher Columbus landed on the island of Haiti and saw people playing with an elastic ball. At that time rubber latex was harvested from the rubber tree *Hevea brasiliensis* as a sticky lump, which had limited applications. However, by 1844 Charles Goodyear discovered and patented a method to sulfur vulcanize rubber, and since then it has been widely used in the tire industry.

The first synthetic polymer was invented by Leo Hendrik Baekeland in 1907. This was a thermosetting phenol-formaldehyde resin called Bakelite. In recent decades, the rapid development of polymers has made a large contribution to technology with the invention of a highly effective catalytic polymerization process. Because commodity polymers—polyethylene, polypropylene, polystyrene, and poly(vinyl chloride) (PVC)—can be produced so cheaply, their use has been exploited for the mass production of disposable packaging. Thus, around the world, polymer pollution has become a serious issue. These petroleum-derived commodity synthetic polymers require hundreds of years to fully degrade into harmless soil components. This, together with the reducing reserves of crude oil, is an encouraging research into the development of renewable sources of raw materials for polymers. Figure 2.1 shows the general trend of polymer development globally.

Although steps have been taken to educate people about the environmental impact caused by the exploitation of plastics, these materials continue to represent the largest proportion of domestic waste. Conventional plastic waste takes a very long time to be broken down into harmless substances compared with organic materials. For instance, a telephone top-up card takes over 100 years to naturally degrade, while an apple core requires just 3 months to be naturally transformed into organic fertilizer. Due to the better degradability of biomass over conventional plastics, polymer–biomass blends were the first step in providing alternatives to help reduce plastic waste

Ebnesajjad: Handbook of Biopolymers and Biodegradable Plastics. http://dx.doi.org/10.1016/B978-1-4557-2834-3.00002-1
© 2012 Elsevier Inc. All rights reserved. Reproduced from a chapter in: Tin Sin, *Polylactic Acid* (2012).

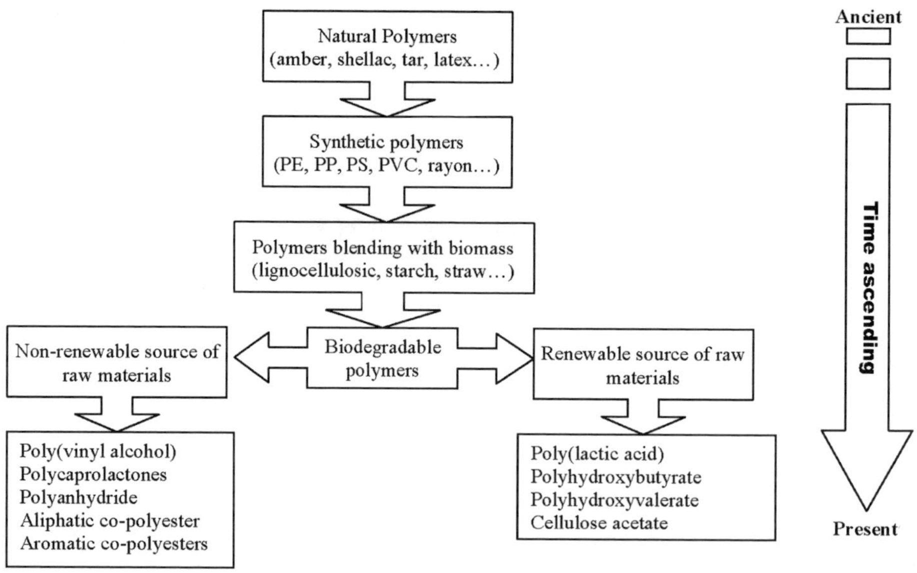

Figure 2.1 Trends in polymer development.

problems. Generally, abundant biomass such as lignocellulosics and starches are blended with synthetic polymers. These polymer compounds are partially degradable by microorganisms. However, after the biomass portion has been consumed, the leftover polymer skeleton will still cause harmful effects to the environment.

These days, the focus is on developing environmentally friendly polymers. These polymers are naturally degradable when disposed in the environment. The carbon footprint of production of these polymers is monitored to ensure sustainable environmental protection.

Biodegradable polymers can be divided into two categories—petroleum-derived and microorganism-derived biodegradable polymers (Fig. 2.1). The petroleum-derived biodegradable polymers, such as poly(vinyl alcohol) (PVOH), use ethylene to produce vinyl acetate for polymerization of poly(vinyl acetate) and is further hydrolyzed into PVOH. The production cost of this polymer is very sensitive to the fluctuation of crude oil prices and it is not environmental friendly, due to the emission of greenhouse gases during production. However, microorganism-derived biodegradable polymers utilize the bio-activity of bacteria to convert plant products, such as starch, into the starting product for polymerization. Poly(lactic acid), also known as polylactide (PLA), is the subject of this chapter, and is produced in this way, utilizing the activity of microorganisms. Polyhydroxyalkanoate (PHA) is also the product of bacterial fermentation. These polymers use renewable feedstock, and the production process possesses carbon credit.

There are also some polymer products on the market called oxo-biodegradable plastics. These so-called "biodegradable" plastics have caused controversy, and disputes with the environmentalists. Oxo-biodegradable plastics are actually degraded using a controlled catalyst to kick-start a chain-scissioning reaction to attack the polymer macromolecules. This catalyst is made of series of active organo transition metals, which are added to the polymer. When oxo-biodegradable polymers are exposed to ultraviolet light and free oxygen attacks, the chain-scissioning reaction occurs extensively, finally reducing the plastic to carbon dioxide. In the market, the oxo-degradation additives are mostly added to polyethylene and polypropylene. The additives are present in very small amounts ($<1\%$) and are highly effective. Nevertheless, controversy has also arisen about these types of "eco-friendly" plastics because they are still derived from petroleum-based products and the degradation still generates carbon dioxide, which is against the principle of carbon credit products. In the short term, these plastics may help to reduce the burden on landfill. However, the use of these oxo-biodegradable materials also causes other environmental problems. The most serious of these is that the plastics take time to be fully degraded into carbon dioxide. During the early breakdown process, fragmentation of the plastic causes pollution to the soil, and this can be accidentally consumed by organisms living off the soil. Again, this has shown that a fully biodegradable polymer with carbon credit is crucial for a sustainable future.

Prior to a more detailed discussion of PLA, several biodegradable polymers will now be examined and compared with PLA, to determine the reasons for which PLA is the most popular among the biodegradable polymers nowadays. PVOH and PLA are the most widely produced biodegradable polymers, while other biodegradable polymers, such as polycaprolactone (PCL) and polyhydroxybutyrate (PHB), are produced in small quantities at the laboratory scale or at pilot plants. In 2006, the world production of PVOH reached over 1 million metric tons (MT) per annum. However, PVOH is a petrochemical-type biodegradable polymer. The major markets for PVOH are textile sizing agents, coatings, and adhesives. Only a limited amount of PVOH is made for packaging applications. The main reason for this is the hydrophilic behavior of PVOH. Prolonged environmental exposure causes PVOH to absorb moisture extensively. There are hydrolyzed and partially hydrolyzed forms of PVOH. Both types of PVOH are soluble in water, and the solubility temperature of hydrolyzed PVOH is higher. The major producer of PVOH is Kuraray, in the United States, which has nearly 16% of the world's production. China is still the country that produces the most PVOH; it accounts for 45% of the global output.

In the early 1800s, PLA was discovered when Pelouze condensed lactic acid through a distillation process of water to form low-molecular-weight PLA. This is the early polycondensation process of lactic acid to produce low-molecular-weight PLA and lactide. Lactide is a prepolymer or an intermediate product used for the transformation to high-molecular-weight PLA. This polycondensation process merely produces low yield and low purity PLA. Almost a century later, DuPont scientist Wallace Carothers found that the heating of lactide in a vacuum produced PLA. Again, for high purity PLA this process is not feasible on an industrial scale due to the high cost of purification, which limits it to the production of medical grade products, such as sutures, implants, and drug carriers. The ambitious company Cargill has been involved in the research and development of PLA production technology since 1987, and first set up a pilot plant in 1992. Later on, in 1997, Cargill and Dow Chemical formed a joint venture named Cargill Dow Polymer LLC to further commercialize PLA. Their efforts have been fruitful, with the introduction of products branded as Ingeo™. As part of this joint venture, Cargill has made efforts to improve the hardening time for products made of PLA, while Dow has focused on the manufacture of PLA [1]. Generally, PLA's monomer, lactic acid, can be obtained from the fermentation of dextrose by bacteria; dextrose is derived from plant starch. Thus, PLA is a polymer made from renewable sources, and has the potential to reduce our dependence on conventional plastics made from fossil-based resources. In recent years, PLA research has developed tremendously, with many inventions and publications globally (Figs 2.2 and 2.3).

In addition to PVOH and PLA, there are some other biodegradable polymers on the market; these are listed in Table 2.1. These polymers are only

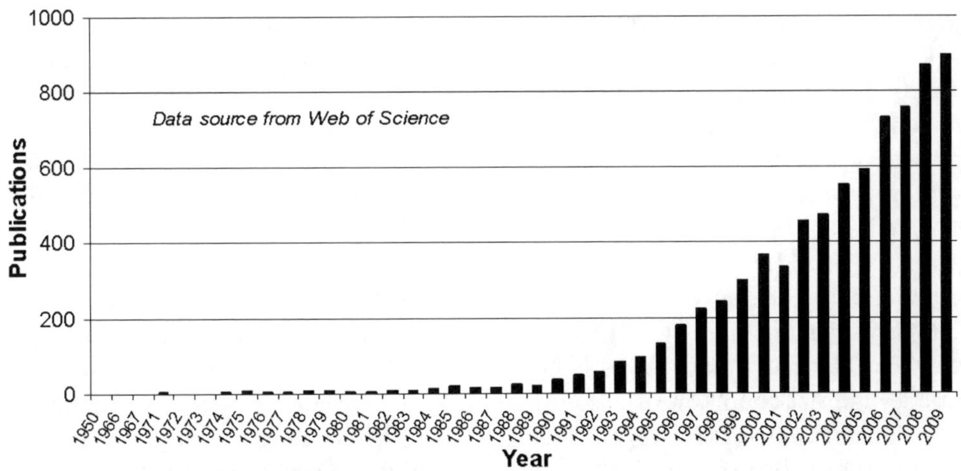

Figure 2.2 Research publications about PLA 1950–2009.

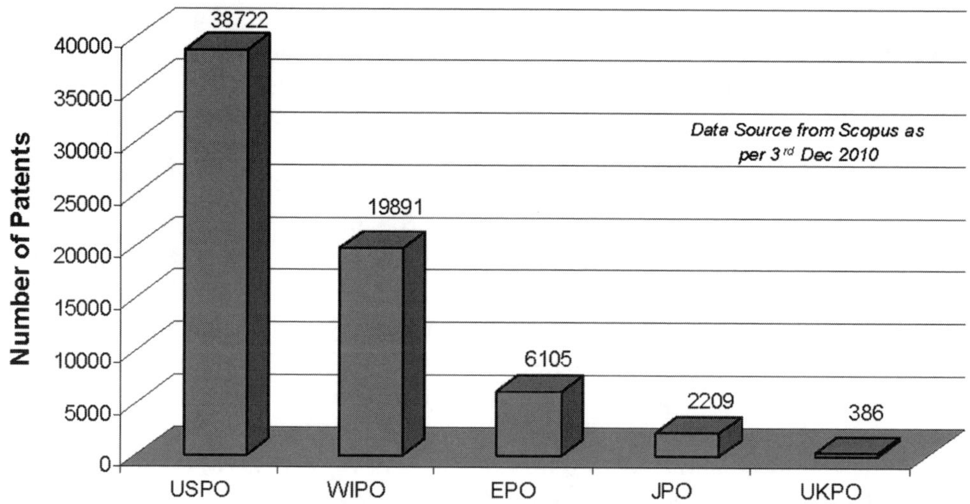

Figure 2.3 Number of patents published about PLA (USPO = United States Patent Office, WIPO = World Intellectual Property Organization, EPO = European Patent Office, JPO = Japanese Patent Office, UKPO = United Kingdom Patent Office).

produced on a small scale, primarily for biological applications, but also for exploration of commercial potential. Most of the biodegradable polymers are in the polyester group. Biodegradable polymers can be derived from renewable and nonrenewable sources (Fig. 2.4). Useful biodegradable polymers are not limited to neat polymers, but also include copolymers (polymerization of different monomers), the latter having improved biodegradability and structural properties. PCL, polyglycolic acid (PGA), and polydioxanone (PDO) are common biodegradable materials for sutures, pins, and drug carrier implants. Generally, PGA and PDO are preferable to PCL in biomedical applications because PCL takes longer time *in vivo* to be resorbed. A clinical study of the PCL-based implantable biodegradable contraceptive Capronor®, containing levonorgestrel, remains intact during the first year of use and is finally degraded and absorbed by the body after 2 years [2].

PHB and poly-3-hydroxybutyrate-co-valerate (PHBV) both belong to the PHA family, being developed using biological fermentation of dextrose as well. A joint venture between Metabolix and ADM, under the name of Telles™, has produced PHB with the trade name Mirel™. Their PHB compost bags take 6–12 months to be naturally degraded. Sanford, the international stationary manufacturer, uses PHB in their famous PaperMate® product range. PHB is not easily degraded under normal condition of usage or storage, even in a humid environment. However, when a PaperMate® pen made of PHB is buried in soil and compost, the pen decomposes in nearly a year.

Cellulose acetate is commonly used for cigarette filters, textiles, spectacle frames, and film media. Since the early part of the twentieth century, cellulose acetate has been a very important base material for the photographic film industry. Over the decades, the application of cellulose acetate has changed. Nowadays, a modified cellulose acetate has been produced that is suitable for injection molding to produce biodegradable plastic articles. Some ranges of sunglasses marketed by Louis Vuitton are made of cellulose acetate. This material comes in a wide variety of colors and textures and has the ability to be adjusted easily, but it tends to become brittle with age. A knitted cellulose acetate fabric treated with a specially formulated petrolatum emulsion is used as a wound dressing—it helps to protect the wound and prevents the dressing from adhering. Prolonged exposure of cellulose acetate to moisture, heat, or acids reduces the acetyl (CH_3C) groups attached to the cellulose. The degradation process causes the release of acetic acid; this is known as "vinegar syndrome." This is why when cellulose acetate film is stored under hot and humid conditions, there is a release of saturation acetic acid resulting in smelt. The release of acetic acid further attacks the polymer chain and deteriorates the cellulose. A study of cellulose acetate reported by Ref. [3] showed that cellulose acetate was biodegraded to cellulose diacetate in a wastewater treatment assay by approximately 70% in 27 days; the rate of degradation also depended on the degree of substitution of acetate. A high degree of substitution of acetate requires longer exposure.

Table 2.1 Some Common Biodegradable Polymers on the Market

Polymer	Chemical Composition	Producer	Applications	Biodegradability
ϵ-Polycaprolactone (PCL)	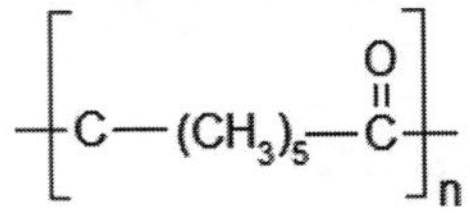	DURECT Corporation: Lactel® Daicel Chemical Industry: Celgreen® Union Carbide Corporation: TONE® Solvay Group: CAPA® Purac: Purasorb® PC 12	Ethicon: Monocryl®—suture Capronor®—contraceptive implant Agrotec: Agrothane®—paint and metal protection film	>12 months
Polyglycolide or polyglycolic acid (PGA)	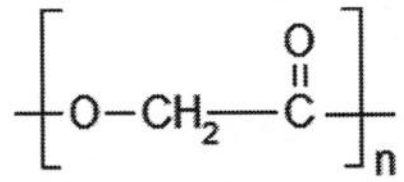	Purac: Purasorb® PG 20 Teleflex Incorporated Kureha Corporation	Dolphin: Petcryl®—sutures Bondek®—sutures Dexon™ S—sutures DemeTech®—sutures	>3 months
Polyhydroxyalkanoate: polyhydroxybutyrate (PHB) and polyhydroxyvalerate (PHV)	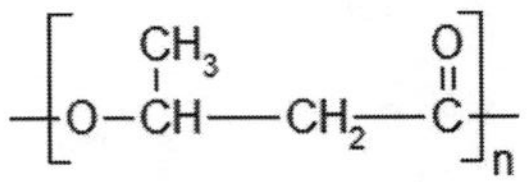 (PHB) 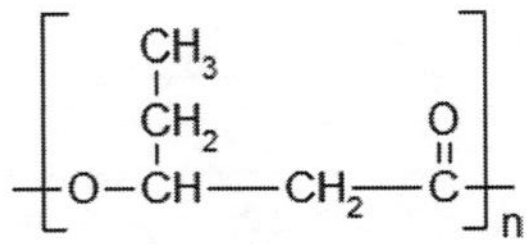 (PHV)	Metabolix/ADM (Telles): Mirel™ Ningbo Tianan Biologic Material: Enmat™ Copersucar: Biocycle® Biomer: Biomer®	Compost bags Consumer packaging Agriculture/horticulture film Rubbermaid®, Calphalon®, PaperMate® BioTuf™ EcoGen™	3–12 months
Polydioxanone (PDO)	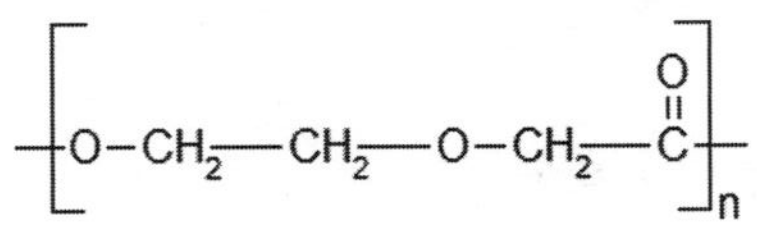	Ethticon Samyang	DemeTech® sutures Duracryl® sutures D-Tek® sutures Surgeasy® sutures	<7 months

(Continued)

Table 2.1 Some Common Biodegradable Polymers on the Market—*Cont'd*

Polymer	Chemical Composition	Producer	Applications	Biodegradability
			Ethicon® PDS* II sutures OrthoSorb® pin	
Cellulose acetate	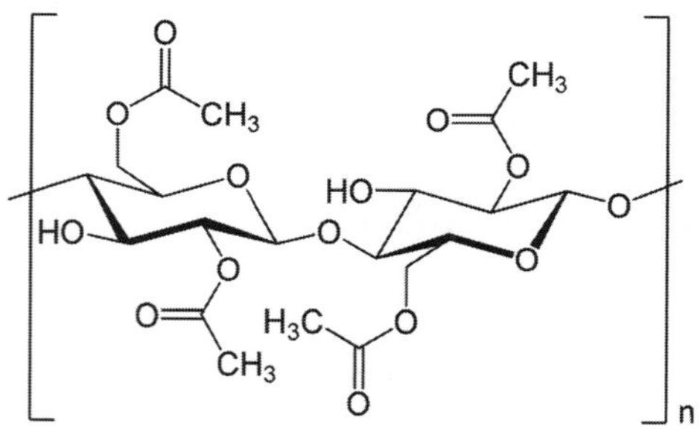	Celanese Rhodia	Cigarette filters Textiles Spectacle frames Film media Wound dressings—ADAPTIC™ Bioceta®: toothbrush	<24 months (depends on acetate content)

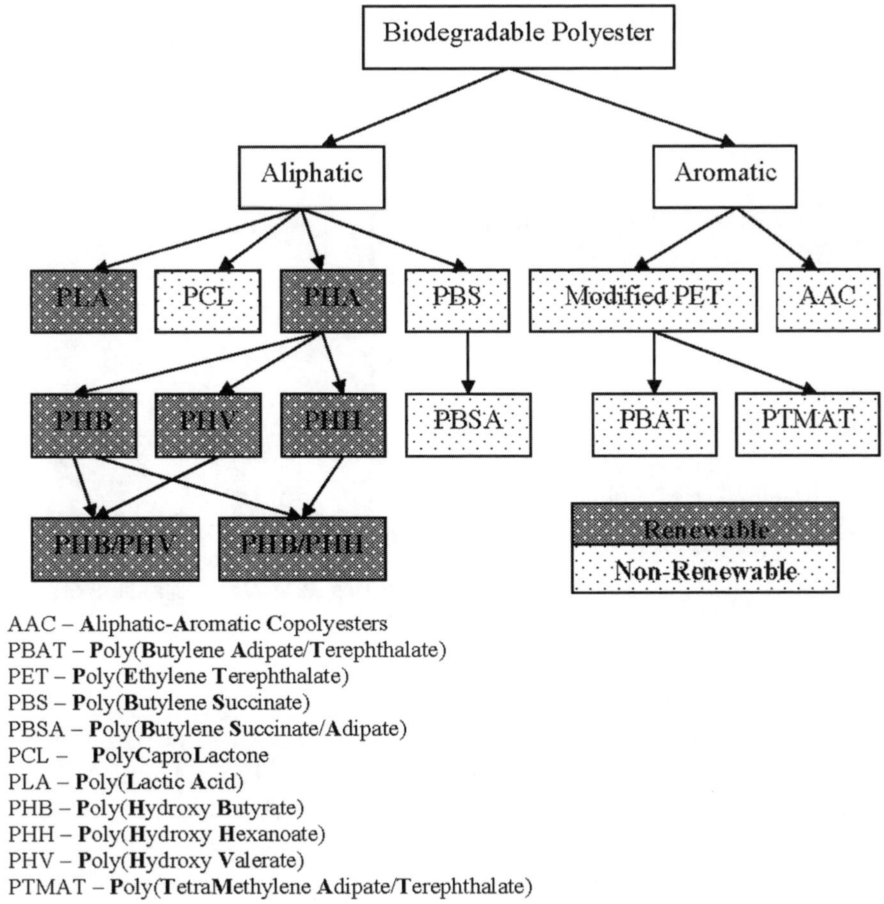

Figure 2.4 Biodegradable polyester family.

As can be seen, most of the biodegradable polymers mentioned belong to the polyester group (Fig. 2.4). This is due to the ester-containing covalent bond with a reactive polar nature. It can be broken down easily by the hydrolysis reaction. The biodegradable polyesters can be divided into aliphatic and aromatic groups, with members of each group being derived from renewable and nonrenewable sources. PLA and PHA are both aliphatic polyesters from renewable agricultural sources, while PCL and PBS/PBSA are aliphatic polyesters produced from nonrenewable feedstock. Most of the PCL on the market is used in biomedical applications. PBS/PBSA as marketed by Showa Denko, under the trade name Bionolle™, is supplied for Japanese local government programs for packing domestic solid waste before collection.

Generally, all the aromatic polyesters are produced from petroleum. Some consider the petroleum-based biodegradable polymers to be more viable than bio-based biodegradable polymers. The reason is that the manufacturing of bio-based polymers has led to the competition between food supply and plastic production, and this continues to be an issue as many people in the third world are still living with food shortage. However, this view should not be an obstacle to the development of bio-based polymers, because a small step in this direction has the potential to lead to a giant leap in reducing our dependence on fossil resources.

BASF has introduced their aliphatic–aromatic copolyesters (AAC) product under the name Ecoflex®. This material is widely used to produce compostable packaging and films. According to the BASF's corporate website, annual production of Ecoflex® has risen to 60,000 MT to keep up with the demand for biodegradable plastics, which is growing at a rate of 20% per year. At the same time, BASF also produces a blend of polyester and PLA—a product called Ecovio®. This high-melt-strength polyester–PLA can be directly processed on conventionally blown film lines without the incorporation of additives. Moreover,

Ecovio® has extraordinary puncture- and tear-resistance and weldability. Another company, Eastman, has also produced AAC, with the trade name Eastar Bio®. Eastar Bio® has a highly linear structure, while Ecoflex® contains long-chain branching. Late in 2004, the Eastar Bio® AAC technology was sold to Novamont S.p.A. Eastar Bio® is marketed in two different grades: Eastar Bio® GP is mainly for extrusion, coating, and cast film applications; Eastar Bio® Ultra is marketed for use in blown films. A study reported by BASF (2009) [4] shows that the AAC of Ecoflex® has comparable biodegradability to cellulose biomass, which is 90% degraded in 180 days as per CEN (European Committee for Standardization) EN 13432. This has shown that a petroleum-based biodegradable polymer can be as good as a natural material in terms of degradability.

The conventional polyethylene terephthalate (PET) takes hundreds of years to naturally degrade. However, the situation is different with PET with appropriate modification, such as comonomer ether, amide, or aliphatic monomer. The irregular weak linkages promote biodegradation through hydrolysis. The weaker linkages are further susceptible to enzymatic attack on the ether and amide bonds [5]. Such modified PET materials include polybutylene adipate/terephthalate (PBAT) and polytetramethylene adiphate/terephthalate (PTMAT). DuPont has commercialized Biomax® PTT 1100 with a plastic melting point of 195 °C for high service-temperature applications. This product is suitable for use as fast-food disposable packaging for hot food and drink. In general, the development of biodegradable polymers is still in the preliminary stages and it is anticipated that this will expand in the near future.

2.2 Market Potential of Biodegradable Polymers and PLA

Plastics manufacturing is the major industry worldwide. Every year, billions of tons of virgin and recycled plastics are produced. The world production of plastics has increased 160 times in a little less than 60 years, from 1.5 million tons in 1950 up to 245 million tons in 2008. Figure 2.5 shows that the production of polymers has increased year on year, with the exception of 2008, which showed a reduction in plastic production due to the global financial crisis. The demand for plastics soon recovered with the rebound of the world economy. This is evidenced

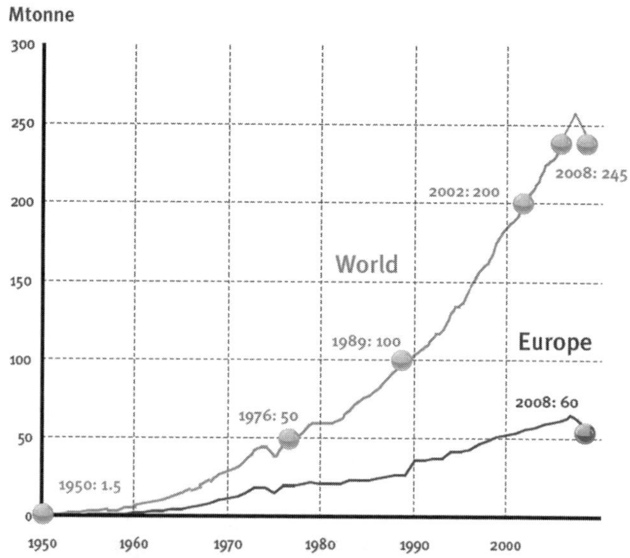

Figure 2.5 World plastics production from 1950 to 2008. *Adapted with permission from Ref. [6].*

by the fact that the giant global producers Dow Chemical, ExxonMobil Chemical, and BASF showed double-digit gains in sales and volumes [7]. Dow's sales were up by 33% in all geographical areas. This was contributed to by the high growth in the automotive industry and the need for elastomer materials for the increased demand for vehicles worldwide. BASF reported an increase in sales by 26% in the first quarter of 2009 due to substantial volume gains from the automotive and electrical/electronic sectors. Sales of the giant chemical company ExxonMobil rose 38%, or US$6.3 billion, in the first quarter of 2009 due to the larger chemical margins, with a large portion contributed by its plastic business.

Overall, the worldwide demand for plastic is forecast to be 45 kg per capita by 2015 [6]. The plastics market is still a big cake to be shared among the existing players, and the newcomers will also have the opportunity to gain a market share. From research data provided by a global management consulting company [8], the highest growth in polymer consumption belongs to the electrical/electronics sector. The highly sophisticated electrical/electronic products on the market, such as smart phones, computers, and entertainment appliances, require durable and lightweight parts, which make polymers crucial for use in their design. A variety of plastic products, both liquid and solid, including packaging, toys, containers, and stationery, remains the sector with the highest polymer

Table 2.2 World Polymer Consumption Forecast (Data: Ref. [8])

Market Sector	2006 (Thousand MT)	2016 (Thousand MT)	2006–2016 Compound Annual Growth Rate (%)
Food	42,025	71,774	5.5
Textiles	32,176	51,630	4.8
Furniture	13,687	22,993	5.3
Printing	780	1,220	4.6
Plastic products	43,500	78,361	6.1
Fabricated metals	1,519	2,259	4.0
Machinery	2,397	3,658	4.3
Electrical/electronic	13,810	25,499	6.3
Other transportation	9,330	16,181	5.7
Vehicles and parts	10,746	15,625	3.8
Other equipment	3,852	6,334	5.1
Other manufacturing	21,238	33,569	4.7
Construction	45,886	72,919	4.7
Total	240,947	402,022	5.3

consumption, with forecasts reaching 78,361 thousand MT per annum (Table 2.2).

These figures provide strong evidence that the demand for plastic products will grow further in future. However, the majority polymers on the market are petroleum-based products. Although the current price of crude oil has returned to an affordable level since the price hike of US$147 in July 2008, the price of many petroleum commodity products, especially polymers, has reached a historical high. Today, many believe that another petroleum price hike is very likely to happen in the next decade, due to the limited crude oil reserves. Continual exploitation of these natural resources has also caused serious global warming. Thus, the search for alternative sources of energy and nonpetroleum-based products is crucial for a sustainable economy and environment.

As mentioned previously, biodegradable polymers can be derived from both petroleum and renewable sources. Both types of biodegradable polymers have attracted attention in the industry. Petroleum-based biodegradable polymers may help to overcome the accumulation of nondegradable plastic waste. However, renewable biodegradable polymers not only possess biodegradability, but the polymers are also derived from sustainable sources with the environmental credit.

Many countries have imposed regulations to reduce or ban the use of nondegradable plastics for environmental protection. For instance, China, the largest polymer-consuming country with a population of 1.3 billion, has banned the usage of plastic bags. Major supermarkets do not provide free plastic bags to their customers. These actions have helped to save at least 37 million barrels of oil per year. In Europe, several regulations have driven forward organic waste management to help reduce soil/water poisoning and the release of greenhouse gases. Recycling of bio-waste is the first measure to reduce the generation of methane (a greenhouse gas) from landfills. Directive 1999/21/EC on the Landfill of Waste requires European Union members to reduce the amount of biodegradable waste to 35% of 1995 levels by 2016. The second measure is to increase the usage of compostable organic materials, so that they become useful in helping to enrich the soil. This can help replace the lost carbon from the soil as emphasized in Directive 2008/98/EC on waste (Waste Framework Directive). Following the introduction of Directive 94/62/EC on Packaging and Packaging Waste, which imposed requirements for plastic and packaging waste, plastic and packaging waste should now fulfill the European standard EN 13432, with these materials to be declared as compostable prior to being marketed to the public [9].

Ireland was one of the first countries to introduce a plastic bag levy. Ireland's Department of Environment, Heritage and Local Government introduced a charge of 15 cents on plastic bags in 2002. This move had an immediate effect, reducing the usage of plastic bags from 328 to 21 bags per capita. After this encouraging outcome, the Irish Government increased the levy to 22 cents, further reducing the usage of plastic bags [10]. Although biodegradable plastic bags degrade more quickly than standard ones, the Irish Government did not distinguish between the two in their laws. However, reusable plastic bags sold in the shops are exempt from the levy, with the condition that they should not be sold for less than 70 cents.

Because the use of plastic bags is not entirely avoidable in modern life, the production of reusable plastic bags made of a compostable material is recommended, so that disposal will not burden the environment. As people have become more aware about using compostable packaging, many companies have tried to make their products at least appear to have such packaging. Consequently, various types of "eco-packaging" are available in the market. Such eco-plastic products need to undergo a series of tests to verify their biodegradability and compostability. In the European Union, compostable packaging must fulfill the requirements of EN 13432, while other countries have their own standard to be met in order to allow the use of a compostable logo (Table 2.3).

The production of biodegradable polymers has increased tremendously over the past few decades. In an overview of the products and market of bio-based plastics by Shen *et al.* in Ref. [11] known as PRO-BIP 2009, the global output of bio-based plastics was 360,000 MT in 2007. This represented only 0.3% of the total amount of plastics produced worldwide. However, the production of bio-based plastics grew rapidly at a rate of 38% annually between 2003 and 2007 [11]. Shen *et al.* [11] have predicted that bio-based plastic production will increase to 3.45 million MT in 2020, and will be primarily made up of starch plastics (1.3 million MT), PLA (800,000 MT), bio-based polyethylene (600,000 MT), and PHA (400,000 MT).

Bio-based polyethylene is produced from the feedstock of ethylene, which is based on the dehydration of bio-ethanol from sugar fermentation. A large number of bio-based projects have been started in the United States, Europe, and Japan, and then production has been transferred to other parts of the world.

Based on the information from PRO-BIP 2009, the production output for different types of biodegradable polymer in 2009 is summarized in Fig. 2.6. Cellulose-based polymers represent the largest proportion of biodegradable polymers globally. Cellulose polymers are mainly used in the manufacture of fiber for textiles, bedding, cushions, filters, etc. Most of the cellulose is harvested from cotton and chemically treated or modified to suit the end use. Starch-based polymers relate to starch—polymer blends and thermoplastic starch. Companies such as Novamont S.p.A, Plantic DuPont, and Cereplast blend starch with other synthetic polymers to improve the processability and mechanical properties of the starch alone. Normally, blending of starch with a biodegradable polymer such as PCL, PLA, and PHB is preferable, to ensure the resulting blends are fully biodegradable. Some starch-based polymer producers also blend starch with polyolefin. These starch—polymer blends are partially degradable, with starch initiating the degradation. However, the left-over polymer skeleton can still cause harmful effects to the environment.

PLA, PHA, and other biodegradable polymers contributed to 14% of the world production in 2009. PLA is the most widely produced of the renewable biodegradable polymers. Currently, most of the renewable biodegradable polymers are still in the developing stages. PLA represents a large portion of the market because of the maturity of its technology for mass production. Technologists prefer PLA due to its renewable feedstock for carbon credit. The establishment of downstream processing and the market by renowned producers, especially NatureWorks, have also contributed to the expansion of the PLA production in a range of countries. In future, the production of PLA may overtake the sum of other biodegradable polymers, such as PBS, PBT, PCL, PBAT, etc. (Fig. 2.7). Future mass production and market competition will also assist the development of economically viable technology to offer cheaper products. Investors are likely to favor bulk production of PLA with its known profitability and long-term low-cost feedstock from agricultural sources. Moreover, the development of starch-based and other bio-plastics will also increase the demand for PLA. This is because fully biodegradable starch blended with PLA helps to improve the properties of the weaker starch structure itself. Similarly, BASF's AAC Ecovio® is blended with PLA for better processability and flexibility of the end product.

Table 2.3 Certification of Compostable Plastic for Different Countries

Certificating Body	Standard of Reference	Logo
Australia Bioplastics Association (Australia) www.bioplastics.org.au	EN 13432: 2000	
Association for Organics Recycling (UK) www.organics-recycling.org.uk	EN 13432: 2000	
Polish Packaging Research and Development Centre (Poland) www.cobro.org.pl/en	EN 13432: 2000	
DIN Certco (Germany) www.dincertco.de/en/	EN 13432: 2000	
Keurmerkinstituut (the Netherlands) www.keurmerk.nl	EN 13432: 2000	
Vincotte (Belgium) www.okcompost.be	EN 13432: 2000	
Jätelaito-syhdistys (Finland) www.jly.fi	EN 13432: 2000	
Certiquality/CIC (Italy) www.compostabile.com	EN 13432: 2000	
Biodegradable Products Institute (USA) www.bpiworld.org	ASTM D 6400-04	

(*Continued*)

Table 2.3 Certification of Compostable Plastic for Different Countries—*Cont'd*

Certificating Body	Standard of Reference	Logo
Bureau de normalisation du Québec (Canada) www.bnq.qc.ca	BNQ 9011-911/2007	COMPOSTABLE www.compostable.info
Japan BioPlastics Association (Japan) www.jbpaweb.net	Green Plastic Certification System	グリーンプラ® 生分解性プラスチック

Figure 2.8 shows the average prices of biodegradable plastics and conventional commodity plastics in 2009. The price of PLA is the lowest of the biodegradable polymers. The nearest competing biodegradable polymer is PVOH, which is produced by hydrolysis of polyvinyl acetate from petroleum sources. PLA and PVOH are very unlike to compete directly in the biodegradable polymer industry due to their respective characteristics. PVOH possesses hydrophilic properties, and is used as a sizing agent,

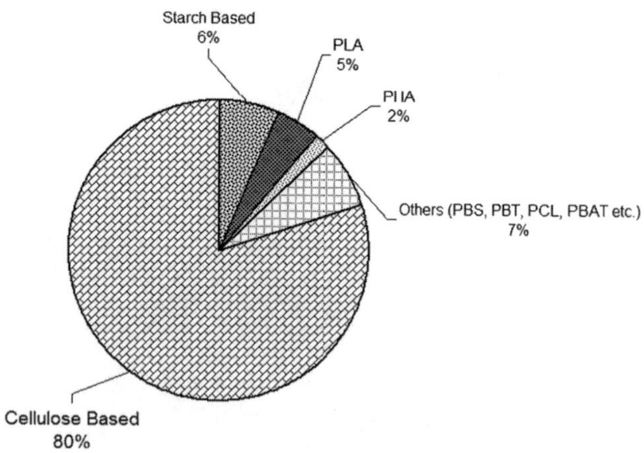

Figure 2.6 World production of renewable biodegradable plastics in 2009.

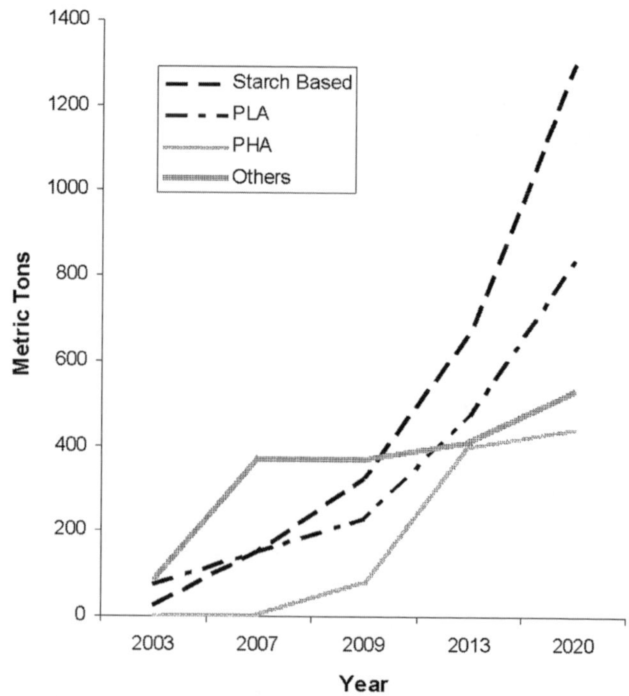

Figure 2.7 World production of renewable biodegradable polymers in 2003–2020 (projected).

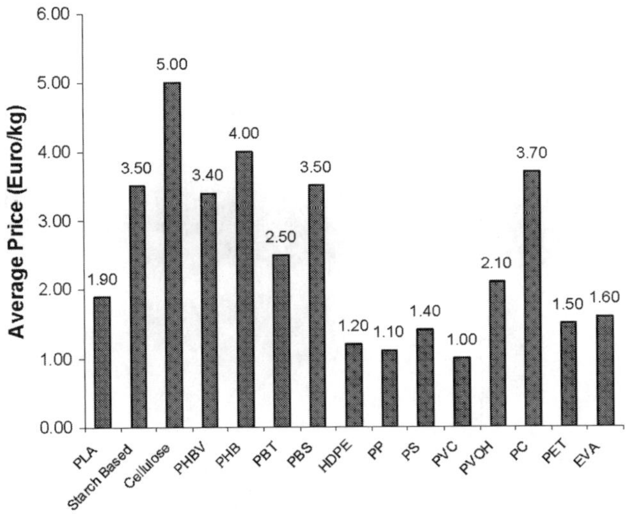

Figure 2.8 Average prices of polymers in 2009.

adhesive, and paper coating. Only a limited amount of PVOH is used for the manufacture of packaging film for food. PVOH tends to be soluble in water at 90 °C. In contrast, PLA is hydrophobic, and has the potential to be used as a substitute for some of the existing polyolefin polymers. The starch-based plastics have a higher price compared to PLA; this can be attributed to the technological processing of starch, which is remarkably complex. Starch needs to be blended with other polymers, such as PP and PLA, and, consequently, this leads to higher costs and extra processing on melt blending of starch with PP or PLA. Although cellulose is the most produced biodegradable plastic, its price remains higher due to its specialty application. The ability of cellulose to be injection molded is also limited. Extra treatment and modification of cellulose is crucial for processability using injection molding.

From the direct comparison in Fig. 2.8, PLA is the nearest competitor to the commodity polymers polyethylene (PE), polypropylene (PP), polystyrene (PS), PET, and ethylene vinyl acetate copolymer (EVA). At the same time, the price of PLA is much less than polycarbonate (PC). The potential of PLA to substitute PC is great, especially in the fabrication of electric/electronic casings. Fujitsu has introduced a laptop casing made of PC and PLA. This PC–PLA laptop casing has a 14.8% lower carbon oxide emission compared to a conventional PC–ABS casing. Overall, the PLA resin price is relatively high compared to commodity plastic. However, increasing of production efficiency and a competitive marketplace are likely to provide better prices in the near future.

Although PLA was first synthesized in the early 1800s, the development of PLA has taken long time to reach production viability. In the early stages of commercialization, the PLA produced was limited to use in biomedical devices, because the cost of synthesis was expensive and was not mass-produced. Direct polycondensation requires critical process control in order to achieve high-molecular-weight PLA. In the 1990s, the market for PLA started to expand, with the first pilot plant being set up in 1992 by Cargill, using the indirect polymerization of lactide monomer for a higher production yield of PLA. In 1997, the Cargill and Dow Chemical joint venture founded the company NatureWorks with their preliminary commercial products coming to market under the name Ingeo™. A plant was built at Blair, in the United States, costing US \$300 million in 2002. Later, in 2007, Dow Chemical sold its 50% stake in NatureWorks to Japan's Teijin. Teijin has been very committed to developing green plastic technologies to expand their existing polymer resins range. During the recent economic downturn, Teijin underwent restructuring, and transferred its 50% ownership to Cargill [12–14]. Teijin is now focusing in the development of their PLA product BIOFRONT™, a heat-resistant type of PLA plastic for the substitution of PET. BIOFRONT™ has 40 °C higher melting temperature than existing poly-L-lactic acid. Teijin's BIOFRONT™ has been produced in collaboration with Mazda, to develop a car-seat fabric made of 100% bio-based fibers [15]. More recently, Teijin has announced the codevelopment of a PLA compound with Panasonic Electric Works; MBA900H has superior moldability, and 1000 MT are set to be produced in 2012. Since the withdrawal of Teijin, NatureWorks has been wholly owned by Cargill. In a March 2009 corporate press release, NatureWorks announced that the company was assessing for a new production plant for Ingeo™ [12,13]. Ingeo™ is used by hundreds of leading brands and retailers in the United States, Europe, and Asia (Table 2.4).

Purac, currently the world's largest lactic acid producer, operates a lactic acid plant in Thailand with an annual output of 100,000 MT in 2007. This entire plant has the capacity of 200,000 MT annually in the future. Currently, Purac supplies over 60% of lactic acid globally from its operation facilities located in the Netherlands, Spain, Brazil, and USA. Purac has been manufacturing PLA and PLA copolymers for biomedical applications such as sutures, pins, screws, and tissue scaffolding materials. In planning for further business expansion and with the maturity of the PLA market, Purac has

Table 2.4 Examples of PLA Product Applications

Company	Area of Application	Market Products
CL Chemical Fibers	Spunbond fabrics	Medical applications, shopping bags, and landscape textiles 
Dyne-a-Pak	Foam meat trays	Dyne-a-Pak Nature™ trays 
Bodin (France)	Foam trays	Trays for meat, fish, and cheese 
CDS srl	Food serviceware	Cutlery 

Table 2.4 Examples of PLA Product Applications—*Cont'd*

Company	Area of Application	Market Products
Cargo Cosmetics	Casings	Casings for cosmetics 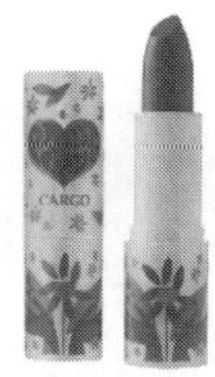
DS Technical Nonwoven	Exhibition grade carpeting	Ecopunch® carpets 
Sant'Anna, Swangold, Cool Change, Good Water, Primo Water	Bottles	Bottles for juice and still water 
Natures Organics	PLA bottles	Shampoo bottles in Australia 

(*Continued*)

Table 2.4 Examples of PLA Product Applications—*Cont'd*

Company	Area of Application	Market Products
Naturally Iowa	EarthFirst® shrink sleeve labels	Bottles for debuted Yogurt 7.0 
Priori	Cosmetic packaging	CoffeeBerry® 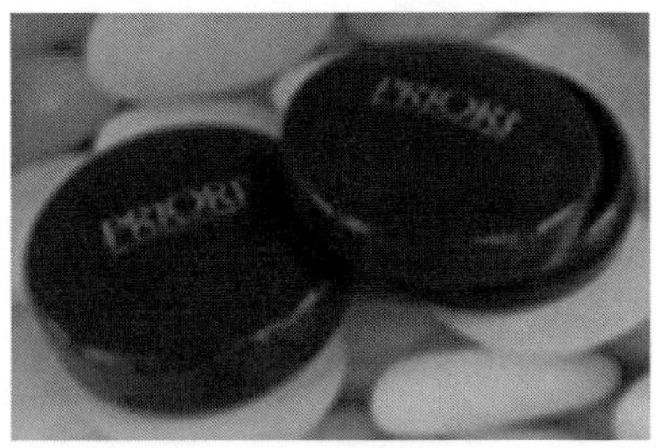
Frito-Lay	Packing bags	SunChips® 

Table 2.4 Examples of PLA Product Applications—*Cont'd*

Company	**Area of Application**	**Market Products**
InnoWare Plastics	Deep hinged trays and lids	ECO OctaView™ and ECO Expressions™ 
Ahlastrom	Nonwoven fabric	Tea bags 
Telecom Italia and MID product design studio	Telecommunication casing	Cordless telephones 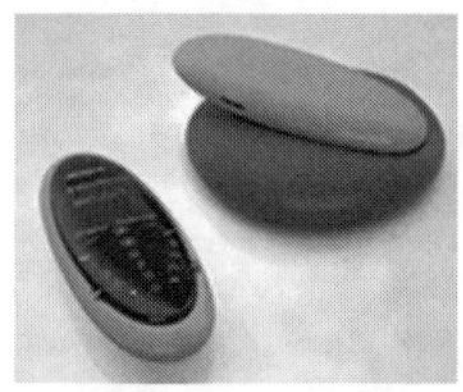
Carrefour Belgium	Film	Clear film overwrap for trays 
Kik & Boo	Fiber	Soft toys filled with PLA fiber 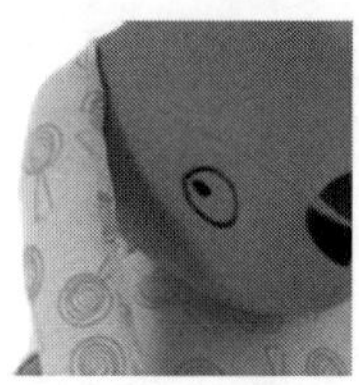

(*Continued*)

Table 2.4 Examples of PLA Product Applications—*Cont'd*

Company	Area of Application	Market Products
Stilolinea	Stationery	Pens 
DDCLAB USA	Fabric	Slim fit men's shirts and trousers 
Pacific Coast Feather Company	Fiberfill	Comforters, pillows 
Method	Fiber cloths	Sweeper cloths for omop™ 
Valor Brands	Fiber	Diapers—Natural Choice™ 

Table 2.4 Examples of PLA Product Applications—*Cont'd*

Company	Area of Application	Market Products
Kimberly-Clark	Fiber	Huggies® Pure and Natural diapers 
Fujitsu	Computer casing	FMV-BIBLO notebook 
Toyota	Automotive	Toyota Eco-Plastic—spare tire covers and floor mats 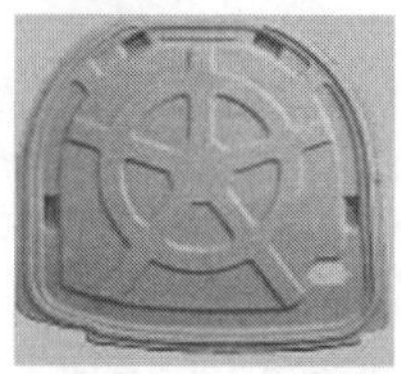
Bioserie	Electronics covers	iPhone covers 

decided to utilize its production of lactic acid for PLA manufacture. With its existing high-volume production of lactic acid, Purac has the opportunity to convert lactic acid into L-lactide and D-lactide under the brand name PURALACT™. Purac has invested 45 million to produce 75,000 MT of PLA at its lactide plant in Thailand. The new plant is scheduled for its first production in the second half of 2011.

Purac in the Netherlands and Sulzer Chemtech AG in Switzerland have joined forces to produce PLA foam. Synbra, a company in Etten-Leur, the Netherlands, has been engaged to set up the PLA foam technology for Purac–Sulzer, expanding their product range, which includes a green polymer foam called BioFoam® (Fig. 2.9). Synbra has been in the Styrofoam manufacturing line for more than 70

(a)

(b)

Figure 2.9 (a) Sulzer's 23 kg/h pilot plant in Switzerland using Purac's new lactide monomer; (b) Purac's 75,000 MT/year lactide monomer plant operating in Thailand from 2011.

years. The expandable PLA of Synbra utilizes the lactide produced by Purac's lactide facility in Spain. Purac's Spanish plant will have the production capacity of 10 million lb per year in the near future once it is fully commissioned. In September 2010, Purac entered into collaboration with Arkema to develop high-purity functional block copolymers, containing PLA segments, using the Purac's lactide. The output of the development is an improvement on the current lactide polymerization process with the absence of metal residues, which ensures safe medical and consumer goods packaging. In addition, Purac is also collaborating with Toyobo, a Japanese film, fiber, and biotechnology firm, to make an amorphous and biodegradable PLA product for the European market under the brand name Vyloecol®. Unlike the production technology used by Purac–Sulzer, Vyloecol® developed by Purac–Toyobo is a patented amorphous PLA for application as coatings or adhesives for packaging films and materials.

Purac is also actively involved in PLA production in the European Union, with Galactic and Total Petrochemicals. They established a 50/50 joint venture—Futerro—in September 2007 to develop PLA technology. The preliminary project was to construct a demonstration plant with a 1500 MT PLA production capacity; this pilot unit costs 15 million. The Galactic production site is located at Escanaffles, Belgium. The monomer, lactide, is obtained from fermenting sugar beet. Another joint venture, known as Pyramid Bioplastics Guben GmbH, is also planning to construct and operate a plant for the production of PLA, this time in Guben, eastern Germany. The company is a partnership between Pyramid Technologies Ltd, of Zug in Switzerland, and the German company Bioplastics GmbH, of Guben. The first construction plant will have a 60,000 MT capacity of PLA per annum by 2012. Hycail, a pilot-plant scale producer, used to produce a small quantity of PLA before selling it to Tate & Lyle in 2006. This plant was shut down 2 years later.

In Asia, many companies have been established to explore PLA technology. Japan is the first country to be involved in the research and development of PLA.

China then followed, as the market for PLA started to grow. Although Japan was involved in PLA technology earlier than other Asian countries, some of the large ambitious companies halted production due to high production costs, lack of availability of raw materials, and an immature market to accept such premium plastics with a higher price. Shizmadu initially operated a pilot plant to produce small commercial quantities of PLA. Since then, production has ceased and the technology sold to the Toyota Motor Corporation. Toyota increased production to 1000 MT per year, mainly for automotive applications. In 2008, the plant was sold to Teijin, and now Teijin is expanding production for its BIOFRONT™ products. The company plans to increase the productivity of BIOFRONT™ to 5000 MT per year in 2011. Unitika Ltd, a 120-year-old textile company, has marketed PLA products under the Teramac® brand. Teramac® resin can be processed using wide range of plastic technologies, including injection, extrusion, blow, foam, and emulsion. The Korean company Toray has launched full-scale commercialization of Ecodear™ PLA films and sheets. Ecodear™ possesses heat and impact resistance as well as flexibility and high transparency equivalent to petroleum-based plastic films.

Since 2007, many projects have been announced in China. However, many of these have seen a lack of further development [16]. Zhejiang Hisun Biomaterial was the first company in China to produce PLA on a commercial scale, with an annual production of 5000 MT per year. Other companies had smaller plants at the time: Shanghai Tong-jie-liang Biomaterial had a pilot plant producing 300 MT per year PLA, and Nantong Jiuding Biological Engineering had a larger facility that could produce up to 1000 MT per year. At the end of 2009, Nantong Jiuding Biological Engineering secured funding of US$1.4 million from the National Development Reform Commission to expand its PLA project [17]. This was followed by an expansion project, involving a total investment of US$19 million, to boost production to 20,000 MT per year. Henan Piaoan Group, a medical equipment and supplies manufacturer, has purchased the patented PLA technology of Japan's Hitachi Plant Technologies Ltd. The Henan Piaoan plant is expected to produce 10,000 MT of PLA annually. Most of the PLA produced in China is for export rather than internal use, because the biodegradable market in China is still in its infancy and there is a lack of local regulation on biodegradable polymer use for environmental protection.

A list of PLA resin producers worldwide is given in Table 2.5.

Table 2.5 Polylactic Acid Resin Producers

Producer	Capacity (MT/year)	Location
NatureWorks	140,000	Nebraska, USA
Purac–Sulzer Chemtech–Synbra Technology	5,000	The Netherlands
Galactic–Total Petrochemicals: Futerro	1,500	Belgium
Zhejiang Hisun Biomaterial	5,000	Zhejiang, China
Shanghai Tong-jie-liang Biomaterial	300	Shanghai, China
Mitsui Chemical—LACEA®	No data	Japan
Unitika–Terramac	5,000	Japan
Nantong Jiuding Biological Engineering	1,000	Jiangsu, China
Piaoan Group	10,000 (in planning)	Henan, China
Purac–Toyobo	No data	Japan
Toray Industries	5,000	Kyungsangbuk-do, South Korea
Pyramid Bioplastics Guben GmBH	60,000 (in planning)	Guben, Germany
Teijin Limited	1,200	Matsuyama, Ehime Prefecture, Japan

2.3 General Properties and Applications of PLA

2.3.1 PLA for Domestic Applications

NatureWorks is the largest PLA producer in the world. Their product range includes injection molding, extrusion, blow molding, thermoforming, films, and fiber applications. Ingeo™, NatureWorks' PLA resin, is produced at a rate of 140,000 MT per year from a facility located in Nebraska, United States. The company has 19 worldwide distribution points from which to sell and promote their products. NatureWorks has initiated a co-branding partnership program for better market positioning of Ingeo™. Currently, there are over 900 companies involved in this partnership program, which has successfully strengthened the Ingeo™ brand worldwide.

Tables 2.6—2.8 give a summary of the properties of Ingeo™. As with commodity plastics such as polyethylene and polypropylene, the selection of Ingeo™ is made according to the processing technique as well as the end use of the product. According to Patrick Gruber, Chief Technology Officer at NatureWorks, and colleagues [18] the variety grades of PLA are formulated using the principle of stereochemical purity, molecular weight, and the incorporation of additive packages. Manipulation of the stereochemical composition of PLA has a significant effect on the melting point, rate of crystallization, and ultimately the extent of crystallization [18]. Pure PLA either fully in L or D stereochemistry has a melting point of 180 °C and a glass transition temperature at 60 °C [19]. Copolymerization of D-lactide or *meso*-lactide affects the stereochemical purity. The crystallinity of PLA is totally destroyed after the incorporation of 15% *meso*-lactide or D-lactide in poly(L-lactide) (PLLA). The copolymerization of L and D stereochemistry induce the formation of an amorphous structure in the resulting polymer. Nevertheless, the higher melting point of the resulting polymer is preferable to avoid heat deflection of the PLA-formed article, typically in hot food serviceware. Purac claims that through the manipulation of the stereo complex and stereo block of lactide during the copolymerization process, the melting temperature can effectively be increased to 230 °C, which is almost as good as polystyrene (melting point of polystyrene is about 240 °C). In spite of that, it is important that the rheological properties of the resulting polymer suit the processing technology. PLA is typical of aliphatic polyesters, having relatively poor strength and lacking in shear sensitivity. The introduction of branching in PLA makes it possible to be able to obtain a longer chain of the resulting polymer for better entanglement, which can result in better melt strength for blow film application [20]. However, the details of such modifications are rarely disclosed by the manufacturers. Further details of research work on the rheological properties of PLA are discussed in Chapter 3.

Unitika Limited and FKuR Kunststoff GmBH have marketed their products based on NatureWorks' Ingeo™ under the trade names of Bio-Flex® and Terramac®, respectively. Although both the manufacturers have stressed that their products are based on Ingeo™, some modifications or additives have been incorporated into the product to improve the original properties of the PLA. It can be seen from Tables 2.9—2.11 that the heat distortion/deflection temperature of the Terramac® series is higher than that for Ingeo™. A higher heat distortion/deflection temperature is crucial for certain products, particularly food serviceware for hot food and drink. Bio-Flex® (Table 2.12) also has different properties to Ingeo™, after converting unit of analysis. The improvements to PLA made by other manufacturers are considered to be positive moves to enable PLA to fulfill a wide range of market needs. In its series of Terramac® products, Unitika has also included a foam and emulsion of PLA. The foam PLA is targeted to replace Styrofoam, while reducing the environmental pollution. The emulsion grade of PLA is suitable as a coating agent. Similarly, Toyobo's PLA under the trade name of Vyloecol® is mainly produced to be used as a general-purpose coating agent (Table 2.13).

In addition to converting and improving Ingeo®, Zhejiang Hisun Biomaterial has produced two other grades, REVODE201 and REVODE101 (Table 2.14), for injection molding and extruded sheet thermoforming applications respectively, from its facility located in China. The Galactic and Total Petrochemical joint venture has introduced Futerro® polylactide consisting of three grades, for thermoforming, fiber, and injection molding applications (Table 2.15). Other manufacturers such as Mitsui, Teijin, Purac, Toray, and some Chinese manufacturers lack data about their product grades. This might be due to the manufacturer's technology still being in the pilot stage and, therefore, yet to produce detailed specifications prior to mass production for the market.

Table 2.6 NatureWorks PLA Grades for Thermoform and Injection Molding (Data: NatureWorks)

Grade	2003D	3001D	3051D	3251D	3801X
Specific gravity	1.24[1]	1.24[1]	1.25[1]	1.24[1]	1.33[1]
Melt index (g/10 min)	5–7[2]	10–30[3]	10–25[2]	70–85[2]	8[3]
Tensile strength at break (MPa)	53[4]	—	—	—	—
Tensile yield Strength (MPa)	60[4]	48[5]	48[5]	48[5]	25.9[5]
Tensile modulus (MPa)	3500[4]	—	—	—	2980[5]
Tensile elongation (%)	6[4]	2.5[5]	2.5[5]	2.5[5]	8.1[5]
Notched Izod impact (J/m)	12.81[6]	0.16[6]	0.16[5]	0.16[5]	144[6]
Flexural strength (MPa)	—	83[7]	83[7]	83[7]	44[7]
Flexural modulus (MPa)	—	3828[7]	3828[7]	—	2850[7]
Crystalline melt temperature (°C)	—	—	150–165[8]	—	160–170[8]
Glass transition temperature (°C)	—	—	55–65[9]		
Applications	General extrusion for thermoform production of food packaging, dairy containers, food serviceware, transparent containers, hinged ware and cold drink cups	Injection molding applications for clear cutlery, cups, plates, etc. with heat deflection temperature <55 °C	Injection molding applications with the requirement for clarity and heat deflection temperature <55 °C	Injection molding applications with higher melt flow capability. High gloss, UV-resistance and stiffness	Injection molding for high-heat and high-impact applications. More rapid crystallization kinetics for shorter cycle time. Application at heat deflection temperatures 65–140 °C without food contact

[1]ASTM D792[2]ASTM D1238 (210 °C/2.16 kg)[3]ASTM D1238 (190 °C/2.16 kg)[4]ASTM D882[5]ASTM D638[6]ASTM D256[7]ASTM D790[8]ASTM D3418[9]ASTM D3417

Table 2.7 NatureWorks PLA Grades for Films and Bottles (Data: NatureWorks)

Grade		4043D	4060D	7001D	7032D
Density (g/cm^3)		1.24[1]	1.24[1]	1.24[2]	1.24[2]
Melt index (g/10 min)		—	—	5–15[3]	5–15[3]
Tensile strength	MD (kpsi)	16[4]	—	—	—
	TD (kpsi)	21[4]	—	—	—
Tensile modulus	MD (kpsi)	480[4]	—	—	—
	TD (kpsi)	560[4]	—	—	—
Elongation at break	MD (%)	160[4]	—	—	—
	TD (%)	100[4]	—	—	—
Elmendorf tear	MD (g/mil or g/25 μm)	15[5]	—	—	—
	TD (g/mil or g/25 μm)	13[5]	—	—	—
Transmission rate	Oxygen (cc mil/m^2/24 h atm or cm^3 25 μm/m^2/24 h atm)	550[6]	550[6]	550[6]	550[6]
	Carbon dioxide (cc mil/m^2/24 h atm or cm^3 25 μm/m^2/24 h atm)	3000[6]	330[6]	3000[6]	3000[6]
	Water vapor (g mil/m^2/24 h atm or g 25 μm/m^2/24 h atm)	325[7]	325[7]	325[7]	325[7]
Optical characteristics	Haze (%)	2.1[8]	2[8]	—	—
	Gloss (20°)	90[8]	90[8]	—	—
Thermal characteristics	Melting point (°C)	135[8]	—	145–155[9]	160[9]
	Glass transition temperature (°C)	—	52–58[9]	52–58[10]	55–60[10]
	Seal initiation temperature (°C)	—	80[11]	—	—
Application		Biaxial oriented film application Excellent optics, twist, and deadfold Barrier to flavor, grease, and superior oil resistance	For heat seal layer in coextruded oriented films Excellent heat seal and hot tack	Injection stretch blow molded bottles Potential for fresh dairy, edible oil, fresh water and liquid hygiene products	Injection stretch blow molded bottles. Ideal for applications requiring heat setting—fruit juices, sports drinks, jams, jellies

[1]ASTM D1505[2]ASTM D792[3]ASTM D1238 (210 °C/2.16 kg)[4]ASTM D882[5]ASTM D1922[6]ASTM D1434[7]ASTM E96[8]ASTM D1003[9]ASTM D3418[10]ASTM D3417

Table 2.8 NatureWorks PLA Grades for Fiber Application (Data: NatureWorks)

Grade	5051X	6060D	6201D	6202D	6204D	6251D	6302D	6550D	6400D	6751D
Specific gravity	1.24[1]	1.24[1]	1.24[1]	1.24[1]	1.24[1]	1.24[1]	1.24[1]	1.24[1]	1.24[1]	1.24[1]
Melt index (g/10 min)	—	10[2]	15–30[2]	15–30[2]	15–30[2]	70–85[2]	20[2]	65[2]	4–8[2]	15[2]
Crystalline melt temperature (°C)	145–155[3]	125–135[3]	160–170[3]	160–170[3]	160–170[3]	160–170[3]	125–135[3]	145–160[3]	160–170[3]	150–160[3]
Glass transition temperature (°C)	55–65[4]	55–60[4]	55–60[4]	55–60[4]	55–65[4]	55–60[4]	55–60[4]	55–60[4]	55–60[4]	55–60[4]
Denier per filament	>1.5	>4	>0.5	>0.5	>0.5	1–2	>4	—	10–20	>1.5
Tenacity (g/d)	2.5–4.0[5]	3.5[5]	2.5–5.0[5]	2.5–5.0[5]	2.5–5.0[5]	—	3.5[5]	—	2.0–2.4[5]	2.5–4.0[5]
Elongation (%)	10–70[5]	50[5]	10–70[5]	10–70[5]	10–70[5]	—	50[5]	—	10–70[5]	10–70[5]
Modulus (g/d)	20–40[5]	—	30–40[5]	30–40[5]	30–40[5]	—	—	—	—	20–40[5]
Hot air shrinkage (%)	<8[6]	—	5–15[6]	<8[7]	5–15[7]	—	—	—	—	8
Application	Nonwoven spunlace wipes	Low melt binder polymer in a sheath-core configuration. Good for thermal bonded nonwovens	Woven and knitted 100% continuous filament apparel, intimate staple blend fabrics, including blends with cotton, wool, or other fibers; for home furnishings and civil engineering applications	Fiberfill, nonwovens, agricultural woven, and nonwoven fabrics, articles for household disposal	Woven and knitted 100% continuous filament apparel, intimate staple blend fabrics, including blends with cotton, wool, or other fibers; for home furnishings and civil engineering applications	Suitable for wipes, geotextiles, hospital garments, absorbent pad liners, and personal hygiene products, agricultural/horticultural products	Low melt binder polymer in a sheath-core configuration. Good for thermal bonded nonwovens	For extrusion into spunbond nonwovens using conventional bi-component PET spunbond equipment, where filament velocities >4000 m/min	For bulk continuous filament, tufted carpet-loop/cut pile, broad loom carpets and carpet mats	Suitable for nonwoven (spunlace wipes) and multifilament twine

[1]ASTM D792[2]ASTM D1238 (210 °C/2.16 kg)[3]ASTM D3418[4]ASTM D3417

Table 2.9 Unitika–Terramac® PLA Grades for Injection Molding

Grade	ISO	Basic Grade TE-2000	High Impact Grade TE-1030	High Impact Grade TE-1070	Heat-Resisting Grade TE-7000	Heat-Resisting Grade TE-7307	Heat-Resisting Grade TE-7300	High-Durability Grade TE-8210	High-Durability Grade TE-8300
Density	1183	1.25	1.24	1.24	1.27	1.42	1.47	1.42	1.47
Melting point (°C)	—	170	170	170	170	170	170	170	170
Breaking strength (MPa)	527	63	51	34	70	54	54	50	56
Tensile elongation (%)	527	4	170	>200	2	2	1	2	1
Blending strength (MPa)	178	106	77	50	110	85	98	90	104
Bending modulus (GPa)	178	4.3	2.6	1.4	4.6	7.5	9.5	6.8	9.3
Charpy impact strength: with notch (kJ/m^2)	179	1.6	2.3	5.6	2.0	2.5	2.4	4.0	2.8
Deflection temperature under load of 0.45 MPa (°C)	75	58	51	54	11-	120	140	120	140
Molding shrinkage (%)	—	0.3–0.5	0.3–0.5	0.3–0.5	1.0–1.2	1.0–1.2	1.0–1.2	1.0–1.2	1.0–1.2

Table 2.10 Unitika–Terramac® PLA Grades for Extrusion, Blow, and Foam Sheet

Grade	ISO	Basic Grade TP-4000	Soft TP-4030	Foam HV-6250H
Density	1183	1.25	1.24	1.27
Melting point (°C)	—	170	170	170
Breaking strength (MPa)	527	66	50	69
Tensile elongation (%)	527	5	44	2
Bending strength (MPa)	178	108	71	111
Bending modulus (GPa)	178	4.6	2.4	4.7
Charpy impact strength: with notch (kJ/m^2)	179	1.6	2.6	1.9
Deflection temperature under load of 0.45 MPa (°C)	75	59	52	120
Molding shrinkage (%)	—	3–5	3–5	1–3

2.3.2 PLA and Copolymers for Biomedical Applications

In addition to the usage of PLA for the production of environmentally friendly domestic articles to substitute existing petrochemical-based plastic products, PLA is also widely used in the biomedical field, for the production of bioresorbable implants and devices. Most of the PLA in biomedical applications is produced from L-lactic acid. The implants made of poly(L-lactide) can be easily degraded and resorbed by the body through the action of enzymes. Unfortunately, the stereoisomer D-lactic acid is not degraded by the body's enzymes. However, prolonged hydrolysis in body fluids eventually breaks down the bulk of poly(D-lactide). This degradation mechanism is discussed in Chapter 3.

A considerable amount of PLA copolymer is synthesized for tissue engineering. The main objective during the synthesizing of such copolymers is to fine-tune the period of degradation from weeks to years [21]. Commonly, the monomer of glycolide acid and ϵ-caprolactone are copolymerized with lactide. As can be seen from Table 2.16, when *in vitro* at 37 °C, the mass of poly(L-lactide) is significantly increased after being copolymerized with glycolide and ϵ-caprolactone. This is very important for the fabrication of scaffold for tissue engineering and for wound dressings. The degradation of the copolymer is designed to couple with the growth of tissue and the loss of mass and strength of the prescribed implants. Eventually, the scaffold structure is substituted by the permanent tissue of the patient.

Table 2.11 Unitika–Terramac® PLA Grade for Emulsion

Grade	Standard Type LAE-013N
Solids content concentration (wt%)	50–55
pH	3.5–5.5
Particle diameter (μm)	<1
Viscosity (mPa s)	300–500
Lowest film-foaming temperature (°C)	60–70

PLA and its copolymers can be used for a wide range of biomedical applications such as sutures, anchors, screws, and scaffolds. They have uses in oral, orthopedic, auricular, and craniofacial augmentations in plastic surgery (Table 2.17). Screws and anchors are produced by the injection molding method, and sutures are manufactured using a fiber spinning process. Bioresorbable scaffolds are prepared using a range of techniques, including phase separation, solvent evaporation, casting/salt leaching, and fiber bonding to form a polymer mesh. PLA copolymers are also widely used as a drug carrier medium (Table 2.18). Such drug carriers contain active drugs, which can be efficiently delivered to the target cells and subsequently released at a controlled rate [23,24]. One of the best-known products on the market, Zoladex®, is a polylactide-co-glycolide with a formulation of goserelin as a controlled release drug for the treatment of breast cancer [25]. Zoladex®

Table 2.12 FKuR Kunststoff GmbH PLA Specification

Grade	Test Method	Bio-Flex® A 4100 CK	Bio-Flex® F 1110	Bio-Flex® F 1130	Bio-Flex® F 2110	Bio-Flex® F 6510	Bio-Flex® S 5630	Bio-Flex® S 6540
Tensile modulus of elasticity (MPa)	ISO 527	1840	230	390	730	2.600	2160	2800
Tensile strength (MPa)	ISO 527	44	16	17	20	47	32	31
Tensile strain at tensile strength (%)	ISO 527	5	>300	>300	>300	4	6	5
Tensile stress at break (MPa)	ISO 527	22	No break	No break	No break	23	29	28
Tensile strain at break (%)	ISO 527	12	No break	No break	No break	19	9	7
Flexural modulus (MPa)	ISO 178	1770	215	370	680	2.650	2400	2890
Flexural strain at break (%)	ISO 178	No break	No break	No break	No break	No break	No break	6
Flexural stress at 3.5% (MPa)	ISO 178	48	6	9	17	64	46	50
Notched impact strength (Charpy), RT (kJ/m^2)	ISO 179-1/1 eA	3	No break	No break	83	7	3	3
Impact strength (Charpy), RT (kJ/m^2)	ISO 179-1/1 eU	44	No break	No break	No break	No break	51	36
Density (g/cm^3)	ISO 1183	1.24	1.28	1.40	1.27	1.30	1.55	1.62
Melt temperature (°C)	ISO 3146-C	>155	>155	>155	145–160	150–170	140–160	110–15-
Vicat A softening temperature (°C)	ISO 306	44	68	89	78	60	105	105

Heat distortion temperature HDT B (°C)	ISO 75	40	n/a	n/a	n/a	n/a	68	n/a
Melt flow rate—190 °C/2.16 g (g/10 min)	ISO 1133	10–12	2–4	2–4	3–5	2.5–4.5	10–12	8–10
Water vapor (g/m^2 d)	ISO 15 106-3	170	—	70	130	130	—	
Oxygen (cm^3/(m^2 d bar))	ISO 15 105-2	130	—	850	1450	1.060	—	
Nitrogen—25 μm film (cm^3/(m^2/d/bar))	DIN 53380-2	65	—	160	230	150	–	
Application		Film extrusion	Film extrusion	Film extrusion	Film extrusion	Film extrusion	Thermoforming and injection molding	Injection molding

Table 2.13 Toyobo PLA Specification

Grade	Vyloecol BE-400	Vyloecol BE-600
Form	Pellet	Sheet
Molecular weight	43,000	25,000
Specific gravity at 30 °C	1.26	1.24
T_g (°C)	50	30
Hydroxyl group value KOH (mg/g)	3	11
Features and applications	General purpose grade, agent for various coatings	Anchor coating for vapor deposition film, anchor coating for printing ink

allows the slow release of the drug, which inhibits the growth of cancer cells that are hormone dependent. U.S. Food and Drug Administration also approved Zoladex® for the treatment of prostate cancer. A couple of other PLA-copolymer-related drug delivery systems are widely available on the market.

Purac is the main global company actively involved in producing biomedical and drug delivery grade PLA and copolymers; they are marketed as Purasorb®. Durect Corporation also markets a bio-absorbable polymer under the trade name Lactel®. As can be seen from the grade specification of both

Table 2.14 Hisun Biomaterial PLA Specification

Grade	Test Method	REVODE201	REVODE101
Specific gravity	GB/T1033-1986	1.25 ± 0.05	1.25 ± 0.05
Melt index—190 °C/2.16 kg (g/10 min)	GB/T3682-2000	10–30	2–10
Melting point (°C)	GB/T19466.3-2004	137–155	140–155
Glass transition temperature (°C)	GB/T19466.2/2004	57–60	57–60
Tensile strength (MPa)	GB/T1040-1992	45	50
Tensile elongation (%)	GB/T1040-1992	3.0	3.0
Impact strength (kJ/m^2, Izod)	GB/T1040-1992	1–3	1–3
Applications		For injection molding, including cutlery, toys, plates, cups, etc.	Easily processed using conventional extrusion equipment for producing sheet ranging between 0.2–10 mm in thickness for thermoforming. Suitable for dairy containers, food serviceware, transparent food containers, and cold drink cups

Table 2.15 Futerro PLA Specification

Grade	Test Method	Futerro Polylactide —Extrusion Grade	Futerro Polylactide —Fiber Melt Spinning Grade	Futerro Polylactide —Injection Grade
Specific gravity at 25 °C	ISO 1183	1.24	1.24	1.24
Melt index—190 °C/2.16 kg (g/10 min)	ISO 1133	2–4	10–15	10–30
Haze—2 mm (%)	ISO 14782	<5	<5	<5
Glass transition temperature (°C)	ISO 11357	52–60	52–60	52–60
Crystalline melt temperature (°C)	ISO 11357	145–175	145–175	145–175
Tensile strength at break (MPa)	ISO 527	55	55	55
Tensile yield strength (MPa)	ISO 527	60	60	60
Tensile modulus (MPa)	ISO 527	3500	3500	3500
Tensile elongation (%)	ISO 527	6.0	6.0	6.0
Notched Izod impact (kJ/m^2)	ISO 180	3.5	3.5	3.5
Flexural yield strength (MPa)	ISO 178	90	90	90
Application		For extrusion and thermoforming applications	For extrusion into mechanically drawn staple fibers or continuous filament. Potential for woven and knitted apparel, fabrics or netting for civil engineering applications	For injection molding applications with deflection temperatures <55 °C

Table 2.16 Physical Properties of Synthetic Biodegradable Polymers Used as Scaffolds in Tissue Engineering [21]

Polymer	**Poly(glycolide)**	**Poly (L-lactide)**	**Poly (ϵ-caprolactone)**	**Copolymer of L-lactide and glycolide (10:90)**	**Copolymer of L-lactide and ϵ-caprolactone (75:25)**	**Copolymer of (L-lactide)**
T_m (°C)[1]	230	170	60	200	130–150	90–120
T_g (°C)[2]	36	56	−60	40	15–30	−17
Shape	Fiber	Fiber, sponge, film	Fiber, sponge, film	Fiber	Fiber, sponge, film	Fiber, sponge, film
Tensile strength (MPa)	890 (fiber)	900 (fiber)	10–80 (fiber)	850 (fiber)	500 (fiber)	12 (film)
Young's modulus (GPa)	8.4 (fiber)	8.5 (fiber)	0.3–0.4 (fiber)	8.6 (fiber)	4.8 (fiber)	0.9 (film)
Elongation at break (%)	30 (fiber)	25 (fiber)	20–120 (fiber)	24 (fiber)	70 (fiber)	600 (fiber)
P_{wo}[3]	2–3 months	3–5 years	>5 years	10 weeks	1 year	6–8 months
P_{t50}[4]	2–3 weeks	6–12 months	—	3 weeks	8–10 weeks	4–6 weeks

[1]Melting point[2]Glass transition temperature[3]Period until the polymer mass becomes zero (in saline at 37 °C)[4]Period until tensile strength of polymers becomes 50% (in saline at 37 °C)

Source: With permission from Marcel Dekker

Table 2.17 PLA in Biomedical Applications

Polymer	Area of Application	Products
Poly(lactide)	Orthopedic surgery, oral, and maxillofacial surgery 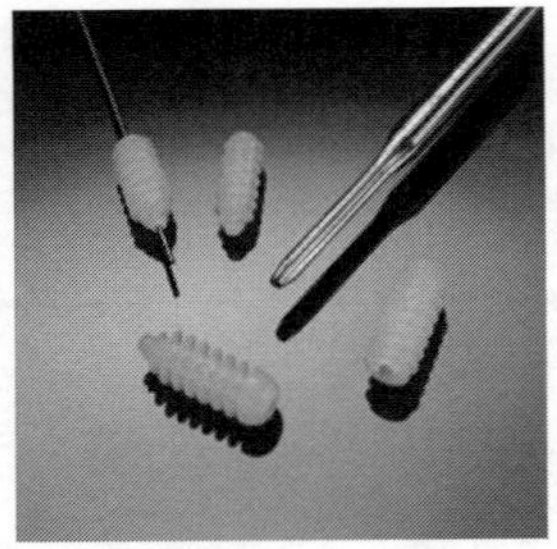	Takiron: Osteotrans™ MX, Fixsorb™ MX (screws, nails, pins) Gunze: Grandfix®, Neofix® (screws, nails, pins) Arthrex: Bio-Tenodesis® (interference screw), Bio-Corkscrew® (suture anchor) Conmed Linvatec: SmartScrew®, SmartNail®, SmartTack®, SmartPin® BioScrew® Stryker: Biosteon®, Biozip® (interference screw, anchor) Zimmer: Bio-Statak® (suture anchor), prostatic stent, suture anchor, bone cement plug Dermik Laboratories: Sculptra® (injectable facial restoration) Kensey Nash: EpiGuide®
Poly(D,L-lactide-co-glycolide) Poly (D,L-lactide-co-glycolide) 85/15 Poly(D,L-lactide-co-glycolide) 82/18 Poly(D,L-lactide-co-glycolide) 10/90	Sutures Drug delivery Oral and maxillofacial surgery General surgery Sutures, periodontal surgery, general surgery 	USS Sport Medicine: Polysorb™ sutures Instrument Makar: Biologically Quiet™ (interference screw) Staple 85/15 Biomet: ALLthread™ LactoSorb®, screw, plates, mesh, surgical clip, pins, anchor Ethicon: Vicryl suture, Vicryl mesh

(*Continued*)

Table 2.17 PLA in Biomedical Applications—*Cont'd*

Polymer	Area of Application	Products
Poly(L-lactide-co-D,L-lactide) 98/2 Poly(L-lactide-co-D-lactide) 96/4 Poly(L-lactide-co-D,L-lactide) 50/50 Poly(L-lactide-co-D,L-lactide) 70/30 Poly(D-lactide-co-D,L-lactide-co-L-lactide)	Orthopedic surgery Oral and maxillofacial surgery 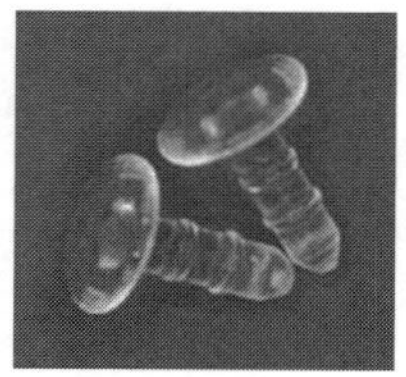	Phusiline® interference screw, Sage ConMed: Bio-Mini Revo® Sulzer: Sysorb® screw (50/50) Geistlich: ResorPin® 70/30 Kensey Nash: Drilac® Surgical dressing
Poly(D,L-lactide-co-caprolactone)	Nerve regeneration 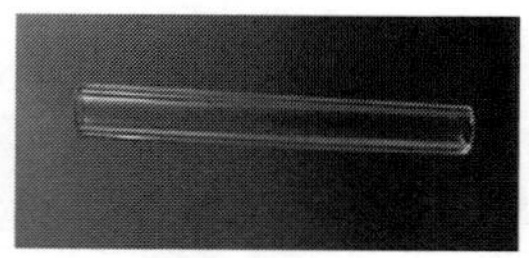	Ascension Orthopedics: Neurolac® Polyganics: Vivosorb®

Table 2.18 List of Commercially Available PLA and Copolymer Delivery Carriers and the Corresponding Therapeutic and Its Indication

Delivery System	Material Composition	Product Name	Therapeutic	Type of Drug: Indications
Microspheres	PLA (poly(lactic acid))	Lupron Depot	Leuprolide acetate	Peptide hormone: cancer and Alzheimer's
	PLGA (polylactide-glycolide)	Eligard	Leuprolide acetate	Peptide hormone: cancer and Alzheimer's
		Risperdal Consta	Risperidone	Peptide: schizophrenia
		Trelstar LA	Triptorelin pamoate	Peptide hormone: prostate cancer
	PLGA-glucose	Sandostatin LAR	Octreotide	Peptide: anti-growth hormone
Implant	PLGA	Durin	Leuprolide	Peptide hormone: cancer and Alzheimer's
		Zoladex	Goserelin acetate	Peptide hormone: prostate/ breast cancer
Gel	PLGA	Oncogel	Paclitaxel	Small molecule: anticancer

Source: Extracted from Ref. [22]

manufacturers (Tables 2.19–2.21), the PLGA copolymer is the most widely produced grade. All grades are tested for their intrinsic viscosities as guidance on the molecular weight of the synthesized polymer. This is very important in biomedical applications, as it ensures the rate of resorption in the body. When the polymer is exposed to aqueous media or tissue, the ester linkages of the polymer react with the absorbed water through a hydrolysis reaction. Over time, the long polymer chains are broken into shorter ones to form water-soluble fragments. Eventually, the water-soluble fragments diffuse away from the initial polymer structure and finally hydrolyze to glycolic and lactic acid for metabolism by the liver. Generally, the rate of degradation is higher at low molecular weights and for higher glycolide content [26]. Overall, PLA and copolymers have contributed significantly to the medical industry.

2.4 Environmental Profile of PLA

PLA is produced from renewable agricultural sources, hence it is known for its eco-friendliness. Though the technology used by NatureWorks for mass production is from corn, sugarcane is also used as the input for producing lactic acid. Sugarcane-based production of lactic acid has been developed by Purac, with the setting up of a commercial lactic acid plant in Thailand. In general, PLA is produced using a direct polycondensation reaction and ring-opening polymerization approaches. The majority of commercial producers find that ring-opening polymerization is preferable for better control of the process and better production quality.

In the environmental credit analysis of PLA, there are two major aspects that need to be considered—the PLA manufacturing process and the postconsumer PLA product disposal. Several research projects on life cycle analysis of PLA mass production have been conducted in recent years. Two of the life cycle analyses of PLA production have been undertaken by NatureWorks and Purac. The objective here is to summarize these studies rather than directly perform life cycle analysis of PLA. More detailed information can be found in the relevant publications [27–30].

2.5 Eco-profile of PLA in Mass Production

PLA is produced from sugar fermentation by bacteria. The source of sugar is starch, and this currently comes mainly from corn and cassava.

Table 2.19 Purac Purasorb® PLA for Medical Devices

Grade	Structure	Inherent Viscosity Midpoint (dl/g)
Purasorb PL 18	Poly(L-lactide)	1.8
Purasorb PL 24		2.4
Purasorb PL 32		3.2
Purasorb PL 38		3.8
Purasorb PL 49		4.9
Purasorb PL 65		6.5
Purasorb PD 24	Poly(D-lactide)	2.4
Purasorb PDL 45	Poly(DL-lactide)	4.5
Purasorb PLDL 8038	80/20 L-lactide/DL-lactide copolymer	3.8
Purasorb PLDL 8058		5.8
Purasorb PLDL 7028	70/30 L-lactide/DL-lactide copolymer	2.8
Purasorb PLDL 7038		3.8
Purasorb PLDL 7060		6.0
Purasorb PLD 9620	96/04 L-lactide/D-lactide copolymer	2.0
Purasorb PLD 9655		5.5
Purasorb PLG 8523	85/15 L-lactide/glycolide copolymer	2.3
Purasorb PLG 8531		3.1
Purasorb PLG 8560		6.0
Purasorb PLG 8218	82/18 L-lactide/glycolide copolymer	1.8
Purasorb PLG 8055	80/20 L-lactide/glycolide copolymer	5.5
Purasorb PLG 1017	10/90 L-lactide/glycolide copolymer	1.7
Purasorb PLC 9517	95/05 L-lactide/caprolactone copolymer	1.7
Purasorb PLC 9538		3.8
Purasorb PLC 8516	85/15 L-lactide/caprolactone copolymer	1.6
Purasorb PLC 7015	70/30 L-lactide/caprolactone copolymer	1.5
Purasorb PDLG 8531	85/15 DL-lactide/glycolide copolymer	3.1
Purasorb PDLG 5010	50/50 DL-lactide/glycolide copolymer	1.0

NatureWorks grows corn to produce starch as the input for their PLA production, while Purac uses cassava to produce PLA, using the Synbra–Sulzer Chemtech technology. Both the technologies utilize the fermentation approach to produce lactic acid. This is followed by transforming lactic acid into lactide and finally undergoing ring-opening polymerization into PLA.

According to Ref. [30], the initial technology of NatureWorks required 54.1 MJ of fossil energy to produce every kilogram of Ingeo™ PLA. Fossil energy is used for running the factory, transportation of corn to the wet mill, wastewater treatment, etc. Although the combustion energy of corn residue is a renewable energy, it merely contributes 34.4% of the overall energy required (82.5 MJ/kg of PLA) in the plant. Figure 2.10 shows the gross energy required to produce PLA by NatureWorks' first generation technology. Energy is required to operate supplies such as fertilizers and pesticides for growing the corn (total 3.8 MJ/kg of PLA) as well as transportation of the corn to the wet mill and related wastewater treatment throughout the production

Table 2.20 Purac Purasorb® PLA for Drug Delivery

Grade	Structure	Intrinsic Viscosity Midpoint (dl/g)
Purasorb PDL 02A—acid terminated	Poly(DL-lactide)	0.2
Purasorb PDL 02		0.2
Purasorb PDL 04		0.4
Purasorb PDL 05		0.5
Purasorb PDL 20		2.0
Purasorb PDLG 7502	75/25 DL-lactide/glycolide copolymer	0.2
Purasorb PDLG 7502A—acid terminated	75/25 DL-lactide/glycolide copolymer	0.2
Purasorb PDLG 7507	75/25 DL-lactide/glycolide copolymer	0.7
Purasorb PDLG 5002	50/50 DL-lactide/glycolide copolymer	0.2
Purasorb PDLG 5002A—acid terminated	50/50 DL-lactide/glycolide copolymer	0.2
Purasorb PDLG 5004	50/50 DL-lactide/glycolide copolymer	0.4
Purasorb PDLG 5004A—acid terminated	50/50 DL-lactide/glycolide copolymer	0.4
Purasorb PDLG 5010	50/50 DL-lactide/glycolide copolymer	1.0

Table 2.21 Durect Lactel® Absorbable Polymer

Grade	Chemical Name	Inherent Viscosity Midpoint (dl/g)
B6017-1	50:50 Poly(DL-lactide-co-glycolide)	0.2
B6010-1	50:50 Poly(DL-lactide-co-glycolide)	0.4
B6010-2	50:50 Poly(DL-lactide-co-glycolide)	0.65
B6010-3	50:50 Poly(DL-lactide-glycolide)	0.85
B6001-1	65:35 Poly(DL-lactide-co-glycolide)	0.65
B6007-1	75:25 Poly(DL-lactide-co-glycolide)	0.65
B6006-1	85:15 Poly(DL-lactide-co-glycolide)	0.65
B6005-1	Poly(DL-lactide)	0.40
B6005-2	Poly(DL-lactide)	0.65
B6002-2	Poly(L-lactide)	1.05
B6013-1	50:50 Poly(DL-lactide-co-glycolide)	0.20
B6013-2	50:50 Poly(DL-lactide-co-glycolide)	0.65
B6015-1	25:75 Poly(DL-lactide-co-ϵ-caprolactone)	0.8
B6016-1	80:20 Poly(DL-lactide-co-ϵ-caprolactone)	0.8

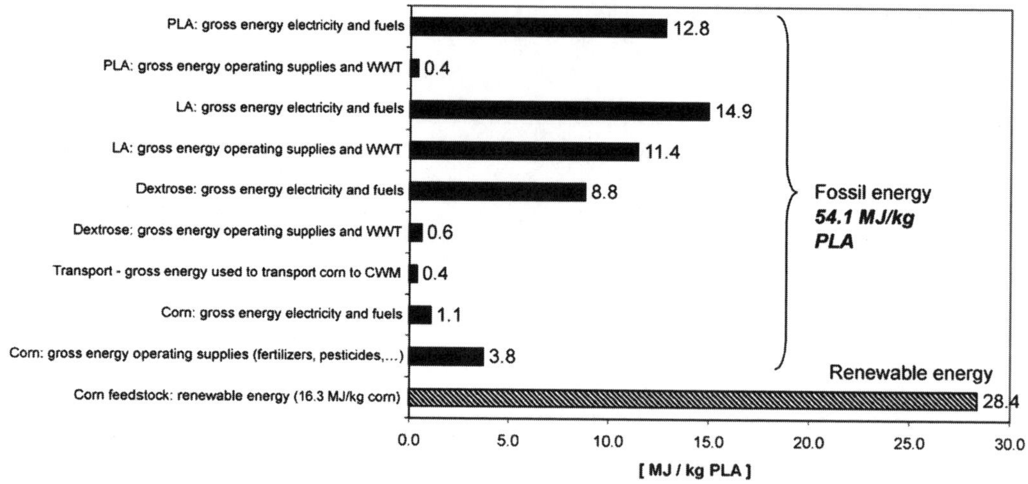

Figure 2.10 Energy requirement for the production of NatureWorks' first generation PLA [30]. LA= lactic acid, WWT= waste water treatment, CWM= corn wet mill. *Published with permission of Elsevier.*

process. All these operations require an external supply of energy, because it is not possible to self-supply using the heat of combustion of the corn residue. Most people think that PLA is a novel environmentally friendly polymer. However, this usage of fossil energy still generates greenhouse gases. Nevertheless, PLA is still worthy of exploration due to its fully biodegradable nature when disposed of in the natural environment. In fact, the biodegradability of PLA is its most important selling point in the domestic market.

Despite the fact that the gross fossil energy consumption is considered high (<50% of the total energy to produce per kilogram of PLA), when NatureWorks first-generation PLA is compared to a petrochemical polymer, PLA remains its outstanding production characteristics [30]. Researchers compared ten commercially available polymers with first generation of Ingeo™ and found that PLA consumed the least fossil energy (Fig. 2.11). Over the years, NatureWorks has shown initiative by maximizing the usage of biomass as well as wind power to reduce dependence on fossil fuel. NatureWorks has highlighted that the advances of second-generation PLA technology manages to capture more free carbon in the air. The production of second generation PLA can achieve a negative emission impact to protect the environment against global warming. The second generation Ingeo™ production system in 2006 emitted 0.27 kg CO_2 eq/kg PLA and used 27.2 MJ/kg PLA of fossil energy. This represents a reduction of 85 and 50% respectively when compared to Ingeo's 2003 eco-profile data [29]. In an announcement in early 2009, NatureWorks claimed that Ingeo™ production had been further improved with greenhouse gas emissions lowered by 36% and nonrenewable energy utilization reduced by 44% compared to data from 2005. The latest Ingeo™ technology generates 2.24 kg CO_2 eq/kg of Ingeo™ (Fig. 2.12) and uses 42 MJ of nonrenewable energy [28]. At the same time, the gross water saving for the production of PLA is encouraging compared to the majority of petrochemical polymers (Fig. 2.13). However, the total gross water required for amorphous PET production is slightly lower than for PLA. This is because the production of PLA uses an agricultural source, which needs water for irrigation. Furthermore, the fermentation and wastewater treatment also require plenty of water. Thus, water is considered an unavoidable input for the production of PLA.

Purac's technology uses sugarcane as the feedstock for lactic acid production. Purac's lactic acid facility in Thailand has been in operation since 2007. The lactic acid is scheduled for conversion into lactide once the new large-scale plant is ready in 2011. During the developmental stage, most of the lactic acid has been exported for conversion at Purac's lactide plant in Spain. Reference [27] in life cycle assessment of lactide and PLA production from sugarcane in Thailand reported that every ton of PLA emits 500 kg CO_2. Although alternative renewable energy can be obtained through the burning of sugarcane bagasse—in the range of 17–95 kWh/MT of sugarcane, Ref. [27] points out that environmental credit varies, depending on the type of byproducts, combustion technology, and the mix of energy in application. In other words, every source of PLA has

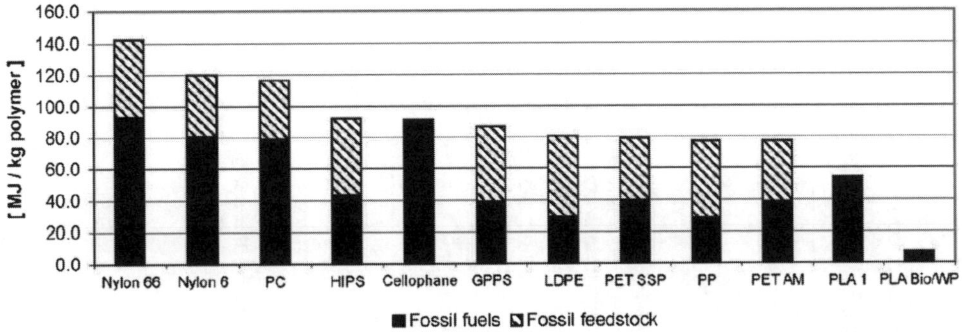

Figure 2.11 Fossil energy requirement for petrochemical polymers and PLA. The cross-hatched area of the bars represents the fossil energy used as chemical feedstock (i.e., fossil resource to build the polymer chain). The solid part of the bars represented the gross fossil energy used for the fuels and operation supplies used to drive the production processes. PC = polycarbonate; HIPS = high-impact polystyrene; GPPS = general purpose polystyrene; LDPE = low-density polyethylene; PET SSP = polyethylene terephthalate, solid-state polymerization (bottle grade); PP = polypropylene; PET AM = polyethylene terephthalate, amorphous (fiber and film grade); PLA = PLA first generation; PLA B/WP (PLA, biomass/wind power scenario). *Adapted from Ref. [30].*

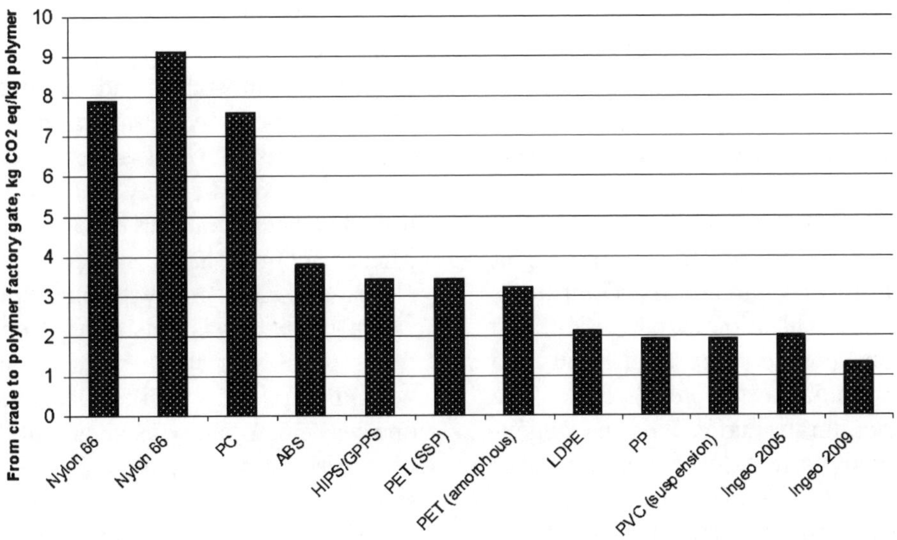

Figure 2.12 Contribution of petrochemical polymers and Ingeo™ PLA to global climatic change. *Adapted from Ref. [28].*

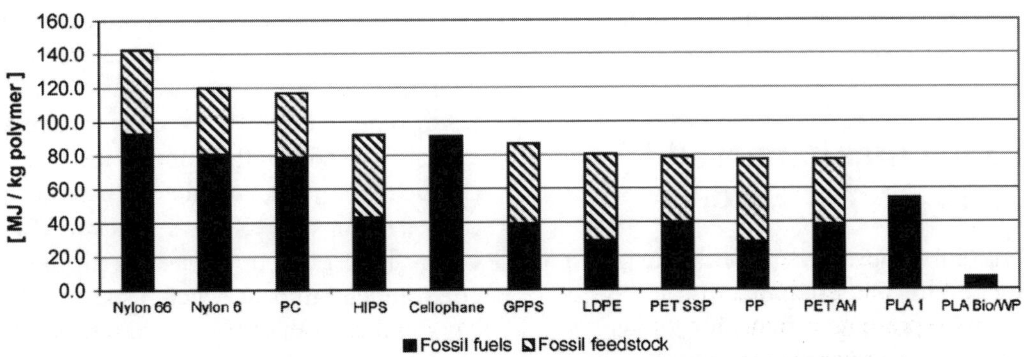

Figure 2.13 Gross water used in the production of petrochemical polymers and PLA. *Adapted from Ref. [30].*

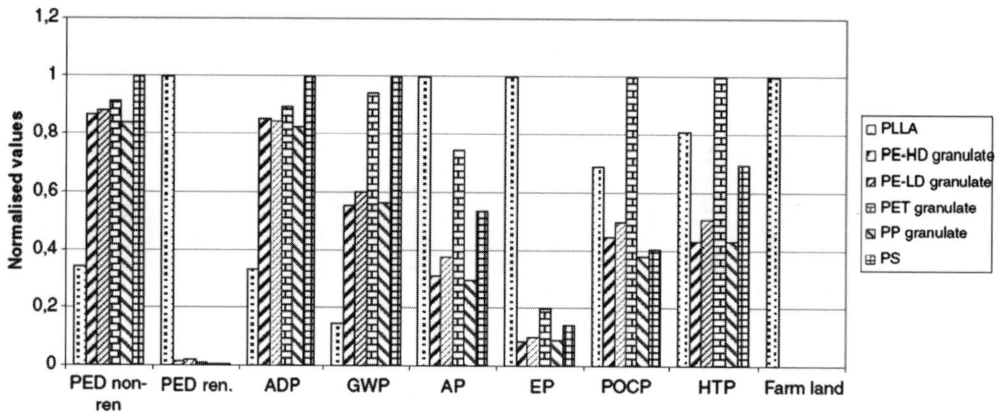

Figure 2.14 Comparison of the most relevant ecological factors involved in the production of PLLA and fossil-based derived polymers. PED = primary renewable energy; PED nonren = primary nonrenewable energy; GWP = global warming potential; AP = acidification potential; EP = eutrophication potential; POCP = photochemical ozone creation potential; ADP = abiotic resource depletion potential; HTP = human toxicity potential. *(Adapted from Ref. [27].)*

a unique eco-profile. Thus, it is of utmost importance to develop a green PLA through the careful selection of processes. The environmental impact of PLA is shown in Fig. 2.14 together with that of some of the petrochemical polymers. It is clear that some ecological aspects of PLA production need improvement to become greener. The most detrimental impact scores of PLA belong to the process of sugarcane cultivation and transformation into sugar. In addition, the farming of sugarcane also contributes significantly to the eutrophication, acidification, and photochemical ozone creation due to the nitrogen emission of ammonia-based fertilizers. The combustion of agricultural residues for the cogeneration operation tends to release greenhouse gases such as NO_x, SO_x, and CO. Some of the related soil activity by microorganism can cause emission of NO_x and methane as well. PLA is the one polymer that causes effects on farmland due to the continuous replanting, resulting in soil erosion and loss of natural nutrients. As a result, precautions and environmental assessment need to be conducted prior to deforestation for the farming of sugarcane.

2.6 Environmental Impact of PLA at the Postconsumer Stage

PLA is a suitable substitute for existing petrochemical polymers in the manufacture of cups, containers, and packaging. PLA is known to degrade well when disposed along with municipal waste, and so is less of a burden to the environment. Unlike petrochemical polymers such as PE, PP, PET, PC, and PS, which require 100 years to breakdown into harmless substances, PLA is fully compostable and is accepted as a green product, especially in Japan, the United States, and EU countries. Several reports have been published about the eco-efficiency of PLA postconsumer, and this has been compared to conventional plastics. These reports have included PLA cups [31], clamshells [32], and wrappings [33].

An eco-analysis was carried out comparing four types of plastic cups—the reusable PC cup, one-way PP cup, one-way PE-coated cardboard cup, and one-way PLA cup—used at public events held in Flanders (Belgium). Vercalsteren *et al.* [31] presented their findings in a report to the Flemish Institute for Technological Research (VITO), which concluded that there was no obvious indication as to which cup system had the highest or lowest environmental impact. There is no decisive formula that makes it possible to use all the impact categories—e.g., carcinogens, ecotoxicity, fossil fuels, etc.—to indicate which cup system is superior (Fig. 2.15a). For instance, the PLA cup uses less fossil fuel than the PP cup; however, the respiratory effects caused by the inorganics of the PE-coated cardboard cup remain the highest. The size of the event also has an effect on the eco-efficiency of the cups. The PC cup appears to have the lowest environment impact when used for a small event. This is due to the reusable nature of PC, which can be washed by hand, meaning that less water and detergent is used in the cleaning process. However,

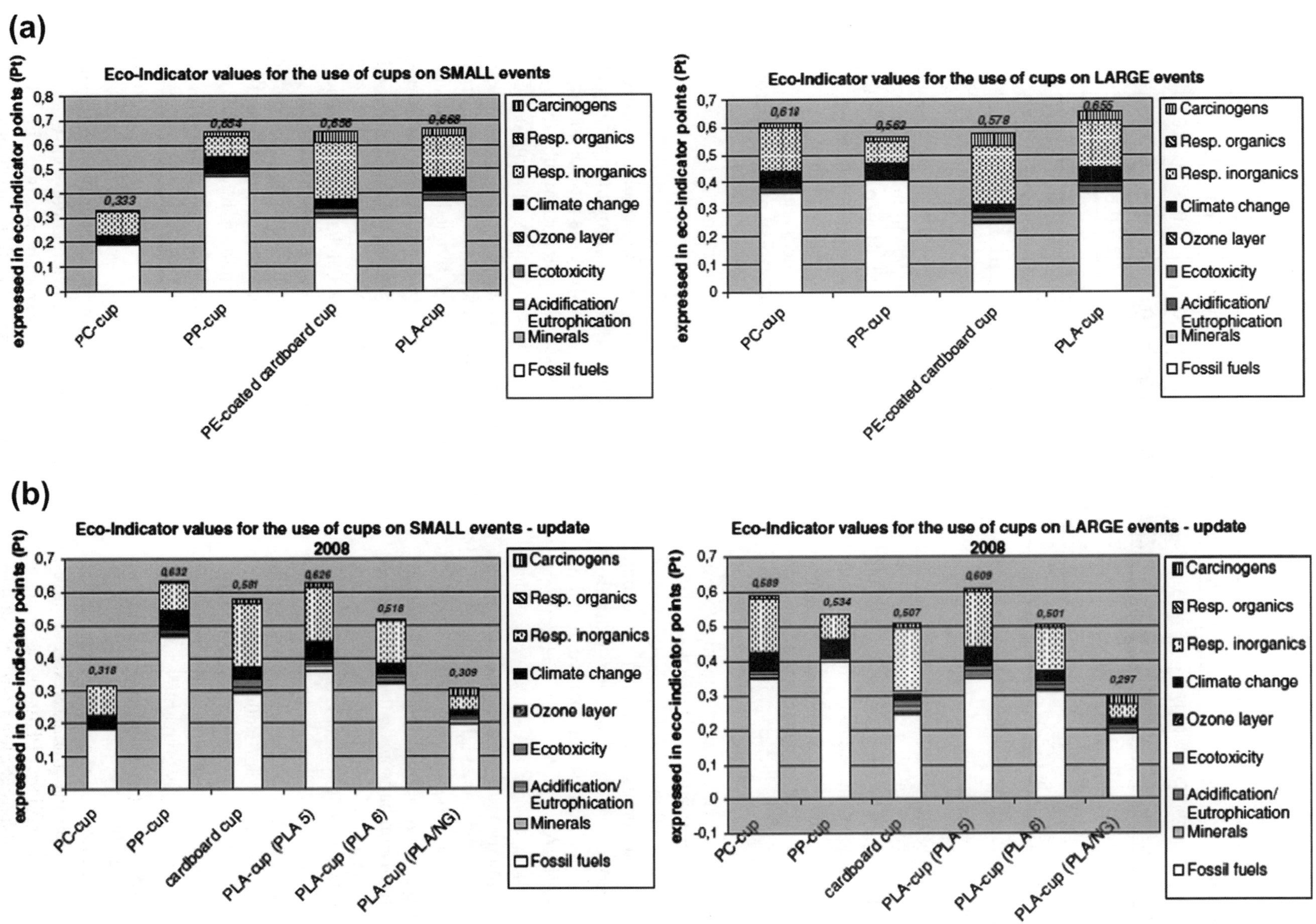

Figure 2.15 (a) Eco-indicator values for the usage of cups at small-scale indoor and large-scale outdoor events. (b) Eco-indicator values for the usage of cups at small-scale indoor and large-scale outdoor events for PLA6 and PLA/NG. *Adapted from Ref. [31].*

Table 2.22 Comparison of the Ecological Aspects of Ingeo™, Virgin PET (vPET), and Recycled PET (rPET) for Different End-of-Cycle Treatment Approaches under European Union Framework

Treatment	**Landfill**			**Incineration**		
Clamshell	**Ingeo™**	**vPET**	**rPET**	**Ingeo™**	**vPET**	**rPET**
Renewable primary energy (GJ)	0.53	0.02	0.02	0.52	0.01	0.02
Nonrenewable primary Energy (GJ)	1.22	1.70	1.04	0.96	1.37	0.88
Aquatic eutrophication (g PO_4)	9.73	3.81	2.20	6.61	0.68	0.62
Acidification (kg SO_2)	0.52	0.34	0.20	0.49	0.33	0.19
Climate change (kg CO_2)	60.6	77.8	49.4	81.8	104	62.7
Fossil resources (kg crude oil)	13.5	26.0	14.6	9.9	21.4	12.3

Source: Data extracted from Ref. [32]

the turnover usage of the PC cup is higher at a large event. Consequently, washing is carried out frequently, and so the PC cups wear out rapidly and require regular replacement. Although the PLA cup has the highest eco-indicator points, PLA is also likely to be competitive in long-term applications. This is because PLA technology is still in its infancy and there will be future improvements to environmental issues such as acidification/eutrophication and the dependence on fossil fuels. Eco-improvement initiatives conducted by NatureWorks have proved fruitful for the production of second generation Ingeo™ (PLA6), the eco-indication points for which are 20% lower than for the first generation PLA (PLA5) (Fig. 2.15b). NatureWorks is currently working on PLA/NG (i.e., next generation Ingeo™), which should be an absolutely green product, for better environmental protection.

The Institute for Energy and Environmental Research (IFEU), Heidelberg, Germany, has carried out a head-to-head comparison of the life cycle of clamshell packaging made of Ingeo™, virgin, and recycled PET. The report by Krüger *et al.* in Ref. [32] compared the environmental impact according to the treatment of the respective clamshells using landfill and incineration approaches. Both methods are commonly used in Europe and the United States. Data from the report are summarized in Table 2.22, and shows that Ingeo™ has numerous advantages compared to virgin PET. The aquatic eutrophication and acidification of Ingeo™ appears to be higher, mainly due to the production stage involving farming and soil activity, which generate greenhouse gases. Although recycled PET seems to be a greener product compared to PLA, recycled PET is actually made of virgin PET; thus, the upstream fabrication process is offset during the virgin PET calculation. It is confidently believed that Ingeo™ can yield a better ecological performance in recycled usages as well. However, this requires a thorough analysis in the near future. In conclusion, the green status of PLA is undoubted for sustainable environmental protection.

2.7 Conclusion

PLA has been around for decades, but it is only in more recent years that the growth in its applications has expanded rapidly. PLA is a biodegradable polymer that possesses the potential to substitute existing petroleum-based commodity polymers, to help overcome the accumulation of plastic waste in landfills. In addition to use in general and packaging products, it also has biomedical applications in surgery, due to its compatibility with living tissue. PLA is favored because it can be mass produced from agricultural sources, which are renewable, allowing society to reduce its dependency on petrochemicals. Continued research and development has made it possible to lower greenhouse emissions associated with the production process. In conclusion, PLA has got great potential and marketability as a biodegradable polymer for a sustainable future.

References

[1] Economic Assessment Office - National Institute of Standards and Technology, Cargill, Inc, Research Center— Improving Biodegradable Plastic Manufactured from Corn, Advance

Technology Program (2007). http://statusreports.atp.nist.gov, 2007 (assessed 4.9.10).
[2] P.D. Darney, S.E. Monroe, C.M. Klaisle, A. Alvarado, Clinical evaluation of the Capronor contraceptive implant: preliminary report, Am. J. Obstet. Gynecol. 160 (1989) 1292−1295.
[3] C.M. Buchanan, R.M. Gardner, R.J. Komarek, Aerobic biodegradation of cellulose acetate, J. Appl. Polym. Sci. 47 (1993) 1709−1719.
[4] BASF Corporation (2009). Totally convincing: Ecoflex® the biodegradable plastic that behaves just like a natural material. Trade Brochure.
[5] R. Leaversuch, Biodegradable Polyester: Packaging goes Green. Feature Article. http://www.ptonline.com/articles/200209fa3.html, 2002 (accessed on 11. 9.2010).
[6] PlasticsEurope, The compelling facts about plastics 2009: An Analysis of European Plastics Production, Demand and Recovery for 2008. www.plasticseurope.org, 2009.
[7] Plastics Today, Q1 earning at Dow, ExxonMobil, and BASF point to global plastics demand growth. www.plasticstoday.com, 2010.
[8] Accenture, Trends in Manufacturing Polymers, Achieving High Performance in a Multi-Polar World (2008).www.accenture.com, 2008.
[9] European Bioplastics, Fact sheet Nov 2009 Industrial Composting (2009). www.european-bioplastics.org, 2009.
[10] IDEHLG − Ireland Department of the Environment, Heritage and Local Government. Waste Management (Environmental Levy) (Plastic Bag) (Amendment) (No.2) Regulations 2007.
[11] L. Shen, J. Haufe, M.K. Patel, Product Overview and Market Projection of Emerging Bio-based Plastics. PRO-BIP 2009, Final Report, Report Commissioned by European Polysaccharide Network of Excellence (EPNOE) and European Bioplastics, Group Science, Technology and Society, Universiteit Utrecht, The Netherlands, 2009.
[12] NatureWorks LLC, Cargill acquires full NatureWorks ownership from Teijin. http://www.natureworksllc.com/news-and-events/press-releases/2009/07-01-09-ownership-change.aspx, 2009a.
[13] NatureWorks LLC, NatureWorks Assesses Second Ingeo Manufacturing Location (2009b). http://www.natureworksllc.com/news-and-events/press-releases/2009/03-12-09-manufacturing-location2.aspx, 2009b.
[14] Teijin Limited, Teijin expands hygrothermal resistance of BioFront Bioplastic upgraded version now offers high durability comparable to PET. http://www.teijin.co.jp/english/news/2009/ebd090708.html, 2009.
[15] Teijin, Teijin launches BioFront Heat-Resistance Bio Plastic − 100% BioFront Car Set Fabrics Developed with Mazda. http://www.teijin.co.jp/english/news/2007/ebd070912.html, 2007.
[16] K.J. Jem, J.F. Pol, S. Vos, Microbial lactic acid, its polymer poly(lactic acid), and their industrial applications, in: G.-Q. Chen (Ed.), Plastics from Bacteria: Natural Functions and Applications, Microbiology Monographs, vol. 14, 2010. http://dx.doi.org/10.1007/978-3-642-03287_5_13.
[17] CCM International Limited, Corn Products China News vol. 3 (Issue 1) (2010).
[18] R.E. Drumright, P.R. Gruber, D.E. Henton, Polylactic acid technology, Adv. Mater. 12 (2000) 1841−1846.
[19] A.J. Nijenhuis, D.W. Grijpma, A.J. Pennings, Highly crystalline as-polymerized poly(L-lactide), Polym. Bull. 26 (1991) 71−77.
[20] D.E. Henton, P. Gruber, J. Lunt, J. Randall, Polylactic acid technology, in: A.K. Mohanty, M. Misra, L.T. Drzal (Eds.), Natural Fibers, Biopolymers, and Biocomposites, Taylor & Francis, Boca Raton, FL, 2005, pp. 527−577.
[21] S.-I. Morita, Y. Ikada, Lactide copolymers for scaffolds in tissue engineering, in: K.-U. Lewandrowski, D.L. Wise, D.J. Trantolo, J.D. Gresser, M.J. Yaszemski, D.E. Altobelli (Eds.), Tissue Engineering and Biodegradable Equivalents Scientific and Clinical Applications, Marcel Dekker, New York, Basel, 2002, pp. 111−122.
[22] M.C. Branco, J.P. Schneider, Self-assembling materials for therapeutic delivery, Acta Biomater 5 (2009) 817−831.
[23] M-H. Seo, I-J. Choi, Y.-H. Cho, Positively charged amphiphilic block copolymer as drug carrier and complex thereof with negatively charged drug. US Patent 7226616, 2007.
[24] H. Yin, S. Yu, P.S. Casey, G.M. Chow, Synthesis and properties of poly (D, L-lactide) drug carrier with maghemite nanoparticles, Mater. Sci. Eng. C 30 (2010) 618−623.

[25] R Jain, K.C. Jindal, S.K. Devarajan, Injectable Depot Compositions and Its Process of Preparation. US Patent 20100015195, 2010.
[26] Durect (2010).www.absorbables.com/biodegradation.htm, 2010 (accessed 22.11.10).
[27] W.J. Groot, T. Borén, Life cycle assessment of the manufacture of lactide and PLA biopolymers from sugarcane in Thailand, Int. J. Life Cycle Assess. 15 (2010) 970–984.
[28] E.T.H. Vink, S. Davies, J.J. Kolstad, The eco-profile for current Ingeo® polylactide production, Ind. Biotechnol. 6 (2010) 212–224.
[29] E.T.H. Vink, D.A. Glassner, J.J. Kolstad, R.J. Wooley, R.P. O'Connor, The eco-profiles for current and near future NatureWorks® polylactide (PLA) production, Ind. Biotechnol. 3 (2007) 58–81.
[30] E.T.H. Vink, K.R. Rábago, D.A. Glassner, P.R. Gruber, Applications of life cycle assessment to NatureWorks™ polylactide (PLA) production, Polym. Degrad. Stab. 80 (2003) 403–419.
[31] A. Vercalsteren, C. Spririnckx, T. Geerken, Life cycle assessment and eco-efficiency analysis of drinking cups used at public events, Int. J. Life Cycle Assess. 15 (2010) 221–230.
[32] M. Krüger, B. Kauertz, A. Detzel, Life Cycle Assessment of Food Packaging Made of IngeoTM Biopolymer and (r)PET. Final Report, IFEU GmbH, Heidelberg, Germany, 2009.
[33] B.G. Hermann, K. Blok, M.K. Patel, Twisting biomaterials around your little finger: environmental impacts of bio-based wrapping, Int J. Life Cycle Assess. 15 (2010) 346–358.
[34] BASF Corporation, BASF Announces Major Bioplastics Production Expansion (2008). http://www.basf.com/group/pressrelease/P-08-229, 2008 (accessed 11.9.10).

3 Applications of Poly(lactic Acid)

Lee Tin Sin, Abdul R. Rahmat and Wan A.W.A. Rahman

OUTLINE

3.1 Introduction

Poly(lactic acid) (PLA) is a biodegradable polymer that has a variety of applications. It has been widely used in the biomedical and pharmaceutical fields for several decades due to its biocompatibility and biodegradability in contact with mammalian bodies. For many years, however, the application of PLA was very limited, due to the high cost of synthesis in the laboratory. For the most part, the direct polycondensation route (Fig. 3.1) was employed to produce PLA from lactic acid. The resultant PLA had a low molecular weight and poor mechanical properties.

The properties of PLA improved tremendously with the development of production using ring-opening polymerization. This route requires an intermediate substance known as lactide. Lactide is the cyclic dimers of lactic acid, and it can be in the form of L-lactide, L,D-lactide (*meso*-lactide) and D-lactide stereocomplex (Fig. 3.2). Nowadays, the synthesis of PLA rarely starts from chemically synthesized lactic acid. The lactic acid used is yielded from the fermentation of carbohydrates such as starch and cellulose. A large proportion is derived from the crops corn and cassava. Microorganism-based fermentation yields mainly L-lactic acid.

Currently, the NatureWorks is the largest producer of PLA for domestic applications. NatureWorks employs lactide ring-opening polymerization for the mass production of 140,000 MT per year of PLA, which is branded as Ingeo™. NatureWorks' PLA is

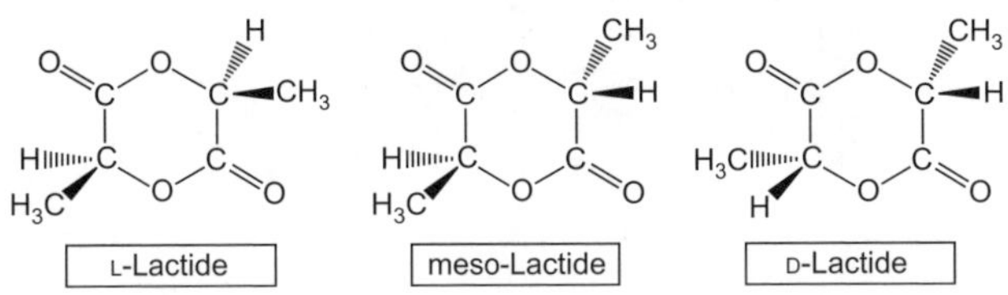

Figure 3.2 D-Lactide stereocomplex.

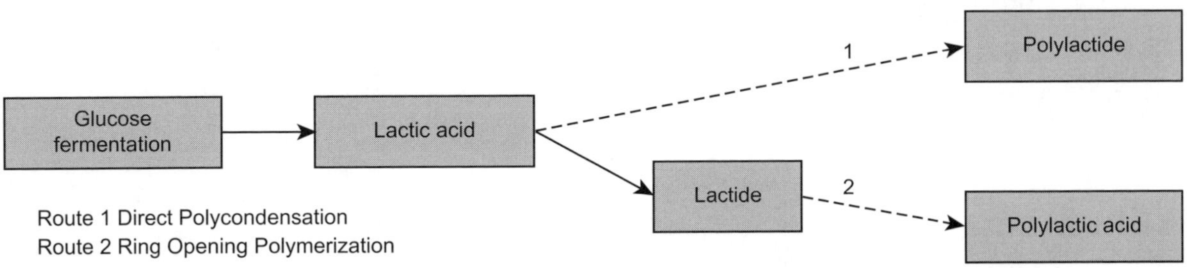

Figure 3.1 General route of PLA production.

Ebnesajjad: Handbook of Biopolymers and Biodegradable Plastics. http://dx.doi.org/10.1016/B978-1-4557-2834-3.00003-3
© 2012 Elsevier Inc. All rights reserved. Reproduced from a chapter in: Tin Sin, *Polylactic Acid* (2012).

Table 3.1 Domestic Applications of PLA

Application	Manufacturer/User (Product)	Description	Illustrations
Apparel	Mill Direct Apparel (jackets, caps, polo shirts), Codiceasbarre (shirts), Gattinoni (wedding dresses), Descente (sportswear), etc.	PLA fiber is used as a material for making garments. According to [1]; substitution of 10,000 polyester performance sports shirts with the usage of Ingeo™ can help to save fossil fuels equating to 540 gal gas/greenhouse gas emissions or 11,500 miles of driving a car. Apparel made of PLA has excellent wicking properties and has low moisture and odor retention. It is hypoallergenic, eliciting no skin irritation. For apparel, Ingeo™ can be blended with a maximum of 67% natural, cellulosic or man-made fiber to achieve a variety of properties.	  
Bottles	Shiseido-Urara (shampoo bottles), Polenghi LAS (lemon juice bottles), Sant'Anna (mineral water bottles), etc.	PLA is known to be suitable for making bottles. Most of the PLA grades are suitable for application at or slightly above the room temperature. This is because PLA bottles tend to deform at temperatures of 50–60 °C [2], i.e., the glass transition temperature (T_g) of PLA. When the temperature reaches T_g, the amorphous chain mobility of the plastic starts to increase significantly. The PLA material, which is glassy and rigid at room temperature, gradually turns mobile and rubbery at T_g. However, PLA bottles have excellent gloss, transparency, and clarity—equal to polyethylene terephthalate (PET). The PLA also has exceptional flavor and aroma barrier properties. The substitution of 100,000 32-oz juice	

Application	Manufacturer/User (Product)	Description	Illustrations
		bottles can save fossil fuels equating to 1160 gal of greenhouse gases or a car traveling for 23,800 miles [3].	 
Cups and food serviceware	Fabri-Kal (cold drink cups and lids), Coca-Cola (lining of paper hot cups), Avianca (in-flight cold drink cups), StalkMarket (cutlery sets), etc.	This is one of the most important applications of PLA. PLA is used for these applications in order to reduce the volume of nondegradable disposable food serviceware, such as cups, plates, utensils, and cutlery going to landfill. Conventionally, polystyrene and polypropylene have been widely used for making food serviceware due to their low cost, light weight, and acceptable properties. PLA is a good alternative; it has excellent gloss, clarity, printability, and rigidity. It has good barrier properties with grease, oil, and moisture, and has the flexibility to adapt with high production plastic technologies, such as injection molding and thermoforming. PLA is also suitable for coating or lining paper cups. The environmentally friendly characteristics of PLA means that it can help to save 5950 gal of gas/greenhouse gas emissions for	 

(Continued)

Table 3.1 Domestic Applications of PLA—*Cont'd*

Application	Manufacturer/User (Product)	Description	Illustrations
		every million of cups, forks, spoons, and knives, when substituting petrochemical polymers [4].	
Food packaging	Lindar (thermoform container), InnoWare Plastics (thermoform container), Carrefour Belgium (grocery bags), etc.	PLA is suitable to be used for light weight and transparent food packaging containers. It is highly glossy and can be easily printed—equal to the existing materials, such as polystyrene, polyethylene, and polyethylene terephthalate. Container lidding made from PLA is compostable and renewable; typical lidding applications include yogurt pots, sandwich containers, and fresh food trays for fruits, pastas, cheeses, and other delicatessen products. The design solution of compostable delicatessen lidding of NatureWorks® PLA is shown. 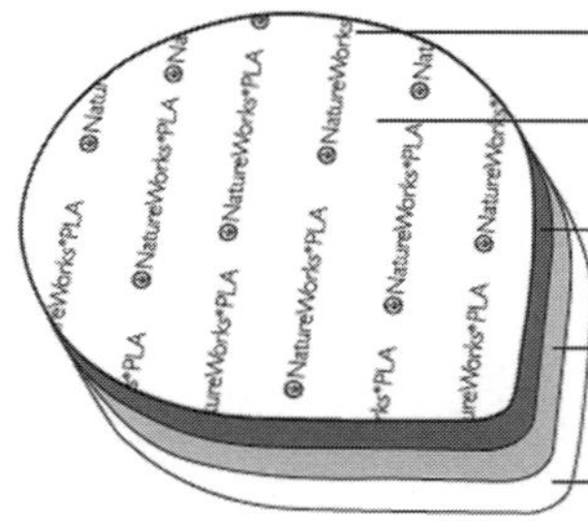 The advantages of this lidding design are: superior flavor and aroma barrier up to 47 °C, with strong resistance to most oils and fats in contact with food products [5]. The heating sealing can be done at temperatures as low as 80 °C with the heat seal strength >1.5 lb/in. PLA has good compatibility with many ink formulations with a natural surface energy of 38 dyne/cm^2. Additional treatment with both corona and flame can further enhance surface energy to over 50 dyne/cm^2. The conversion of 250,000 medium-sized deli	  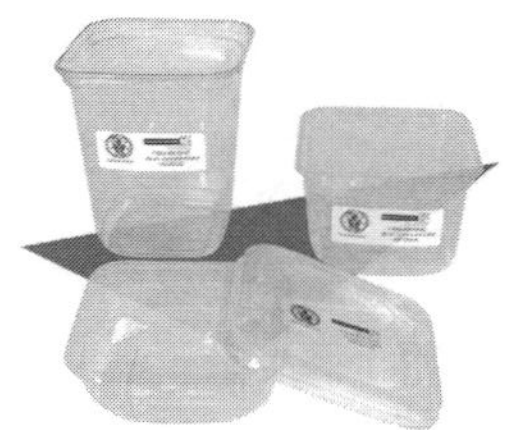

Application	Manufacturer/User (Product)	Description	Illustrations
		containers to PLA can save 3000 gal of gas/greenhouse gas emissions progressively [6].	
Films	Frito-Lay (SunChip), Walmart (salad packaging), Naturally Iowa (EarthFirst® shrink sleeve label), etc.	PLA films are made for bakery goods, confectionery, salads, shrink wrap, envelope windows, laminated coatings, multilayer performance packaging, etc. PLA can be made into biaxially oriented plastic film for packaging bags. PLA plastic bags take a few months to fully degrade when buried in compost. The thickness of the film affects the rate of degradation and mass losses. PLA marketed by NatureWorks is specially made for processing using the blown film equipment for low-density polyethylene film. It can be also processed using the oriented polypropylene facility with minor modifications to setting. Every year, millions of plastic bags are disposed of, causing white pollution to the ground and water. The substitution of petroleum-based plastic bags for PLA bags can make significant environmental savings. The replacement of 20 million medium salad package bags can help to save fossil fuel equal to 29,200 gal of greenhouse gas emissions [7].	  
Cards for transactions	Apple Store (iTunes), The Plastic Card Shop® (gift card), etc.	Transaction cards made of PLA are as durable as polyethylene, polyvinyl chloride (PVC), or polyethylene terephthalate. Most of the existing plastic cards are made for single use, such as gift cards or prepaid top-up cards. There are millions of regular hotel key cards, and loyalty and transaction cards produced every year. PLA cards have good adaptability to cope with security features and magnetic strips. They have durable characteristics and can be film-laminated. Water-based acrylic and solvent-based nitrocellulose and polyamide are the suitable inks for printing onto PLA cards. By converting 40 million plastic cards to PLA, this can make an	

(Continued)

Table 3.1 Domestic Applications of PLA—*Cont'd*

Application	Manufacturer/User (Product)	Description	Illustrations
		environmental saving equivalent to 20,800 gal of gas/greenhouse gas emissions or a car traveling 691,700 miles [8].	
Rigid consumer goods	Bioserie (iPod and iPad covers), Henkel (correction roller and stationery), NEC (Nucycle desktop computer), Cargo (lipstick case)	PLA is widely used as the casing for electronic devices, cosmetics, and stationary. The rigid character of PLA can provide protection to enclosures for highly sensitive products, such as electronics and cosmetics. There are a few grades of PLA on the market specially designed for high-impact and heat-stable applications. PLA is readily coupled with fibers to form composites for extreme applications. Potential applications for PLA composite include computer casings with good stiffness. PLA is very important for electronics industry nowadays, because the development and turnover of electronic appliances is tremendous. A handheld device can become outdated because of embedded software in a single year. Every year, millions of mobile phone casings are disposed. Every 1,000,000 casings generate 6,400 gal of greenhouse gas emissions. Laptop cases, disposable razors, pens, cosmetic containers, etc. all put burden on landfill. Substitution of petrochemical-based plastics with PLA can reduce the volume of waste in landfill sites due to the biodegradability of PLA. Life cycle analysis demonstrates that a desktop computer with PLA content (~75% plant based) offers a significant carbon footprint reduction, lowering CO_2 emissions by around 50% compared to the petroleum-based polycarbonate/ABS blends.	  

Application	Manufacturer/User (Product)	Description	Illustrations
Home textiles	Eco-centric (cushion), Ahlstrom (tea bag), Natural Living® (mattress topper), etc.	PLA can be transformed into fiber to substitute existing PET products, such as fabrics. PLA in this form has equally good breathability and comfort. It has outstanding moisture management properties and good thermoregulating characteristics. PLA fabric is easy to care for quick drying and requires no ironing. In a comparison of PLA fiber with soy and bamboo fibers to determine the percentage of shrinkage after washing and tumble drying following the AATCC 135-2004 IIIA [9], PLA fiber showed a reduction of 2.2% in length after three washes, while soy and bamboo fibers reduced by 15.0% and 17.2 %, respectively [10]. Although bamboo, soy, and PLA are all biodegradable and agriculturally derived, PLA fiber tends to show superior properties.	  
Nonwoven products	GroVia (diapers), Elements Naturals® (baby wipes), Renewable Fiber LLC (shopping bags), etc.	Many nonwoven products can be made from PLA instead of PET and polypropylene. Existing synthetic nonwoven products, such as diapers, baby wipes, sanitary pads, and shopping bags, require hundreds of years to degrade after landfill burial. PLA is favorable because it can be spun into fibers. It has low flammability, with a limiting oxygen index of 26, high resilience and excellent wicking. It has also been found that PLA fibers exhibit 20% and 45% higher extension than wool and cotton, respectively [11]. It has been shown in tests that PLA does not cause irritation to the mammalian body [12]. When 1 million diapers are converted from PET and	

(*Continued*)

Table 3.1 Domestic Applications of PLA—*Cont'd*

Application	Manufacturer/User (Product)	Description	Illustrations
		polypropylene to PLA, it can help to save fossil fuel equivalent to 1,000 gal of gas/greenhouse gas emissions or driving a car for 12,800 miles.	 
Foam trays	Sealed Air (Cryovac® NatureTRAY food tray), Dyne-a-pak Inc. (Dyne-a-pak Nature™ meat foam tray), etc.	Foam trays are important in packaging, especially for fresh food. "Styrofoam" is the well-known foam tray made from polystyrene. This type of polystyrene is cheap but nondegradable. Recycling of foam trays is not a profitable business because the collection volume is large in order to rework it into a small amount of dense resin. The density of Styrofoam is 0.025 g/cm^3 compared to virgin polystyrene resin, which is 1.05 g/cm^3. This means that 42 foam trays are needed to revert to the original dense polystyrene. PLA is a good replacement because the disposed PLA foam tray can be composted easily without causing adverse effects to the environment. Moreover, the compostable nature of PLA provides enriching nutrients when buried in soil.	

Application	Manufacturer/User (Product)	Description	Illustrations
Expanded foam	Foam Fabricator, Inc. (expandable foam for cushioning)	The technology was developed by Biopolymers Network, and the work received the "Best Innovations in Bioplastics Award" at the Annual European Bioplastics Conference. The technology relies on the application of an expansion agent of CO_2, which is a safer substance compared to expandable polystyrene using pentane as the expansion agent. The compostability of the expanded PLA foam provides an environmentally friendly solution to the electrical and electronics industry, which uses expanded foam as a cushioning material during shipping.	
Children's toys	Kik&Boo (soft toy filled with PLA fiber)	PLA can be used to make both rigid and soft toys for children. In one example, the fabric of the soft toy is produced from woven PLA fiber, while the soft toy is filled with PLA fiber padding. Both soft and rigid toys made of PLA are washable and hygienic. The production of PLA does not involve toxic petrochemicals; thus, it reduces the exposure of the children to toxins.	 
Fashion products	Fashion Helmet (designer helmet), Rizieri (ladies shoes), etc.	Environmentally friendly PLA can be used to produce typical parts of the helmet. This is only limited by the artistic design; the outer part of the helmet is covered with PLA-calendered cloth. Similarly, the ladies fashion brand, Rizieri, of Milan, Italy, has created an innovation known as "Zero Impact," involving models of "handmade" products based on PLA or Ingeo® fabric. These products have all the delicacy of silk to the touch.	 

Table 3.2 Engineering and Agricultural Applications of PLA

Application	Manufacturer/User (Product)	Description	Illustrations
Engineering materials	Singoshu (Lactboard® for draining plate)	Drainage material is used in construction ground works to reduce or eliminate hydrostatic pressure while improving the stability of the enclosed materials. PLA drainage material has good workability for soft ground with sufficient permeability and tensile strength. The favorable biodegradability of PLA enables the drainage material to return to nature safely. In other words, after the consolidation period, PLA can reduce the load on the surrounding environment and be detoxified. The PLA material can become impaired after completion of the shield for excavation and underground construction consolidation settlement.	
Automotives	Toyota (floor mat of Toyota Prius and spare tire cover), Toray (fiber for car mat), etc.	The automotive industry uses large quantities of plastics, especially polyethylene, polyvinyl chloride (PVC), and acrylonitrile-butadiene-styrene (ABS), which are derived from nonrenewable petroleum sources. The levels of recycled plastics in use are as low as 30% (by weight); the remaining are virgin polymers. When the car is disposed of, the percentage of plastic recycled from it can be as low as 20%. This means that a large volume of automotive plastics eventually end up polluting the environment. PLA is an environmentally friendly material for automotive applications. This is particularly important for those parts that cannot be recycled, such as car mats and cushion fabrics. The rigidity of PLA is an advantage for external cover applications. Although PLA is biodegradable, the rate of degradation is low and requires high moisture conditions to initiate the hydrolysis process (the depolymerization reaction). The involvement of microorganisms takes part only after the depolymerization reaction transforms the material to low-molecular-weight oligomer lactate. Normally,	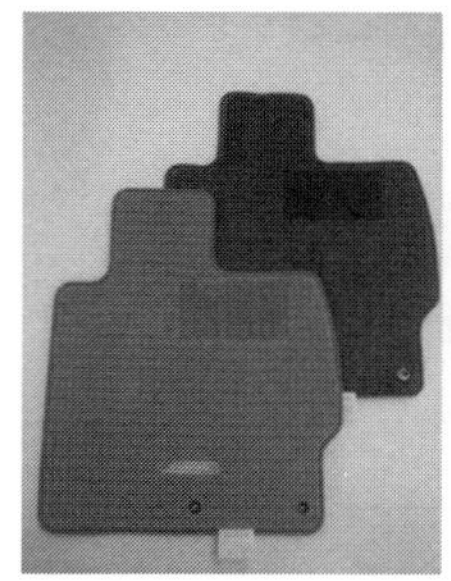 

Application	Manufacturer/User (Product)	Description	Illustrations
		this process takes time, and this exceeds the lifetime of the products.	
Building materials	LG Hausys (laminated flooring and wallpapers), Saint Maclou (carpets), Sommer Needlepunch (Eco2punch® carpets), etc.	Most PLA products in the construction industry are related to flooring. Products include carpet, laminated flooring materials, and wallpapers. PLA in this area is aimed at substituting PVC, which dominates as a building material. One of the problems of PVC is that its processing requires plasticizers, which increases flammability. Consequently, halogen flame retardants are added to achieve better fire resistance. In contrast, PLA is derived from agricultural sources, and involves less toxic substances during processing stage. Most of the building materials made of PLA can last well when well maintained. These PLA products can be disposed without causing serious pollution to the environment at the end of life.	 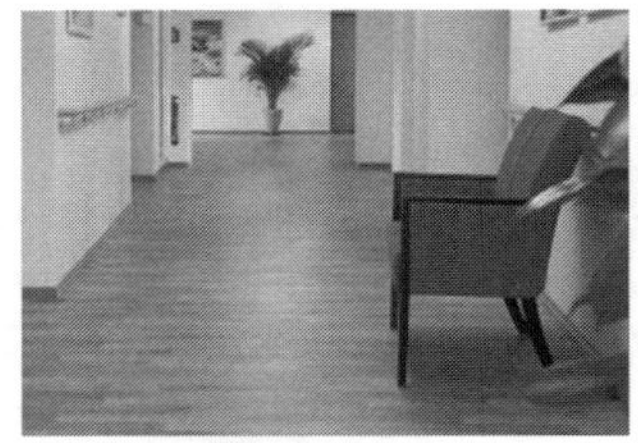
Electrical and electronics	Fujikura (conductor cable coating), Renesas (computer network device casing), ABB (socket casing), etc.	The usage of PLA in the electrical industry is still in the developing stage. PLA can be used as the coating agent for conductor wire. It can also be easily formed into rigid casing for socket and plug applications. Reference [13] compared PLA with polyethylene and polyvinyl chloride (PVC) and found that the resistivity of PLA (4.3 × 1017 Ωcm) is higher than polyethylene (>1016 Ωcm) and PVC (1011 −1014 Ωcm). The dielectric dissipation factor of the three polymers are PLA = 0.01%, polyethylene = 0.01%, and PVC = 0.10%. Generally, PLA has equally good electrical properties as other commodity polymers used in the electric and electronics industries. (See Table 3.3 for a comparison of PLA and PVC cable.)	 

(*Continued*)

Table 3.2 Engineering and Agricultural Applications of PLA—*Cont'd*

Application	Manufacturer/User (Product)	Description	Illustrations
Agricultural	FKuR Kunststoff GmbH (Bio-Flex mulch film), Desch Plantpak B.V. (D-Grade® Bio thermoformed flower pot, trays and packs), BASF (Ecoflex® mulch film)	The biodegradable characteristic of PLA is favorable in agricultural applications. This is because PLA can be well composted without leaving harmful residues in the soil. PLA mulch film can provide soil protection, weed management, fertilizer retention, etc. Over time, the mulch films slowly degrade and finally decompose when the crops reach the harvest period. This eliminates the need for farmers to collect and dispose the used mulch film. The composted PLA mulch film also provides soil nutrients. Flower pots made of PLA can be buried in soil and left there to degrade when the plant is ready to be planted in the ground.	 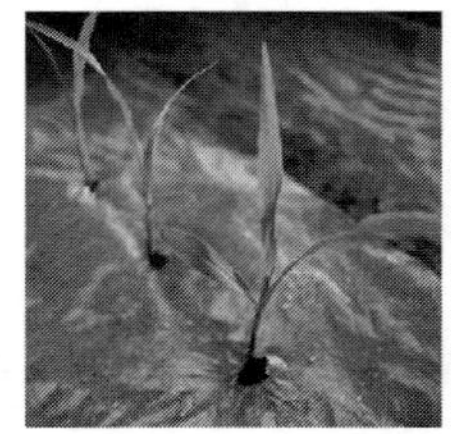

produced mainly for biodegradable packaging, containers, clothing, fibers, etc. Purac is the major producer of PLA for the biomedical and pharmaceutical industries.

In this section the product applications of PLA are summarized. The applications of PLA can be grouped into three main categories: domestic, pharmaceutical/biomedical, and engineering. Products, trade names, and producers have been included where useful. The intention is not to advertise but rather to provide supportive information and references.

3.2 Poly(lactic Acid) for Domestic Applications

Most of the PLA produced worldwide is made for domestic applications, such as apparel, bottles, cups, and food serviceware (Table 3.1). All these PLA products are targeted to substitute the existing petrochemical polymers, with the advantage that the PLA products have environmentally friendly production and are biodegradable upon disposal.

3.3 Poly(lactic Acid) for Engineering and Agricultural Applications

PLA is suitable for typical engineering applications that impose environmental burdens at the end of life. The rigidity of PLA can ensure good mechanical properties during applications, and yet it can easily undergo biodegradation after disposal. The use of PLA for essential engineering parts is limited. The use of PLA is mostly focused on secondary applications as listed in Table 3.2. In relation to its use in electronics and electrical applications, Table 3.3 sets out a comparison of PLA and PVC-coated cables.

3.4 Poly(lactic Acid) for Biomedical Applications

In the early days of PLA development, most of its applications were in the biomedical field. PLA continues to be used in this arena (Table 3.4). It is widely used in scaffolds to provide temporary

Table 3.3 Evaluation of PLA-Coated Cable in Comparison with PVC-Coated Cable [1]

Item/Cable	Pure PLA	Plasticized PLA with Flexibility	600 V PVC Cable (IV) JIS C 3307
Extrusion	■: Excellent appearance ▲: Void in surface between conductor and insulation	▲: Excellent appearance ■: Analogous with pure PLA	-
Bending	▼: Whitening at 10 times bending and cracking at 4 times bending	▲: Whitening at 2 times bending ■: No cracking at self-diameter bending	-
Tensile	■: Strength = 59 MPa ▼: Elongation = 12%	■: Strength = 43 MPa ▼: Elongation = 25%	Strength > 10 MPa Elongation > 100%
Heat deformation	☺: 60–120 °C = reduction < 10%	☺: 60–90 °C = reduction < 10% ▼: 120 °C = reduction 58%	Thickness reduction less than 50%
Electrical	☺: tan δ = 0.35%, □ = 3.2 ☺: $\rho = 2.7 \times 10^{16}$ Ωcm	▼: tan δ = 2.31 %, □ = 4.1 ■: $\rho = 4.6 \times 10^{12}$ Ωcm	$\rho = 5 \times 10^{12}$ Ωcm
Dielectric breakdown	☺: 35–45 kV (0.7 mm thickness)	☺: 45–50 kV (0.7 mm thickness)	Withstand voltage test 1.5 kV × 1 min
Dielectric breakdown with bending	▼: Cracking at 4 times bending	■: 25 kV at self-diameter bending	—

Table 3.4 Biomedical Applications of PLA

Application	Manufacturer/User (Product)	Description	Illustrations
Surgical implants	Zimmer (Bio-Statak® suture anchor and bone cement plug), Ethicon (Vicryl suture and Vicryl mesh) and Sulzer (Sysorb® screw), etc.	PLA and its copolymer PLGA (polylactide-co-glycolide) are compatible with living tissue. However, this is limited to the L stereoisomer of PLA because mammalian bodies only produce an enzyme that breaks down this one. PLA and PLGA are used to fabricate screws, pins, scaffolds, etc., to provide a temporary structure for the growth of tissue, eventually breaking down after a certain period. The purpose of copolymerizing with comonomer glycolide is to control the rate of degradation through the modification of crystallization. Sometimes, L and D isomers of lactides are copolymerized for this purpose. Although poly(D-lactic acid) cannot be consumed by the body's enzymes prolonged exposure to body fluid tends to initiate hydrolysis, which eventually breaks down the macromolecules. Orthopedic surgery often uses PLA and copolymers to fabricate artificial bones and joints. PLA has been used to make surgical sutures for decades. In short, PLA is an important material for biomedical surgical applications.	 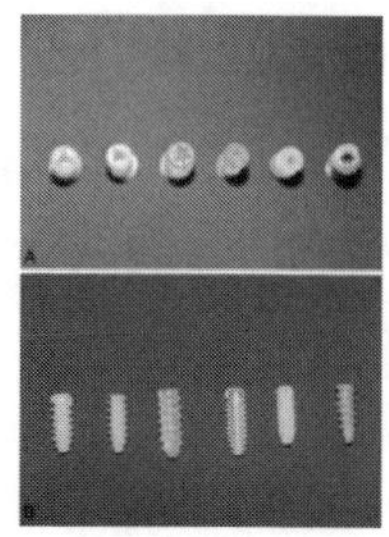 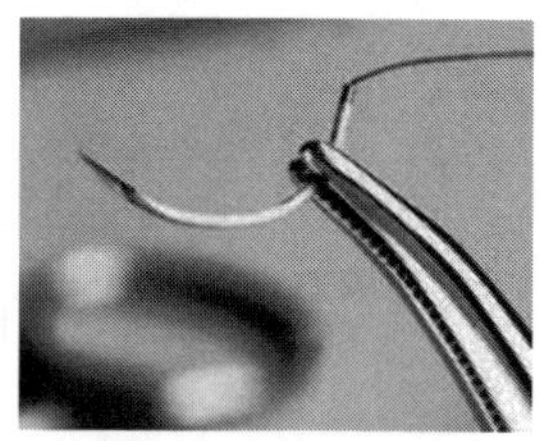
Drug carrier	Abbott (Lupron Depot® for palliative treatment of advanced prostate cancer), AstraZeneca UK Limited (Zoladex®, an injectable hormonal treatment for men with certain types of prostate cancer), Janssen Pharmaceuticals (Risperdal® Consta®, for treatment of schizophrenia and for the long-term treatment of bipolar I disorder), etc.	Most of the PLA drug carriers on the market are available in the copolymer form. This is due to the fact that high purity PLA possesses high crystallinity and takes a longer time to degrade while releasing active drugs. The majority of PLA drug carriers are copolymerized with different percentages of polyglycolic acid (PGA). Normally, such drug carriers slowly release the medication for long-term treatments. For instance, leuprolide acetate applied with a miscrosphere delivery system of PLA and PLGA is used for the treatment of cancer and fibroids. PLGA (polylactide-co-glycolide) can	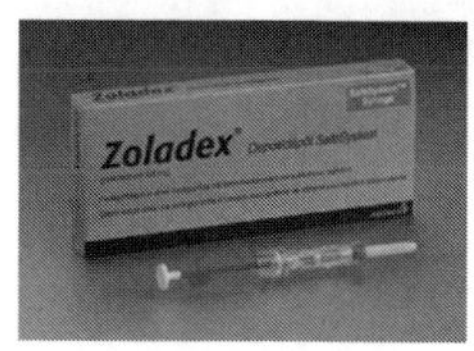 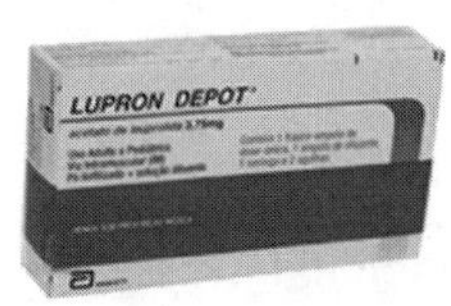

Table 3.4 Biomedical Applications of PLA—*Cont'd*

Application	Manufacturer/User (Product)	Description	Illustrations
		be used in the form of implants and gels with the therapeutics goserelin acetate and paclitaxel for the treatment of prostate/breast cancer, or other anticancer drugs.	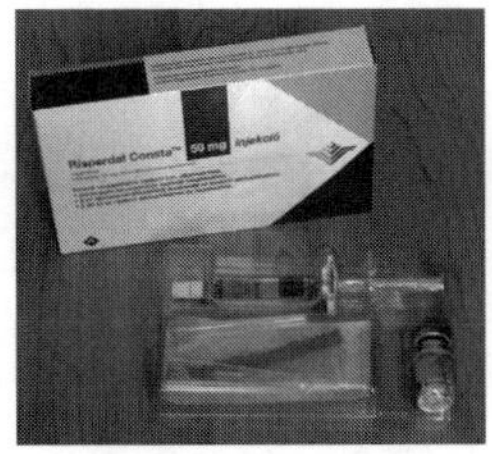

structural support for the attachment and growth of tissues in surgery. It is also used as a drug carrier, containing controlled release active agents for long-term treatments, including for cancer.

3.5 Conclusion

PLA is a very useful polymer that has found applications in a wide range of industries. PLA is well positioned in a niche market because of its biodegradable and environmentally friendly characteristics. Its applications in the biomedical and pharmaceutical field can be traced back several decades. The development of PLA applications in recent years mainly relates to environmental concerns and the adverse effects of using nondegradable petrochemical-based polymers. The use of PLA has grown well in the domestic market for general consumer goods and, importantly, in biodegradable packaging. The development of PLA is forecast to grow tremendously in future, making the price of PLA as economical as commodity plastics, but with benefit of being kinder to the environment.

References

[1] Natureworks, Can a t-shirt help change the world? <http://www.natureworksllc.com/Product-and-Applications/Apparel.aspx>, 2011a.

[2] NatureWorks, Thermal stability of PLA preform. <www.natureworksllc.com>, 2011b.

[3] NatureWorks, Choosing a bottle to make a difference. <http://www.natureworksllc.com/Product-and-Applications/Bottles.aspx>, 2011c.

[4] NatureWorks, Can plastic dinnerware make a difference? <http://www.natureworksllc.com/Product-and-Applications/Serviceware.aspx>, 2011d.

[5] NatureWorks, Top if off with NatureWork® PLA Dairy and Delicatessen Container Lidding Solutions. <www.natureworkllc.com>, 2011e.

[6] NatureWorks, Can fresh food packaging help change anything? <http://www.natureworksllc.com/Product-and-Applications/Fresh-Food-Packaging.aspx>, 2011f.

[7] NatureWorks, Can a simple plastic film wrap really make a difference? <http://www.natureworksllc.com/Product-and-Applications/Films.aspx>, 2011g.

[8] NatureWorks, Can your next plastic card really make a difference? <http://www.natureworksllc.com/Product-and-Applications/Cards.aspx>, 2011h.

[9] American Association of Textile Chemists and Colorists, AATCC Test Method 135-2004 Dimensional Changes of Fabrics after Home Laundering, 2006.

[10] NatureWorks, Ingeo™ fibers comparison with soy and bamboo fibers. <www.natureworksllc.com>, 2011i.

[11] NatureWorks, Basic fiber properties. <www.natureworksllc.com>, 2011j.

[12] NatureWorks, Wipes toxicology study/ regulatory information.<www.natureworksllc.com>, 2011k.

[13] T. Nakatsuka, Polylactic acid-coated cable, Fujikura Tech. Rev. 40 (2011) 39–45.

4 The State of the Art of Polymers from Renewable Resources

Alessandro Gandini and Mohamed Naceur Belgacem

OUTLINE

4.1 The Context

The biosynthesis of macromolecules through enzymatic bacterial and chemical polymerizations of specific molecular structures constitutes a key step in the evolution of living organisms. Natural polymers have therefore been around for a very long time and always constituted one of the essential ingredients of sustainability, first and foremost as food, but also as shelter, clothing, and source of energy. These renewable resources have also played an increasingly important role as materials for humanity through their exploitation in a progressively more elaborated fashion. The ever improving technologies associated with papermaking, textile and wood processing, vegetable oils, starch and gelatin utilization, the manufacture of adhesives, etc. represent clear examples of the progressive sophistication with which man has made good use of these natural polymers throughout the millennia. Concurrently, natural monomers have been polymerized empirically for equally long periods for applications such as coatings, paint and ink setting, leather tanning, etc.

The progress of chemistry, associated with the industrial revolution, created a new scope for the preparation of novel polymeric materials based on renewable resources, first through the chemical modification of natural polymers from the mid-nineteenth century, which gave rise to the first commercial thermoplastic materials, like cellulose acetate and nitrate, and the first elastomers, through the vulcanization of natural rubber. Later, these processes were complemented by approaches based on the controlled polymerization of a variety of natural monomers and oligomers, including terpenes, polyphenols, and rosins. A further development called upon chemical technologies that transformed renewable resources to produce novel monomeric species like furfuryl alcohol.

The beginning of the twentieth century witnessed the birth of a novel class of materials, the synthetic

Ebnesajjad: Handbook of Biopolymers and Biodegradable Plastics. http://dx.doi.org/10.1016/B978-1-4557-2834-3.00004-5
© 2013 Elsevier Inc. All rights reserved. Adapted from a chapter in: Belgacem, *Monomers, Polymers and Composites from Renewable Resources* (2008).

polymers based on monomers derived from fossil resources, but the progress associated with them was relatively slow up to the Second World War and did not affect substantially the production and scope of the naturally based counterparts. Some hybrid materials, arising from the copolymerization between both types of monomers, were also developed at this stage as in the case of the first alkyd resins. Interestingly, both monomers used in the first process to synthesize Nylon in the late thirties were prepared from furfural, an industrial commodity obtained from renewable resources, in a joint venture between Quaker Oats and DuPont.

The petrochemical boom of the second half of the last century produced a spectacular diversification in the structures available through industrial organic chemistry. Among these, monomers played a very significant role, as is clear from the high percentage of such structures represented in the list of the most important chemical commodities in world production. The availability of a growing number of cheap chemicals suitable for the production of macromolecular materials gave birth to "the plastic age," in which we still live today, with of course greatly enhanced quantitative and qualitative features.

This prodigious scientific and technical upsurge went to the detriment of any substantial progress in the realm of polymers from renewable resources. In other words, although these materials never ceased to exist, very modest investments were devoted to their development, compared with the astronomical sums invested in petrochemistry. As a consequence, although cotton, wool, and silk textiles are still plentiful, the competition of synthetic fibers has not stopped growing. Likewise, the application of natural rubber is very modest today, compared with its synthetic counterparts, not only in relative tonnage, but also in the continuously widening degree of sophistication associated with the properties of the latter materials. In virtually all other domains associated with polymeric materials, the present contribution of structures derived from renewable resources is very modest and has not played an appreciable role in terms of bringing about specific functional properties. On the other hand, paper has resisted all attempts to be replaced by synthetic polymers, although these have been playing a growing role as bulk and surface additives, albeit without modifying the essential constitution of this material, which still relies on the random assembly of cellulose fibers.

We are deeply convinced that this state of affairs has nothing to do with any consideration of relative merits associated with the different structures and chemical processes involved in either context. Its origin is instead to be found in a purely economic aspect, i.e., in the enormous difference in investment that favored petrochemistry for the last half century, which was also the period when the chemical industry witnessed its fastest progress ever. The choice to finance R&D activities in polymers derived from fossil resources in a massive way was to the benefit of the corresponding materials as we know them today. This objective situation, however, does not prove anything against the potential interest of alternative counterparts made from renewable resources, simply because no such investments were ever made to that effect.

Should fossil resources be available for us to exploit for centuries to come, the above arguments would sound like a futile exercise in style. Their validity stems precisely and primarily from the very fact that fossil resources are dwindling and becoming progressively more expensive. Furthermore, they are not a commonly shared richness, since their global distribution is totally uneven, which implies that certain countries are heavily dependent on others in this respect. These problems are of course affecting the energy outlook in the first instance, since some 95% of the fossil resources are used as fuel, but the looming crisis will inevitably affect the corresponding chemical industry as well.

The purpose here is to show succinctly through the wide spectrum of materials already potentially available, that renewable resources are perfectly apt to provide as rich a variety of monomers and polymers as that currently available from petrochemistry. Implicit in this statement is the condition that substantial investments should be placed in the future to carry out the required research.

If, on the one hand, it is encouraging to see a very impressive increase in this type of activity worldwide, the situation relative to petrochemistry, on the other hand, is still at a very low level of competitiveness. Qualitatively, a change in awareness has indeed taken place, parallel to the preoccupation surrounding the energy issues. This chapter is intended to amplify these promising initial stirrings by providing very sound examples of what can be achieved thanks to this alternative strategy.

Renewable resources are intrinsically valuable in this realm because of their ubiquitous character, which gives any society precious elements of sustainability, including with respect to polymeric materials. In all the topics, emphasis is made, explicitly or implicitly, to the essential fact that the specific sources utilized for the purpose of producing new polymers, are taken neither from food nor from natural materials, but instead from *by-products* of agricultural, forestry, husbandry, and marine activities. One of the best examples of this strategy is the production of furfural, since it can be carried out industrially virtually anywhere in the world, given the fact that any vegetable by-product containing pentoses represents an excellent raw material for its synthesis.

The term "renewable resource" is defined as any animal or vegetable species which is exploited without endangering its survival and which is renewed by biological (short term) instead of geochemical (very long term) activities.

4.2 Vegetable Resources

It is estimated that the world vegetable biomass amounts to about 10^{13} tons and that solar energy renews about 3% of it per annum. Given its fundamental role in the maintenance of the oxygen level, the principle of sustainability limits its exploitation at most to that renewed percentage. Vegetable biomass can be divided into wood, annual plants, and algae.

4.2.1 Wood

Wood is the most abundant representative of the vegetable realm and constitutes the paradigm of a composite material. It displays, on the one hand, a basic universal qualitative composition in terms of its major constituents (cellulose, lignin, hemicelluloses, and polyphenols) and, on the other hand, species-specific components, which can be polymeric, like polyisoprene (natural rubber) and suberin, or small molecules, like terpenes, steroids, etc. An example of its morphology is shown in Fig. 4.1, which illustrates the role of the three basic components respectively as the matrix (lignin), the reinforcing elements (cellulose fibers), and the interfacial compatibilizer (hemicelluloses). The middle lamella (0.5−2 μm) is mainly composed of lignin (70%), associated with small amounts of hemicelluloses, pectins, and cellulose. The primary wall, often hard to distinguish from the middle lamella, is very thin (30−100 nm) and is composed of lignins (50%),

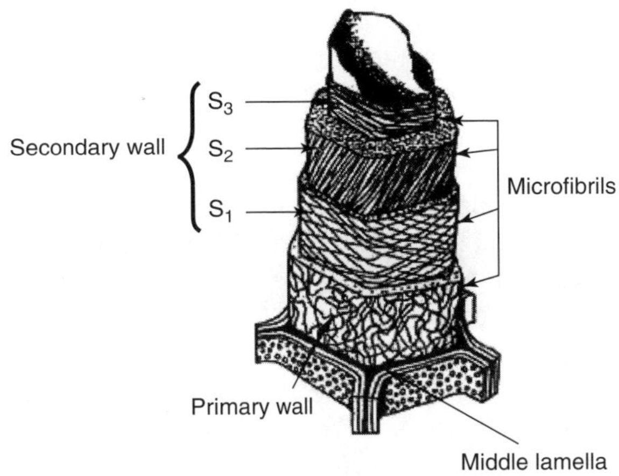

Figure 4.1 Typical morphology of a wood fiber.

pectins, and hemicelluloses. The secondary wall is the main part of the vegetal fibers. Its essential component is cellulose and it bears three layers, namely the external S_1 (100−200 nm), the central S_2 (the thickest layer of 0.5−8 μm), and the internal or tertiary layer S_3 (70−100 nm) situated close to the lumen.

Wood is the structural aerial component of trees. The rest of their anatomy, namely roots, leaves, flowers, and fruits, are not relevant to the aim of this book, and will therefore not be dealt with.

Cellulose dominates the wood composition, although its proportion with respect to the other main components can vary appreciably from species to species. Conversely, polyphenols are the least-abundant components and moreover can exhibit quite different structures. As for lignins and hemicelluloses, their relative abundance and their detailed structures are essentially determined by the wood family: softwoods are richer in lignins, whereas, hardwoods are richer in hemicelluloses. These three basic polymeric components represent fundamental sources of interesting materials and are thoroughly examined in this context, in specific chapters of this book, which focus their attention on the exploitation of these natural polymers after appropriate chemical treatment, with the aim of obtaining novel polymeric materials. The uses of wood itself as a structural material, as a source of furniture or flake boards, and as a raw material in pulping will not be treated here, because these applications call upon the exploitation of this fundamental natural resource through well established technologies. Obviously, all these processes undergo regular improvement in their chemistry and engineering, but we deemed that their inclusion in the present treaty would have unduly overcharged its contents. The interested reader will find excellent monographs on each topic, going from introductory texts to highly specialized books [1–6].

The only area in which wood-related materials are witnessing important research contributions concerns their physical and chemical modification, in order to protect them against degradation by various reagents or to obtain novel properties such as a thermoplastic behavior.

The traditional uses of cellulose, like papermaking and cotton textiles, whose technologies are very thoroughly documented [7–10], are not relevant here. As for those cellulose derivatives which have been exploited for a very long time, like some cellulose esters and ethers, again a systematic treatment of the corresponding processes and properties are available [11–13].

4.2.1.1 Cellulose

Virtually all the natural manifestations of cellulose are in the form of semicrystalline fibers whose morphology and aspect ratio can vary greatly from species to species, as shown in Fig. 4.2. The subunits of each individual fiber are the microfibrils, which in turn are made up of highly regular macromolecular strands bearing the cellobiose monomer unit, as shown in Fig. 4.3.

The interest of cellulose as a source of novel materials is reflected in this book through chapters dealing, respectively, with (i) the chemical bulk modification for the preparation of original macromolecular derivatives with specific functional properties; (ii) the surface modification of cellulose fibers in view of their use as reinforcing elements in composite materials and as high-tech components; (iii) the processing and characterization of these composites, including the use of nanofibers; and (iv) the technology and applications associated with bacterial cellulose. These contributions clearly show that cellulose, the most abundant and historically the most thoroughly exploited natural polymer, still provides new stimulating avenues of valorization in materials science and technology.

4.2.1.2 Lignins

Lignin, the amorphous matrix of wood, is characterized by a highly irregular structure compared with that of cellulose, and is moreover known to vary considerably as a function of wood family (*in situ*) and of the isolation process, which always involves a depolymerization mechanism. Figure 4.4 gives a typical example of the structure of a lignin macromolecule with its most characteristic building blocks. The pulping technology, which calls upon a delignification mechanism based on the use of sulfites, yields lignin fragments bearing sulfonate moieties, i.e., polyelectrolytes.

Traditionally, the aim of separating the wood components has been associated with papermaking, in which delignification isolates the cellulose fibers. In this context, the dissolved lignins have been utilized as fuel, which provides not only the energy required by the process but also a convenient way of recovering its inorganic catalysts.

The idea of using these lignin fragments as macromonomers for the synthesis of polymers, by introducing them into formaldehyde-based wood resins, or by exploiting their ubiquitous aliphatic and phenolic hydroxyl groups, began to be explored only in the last quarter of the twentieth century. Given the fact that these industrial oligomers are produced in colossal amounts, it seems reasonable to envisage that a small proportion could be isolated for the purpose of producing new polymers, without affecting their basic use as fuel. Additionally, novel papermaking

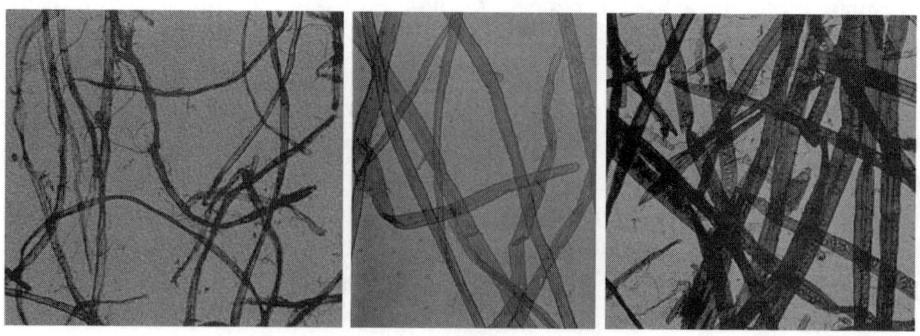

Figure 4.2 Cotton, pine, and fir fibers.

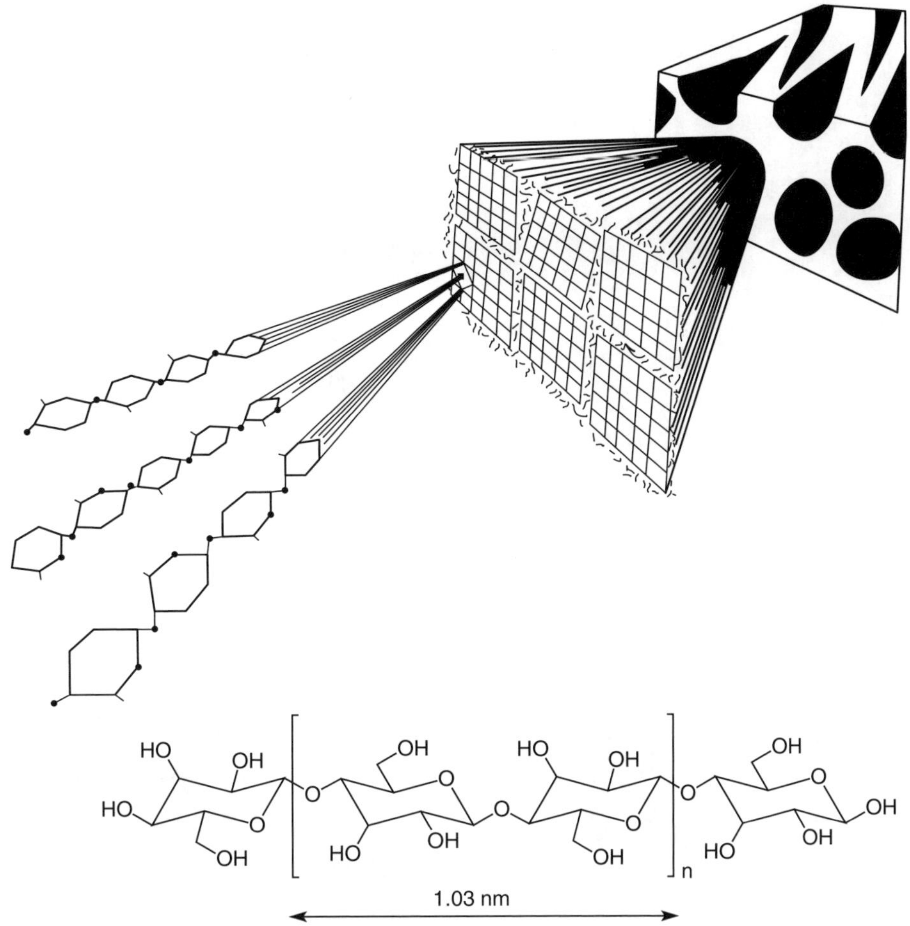

Figure 4.3 Schematic view of the components of cellulose fiber.

technologies, like the organosolv processes and biomass refinery approaches such as steam explosion, provide lignin fragments without the need of their use as a source of energy and with more accessible structures, in terms of lower molecular weights and higher solubility. Therefore, lignin macromonomers represent today a particularly promising source of novel materials based on renewable resources.

4.2.1.3 Hemicelluloses

Wood hemicelluloses are polysaccharides characterized by a relative macromolecular irregularity, compared with the structure of cellulose, both in terms of the presence of more than one monomer unit and by the possibility of chain branching. Figure 4.5 gives typical examples of such structures.

In papermaking processes, part of the wood hemicelluloses remain associated with the cellulose fibers, which results in the improvement of certain properties of the final material. The rest of these polysaccharides is dissolved together with lignin and in most processes it is burned with it. In some instances, however, particularly in the case of organosolv pulping or steam explosion technologies, the hemicelluloses can be recuperated as such. The utilization of wood hemicelluloses, but also of counterparts extracted from annual plants, has interested several industrial sectors for a long time, in particular that of food additives. In recent years, new possible outlets for hemicelluloses have been and are being explored.

4.2.1.4 Natural Rubber

Turning now to more species-specific components, natural rubber is certainly one of the most important representatives. Different tropical trees

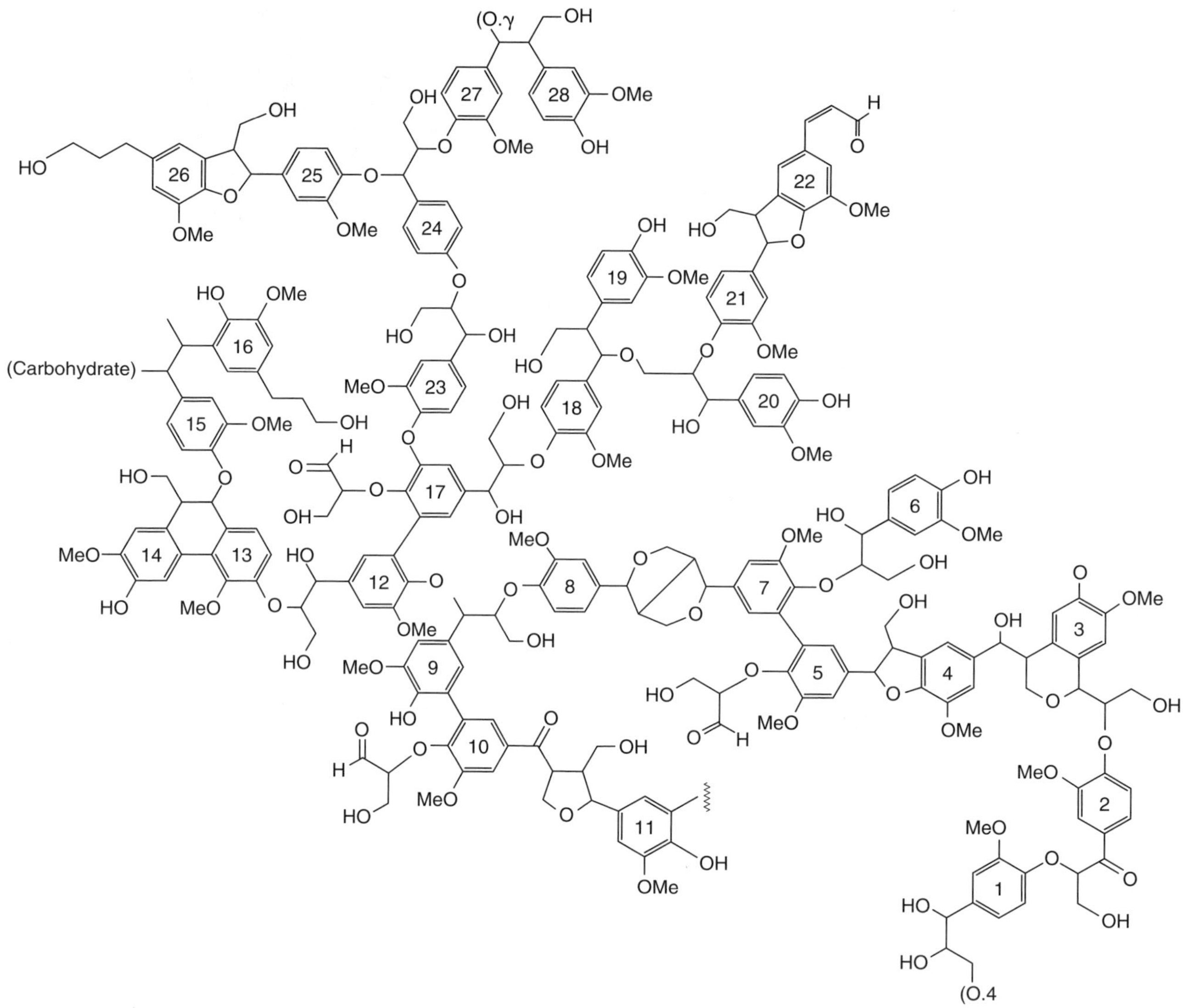

Figure 4.4 Lignin main moieties in a typical macromolecular assembly.

produce different forms of poly(1,4-isoprene), which are exuded or extracted as an aqueous emulsion (latex) or as a sap-like dispersion, before coagulation. The *cis*-form of the polymer (Fig. 4.6a) tends to be amorphous and has a glass transition temperature of about −70 °C, which makes it ideally suitable for application as elastomers, following chemical crosslinking (vulcanization), which involves some of its C=C unsaturations. Its world production in 2004 was estimated at about 8 million tons. The *trans*-form (Fig. 4.6b), called *gutta percha* or *balata*, readily crystallizes forming rigid materials melting at about 70 °C. As in the case of papermaking and cotton textile, the extraction and processing technology of these valuable natural polymers, as well as the preparation and optimization of their corresponding materials, represent a well-established and well-documented know-how [14]. Examples of interesting recent contributions to the biosynthesis [15] and chemical modification [16−18] of natural rubber are available. Moreover, the use of natural rubber in blends with other biopolymers like starch and lignins are discussed in the corresponding chapters.

4.2.1.5 Suberin

The other macromolecule found only in certain wood species is suberin. This nonlinear polyester contains very long aliphatic moieties which impart a characteristic hydrophobic feature to the natural

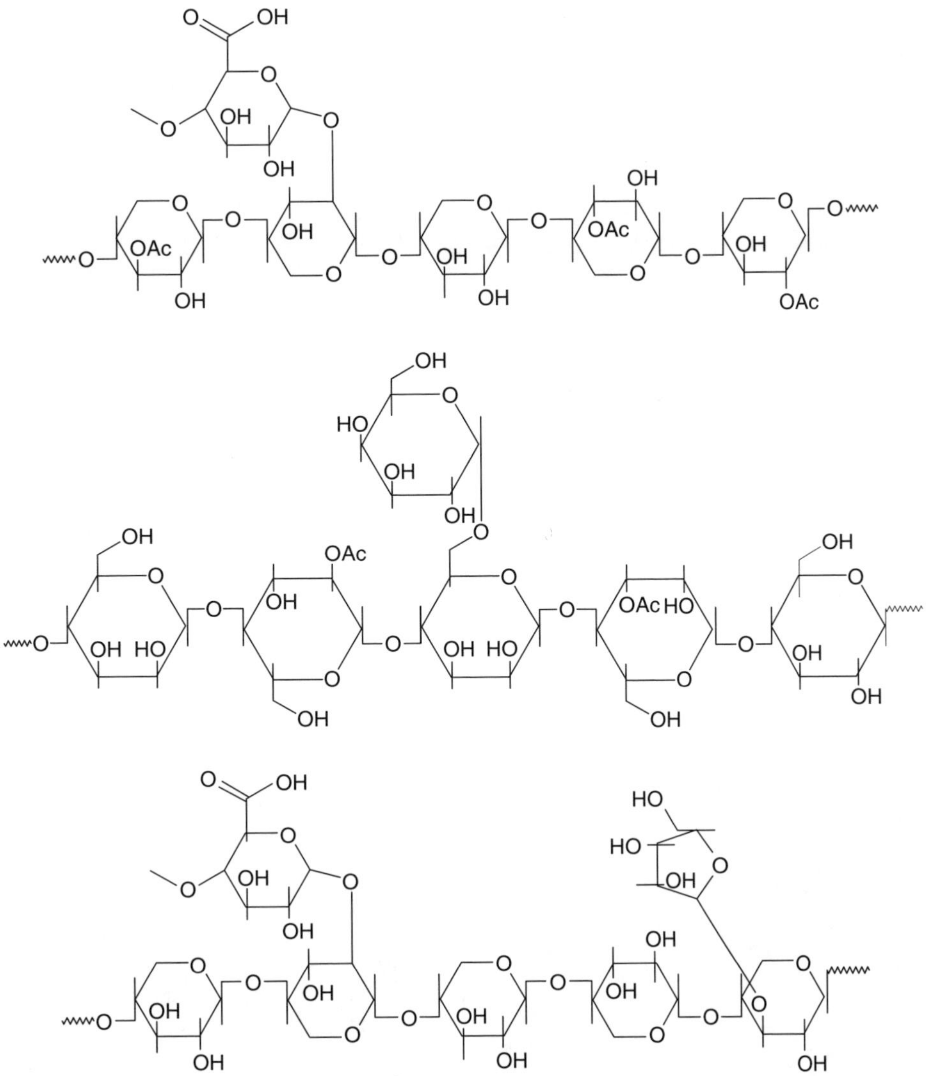

Figure 4.5 Three typical hemicellulose structures.

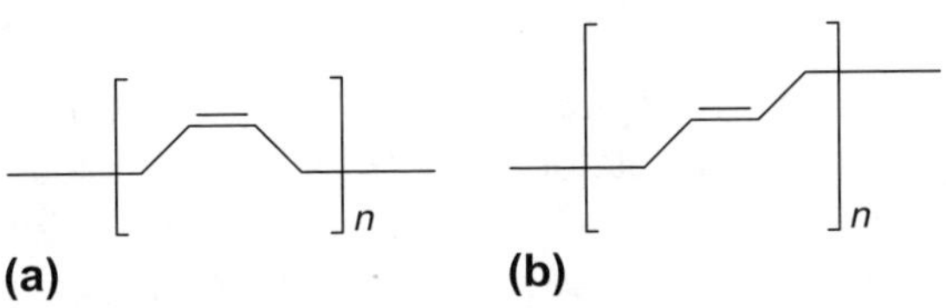

Figure 4.6 The two main structures of poly(1,4-isoprene) in natural rubber: (a) the *cis*-form and (b) the *trans*-form.

material that contains it. Figure 4.7 shows a schematic structure of suberin. By far the most representative species containing this polymer in its very thick bark (the well-known cork) is *Quercus suber*, which grows in the Mediterranean area, but Nordic woods like birch, also have a thin film of suberin coating their trunks. The sources of suberin, as well as the corresponding structure and composition, are now an important research area, together with the use of suberin fragments, obtained from its hydrolytic splicing, as monomers for the synthesis of biodegradable aliphatic polyesters.

4.2.1.6 Tannins

Among the polyphenols present in the tree barks, tannins are by far the most interesting oligomers (molecular weights of 1000–4000) in terms of their utilization as macromonomers for the crosslinking of proteins in leather (tanning) and for macromolecular syntheses. The two representative structures of the

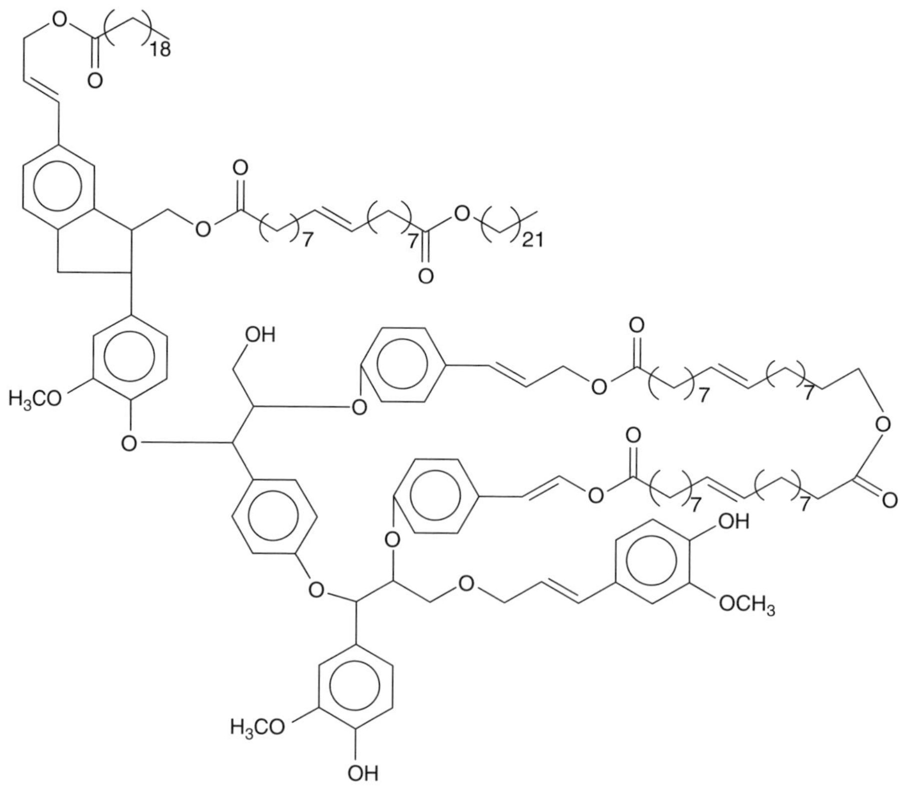

Figure 4.7 A schematic representation of the structure of suberin.

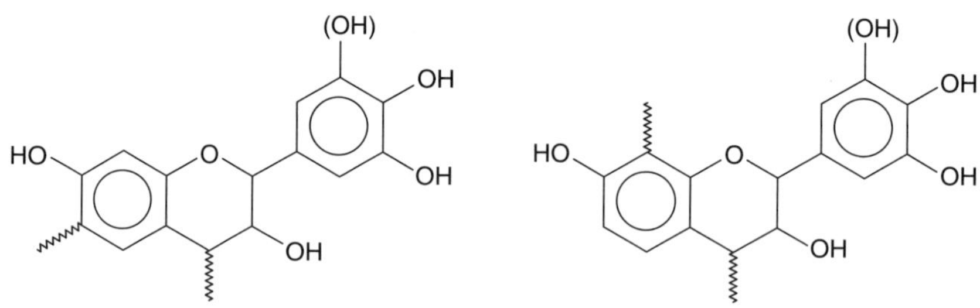

Figure 4.8 Two typical monomer units found in tannins.

flavonoid units in tannins are shown in Fig. 4.8. The most salient aspects are related to the sources, structures, and production of tannins and to their exploitation in polymer modification and manufacture.

4.2.1.7 Wood Resins

A number of resinous materials are secreted by trees. Their molecular weights are low (a few hundreds to a few thousands), hence a low melt viscosity, but their glass transition temperature can be as high as 100 °C. Rosins (extracted from pine trees) are the most important representative of this family. They are made up of a mixture of unsaturated polycyclic carboxylic acids of which abietic acid (Fig. 4.9) is the major representative. The sources, extraction, structures, and chemical modification of these substances, together with their use as sources of polymeric materials, constitute a realm that is currently being revived.

4.2.1.8 Terpenes

Apart from all these natural polymers and oligomers, either general or specific wood components, certain trees also produce monomers in the form of terpenes, which are unsaturated cyclic hydrocarbons of general formula $(C_5H_8)_n$, with n mostly equal to 2.

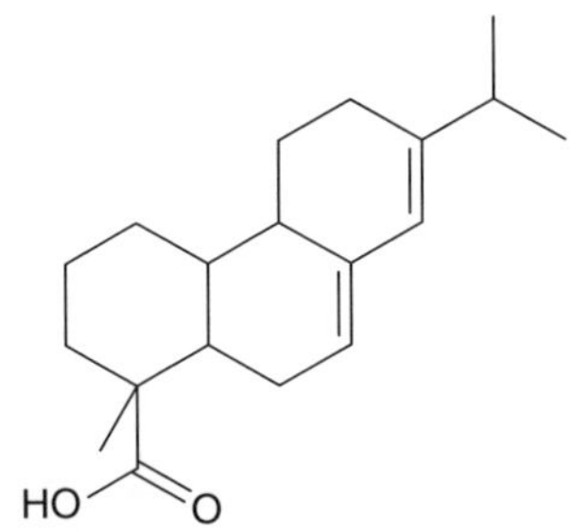

Figure 4.9 Abietic acid.

Figure 4.10 shows typical representatives of such compounds. The description of the sources of these monomers, their relative abundance, and their polymerization represent important traditional topics, with novel insight arising from the application of more efficient types of macromolecular syntheses.

Numerous other interesting molecules are found in the different elements of the tree anatomy, which find specific uses as a function of their structure, e.g., as medicines, cosmetics, dyestuffs, etc., but which are not exploitable in polymer synthesis. These valuable compounds have been the object of much (still ongoing) research and development [19].

4.2.2 Annual Plants

The term annual plant is used here to define plants and crops with a typical yearly turnover, but also includes species with shorter or longer cycles. The primary interest of annual plants, which have been optimized by human selection throughout the ages, is the production of food. Nevertheless, man has also exploited their residues for different purposes, including shelter, clothing, etc. In a complementary vein, annual plants have also been grown for the production of medicines, dyestuffs, and cosmetics. In the framework of this book, attention will be focused on two different types of raw materials, namely the fibrous morphologies of their basic structure and the substances they produce.

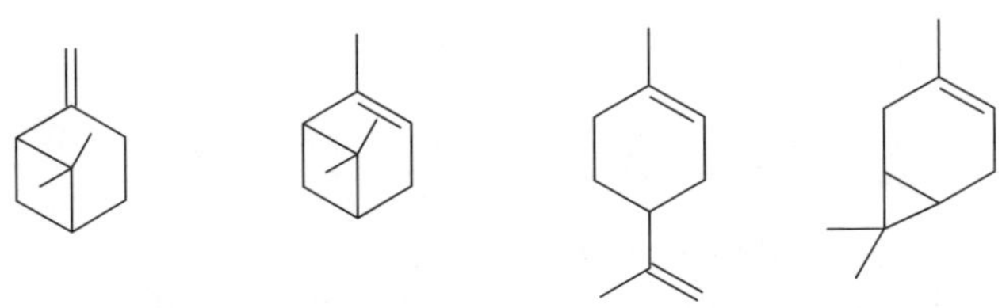

Figure 4.10 Four common terpenes.

Concerning the former, the lignocellulosic fibers of most plant stems fall into the same category as their wood counterpart and are discussed in the chapters devoted to cellulose. As for cotton, i.e., pure cellulose fibers annually produced by the corresponding plants, its traditional uses in textile, pharmaceutical aids, etc. reflect well-established and well-documented technologies, with little relevance to the primary scope of this book, as already pointed out in the case of papermaking and natural rubber technologies.

The relevant contribution of the output of annual plants to the realm of polymer synthesis and applications stems, instead, from some specific products, namely starch as a polymer, vegetable oils as triglyceride oligomers, and hemicelluloses and monosaccharides as potential monomers or precursors to furan derivatives.

4.2.2.1 Starch

Starch is an extremely abundant edible polysaccharide present in a wide variety of tubers and cereal grains. In most of its manifestations, it is composed of two macromolecules bearing the same structural units, 1,4-D-glucopyranose, in linear (amylose, Fig. 4.11a) and highly branched (amylopectine, Fig. 4.11b) architectures, present in different proportions according to the species that produces it. The utilization of starch or its derivatives for the production of adhesives, or as wet-end additives in papermaking, constitutes traditional applications to which a variety of novel materials, including plasticized starch, blends, and composites, have been recently added from a worldwide flurry of fundamental and technological investigations.

4.2.2.2 Vegetable Oils

Vegetable triglycerides are among the first renewable resources exploited by man primarily in coating applications ("drying oils"), because their unsaturated varieties polymerize as thin films in the presence of atmospheric oxygen. This property has been exploited empirically for millennia and has received much scientific and technological attention in the last few decades. These oils are extracted from the seeds or fruits of a variety of annual plants, mostly for human consumption. Within their general structure, consisting of glycerol esterified by three

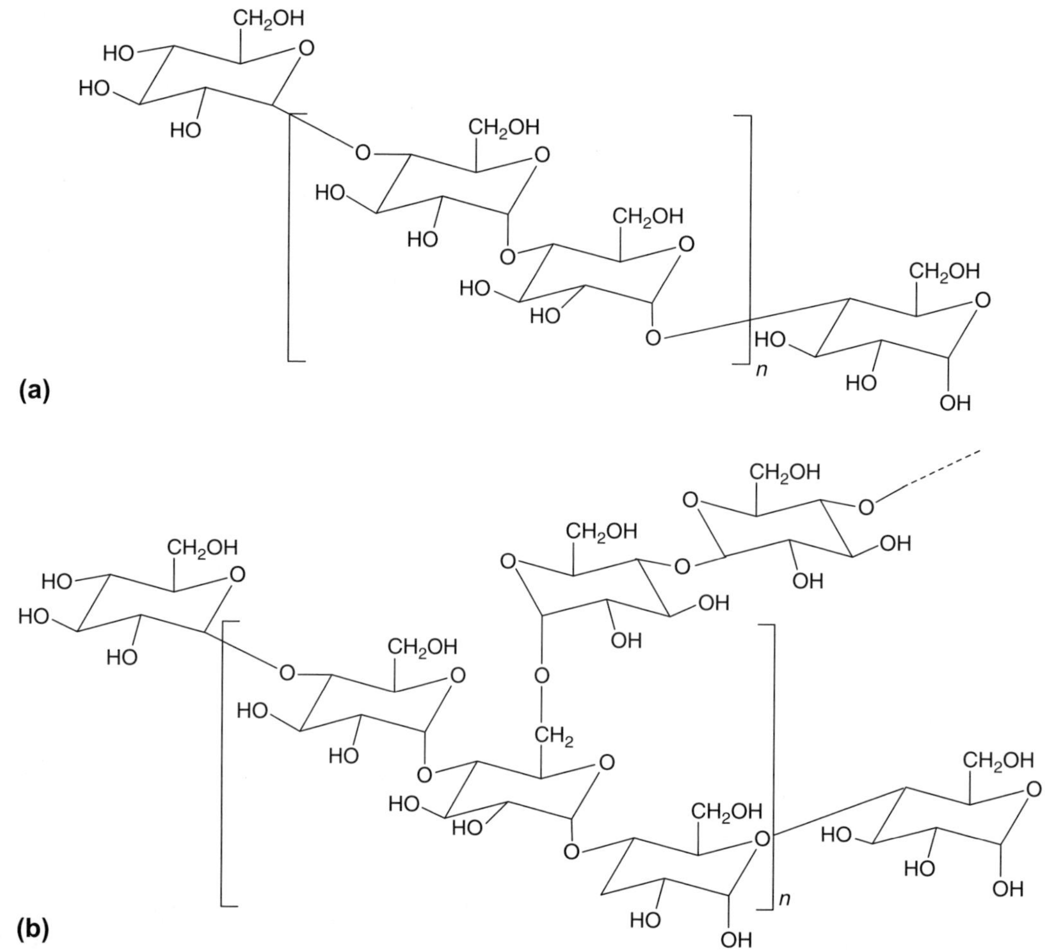

Figure 4.11 The two macromolecular components of starch: (a) amylose and (b) amylopectin.

long-chain aliphatic acids (Fig. 4.12) bearing variable number of carbon atoms, the most relevant difference is undoubtedly the number of C=C unsaturations borne by the chains, but other more peculiar features are also encountered, e.g., hydroxyl moieties. Their essential role as components of paints and inks constitutes the most important application for the elaboration of materials. This traditional

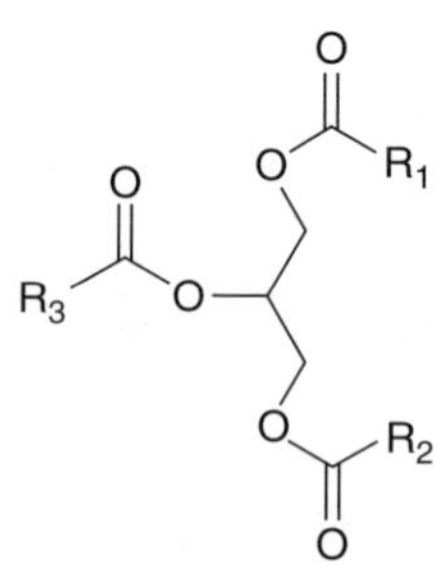

Where R_1, R_2 and R_3 are fatty acid chains.

Figure 4.12 The basic structure of triglycerides.

technology is presently being updated through research aimed at modifying the pristine structure of the oils in order to enhance their reactivity, in such realms as photosensitive coatings, polyurethanes, and other macromolecular materials, to render them competitive with respect to petroleum-based counterparts.

The explosion of activities related to the synthesis of biodiesel through transesterification reactions of different triglycerides has generated a growing interest in glycerol (a by-product of these syntheses) and its chemical transformation into other useful chemicals. This topic has been recently reviewed in Refs. [20–22].

4.2.2.3 Hemicelluloses

Annual plants produce a rich selection of hemicelluloses, often with quite different structures compared with those found in woods, although the basic chemical features are always those of polysaccharides. It follows that specific applications are associated with these different structural features, notably as food additives. The presence of charged monomer units is one of the most exploited characteristics, because of the ensuing rheological sensitivity to physical parameters. The properties and applications of plant and seaweed hemicelluloses is another hot topic within the present scenario, particularly within the specific features related to polyelectrolytes.

In a different vein, plants rich in C5 hemicelluloses, and more specifically xylans, are excellent raw materials for the production of furfural (Fig. 4.13). This simple technology, developed more than a century ago, has been applied to a whole host of annual plant residues, after the extraction of their food component, ranging from corn cobs, rice hulls, and sugar cane bagasse to olive husks. The industrial production of furfural is therefore possible in any country and indeed implemented in many of them, because of the wide variety of biomass containing its precursor, and it represents a beautiful example of the exploitation of renewable resources using a readily implemented and cheap process.

The use of furfural as such, as well as its transformation into a variety of furan monomers, together with the synthesis and properties of the corresponding macromolecular materials, is perceived today as a "sleeping giant" awaiting to irrupt on the scene of polymer science and technology.

Another possible exploitation of annual plant residues, after the separation of their foodstuff, is their conversion into polyols by oxypropylation. In this context, the whole of the residue is involved in the transformation, providing a convenient and ecological source of macromonomers for novel materials like polyurethanes, polyethers, and polyesters.

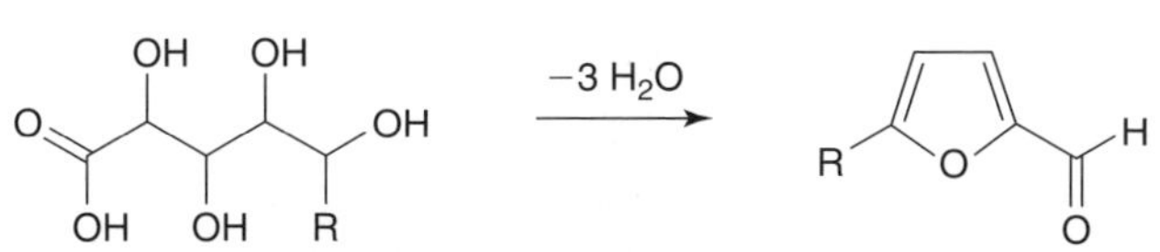

R = H (furfural) or CH_3 (5-methyl furfural)

Figure 4.13 Schematic conversion of aldopentoses into furfural and 5-methyl furfural.

4.2.2.4 Mono- and Disaccharides

The traditional use of some of the most important mono- and disaccharides as sweeteners, whether energetic or not, is of course outside the scope of this book. The interest in using this family of compounds, produced by different annual plants, as precursors to novel materials has increased considerably in recent years, mostly in three different directions, viz., (i) the conversion of fructose to hydroxymethyl furfural (Fig. 4.14), (ii) the synthesis of polycondenzation materials using sugars as comonomers, and (iii) the preparation of surfactants based on renewable resources.

Hydroxymethyl furfural is the other first generation furan derivative readily obtained from hexoses (C6 saccharides) [22]. Its role as a precursor to a large spectrum of monomers and the interest of the ensuing polymers, complementary to those relative to furfural, represent the second face of the "sleeping giant."

The synthesis of novel polymers, mostly polyesters and polyurethanes, based on mono- and disaccharides (Fig. 4.15), together with their properties and possible applications, has attracted much attention in recent years, with very convincing results in terms of materials alternative to counterparts derived from petrochemical monomers.

The fermentation of glucose has opened new avenues in the synthesis of polymers derived from renewable resources, with particular emphasis on the exploitation of lactic acid as a monomer. This topic is dealt with in Chapters 2 and 3.

4.2.3 Algae

Marine biomass is also a very interesting source of precursors to materials, both in terms of vegetable and animal resources. Polysaccharides derived from

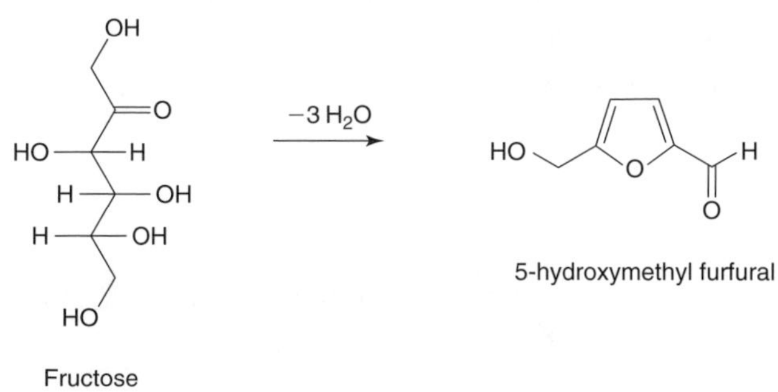

Figure 4.14 Schematic conversion of fructose into 5-hydroxymethyl furfural.

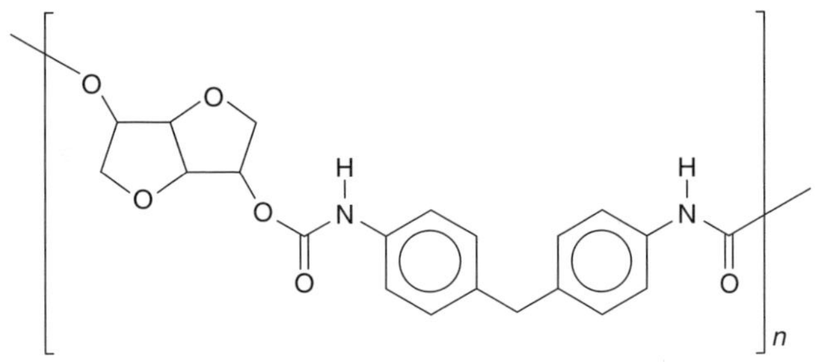

Figure 4.15 A sugar-based polyurethane.

certain algae, like alginates, have been exploited for a long time as polyelectrolyte materials.

4.3 Animal Resources

As in the case of vegetable resources, all the traditional technologies of exploitation of materials derived from the animal realm will not be discussed here. Thus, readers interested in leather [23], wool [24], silk [25], gelatin [26], animal fats and waxes [27], and carbon black [28], as well as animal-based resins like shellac [29], are invited to consult the corresponding monographs quoted here. The reason for these exclusions stems from the fact that the processes associated with the production of these materials have not been the object of any major qualitative improvement in recent times.

4.3.1 Chitin and Chitosan

Chitin is undoubtedly the most abundant animal polysaccharide on earth. It constitutes the basic element of the *exo*-skeleton of insects and crustaceans, but it is also found in the outer skin of fungi. Chitin is a regular linear polymer whose structure differs from that of cellulose by the presence of *N*-methylamide moieties instead of the hydroxyl groups at C2 (Fig. 4.16). Given the susceptibility of this function to hydrolysis, chitin often bears a small fraction of monomer units in the form of primary amino groups resulting from that chemical modification.

Chitin is sparingly soluble even in very polar solvents, because of its high cohesive energy associated with strong intermolecular hydrogen bonds (NH−CO), which is also the cause of its lack of melting, because the temperature at which this phase change would occur is higher than that of the onset of its chemical degradation, just like with cellulose. It follows that the potential uses of chitin are strongly limited by these obstacles to processing. The possibility of exploiting chitin is therefore dependant on its transformation into its deacetylated derivatives through hydrolysis. As the proportion of the amide function converted into primary amino groups increases along the macromolecule, the corresponding material becomes progressively more soluble in such simple media as weak aqueous acids or polar protic solvents. The ideal fully hydrolyzed polymer takes the name of chitosan (Fig. 4.17), a term which is in fact also attributed to all random copolymers bearing more than about 50% of amino monomer units.

Chitosan has become one of the most attractive polymers derived from renewable resources, because it possesses remarkable properties that find applications in many areas of material science and technology, particularly related to biomaterials and medical aids. It is not exaggerated to talk about the boom of chitosan-related research, considering the explosion of scientific and technical literature on this polymer, accompanied by the creation of learned societies and the frequent international meetings covering its progress. Industrial units devoted to the extraction of chitin followed by the production of chitosan are springing up throughout the world because the materials derived from chitin and chitosan are steadily gaining in importance thanks to their widening variety of high-tech applications.

4.3.2 Proteins

Because of their highly polar and reactive macromolecular structure, proteins have attracted

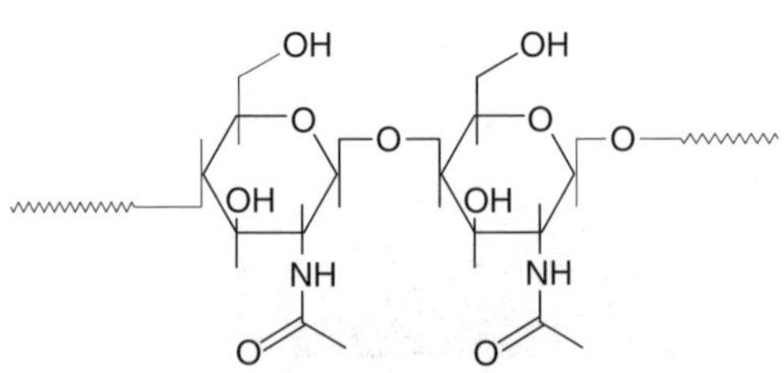

Figure 4.16 Chitin.

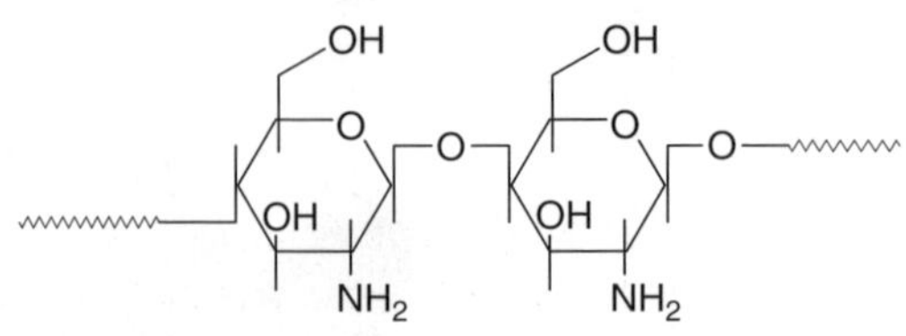

Figure 4.17 The main monomer units in chitosan.

much attention in the last few decades as possible sources of novel polymeric materials. A particularly interesting natural proteinic material is undoubtedly the spider dragline silk, because of its extraordinary mechanical properties. Given the obvious difficulties related to gathering viable amounts of this biopolymer, much research is being devoted to its bioengineering production [30].

4.3.3 Cellulose Whiskers and Nanofibrils

Although cellulose is the supreme example of a *predominantly vegetable* natural polymer, exotic animal species are known to produce this polysaccharide and, more particularly, some of its most regular manifestations. Thus, the tunicate mollusk has become the very symbol associated with cellulose whiskers, i.e., extremely regular nanorods with remarkable mechanical and rheo-modifying properties. Researchers have been inspired by this natural manifestation in the last decade and a whole realm has flourished, based on the production, characterization, and exploitation of whiskers and nanofibrils from conventional cellulose fibers.

4.4 Bacterial Polymers

Although the polymerization induced by bacteria has been known and studied for a long time, the strategy based on using this biological activity to actually harvest commercial materials is a relatively recent endeavor. Two specific instances are prominent in this context, namely the production of poly(hydroxyalkanoates) and the synthesis of bacterial cellulose.

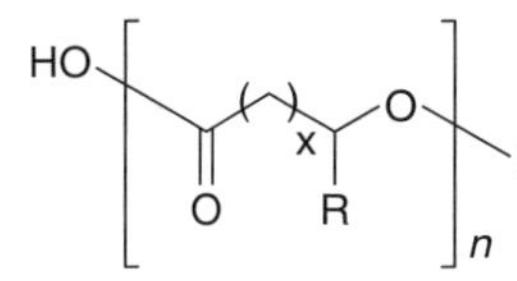

Figure 4.18 The general formula of poly (hydroxyalkanoates).

4.4.1 Poly(Hydroxyalkanoates)

This family of polyesters and copolyesters (Fig. 4.18) has interested the polymer community both because of their remarkable physical properties and biodegradability. Efforts have been actively implemented to improve the economy of the biotechnological processes used to prepare these materials, so that they can become commercially competitive compared with petroleum-based polymers with similar properties.

4.4.2 Bacterial Cellulose

Although the chemical structure of bacterial cellulose is identical to that of any other vegetable-based counterpart, its fibrous nanomorphology (Fig. 4.19), as obtained directly in its biotechnological production, is unique and consequently the properties associated with this original material are also peculiar and promise very interesting applications, notably in the area of biomedical artifacts.

4.5 Conclusions

Several monographs have been published in recent years [31–40], including two excellent reviews by Corma *et al.* covering the transformation of biomass for the production of fuel [41], chemicals [21] and other recent contributions [42,43]. The concept of the bio-refinery [31,44] is explicitly or implicitly at the

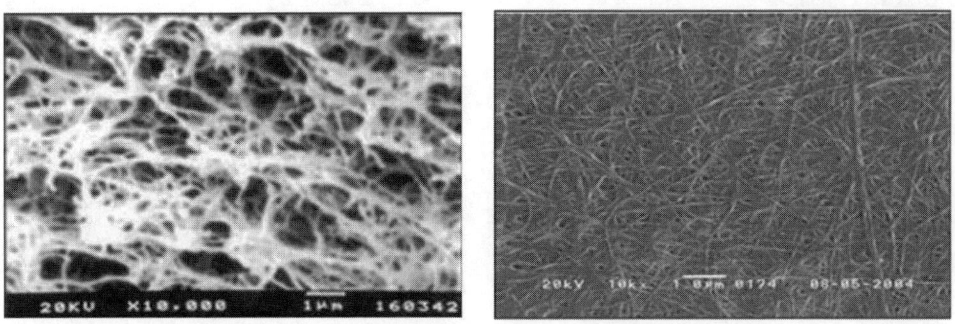

Figure 4.19 The unique morphology of bacterial cellulose.

basis of all these treaties, i.e., the working hypothesis proposing a rational separation and exploitation of all the components of a given natural resource. The other common denominator to many of these collective overviews is the biodegradable character of the ensuing material.

A recent book provided a thorough approach to all the aspects sketched above [45], with the materials science elements as the predominant feature. This was the first attempt at a comprehensive treatment of the topic of monomers, polymers, and composites derived from renewable resources. Given the fast pace of progress in the area, updates have been published [46–48] to keep abreast of novel contributions.

References

[1] D. Fengel, G. Wegener, Wood Chemistry Ultrastructure Reactions, Walter de Gruyter, Berlin, 1989;
J.F. Kennedy, G.O. Philips, P.A. Williams, The Chemistry and Processing of Wood and Plant Fibrous Materials, Woodhead, Cambridge, UK, 1996.

[2] C.A.S. Hill, Wood Modification: Chemical, Thermal and Other Processes, John Wiley & Sons, Ltd, Chichester, 2006.

[3] Pentti O. Kettunen, Wood: structure and properties, in: Mater. Sci. Found, Trans Tech Publications Ltd., Uetikon-Zuerich, 2006, pp. 29–30.

[4] R.M. Rowell, Handbook of Wood Chemistry and Wood Composites, CRC Press, Boca Raton, 2005.

[5] J. Gullichsen, H. Paulapuro, Papermaking Science and Technology, vol. 2 & 3, Fapet Oy, Helsinki, 1999.

[6] J.C. Roberts, Paper Chemistry, third ed. Chapman & Hall, United Kingdom, 1996;
D. Eklund, T. Lindström, Paper Chemistry: An Introduction, DT Paper Science Publications, Grankulla, Finland, 1991.

[7] J.P. Casey, Pulp and Paper Chemistry and Chemical Technology, third ed., vol 1–4, John Wiley and Sons, New York, 1980.

[8] J. Gullichsen, H. Paulapuro, Papermaking Science and Technology, vol 4–10, Fapet Oy, Helsinki, 1999.

[9] P.J. Wakelyn, N.R. Bertoniere, A.D. French, D.P. Thibodeaux, B.A. Triplett, M.A. Rousselle, W.R. Goynes Jr., J.V. Edwards, L. Hunter, D.D. McAlister, G.R. Gamble, Cotton Fibre Chemistry and Technology,, CRC Press, Boca Raton, 2007.

[10] M. Ash, I. Ash, Handbook of Textile Processing Chemicals, Synapse Information Resources, Endicott, NY, 2001;
A.R. Horrocks, S.C. Anand, Handbook of Technical Textiles, Woodhead Publ., Cambridge, UK, 2000.

[11] T. Heinze, T. Liebert, A. Koschella, Esterification of Polysaccharides, Springer, Berlin, 2006.

[12] D.N.S. Hon, N. Shiraishi, Marcel Dekker Inc., Wood and Cellulosic Chemistry New York, 1992.

[13] K. Goetze, Synthetic Fibers by the Viscose Process, Springer-Verlag, 1967;
T.P. Burt, A.L. Heathwaite, T.S. Trudgill, Nitrate: Processes, Patterns and Management, Wiley, Chichester, UK, 1993.

[14] R. Brendan, Rubber Compounding: Chemistry and Applications, Marcel Dekker Inc., New York, 2004;
D.A.D. Parry, J.M. Squire, Fibrous Proteins: Coiled-Coils, Collagen and Elastomers [Adv. Protein Chem., 70, 2005], Elsevier, San Diego, 2005.

[15] J.E Puskas, E. Gautriaud, A. Deffieux, J.P. Kennedy, Prog. Polym. Sci. 31 (2006) 533.

[16] S. Kawahara, T. Saito, J. Polym. Sci. Part A: Polym. Chem. 44 (2006) 1561.

[17] W. Kangwansupamonkon, R.G. Gilbert, S. Kiatkamjornwong, Macromol. Chem. Phys 206 (2005) 2450.

[18] M. Jacob, B. Francis, K.T. Varughese, S. Thomas, Macromol. Mater. Eng 291 (2006) 1119.

[19] D.Y. Murzin, P. Maki-Arvela, T. Salmi, B. Holmbom, Chem. Eng. Technol 30 (2007) 569.

[20] M. Pagliaro, R. Ciriminna, H. Kimura, M. Rossi, C. Della Pina, Angew. Chem. Int. Ed 46 (2007) 4434.

[21] A. Corma, S. Iborra, A. Velty, Chem. Rev 107 (2007) 2411.

[22] J.N. Chheda, G.W. Huber, J.A. Dumesic, Angew. Chem. Int. Ed 46 (2007) 7164.

[23] K.J. Bienkiewicz, Physical Chemistry of Leather Making, Krieger, Melbourne, FL, 1983.

[24] W.S. Simpson, G.H. Crawshaw, Wool: Science and Technology, Woodhead Publ., Cambridge, UK, 2002.

[25] T. Scheibel, Silk. [Special Issue of Appl. Phys. A: Mater. Sci. Process., 82, 2006], Springer, Heidelberg, 2006.

[26] R. Schreiber, H. Gareis, Gelatine Handbook, Wiley-VCH, Weinheim, 2007.

[27] R.J. Hamilton, Waxes: Chemistry, Molecular Biology and Functions, The Oily Press, Dundee, UK, 1995.;
K.G. Berger, Animal fats-BSE and after, in: P.J. Barnes, Associates (Eds.), Proceedings of a Joint Meeting of the SCI Oils & Fats Group and the SCI Food Commodities & Ingredients Group, held in London, UK, 10 June 1997, Bridgewater, UK, 1997.

[28] J.B. Donnet, R.C. Bansai, M.J. Wang, Carbon Black: Science and Technology, second ed. Marcel Dekker, New York, 1993.

[29] E. Hicks, Shellac: Its Origin and Applications, MacDonald, London, 1962.

[30] E. Bini, C.W. Po Foo, J. Huang, V. Karageorgiou, B. Kitchel, D.L. Kaplan, Biomacromolecules 7 (2006) 3139.

[31] B. Kamm, P.R. Gruber, M. Kamm, Biorefineries—Industrial Processes and Products, vol 1 & 2, Wiley VCH, Weinheim, 2006.

[32] K. Khemani, C. Scholz (Eds.), Degradable Polymers and Materials: Principles and Practice. ACS Symp. Ser. 939, ACS, Washington DC, 2006.

[33] M. Fingerman, R. Nagabhushanam, Biomaterials from Aquatic and Terrestrial Organisms, Science Publishers, Enfield, NH, 2006.

[34] S.S. Im, Y.H. Kim, J.S. Yoon, I.J. Chin (Eds.), Bio-Based Polymers: Recent Progress, Wiley, New York, 2005.

[35] G. Scott, Degradable Polymers, In: Principles and Applications Kluwer Academic, New York, 2003.

[36] A. Steinbüchel, R.H. Marchessault, Biopolymers for Medical and Pharmaceutical Applications, Wiley, Weinheim, 2005.

[37] C.V. Stevens, R.G. Verhé, Renewable Bioresources: Scope and Modification for Non-Food Applications, John Wiley & Sons, Ltd, Chichester, 2004.

[38] R.P. Wool, X.S. Sun, Bio-Based Polymers and Composites, Elsevier, Amesterdam, 2005.

[39] Wolf O., Techno-Economic Feasibility of Large-scale Production of Bio-Based Polymers in Europe, Technical Report EUR 22103, European Commission, 2005.

[40] ACS Symposium Series 954D.S. Argyropoulos (Ed.), Materials, Chemcals and Energy from Forest Biomass, ACS, Washingon DC, 2007.

[41] G.W. Huber, S. Iborra, A. Corma, Chem. Rev 106 (2006) 4044.

[42] P. Gallezot, Green Chem 9 (2007) 295.

[43] M. Darder, P. Aranda, E. Ruiz-Hitzky, Adv. Mater. 19 (2007) 1309.

[44] J. Sanders, E. Scott, R. Weusthuis, H. Mooibroek, Macromol. Biosci. 7 (2007) 105.

[45] M.N. Belgacem, A. Gandini (Eds.), Monomers, Polymers and Composites from Renewable Resources, Elsevier, Amsterdam, 2008.

[46] A. Gandini, Macromolecules 41 (2008) 9491.

[47] A. Gandini, Green Chem. 13 (2011) 1061. and refs. therein.

[48] P. Gazellot, Chem. Soc. Rev. 41 (2012) 1538.

5 Polymeric Biomaterials

Wei He and Roberto Benson

OUTLINE

5.1 Introduction

Biomaterials is an exciting and highly multidisciplinary field. These materials have matured into an indispensable element in improving human health and quality of life in the modern era. Applications of biomaterials range from diagnostics such as gene arrays and biosensors, to medical supplies such as blood bags and surgical tools, to therapeutic treatments such as medical implants and devices, to emerging regenerative medicine involving tissue-engineered skin and cartilage, etc. A general classification divides biomaterials into three main categories: metals, ceramics, and polymers. Polymers, being organic in nature, offer a versatility that is unmatched by metals and ceramics. The wide spectrum of physical, mechanical, and chemical properties that polymers can provide has fueled the extensive research, development, and applications of polymeric biomaterials. Furthermore, the significance of polymers in the field of biomaterials is clearly reflected in the staggering market size of medical polymers, estimated to be roughly $1 billion with yearly growth of 10–20% [1].

This chapter provides a brief overview of several medical applications that polymers have made seminal contributions to over the years. Many of the polymers discussed here are initially developed as plastics, elastomers, and fibers for nonmedical industrial applications. They were "borrowed" by the surgeons post–World War II to address medical problems. Since then, they have led to the development of biomedical-specific materials. Currently, with the rapid growth in modern biology and collaborative effort, cross-discipline work involving materials science, engineering, chemistry, biology, and medicine, is resulting in polymeric biomaterials that are bioactive, biomimetic, and, most importantly, have excellent biocompatibility. Examples of this newer generation of polymeric biomaterials are also included in this chapter.

5.2 Polymeric Biomaterials in Ophthalmology

Ophthalmology focuses on the diseases of the eye, which is a complex and vital organ for daily life.

Ebnesajjad: Handbook of Biopolymers and Biodegradable Plastics. http://dx.doi.org/10.1016/B978-1-4557-2834-3.00005-7
© 2011 Elsevier Inc. All rights reserved. Reproduced from a chapter in: Kutz, *Applied Plastics Engineering Handbook* (2011).

Application of biomaterials in ophthalmology can be dated back to the mid-nineteenth century, when Adolf Fick successfully invented the glass contact lens. Since then, a wide variety of ophthalmological biomaterials have been developed and some are finding overwhelming success in clinical applications. Applications of biomaterials in ophthalmology include contact lenses [2], intraocular lenses (IOLs) [3], artificial orbital walls [4], artificial corneas [5], artificial lacrimal ducts [6], glaucoma filtration implants [7], viscoelastic replacements [8], drug delivery systems [9], scleral buckles [10], retinal tacks and adhesives [11], and ocular endotamponades [12]. Although ceramics and metals have also been used in ophthalmology, modern ophthalmic implants are mainly made of polymers. The focus of this section will be on polymers used for contact lens, IOL, and artificial corneas.

5.2.1 Polymeric Contact Lens

A contact lens is an optical device placed on the cornea of the eye for corrective, therapeutic, or cosmetic effects. It is estimated that there are approximately 125 million contact lens wearers worldwide. A myriad of principle properties have been sought in high performance contact lens materials, including (1) good transmission of visible light; (2) high oxygen permeability; (3) tear-film wettability; (4) resistance to deposition of components from tear-film, such as lipid, protein, and mucus; (5) ion permeability; (6) chemical stability; (7) good thermal conductivity; and (8) amenability to manufacture [13]. A wide variety of polymers have been used in contact lenses, and their modulus of elasticity defines contact lenses to be either hard or soft. Structures of the various monomers commonly used in contact lenses are shown in Fig. 5.1. The first generation of polymeric contact lenses was made of poly(methyl methacrylate) (PMMA), a polymer commercially known as Plexiglas® and a classical example of hard or rigid lens material. PMMA can be prepared using bulk free radical polymerization and lathed into lens shape. It has excellent optical properties such as index of refraction with greater clarity than glass, remarkable durability, and good resistance against deposition of components from the tear-film due to its hydrophobicity. However, major drawbacks such as lack of oxygen permeability and tendency to change the shape of the eye have limited the usage of PMMA contact lenses. In order to improve the permeability of oxygen, rigid gas-permeable (RGP) contact lenses were developed in the late 1970s. Materials used for RGP contact lenses are typically copolymers of methyl methacrylate (MMA) with a monomer that imparts high oxygen permeability, e.g., methacryloxypropyl tris(trimethylsiloxy silane) (TRIS), hexafluoroisopropyl methacrylate (HFIM), and 2,2,2-trifluoroethyl methacrylate (TFEMA). The incorporation of highly hydrophobic siloxane into the copolymer reduces the lens wettability, which leads to undesired increase of lipid deposition. Therefore, hydrophilic monomers, such as methacrylic acid (MAA), 2-hydroxyethyl methacrylate (HEMA), or *N*-vinyl-2-pyrrolidone (NVP), are commonly used as wetting agents in RGP lens formulation to compensate for reduction in wettability.

Soft contact lenses emerged in the 1960s when Otto Wichterle developed poly(2-hydroxyethyl methacrylate) (PHEMA) [14] and forever changed the contact lens industry. Generally, soft contact lenses are made from hydrogel, a cross-linked network capable of retaining a significant amount of water. The first PHEMA soft lens contained 40% water of hydration. Despite its improvement in wearer comfort over rigid lens, the low oxygen permeability of PHEMA was interfering with the normal corneal metabolism. Since the extent of hydration directly affects the permeability of oxygen, hydrogels with high water content (>50%) have been developed by copolymerizing HEMA with highly hydrophilic monomers such as NVP, MAA, and glyceryl methacrylate (GMA). A drawback associated with increased hydrophilicity is the higher protein binding to the lens, which could cause discomfort and complications such as increased bacterial adhesion [15]. High water content hydrogels also tend to cause corneal desiccation. In the quest for high oxygen permeability, researchers have developed a new type of siloxane-containing hydrogels for soft contact lenses. It is well known that due to the bulkiness of the siloxane groups ($-Si(CH_3)_2-O-$) and the chain mobility, siloxane-containing materials typically have high diffusivity of oxygen. On the other hand, siloxane materials are highly hydrophobic and, therefore, prone to lipid deposition and are less comfortable with rubbery-like behavior. To offset these shortcomings, functionalized siloxane macromer (shown in Fig. 5.1) was copolymerized with hydrophilic monomers (e.g., NVP and HEMA) into hydrogels that offer

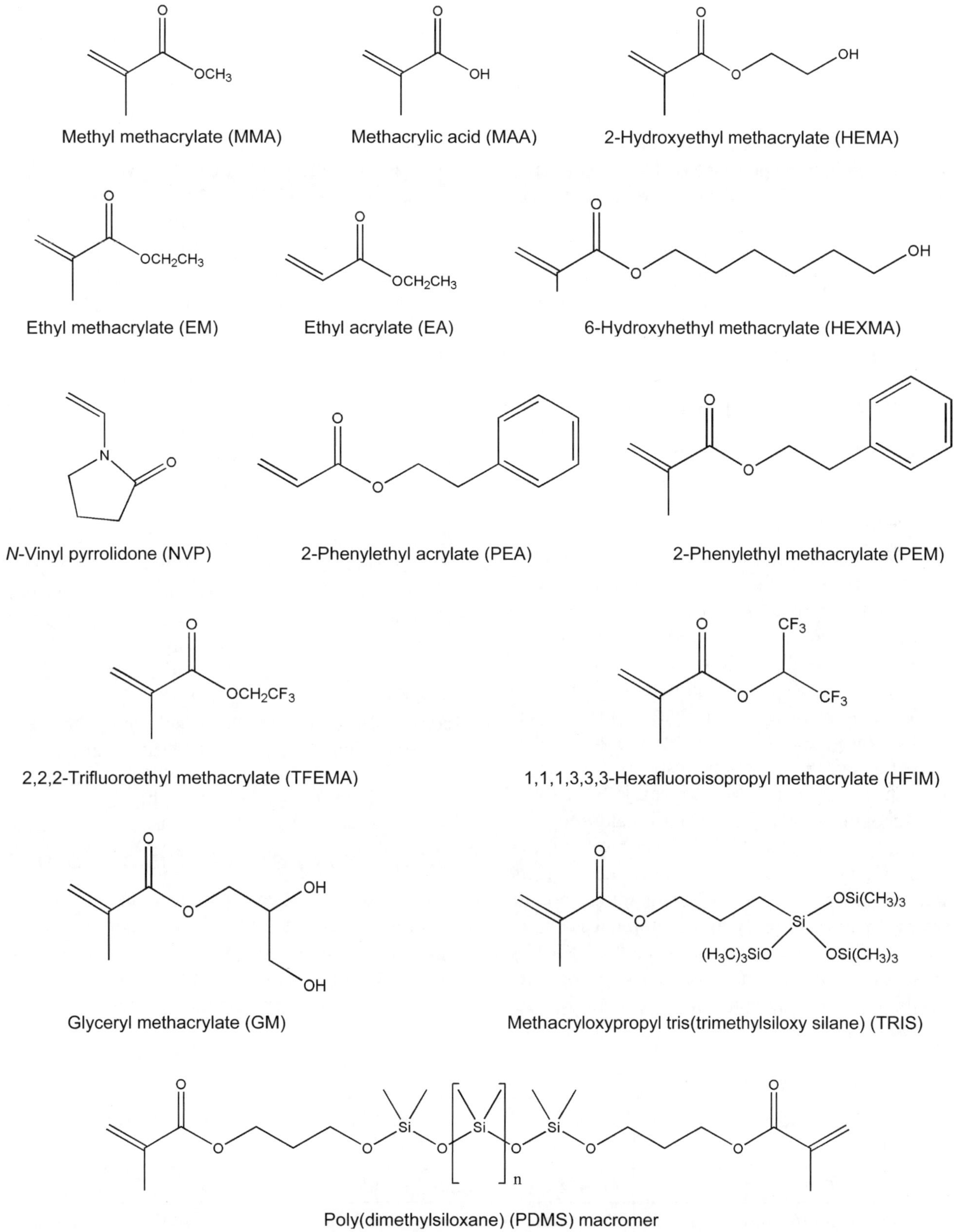

Figure 5.1 Chemical structures of common monomers used in contact lenses and intraocular lenses.

sufficiently high oxygen transmission required by the cornea as well as the softness for comfortable extended wear. Currently, commercialized siloxane hydrogel contact lenses include Focus® Night & Day™ (lotrafilcon A by CIBA Vision Corp.) and PureVision™ (balafilcon A by Bausch and Lomb). It is worth noting that the presence of siloxane moieties on the surface of these hydrogels demands further treatments in order for the lens to be tolerated on the eye. Examples of surface treatment for siloxane hydrogels include radiofrequency glow discharge (RFGD) [16] and graft polymerization of hydrophilic monomers (e.g., acrylamide [17]) on the lens surface to improve surface hydrophilicity.

5.2.2 Polymeric Intraocular Lens

IOLs are commonly used to replace natural lenses and provide clear optical imaging for patients undergoing cataract surgery. IOL is a major area in ocular biomaterials research for its critical role in treating cataract-induced blindness, which was predicted to reach 40 million cases by the year 2020 [18]. IOL also holds a special place in the biomaterials history, where its invention was originated from Sir Harold Ridley's accurate observations of the biological reaction to accidentally implanted pieces of canopy in a World War II pilot's eyes. Since the canopy material, PMMA, was well tolerated by the eye, Ridley was inspired to use the material to invent the first biocompatible IOL, and it is well recognized as a pioneering breakthrough in biomaterial science. The key material requirements for IOLs include the optical property, i.e., able to maintain a clear path for optical imaging and the long-term biocompatibility as the implant is intended to reside in the eye permanently. PMMA dominated the IOL market for 40 years before other materials emerged. Despite its excellent optical properties and relative tolerance by the eye, PMMA still induces damage to the tissues around the IOL implant. Of primary concern are the injury of the corneal endothelium associated with the lens rigidity, and the accumulation of inflammatory cells to the IOL surface, which can lead to complications such as iris adhesion to the IOL, uveitis, and loss of vision [19]. Such issues have led to newer IOL designs and materials selection. In contrast to the original hard and bulky PMMA IOL, common IOLs nowadays are featured as soft and foldable.

The most widely used foldable IOL, AcrySof®, is fabricated from a copolymer of phenylethyl acrylate and phenylethyl methacrylate with a cross-linking reagent and a UV-absorbing chromophore. Its improved optical property, i.e., higher refractive index (n = 1.55) compared to PMMA (n = 1.49), allows a thinner IOL configuration. The mechanical characteristic of the copolymer results in a slow and better controlled unfolding of the IOL, which contributes to the significant reduction in posterior capsular opacification (PCO). Other materials used in foldable IOL fabrication include silicone elastomers, hydrophilic acrylics (with water content higher than 18%), and collagen copolymers (Table 5.1). Although hydrophilic acrylic IOLs have shown good uveal biocompatibility due to the reduction in protein adsorption and macrophage adhesion, they tend to present higher rate of PCO and cause anterior capsular opacification, which has reduced their application in the market [22–25].

5.2.3 Polymeric Artificial Cornea

The cornea is a transparent tissue situated at the front of the eye. It is the main element in the ocular optical system, and plays various roles from refracting light onto the retina to form an image, to acting as a protective barrier for the delicate internal eye

Table 5.1 Examples of Biomaterials for IOLs

Manufacturer	Lens Type	Material	Refractive Index
Advanced Medical Optics	Rigid	PMMA	1.49
ALCON	ACRYSOF® foldable	PEA/PEMA	1.55
Bausch & Lomb	Hydroview® foldable	HEMA/HEXMA	1.47
Calhoun Vision	Multifocal foldable	PDMS	1.41
STAAR Surgical	Collamer® foldable	Collagen/HEMA	1.45

(Modified from Lloyd and Patel [20,21])

tissue. Damage to the cornea can result in loss of vision, which accounts for the second most common cause of blindness worldwide after cataract [26]. The most widely accepted treatment of corneal blindness is transplantation of human donor corneas. However, limitations in the availability of donor cornea tissues have called for design and development of artificial cornea substitutes. Artificial corneas, also known as keratoprostheses, come in a variety of forms, from fully synthetic to tissue-engineered. The focus of this discussion will be on polymer-based synthetic keratoprostheses. Several excellent comprehensive reviews on artificial corneas are available for further reading [27–29].

The cornea tissue is complex, avascular, highly innervated, and immune privileged. It is arranged in three major cellular layers: an outer stratified epithelium, an inner single-layered endothelium, and sandwiched in between a stromal compartment, which is responsible for the optical properties of the cornea. Although it is challenging to duplicate the complex structure of the natural cornea, it is possible to construct an artificial cornea that can simulate the physical features of the natural cornea and restore some functional level of vision. An ideal artificial cornea should meet the following specific requirements: (1) transparent with a smooth anterior surface of appropriate curvature, (2) suitable refractive index, (3) flexible and sufficient tensile strength for surgical handling, (4) ability to heal with the host cornea, (5) ability to promote and sustain the growth of epithelium over the anterior surface of the artificial cornea, (6) ability to avoid the formation of a retrocorneal fibroblastic membrane, and (7) biocompatibility [30]. Early generations of artificial corneas were made from a number of different hydrophobic polymers, such as PMMA, nylon, poly(tetrafluoroethylene) (PTFE), polyurethane (PU), and poly(ethylene terephthalate) (Dacron®) [20,31–33]. The design evolved from one material button-like full piece to the more widely used "core-and-skirt" configuration, where the core is made from transparent material with good optical properties and the skirt is made either from the same or different material to ensure host integration. Among these polymers, PMMA is arguably the most extensively used due to its remarkable optical properties, as discussed in the above IOL section. Even though the application of PMMA in artificial cornea continues, the associated complications such as retroprosthetic membrane formation, glaucoma, extrusion, endophthalmitis, and rejection [34–36] have led to the development of soft, hydrogel-based artificial corneas. Most of the research has been directed toward HEMA-based hydrogel. An interesting observation with HEMA is that when the monomer is polymerized with the presence of 40% or less of water, it forms a homogeneous transparent hydrogel; when the water content is higher, phase separation occurs during polymerization, and the resulting hydrogel is heterogeneous and opaque. Taking advantage of such characteristics, the first "core-and-skirt" hydrogel-based artificial cornea was created using HEMA, and the device is commercially known as AlphaCor [37]. The core is the transparent, lower water content PHEMA, and the skirt is the phase-separated, macroporous opaque PHEMA. Even though PHEMA is considered a hydrophilic polymer, its water content remains far below the water level found in the natural cornea (78%). Such high water content is essential for the stability and survival of the epithelium as it facilitates nutrient diffusion. In order to increase the water content of the artificial cornea, various strategies have been explored. Examples include copolymerization of HEMA with an ionic acrylate MAA [38], and hydrogels made from homopolymer of poly(vinyl alcohol) (PVA), which can contain over 80% water at equilibrium [39,40]. Several groups have also reported making biomimetic hydrogels for artificial corneas. As the extracellular matrix of the cornea is dominated by type I collagen, it has been used in the preparation of a copolymeric hydrogel based on *N*-isopropylacrylamide (NIPAAm), acrylic acid, *N*-acryloxysuccinimide, and collagen [41]. The engineered hydrogel is essentially a network comprising of collagen cross-linked to the copolymers of acrylic acid and NIPAAm using the succinimide pendant groups. This material has demonstrated the biomechanical properties and the required optical clarity to be used for corneal transplantation. *In vivo* animal studies have shown successful regeneration of host corneal epithelium, stroma, and nerves [41]. Clinical trials are currently underway to evaluate this material for therapeutic use in humans.

Interpenetrating polymer networks (IPNs) have also been used for artificial cornea applications. IPN represents a mixture of polymer networks where one polymer is cross-linked in the presence of another polymer network to form a mesh of two different polymers. The major advantage with IPN is that it combines the beneficial properties of both polymers into the final material. Early application of IPNs in artificial corneas was at the connection between the optical core and the peripheral skirt, where an interdiffusion zone of IPN provides a permanent and reliable union of the PHEMA sponge skirt with the PHEMA core [42]. More recent efforts focus on incorporating IPNs in the entire artificial cornea construct. One design is based on IPNs of poly(dimethylsiloxane) (PDMS) and PNIPAAm [43], where the mechanical strength, transparency, and oxygen permeability of PDMS is combined with the hydrophilicity and nutrient permeability of PNIPAAm to form a functional artificial cornea. Another example is IPN of a neutral cross-linked poly(ethylene glycol) (PEG) and a charged, loosely cross-linked polyacrylic acid (PAA) [44,45]. Such IPN has displayed optical transparency with good mechanical properties and glucose diffusion coefficients comparable to that of the natural cornea [46]. Although most of the artificial corneas have shown satisfying biocompatibility in animal models, it is critical to ensure that the materials are nontoxic, nonimmunogenic, and nonmutagenic, and do not result in corneal opacification.

5.3 Polymeric Biomaterials in Orthopedics

Traditionally, orthopedic biomaterials are mainly metallic, largely due to the close property resemblance to that of bone tissue such as high strength, hardness, and fracture toughness. Polymers have also been used in orthopedics over the years, and they are receiving increasing interest for bone tissue engineering. Historically, the use of polymers in orthopedics for the most part is reserved for those capable of performing well for fixation of structural devices and under cyclic load-bearing conditions such as in knee and hip arthroplasty. Despite hundreds of orthopedics applications available in the market, they are dominated by only a few types of polymers, including ultrahigh-molecular-weight polyethylene (UHMWPE) and PMMA.

5.3.1 Polyethylene

UHMWPE is a linear polyethylene with molecular weight usually between 2 and 6 million. The fracture toughness, low friction coefficient, high impact strength, and low density of UHMWPE have made it a popular choice as the articulating surfaces of joint replacements, such as hip, knee, ankle, and shoulder. Although UHMWPE possesses numerous attractive bulk and surface properties, these properties can be compromised by the presence of long-term radicals in the bulk resulting from the ionizing radiation employed in the sterilization process [47]. These radicals can interact with oxygen, leading to the generation of oxygen containing functional groups and deterioration of the surface and bulk properties, particularly the rate of production of particles during the wear process. An overproduction of wear debris has been linked to the inflammatory reaction in the tissues adjacent to the implant. This adverse tissue response will lead to granulomatous lesions, osteolysis, bone resorption, and implant failure [48]. In an effort to overcome the oxidation, a number of additives, such as antioxidant α-tocoferol and vitamin C, are currently used to retard oxidation and enhance surface properties [49]. UHMWPE has been considered the weak link in any total joint replacement because of the wear issue. To improve wear resistance, highly cross-linked UHMWPE has been produced and used in joint replacement. Cross-linking is achieved by irradiating UHMWPE with electron beam or gamma irradiation, followed by a melting step to eliminate the free radicals produced during irradiation. Currently, there is a debate on cross-linking and the clinical performance of UHMWPE. Those in favor have shown evidence of the efficacy of highly cross-linked UHMWPE in reducing the wear in total joint arthroplasties and the associated periprosthetic osteolysis [50]. The opposition states that improvement of wear resistance by cross-linking is at the expense of reduction in the static mechanical properties, such as tensile and yield strength as well as fatigue crack propagation resistance, which could affect the implant longevity, especially in total knee arthroplasty [51]. Complete data regarding the ultimate long-term performance of highly cross-linked UHMWPE will help settle the scientific debate.

5.3.2 Polyacrylates

Application of PMMA as fixative for bone was first demonstrated by Charnley [52]. The PMMA bone cement is composed of the liquid monomer MMA, a partially polymerized PMMA powder, an initiator (commonly used dibenzoyl peroxide), an activator (*N*,*N*-dimethyl-*p*-toluidine), a radiopacifier (visible to X-rays) such as barium sulfate or zirconium oxide, and a copolymer to influence the mixing and handling of the cement [53]. In some cases, an antibiotic (e.g., gentamicin) is included in the formulation to minimize infection during implantation. The polymerization is initiated by the interaction between the activator and the initiator, yielding a free radical that reacts with the monomer. The solidified polymer is able to secure a firm fixation of the prosthesis in the bones. Although acrylic bone cements are widely used in orthopedics, several drawbacks are related with their use. The residual monomer could leak into the body and cause fat embolism [54]. The exothermic nature of the polymerization process can be a potential cause for necrosis of the surrounding tissue. The most critical drawback is aseptic loosening, i.e., loosening of the implant within the cement. The cause of aseptic loosening could be mechanical and/or biochemical. Mechanically, cyclic loading of the implant could lead to fatigue fracture of the cement [55]. Biochemically, wear debris of the polyethylene component could migrate to the bone—cement interface and trigger inflammatory response, leading to osteolysis and weakening the implant interface [56]. In order to improve upon PMMA fixation, a possible strategy is to avoid cement fracture by increasing the mechanical strength of the cement. Researchers have developed bone cement with higher bonding strength and compressive modulus than conventional PMMA, using a bisphenol-A-glycidyl dimethacrylate (Bis-GMA)-based resin impregnated with bioactive glass ceramics [57,58]. Another approach takes advantage of composites by reinforcing PMMA with hydroxyapatite (HA) [59] and bioactive glass [60], which combines strength and elasticity with bioactivity.

The other acrylate bone cement is based on polyethylmethacrylate (PEMA) and *n*-butylmethacrylate (*n*-BMA) monomer [61]. Comparing to PMMA cement, less heat is produced during polymerization of the PEMA-*n*-BMA cement, and the polymer has a relatively low modulus and high ductility to reduce the issue of fracture. The biocompatibility of the PEMA-*n*-BMA cement has been excellent [62]. But these bone cements have been found to be susceptible to creep. To improve creep resistance, bioactive HA particles were incorporated [63]. Although HA improved bioactivity and creep behavior of the cement, the cement failed at lower number of cycles.

5.3.3 Natural Polymers

Natural polymers are finding increasing applications in the area of bone replacement and hard tissue augmentation. Ideally, materials used for such purpose should be biocompatible; able to mimic the three-dimensional characteristics, physical, and mechanical nature of the bone and hard tissue; able to support appropriate cellular functions; and able to be replaced gradually by the regenerating new tissue. A variety of natural polymers have been used, including extracellular matrix proteins such as collagen [64]; polysaccharides such as chitosan [65], alginate [66], starch [67], and cellulose [68]; as well as glycosaminoglycans such as hyaluronic acid [69]. Some of the natural polymers can provide a template for biomimetic apatite formation, which is highly desirable to induce rapid bone colonization. Recent studies by Hutchens *et al.* [70] revealed the formation and characterization of bacterial cellulose/hydroxyapatite composites with the potential for bone replacement. Both degradable and nondegradable bacterial cellulose were used to form the composite. The hydroxyapatite present in the composite has ordered nanometer needle-like particles with nonstoichiometric composition similar to that observed in human bone. The combined bioactivity and biocompatibility substantiates the potential of this composite for orthopedic application.

5.4 Polymeric Biomaterials in Cardiovascular Diseases

Biomaterials have played a vital role in the treatment of cardiovascular diseases; examples of applications include heart valve prostheses, vascular grafts, stents, indwelling catheters, ventricular assist devices, total implantable artificial hearts, pacemakers, automatic internal cardioverter defibrillators, intra-aortic balloon pumps, etc. A key requirement for materials in cardiovascular applications, particularly blood-contacting devices, is blood compatibility, i.e., nonthrombogenic. Additional requirements include mechanical and surface properties that are application-specific. Surveying the field of polymers used in cardiovascular applications reveals that PUs, polyethylene terephthalate (PET), and expanded PTFE (ePTFE) are the most commonly used. This section will review each of the three polymers followed by a brief introduction of other emerging polymers for use in the cardiovascular area.

5.4.1 Polyurethanes

PUs are among the most commonly selected biomedical polymers for blood-contacting medical devices. They can be found in hemodialysis bloodlines, catheters, stents, insulation for pacemaker leads, heart valves, vascular grafts and patches, left ventricular assist devices (LVADs), etc. PUs are characterized as segmented block copolymers with a wide range of mechanical and blood contact properties, simply by varying the type and/or molecular weight of the soft segment and coupling agents. The urethane linkage, −NH−C(=O)−O−, in biomedical PUs can be formed through a two-step process. The initial step is a reaction involving the end-capping of the macrodiol soft segments (e.g., polyether, polyester, polycarbonate, and polysiloxane) with diisocyanate to form a prepolymer. The second reaction is the coupling of the prepolymer with a low-molecular-weight chain extender—generally a diol or a diamine [71]. The hard segment usually refers to the combination of the chain extender and the diisocyanate components.

Due to the chemical incompatibility between the soft and hard segments, the morphology of PUs consists of hard segments aggregation to form domains that are dispersed in a matrix formed by the soft segments [72,73]. Such unique morphology is responsible for the exceptional mechanical properties and biocompatibility of the biomedical PUs. For example, depending on the relative molecular weights and amounts of the hard and soft segments, the obtained PU can be elastomeric or rigid. The mechanical properties of PU can also be tailored by changing the chemical nature of the chain extender. Generally, PUs prepared with aliphatic chain extender are softer than those with aromatic chain extender. Biocompatibility of PU is also closely related to the chemical nature of the chain extender and the soft segment. Early studies by Lyman *et al.* [74] showed that changes in the molecular weight of the polypropylene soft segments affected protein adsorption. Lysine diisocyanate and hexamethylene diisocyanate are preferred over aromatic diisocyanates in the synthesis of biodegradable PUs, partly because of the putative carcinogenic nature of aromatic diisocyanates [75]. Recent studies have reported using natural polymers, such as chitin [76] and chitosan [77] as chain extender to improve the biocompatibility of PUs.

Biostability has been and continues to be a main research focus of PUs. Depending on the intended medical applications, the desired biostability of PUs varies. For example, PUs used as a pacemaker lead covering should have superior long-term stability, whereas PUs used as a scaffold to build engineered tissue construct for the replacement of diseased cardiovascular tissues should be biodegradable. The challenge to maintain long-term *in vivo* biostability of PUs lies in the fact that biodegradation of PUs is a complicated and multifactor-mediated process. Mechanisms responsible for PU biodegradation include (1) hydrolysis, (2) oxidative degradation, metal or cell catalyzed, (3) enzymatic degradation, (4) surface cracking, (5) environmental stress cracking, and (6) calcification [75]. It is well known that PUs containing polyester soft segments have poor hydrolytic stability, and PUs with polyether soft segments are prone to oxidative degradation. Guided with valuable information collected from extensive investigation of molecular pathways leading to the biodegradation of PUs, more bioresistant PUs have been designed over the years. These strategies include using polycarbonate macrodiols [78,79], polyether macrodiols with larger hydrocarbon segments between ether groups [80], and siloxane-based macrodiols [81−83]. On the other end of the spectrum, bioresorbable PUs are attracting increasing attention as elastomeric tissue engineering scaffolds. For this class of PUs, soft segments such as polylactide or polyglycolide, polycaprolactone, and polyethylene oxide are most commonly used [84]. Furthermore, degradation is engineered into the hard segments. Enzyme-sensitive linkages have been incorporated into the hard segment, leading to specific enzymatic degradation in contrast to nonspecific hydrolytic degradation [85−87]. Another interesting addition to the hard segments is bioactive molecule such as antimicrobial drug [88]. Polymer degradation will thus lead to free drug release, making this class of PUs very attractive for biomedical applications.

5.4.2 Polyethylene Terephthalate

PET is a member of the engineering polyester family. It is a semicrystalline polymer with industrial applications as synthetic fibers and beverage and food containers. In the medical field, PET is widely used as prosthetic vascular grafts, sutures, and wound dressings in either fiber or fabric form (commercially known as Dacron®). Despite the presence of hydrolytically cleavable ester linkage, PET is relatively stable *in vivo* largely due to the high crystallinity and hydrophobicity. It is one of the two standard biomaterials of prosthetic vascular grafts used clinically. It is widely used for larger vessel (diameter >6 mm) applications. PET for vascular applications can be prepared either woven or knitted, which will determine the porosity and mechanical property of the graft. Generally, a woven finish has less porosity than a knitted graft, therefore, reducing the chance of transmural blood extravasation. Dacron® vascular graft is strong and stiff, much less compliant than natural arteries [89]. Such compliance mismatch has been considered the cause of patency loss of the graft over a long time frame (>6 months) [90]. The other major complication related to the PET graft is its thrombogenicity. When the graft comes in contact with blood, plasma protein is adsorbed to the luminal and capsular surfaces, leading to thrombus formation and inflammatory response. Various strategies have been explored to make the graft surface thromboresistant, including passivating the surface with albumin [91], coating with fluoropolymer [92], coating with hydrophilic polymer [93], covalent or ionic binding of the anticoagulant heparin−albumin [94−96], covalent linkage of antithrombotic agent thrombomodulin [97], etc. Although some improvement has been reported in terms of acute thrombosis, there is still a long way to go to achieve satisfying long-term functionality of PET-based vascular grafts.

5.4.3 Expanded PTFE

ePTFE, commercially also known as Gore-Tex®, is one of the two standard biomaterials of prosthetic vascular grafts used clinically. Besides vascular uses, ePTFE is also used as patches for soft tissue regeneration, such as hernia repair, and surgical sutures. It is produced by a series of extrusion, stretching, and heating processes to create a microporous material with pore size ranging from 30 to about 100 μm. Similar to PET, ePTFE is highly crystalline, hydrophobic, and highly stable. It has an extremely low coefficient of friction, making it easy for handling. Its tensile strength and tensile modulus are lower than those of PET. Even though the compliance of ePTFE grafts is relatively lower than that of PET grafts, it is still too high compared to natural arteries. Generally, ePTFE is the choice over PET to bypass smaller vessels. However, it still faces a patency issue. Femoropopliteal reconstruction using ePTFE has a 5-year patency rate of 40−50%, compared to the 70−80% achieved by using autogenous vein grafts [98]. Similar to PET, the cause of low patency is the thrombogenicity of the material. It has been reported that the graft fails to develop a full coverage of endothelial cells on the lumen side of the graft [99,100]. To address this issue, one approach is to increase the porosity to promote tissue ingrowth. But it requires a careful balance to prevent leakage of blood elements as mentioned earlier. Other approaches focus on reducing surface thrombogenicity, including carbon coating to increase surface electronegativity [101], attachment of anticoagulant or antithrombotic agents [102,103], and impregnation with fibrin glue to deliver growth factors that can promote endothelialization [104,105]. The actual benefits of these treatments are yet to be determined through longer-term *in vivo* investigations.

The challenge posed by small diameter vascular repair has spurred research for alternative biomaterials that would match or surpass the autograft. A notable effort is to build a tissue-engineered graft *ex vivo* using a synthetic biodegradable scaffold. Conceptually, such graft will have mechanical properties closely mimicking those of the native tissues without the concern of chronic inflammatory responses commonly induced by the presence of synthetic materials. Till date, a wide variety of biodegradable polymers have been used to build such constructs, including poly(α-hydroxyesters); poly(glycolic acid) (PGA); poly(lactic acid) (PLA); and their copolymers poly(lactic-*co*-glycolic acid) (PLGA), polycaprolactone, polyanhydride, polyhydroxyalkanoate, and polypeptide. Several excellent reviews are available discussing the current status of materials as scaffolding for vascular tissue engineering [106−108].

The other cardiovascular application in which polymers are poised to make a significant impact is biodegradable stents. Current stents are mainly made of metallic materials, such as stainless steel, cobalt−chromium, or Nitinol. However, long-term complications associated with metal stents have prompted research of fully degradable replacement. Several key requirements have to be satisfied by the polymeric stent, with the top two being mechanical properties and degradation characteristics. In terms of degradation, the products of degradation should be biocompatible, and the degradation process should not compromise the structural integrity of the device up to 6 months [90]. As for mechanical properties, the polymer should withstand the deployment and the blood vessel contractions. Both requirements are challenging, but with a good appreciation of the underlying biology and the versatility of polymer structure−property relationship, newer materials are likely to emerge in the near future. For example, researchers are imparting degradation and shape memory capabilities into polymers that can self-expand and degrade over time [109,110].

5.5 Polymeric Biomaterials for Wound Closure

Surgical wounds can be closed by various means, including sutures [111], adhesives [111], tapes [111], staples [111], and laser tissue welding [112]. Among these methods, sutures are the most frequently used. The sutures are sterile filaments used to approximate and maintain tissue until the healing has provided the wound with appropriate strength to withstand mechanical stresses. Sutures can be classified based on the origin of the materials as natural or synthetic; performance of the materials as absorbable or nonabsorbable; and physical configurations as monofilament, multifilament, braided, or twisted. In general, polymers selected for sutures should elicit minimal adverse biological response in addition to having fiber-forming rheological properties. The sutures must have minimum tissue drag, good strength retention, and knot security. To improve the lubricity and reduce tissue drag, coatings such as tetrafluoroethylene and silicones are normally applied to the suture. The following sections will discuss some of the common nonabsorbable and absorbable polymeric sutures currently commercially available.

In general, nonabsorbable sutures can retain their tensile strength longer than 2 months [113]. The synthetic polymers used to make nondegradable sutures include polypropylene (PP), polyamides, polyesters such as PET and polybutylene terephthalate (PBT), and polyether−ester based on poly(tetramethylene glycol), 1,4-butanediol, and dimethyl terephthalic acid [114]. The base polymer and filament configurations for common nonabsorbable sutures are summarized in Table 5.2.

The PP monofilament sutures are made from isotactic polypropylene [115]. During preparation, the PP monofilament is subjected to a series of post-spinning operations including annealing, designed to increase crystallinity [116]. Although PP sutures are highly resistant to hydrolytic degradation, it can undergo thermo-oxidative degradation [117]. PP sutures are usually sterilized by ethylene oxide or autoclave due to their susceptibility to ionizing radiation such as γ-radiation from cobalt-60 source that is normally used for radiation sterilization. In terms of performance, PP sutures cause one of the lowest tissue responses.

Polyamide sutures are commonly made out of nylon-6 and nylon-6,6. Nylon-6 is synthesized by

Table 5.2 A List of Commercially Available Nonabsorbable Suture Materials

Generic Name	Polymer	Configuration	Trade Name
Polyamide	Nylon-6, nylon-6,6	Monofilament	Dermalon
		Multifilament, braided	Nurolon
		Multifilament, braided and silicone coating	
Polypropylene	PP	Monofilament	Prolene
		Monofilament	Surgipro
Polyethylene	Polyethylene	Monofilament	Dermalene
Polyester	PET	Braided	Dacron
		Braided with silicone coating	Ti-Cron
		Braided with polybutylate coating	Ethibond
		Braided with PTFE coating	Polydek

ring-opening polymerization of caprolactam, while nylon-6,6 is prepared by condensation polymerization of adipic acid and hexamethylene diamine. These polyamide sutures are processed into monofilament, braided multifilament, and core–sheath configurations. The braided multifilament nylon sutures are often coated (e.g., silicone coating) to reduce tissue drag. The observed decrease in strength retention over time is associated with the susceptibility of the amide bond to hydrolytic degradation in the nylon structure. The tensile strength of nylon sutures decreases at a yearly rate of approximately 15–25% [118]. The tissue reaction to nylon sutures appears to be independent of configuration, with both braided and monofilament eliciting low reactivity.

The need to suture very delicate and complicated tissues have led to the development and use of sutures based on fluoropolymers such as PTFE and polyvinylidene fluoride (PVDF), and copolymers of PVDF and hexafluoropropylene (HFP) [119]. PTFE is a stable (T_m = 327 °C) semicrystalline linear polymer. ePTFE sutures are highly crystalline microporous fibers prepared by wet spinning an aqueous mixture of PTFE powder and cellulose xanthate. The morphology of ePTFE fibers consists of nodules connected by thin crystalline fibers that control tensile strength. The mechanical properties, biological response, and handling can be directly correlated with the porosity of the PTFE fibers [120]. The bending stiffness of ePTFE sutures is low due to the microporous structure [121], but the porous structure also contributes to the decrease in strength. PVDF is also highly crystalline (T_m = 175 °C). Sutures prepared from PVDF exhibit good creep resistance and tensile strength retention. Morphological studies have demonstrated high surface stability, i.e., no visible signs of bulk or surface fracture [122]. PVDF sutures are susceptible to thermo-oxidative degradation, but can be readily sterilized with γ-radiation. PVDF elicits moderate tissue and cell response—a behavior similar to PP sutures. Sutures derived from copolymers of PVDF and HFP were originally designed to combine the beneficial handling properties and biological response of PVDF and PP into one material. In addition, PVDF/HFP sutures were also designed to emulate the durability of polyester sutures. The tensile strength, size, biological response, and handling of the PVDF/HFP sutures can be tailored by manipulating the copolymer compositions. The major target areas for usage of PVDF/HFP sutures are wound closure during cardiovascular, neurological, and ophthalmic surgeries [119]. These PVDF/HFP sutures are normally used as uncoated monofilaments.

Among the most commonly used polyester-based nonabsorbable sutures are PET and PBT. In addition, there are polyester-based sutures made from copolymers of poly(tetramethylene ether terephthalate) and poly(tetramethylene terephthalate) called polyetheresters. PET is synthesized by condensation polymerization of ethylene glycol and terephthalic acid. PET is a polymer with a melting temperature of approximately 265 °C. The thermal stability of PET enables melt spinning to form monofilament fibers with variable profiles. During processing, the fibers are subjected to hot drawing that enhances molecular orientation, crystallinity, and tensile strength. The PET sutures are commercially available as coated or uncoated monofilament or braided multifilament configurations. The surface treatments of PET sutures include coatings of PTFE and silicone. PET sutures are very stable in the biological environment with no evidence of hydrolytic degradation. The strength retention of PET sutures remains for an extended period of time. The tissue response to PET sutures is dependent on the configuration with braided multifilament and monofilament having moderate and low tissue reactivity, respectively. Compared to PET, PBT sutures are generally less brittle and stiff, due to the longer aliphatic segment in the polymer structure. Polybutester sutures are obtained from block copolymers of PBT and poly(tetramethylene ether) glycol terephthalate (PTMG). In the copolymer, the PBT is the hard segment and PTMG is the flexible segment. Chemical incompatibility between the hard PBT and soft PTMG blocks renders these copolymers elastomeric. Such unique mechanical behavior makes the polybutester sutures ideal for wounds prone to edema formation.

The synthetic absorbable sutures are made from polymers capable of degradation in the biological environment without adverse effects. One overall advantage of absorbable sutures is the elimination of clinical visits for their removal. These sutures are either homopolymers or copolymers based on degradable polymeric units such as polyglycolic acid, polylactic acid, or poly-*p*-dioxanone.

Polyglycolic acid (PGA) can be synthesized by condensation or ring-opening polymerization. Sutures based on PGA were the first absorbable sutures made [123]. PGA sutures are commercially available coated or uncoated in a braided configuration. Glycolide has been copolymerized with lactic acid, trimethylene carbonate, and ε-caprolactone [119]. Glycolic acid was copolymerized with L- or DL-lactic acid to form random copolymer. The performance of the glycolide-L-lactide sutures is dependent on composition. The initial tensile strength and retention through the healing process of the glycolide-L-lactide sutured wound is directly dependent on the concentration of the crystallizable glycolide monomers [124]. Copolymers based on the DL-lactide do not exhibit the same composition dependence as observed for the L-lactide copolymers [119]. Glycolide has been copolymerized with trimethylene carbonate to form a triblock copolymer where the middle block is a random copolymer of glycolide and trimethylene carbonate and the terminal blocks based on glycolide. These sutures are available as uncoated monofilaments. The copolymerization of glycolide and ε-caprolactone leads to formation of segmented copolymers. In these copolymers, the glycolide and ε-caprolactone form the soft and hard segments, respectively.

Poly-*p*-dioxanone (PDS) is synthesized by ring-opening polymerization of 1,4-dioxanone-2,5-dione. The monofilament sutures are produced by melt spinning. The fibers are subjected to a drawing process to improve tensile strength and performance. Recently, attempts have been made to copolymerize PDS with PGA and PLLA to produce sutures with different properties [125].

Current research focus in wound closure suture is to incorporate extra functionality to the suture besides closing the wound. These efforts include to control wound infection by developing antimicrobial sutures, and to accelerate the wound healing process by using bioactive material such as chitin, or to deliver therapeutics that can impact the wound healing response.

5.6 Polymeric Biomaterials in Extracorporeal Artificial Organs

Extracorporeal artificial organs provide mass-transfer operations to support failing or impaired organ systems [126]. Common examples include kidney substitute, hemodialysis, cardiopulmonary bypass (CPB), apheresis therapy, peritoneal dialysis, lung substitute and assist, and plasma separation. A critical component involved in the extracorporeal artificial organ is the membrane that serves to separate the undesired substance from the blood or plasma. Ideally, materials used as the membrane in these particular applications should have appropriate cellular and molecular permeability, as well as blood compatibility (i.e., hemocompatibility). Over the years, both natural and synthetic polymers have been used as membrane materials.

The most widely used natural membrane is cellulosic. Taking hemodialysis as an example, early applications of cellulose membrane in the dialyzer used regenerated cellulose, i.e., unsubstituted with rich hydroxyl groups along the repeating saccharide units. Studies have found that regenerated cellulose has poor hemocompatibility. It activates the complement system, which leads to inflammation and other serious immune responses. The complement activation has been attributed to the high concentration of hydroxyl groups on the membrane rendering it nucleophilic and susceptible to protein deposition, particularly C3b. Such observation spurred later research of using substituted cellulose for dialysis membrane, examples include cellulose acetate and cellulose triacetate, where in both cases a fraction of the hydroxyl groups are replaced with acetate functionality. These modified cellulose materials greatly limit complement activation by eliminating the active surface sites for complement protein interaction. Besides chemically blocking complement interaction, approaches using steric hindrance effect have also been explored. A bulky chemical group such as benzyl substitution group or tertiary amine group has been used to replace the hydroxyl group to sterically minimize the complement protein interaction with the membrane [127,128].

Current dialysis membranes are mostly made from synthetic polymers, including polysulfone, polyethersulfone, polyacrylonitrile, PMMA, polyamide, and polypropylene hollow fibers. Compared with natural cellulosic membrane, synthetic membranes are less prone to complement activation. The reason behind the improved complement compatibility is the diminished level of surface nucleophiles for C3b deposition. Furthermore, some of the synthetic membranes are rich in negative charges on the surface, which can absorb the activated cationic complement peptide (e.g., C5a) and minimize the subsequent cascade of inflammation. Synthetic membranes generally have significantly larger pore sizes and higher hydraulic permeability than cellulosic membranes [129]. Therefore, synthetic membranes are the choice for high-flux applications. The larger pore size also allows for removal of middle molecules with molecular weight between 500 and 2000 Da, which have been deemed bioactive and may have a potential biological impact [130]. The hydrophobic nature of most synthetic membranes contributes to the adsorptive capacity toward noxious compounds such as interleukin-1, tumor necrosis factor, interleukin-6, and β_2-microglobulin [131]. PMMA and polyacrylonitrile usually exhibit the most pronounced adsorption capacity. Regardless of their origin, the membranes have been used either in hollow-fiber design, which is most common, or as sheet films in parallel-plate design.

5.7 Polymeric Biomaterials for Nerve Regeneration

Repair of the damaged nerves presents enormous challenge due to the physiology complexity of the nervous system. Even though progress has been made over the past decades, it is still not possible to fully repair the damage so that lost functions of the nervous systems can be restored. The nervous system is generally classified into the central nervous system (CNS) and the peripheral nervous system (PNS). Various strategies have been explored for nerve repair in both the CNS and the PNS, including guidance conduits, scaffolds with cell transplantation, and delivery of therapeutics. This section will mainly focus on polymers used in the nerve guidance conduit approach.

It has been widely accepted that physical guidance of axons, the long processes extending from the neuron cell body and conducting electrical signals, plays a critical role in nerve repair. The nerve guidance conduit is designed to (1) direct the outgrowth of axons from the proximal nerve end bridging across the lesion, (2) provide a channel for the diffusion of biomolecules secreted by the injured nerve ends, and (3) reduce the scar tissue invasion to the regeneration zone [132]. To fulfill these functions, an ideal nerve guidance conduit should be semipermeable with oriented topographical features inside the conduit, supportive of electrical activity, able to deliver bioactive factors, and able to support cell adhesion and migration. The versatility of polymers makes them the top choice in engineering of nerve guidance conduits. Early research has used nondegradable synthetic polymers including silicone [133] and ePTFE [134]. Although silicone nerve guidance conduits have shown success in bridging gaps up to 10 mm, they have failed to support regeneration across larger defects. Therefore, effort has been shifted to develop biodegradable guidance conduits. The advantage of using a degradable material lies in the fact that long-term complications such as fibrotic reaction and nerve compression can be minimized. The degradation characteristics of the material should meet the following requirements: (1) the degradation profile should match with the axonal outgrowth profile, so that the guidance conduit will maintain sufficient mechanical support during the regeneration process and (2) the degradation product(s) should induce minimum to zero tissue reaction. A series of degradable polymers have been used, including biodegradable poly(esters) such as PGA [135], PLA [136], PLGA [137], and poly(caprolactones) [138]; polyphosphazenes [139]; polyurethanes [140]; and poly(3-hydroxybutyrate) [141].

Since the emergence of studies showing that electrical charge affects neurite extension *in vitro* [142,143] and improves nerve regeneration *in vivo* [144], polymers that can provide electrical stimulus have been included in guidance conduit development. These polymers include piezoelectric polymers such as PVDF and its copolymer [144], and conducting polymers like polypyrrole and its biologically modified derivatives [143,145]. Other electroactive polymers, such as polyaniline, may also provide support for nerve growth, as studies have shown encouraging results with cardiac myoblast cells [146].

Nerve guidance conduits can be hollow or filled with matrix to support axonal elongation. A popular filler choice is natural polymeric gel. Ideally, the gel should be soft with mechanical properties matching those of the nervous tissue, porous to allow axonal ingrowth, biodegradable, and biocompatible. A number of natural polymers have been investigated, including agarose [147], chitosan [148], methylcellulose [149], hyaluronic acid [150], alginate [151], fibrin gels [152], collagen [153], keratin [154], and self-assembling peptide scaffolds [155]. Agarose is a thermally reversible polysaccharide hydrogel. Its gelling temperature can be modified by changing the functional groups attached to the sugar residues. It can also be functionalized with various biological motifs, such as laminin-derived peptide sequences RGD, YIGSR, and IKVAV, to enhance neurite extension [156]. Fibrin is a natural wound-healing matrix that can be found in the early stages of regeneration. It is formed from the blood coagulation cascade to restore hemostasis and initiate tissue repair. Using fibrin gels as the filler can closely mimic the natural matrix formed in the guidance conduit bridging short nerve gaps, where a fibrin cable is usually formed from the exuding serum by the damaged blood vessels in the nerve ends [157]. Peptide sequences have also been cross-linked into the filling fibrin matrix to further induce neurite extension [158]. In addition to gel filler, longitudinal filaments, either synthetic or natural, have been used in the conduit to align the growing axons in the direction of regeneration. Materials used in filament preparation include polyamide, catgut, polydioxanone, polyglactin, poly(acrylonitrile-*co*-methylacrylate), collagen, PLA, PGA, etc. [159–162].

Recently, materials research on nerve guidance conduits has been taken to a new level, where the old paradigm of passive material has shifted to new bioactive material design. Chemical messengers such as neurotransmitters have been polymerized into the polymer backbone to impart neuroactivity for the resulting biomaterial [163]. The first example of this new class of polymer is dopamine polymerized with a diglycidyl ester to form a biodegradable material that has shown vigorous neurite outgrowth *in vitro* and good tissue compatibility *in vivo*. Another example of new bioactive polymer is polysialic acid and its hydrogel. Polysialic acid is a dynamically regulated post-translational modification of the neural cell adhesion molecule [164]. It has been shown to significantly improve cell adhesion and viability *in vitro*. With the increasing understanding of the biology behind nerve regeneration, it is expected that more bioactive materials will be developed in the future to achieve timely functional recovery from nerve damage.

5.8 Conclusions and Future Outlook

Polymers have made significant impact on biomedical research and medical practice, and will continue to be the major workforce for biomaterials in the twenty-first century. The polymeric biomaterials and their applications presented here are only the tip of an iceberg. With the growing understanding of the biological response to existing biomaterials and a better grasp of human organ composition, function, biomechanics, and disease etiology, chemists and polymer scientists should continue working collaboratively with biologists, physicians, and engineers to develop tailor-made polymers for biomedical applications. In contrast to the old inert synthetic polymers, bioactive, biomimetic, and smart polymers will be at center stage. Furthermore, as the interactions of the biological system with polymers occur at the interface, surface-related research will continue to thrive, especially surface characterization and surface modification. One can be hopeful to foresee a better management of diseases with the help of a new generation of biomaterials, and a seamless integration of the biomaterials into the body.

References

[1] M.S. Reisch, Medical polymers renaissance, Chem. Eng. News 85 (2007) 14–17.

[2] I. Mann, A brief review of contact lens work, Trans. Ophthalmol. Soc. Aust 1 (1939) 107–115.

[3] R.J. Schillinger, R.V. Shearer, O.R. Levy, Animal experiments with a new type of intraocular acrylic lens, Arch. Ophthalmol 59 (1958) 423–434.

[4] A.C. Perry, Advances in enucleation, Ophthalmol. Clin. North Am. 4 (1991) 173–182.

[5] C.W. Flowers, P.J. McDonnell, Mechanical methods in refractive corneal surgery, Curr. Opin. Ophthalmol 5 (1994) 81–89.

[6] M.E. Migliori, A.M. Putterman, Silicone intubation for the treatment of congenital lacrimal duct obstruction – successful results removing the tubes after 6 weeks, Ophthalmology 95 (1988) 792–795.

[7] A.C.B. Molteno, New implant for drainage in glaucoma, Animal trial, Brit. J. Ophthalmol 53 (1969) 161–168.

[8] T.J. Liesegang, Viscoelastic substances in ophthalmology, Surv. Ophthalmol 34 (1990) 268–293.

[9] R. Bawa, M. Nandu, Physicochemical considerations in the development of an ocular polymeric drug delivery system, Biomaterials 11 (1990) 724–728.

[10] C.L. Schepens, F. Acosta, Scleral implants: an historical perspective, Surv. Ophthalmol 35 (1991) 447–453.

[11] C.E. Gilbert, Adhesives in retinal-detachment surgery, Br. J. Ophthalmol 75 (1991) 309–310.

[12] J.B. Jonas, H.L. Knorr, R.M. Rank, W.M. Budde, Intraocular pressure and silicone oil endotamponade, J. Glaucoma 10 (2001) 102–108.

[13] S.M. Mc Glinchey, C.P. McCoy, S.P. Gorman, D.S. Jones, Key biological issues in contact lens development, Expert Rev. Med. Devices 5 (2008) 581–590.
[14] O. Wichterle, D. Lim, Hydrophilic gels for biological use, Nature 185 (1960) 117–118.
[15] R.L. Taylor, M.D. Willcox, T.J. Williams, J. Verran, Modulation of bacterial adhesion to hydrogel contact lenses by albumin, Optom. Vis. Sci. 75 (1998) 23–29.
[16] R.M. Hesby, C.R. Haganma, C.M. Standford, Effects of radiofrequency glow discharge on impression material surface wettability, J. Prosthet. Dent 77 (1997) 414–422.
[17] T. Okada, Y. Ikada, Modification of silicone surface by graft polymerization of acrylamide with corona discharge, Makromol. Chem. 192 (1991) 1705–1713.
[18] G. Brian, H. Taylor, Cataract blindness: challenges for the 21st century, Bull. World Health Organ 79 (3) (2001) 249–256.
[19] A.S. Obstbaum, Biologic relationship between poly-(methyl methacrylate) intraocular lenses and uveal tissue, J. Cataract Refract. Surg 18 (1992) 219–231.
[20] A.W. Lloyd, R.G.A. Faragher, S.P. Denyer, Ocular biomaterials and implants, Biomaterials 22 (2001) 769–785.
[21] A.S. Patel, Intraocular lens implants: a scientific perspective, in: B.D. Ratner, A.S. Hoffman, F.J. Schoen, J.E. Lemons (Eds.), Biomaterials Science: An Introduction to Materials in Medicine, Elsevier, San Diego, CA, 2004. Chapter 7.11.
[22] M.U. Koch, D. Kalicharan, J.J.L. Vanderwant, Lens epithelial cell formation related to hydrogel foldable intraocular lenses, J. Cataract. Refract. Surg 25 (1999) 1637–1640.
[23] L. Werner, Biocompatibility of intraocular lens materials, Curr. Opin. Ophthalmol 19 (2008) 41–49.
[24] L. Werner, D.J. Apple, M. Kaskaloglu, S.K. Pandey, Dense opacification of the optical component of a hydrophilic acrylic intraocular lens: a clinicopathologic analysis of 9 explanted lenses, J. Cataract Refract. Surg 27 (2001) 1485–1492.
[25] A.M. Izak, L. Werner, S.K. Pardey, D.J. Apple, Calcification of modern foldable hydrogel intraocular lens designs, Eye 17 (2003) 393–406.
[26] J.P. Whitcher, M. Srinivasan, M.P. Upadhyay, Corneal blindness: a global perspective, Bull. World Health Organ 79 (2001) 214–221.
[27] D. Myung, P.E. Duhamel, J.R. Cochran, J. Noolandi, C.N. Ta, C.W. Frank, Development of hydrogel-based keratoprostheses: a materials perspective, Biotechnol. Prog 24 (2008) 735–741.
[28] M. Griffith, W.B. Jackson, N. Lagali, K. Merrett, F. Li, P. Fagerholm, Artificial corneas: a regenerative medicine approach, Eye (in press).
[29] H. Sheardown, M. Griffith, Regenerative medicine in the cornea, in: A. Atala, R. Lanza, J. Thompson, R. Nerem (Eds.), Principles of Regenerative Medicine, Elsevier, Boston, 2008, pp. 1060–1071.
[30] T.V. Chirila, C.R. Hichs, P.D. Dalton, S. Vijayasekaran, X. Lou, Y. Hong, A.B. Clayton, B.W. Ziegelaar, J.H. Fitton, S. Platten, G.J. Crawford, I.J. Constable., Artificial cornea, Prog. Polym. Sci. 23 (1998) 447–473.
[31] D.R. Caldwell, The soft keratoprosthesis, Trans. Am. Ophthalmol. Soc. 95 (1997) 751–802.
[32] S. Pintucci, F. Pintucci, S. Caiazza, M. Cecconi, The Dacron felt colonizable keratoprosthesis, after 15 years, Eur. J. Ophthalmol 6 (1996) 125–130.
[33] J.C. Barber, Keratoprosthesis: past and present, Int. Ophthalmol. Clin 28 (1988) 103–109.
[34] F. Yaghouti, C.H. Dohlman, Innovations in keratoprosthesis, proved and unproved, Int. Ophthalmol. Clin 39 (1999) 27–36.
[35] B.E. Khan, J. Dudenhoefer, C.H. Dohlman, Keratoprosthesis, an update, Curr. Opin. Ophthalmol 12 (2001) 282–287.
36] M. Nouri, H. Terada, E.C. Alfonso, C.S. Foster, M.L. Durand, C.H. Dohlman, Endophthalmitis after keratoprosthesis, incidence, bacterial causes risk factors, Arch. Ophthalmol 11 (2001) 484–489.
[37] T.V. Chirila, An overview of the development of artificial corneas with porous skirts and the use of PHEMA for such an application, Biomaterials 22 (2001) 3311–3317.
[38] J.T. Jacob, C. Wallace, J. Bi, Characterization of corneal epithelial cell adhesion on novel hydrogels, Invest. Ophthalmol. Vis. Sci. 45 (2004). U564–U564.
[39] N.A. Peppas, E.W. Merrill, Development of semicrystalline poly(vinyl alcohol) hydrogels for

biomedical applications, J. Biomed. Mater. Res. 11 (1977) 423–434.

[40] H. Miyashita, S. Shimmura, H. Kobayashi, T. Taguchi, N. Asano-Kato, Y. Uchino, M. Kato, J. Shimazaki, J. Tanaka, K. Tsubota, Collagen-immobilized poly(vinyl alcohol) as an artificial cornea scaffold that supports a stratified corneal epithelium, J. Biomed. Mater. Res. 76B (2005) 56–63.

[41] F. Li, D. Carlsson, C. Lohmann, E. Suuronen, S. Vascotto, K. Kobuch, H. Sheardown, R. Munger, M. Nakamura, M. Griffith, Cellular and nerve regeneration within a biosynthetic extracellular matrix for corneal transplantation, Proc. Natl. Acad. Sci. USA 100 (2003) 15346–15351.

[42] T.V. Chirila, S. Vijayasekaran, R. Horne, Y.C. Chen, P.D. Dalton, I.J. Constable, G.J. Crawford, Interpenetrating polymer network (IPN) as a permanent joint between the elements of a new type of artificial cornea, J. Biomed. Mater. Res. 28 (1994) 745–753.

[43] L. Liu, H. Sheardown, Sheardown Glucose permeable poly(dimethyl siloxane) poly(*N*-isopropylacrylamide) interpenetrating networks as ophthalmic biomaterials, Biomaterials 26 (2005) 233–244.

[44] D. Myung, W. Koh, J. Ko, Y. Hu, M. Carrasco, J. Noolandi, C.N. Ta, C.W. Frank, Biomimetic strain hardening in interpenetrating polymer network hydrogels, Polymer 48 (2007) 5376–5387.

[45] D. Myung, W. Koh, A. Bakri, F. Zhang, A. Marshall, J. Ko, J. Noolandi, M. Carrasco, J.R. Cochran, C.W. Frank, C.N. Ta, Design and fabrication of an artificial cornea based on a photolithographically patterned hydrogel construct, Biomed Microdev 9 (2007) 911–922.

[46] D. Myung, N. Farooqui, D. Waters, S. Schaber, W. Koh, M. Carrasco, J. Noolandi, C.W. Frank, C.N. Ta, Glucose-permeable interpenetrating polymer network hydrogels for corneal implant applications, a pilot study, Curr. Eye. Res. 9 (2008) 29–43.

[47] V. Premnath, W.H. Harris, M. Jasty, E.W. Merrill, Gamma sterilization of UHMWPE articular implants: an analysis of the oxidation problem, Biomaterials 17 (1996) 1741–1753.

[48] W.J. Maloney, R.L. Smith, Periprosthetic osteolysis in total hip arthroplasty: the role of particulate debris, J. Bone Joint Surg 77A (1995) 1448–1461.

[49] N. Tomita, T. Kitakura, N. Onmori, Y. Ikada, E. Aoyama, Prevention of fatigue cracks in ultrahigh molecular weight polyethylene joint components by the addition of vitamin E, J. Biomed. Mater. Res. 48 (1999) 474–478.

[50] M. Jasty, H.E. Rubash, O. Muratoglu, Highly cross-linked polyethylene: the debate is over—in the affirmative, J. Arthroplasty 20 (2005) 55–58.

[51] M.D. Ries, Highly cross-linked polyethylene: the debate is over—in opposition, J. Arthroplasty 20 (2005) 55–58.

[52] J. Charnley, The bonding of prosthesis to bone by cement, J. Bone Joint Surg 46 (1964) 518–529.

[53] M. Navarro, A. Michiardi, O. Castano, J.A. Planell, Biomaterials in orthopaedics, J.R. Soc. Interface 5 (2008) 1137–1158.

54] M.J. Koessler, R.P. Pitto, Fat and bone marrow embolism in total hip arthroplasty, Acta Orthop. Belg 67 (2001) 97–109.

[55] W.J. Maloney, M. Jasty, A. Rosenberg, W.H. Harris, Bone lysis in well-fixed cemented femoral components, J. Bone Joint Surg. Br. 72 (1990) 966–970.

[56] M.A. Freeman, G.W. Bradley, P.A. Revell, Observations upon the interface between bone and polymethylmethacrylate cement, J. Bone Joint Surg. Br. 64 (1982) 489–493.

[57] K. Kawanabe, J. Tamura, T. Yamamuro, T. Nakamura, T. Kokubo, S. Yoshihara, New bioactive bone cement consisting of bis-GMA resin and bioactive glass powder, J. Appl. Biomater. 4 (1993) 135–141.

[58] J. Tamura, T. Kitsugi, H. Iida, H. Fujita, T. Nakamura, T. Kokubo, S. Yoshihara, Bone bonding ability of bioactive cements, Clin. Orthop 343 (1997) 183–191.

[59] M.J. Dalby, L. Disilvio, E.J. Harper, W. Bonfield, In vitro evaluation of a new polymethylmethacrylate cement reinforced with hydroxyapatite, J. Mater. Sci. Mater. Med 10 (1999) 793–796.

[60] J.T. Heikkila, A.J. Aho, I. Kangasniemi, A. Yli-Urpo, Polymethylmethacrylate composites: disturbed bone formation at the surface of

bioactive glass and hydroxyapatite, Biomaterials 17 (1996) 1755–1760.

[61] B. Weightman, M.A.R. Freeman, P.A. Revell, M. Braden, B.E.J. Alberkttsson, L.V. Carlson, The mechanical properties of cement and loosening of the femoral component of hip replacements, J. Bone Joint Surg 69B (1987) 558–564.

[62] P. Revell, M. Braden, B. Weightman, M. Freeman, Experimental studies of the biological response to a new bone cement: II soft tissue reactions in the rat, Clin. Mater 10 (1992) 233–238.

[63] E.J. Harper, J.C. Behiri, W. Bonfield, Flexural and fatigue properties of a bone cement based upon polyethylmethacrylate and hydroxyapatite, J. Mater. Sci. Mater. Med 6 (1995) 799–803.

[64] T. Uemura, J. Dong, Y. Wang, H. Kojima, T. Saito, Iejima, M. Kikuchi, J. Tanaka, T. Tateishi, Transplantation of cultured bone cells using combinations of scaffolds and culture techniques, Biomaterials 24 (2003) 2277–2286.

[65] T. Jiang, W.I. Abdel-Fattah, C.T. Laurencin, In vitro evaluation of chitosan/poly(lactic acid-glycolic acid) sintered microsphere scaffolds for bone tissue engineering, Biomaterials 27 (2006) 4894–4903.

[66] E. Fragonas, M. Valente, M. Pozzi-Mucelli, R. Toffanin, R. Rizzo, F. Silvestri, F. Vittur, Articular cartilage repair in rabbits by using suspensions of allogenic chondrocytes in alginate, Biomaterials 21 (2000) 795–801.

[67] B. Ongpipattanakul, T. Nguyen, T.F. Zioncheck, R. Wong, G. Osaka, L. DeGuzman, W.P. Lee, L.S. Beck, Development of tricalcium phosphate/amylopectin paste combined with recombinant human transforming growth factor beta 1 as a bone defect filler, J. Biomed. Mater. Res. 36 (1997) 295–305.

[68] G.J. Dias, P.V. Peplow, F. Teixeira, Osseous regeneration in the presence of oxidized cellulose and collagen, J. Mater. Sci. Mater. Med 14 (2003) 739–745.

[69] E. Vögelin, N.F. Jones, J.I. Huang, J.H. Brekke, J.R. Lieberrman, Healing of a critical-sized defect in the rat femur with use of a vascularized periosteal flap, a biodegradable matrix, and bone morphogenetic protein, J. Bone Joint Surg. Am. 87 (2005) 1323–1331.

[70] S.A. Hutchens, R.S. Benson, B.R. Evans, H.M. O'Neill, C.J. Rawn, Biomimetic synthesis of calcium-deficient hydroxyapatite in a natural hydrogel, Biomaterials 27 (2006) 4661–4670.

[71] D.J. Lyman, Polyurethanes. 1. The solution polymerization of diisocyanates with ethylene glycol, J. Polym. Sci. 45 (1960) 49–59.

[72] J. Blackwell, K.H. Gardner, Structure of the hard segments in polyurethane elastomers, Polymer 20 (1979) 13–17.

[73] J. Blackwell, C.D. Lee, Hard-segment polymorphism in MDI diol-based polyurethane elastomers, J. Polym. Sci. Polym. Phys. 22 (1984) 759–772.

[74] D.J. Lyman, J.L. Brash, S.W. Chaikin, K.G. Klein, M. Carini, Effects of chemical structure and surface properties of synthetic polymers on coagulation of blood. 2. Protein and platelet interaction with polymer surfaces, Trans. Am. Soc. Artif. Int. Org 14 (1968) 250–255.

[75] J.P. Santerre, K. Woodhouse, G. Laroche, R.S. Labow, Understanding the biodegradation of polyurethanes: from classical implants to tissue engineering materials, Biomaterials 26 (2005) 7457–7470.

[76] K.M. Zia, M. Barikani, I.A. Bhatti, M. Zuber, H.N. Bhatti, Synthesis and characterization of novel, biodegradable, thermally stable chitin-based polyurethane elastomers, J. Appl. Polym. Sci. 110 (2008) 769–776.

[77] D. Xu, Z. Meng, M. Han, K. Xi, X. Jia, X. Yu, Q. Chen, Novel blood-compatible waterborne polyurethane using chitosan as an extender, J. Appl. Polym. Sci. 109 (2008) 240–246.

[78] M. Szycher, V.L. Poirier, D.J. Dempsey, Development of an aliphatic biomedical grade polyurethane elastomer, J. Elastom. Plast 15 (1982) 81–95.

[79] L. Pinchuk, A review of the biostability and carcinogenicity of polyurethanes in medicine and the new generation of "biostable" polyurethanes, J. Biomater. Sci. Polym. Ed. 6 (1994) 225–267.

[80] P.A. Gunatillake, G.F. Meijs, E. Rizzardo, R.C. Chatelier, S.J. McCarthy, A. Brandwood, K. Schindhelm, Polyurethane elastomers based on a novel macrodiols and MDI: synthesis, mechanical properties and resistance to hydrolysis and oxidation, J. Appl. Polym. Sci. 46 (1992) 319–328.

[81] P.A. Gunatillake, G.F. Meijs, S.J. McCarthy, Polysiloxane-containing polyurethane elastomeric compositions, International Patent Application PCT/AU97/00619, 1996.
[82] A. Thakahara, R.W. Hergenrother, A.J. Coury, S.L. Cooper, Effect of soft segment chemistry on the biostability of segmented polyurethanes. I. In vitro oxidation, J. Biomed. Mater. Res. 25 (1991) 341–356.
[83] A. Thakahara, R.W. Hergenrother, A.J. Coury, S.L. Cooper, Effect of soft segment chemistry on the biostability of segmented polyurethanes. II. In vitro hydrolytic stability, J. Biomed. Mater. Res. 26 (1992) 801–818.
[84] S.A. Guelcher, Biodegradable polyurethanes: synthesis and applications in regenerative medicine, Tissue Eng. PT B Rev. 14 (2008) 3–17.
[85] G.A. Skarja, K.A. Woodhouse, Structure-property relationships of degradable polyurethane elastomers containing an amino acid-based chain extender, J. Appl. Polym. Sci. 75 (2000) 1522–1534.
[86] G.A. Skarja, K.A. Woodhouse, In vitro degradation and erosion of degradable, segments polyurethanes containing an amino acid-based chain extender, J. Biomater. Sci. Polym. Ed. 12 (2001) 851–873.
[87] J. Guan, W.R. Wagner, Synthesis, characterization and cytocompatibility of polyurethaneurea elastomers with designed elastase sensitivity, Biomacromolecules 6 (2005) 2833–2842.
[88] G.L.Y. Woo, M.W. Mittelman, J.P. Santerre, Synthesis and characterization of a novel biodegradable antimicrobial polymer, Biomaterials 21 (2000) 1235–1246.
[89] A.M. Seifalian, A. Giudiceandrea, T. Schmitz-Rixen, G. Hamilton, Noncompliance: the silent acceptance of a villain, in: P. Zille, H.P. Greisler (Eds.), Tissue Engineering of Vascular Prosthetic Grafts, Landes, Georgetown, 1999. Chapter 2.
[90] S. Venkatraman, F. Boey, L.L. Lao, Implanted cardiovascular polymers: natural, synthetic and bio-inspired, Prog. Polym. Sci. 33 (2008) 853–874.
[91] J. Rumisek, C. Wade, K. Kaplan, C. Okerberg, J. Corley, M. Barry, J. Clarke, The influence of early surface thromboreactivity on long-term arterial graft patency, Surgery 105 (1989) 654–661.
[92] J.P. Eiderg, O. Roder, M. Stahl-Madsen, N. Eldrup, P. Qvarfordt, A. Laursen, et al., Fluropolymer-coated Dacron graft versus PTFE grafts for femorofemoral crossover by pass, Eur. J. Vasc. Endovasc. Surg 32 (2006) 431–438.
[93] J. San Román, J. Buján, J.M. Bellón, A. Gallardo, M.C. Escudero, E. Jorge, J. de Haro, L. Alvarez, J.L. Castillo-Olivares, Experimental study of the antithrombogenic behavior of Dacron vascular grafts coated with hydrophilic acrylic copolymers bearing salicylic acid residues, J. Biomed. Mater. Res. 32 (1996) 19–27.
[94] K. Kottke-Marchant, J. Anderson, Y. Umemura, R. Marchant, Effect of albumin coating on the in vitro blood compatibility of Dacron arterial prostheses, Biomaterials 10 (1989) 147–155.
[95] Y. Merhi, R. Roy, R. Guidoin, J. Hebert, W. Mourad, S.B. Slimane, Cellular reactions to polyester arterial prostheses impregnated with cross-linked albumin: in vivo studies in mice, Biomaterials 10 (1989) 56–58.
[96] H. Parsson, W. Jundzill, K. Johansson, T. Jonung, L. Norgren, Healing characteristics of polymer-coated or collagen-treated Dacron grafts: an experimental porcine study, Cardiovasc. Surg 2 (1994) 242–248.
[97] A. Kishida, Y. Ueno, N. Fukudome, E. Yashima, I. Maruyama, M. Akashi, Immobilization of human thrombomodulin onto poly(ether urethane urea) for developing antithrombogenic blood-contacting materials, Biomaterials 15 (1994) 848–852.
[98] F.J. Veith, S.K. Gupta, E. Ascer, S. White-Flores, R.H. Samson, L.A. Scher, J.B. Towne, V.M. Bernhard, P. Bonier, W.R. Flinn, P. Astleford, J.S.T. Yao, J.J. Bergan, Six-year prospective multicenter randomized comparison of autologous saphenous vein and expanded polytetrafluoroethylene grafts in infringuinal arterial reconstruction, J. Vasc. Surg 3 (1986) 104–114.
[99] A.W. Clowes, A.M. Gown, S.R. Hanson, M.A. Reidy, Mechanisms of arterial graft failure. 1. Role of cellular proliferation in early healing of PTFE prostheses, Am. J. Pathol 118 (1985) 43–54.
[100] J.M. Bellon, J. Bujan, L.A. Contreras, A. Hernando, F. Jurado, Similarity in behavior of polytetrafluoroethylene (ePTFE) prostheses

implanted into different interfaces, J. Biomed. Mater. Res. 31 (1996) 1–9.
[101] D.L. Akers, Y.H. Du, R.F. Kempscinski, The effect of carbon coating and porosity on early patency of expanded polytetrafluoroethylene grafts: an experimental study, J. Vasc. Surg 18 (1993) 10–15.
[102] B.H. Walpoth, R. Rogulenko, E. Tikhvinskaia, S. Gogolewski, T. Schaffner, O.M. Hess, U. Althaus, Improvement of patency rate in heparin-coated small synthetic vascular grafts, Circulation 98 (1998) II319–II323.
[103] J.L. Fisher, R.C. Thomson, J.W. Moore, P.C. Begovac, Functional parameters of thromboresistant heparinized e-PTFE vascular grafts, Cardiovasc. Pathol 11 (2002). 42–42.
[104] H.P. Greisler, D.J. Cziperle, D.U. Kim, J.D. Garfield, D. Petsikas, P.M. Murchan, E.O. Applegren, W. Drohan, W.H. Burgess, Enhanced endotheliazation of expanded polytetrafluoroethylene grafts by fibroblast growth factor type 1 pretreatment, Surgery 112 (1992) 244–254.
[105] B.H. Walpoth, P. Zammaretti, M. Cikirikcioglu, E. Khabiri, M.K. Djebaili, J.C. Pache, J.C. Tille, Y. Aggoun, D. Morel, A. Kalangos, J.A. Hubbell, A.H. Zisch, Enhanced thickening of expanded polytetrafluoroethylene grafts coated with fibrin or fibrin-releasing vascular endothelial growth factor in the pig carotid artery interposition model, J. Thorac. Cardiovasc. Surg 133 (2007) 1163–1170.
[106] F. Couet, N. Rajan, D. Mantovani, Macromolecular biomaterials for scaffold-based vascular tissue engineering, Macromol. Biosci 7 (2007) 701–718.
[107] L. Xue, H.P. Greisler, Biomaterials in the development and future of vascular grafts, J. Vasc. Surg 37 (2003) 472–480.
[108] E. Rabkin, F.J. Schoen, Cardiovascular tissue engineering, Cardiovasc, Pathol 11 (2002) 305–317.
[109] M.C. Chen, H.W. Tsai, Y. Chang, W.Y. Lai, F.L. Mi, C.T. Liu, H.S. Wong, H.W. Sung, Rapidly self-expandable polymeric stents with a shape-memory property, Biomacromolecules 8 (2007) 2774–2780.
[110] Y.S. Wong, Y. Xiong, S.S. Venkatraman, F.Y. Boey, Shape memory in un-cross-linked biodegradable polymers, J. Biomater. Sci. Polym. Ed. 19 (2008) 175–191.
[111] D. Reiter, Methods and materials for wound closure, Otolaryngol. Clin. North Am. 28 (1995) 1069–1080.
[112] D.K. Drew, L. Supik, C.R. Darrow, G.F. Price, Tissue repair using laser: a review, Orthopaedics 16 (1993) 581–587.
[113] N.A. Swanson, T.A. Tromovitch, Suture materials, 1980s: properties, uses, and abuses, Int. J. Dermatol 21 (1982) 373–378.
[114] R.B. Seyomour, C.E. Carraher (Eds.), Structure–Property Relationships in Polymers, Plenum Press, New York, 1984.
[115] G.L. Listner, Polypropylene (PP) sutures, Patent 3 630 (1971) 205.
[116] M. Wishman, G.E. Hagler, Polypropylene fibers, in: M. Lewin, E.M. Pearce (Eds.), Handbook of Fiber Science and Technology, vol. 4, Marcel Dekker, New York, 1985.
[117] D.J. Apple, N. Mamalis, S.E. Brady, K. Loftfield, D. Kavka-Van Norman, R.J. Olson, Biocompatibility of implant materials: a review and scanning electron microscopic study, J. Am. Intraocul. Implant Soc. 10 (1984) 53–66.
[118] C.C. Chu, Chemical structure and manufacturing processes, in: C.C. Chu, J. von Fraunhofer, H.P. Greisler (Eds.), Wound Closure Biomaterials and Devices, CRC Press, Boca Raton, FL, 1997.
[119] C.C. Chu, Textile-based biomaterials for surgical applications, in: S. Dumitriu (Ed.), Polymeric Biomaterials, Marcel Dekker, New York, 2003.
[120] C.C. Chu, Z. Kizil, Qualitative-evaluation of stiffness of commercial suture materials, Surg. Gynecol. Obstet 168 (1989) 233–238.
[121] M.C. Dang, J.G. Thacker, J.C.S. Hwang, G.T. Rodeheaver, S.M. Melton, R.F. Edlich, Some biomechanical considerations of polytetrafluoroethylene sutures, Arch. Surg 125 (1990) 647–650.
[122] E. Urban, M.W. King, R. Guidoin, G. Laroche, Y. Marois, L. Martin, A. Cardou, Y. Douville, Why make monofilament sutures out of polyvinylidene fluoride? ASAIO 40 (1994) 145–156.
[123] E.J. Frazza, E.E. Schmitt, A new absorbable suture, J. Biomed. Mater. Res. 5 (1971) 43–58.

[124] R.A. Miller, J.M. Brady, D.E. Cutright, Degradation rates of oral resorbable implants (polylactates and polyglycolates): rate modification with changes in PLA/PGA copolymer ratios, J. Biomed. Mater. Res. 11 (1977) 711–719.
[125] S.W. Shalaby, Synthetic absorbable polyesters, in: S.W. Shalaby (Ed.), Biomedical Polymers: Designed to Degrade Systems, Hanser Press, New York, 1994.
[126] P.S. Malchesky, Extracorporeal artificial organs, in: B.D. Ratner, A.S. Hoffman, F.J. Schoen, J.E. Lemons (Eds.), Biomaterials Science: An Introduction to Materials in Medicine, Elsevier, San Diego, CA, 2004. Chapter 7.6.
[127] R. Schaefer, W. Horl, K. Kokot, A. Heidland, Enhanced biocompatibility with a new cellulosic membrane: cuprophan vs hemophan, Blood Purif 5 (1987) 262–267.
[128] S. Bowry, T. Rintelen, Synthetically modified cellulose: a cellulosic hemodialysis membrane with minimized complement activation, ASAIO J 44 (1998) M579–M583.
[129] W.R. Clark, R.J. Hamburger, M.J. Lysaght, Effect of membrane composition and structure on solute removal and biocompatibility in hemodialysis, Kidney Int 56 (1999) 2005–2015.
[130] R. Vanholder, R. De Smet, G. Glorieux, et al., Review on uremic toxins: classification, concentration, and interindividual variability, Kidney Int 63 (2003) 1934–1943.
[131] C.S. Bouman, R.W. van Olden, C.P. Stoutenbeek, Cytokine filtration and adsorption during pre- and postdilution hemofiltration in four different membranes, Blood Purif 16 (1998) 261–268.
[132] C. Schmidt, J.B. Leach, Neural tissue engineering: strategies for repair and regeneration, Annu. Rev. Biomed. Eng 5 (2003) 293–347.
[133] L. Dahlin, G. Lundborg, The use of silicone tubing in the late repair of the median and ulnar nerves in the forearm, J. Hand Surg. (Br) 26 (2001) 393–394.
[134] B.C. Vasconcelos, C. Gay-Escoda, Facial nerve repair with expanded polytetrafluoroethylene and collagen conduits: an experimental study in the rabbit, J. Oral Maxillofac. Surg 58 (2000) 1257–1262.
[135] H. Molander, Y. Olsson, O. Engkvist, S. Bowald, I. Eriksson, Regeneration of peripheral nerve through a polyglactin tube, Muscle Nerve 5 (1982) 54–57.
[136] G.R. Evans, K. Brandt, M.S. Widmer, L. Lu, R.K. Meszlenyi, P.K. Gupta, A.G. Mikos, J. Hodges, J. Williams, A. Gürlek, A. Nabawi, R. Lohman, C.W. Patrick Jr., In vivo evaluation of poly(L-lactic acid) porous conduits for peripheral nerve regeneration, Biomaterials 20 (1999) 1109–1115.
[137] E. Nyilas, T.H. Chiu, R.L. Sidman, E.W. Henry, T.M. Brushart, P. Dikkes, R. Madison, Peripheral nerve repair with bioresorbable prosthesis, Trans. Am. Soc. Artif. Int. Org 29 (1983) 307–313.
[138] A. Valero-Cabré, K. Tsironis, E. Skouras, G. Perego, X. Navarro, W.F. Neiss, Superior muscle reinnervation after autologous nerve graft or poly-L-lactide-epsilon-caprolactone (PLC) tube implantation in comparison to silicone tube repair, J. Neurosci. Res. 63 (2001) 214–223.
[139] N. Nicoli Aldini, M. Fini, M. Rocca, G. Giavaresi, R. Giardino, Guided regeneration with resorbable conduits in experimental peripheral nerve injuries, Int. Orthop 24 (2000) 121–125.
[140] G. Soldani, G. Varelli, A. Minnocci, P. Dario, Manufacturing and microscopical characterization of polyurethane nerve guidance channel featuring a highly smooth internal surface, Biomaterials 19 (1998) 1919–1924.
[141] R.C. Young, M. Wiberg, G. Terenghi, Poly-3-hydroxybutyrate (PHB): a resorbable conduit for long-gap repair in peripheral nerves, Br. J. Plast. Surg 55 (2002) 235–240.
[142] R.F. Valentini, T.G. Vargo, J.A. Gardella Jr., P. Aebischer, Electrically charged polymeric substrates enhance nerve fiber outgrowth in vitro, Biomaterials 13 (1992) 183–190.
[143] C.E. Schmidt, V.R. Shastri, J.P. Vacanti, R. Langer, Stimulation of neurite outgrowth using an electrically conducting polymer, Proc. Natl. Acad. Sci. USA 94 (1997) 8948–8953.
[144] E.G. Fine, R.F. Valentini, R. Bellamkonda, P. Aebischer, Improved nerve regeneration through piezoelectric vinylidenefluoride-trifluoroethylene copolymer guidance channels, Biomaterials 12 (1991) 775–780.
[145] J.H. Collier, J.P. Camp, T.W. Hudson, C.E. Schmidt, Synthesis and characterization of

polypyrrole-hyaluronic acid composite biomaterials for tissue engineering applications, J. Biomed. Mater. Res. 50 (2000) 574–584.

[146] P.R. Bidez III, S. Li, A.G. Macdiarmid, E.C. Venancio, Y. Wei, P.I. Lelkes, Polyaniline, an electroactive polymer, supports adhesion and proliferation of cardiac myoblasts, J. Biomater. Sci. Polym. Ed. 17 (2006) 199–212.

[147] A.P. Balgude, X. Yu, A. Szymanski, R.V. Bellamkonda, Agarose gel stiffness determines rate of DRG neurite extension in 3D cultures, Biomaterials 22 (2001) 1077–1084.

[148] G. Haipeng, Z. Yinghui, L. Jianchun, G. Yandao, Z. Nanming, Z. Xiufang, Studies on nerve cell affinity of chitosan-derived materials, J. Biomed. Mater. Res. 52 (2000) 285–295.

[149] M.R. Wells, K. Kraus, D.K. Batter, D.G. Blunt, J. Weremowitz, S.E. Lynch, H.N. Antoniades, H.A. Hansson, Gel matrix vehicles for growth factor application in nerve gap injuries repaired with tubes: a comparison of biomatrix, collagen, and methylcellulose, Exp. Neurol 146 (1997) 395–402.

[150] B.R. Seckel, D. Jones, K.J. Hekimian, K.K. Wang, D.P. Chakalis, P.D. Costas, Hyaluronic acid through a new injectable nerve guide delivery system enhances peripheral nerve regeneration in the rat, J. Neurosci. Res. 40 (1995) 318–324.

[151] T. Hashimoto, Y. Suzuki, M. Kitada, K. Kataoka, S. Wu, K. Suzuki, K. Endo, Y. Nishimura, C. Ide, Peripheral nerve regeneration through alginate gel: analysis of early outgrowth and late increase in diameter of regenerating axons, Exp. Brain Res. 146 (2002) 356–368.

[152] C.B. Herbert, C. Nagaswami, G.D. Bittner, J.A. Hubbell, J.W. Weisel, Effects of fibrin micromorphology on neurite growth from dorsal root ganglia cultures in three-dimensional fibrin gels, J. Biomed. Mater. Res. 40 (1998) 551–559.

[153] T. Satou, S. Nishida, S. Hiruma, K. Tanji, M. Takahashi, S. Fujita, Y. Mizuhara, F. Akai, S. Hashimoto, A morphological study on the effects of collagen gel matrix on regeneration of severed rat sciatic nerve in silicone tubes, Acta Pathol. Jpn 36 (1986) 199–208.

[154] P. Sierpinski, J. Garrett, J. Ma, P. Apel, D. Klorig, T. Smith, L.A. Koman, A. Atala, Van M. Dyke, The use of keratin biomaterials derived from human hair for the promotion of rapid regeneration of peripheral nerves, Biomaterials 29 (2008) 118–128.

[155] T.C. Holmes, S. de Lacalle, X. Su, G. Liu, A. Rich, S. Zhang, Extensive neurite outgrowth and active synapse formation on self-assembling peptide scaffolds, Proc. Natl. Acad. Sci. USA 97 (2000) 6728–6733.

[156] X. Yu, G.P. Dillon, R.V. Bellamkonda, Tissue-engineered scaffolds are effective alternatives to autografts for bridging peripheral nerve gaps, Tissue Eng 9 (1999) 421–430.

[157] L.R. Williams, S. Varon, Modification of fibrin matrix formation in situ enhances nerve regeneration in silicone chambers, J. Comp. Neurol 231 (1985) 209–220.

[158] J.C. Schense, J.A. Hubbell, Cross-linking exogenous bifunctional peptides into fibrin gels with factor XIIIa, Bioconjug. Chem. 10 (1999) 75–81.

[159] M.B. Chen, F. Zhang, W.C. Lineaweaver, Luminal fillers in nerve conduits for peripheral nerve repair, Ann. Plast. Surg 57 (2006) 462–471.

[160] T.T. Ngo, P.J. Waggoner, A.A. Romero, K.D. Nelson, R.C. Eberhart, G.M. Smith, Poly(L-Lactide) microfilaments enhance peripheral nerve regeneration across extended nerve lesions, J. Neurosci. Res. 72 (2003) 227–238.

[161] X. Wang, W. Hu, Y. Cao, J. Yao, J. Wu, X. Gu, Dog sciatic nerve regeneration across a 30-mm defect bridged by a chitosan/PGA artificial nerve graft, Brain 128 (2005) 1897–1910.

[162] D. Ceballos, X. Navarro, N. Dubey, G. Wendelschafer-Crabb, W.R. Kennedy, R.T. Tranquillo, Magnetically aligned collagen gel filling a collagen nerve guide improves peripheral nerve regeneration, Exp. Neurol 158 (1999) 290–300.

[163] J. Gao, Y.M. Kim, H. Coe, B. Zern, B. Sheppard, Y. Wang, A neuroinductive biomaterial based on dopamine, Proc. Natl. Acad. Sci. USA 103 (2006) 16681–16686.

[164] Y. Haile, K. Haastert, K. Cesnulevicius, K. Stummeyer, M. Timmer, S. Berski, G. Drager, R. Gerardy-Schahn, C. Grothe, Culturing of glial and neuronal cells on polysialic acid, Biomaterials 28 (2007) 1163–1173.

6 Biodegradable Polymers and Polymer Blends

Long Jiang and Jinwen Zhang

OUTLINE

6.1 Introduction

Since the subject of biodegradable polymers received wide attention in the early 1970s, biodegradable polymers have undergone extensive investigations in academia and industry and experienced several important stages of development. Since plastics account for a significant portion (~21% by volume in the United States) of municipal waste, degradable or biodegradable plastics were initially intended to address the issue of "landfill crisis," with the anticipation that some landfill space would be freed up if the waste plastic materials could be biodegraded. Therefore, the first generation of degradable plastics did not place biodegradability and environmental footprint as a priority but focused only on landfill space saving. Most of these products are based on the compounds of conventional resins such as polyolefins filled with starch or activated with metal oxide or transition melt salt, which only disintegrate into small pieces over time due to the biodegradation of the starch ingredient or catalyzed photodegradation of the polyolefins. Later on, a series of petroleum-based synthetic polymers, which can be termed second-generation degradable plastics and are truly biodegradable, were developed and entered the marketplace with an insignificant share. These biodegradable polymers mainly include aliphatic polyesters such as polycaprolactone (PCL), poly(butylene succinate) (PBS), poly(butylene succinate-*co*-adipate) (PBSA), and other aliphatic copolyesters and aliphatic–aromatic copolyesters such as poly(butylene adipate-*co*-terephthalate) (PBAT). Meanwhile, researchers also devoted tremendous effort to developing biodegradable polymers and plastic materials from renewable resources such as starch, soy protein (SP), cellulose, and plant oil. Starch and SP can be thermoplasticized under heat and mechanical agitation in the presence of appropriate processing agents. Thermoplasticized starch and SP can be effectively blended with other thermoplastic

Ebnesajjad: Handbook of Biopolymers and Biodegradable Plastics. http://dx.doi.org/10.1016/B978-1-4557-2834-3.00006-9
© 2013 Elsevier Inc. All rights reserved. Adapted from a chapter in: Kutz, *Applied Plastics Engineering Handbook* (2011).

polymers to form biodegradable polymer composites. Cellulosic fiber can be directly used as reinforcement fiber in fiber-reinforced polymer composites. It can also be dissolved using appropriate solvents and then shaped into "regenerated cellulose" products such as fibers and sheets. Polylactic acid (PLA) and polyhydroxyalkanoates (PHAs) represent the two most important biodegradable polymers derived from renewable resources. They are thermoplastics and show mechanical properties and processability similar to that of some petroleum-based polymers. The advent of PLA and PHAs is a great leap forward in the development of biodegradable polymers.

Historically, the research interest and effort on biodegradable and biobased polymers has been up and down, in accordance with the cycle of oil price. The most recent oil price spike and national energy policy shift will definitely promote the already intensive research on alternative energy and renewable materials. With the tremendous interests and efforts being put in this area, new progress and achievements are made continuously as evidenced by the increasing numbers of publications. This chapter summarizes contemporary research achievements and situations in biodegradable and biobased polymers. In the following sections, we first discuss the naturally occurring biodegradable polymers, and then the biodegradable polymers derived from renewable resources and the biodegradable polymers based on petroleum. Finally, we briefly discuss several biobased polymers that may not be biodegradable.

6.2 Naturally Occurring Biodegradable Polymers

The utilization of natural polymers for nonfood uses can be traced back to ancient times. Skin and bone parts of animals, plant fibers, starch, silk, etc., are typical examples of the natural polymers used in different periods of the human history. In the last century, the development of natural polymers was significantly hindered due to the advent of low-cost petrochemical polymers. It was only about two decades ago that intensive research on natural polymers was revived, primarily due to the issues of environmental pollution and the depletion of fossil oils. Modern technologies provide new insights of the synthesis, structures, and properties of the natural polymers. These new findings have enabled developments of natural polymers with novel processing characteristics and properties, which can be used for many more advanced applications. This section deals with three major natural polymers: starch, cellulose, and SP. All of them have primarily been used as human and animal foods through history. New developments have allowed them to be used as a material component in polymer blends and composites to make biodegradable products.

6.2.1 Starch

Starch is traditionally the largest source of carbohydrates in human diet. Being a polysaccharide polymer, starch has been intensively studied in order to process it into a thermoplastic polymer in the hope of partially replacing some petrochemical polymers. Starch is a mixture of linear (amylose) and branched (amylopectin) poly-(1,4)-α-glucose (Fig. 6.1) and exists in the form of discrete granules. Amylose has a typical molecular weight of several hundred thousand, whereas the molecular weight of amylopectin is much higher and is in the order of tens of millions. Depending on the botanic origin of starch, the ratio of amylose is typically around 20–35%. Some so-called "waxy" starches have very low amylose content. For example, waxy maize starch contains less than 2% amylose. Starch granules are semicrystalline, containing both ordered structure (mainly double helices of amylopectin short chains) and amorphous structure (amylopectin long chains and branch points and amylose) [1].

In its natural form, starch is not meltable and therefore cannot be processed as a thermoplastic. However, starch granules can be thermoplasticized through a gelatinization process. In this process, the granules are disrupted and the ordered crystalline structure is lost under the influence of plasticizers (e.g., water and glycerol), heat, and shear. The resultant melt-processable starch is often termed

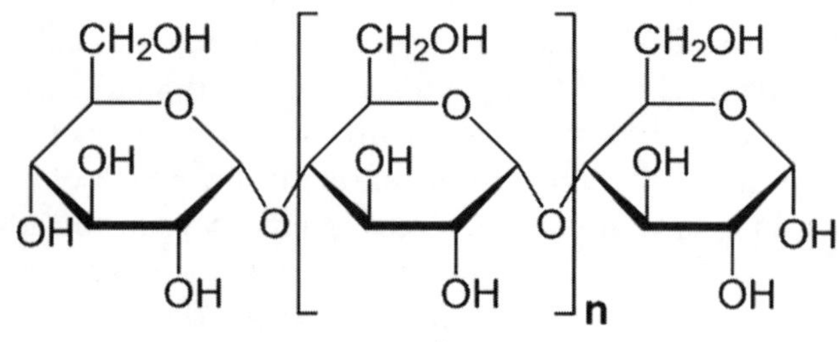

Figure 6.1 Chemical structure of starch.

thermoplastic starch (TPS). Since the advent of TPS, numerous studies have been conducted to explore its use as a thermoplastic polymer by overcoming its inherent drawbacks including low strength, high moisture sensitivity, and brittleness caused by starch retrogradation and gradual loss of the plasticizers.

To destruct the crystalline structure of starch and allow flowability, large contents of plasticizers are used in the preparation of TPS. Depending on the amount of plasticizers used, TPS materials range from glassy to rubbery state. Their stress–strain behaviors are dependent on the content of the plasticizers. Being hydrophilic, TPS is susceptible to moisture attack during storage or service. The increase in water content decreases glass transition temperature (T_g) of TPS and subsequently reduces its tensile strength. It was found that T_g of starch–glycerol–water blends depended linearly on the water content [2]. Glass transition temperature decreased from −53 °C to −105 °C when the water content increased from 2 to 30%. A small amount of glycerol also caused a large decrease in T_g of the blends, but further addition of glycerol only slightly affected T_g [2]. Besides the strength and T_g, TPS's susceptibility to water also leads to poor dimensional stability of its final products.

On the other hand, when humidity and temperature are constant, mechanical properties of TPS depend on the storage or service time of the products [3]. The tensile strength of TPS increased and the elongation degreased after the product was stored for 5 weeks at constant humidity and temperature [2]. This was due to time-dependent retrogradation (postcrystallization) of starch in the presence of water and glycerol.

The time-dependent properties of TPS are a combined result of starch retrogradation, water content fluctuation, and plasticizer (e.g., glycerol) diffusion. These factors are difficult to control during the storage and service life of TPS. As a result, TPS is rarely used alone but is often blended with hydrophobic thermoplastic polymers to form starch-containing polymer blends so that the mechanical performance, moisture resistance, and dimensional stability of TPS can be improved.

In the early years of starch-containing polymer blends, dry starch granules were directly used in the blends as a filler [4]. Since most polymers are hydrophobic, hydrophilic starch is thermodynamically immiscible with these polymers and consequently resulted in weak interfacial bonding between the starch and the polymer matrix. This in turn led to poor mechanical properties of the blends, e.g., low tensile strength, low elongation, and brittleness. In view of this, compatibilization between the starch granules and the polymer matrix was carried out. Maleic anhydride (MA) is most commonly used and is also one of the most effective coupling agents for the starch-containing blends. In the literature, both biodegradable polymers, e.g., ethylene vinyl acetate (EVA), low-density polyethylene (LDPE), and high-density polyethylene (HDPE), and non-biodegradable polymers, e.g., PBS, PCL, and PLA, have been functionalized by MA and used as compatibilizers in the corresponding starch–polymer blends. Remarkable strength increase was realized after the addition of the compatibilizers. The modulus and elongation of the blends were relatively less affected [5–7].

Compared to granular starch, TPS offers a great advantage in material processability and morphology control as the TPS can be deformed and dispersed to a much finer state than the dry native starch. Property-enhancing microstructures such as co-continuous structure can be formed during melt blending. The polymers used to blend with TPS included LDPE [8], polystyrene (PS) [9], and most often biodegradable polymers such as poly(hydroxyl ester ether) [10], castor oil-based polyurethane (PU) [11], poly(ester amide) [12], PCL [13], and poly(3-hydroxybutyrate-*co*-3-hydroxyvalerate) (PHBV) [14–16]. In general, the typical disadvantages of TPS such as moisture susceptibility, brittleness, and low strength have been reduced to various degrees by blending with these polymers. However, mechanical properties of these blends still decreased as the TPS content increased. This limited the content of TPS in the blends if the strength of the matrix needed to be maximally maintained. The reports on compatibilization between TPS and polymer matrix are surprisingly scarce. Using a twin-screw extruder, Huneault and Li [16] first grafted MA to PLA by free radical grafting and then allowed the resultant MA-*g*-PLA to react with TPS. In the PLA-*g*-MA TPS extrudates, TPS was shown to be dispersed in a much finer state than in the PLA–TPS blend without MA grafting. The tensile strength and modulus showed no obvious variation with or without MA grafting. However, the elongation was significantly increased when PLA-*g*-MA was used.

Besides being used as a dispersed component in polymer blends, TPS has also been used as a matrix polymer and has been reinforced by natural fibers. TPS—sisal fiber composite prepared by compression molding exhibited improved tensile strength. The strength increased from 4 MPa (neat TPS) to 8 MPa at 10% fiber content [17]. Ma *et al.* [18] prepared TPS—winceyette fiber composite by extrusion. The composite exhibited a tensile strength of 15 MPa at 20% fiber content, tripling the strength of the unreinforced TPS. Using a starch-based emulsion-type resin, Ochi prepared unidirectional continuous hemp fiber-reinforced starch composites [19]. A tensile strength of 365 MPa was obtained at the highest fiber content of 75%.

In recent years, nanoclay has also been studied for its effects on mechanical and barrier properties of TPS and TPS—polymer blends. For example, Wilhelm *et al.* reported a 70% increase in tensile strength of TPS—hectorite nanocomposite films at a 30% clay level [20]. Avella *et al.* also reported increased mechanical properties of potato starch—montmorillonite (MMT) nanocomposite films [21]. Especially significant, Huang *et al.* observed an increase of 450% and 20% in tensile strength and strain, respectively, after the addition of 5% clay to corn starch—MMT nanocomposites [22]. Most recently, Tang *et al.* reported significantly increased tensile strength (up to 92% higher) and reduced water vapor permeability (up to 67% lower) of TPS—MMT films prepared by melt extrusion and subsequent casting [23].

Starch can also be foamed by water vapor to make compostable packaging foams [24]. To provide water resistance, acetylated starch, which is a less polar material and is more water resistant, can be used as the foaming material [25]. For the same purpose, TPS was also first blended with hydrophobic polymers (e.g., PHBV, PCL, PBS, PVA, and PLA) and subsequently foamed [26—28]. Besides water vapor, CO_2 has been also used as the foaming agent in TPS—PLA foams [29].

6.2.2 Cellulose

Cellulose is the most abundant renewable biopolymer on earth. About 33% of all plant matter is cellulose. The purest natural cellulose form is cotton (~90%). Wood contains about 50% of cellulose [30]. Cellulose can also be synthesized by some bacteria. Cellulose is a polysaccharide having a molecular structure similar to starch. However, the D-glucose units are linked by β-glycosidic bonds in cellulose

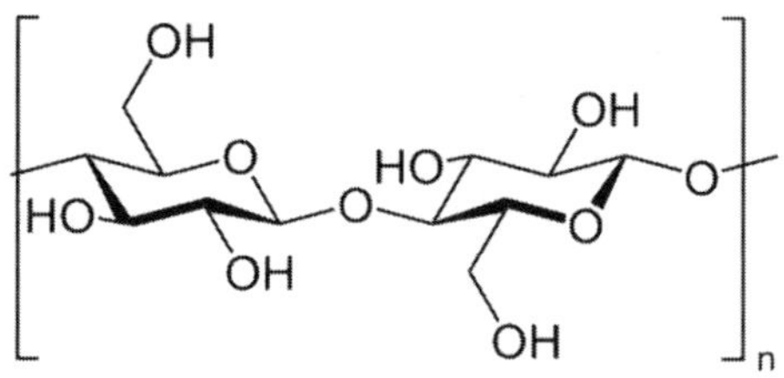

Figure 6.2 Chemical structure of cellulose.

(Fig. 6.2) instead of α-glycosidic bonds in starch. Due to this β-glycosidic bond, cellulose molecules adopt an extended and stiff rodlike conformation. The multiple hydroxyl groups from one chain form hydrogen bonds with oxygen molecules on another chain, holding the chains firmly together side by side and forming elementary crystallites (cellulose nanowhiskers (CNWs)) with exceptional high tensile strength. These nanowhiskers, embedded in amorphous hemicellulose and lignin, form microfibrils and further the cell wall of plant cells.

Cellulosic natural fibers (e.g., abaca, bamboo, jute, flax, and hemp) have long been used as load-bearing materials to reinforce polymer matrix. Compared to traditional reinforcement fibers, e.g., glass fibers and carbon fibers, cellulosic fibers show the advantages of low material cost, low environmental impact (renewability and carbon dioxide neutral, i.e., no excess carbon dioxide is returned to the environment when composted or combusted), and competitive strength—density ratio [31]. Cellulosic fibers are almost nonabrasive to processing equipment, which contributes to substantial reduction in production cost. They are also safer to handle compared to man-made fibers.

A major disadvantage of cellulosic fibers is their hydrophilicity due to the existence of large amounts of hydroxyl groups in polysaccharide molecules. While blending with hydrophobic matrix polymers, this leads to serious dispersion and interface problems which often result in poor mechanical properties. Other drawbacks of using cellulosic fibers include limited processing temperature (<200 °C), high moisture absorption and swelling, nonuniform dimensions and properties, and low microbial resistance and product durability (if this is needed). However, many of these disadvantages can be reduced or even eliminated by appropriate fiber treatment and composite processing.

One of the most promising uses of cellulosic fibers is the development of fully biodegradable "green" composites using biopolymers as the matrix. The "green" composites should be environmentally friendly, biodegradable, and sustainable. The disposal of the composites at the end of their service poses no harm to the environment. For example, Plackett *et al.* prepared PLA–jute fiber mat composite by film stacking technique [32]. Tensile strength of the composite was significantly increased when the composite was pressed within the temperature range of 180–220 °C. Interfacial bonding between the hydrophilic fibers and the hydrophobic polymers can be improved by compatibilization. Lee and Wang studied the effects of coupling agent (lysine-based diisocyanate) on the properties of bamboo fiber (BF) composites [33]. They found that tensile strength, water resistance, and interfacial bonding were all improved for the PLA–BF and PBS–BF composites after the addition of coupling agent. Jiang *et al.* prepared PHBV–bamboo pulp fiber (BPF) composites by melt compounding and injection molding [34]. Tensile strength and modulus, flexural strength and modulus, impact strength, and crystallization rate were all substantially increased by the addition of BPF. Tensile and flexural elongations were also moderately increased at low fiber contents (<20 wt%). When PHBV8-*g*-MA was used as the compatibilizer, the strength and modulus were further increased due to improved polymer–fiber interfacial bonding. However, the toughness of the composites was substantially reduced due to the hindrance to fiber pullout, a major energy dissipation source during the composite deformation. Cellulosic fibers were also used in SP plastics as reinforcing agents. Lodha and Netravali investigated ramie fiber [35] and flax yarn [36] reinforced SP isolate (SPI) resin. They found that stearic acid modified SPI (MSPI)–ramie fiber composites showed significantly higher mechanical properties compared to SPI–ramie fiber composites. A polycarboxylic acid-based modifier (Phytagel®) also considerably improved the mechanical and moisture properties as well as thermal stability of the SPI–flax yarn composites.

The elementary crystallites of cellulose, CNWs, exhibit a Young's modulus of over 100 GPa and a surface area of several hundred square meters per gram [37]. They have the potential to significantly reinforce polymers at low volume ratios as is being realized by other nanomaterials such as carbon nanotubes (CNTs) and nanoclays. CNWs can be separated from cellulosic fibers by acid hydrolysis. Oksman's group has performed extensive research on the production of CNWs and the processing of various polymer–CNW nanocomposites by both solution casting and extrusion blending [38–45]. Homogeneous dispersion of the whiskers posed a great challenge in CNW nanocomposite processing due to hydrogen bonding-induced agglomeration of the whiskers. This was especially true when the freeze-dried whisker powder and polymers were compounded by extrusion. Good dispersion of CNW in the polymer matrix was obtained by solution casting [43] or by directly pumping whisker suspension into the extruder during the extrusion compounding process [39,40,45]. Dispersion agents/compatibilizers were found to improve the dispersion of CNW [40,41,43,45]. Composites obtained without good dispersion showed insignificant effects on the properties of the composites. Various degrees of success have also been made by other researchers on cellulose whisker-reinforced polymer composites prepared by solution casting [46–48].

Besides being directly used as reinforcement fiber, cellulose has also been chemically treated to form cellulose derivatives and then dissolved in appropriate solvents (or directly dissolved in suitable solvents) to produce highly viscous cellulose derivative (or cellulose) solutions. This process imparts flowability to cellulose and thus enables its processing using traditional polymer-processing equipment. Cellulose is regenerated when the solutions are passed through a coagulation bath. The most widely known "regenerated cellulose" materials are cellophane, a thin transparent film, and rayon and lyocell, which are both cellulose fibers. The production of lyocell is more eco-friendly than that of rayon and cellophane because the former does not use hazardous CS_2 to form cellulose derivative and its solvent can also be fully recycled and reused.

6.2.3 Soy Protein Plastic

Similar to starch and cellulose, SP is an abundant, low-cost, and renewable biopolymer that shows great potential in the polymer industry as a replacement for petrochemical polymers in many applications. SP is commercially available in three different SP concentrations: soy flour (SP concentration ~54%), SP concentrate (65–72%), and SPI (~90%). The rest of SP is primarily carbohydrates. SP is made from dehulled, defatted soybean meal. The concentration of protein is achieved by removing (e.g., water/acid/aqueous alcohol wash, precipitation, and centrifuge) most of the soluble nonprotein compounds, including mainly soluble carbohydrates, some nitrogenous substances, and minerals. SP molecules comprise 20 different amino acids with strong inter- and intramolecular interactions. These interactions make SP unmeltable, and therefore it is impossible to process SP as a thermoplastic polymer, unless enough amount of plasticizers, e.g., water, glycerol, ethylene glycol, sorbitol, etc., are applied [49]. Other processing agents, such as sodium tripolyphosphate for interrupting SP ionic interactions [50], or sodium sulfite as a reducing agent to break the disulfide bonds [51], are also used. The use of significant amounts of plasticizers results in low mechanical properties of SP plastics. On the other hand, when the plasticizers migrate away from the SP plastics during storage or service, the materials become very brittle. Moreover, the hydrophilicity of SP and the plasticizers leads to low moisture resistance of SP plastics.

Blending SP plastics with biodegradable polymers is a natural choice to overcome the aforementioned drawbacks of SP plastics. PCL [52,53], PLA [54], PBSA [55], and poly(tetramethylene adipate-*co*-terephthalate) (PTAT) [56] have been used to blend with SP. These polymers are hydrophobic and therefore can not establish strong interfacial bonding with SP. As a result, their blends with SP showed inferior properties. On the other hand, poly(hydroxyl ester ether) can form strong hydrogen bonding with SP. Therefore, its blends with SP exhibited better properties [57]. John and Bhattacharya [55] showed that using a small amount of MA-grafted biopolyesters, the mechanical properties, moisture resistance, and processing conditions of SP–biopolyester composites can be improved due to enhanced interfacial interactions. Zhong and Sun found that methylene diphenyl diisocyanate (MDI) was an effective compatibilizer to increase the tensile strength of SP–PCL blends [58]. Zhang *et al.* compared the different morphologies and properties between SPI–PLA and SPC–PLA blends [54]. SPC–PLA blend showed finer phase structures and higher mechanical properties than SPI–PLA due to SPC's higher compatibility with PLA. Co-continuous structure was realized in the SPC–PLA blends in a broad composition range. Moreover, after applying poly(2-ethyl-2-oxazoline) (PEOX) as a compatibilizer to both SP blends, the phase structure, mechanical properties, and water resistance of both blends were all improved. Very recently, by introducing urethane and isocyanate groups to PBS, Li *et al.* obtained SPI–PBS blends with substantially improved phase structures and mechanical properties [59]. The authors attributed the compatibility to the hydrogen bonding between the urethane groups (–OCONH–) in PBS and the amide groups (–CONH–) in SPI. The residual NCO groups in isocyanate-containing PBS can also react with NH_2 groups in SPI, which further strengthens interfacial bonding between the two phases.

Jinwen Zhang and his group at Washington State University have conducted intensive research on the effects of processing techniques, plasticizers, compatibilizers, and other additives on the microstructure and properties of SP blends. Chen and Zhang prepared SP–PBAT blends using SP with different water contents [60]. They found that with increasing water content, SP's behavior ranged from resembling a rigid filler to a deformable filler and eventually to a thermoplastic. A percolate SP thread structure was formed in the blends at high water content. As a result, mechanical properties of the blends were significantly improved. The authors also investigated the effects of SP content and a maleic anhydride-grafted PBAT compatibilizer on the morphology, mechanical properties, rheological properties, and dynamic properties of the blends [61]. The compatibilizer refined phase structure of the blends and an SP content over 25% led to a percolated SP network structure. Both of them contributed to the greatly improved mechanical properties of the blends. The effects of SP plasticization and shear stress on the phase structure development and mechanical properties of SP blends were further studied and an empirical percolation model was shown to be able to predict the formation of the percolated SP structure [62].

Liu *et al.* on the other hand prepared SP–PLA blends using the same method and studied the effects of plasticizers, compatibilizers, and other processing aids on the properties of the blends. They compared the plasticizing effects of water and glycerol on soy protein and their influences on final morphology of the blends [63]. They showed that PEOX and polymeric methylene diphenyl diisocyanate (pMDI) were effective compatibilizers for the blends. When used together, they show significant synergy in improving the mechanical properties of the blends [64]. Acetyl tri-*n*-butyl citrate and an alkene bis fatty amide were used as processing aids to improve the processability of the blends. Their lubricating mechanisms and effects on the properties of the resultant blends were discussed [65]. Finally, the group extruded SP/PLA foams using 4-methylbenzene-1-sulfonohydrazide as the chemical blowing agent. The morphology of the foams was studied and correlated to extrusion temperature, foaming agent content, and pMDI content [66].

The low strength of SP plastics can also be remedied by cellulosic fibers and nanosized reinforcing fillers. Liu *et al.* found that raw grass fibers could improve mechanical (tensile, flexural, and impact) and thermal properties of SP plastics [67,68]. Alkali pretreatment of the raw grass fibers removed hemicelluloses and lignin, which resulted in a larger length–diameter ratio of the fibers and better fiber dispersion. This in turn led to even higher mechanical and thermal properties of the SP-treated grass composites. SP plastics were also reinforced by CNWs. Wang *et al.* prepared SP–nanowhisker composites by solution dispersion, freeze drying, and hot pressing [69]. The composites showed increased tensile strength and modulus and improved water resistance and thermal stability. The authors ascribed these properties to the crosslink network caused by intermolecular hydrogen bonds between the cellulose whiskers and the SPI matrix. Chen *et al.* prepared SP–MMT nanocomposites by combined aqueous dispersion and melt extrusion method [70]. MMT was shown to be highly exfoliated in the SP matrix at low MMT concentrations (<12 wt%). Above this range, MMT was intercalated. The mechanical strength and thermal stability of the SPI–MMT composites were significantly improved due to the fine dispersion of the MMT layers and the strong restriction effects on the interfaces, which was created by the surface electrostatic interaction between the positive charge-rich domains of SP and the negatively charged MMT layers as well as the hydrogen bonding between the –NH and Si–O groups. SP was also blended with another important category of nano-reinforcement agents, CNTs, for property improvement [71]. Multiwalled CNTs (MWCNTs) of different sizes were compounded with SP in solution, freeze dried, and pressed into sheets. Various degrees of improvement on tensile strength, modulus, elongation, and water resistance were observed for the composites with different sizes and concentrations of MWCNTs. Optimal nanotube size and concentration were identified. Depending on the size (internal and external diameters of MWCNTs), the authors hypothesized two SP–MWCNT microstructures: SP molecules wrapping on MWCNTs (small nanotubes) and SP molecules penetrating the internal channels of MWCNTs (large nanotubes).

6.2.4 Sugar Beet Pulp (SBP) Plastics and Composites

SBP is the residual after sugar extraction from sugar beets. Its main ingredients are polysaccharides (75%) including pectin, cellulose, and hemicellulose. Other ingredients include lignin, protein, residual sugar, etc. The United States is the largest producer of sugar beets in the world with an annual dry SBP production of 1.5 million tons. SBP is traditionally used as an animal feed. Extracting pectin and cellulose fibers from SBP has been explored to increase the value of SBP. Using SBP as a thermoplastic or a component in polymer composites is relatively new.

Pectin is a group of water-soluble 1,4-linked α-D-galacturonic acid residues with variable numbers of methyl ester groups. It accounts for approximately 25% of the dry mass of SBP. In the presence of sufficient water and/or other plasticizers, SBP can be converted into a thermoplastic with pectin as the matrix material by twin screw extrusion [72,73]. In this process, the cellular structure of SBP is destructed under the influences of shear stress, heat, and plasticizers. Pectin is released and cellulose fibers and other ingredients of SBP are redistributed in the pectin matrix to virtually form a thermoplastic composite. Liu *et al.* showed that the content of plasticizers had a strong effect on mechanical properties of the SBP plastics [73]. High content of water and other plasticizers resulted in low tensile strength and modulus but high elongation of the plastics.

SBP can also be used as a dry filler to be directly blended with other thermoplastic polymers [74–76]. Embedding SBP in a hydrophobic polymer matrix increases its moisture resistance. Modulus of the polymer was increased after the addition of SBP, whereas tensile strength and elongation were decreased. Chen *et al.* studied the effects of pMDI as a compatibilizer for PLA–SBP composites to increase its tensile strength [74]. The strength of PLA/SBP (70/30) was increased from 37 MPa (without pMDI) to 61 MPa (with 2% pMDI). The dispersion of SBP in the PLA matrix and the water resistance of the composites were also improved due to enhanced interfacial adhesion.

Another method to use SBP in polymer composites is to convert SBP into thermoplastic SBP (TSBP) first and then blend it with other polymers. This method can improve SBP dispersion and surface quality of the composites. Glycidyl methacrylate (GMA), PLA, and PBAT have been used to blend with TSBP [77–79]. The effects of TSBP on the mechanical properties of the composites varied with different polymer matrixes. Liu *et al.* compared the mechanical properties of PBAT–TSBP and PBAT–SBP (i.e., SBP as dry filler) composites and showed that the former had higher properties at equal SBP contents [79]. This high performance was attributed to good dispersion of TSBP and its fiber-like structure that was formed during melt blending. The authors further showed that after adding pMDI as a compatibilizer for PBAT and TSBP, the tensile strength, modulus, and elongation of the composites were all significantly increased.

6.3 Biodegradable Polymers Derived from Renewable Resources

Unlike the aforementioned natural polymers, which can be harvested directly from the nature, some polymers are not available (or available in meaningful quantity) from the nature but can be produced with human intervention from naturally occurring biosources. PLAs and PHAs are the two most important polymers within this category. They have received intensive research interests in the past two decades and are finding more and more applications due to their unique combinations of properties.

6.3.1 Polylactic Acid

PLA is a synthetic biodegradable polyester with its monomer, lactic acid (LA), derived from natural resource. Lactic acid is made by bacterial fermentation of carbohydrates such as corn, sugarcane, potatoes, and other biomass. High-molecular-weight PLA can be synthesized using three different routes: direct condensation polymerization, azeotropic dehydrative condensation, and ring-opening polymerization of lactide. The last route was patented by Cargill in 1992 [80] and is the most commonly used method (Fig. 6.3). Direct condensation polymerization is the least expensive method, but it can only obtain low-molecular-weight PLA because it is difficult to remove water completely from the reaction mixture.

PLA is well known for its biocompatibility and biodegradability. Moreover, PLA is a thermoplastic polymer and can be conveniently processed using existing polymer-processing equipment and techniques. PLA can be processed into fiber, film, sheet, and 3D articles by fiber drawing, film blowing, extrusion, and injection molding. With the continual drop of resin prices, PLA is gradually gaining the market share. Its clarity makes it suitable for biodegradable packaging, such as bottles, food containers, and wrappers. It has also been used for food service ware, lawn and food waste bags, coatings for paper and cardboard, and fibers for clothing, carpets, sheets and towels, and wall coverings. In biomedical applications, it is used for sutures, stents, prosthetic materials, dialysis media, and drug delivery devices. PLA degrades primarily by hydrolysis through a two-stage process. First, random chain scission of the ester groups of PLA reduces its molecular weight. The speed of chain scission depends on the pH value, temperature, and moisture levels of the environment [81]. Embrittlement of the polymers occurs with the reduction of its molecular weight. Second, low-molecular-weight PLA is metabolized by microorganisms, yielding carbon dioxide, water, and humus [82].

The properties of PLA can be quite different because of the presence of the pendent methyl group on the alpha carbon atom. L-, D-, and DL-lactide isomers exist due to this structure. L-Lactide is produced by most microorganisms. DL-Lactide is the

Figure 6.3 Synthesis of PLA.

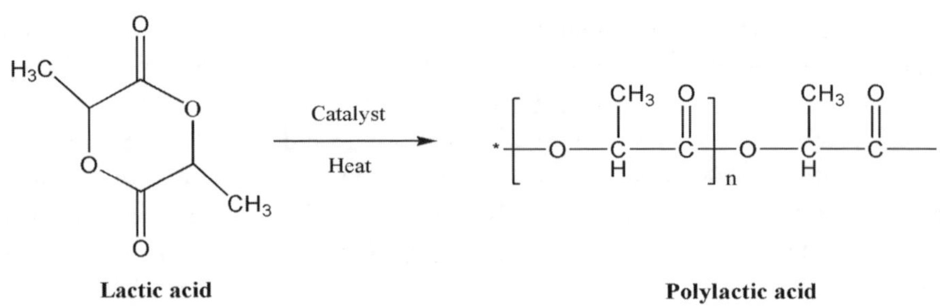

synthetic blend of D-lactide and L-lactide. The homopolymer of L-lactide (PLLA) is a semi-crystalline polymer with a typical melting point (T_m) of 160–180 °C and a glass transition temperature (T_g) of 55–65 °C. It possesses high tensile strength/modulus and low elongation, making it suitable for load-bearing applications such as in orthopedic fixation and sutures. Poly(DL-lactide) (PDLLA) is an amorphous random copolymer of L-lactide and D-lactide. It has lower tensile strength, higher elongation, and a much more rapid degradation time, making it more attractive as a drug delivery system. PLLA is a slow-crystallizing material, and its crystallinity significantly depends on processing conditions such as cooling rate and annealing status. For instance, PLLA products produced by injection molding show minimum crystallinity due to the fast cooling in the mold [83]. Annealing the products above its T_g considerably increases their crystallinity. Higher annealing temperature and time resulted in more perfect, higher melting crystals [84]. General purpose PLLA exhibits much higher strength (~60 MPa) and modulus (~3 GPa) than many other commodity fossil oil-based plastics. For instance, HDPE has a typical strength and modulus of 20 MPa and 1 GPa, while polypropylene (PP) shows a typical strength and modulus of 30 MPa and 1.5 GPa. On the other hand, PLA is a brittle polymer, exhibiting a typical tensile strain at break of less than 6%. This brittleness significantly limits PLA's use in many applications. As a result, PLA toughening has received intensive research interests and a number of toughening strategies have since been developed.

First, PLA can be toughened by copolymerizing lactides with other monomers. For instance, the copolymers of lactide/caprolactone were found to be increasingly rubbery when the caproyl units in the copolymers increased from 5% to 20% [85]. When 50% of trimethylene carbonate (TMC) was copolymerized with lactide, elongation of the copolymer increased to 900%, whereas tensile strength decreased 10-fold to 5 MPa [86]. Second, lowering T_g of PLA by adding miscible plasticizers is another important method to toughen PLA. Lactide monomer and oligomer are natural PLA plasticizers. At ~20 wt% of LA monomer or oligomer, the elongation at break of PLA was increased to over 200% [87,88]. Other plasticizers such as poly(ethylene glycol) (PEG) [89] and triethyl citrate [90] were also used.

Blending PLA with ductile polymers is probably the most convenient and therefore the most studied route for PLA toughening. To maintain biodegradability of the final blends, ductile biodegradable polymers are often used to blend with PLA. Among these polymers, PCL has received the most research interest. PCL is a biodegradable polymer, which is flexible at room temperature due to its low T_g. Blending PLA with PCL resulted in significantly improved ductility and toughness [91,92]. PLA was also toughened by other biodegradable polymers such as PBAT [83,93] and PBS [94].

The aforementioned three toughening routes, e.g., copolymerization, plasticization, and blending, can significantly increase the elongation of PLA. However, the strength and modulus of PLA are substantially decreased at the same time. Polymers toughened by rigid particles were shown to be able to maintain or even increase their strength and modulus. A recent study on MMT toughening of PLA showed that PLA could be reinforced and toughened simultaneously when MMT concentration was low (<2.5 wt%) [95]. Furthermore, Chen *et al.* prepared a PLA–PBS–MMT ternary blend by melt blending [96]. When MMT was epoxy-functionalized, the ternary blend showed similar modulus to the neat PLA (1990 vs. 2215 MPa) and significantly increased the elongation at break (118.1 vs. 6.9%).

In toughening PLA, it is easier to achieve elongation or ductility improvement than impact strength improvement. This is because under high rate impact loading conditions materials tend to be more brittle. Nevertheless, high impact strength is desired in many important applications. Researchers have been looking for ways to significantly increase the impact strength of PLA. Recently, reactive blending of PLA with selected polymers/chemicals under certain conditions was discovered to be an effective method to achieve this goal. Several super-tough PLA blends were obtained using this method. Oyama reported a super-tough PLA–ethylene-*co*-glycidyl methacrylate (E-GMA) (80/20, w/w) blend [97]. The injection molded samples exhibited only a 2–3 fold increase in impact strength compared to neat PLA. However, the impact strength was drastically increased to 50 times of that of the neat PLA after annealing the samples at 90 °C for 2.5 h. The author postulated that the high impact strength was due to the crystalline structure changes of the PLA matrix. Elongation of the blends was reduced to below 35% after the annealing.

Liu *et al.* prepared a novel PLA ternary blend consisting of an ethylene/*n*-butyl acrylate/glycidyl methacrylate terpolymer rubber (EBA-GMA) and a zinc-containing ionomer of ethylene/methacrylic acid copolymer (EMAA-Zn) by twin screw extrusion [98,99]. Simultaneous vulcanization (crosslinking) of EBA-GMA and reactive compatibilization between PLA and EBA-GMA occurred during the extrusion. The blend showed a "salami"-like phase structure (i.e., subdomains within large domains). As a result, the impact strength of the blend was increased up to 35 times of that of neat PLA and the elongation of the blend was greater than 200%. The authors also showed that blending temperature and EBA-GMA—EMAA-Zn ratio of the blend played critical roles in impact strength improvement. High blending temperatures resulted in high extent of rubber crosslink reaction and reactive compatibilization between PLA and EBA-GMA. Large EBA-GMA—EMAA-Zn ratio (i.e., >1) caused an "EMAA-Zn in EBA-GMA" salami structure. The two aspects contributed to the creation of super-tough PLA blends.

6.3.2 Polyhydroxyalkanoates

Unlike PLA whose production involves designed chemical reactions, PHAs are biodegradable polyesters directly produced by bacterial metabolism. PHAs are synthesized and accumulated by bacteria as carbon and energy storage materials under the condition of limiting nutrients in the presence of excess carbon source [100,101]. More than 250 species of bacteria have been reported to produce PHAs. The polymers are stored in the cells as discrete granules with sizes between 0.2 and 0.5 μm. The stored PHAs are degraded by depolymerases and metabolized as carbon and energy source as soon as the supply of the limiting nutrient is restored [100]. Depending on the species of bacteria and their

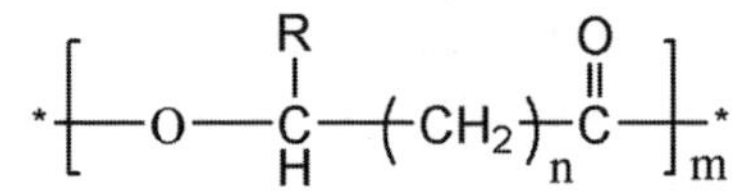

n = 1	R =	hydrogen	poly(-3-hydroxypropionate)
		methyl	poly(-3-hydroxybutyrate)
		ethyl	poly(-3-horoxyvalerate)
		propyl	poly(-3-hydroxyhexanoate)
		pentyl	poly(-3-hydroxyoctanoate)
		nonyl	poly(-3-hydroxydodecanoate)
n = 2	R =	hydrogen	poly(-4-hydroxybutyrate)
n = 3	R =	hydrogen	poly(-5-hydroxyvalerate)

Figure 6.4 General structure of PHAs.

growth conditions, molecular weight of PHAs is in the range of 2×10^5 to 3×10^6 Da.

PHAs include a family of polyesters with different side groups and different numbers of carbon atoms in the repeating units (Fig. 6.4). The most studied PHAs are poly(3-hydroxybutyrate) (PHB) and its copolymer PHBV. The homopolymer PHB is a highly crystalline thermoplastic with a T_m around 175 °C. It possesses several physical properties, e.g., T_m, T_g (15 °C), crystallinity (80%), and tensile strength (40 MPa), similar to those of PP. However, PHB is significantly more brittle than PP (strain at break 6 vs. 400%). With the introduction of 3-hydroxyvalerate (3HV) units to PHB, the regular structure of PHB is disrupted and therefore its crystallinity, crystallization rate, T_g, and T_m decrease as the content of HV increases [102]. As a result, PHBV becomes tougher and more flexible at higher HV percentages. Table 6.1 compares several physical properties of PHB and PHBV with different mole ratios of HV.

PHAs can be consumed by microorganisms as an energy source. Therefore, they are readily biodegradable in microbially active environments such as

Table 6.1 Thermal and Mechanical Properties of PHB and PHBV [103]

Polymer	T_g (°C)	T_m (°C)	Modulus (GPa)	Strength (MPa)	Strain at Break (%)
PHB	9	175	3.8	45	4
PHBV (11% HV)	2	157	3.7	38	5
PHBV (20% HV)	−5	114	1.9	26	27
PHBV (28% HV)	−8	102	1.5	21	700
PHBV (34% HV)	−9	97	1.2	18	970

compost [104]. The enzymes secreted by the microorganisms break down PHAs into monomers. The monomers are then used up by the cell as a carbon source for biomass growth. Many factors, e.g., surface area, microbial activity, pH, temperature, and moisture, affect the degradation rate of PHAs. The end products of PHA degradation in aerobic environments are carbon dioxide and water, while methane is also produced in anaerobic conditions. The degradation rate of PHAs varies with environmental conditions. For instance, PHBV completely degrades after 6, 75, and 350 weeks in anaerobic sewage, soil, and sea water, respectively [102]. However, PHAs do not degrade under normal conditions of storage [105].

With their inherent biocompatibility and biodegradability, PHAs have found important applications in medical and pharmaceutical areas, including wound management (e.g., sutures, skin substitutes, nerve cuffs, and staples), vascular system materials (e.g., heart valves, cardiovascular fabrics, and vascular grafts), orthopedics (e.g., scaffolds, spinal cages, bone graft substitutes, and internal fixation devices), and drug delivery systems [106]. PHAs are also finding more and more applications in packaging, single-use and disposable items, housewares, appliances, electrical and electronics, consumer durables, agriculture and soil stabilization, adhesives, paints and coatings, and automotive parts. In the United States, PHAs are commercially produced by Metabolix through the fermentation of enzyme-thinned starch, plant sugars, and oils using microbial biofactories. A series of PHAs, including homopolymers, copolymers, and terpolymers, are produced under the brand name of Mirel™. This wide range of PHA structures has allowed a broad property envelope of PHAs, from rigid thermoplastics to thermoplastic elastomers and adhesives.

PHAs have been blended with many biodegradable and non-biodegradable polymers to improve their properties and lower material costs. Miscibility, crystallization behavior, and biodegradability of the blends are the main topics of the published articles on PHA blending. PHB was found to be miscible with poly(ethylene oxide) (PEO), poly(vinyl acetate) (PVAc), poly(*p*-vinyl phenol), poly(vinylidene fluoride), and poly(methyl methacrylate) (PMMA) to various degrees under different component ratios and temperatures. PHB is not miscible or only partially miscible with poly(vinyl acetate-*co*-vinyl alcohol), PCL, PLA, poly(oxymethylene), ethylene−propylene rubber, EVA copolymer, and epichloridrin elastomers. The reports on mechanical properties of the blends are limited. Parulekar and Mohanty toughened PHB using epoxidized natural rubber with maleated polybutadiene as the compatibilizer [107]. The toughness of PHB was increased by 440%. PHB was also toughened by poly(*cis*-1,4-isoprene) (PIP) and PIP-*g*-PVAc, respectively [108]. It was found that the tensile properties and impact strength of the PHB−PIP-*g*-PVAc blends were superior to the PHB−PIP blends due to the former's higher compatibility. Li *et al.* prepared PHB−PEO blends by solvent casting [109]. They found that when the molecular weight of PEO was low (0.3×10^6), the blends exhibited very poor properties. When the PEO with a molecular weight of 5×10^6 was used, synergism occurred and the tensile strength, modulus, and elongation of the blends all increased to be significantly higher than those of the component materials.

Properties of PHAs were also modified by natural fibers. It has been shown that the addition of natural fibers increased modulus, T_g, and heat distortion temperature (HDT) of PHB (or PHBV) composites [110,111]. Nevertheless, the improvements in tensile strength and toughness were found to be difficult and dependent on many factors such as fiber length and aspect ratio, interfacial bonding, fiber sources, fiber treatments, and fiber forms (single fiber/fabrics). The studies on the composites of PHB/PHBV with flax [110], recycled cellulose fiber [111], wood fiber [112], and pineapple fiber [113] have shown that the tensile strength and toughness were not improved or were even decreased by the addition of short or flour-type natural fibers. Very recently, Jiang *et al.* reported a PHBV−BPF composite with substantially increased tensile strength, modulus, and toughness [34]. They also found that by using PHBV-*g*-MA as a compatibilizer, the strength and modulus were further improved and the impact strength was decreased, due to the increased interfacial bonding between PHBV and BPF. CNWs have also been used to reinforce PHBV [114]. PHBV−CNW composites were prepared by solvent casting and melt compounding, respectively. Homogeneous dispersion of CNWs was achieved and the composites exhibited improved tensile strength and modulus and increased T_g in the solvent-prepared composites. By contrast, the composites prepared through melt process showed decreased strength and constant T_g due to CNW agglomeration.

PHAs have also been spun into fibers for use in textile products. Commercially available Biopol® PHBV fibers were reported to have a tensile strength of 183 MPa [115]. Iwata *et al.* spun ultra-high-molecular-weight PHB fibers with a tensile strength of 1.3 GPa by a combination of cold-drawing and two-step-drawing methods [116]. Using commercial PHBV, Tanaka *et al.* produced high strength fibers (1.1 GPa) using room temperature drawing after isothermal crystallization at T_g of PHBV [117]. This strength is comparable to that of common polymer fibers such as polyethylene (PE) and poly(ethylene terephthalate) (PET).

6.4 Biodegradable Polymers Derived from Petroleum

Biodegradable polymers can be derived not only from renewable bioresources but also from petroleum. Some synthetic aliphatic polyesters have been known to be biodegradable for decades. Petroleum-based biodegradable polyesters are synthesized by polycondensation reaction between aliphatic diacids and aliphatic diols or by ring-opening polymerization of lactones. Aliphatic acids and terephthalic acids can also be used together to react with aliphatic diols to produce biodegradable aliphatic–aromatic co-polyesters. Typical synthetic aliphatic polyesters include PCL, PBS, and their copolymers. The most widely used aliphatic–aromatic copolyester is PBAT produced by BASF (Ecoflex®), DuPont (Biomax®), and Eastman Chemical (Eastar Bio®).

6.4.1 Polycaprolactone

PCL is a semicrystalline aliphatic polyester synthesized by ring-opening polymerization of caprolactone (Fig. 6.5). It is completely degradable through enzyme activities [118].

Depending on its molecular weight (M_w), PCL can be a waxy solid (M_w below several thousands) or a solid polymer (M_w above 20,000). The waxy PCL is usually used as additive or co-component. High molecular weight PCL polymer has mechanical properties similar to PE, possessing a tensile stress of 12–30 MPa and a break extension of 400–900%. Table 6.2 provides several properties of PCL with different molecular weights [119]. It is obvious that all the properties are strongly molecular weight dependent.

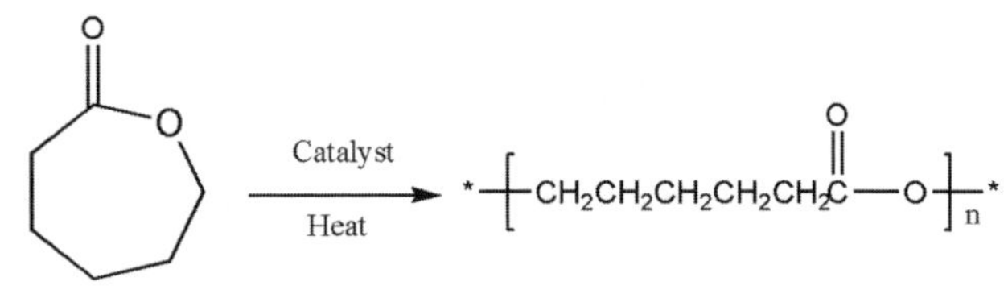

Figure 6.5 Synthesis and structure of PCL.

PCL shows high-molecular-chain flexibility and outstanding processability. It can be spun into fibers or blown films under 200 °C without thermal degradation. The drawback of PCL is its low melting point (~60 °C) and low glass transition temperature (~−60 °C), which prohibits its application at elevated temperatures [120]. Therefore, PCL is often blended with other polymers, such as PP, polycarbonate (PC), polyethylene oxide (PEO), and starch, to produce composites with desired properties [121,122]. Major PCL producers include Dow Chemical in the United States, Solvay in Europe, and Daicel Chemical Industries in Japan. Commercialized with the trade name "TONE" and "CAPA," PCL is extensively used in food packaging and tissue engineering. For instance, microcellular PCL foams are used for tissue regeneration and cell transplantation.

6.4.2 Poly(butylene succinate)

PBS is one of the most important biodegradable polyesters synthesized by polycondensation between succinic acid and butanediol. The reaction proceeds in two steps. First, esterification occurs between the diacid and the diol, and then polycondensation takes place under high temperature to form high-molecular-weight PBS (Fig. 6.6).

Showa Highpolymer (Japan) synthesized various aliphatic copolyesters (PBS and PBSA) based on succinate, adipate, ethylene glycol, and 1,4-

Table 6.2 Properties of PCL with Different Molecular Weight

Properties	PCL 1	PCL 2	PCL 3
Molecular weight	37,000	50,000	80,000
Melting point (°C)	58–60	58–60	60–62
Tensile stress (kg/cm^2)	140	360	580
Elongation at break (%)	660	800	900

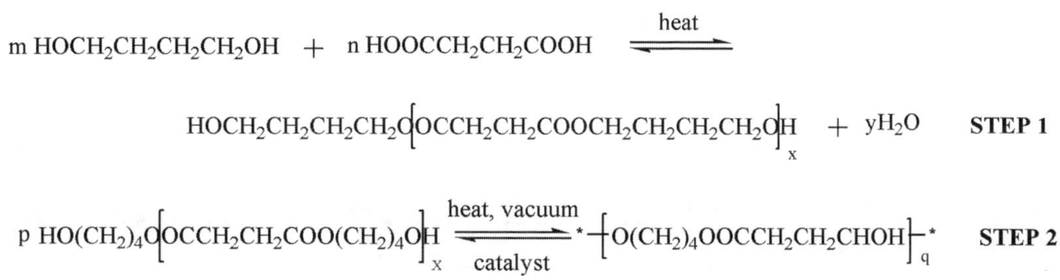

Figure 6.6 Synthesis and structure of PBS.

Table 6.3 Comparison of the Properties of Bionolle and LDPE, HDPE, and PP

Properties	Bionolle (#1001)	Bionolle (#3001)	LDPE	HDPE	PP
Glass transition (°C)	−32	−45	−120	−120	5
Melting point (°C)	114	94	110	129	163
HDT (°C)	97	69	88	110	145
Tensile strength (MPa)	57	47	35	39	44
Yield strength (MPa)	32	19	12	27	31
Flexural modulus (MPa)	656	323	176	1070	1370
Strain at break (%)	700	900	400	650	800
MFR at 190 °C (g/10 min)	1.5	1.4	2	2	4

Source: Data adapted from http://www.shp.co.jp/en/bionolle_data1.htm [125]

butanediol, which were synthesized and commercialized under the trade name "Bionolle" [123]. These polyesters exhibit melting temperature >100 °C and thermal degradable temperature >300 °C and other properties similar to LDPE, HDPE, and PP (Table 6.3) [124]. They can be processed by injection molding, extrusion, and film blowing using conventional equipment. Therefore, they are considered to be potential alternatives to petrochemical polyolefins.

6.4.3 Poly(butylene adipate-co-terephthalate)

PBAT is an aliphatic–aromatic copolyester, which shows higher chain stiffness than entirely aliphatic polyesters such as PCL and PBS due to the inclusion of terephthalic groups in the molecules. The chemical structure of PBAT is shown in Fig. 6.7.

PBAT can be synthesized by conventional bulk polycondensation techniques. It degrades completely in soil, in aqueous environment, and under composting and anaerobic conditions within varying time periods, depending on the average chain length of the aromatic blocks [126–129]. Table 6.4 lists the properties of three commercial biodegradable aliphatic–aromatic polyesters. Currently, these materials are mainly used in packaging and agricultural applications.

6.5 Biobased Polymers Derived from Plant Oil

Plant oils, including soybean oil, vegetable oil, corn oil, etc., are important raw materials for biobased polymers due to their triglyceride structure and fatty acid chains. Triglyceride contains reactive groups, such as carbon–carbon double bond, and

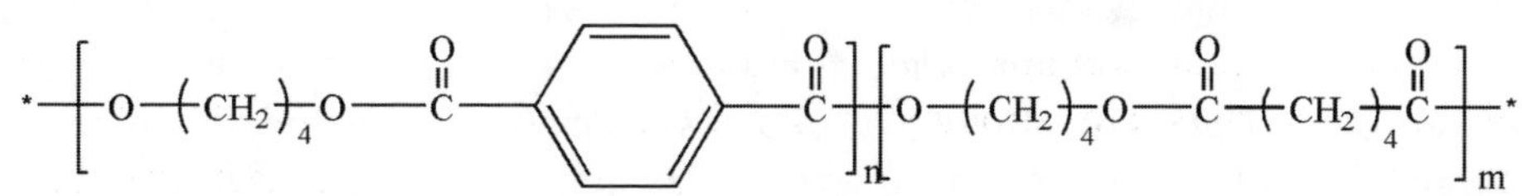

Figure 6.7 Chemical structure of PBAT.

Table 6.4 Properties of Three Biodegradable Aliphatic–Aromatic Polyesters

Trade name	Ecoflex	Eastar	Biomax
Producer	BASF, Germany	Eastman, USA	DuPont, USA
Raw materials	1,4-Butanediol, adipic acid, terephthalic acid	1,4-Butanediol, adipic acid, terephthalic acid	PET with aliphatic dicarboxylic acid
Density (g/cm^3)	1.25–1.27	1.22	1.35
Melting point (°C)	110–115	108	200
Tensile strength (MPa)	32–36	20–22	15–50
Strain at break (%)	580–800	700–730	40–500
Modulus (MPa)	–	106–107	60–2100

Source: Data adapted from R.J. Müller, Handbook of Biodegradable Polymers, Rapra Technology Limited, Shawburry, UK, 2005, Chapter 10, p. 303 [130]

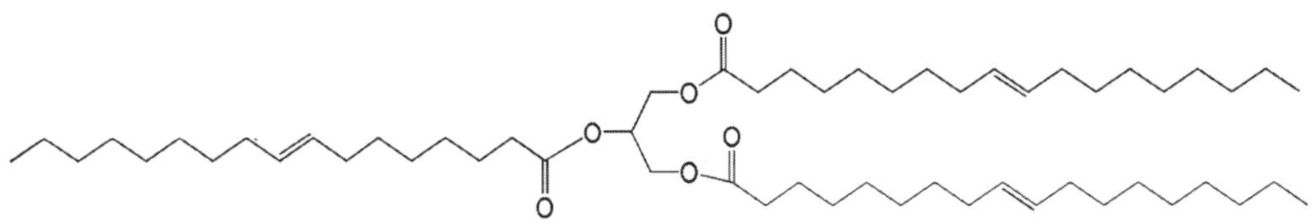

Figure 6.8 Typical structure of plant oil.

allylic and ester groups, which make it possible to introduce polymerizable groups into triglyceride molecules using common synthetic techniques (Fig. 6.8). Before they can be used as monomers for high-molecular-weight polymeric materials, all plant oils need to be functionalized by modifying these active sites.

6.5.1 Functionalization of the Carbon–Carbon Double Bond in Triglycerides

The double bonds in triglycerides cannot be used for polycondensation reaction to produce high-molecular-weight polymers without proper modification. Four types of modified triglycerides are shown in Fig. 6.9 [131–134]. For compounds 1, 2, and 3, free radical polymerization can proceed through the introduced double bonds. For compound 4, the introduced epoxy group can go through ring-opening or condensation polymerization. Wool *et al.* have conducted intensive research on synthesis of biobased polymers using modified plant oils [134–138]. From compound 1, they synthesized a thermosetting resin by free radical polymerization or copolymerization with reactive diluents such as styrene [135]. The resulting resin exhibited mechanical properties similar to that of commercial polyester and vinyl ester resins. The epoxidized plant oil has been extensively used in surface coating, in ink, and as the major composition of several resins [132].

6.5.2 Modification of the Ester Group

Another important modification to plant oils is to convert triglyceride to monoglyceride or diglyceride through glycerolysis reaction. The conversion reaction can be carried out by heating the triglycerides and glycerol at 220–230 °C with $Ca(OH)_2$ as a catalyst, and the resulting mixture product (containing diglyceride and monoglyceride) was reacted with methacrylic anhydride [138] or MA [139] to obtain other triglyceride-based monomers. By this method, several plant oil-based unsaturated polyesters with the melting points in the range of 60–70 °C have been synthesized by Wool's group. On the other hand, using the monoglycerides and diglycerides derived from plant oils, some biobased PUs with excellent chemical and physical properties including increased thermal stability have been synthesized [140,141]. In addition, waterborne PUs

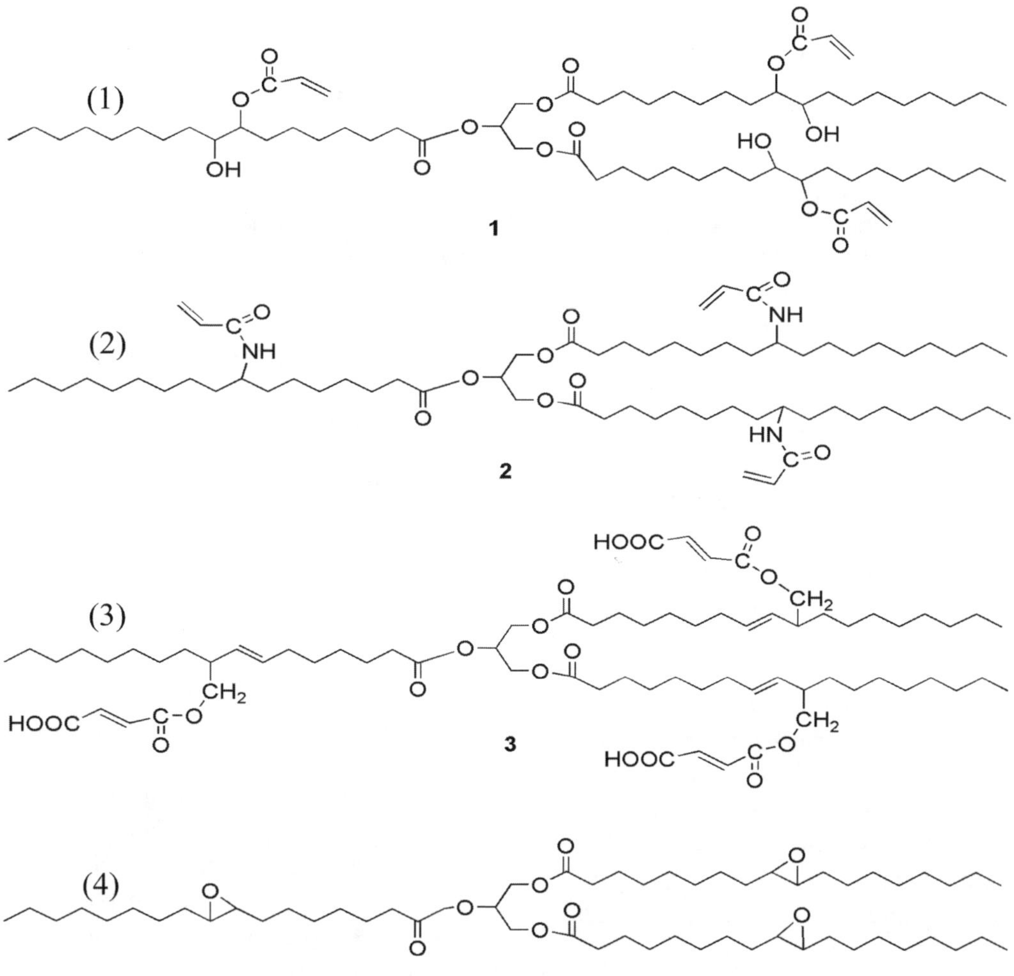

Figure 6.9 Several typical compounds derived from triglyceride [131–134].

and PU/acrylic hybrid latex have been made from soybean oil-based polyols [142].

6.6 Rosin-Based Epoxy Curing Agents

Epoxy can exhibit a wide spectrum of properties by having different combinations of epoxy resins and curing agents. The high mechanical and physical properties of properly cured epoxies are mainly attributed to high content of rigid aromatic or cycloaliphatic moieties in the epoxy or curing agent molecules or both. The majority of the epoxy resins on the market are based on the glycidyl ethers of bisphenol A. Tertiary amines, polyamines, and acid anhydrides are the three main types of epoxy curing agents. Cyclic acid anhydrides (aromatic or cyclic aliphatic), which impart high mechanical and physical properties to the cured resins, are the preferred curing agents because they show low toxicity and low curing exotherms and shrinkages. Important cyclic anhydride curing agents include 1,2-benzenedicarboxylic anhydride (also called phthalic anhydride, PA), 1,2-cyclohexanedicarboxylic anhydride (CHDA), and 1,2,4-benzenetricarboxylic anhydride (BTCA or trimellitic anhydride). These chemicals are all petroleum based and are generally synthesized through complicated chemical processes.

Rosin acids have a characteristic hydrogenated phenanthrene structure that exhibits rigidity similar to that of those petroleum-based cyclic compounds mentioned above. The carboxyl groups and carbon–carbon double bonds in the molecules of rosin acids are reactive. Through their reactions, rosin derivatives suitable for replacing petroleum-based cyclic monomers can be synthesized.

Wang *et al.* synthesized abietyl glycidyl ether and methyl maleopimarate using one of the rosin acids [143]. The two compounds were used as model compounds representing rosin-based epoxies and rosin-based anhydride curing agents, respectively. Nonisothermal curing studies by differential scanning calorimetry (DSC) have suggested that both the curing reaction of epoxide with the rosin anhydride and the curing reaction of rosin epoxide with aniline were autocatalytic, and the cure reactions were similar to the respective conventional epoxy resin systems.

Liu *et al.* synthesized two rosin-based epoxy curing agents: maleopimaric acid and methyl maleopimarate acid [144]. They performed comparison studies on the curing reactions and the properties of cured epoxies between the two rosin-based curing agents and two commercial curing agents with similar functionalities and structures. The results showed that the curing behaviors of the rosin-based curing agents were similar to those of the commercial curing agents. The epoxies cured by the rosin-based curing agents also demonstrated thermal/mechanical properties and thermal stability similar to the epoxies cured by the commercial curing agents. However, the rosin-based curing agents are advantageous because they can be prepared by much simpler and more environmentally friendly synthesis routes.

Liu *et al.* also synthesized rosin-based imide-diacids as epoxy curing agents [145]. A similar imide-diacid based on trimellitic anhydride was also synthesized for comparison. The two types of imide-diacids were used to cure a commercial epoxy resin. The epoxies cured with rosin-based imide-diacids were shown to have significantly higher glass transition temperature, tensile, and dynamic mechanical properties.

The rigid cyclic structures in epoxy resins and curing agents impart high glass transition temperature, high strength, and modulus, but low toughness to the cured epoxies. Flexibility and toughness of the cured epoxies can be significantly improved by incorporating a soft chain segment into either the epoxy or curing agent molecules. Wang *et al.* incorporated a low-molecular-weight PCL diol (caprolactone oligomer) segment between two rosin-derived anhydride moieties to produce a maleopimarate-terminated oligocaprolactone and used it to cure commercial epoxy resins [146]. The test results of the cured epoxies showed that both the epoxide–anhydride equivalent ratio and the molecular length of the rosin anhydride-terminated oligocaprolactone could influence the crosslink density of the epoxies and therefore their mechanical and thermal properties. These results suggest that varying the length of the oligocaprolactone segment and the epoxy-curing agent ratio can result in a broad spectrum of mechanical and thermal properties of the cured epoxies.

6.7 Concluding Remarks

Biodegradable polymers can be either natural or synthetic polymers and they can be derived from either renewable or nonrenewable resources. Non-biodegradable polymers can also be derived from renewable feedstock, e.g., Dupont's Sorona®, which is poly(trimethylene terephthalate), using the corn-derived 1,3-propanediol as the diol monomer. Developing biodegradable polymers from renewable resources appears to be the best scenario in the development of "green" materials and processing techniques. However, this can only be certain after carrying out a life-cycle environmental impact analysis (mainly on energy consumption and CO_2 balance) on individual products. For example, to produce PLA using corn starch, the application of fertilizers, herbicides, and pesticides during the growth of corn may leave a deep environmental footprint. The chemical or biochemical processes such as extraction and purification of lactic acid require water, energy, and chemical or biological additives. The whole production process also produces various wastes, which require energy and material-consuming treatment and disposal. Furthermore, CO_2 is released back to the environment during the degradation of PLA. Therefore, biodegradable polymers derived from renewable resources may not be as "green" as they appear to be at first sight. However, if a significant part of the energy required to produce and process biodegradable polymers is from nonpetroleum sources, e.g., solar power, wind energy, water/tide power, etc., biodegradable polymers based on renewable resources could still possess a substantially better CO_2 balance than petroleum-based polymers.

References

[1] D. French, Organisation of starch granules, in: R.L. Whistler, J.N. BeMiller, E.F. Paschall (Eds.), Starch Chemistry and Technology, Academic Press Inc., Orlando, FL, 1984.

[2] P. Forssell, J. Mikkila, T. Suortti, Plasticization of barley starch with glycol and water, J.M.S.—Pure Appl. Chem. A33 (5) (1996) 703–715.
[3] J.J.G. Van Soest, D.D. Wit, J.F.G. Vliegenthart, Mechanical properties of thermoplastic waxy maize starch, J. Appl. Polym. Sci. 61 (1996) 1927–1937.
[4] S. Jacobsen, H.G. Fritz, Polym. Eng. Sci. 36 (22) (1996) 2799–2804.
[5] R. Mani, M. Bhattacharya, Eur. Polym. J 34 (10) (1998) 1467–1475.
[6] R. Mani, M. Bhattacharya, Eur. Polym. J 37 (3) (2001) 515–526.
]7] J.F. Zhang, X. Sun, Biomacromolecules 5 (2004) 1446–1451.
[8] F.J. Rodriguez-Gonzalez, B.A. Ramsay, B.D. Favis, Polymer 44 (2003) 1517–1526.
[9] D. Schlemmer, E.R. de Oliveira, M.J. Araujo Sales, J. Therm. Analy. Calorim. 87 (2007) 635–638.
[10] P.S. Walia, J.W. Lawton, R.L. Shogren, F.C. Felker, Polymer 41 (2000) 8083–8093.
[11] Y. Lu, L. Tighzert, P. Dole, D. Erre, Polymer 46 (2005) 9863–9870.
[12] J.L. Willett, F.C. Felker, Polymer 46 (2005) 3035–3042.
[13] B.Y. Shin, S.I. Lee, Y.S. Shin, S. Balakrishnan, R. Narayan, Polym. Eng. Sci. 44 (2004) 1429–1438.
[14] Y. Parulekar, A.K. Mohanty, Macromol. Mater. Eng. 292 (2007) 1218–1228.
[15] O. Martin, L. Averous, Polymer 42 (2001) 6209–6219.
[16] M.A. Huneault, H. Li, Polymer 48 (2007) 270–280.
[17] F.G. Torres, O.H. Arroyo, C. Gomez, J. Theromplast, Comp. Mater. 20 (2007) 207–223.
[18] X. Ma, J. Yu, J.F. Kennedy, Carbohydr. Polym. 62 (2005) 19–24.
[19] S. Ochi, Composites Part A 37 (2006) 1879–1883.
[20] H.M. Wilhelm, M.R. Sierakowski, G.P. Souza, F. Wypych, Carbohydr. Polym. (2003) 101–110.
[21] M. Avella, J.J. De Vlieger, M.E. Errico, S. Fischer, P. Vacca, M.G. Volpe, Food Chem. 93 (2005) 467–474.
[22] M. Huang, J. Yu, X. Ma, Carbohydr. Polym. 63 (2006) 393–399.
[23] X. Tang, S. Alavi, T.J. Herald, Cereal Chem. 85 (3) (2008) 433–439.
[24] P.D. Tatarka, R.L. Cunningham, J. Appl. Polym. Sci. 67 (1996) 1157.
[25] G.M. Ganjyal, N. Reddy, Y.Q. Yang, M.A. Hanna, J. Appl. Polym. Sci. (2004) 2627.
[26] Q. Fang, M.A. Hanna, Bioresour. Technol. 78 (2001) 115.
[27] Q. Fang, M.A. Hanna, Cereal Chem. 77 (2002) 779.
[28] J.L. Willett, R.L. Shrongen, Polymer 43 (2002) 5935.
[29] M. Mihai, M.A. Huneault, B.D. Favis, H. Li, Macromol. Biosci. 7 (2007) 907–920.
[30] Encyclopedia Britannica Online. Retrieved 16 July, 2012, from http://www.britannica.com/EBchecked/topic/101633/cellulose, (2008)
[31] A.K. Bledzki, J. Gassan, Prog. Polym. Sci. 24 (1999) 221.
[32] D. Plackett, T.L. Andersen, W.B. Pedersen, L. Nielsen, Compos. Sci. Tech. 63 (2003) 1287–1296.
[33] S.-H. Lee, S. Wang, Composites Part A 37 (2006) 80–91.
[34] L. Jiang, J. Huang, J. Qian, F. Chen, J. Zhang, M.P. Wolcott, Y. Zhu, J. Polym. Environ. 16 (2008) 83–93.
[35] P. Lodha, A.N. Netravali, Compos. Sci. Tech. 65 (2005) 1211–1215.
[36] P. Lodha, A.N. Netravali, Polym. Compos. 26 (2005) 647–659.
[37] A. Sturcova, G.R. Davies, S.J. Eichhorn, Biomacromolecules 6 (2) (2005) 1055–1061.
[38] D. Bondeson, A. Mathew, K. Oksman, Cellulose 13 (2) (2006) 171–180.
[39] D. Bondeson, P. Syre, K. Oksman, J. Biomater, Bioenergy 1 (3) (2007) 367–371.
[40] D. Bondeson, K. Oksman, Composites Part A 38 (2007) 2486–2492.
[41] D. Bondeson, K. Oksman, Compos. Interfaces 14 (2007) 617–630.
[42] I. Kvien, B.S. Tanem, K. Oksman, Biomacromolecules 6 (6) (2005) 3160–3165.
[43] L. Petersson, I. Kvien, K. Oksman, Compos. Sci. Tech. 67 (2007) 2535–2544.
[44] L. Petersson, K. Oksman, in: K. Oksman, M. Sain (Eds.), Cellulose Nanocomposites:

Processing, Characterization and Properties, Preparation and Properties of Biopolymer Based Nanocomposites Films Using Microcrystalline Cellulose (MCC), ACS Symposium Series, vol. 938, Oxford Press, 2006.
[45] K. Oksman, A.P. Mathew, D. Bondeson, I. Kvien, Compos. Sci. Tech. 66 (15) (2006) 2776–2784.
[46] N.L.G. de Rodriguez, W. Thielemans, A. Dufresne, Cellulose 13 (3) (2006) 261–270.
[47] Y.X. Wang, X.D. Cao, L. Zhang, Macromol. Biosci. 6 (7) (2006) 524–531.
[48] J. Sriupayoa, P. Supaphola, J. Blackwellb, R. Rujiravanit, Polymer 46 (15) (2005) 5637–5644.
[49] J. Zhang, P. Mungara, J. Jane, Polymer 42 (2001) 2569.
[50] P. Mungara, J. Zhang, S. Zhang, J. Jane, in: A. Gennadios (Ed.), Protein-Based Films and Coatings, CRC Press, Boca Raton, FL, 2002, pp. 621–638.
[51] J. Jane, S. Wang, US Patent 5 523 293 (1996).
[52] P. Mungara, T. Chang, J. Zhu, J. Jane, J. Polym. Environ. 10 (2002) 31.
[53] R. Deng, Y. Chen, P. Chen, L. Zhang, B. Liao, Polym. Degrad. Stab. 91 (2006) 2189.
[54] J. Zhang, L. Jiang, L. Zhu, J.-L. Jane, J.-L.P. Mungara, Biomacromolecules 7 (2006) 1551.
[55] J. John, M. Bhattacharya, Polym. Int. 48 (1999) 1165.
[56] D. Graiver, L.H. Waikul, C. Berger, R. Narayan, J. Appl, Polym. Sci. 92 (2004) 3231.
[57] C. Wang, J. Carriere, L. Willett, J. Polym, Sci., Part B: Polym. Phys. 40 (2002) 2324.
[58] Z. Zhong, X. Sun, Polymer 42 (2001) 6961.
[59] Y.D. Li, J.B. Zeng, X.L. Wang, K.K. Yang, Y.Z. Wang, Biomacromolecules 9 (2008) 3157–3164.
[60] F. Chen, J.W. Zhang, Polymer 50 (2009) 3770–3777.
[61] F. Chen, J.W. Zhang, Polymer 51 (2010) 1812–1819.
[62] F. Chen, J.W. Zhang, ACS Appl. Mater. Interfaces 2 (2010) 3324–3332.
[63] B. Liu, L. Jiang, H.Z. Liu, L.L. Sun, J.W. Zhang, Macromol. Mater. Eng. 259 (2010) 123–129.
[64] B. Liu, L. Jiang, H.Z. Liu, J.W. Zhang, Ind. Eng. Chem. Res. 49 (2010) 6399–6406.
[65] B. Liu, L. Jiang, J.W. Zhang, J. Polym. Environ. 19 (2011) 239–247.
[66] B. Liu, L. Jiang, J.W. Zhang, Macromol. Mater. Eng. 256 (2011) 835–842.
[67] W. Liu, A.K. Mohanty, P. Askeland, L.T. Drzal, M. Misra, Polymer 45 (2004) 7589–7596.
[68] W. Liu, A.K. Mohanty, L.T. Drzal, M. Misra, Ind. Eng. Chem. Res. 44 (2005) 7105–7112.
[69] Y. Wang, X. Cao, L. Zhang, Macromol. Biosci. 6 (2006) 524–531.
[70] P. Chen, L. Zhang, Biomacromolecules 7 (2006) 6.
[71] H. Zheng, F. Ai, M. Wei, J. Huang, P.R. Chang, Macromol. Mater. Eng. 292 (2007) 780–788.
[72] A. Rouilly, J. Jorda, L. Rigal, Carbohydr. Polym. 66 (2006) 81–87.
[73] B. Liu, J.W. Zhang, L.S. Liu, A.T. Hotchkiss, J. Polym. Environ. 19 (2011) 559–567.
[74] F. Chen, L.S. Liu, P.H. Cooke, K.B. Hicks, J.W. Zhang, Indus. Eng. Chem. Res. 47 (2008) 8667–8675.
[75] V.L. Finkenstadt, L.S. Liu, J. Willett, J. Polym. Environ. 15 (2007) 1–6.
[76] L.S. Liu, V.L. Finkenstadt, C.K. Liu, D.R. Coffin, J.L. Willett, M.L. Fishman, K.B. Hicks, J. Biobased Mater. Bioenergy 1 (2007) 323–330.
[77] A. Rouilly, C. Geneau-Sbarta, L. Rigal, Bioresour. Tech. 100 (2009) 3076–3081.
[78] V. Finkenstadt, C.K. Liu, P.H. Cooke, L.S. Liu, J.L. Willett, J. Polym. Environ. 16 (2008) 19–26.
[79] B. Liu, S. Bhaladhare, P. Zhan, L. Jiang, J.w. Zhang, L.S. Liu, A. Hotchkiss, Indus. Eng. Chem. Res. 50 (2011) 13859–13865.
[80] P.R. Gruber, E.S. Hall, J.H. Kolstad, M.L. Iwen, R.D. Benson, R.L. Borchardt, US Patent 5 142 023 (1992).
[81] R.E. Drumright, P.R. Gruber, D.E. Henton, Adv. Mater. 12 (2000) 1841.
[82] A. Sodergard, M. Stolt, Progr. Polym. Sci. 27 (2002) 1123–1163.
[83] L. Jiang, M.P. Wolcott, J. Zhang, Biomacromolecules 7 (2006) 199–207.
[84] M. Yasuniwa, S. Tsubakihara, Y. Sugimoto, C. Nakafuku, J. Polym, Sci. Part B: Polym. Phys. 42 (2004) 23.
[85] M. Hiljanen-Vainio, P.A. Orava, J.V. Seppala, J. Biomed. Mater. Res. 34 (1997) 39–46.

[86] B. Buchholz, J. Mater. Sci. Mater. Med. 4 (1993) 381–388.
[87] R.G. Sinclair, J. Macromol. Sci. – Pure Appl. Chem. A33 (1996) 585–597.
[88] O. Martin, L. Averous, Polymer 42 (2001) 6206–6219.
[89] M. Baiardo, G. Frisoni, M. Scandola, M. Rimelen, D. Lips, K. Ruffieux, E. Wintermantel, J. Appl. Polym. Sci. 90 (2003) 1731–1738.
[90] L.V. Labrecque, R.A. Kumar, V. Dave, R.A. Gross, S.P. McCarthy, J. Appl. Polym. Sci. 66 (1997) 1507–1513.
[91] S. Aslan, L. Calandrelli, P. Laurienzo, M. Malinconico, C. Migliaresi, J. Mater. Sci. Mater. Med. 35 (2000) 1615–1622.
[92] G. Maglio, A. Migliozzi, R. Palumbo, B. Immirzi, M.G. Volpe, Macromol. Rapid Commun. 20 (1999) 236–238.
[93] T.Y. Liu, Polymer 46 (2005) 12586–12594.
[94] M. Shibata, Polymer 47 (2006) 3557–3564.
[95] L. Jiang, J. Zhang, M.P. Wolcott, Polymer 48 (2007) 7632–7644.
[96] G. Chen, H. Kim, E. Kim, J. Yoon, Polymer 46 (2005) 11829–11836.
[97] H. Oyama, Polymer 50 (2009) 747–751.
[98] H. Liu, F. Chen, B. Liu, G. Estep, J.W. Zhang, Macromolecules 43 (2010) 6058–6066.
[99] H. Liu, W. Song, F. Chen, L. Guo, J.W. Zhang, Macromolecules 44 (2011) 1513–1522.
[100] D. Byrom, Polyhydroxyalkanoates, in: D.P. Mobley (Ed.), Plastic from Microbes: Microbial Synthesis of Polymers and Polymer Precursors, Hanser, Munich, 1994, pp. 5–33.
[101] A. Steinbüchel, Polyhydroxyalkanoic acids, in: D. Byrom (Ed.), Biomaterials, Novel Materials from Biological Sources, Stockton, New York, 1991, pp. 124–213.
[102] S.Y. Lee, Biotechnol. Bioeng. 49 (1996) 1–14.
[103] M. Avella, E. Martuscelli, M. Raimo, J. Mater. Sci. 35 (2000) 523.
[104] Y. Poirier, C. Nawrath, C. Somerville, Biotechnology 13 (1995) 142–150.
[105] J. Mergaert, A. Webb, C. Anderson, A. Wouters, J. Swings, Appl. Environ. Microbiol. 59 (1993) 3233–3238.
[106] S.F. Williams, D.P. Martin, in: Y. Doi, A. Steinbüchel (Eds.), Biopolymers Polyesters III, Applications and Commercial Products, 4 Wiley, Weinheim, 2002, p. 91.
[107] Y. Parulekar, A.K. Mohanty, Green. Chem. 8 (2006) 206.
[108] J.S. Yoon, W.S. Lee, H.J. Jin, I.J. Chin, M.N. Kim, J.H. Go, Eur. Polym. J. 35 (1999) 781.
[109] R.Q. Li, Y.X. An, Y.G. Zhuang, L.S. Dong, F.G. Teng, Z.L. Feng, ACTA Polym. Sinica 2 (2001) 143–146.
[110] S. Wong, R. Shanks, A. Hodzic, Macromol. Mater. Eng. 287 (2002) 647–655.
[111] R. Bhardwaj, A. Mohanty, L.T. Drzal, F. Pourboghrat, M. Misra, Biomacromolecules 287 (2006) 647–655.
[112] V. Reinsch, S. Kelley, J. Appl. Polym. Sci. 64 (1997) 1785–1796.
[113] S. Luo, A.N. Netravalli, J. Mater. Sci. 34 (1999) 3709–3719.
[114] L. Jiang, E. Morelius, J. Zhang, M.P. Wolcott, J. Holbery, J. Compos. Mater. 42 (2008) 2629–2645.
[115] T. Ohura, Y. Aoyagi, K. Takagi, Y. Yoshida, K. Kasuya, Y. Doi, Polym. Degrad. Stab. 63 (1999) 23.
[116] T. Iwata, Y. Aoyagi, M. Fujita, H. Yamane, Y. Doi, Y. Suzuki, A. Takeuchi, K. Uesugi, Macromol. Rapid Commun. 25 (2004) 1100.
[117] T. Toshihisa, F. Masahiro, T. Akihisa, S. Yoshio, U. Kentaro, I. Kazuki, F. Tetsuro, D. Yoshiharu, I. Tadahisa, Macromolecules 39 (2006) 2940–2946.
[118] Y. Tokiwa, T. Suzuki, Nature 270 (1977) 76.
[119] M.B. Gregory, Handbook of Biodegradable Polymers, Rapra Technology Limited, 2005 (Chapter 6) 183.
[120] M. Vert, J. Feijen, A. Albertsson, G. Scott, E. Chiellini (Eds.), Biodegradable Polymers and Plastics, Royal Society of Chemistry, 1992, pp. 56–78.
[121] H. Tsuji, K. Suzuyoshi, Polym. Degrad. Stab. 75 (2002) 347.
[122] P.H. Debois, C. Jacobs, R. Jerome, P.H. Teyssie, Macromolecules 24 (1991) 22–66.
[123] Y. Doi, A Steinbuchel, Biopolymers 4 (2002) 275.
[124] T. Fujimaki, Polym. Degrad. Stab. 59 (1998) 209.
[125] http://www.shp.co.jp/en/bionolle_data1.htm

[126] U. Witt, R.J. Müller, J. Augusta, H. Widdecke, W.D. Deckwer, Macromol. Chem. Phy. 195 (1994) 793.
[127] J.K. Hye, S.P. Sang, J. Appl. Polym. Sci. 72 (1999) 593.
[128] C.H. Ki, O.O. Park, Polymer 42 (2001) 18–49.
[129] S.H. Lee, S.W. Lim, K.H.P. Lee, Polym. Int. 48 (1998) 861.
[130] R.J. Müller, Handbook of Biodegradable Polymers, Rapra Technology Limited, 2005, (Chapter 10) 303.
[131] A. Friedman, S.B. Polovsky, J.P. Pavlichko, L.S. Moral, US Patent 5, 576, 027 (1996).
[132] W.R. Likavec, C.R. Bradley, US Patent 5, 866, 628 (1999).
[133] T. Eren, S.H. Kusefoglu, J. Appl. Polym. Sci. 91 (2004) 4037.
[134] T. Eren, S.H. Kusefoglu, J. Appl. Polym. Sci. 97 (2005) 2264.
[135] R.P. Wool, S.H. Kusefoglu, G.R. Palmese, R. Zhao, US Patent 6, 121, 398 (2000).
[136] R.P. Wool, Chem. Tech. 29 (1999) 44.
[137] R.P. Wool, S.N. Khot, in: Proceedings ACUN-2 Sydney, University of New South Wales Kensington, Australia, 2000.
[138] E. Can, S.H. Kusefoglu, R.P. Wool, J. Appl. Polym. Sci. 69 (2001) 81.
[139] E. Can, Ph.D. Thesis, University of Delaware, Newark, Delaware (2004).
[140] G. Lligadas, J.C. Ronda, M. Galia, U. Biermann, Biomacromolecules 8 (2007) 686.
[141] A. Zlatanic, Z.S. Petrovic, K. Dusek, Biomacromolecules 3 (2002) 1048.
[142] Y.S. Lu, R.C. Larock, Biomacromolecules 8 (2007) 3019.
[143] H.H. Wang, B. Liu, J.W. Zhang, Green Chem. 10 (2008) 1190–1196.
[144] X.Q. Liu, W.B. Xin, J.W. Zhang, Green Chem. 11 (2009) 1018–1025.
[145] X.Q. Liu, W.B. Xin, J.W. Zhang, Bioresour. Tech. 101 (2010) 2520–2524.
[146] H.H. Wang, X.Q. Liu, B. Liu, J.W. Zhang, M. Xian, Polym. Intern. 58 (2009) 1435–1441.

7 Starch: Major Sources, Properties and Applications as Thermoplastic Materials

Antonio J.F. Carvalho

OUTLINE

7.1 Introduction

Starch is the major carbohydrate reserve in higher plants. In contrast with cellulose that is present in dietary fibers, starch is digested by humans and represents one of the main sources of energy to sustain life. Bread, potato, rice, and pasta are examples of the importance of starch in our society. Starch has also been extremely important for centuries in numerous nonfood applications, e.g., as glue for paper and wood [1] and as gum for the textile industry [2,3]. Together with wood, natural fibers, and leather, starch has been one of the choice materials since the inception of human technology.

Polysaccharides represent by far the most abundant biopolymers on earth, with cellulose, chitin, and starch dominating. Starch is certainly one of the most versatile materials for potential use in polymer technology. It can be converted, on the one hand, into chemicals like ethanol, acetone, and organic acids, used in the production of synthetic polymers and, on the other hand, it can produce biopolymer through fermentative processes or be hydrolyzed and employed as a monomer or oligomer. Finally, it can be grafted with a variety of reagents to produce new polymeric materials, used as such or as fillers for other polymers.

The conversion into small molecules is chemically easier for starch than for cellulose, making it an economic option to produce hydroxyl-containing compounds, which can be exploited as monomers or as raw material in the production of other biopolymers like polylactic acid (PLA). This approach is in competition with other renewable resources, namely saccharose from sugar cane, used for the production of ethanol and biopolymers such as poly-β-hydroxybutyrate (PHB) [4,5], and lactic acid [6] as a source of its polymer [7]. Despite the importance of starch as a raw material for the production of chemicals and other polymers, its direct use as a renewable resource commodity is undoubtedly more economical and has

Ebnesajjad: Handbook of Biopolymers and Biodegradable Plastics. http://dx.doi.org/10.1016/B978-1-4557-2834-3.00007-0
© 2013 Elsevier Inc. All rights reserved. Adapted from a chapter in: Belgacem, *Monomers, Polymers and Composites from Renewable Resources* (2008).

been a major area of research in material science over the last few decades.

The chemical modification of starch can provide tailor-made materials and has been reviewed recently [8].

The literature concerning starch is extremely vast and its chemistry and technology have been comprehensively reviewed [2,3,8,9]. More specifically, a renewed interest has arisen in the last decade, to convert starch into a plastic material capable of replacing petroleum-based counterparts. The main aim of this chapter is to review the applications of starch in the development of new polymeric materials in which its main macromolecular structure is preserved. The preparation of monomers and oligomers will also be briefly discussed.

7.2 Main Sources of Starch

Several plants are commercially used for the production of starch. The choice of plant depends mainly on geographic and climatic factors and on the desired functional properties of the corresponding starch [10]. It is always possible to find a highly productive plant to produce starch whatever the climate and agricultural conditions: maize in tempered and subtropical zones, cassava (manioc or tapioca) and banana in tropical environments, rice in inundated areas, and potatoes in cold climates. The main plant sources are listed in Table 7.1, together with their production for 2005 [12].

Apart from the traditional crops, cassava shows great potential because it adapts to tropical zones and constitutes, therefore, an important crop in developing areas of the world. New sources of starch are also arising, such as banana [13], which yields a starch of excellent quality.

The development of new uses for starch, and for materials based on starch, within the broader context of the increasing demand for materials based on renewable resources, will certainly increase the demand for starch production and hence the development of new commercial sources of starch.

Table 7.1 World Production of the Main Starch Crops in 2005 (1 × 1000 Metric Tons)

Crops	World Production in 2005 (1 × 1000 Metric Tons)
Maize	711 762.87
Rice (paddy)	621 588.53
Wheat	630 556.61
Potatoes	324 491.14
Cassava	213 024.81
Bananas	74 236.88
Yams	48 891.21
Sorghum	59 722.09

Source: FAO, 2005 [12].

7.3 Structure of Starch Granules

Starch can be found in various parts of a plant, such as the endosperm, the root, the leaf, and the fruit pulp. It is deposited in the form of semicrystalline granules that are insoluble in cold water and resemble spherulites [14] alternating amorphous and crystalline (or semicrystalline) lamellae. Native starch is composed of two main macromolecular components, namely amylose and amylopectin [15–22]. The monomer units of these natural polymers are α-D-glucopyranosyl moieties linked by (1→4) and (1→6) bonds [23]. Amylose is a predominantly α-(1→4)-D-glucopyranosyl linear macromolecule. Amylopectin is a highly branched and high molecular weight macromolecule composed mostly of α-(1→4)-D-glucopyranose units, with α-(1→6)-linkages at intervals of approximately 20 units (Fig. 7.1) [16,24].

7.3.1 Granule Structure

The starch granule morphology, as well as the structure of its main macromolecular constituents, has been the focus of intense research, which is still ongoing because of the complexity of the problems involved. The granules have been examined using several techniques, like light and electron microscopy, X-ray and neutron scattering, and more recently, atomic force microscopy [25–29]. Starch granules from different plant species are significantly different and can be, in the majority of cases, identified by light microscopy inspection. The most obvious differences in starch granules are in their shape and size which can vary considerably [1,30,31], as reported in Table 7.2 for some common granules.

The morphology of the starch granule varies not only according to the source plant, but also to the different parts of the same plant [17]. Other important factors influencing these aspects are the degree

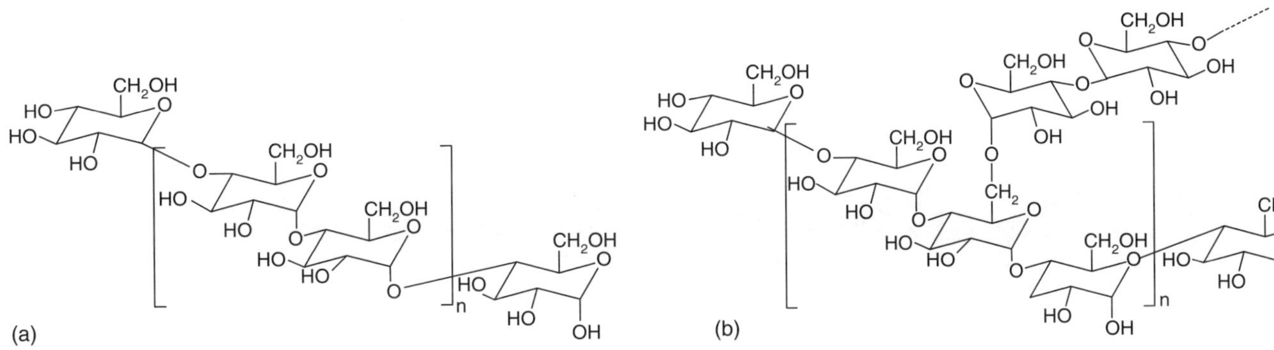

Figure 7.1 Amylose (a) and amylopectin (b) structures.

of polymerization (DP) of amylose and amylopectin and the possible presence of other components in the granule such as lipids, proteins, and inorganic compounds [19].

The semicrystalline nature of starch is well-known and its most important feature is the alternation of long-range molecular order and amorphous regions, defining the corresponding alternation of crystalline and amorphous lamellae [15,17,21,32].

An idealized structure for the starch granule was proposed recently by Gallant *et al.* [22]. This model describes the crystalline and amorphous amylopectin lamellae into effectively spherical blocklets, whose diameter ranges from 20 to 500 nm. These authors also propose the existence of short radial channels of amorphous material. The granule grows alternating radial amorphous and semicrystalline rings from the *hilum*, forming a lamellar structure. The amylopectin, which comprises around 75% of the granule, is mainly responsible for the granule crystallinity. The crystalline and semicrystalline lamellae are composed of amylopectin blocklets, forming a crystalline hard shell composed of large blocklets and a semicrystalline soft shell composed of small blocklets. The crystalline lamellae are around 9–10 nm thick on an average and consist of ordered double-helical amylopectin side chain clusters, interfacing with the more amorphous lamellae of amylopectin branching regions. The size of the semicrystalline soft-shell blocklets ranges from 20 to 50 nm. Figure 7.2 depicts this structure as proposed by Gallant *et al.*

7.3.2 Molecular Structure and Crystallinity

The main technique used to study starch crystallinity is X-ray diffraction, from which starch can be classified to A, B, and C crystallites or polymorph forms [18,33–35]. Starches with these polymorphisms are called A-, B-, and C-type, each type presenting its characteristic diffraction patterns. The most commonly observed structures in native starch are A and B, the former being associated mainly with cereal starches, while the latter dominates generally in tuber starches but also occurs in maize starches with more than 30–40% amylase [18]. The C pattern

Table 7.2 Size, Shape, and Amylose Content of some Starch Granules [2,13,18,30,31]

Source	Diameter (mm)	Amylose Content (wt%)	Shape
Maize	5–25	28	Polyhedric
Waxy maize	5–25	~0	Polyhedric
High-amylose maize	5–35	55–85	Varied smooth spherical to elongated
Cassava	5–35	16	Semi–spherical
Potato	15–100	20	Ellipsoidal
Wheat	20–22	30	Lenticular, polyhedric
Rice (normal)	5/3–8	20–30	Polyhedric
Banana	26–35	9–13	Elongated oval with ridges

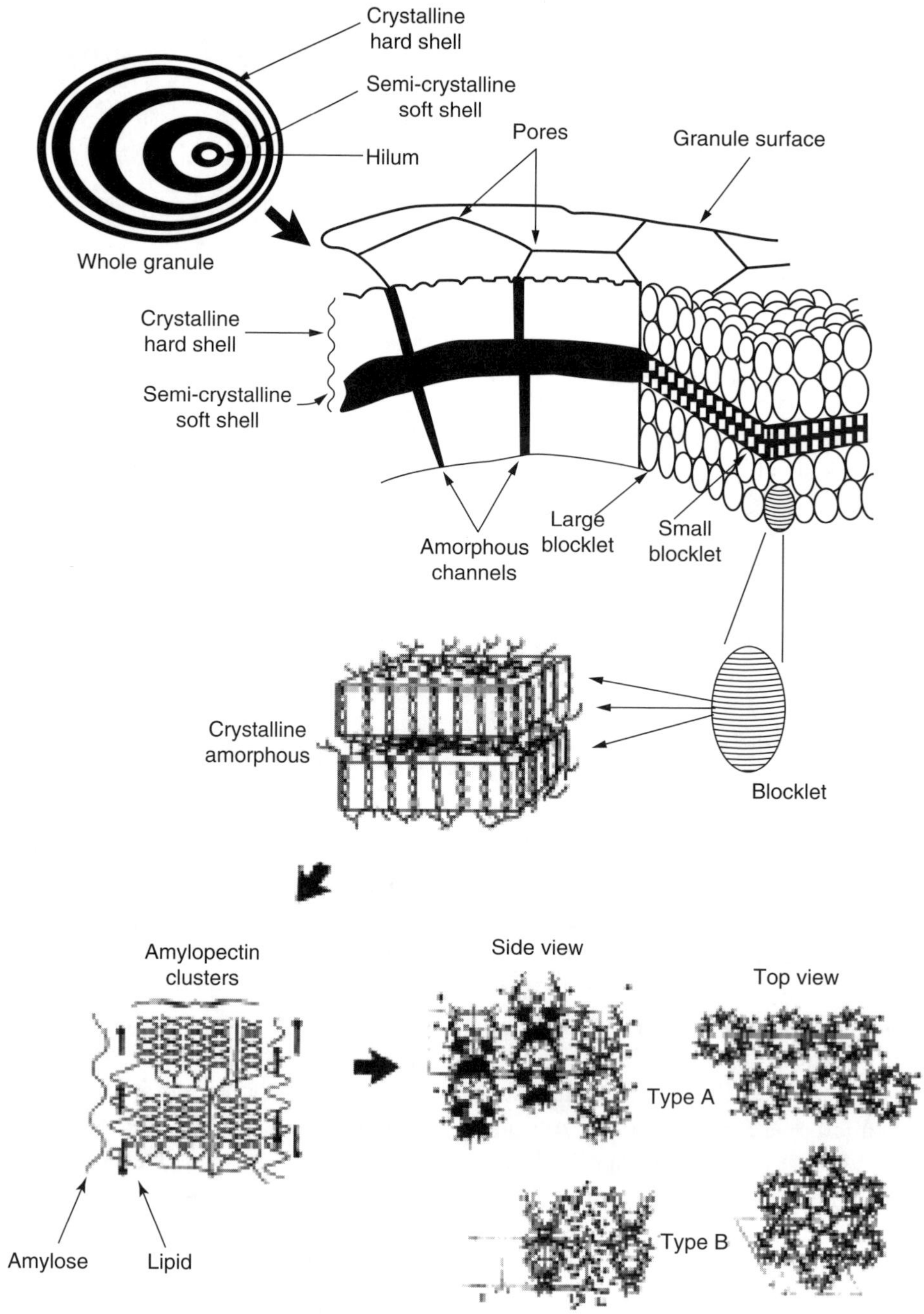

Figure 7.2 Starch granule structure. *Reproduced with permission from Reference [22].*

is a form intermediate between A and B and is usually associated with pea and various bean starches together with other root starches [35,36]. A-type crystallites are denser and less hydrated [34] than B-type counterparts, whereas the C-types arise from the joint presence of the other two homologues [11,18,20,35–40]. Figure 7.3 shows the X-ray diffraction patterns of maize (A-type), potato (B-type), and cassava (C-type) starches.

Although amylose is predominantly made up of linear macromolecules, it has been suggested that amyloses from some starch varieties may contain very long branches [16]. Amylose molecular weights [41] range from 1.5×10^5 to 2.6×10^6, with some variability as a function of the plant variety in the case of cassava [42]. Linear amylose molecules form double helices and can crystallize in a similar way as in starch granule, rendering the A and B structures

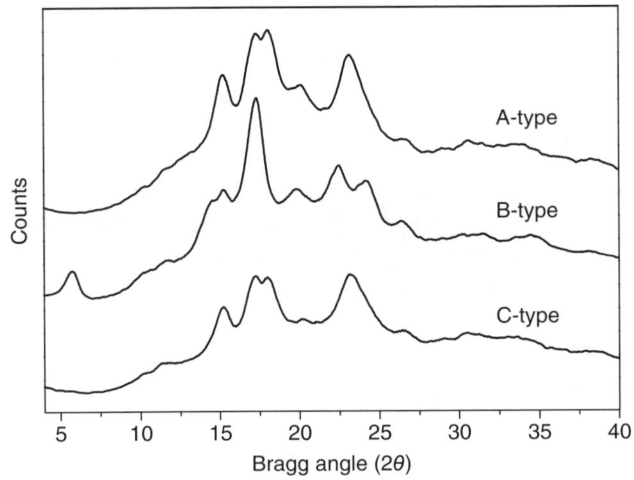

Figure 7.3 X-ray diffraction patterns of maize (A-type), potato (B-type), and cassava (C-type) starches.

[16]. Single crystals of low DP amylose in the polymorphisms A and B were also prepared and characterized [43].

Amylose also form complexes with other materials called V-structures (Verkleisterung) [2], which can be isolated in single crystals [43]. The amylose V-complexes are commonly formed by a left-handed amylose helix with six residues per turn enclosing the aliphatic tail of a lipid in its center [19,44,45], each revolution around the lipid taking 8 Å [46,47]. Depending on the condition of the V-complex formation, they can be crystalline or amorphous [48]. V-complexes are insoluble in water, even in drastic pressure and temperature conditions, a fact that makes them very important in polymer blending when water resistance is a desired property [49,50].

Amylopectin is a highly branched molecule, which is responsible for the main crystalline character of the starch granule. Its structure was modeled as a hyperbranched molecule [18,51,52], as proposed initially by Nikuni [53] and French [17] and later improved by Robin [11] (Fig. 7.4). In this model, short chains with 15 D-glucopyranosyl units branch out at almost regular intervals of 25 units to form either external A-chains or internal chains of amylopectin [11]. Starch crystallites are thus formed in compact areas made up of A-chains with DP 15. Less compact areas mainly occupied by B-chains, where the (1,6)-α-D-branching points are located, are placed between these compact areas.

A parallel with synthetic polymers can be made in which amylopectin is a clear-cut example of a natural dendrimer [51] with a high degree of branching and a spherical shape, each generation being fully

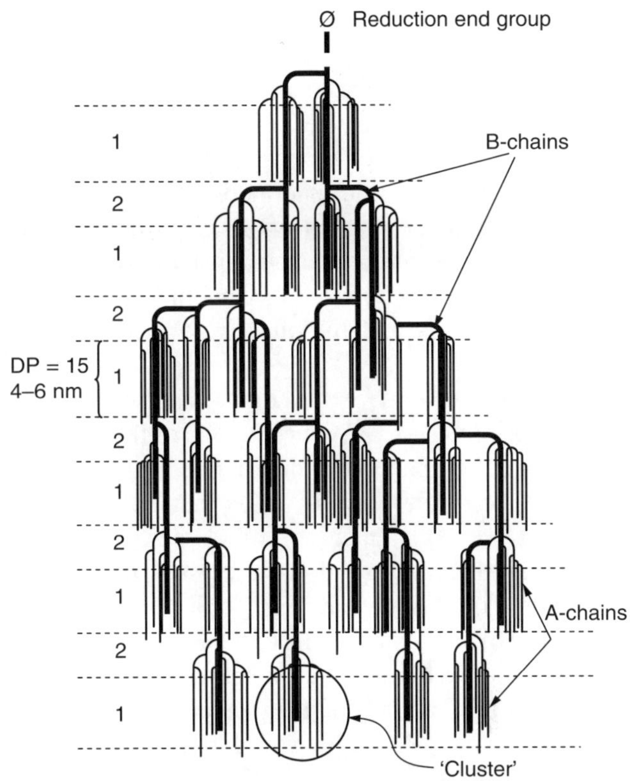

Figure 7.4 Amylopectin model, as proposed in Reference [11]. 1 = compact area; 2 = less compact area where the branch points are located, and Ø = reducing moieties. *Reproduced with permission from Reference [11].*

generated from branching sites with a minimum functionality of three.

7.4 Disruption of Starch Granules

When starch granules are heated in an excess of water (90% w/w) or of another solvent able to form hydrogen bonding (e.g., liquid ammonia, formamide, formic acid, chloroacetic acids, and dimethyl sulfoxide) starch undergoes an irreversible order–disorder transition known as gelatinization or destructuration. This phenomenon occurs above a characteristic temperature known as the gelatinization temperature. During gelatinization, the amylose is dissolved and progressively leached from the granules. The process can be decomposed into two steps: hydration or diffusion of the solvent through the granule takes place during the first, followed by the melting of the starch crystallites in the second. This process can be studied by a calorimetric technique such as differential scanning calorimetry (DSC) or differential thermal analysis (DTA) [14,54,55]. The melting temperature depends on the starch–water ratio. According to Donovan [14], with a large excess of water, only one endotherm appears in the DSC curve, corresponding to the gelatinization temperature, whereas with a modest water content, only one endotherm again appears but at a higher temperature. With intermediate water concentration, two endothermic transitions are observed. This complex behavior has been treated quantitatively using Flory's relationship between the melting point of the crystalline phases and the quantity of added water [14,56,57].

There are several explanations for the multiple peaks in the DSC or DTA curves. One of them is that water is not homogenously distributed in the granule and diffusion can play an important role in that process [14]. This lack of homogeneity leads to partial melting, followed by recrystallization and remelting, implying that this is a nonreversible process [56] and that the Flory–Huggins theory is not fully applicable to starch gelatinization [14,56].

Besides DSC and DTA, many other techniques can also be employed to study the order–disorder transition occurring during gelatinization, including X-ray scattering, light scattering, optical microscopy (birefringence using crossed polarizers), thermomechanical analysis and nuclear magnetic resonance (NMR) spectroscopy and, more recently, small-angle X-ray scattering (SAXS), wide-angle X-ray scattering (WAXS), and small-angle neutron scattering (SANS) [55]. On the bases of evidence gathered using the latter techniques, Jenkins and Donald [55] concluded that the final loss of crystallinity only occurred when the gelatinization was almost complete.

In an extrusion process, where shear forces and high pressures are applied, the entire process is obviously much more complicated. However, both with limited amounts and excess quantities of water, the main step, associated with the melting of crystallites, is the same and the final mass is an amorphous entanglement of amylose and amylopectin macromolecules.

7.5 Applications of Starch as a Raw Material for Plastic Production

7.5.1 Strategies for the Use of Starch as a Source of Polymers

The exploitation of starch as a precursor to macromolecular materials can follow two strategies, namely as a raw material for the production of chemicals used in the synthesis of other polymers or used directly as a high molecular weight polymer by keeping its molecular structure as unchanged as possible.

The first strategy, based on the use of starch for the production of other chemicals, was recently reviewed by Robertson *et al.* [58], Koutinas *et al.* [59], Kennedy *et al.* [60], and Otey and Doane [61]. Three different approaches are applied in this context: (i) starch as a raw material for the production of monomers used in the synthesis of polymers, which can be nonbiodegradable, such as polyethylene, or biodegradable, such as PLA (the main biodegradable commercial polymer whose monomer, lactic acid, can be obtained from the fermentation of starch [62]); (ii) as a raw material for the production of biopolymers like polyhydroxyalkanoates (of which PHB is the main member); and (iii) as a raw material for the production of glucose, dextrin, and other hydroxyl-containing monomers used in the production of mixed compositions based on starch and other monomers.

In all the above processes, the macromolecular structure of starch is destroyed and the polymers derived from the ensuing monomers are totally different from it. It is important to emphasize that the same monomers (e.g., ethylene, sugars, and dextrin) can be produced from other sources, both renewable, such as cellulose and sugar cane, and nonrenewable such as oil.

The second strategy calls upon the use of starch as such or in combination with other materials and is therefore more interesting than the first one, if anything, in terms of cost and yield. Considering, e.g., the conversion of starch into polyethylene through its fermentation to ethyl alcohol and dehydration of the latter, the maximum yield of ethylene produced from starch is close to 35% [61].

In order to adjust the properties of these starch-based materials to the desired application, it is necessary to combine starch with other polymers, as frequently done in the plastic industry. The need for tunable properties may also require starch modifications, such as esterification or etherification, grafting, and reactive or melting extrusion of thermoplastic starch (TPS). The main forms of starch utilization as a polymer are (i) starch grafted with vinyl monomers, (ii) starch as a filler of other polymers, and (iii) plasticized starch (PLS), commonly known as TPS.

Of the two major strategies outlined above, the second constitutes the main subject of this chapter.

7.5.2 Use of Starch in Plastic Production

In the 1960s and 1970s, oxidized starch was used successfully in rubber and other polymer formulations [63,64] and several technologies were developed to optimize its combination with plasticized polyvinyl chloride (PVC) [61,65,66].

In 1972, Griffin [66] described the use of starch as particulate filler for low density polyethylene (LDPE) with the aim of giving a paper-like texture and appearance to extrusion-blown LDPE films. The necessity for highly dried starch to avoid defects caused by water volatilization was a financial limitation to the process since it required appropriate storage of the dried starch prior to its use. Another problem was the poor adhesion of the hydrophilic starch granules to the highly hydrophobic polyethylene, which was improved by treating starch with reactive silanes, but this added an extra cost to the process.

In the 1970s and 1980s, the pollution caused by plastic packages considered "indestructible" had become a serious issue and discussions about possible solutions started based on the search for materials that can degrade in landfills, hence research on biodegradable polymers became an important topic.

LDPE–starch blends seemed an interesting approach, but starch granules were encapsulated into the LDPE matrix and thus became, in principle, inaccessible to biodegradation. Later studies carried out by Griffin demonstrated that even when encapsulated, starch could be degraded in an appropriate environment [66]. A further development called upon the use of a pro-oxidant (Fe^{3+}, Mn^{3+}) in the LDPE matrix [67].

Otey *et al.* [61,68–71] also described the blends of starch with synthetic polymers, in which gelatinized starch was used instead of starch granules. The initial films, composed of 90% of starch blended with poly(ethylene-*co*-acrylic acid) (EAA), were obtained by casting from aqueous dispersions or by dry milling in a rubber mill, followed by a hot roll treatment, but only the cast films attained acceptable characteristics [71]. These materials were intended for application as mulch films for agriculture.

The next generation consisted in compositions of starch and poly(EAA) films, produced by extrusion-blowing [70]. The starch concentration varied between 10 and 40%, that of EAA between 10 and 90%, with some composition also containing between 0 and 50% of PE. Sorbitol and glycerol were also added as plasticizers to some of these mixtures. In a further investigation, Otey *et al.* described a similar composition to which urea, starch-based polyols or glycerol, or mixtures of these materials were added [72]. Urea was used to improve the gelatinization of starch at lower levels of moisture and the polyols were added to increase the levels of biodegradable materials in the final mixture. A typical composition comprised 40% of starch, 20% of EAA, 15% of urea, and 25% of LDPE. These materials were extrusion-blown into films for mulch applications. The critical point for this application is the balance between biodegradability and resistance to it, since the film must neither disintegrate before its estimated lifetime, nor must it offer excessive resistance to biodegradation.

In the 1990s, compositions of starch processed directly in melting equipment such as extruders were described as a new material named destructurized starch or TPS [57,73–75]. This material was patented by the Warner-Lambert Company [76] and was described as being prepared with starch that had been heated to a high enough temperature and for enough time for the melting to occur prior to starch degradation. In this process, the starch processed in extruders contained between 5 and 40% of water. It was also claimed that when starch was heated in a closed volume in appropriate moisture and temperature conditions, it became substantially compatible with hydrophobic thermoplastic synthetic polymers.

Lay *et al.* [76] also described destructurized starch compositions with one or two other polymers, which included a whole variety of both natural and synthetic macromolecules.

In another patented process, starch was destructurized in the presence of low-molecular weight polymers such as polyethylene–vinyl alcohol (EVOH), EAA, poly-ϵ-caprolactone (PCL), and small amounts of moisture and of a high boiling point plasticizer, using a high shear equipment like a twin-screw extruder [77]. From these materials emerged one of the most successful commercial thermoplastic materials based on starch, which took the name of Mater-Bi®.

It is also important to mention materials with high starch content, or based exclusively on starch. The main examples of this family of products are expanded TPS compositions such as those patented by the National Starch and Chemical Investment

Holding Corporation [78] used in place of expanded polystyrene. This was one of the first commercial TPSs placed on the market and is still sold today in growing quantities. Its success is due not only to the replacement of a synthetic polymer by a biodegradable one, but also to its good performance and because its production cost is lower than that of expanded polystyrene.

The grafting of starch with more than 60% of vinyl or acrylic monomers gave materials, which showed excellent mechanical and processing properties, being virtually insoluble in water [61]. However, this process was deemed too expensive and thus limited the use of the ensuing grafted starch.

7.6 Thermoplastic Starch

7.6.1 Definition and Properties

The term TPS describes an amorphous or semicrystalline material composed of gelatinized or destructurized starch containing one or a mixture of plasticizers. TPS can be repeatedly softened and hardened so that it can be molded or shaped by the action of heat and shear forces, allowing its processing to be conducted with the techniques commonly used in the plastic industry. The following sections are devoted to a brief description of the basics of starch extrusion and processing and to the more relevant applications of TPS [50,57,73–75,79–81]. TPS or destructurized starch are also known as PLS [82], because of the inevitable presence of nonvolatile plasticizers in their composition. TPS is, however, the predominant term used for these materials.

TPS is generally produced by processing a starch/plasticizer(s) mixture in an extruder at temperatures between 140 °C and 160 °C at high pressure and high shear. Batch mixers operating in conditions similar to those of extrusion can also be used.

If the final composition only contains water as the plasticizer, in levels greater than 15–20%, it keeps its thermoplastic properties. However, if the processing temperature is higher than 100 °C, water volatilizes and the melted material expands. If controlled, this is a desired effect, exploited in the production of expanded starch used in packaging as a shock absorber. A common feature of these processing techniques is the limited amount of plasticizer and the consequent high viscosity of the melt, but TPS can also be processed in the presence of large amounts of water, as in the technology developed by Otey *et al.* [71,72].

The destructuration temperature profile as demonstrated by Donovan [14] in his classical paper depends on the water content of the sample. Hence, the process in the presence of limited amounts of plasticizer appears to be different from that associated with an excess of it. In general, TPS is produced in the former conditions and the shear forces play a fundamental role in its processing [57].

7.6.2 Plasticizers for TPS

The type and quantity of plasticizer employed determine the preparation/processing conditions and the mechanical and thermal properties of the final material, as discussed in several studies [83–93]. Various authors [83–85] extended existing theories related to the glass transition and the effect of plasticizers on it to the glass transition temperature of TPSs. Thus, Kalichevsy and Blanshard [85] applied Couchman and Karasz's approach despite the difficulties associated with a reliable determination of the glass transition temperature of TPSs. Orford *et al.* [83], on the other hand, estimated the T_g of pure amylose and amylopectin at 500 ± 10 K, by measuring the T_g of a series of monodisperse oligomers of increasing DP. This value is above the thermal degradation temperature of these polymers.

Numerous laboratories have tackled the effect of a series of nonvolatile plasticizers, such as glycerol [84,87–90,93,94], urea [84,87], fructose [85], xylitol, sorbitol, maltitol [87,90,94], glycols (EG, TEG, PG, PEG) [84,87,89,94], ethanolamine [92], and formamide [91]. Several criteria concerning the most appropriate structures for this key role have been put forward [94], although a rough first principle simply predicts that any substance capable of forming hydrogen bonds would be able to plasticize starch [95].

The partial or total replacement of water by nonvolatile organic solvents (plasticizer) such as glycerol and sugars leads to an increase in the gelatinization temperature, as showed by Perry and Donald [96], a feature that needs to be considered when processing TPS. The reason for this effect is not completely understood, but Perry and Donald attributed it to two main causes, namely the higher molecular weight plasticizers are less able to penetrate the starch granule and less able to increase the free volume of the amorphous regions, thus being less effective than water as plasticizers. This effect has alternatively been attributed to a reduction in the activity and the volume fraction of water [97]. However, Perry and Donald [96] showed that even glycerol alone can completely gelatinize starch and induce an increase in the gelatinization temperature of approximately 60 °C, compared with the water-plasticized counterpart. Tan *et al.* [95] recently studied this effect and concluded that the parameters determining solvent transport through the granule, such as viscosity, diffusion, and ingress rate, play an important role in determining the gelatinization temperature. Other solvent properties were also considered relevant, such as molecular size and number of possible hydrogen bonds.

7.6.3 Crystallinity in TPS

The crystalline order observed in the native starch granules is completely destroyed in TPS but, because of the mobility of the starch chains, recrystallization occurs, leading generally to the formation of B-, V-, and E-type crystalline structures. B-type crystallinity appears after TPS is stored above its glass transition temperature or at high plasticizer contents [81]. V- and E-types are observed just after extrusion and are generated during processing [35,98,99]. V-type structures can be observed in two forms, namely Va-type (anhydrous) for materials containing low moisture contents and Vh-type (hydrated) for materials containing higher moisture contents. E-type occurs only in low moisture compositions, is unstable, and rearranges to V-type when the sample is conditioned at ambient temperature with more than 30% relative humidity [35].

As a consequence of their semicrystalline character, the mechanical properties of TPSs are characteristic of partially crystalline polymers [81] with B-type crystallinity being the major factor influencing the mechanical behavior of glycerol–PLS [100]. This recrystallization can also be a problem because after long-term storage TPSs can become rigid and brittle.

Despite the fact that TPS is considered a new material in technological terms, its basic features and processes are in fact the same as those relative to extrusion-cooking starch used in the food industry since the 1960s. This kind of processing is therefore briefly described under the following heading, given its importance in the development of TPS.

7.6.4 Extrusion-Cooking as the Basis for TPS

Extrusion has been applied to pasta processing in the food industry since the mid-1930s, but the low temperatures used (40 °C) were insufficient to cook the extruded material. In the 1960s, extrusions at higher temperatures, sufficient to cook the materials, started, and extrusion-cooking became a process of great importance [35]. This process is conducted in the presence of a limited quantity of water (10–25%) at temperatures that can reach 200 °C but with very short residence times, so that starch decomposition is minimized. This treatment is known as high-temperature short-time (HTST) [35,101,102]. The molten starch shows higher viscosities than common plastics during extrusion, and under isothermal conditions, it can be described as a pseudoplastic material [35]. The main structural modification associated with extrusion-cooking is the destruction of the starch granule morphology (destructurized starch). However, the process is much more complex and chemical reactions that lead to depolymerization and/or other degradation also take place. The expansion of the melted starch occurs at the extrusion head because of the fast evaporation of the moisture present in the melted starch. This expansion is also an important factor in determining the properties of the extruded material. The overall process was extensively studied by Mercier and Feillet [101] who investigated such variables as the temperature of extrusion (170–200 °C), the moisture content of the product before extrusion, and its amylose content.

Extrusion-cooking can be considered the precursor of the modern technology of TPS, the main principles and the changes occurring in starch being the same [63,75]. Wiedmann and Strobel [80] proposed that the compounding of TPS is a combination of extrusion-cooking and plastic compounding.

7.6.5 Macromolecular Scission and Starch Degradation During Destructuring/Plasticization

The changes in molecular weight and its distribution play an important role on the rheological and mechanical properties of starch and have therefore received considerable attention [103–106]. The main factors affecting the molecular weight degradation during TPS preparation and processing are the specific mechanical energy applied [103], the temperature, and plasticizer content [101,104].

Gomez and Aguilera [73] studied the effects of water concentration during the extrusion of maize starch on the properties of the ensuing materials and proposed a model for starch degradation during extrusion. In this model, starch granules are converted into gelatinized starch, then into free polymer chains that, depending on the extrusion conditions, can degrade into dextrinized starch and/or oligosaccharides and sugars. One important conclusion of this work was that, when extruding starch with a moisture level below 20%, a product distinct from gelatinized starch is obtained, since it has been partially dextrinized. Dextrins are dextrorotatory products of partially hydrolyzed starch that can be precipitated with alcohol from their aqueous solution [16].

Willett *et al.* [103] studied the melt rheology and molecular weight degradation of waxy maize (amylopectin) in a corotating twin-screw extruder by processing the native starch and re-extruding the destructurized material. The moisture content of the first extrusion was 35% and the product was re-extruded with 18 or 23% of moisture. Starch degradation was evaluated by multiangle light scattering in dimethyl sulfoxide/water. The weight-average molecular weight decreased moderately with increasing specific mechanical energy, which was considered an important parameter for the prediction of the molecular weight degradation during extrusion.

Myllymäki *et al.* studied the depolymerization of barley starch during its extrusion in the presence of a mixture of water and glycerol [107]. They observed a correlation between the depolymerization of the starch chains and both the water/glycerol concentration and, to a minor extent, the screw speed.

Carvalho *et al.* [108] studied chain degradation in TPS composites of maize starch plasticized with glycerol and reinforced with wood pulp. The product was characterized by high performance size-exclusion chromatography (HPSEC). The matrix (starch/glycerol) and the composites were prepared in an intensive batch mixer at 150–160 °C, with glycerol and fiber contents in the range of 30–50% and 5–15%, respectively. The HPSEC curves obtained for different levels of plasticization without fiber are shown in Fig. 7.5.

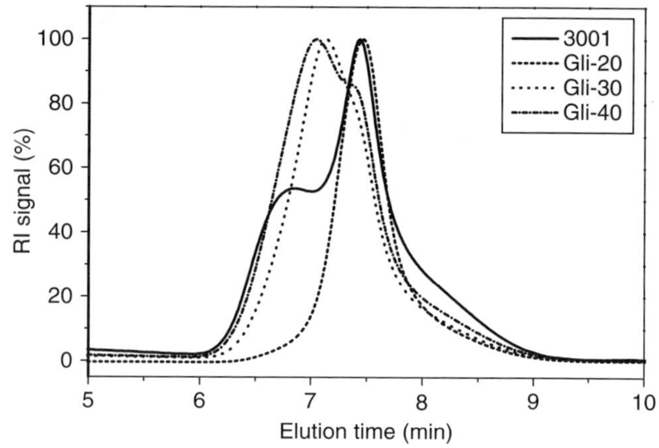

Figure 7.5 High performance size-exclusion chromatography (HPSEC) of native cornstarch, and TPS with 20% (Gli-20), 30% (Gli-30), and 40% glycerol (Gli-40). *Reproduced with permission from Reference [108].*

The results showed that an increase in the glycerol content reduced the starch degradation, whereas an increase in fiber content lead to its increase. The changes in the chromatographic profiles were more pronounced in the high molecular weight fraction, corresponding to amylopectin. The polydispersity index of the matrix was lower than that of native starch due to the selective breakage by shear-induced fragmentation of large amylopectin molecules. The effects of both glycerol and fibers on the molecular weight was of a similar magnitude, in the studied range, for both weight-average molecular weight, M_w, and z-weight-average molecular weight, M_z, and these results were expressed in terms of two equations:

$$M_w = 222\,833 + 13\,500G - 13\,000F \quad (7.1)$$

$$M_z = 267\,500 + 19\,250G - 16\,250F \quad (7.2)$$

where G and F are normalized glycerol and fiber contents, respectively (30% of glycerol is equal to 21 and 50% of glycerol is equal to 11, and likewise for fibers). Glycerol and fibers showed opposite effects, as can be observed in Eqns (7.1) and (7.2). This behavior is related to the shear-induced fragmentation, namely a process in which the largest molecules are the most susceptible. As this process is highly dependent on the melt viscosity, higher viscosities induce correspondingly higher extents of degradation.

Carvalho *et al.* [109] used ascorbic and citric acid as catalysts for the controlled hydrolytic cleavage of starch macromolecules carried out by melt processing in the presence of glycerol and residual moisture. The process proved effective in providing a means for the controlled tuning of the molecular weight of starch in TPS compositions. Several studies have been published since then where citric acid was used for TPS modification intended to improve its performance in polymer blending and in composites and nanocomposites.

7.6.6 TPS Blends

Blending is an important operation for the modification of polymer properties at low cost and without the need of special equipment and techniques. The possibility of blending different polymers increases considerably their range of applications and can also be used to decrease their costs. From a technological point of view, a compatible blend is achieved when the desired properties are improved as a consequence of mixing two or more polymers. In the majority of cases, this mixture is not thermodynamically compatible and thus constitutes a multiphase system [82,110,111].

TPS is blended mainly for two purposes. The first, and probably the most important, is to improve such properties as its water resistance and mechanical performances, whereas the second is to use it as a modifier for other polymers with the purpose of increasing the biodegradability and/or decreasing the cost of the ensuing blends. In fact, starch is cheaper than any other polymer, is readily available and renewable, so that major efforts are employed to maximize the starch content in a blend.

Blends of starch with polar polymers containing hydroxyl groups, such as poly(vinyl alcohol), copolymers of ethylene and partially hydrolyzed vinyl acetate have been prepared since the 1970s, as described by Otey *et al.* [61,68–72]. Since starch and other natural polymers are hydrophilic, water has been commonly used as a plasticizer for these materials. The possibility of using water as plasticizer makes it possible to add the polymer to be blended as an aqueous emulsion, e.g., in the case of natural rubber latex [112], poly(vinyl acetate), and other synthetic polymer latices [71,113,114]. Blends of starch and biodegradable polymers and polymers from renewable resources have been reviewed recently due to their growing importance [82,110,111,115,116]. Table 7.3 gives some polymers commonly used in blends with starch.

Table 7.3 Most Common Polymers Used in Blending with Thermoplastic Starch

Polymer	Reference
Poly(vinyl acetate) PVAc	[71,113]
Poly(methacrylic acid-*co*-methyl methacrylate) MAA/MMA	[114]
Poly(vinyl alcohol) PVA	[119,141]
Poly(acrylic acid) PAA	[125,127]
Poly(ethylene-*co*-acrylic acid) EAA	[61,71]
Poly(ethylene-*co*-vinyl alcohol) EVOH	[49,50,121]
Poly(ϵ-caprolactone) PCL	[50,109,121,122,142]
Poly(ethylene–vinyl acetate)	[123]
Polyethylene	[117,118,123,124,143]
Poly(ester–urethanes)	[144]
Poly(D,L-lactic acid) PLA	[120,145,146]
Poly(3-hydroxybutyrate) PHB	[115]
Poly(3-hydroxybutyrate-*co*-3-hydroxyvalerate-) PHBV	[122,147]
Poly(butylene succinate adipate) PBSA	[148]
Polyesteramide PEA	[142,149]
Zein	[128,129]
Lignin	[150,151]
Cellulose and its derivatives	[121]
Natural rubber	[112,126,127]

Starch blends can be divided into two main categories according to (1) the source and biodegradation properties of the polymer to be blended with starch and (2) the process used for its preparation. As for the first category, the sources can be obtained directly from renewable resources (biodegradable biopolymers), can be synthetic polymers from either oil or renewable resources, and in this latter case, they can be biodegradable or not depending on their structure.

As for the second category, two main processing techniques are used for blending starch, namely melting and solution/dispersion. In *melting processing*, starch blends are obtained during the plasticization of starch granules in an extruder or in a batch mixer. Alternatively, two extruders are used, starch being gelatinized in a first single-screw extruder and then fed into a twin-screw extruder that processes the other component [117,118]. In *solution or dispersion processing*, the product is often obtained by casting. The first commercial materials produced by solution or dispersion were cast films of starch/EAA blends [61,71]. A large number of starch blends have been obtained using this procedure, especially when other natural polymers are involved. As in the case of starch, many natural polymers and also biodegradable synthetic polymers are soluble or dispersible in water. Therefore, solution/dispersion blending is an interesting option for the production of these mixed materials. In some cases, e.g., when using chitosan, melting is not possible because this polymer decomposes before melting and solution blending is the only viable alternative.

Blending is one of the most promising alternatives to make starch useful as a polymer in replacement of other plastics, and the fast progress occurring in this field is attested by several reviews published recently [82,110,111,115]. Indeed, the commercial plastics based on starch presently available are in the form of blends with other polymers [50,116].

In TPS blends, starch can be the continuous or the dispersed phase, depending on the starch/second-polymer ratio and on the processing conditions. As a consequence, the interfacial interactions between starch and the other components will determine the properties of the blend. Several approaches have been investigated in order to enhance the compatibility among the components of these blends:

1. The use of polymers bearing polar groups, particularly those able to form hydrogen bonds (e.g., PVA, EAA, EVOH, and natural polymers like cellulose and its derivatives, gelatin and zein [49,50,110,115]).
2. The use of mixtures of polymers where one of them acts as a compatibilizer between starch and less hydrophilic components (e.g., PVA in TPS/polyethylene blends or a low molecular

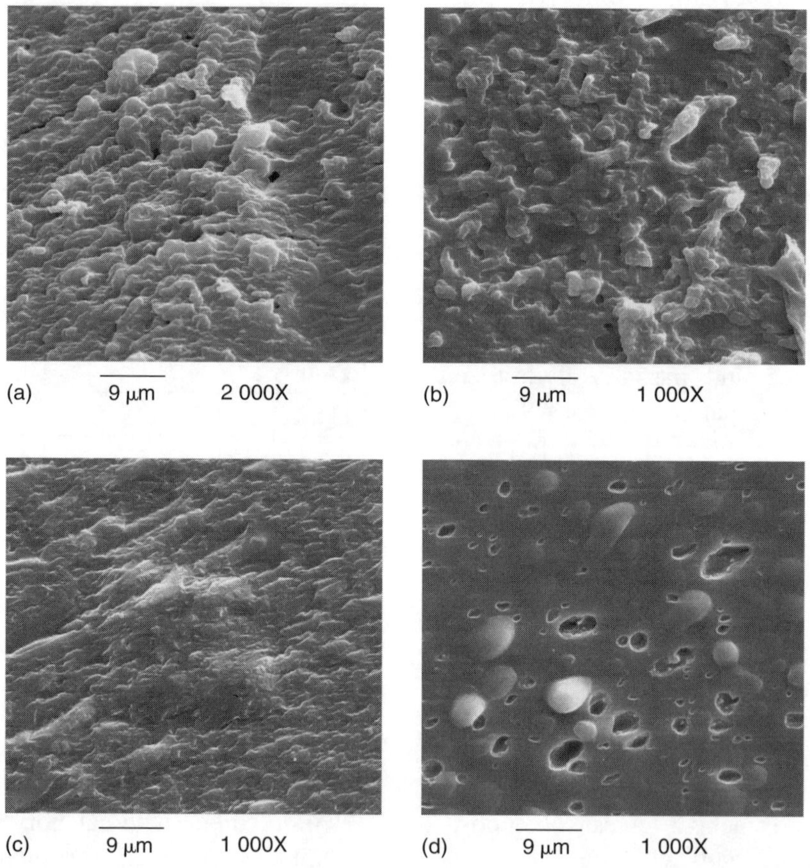

Figure 7.6 Scanning electron micrographs of fragile fractures of starch/natural rubber blends. (a) 20% glycerol and 5% rubber, (b) 30% glycerol and 20% rubber, (c) 40% glycerol and 5% rubber, and (d) 40% glycerol and 20% rubber. All quantities are in w/w based on dry matter. *Reproduced with permission from Reference [112].*

weight polymer like poly(ethylene glycol) in TPS/PLA blends [50,110,115,119]).

3. The use of reactive compatibilizers, which can promote a better interface by polymer–polymer chemical interlinking (e.g., methylenediphenyl diisocyanate (MDI), pyromellitic anhydride, or glycidyl methacrylate [110,115, 120–124]).
4. The formation of complexes between starch and other polymers (e.g., V-type complexes [50,125]).

Arvanitoyannis *et al.* [126] studied blends of starch with 1,4-transpolyisoprene (*gutta percha*) compatibilized with EAA by melt processing and observed that they were biodegradable because of the presence of starch. Their mechanical properties were improved by the addition of glycerol as plasticizer. Rouilly *et al.* [127] also prepared blends of starch and natural rubber by casting mixtures of aqueous starch with glycerol and latex. Carvalho *et al.* [112] blended native starch granules and a natural rubber latex by melt processing calling upon water as a plasticizer for starch. The stable dispersion and the good adhesion between the two natural polymers were attributed in part to the natural nonrubber constituents present in the latex. As little as 2.5% w/w of rubber was sufficient to decrease drastically the brittle character of TPS. Figure 7.6 shows SEM pictures of starch/rubber blends fractured in liquid nitrogen depicting the good dispersion of rubber in the starch matrix.

Zein is a protein obtained from maize as a by-product of maize starch production. It is completely amorphous and, despite the fact of being more hydrophobic than starch, it can also be plasticized by glycerol. It is, therefore, an interesting material for use in blends with starch because it shows some compatibility with starch while conferring to the blends a more hydrophobic character [128,129]. Corradini *et al.* [129] described starch/zein blends prepared by melt processing. The plasticization of zein by glycerol was studied and Fig. 7.7a and 7.7b

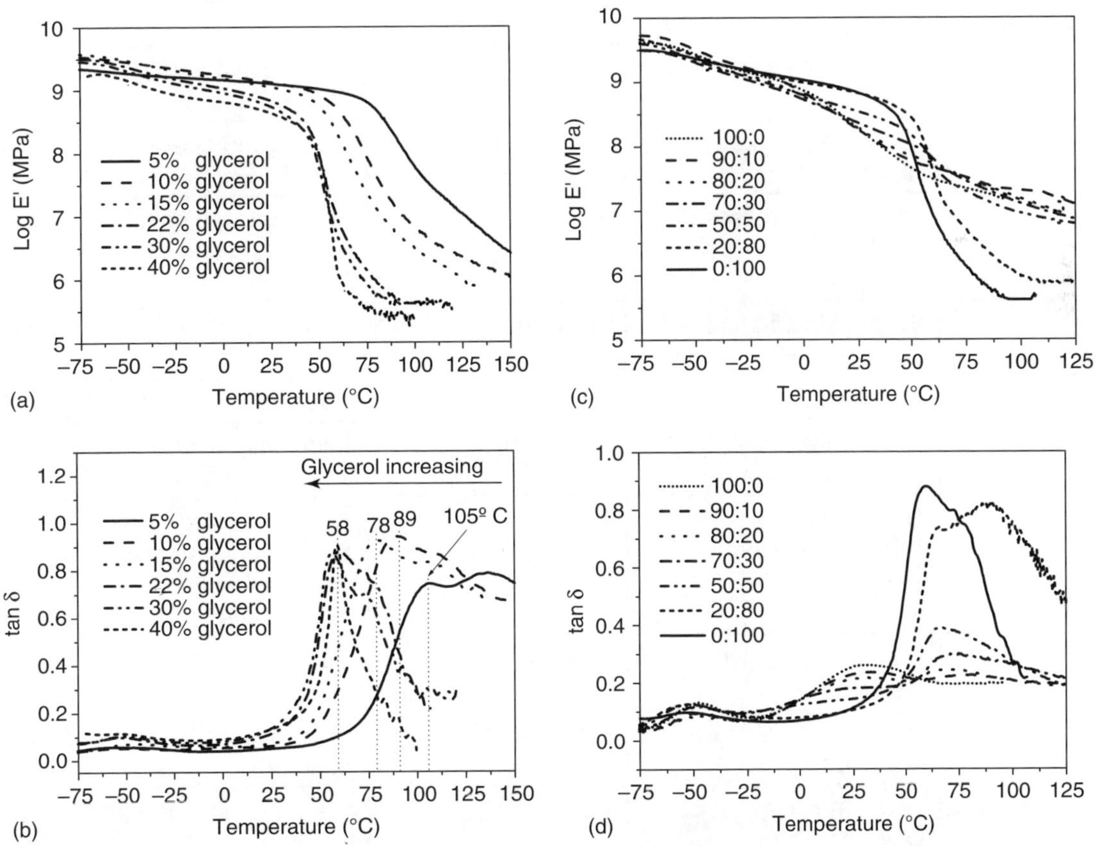

Figure 7.7 Storage modulus (E') and tan δ as a function of temperature for (a) and (b) plasticized zein with 5, 10, 15, 22, 30, and 40% w/w of glycerol, and (c) and (d) starch/zein blends with 22% w/w of glycerol; the proportions given in the key are starch:zein. *Reproduced with permission from Reference [129].*

Table 7.4 Composites and Nanocomposites Based on Thermoplastic Starch

Matrix				
Commercial or Common Name	**Description**	**Source**	**Filler/Reinforcement**	**Reference**
Bioplast GS902	Potato starch blends with cellulose derivatives and synthetic polymers	Biotec Emmerich, Germany	Flax, jute, ramie, oil palm fiber	[152]
TPS/PCL-Tone P-787	Wheat starch/40% poly-ϵ-caprolactone (PCL)	Union carbide Antwerp, Belgium	Flax and ramie	[152]
Mater-Bi ZI01U	Blends of corn starch and poly-ϵ- caprolactone	Montedison, Deutchland	Flax and ramie	[152]
TPS/TPU	Blends of TPS/ thermoplastic polyurethane	Research[1]	Flax	[153]
TPS/PCL	Blends TPS/poly-ϵ-caprolactone (PCL)	Research[1]	Flax	[153]
Mater Bi	TPS blends with PCL, EVOH, etc. with at least 85% of starch	Novamont/ Montedison	Non-woven of flax, hemp, ramie fibers	[154]
SCONACELL A	Modified starch blends	BSL		
Mater-Bi	TPS blends with PCL, etc.	Novamont	Flax cellulose pulp (10–40%)	[155]
TPS/PVA	TPS–polyvinyl alcohol blends	Research	Softowood fiber	[141]
TPS	Starch/glycerol/sorbitol –TPS	Research[1]	Regenerated cellulose fibers: Cellunier F and Temming 500	[156]
TPS	Wheat starch/glycerol/ sorbitol–TPS	Research[1]	Flax and ramie	[152]
TPS	Maize starch–TPS	Research[1]	10–20% of flax fiber	[153]
TPS	Starch/glycerol/ formamide/urea	Research[1]	Micro winceyette fiber	[157]
TPS	Starch/glycerol (30–50% glycerol)	Research[1]	Kraft bleached and thermomechanical wood pulps from *Eucalyptus*	[130, 131]
TPS	Potato starch/glycerol cast film–TPS	Research[1]	Potatoes microfibrils[2]	[132]
TPS	Waxy maize/glycerol cast film	Research[1]	Cellulose whiskers (from tunicate) [2]	[158]
TPS	Waxy-maize/glycerol/ water cast film	Reseach[1]	Starch nanocrystals[2]	[159]
TPS	Wheat starch/glycerol-TPS	Research[1]	Leafwood fibers	[133]

Table 7.4 Composites and Nanocomposites Based on Thermoplastic Starch—*Cont'd*

Matrix				
Commercial or Common Name	**Description**	**Source**	**Filler/Reinforcement**	**Reference**
Mater-Bi	Starch/EVOH	Novamont	Hydroxylapatite-reinforced	[160]
TPS	Maize starch/30% glycerol	Research[1]	Kaolin	[134]
TPS	Potato and wheat starch/ 36.5% glycerol	Research[1]	Up to 10 wt% of montmorillonite (Cloisite Na1, Cloisite 30B-ammonium)[2]	[161, 162]

[1]*Described in research papers.*[2]*Filler or reinforcement with nanometric dimensions*

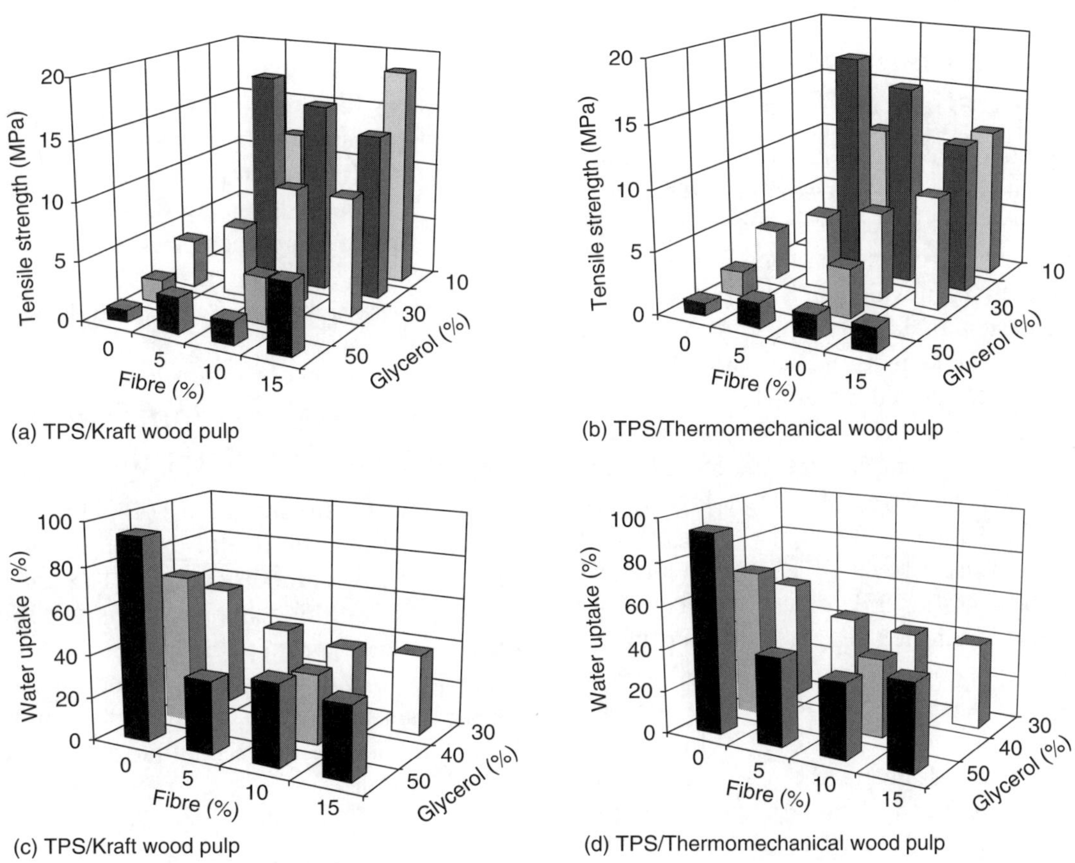

Figure 7.8 Tensile strength and equilibrium water uptake of TPS reinforced with *Eucalyptus* wood pulps: (a) and (c) Kraft pulp; (b) and (d) thermomechanical pulp. *Reproduced with permission from Reference [130].*

shows the DMA plots of the storage modulus and tan δ as a function of temperature. Above 22% w/w of glycerol, its influence on the glass–rubber transition temperature seized. Consequently, the exceeding plasticizer phase separated and, in the presence of starch, migrated into the TPS-phase. Starch/zein blends are immiscible and, depending on the ratio of these two natural polymers, one or the other generated

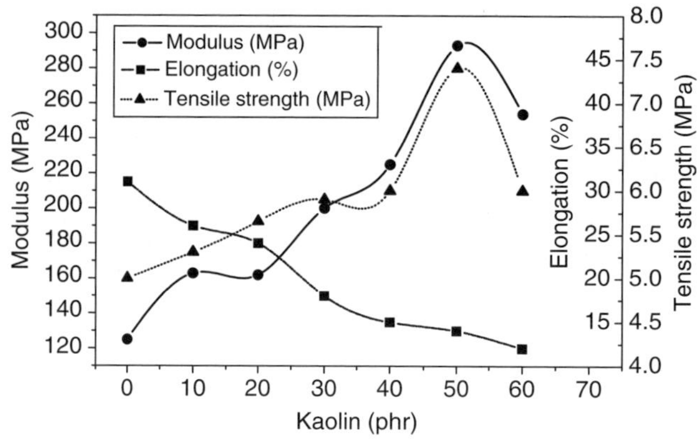

Figure 7.9 Modulus, tensile strength, and elongation of TPS/kaolin composites. *Reproduced with permission from Reference [134].*

the continuous phase. Figures 7.7c and 7.7 d show the dynamic mechanical properties of these blends with the two characteristic tan δ peaks corresponding to the starch/glycerol and zein/glycerol phases.

7.6.7 Composites and Nanocomposites of TPS

As discussed earlier, TPS has two main drawbacks, namely it is mostly water-soluble and has poor mechanical properties; its reinforcement is one of the available options to overcome these weaknesses. Composite materials in which TPS plays the role of the matrix represent a relatively new topic and hence the literature available is rather modest. Table 7.4 lists some of these reported materials.

Reinforcement has proved to be very effective in improving water resistance, and particularly the mechanical properties of TPS. Increases higher than 100% in tensile strength and higher than 50% in modulus were measured when TPS was reinforced with wood fibers [130–133]. Figure 7.8 shows the notable increase in tensile strength and water uptake from TPS to its composites containing 10–50 wt% of glycerol and 0–15 wt% of Kraft and thermomechanical *Eucalyptus* wood pulp [130,131]. With the addition of wood fibers, the water uptake at equilibrium decreased considerably, becoming almost independent of both fiber and glycerol contents. This behavior was attributed mainly to the constraining effect of the reinforcement to the material expansion as water is being absorbed. The effect of lignin, present in thermomechanical fibers, and virtually absent in Kraft counterparts, was negligible.

Kaolin also played a very good role as a reinforcement for TPS matrices as shown by Carvalho *et al.* [134]. The compositions studied in this work were based on TPS containing 30% of glycerol and kaolin in proportions of 10, 20, 30, 40, 50, and 60 phr. Figure 7.9 shows the variation of modulus and tensile strength of these composites as a function of kaolin content with a maximum for both at 50 phr and an increase of 135 and 50%, respectively, relative to the sample prepared without kaolin. The water uptake was reduced considerably for all these kaolin-reinforced TPS composites [134].

A different approach was recently applied to cellulose fibers and starch granules to prepare single-component composites by the partial oxypropylation of these substrates [135].

7.6.8 Chemical Modification of TPS by Reactive Extrusion

Chemical modification of TPS by reactive extrusion has been one of the main techniques used for the chemical modification of TPS and will not be discussed in this chapter since excellent recent reviews have been published [136–140]. Several reactions can be performed including esterification, acidolysis, glycolysis, urethane formation epoxidation, and others. This technique allows the chemical modification of starch in TPS without the use of solvent and complex process and is in general conduced in single-screw and twin corotating screw extruders.

7.7 Conclusions

The necessity to replace materials based on dwindling fossil resources by homologs prepared from renewable counterparts has become an urgent matter for environmental and economical reasons. Within this context, starch will certainly play a very important role as a source of viable alternatives. Its exploitation in nonfood applications is not new, but had declined considerably after the Second World War because of the boom of petrochemistry and the development of polymer materials associated with it. The revival of interest in starch-based plastics began at the end of the last millennium with the emergence of new successful commercial products, and is witnessing a vigorous pursuit. The reasons for this success are to be found in the fact that starch is a cheap raw material, readily available ubiquitously (albeit from different species) and very versatile in terms of chemical and physical modifications.

Acknowledgments

I would like to thank all those who read the manuscript and provided suggestions and comments for its improvement: Dr. Debora T. Balogh, Mrs. Joan Gandini, and Prof. Alessandro Gandini.

References

[1] W. Jarowenko, Starch based adhesives, in: I. Skeist (Ed.), Handbook of Adhesives, second ed., Van Nostrand Reinhold Co., New York, 1977, pp. 192–211 (Chapter 12).

[2] J. Daniel, R.L. Whistler, A.C.J. Voragen, W. Pilnik, Starch and other polysaccharides, in: B. Elvers, S. Hawkins, W. Russey (Eds.), fifth ed., Ullmann's Encyclopedia of Industrial Chemistry, vol. A25, VCH Verlagsgesellschaft mbH, Weinheim, 1994, pp. 1–62.

[3] R.L. Whistler, J.N. Bemiller, E.F. Paschall, Starch: Chemistry and Technology, second ed., Academic Press, San Diego, CA, 1984, p. 718.

[4] J.N. Baptist, Process for preparing poly-b-hydroxybutyric acid, US Patent # US 3036959, 1962; J.N. Baptist, Process for preparing poly-b-hydroxybutyric acid, US Patent # US 3044942, 1962.

[5] J.N. Baptist, F.X. Werber, Plasticized poly-b-hydroxybutyric acid and process, US Patent # US 3182036, 1965.

[6] S.C. Prescott, C.G. Dunn, Industrial Microbiology, McGraw-Hill, New York, 1959.

[7] R.E. Drumright, P.R. Gruber, D.E. Henton, Polylactic acid technology, Adv. Mater. 12 (2000) 1841–1846.

[8] P. Tomasik, C.H. Schilling, Chemical modification of starch, Adv. Carbohydr. Chem. Biochem. 59 (2004) 175–403.

[9] T. Galliard, Starch: Properties and Potential, first ed., John Wiley & Sons, New York, 1987, p 151.

[10] T. Galliard, Starch availability and utilization, in: T. Galliard (Ed.), Starch: Properties and Potential, first ed., John Wiley & Sons, New York, 1987, pp. 1–15 (Chapter 1).

[11] J.P. Robin, C. Mercier, R. Charbonniere, A. Guilbot, Lintnerized starches. Gel filtration and enzymatic studies of soluble residues from prolonged acid treatment if potato starch, Cereal Chem. 51 (1974) 389–406.

[12] FAO (Food and Agriculture Organization of the United Nations), 2007. FAOSTAT Statistical database Agriculture, Rome, Italy, data collected on January 2007.

[13] P. Zhang, R.L. Whistler, J.N. BeMiller, B.R. Hamaker, Banana starch: production, physicochemical properties, and digestibility – a review, Carbohydr. Polym. 59 (2005) 443–458.

[14] J.W. Donovan, Phase transitions of the starch–water system, Biopolymers 18 (1979) 263–275.

[15] P.J. Jenkins, R.E. Cameron, A.M. Donald, A universal feature in the structure of starch granules from different botanical sources, Starch/Stärke 45 (1993) 417–420.

[16] R.L. Whistler, J.R. Daniel, Molecular structure of starch, in: R.L. Whistler, J.N. Bemiller, E.F. Paschall (Eds.), Starch: Chemistry and Technology, second ed., Academic Press, San Diego, CA, 1984, pp. 153–182 (Chapter 6).

[17] D. French, Organization of starch granules, in: R.L. Whistler, J.N. Bemiller, E.F. Paschall (Eds.), Starch: Chemistry and Technology, second ed., Academic Press, San Diego, CA, 1984, pp. 183–247 (Chapter 7).

[18] J.M.V. Blanshard, Starch granule structure and function: a physicochemical approach, in: T. Galliard (Ed.), Starch: Properties and Potential, first ed., John Wiley & Sons, New York, 1987, pp. 14–54 (Chapter 2).

[19] T. Galliard, P. Bowler, Morphology and composition of starch, in: T. Galliard (Ed.), Starch: Properties and Potential, first ed., John Wiley & Sons, New York, 1987, pp. 55–78 (Chapter 3).
[20] M.J. Gidley, S.M. Bociek, Molecular organization in starches: a ^{13}C CP/MAS NMR study, J. Am. Chem. Soc. 107 (1985) 7040–7044.
[21] H.R. Tang, J. Godward, B. Hills, The distribution of water in native starch granules – a multinuclear NMR study, Carbohydr. Polym. 43 (2000) 375–387.
[22] D.J. Gallant, B. Bouchet, P.M. Baldwin, Microscopy of starch: evidence of a new level of granule organization, Carbohydr. Polym. 32 (1997) 177–191.
[23] K. Kainuma, Starch oligosaccharides: linear, branched, and cyclic, in: R.L. Whistler, J.N. Bemiller, E.F. Paschall (Eds.), Starch: Chemistry and Technology, second ed., Academic Press, San Diego, CA, 1984, pp. 125–152 (Chapter 5).
[24] N.W.H. Cheetham, L. Tao, Variation in crystalline type with amylose content in maize starch granules: an X-ray powder diffraction study, Carbohydr. Polym. 36 (1998) 277–284.
[25] A. Ayoub, T. Ohtani, S. Sugiyama, Atomic force microscopy investigation of disorder process in rice starch granule surface, Starch/Stärke 58 (2006) 475–479.
[26] P.M. Baldwin, J. Adler, M.C. Davies, C.D. Melia, High-resolution imaging of starch granule surface by atomic force microscopy, J. Cereal Sci. 27 (1998) 255–265.
[27] T. Ohtani, T. Yoshino, S. Hagiwara, T. Maekawa, High-resolution imaging of starch granule structure using atomic force microscopy, Starch/Stärke 52 (2000) 153–155.
[28] M.J. Ridout, A.P. Gunning, M.L. Parker, R.H. Wilson, V.J. Morris, Using atomic force microscopy to image the internal structure of starch granules, Carbohydr. Polym. 50 (2002) 123–132.
[29] M.J. Ridout, M.L. Parker, C.L. Hedley, T.Y. Bogracheva, V.J. Morris, Atomic force microscopy of pea starch: granule architecture of the rug3-a, rug4-b, rug5-a and lam-c mutants, Carbohydr. Polym. 65 (2006) 64–74.
[30] E.M. Snyder, Industrial microscopy of starches, in: R.L. Whistler, J.N. Bemiller, E.F. Paschall (Eds.), Starch: Chemistry and Technology, second ed., Academic Press, San Diego, CA, 1984, pp. 661–673 (Chapter 22).
[31] L.E. Fitt, E.M. Snyder, Photomicrographs of starches, in: R.L. Whistler, J.N. Bemiller, E.F. Paschall (Eds.), Starch: Chemistry and Technology, second ed., Academic Press, San Diego, CA, 1984, pp. 675–689 (Chapter 23).
[32] R.E. Cameron, A.M. Donald, A small-angle X-ray scattering study of the annealing and gelatinization of starch, Polymer 33 (1992) 2628–2635.
[33] H.C.H. Wu, A. Sarko, The double-helical molecular structure of crystalline B-amylose, Carbohydr. Res. 61 (1978) 7–25.
[34] H.C.H. Wu, A. Sarko, The double-helical molecular structure of crystalline A-amylose, Carbohydr. Res. 61 (1978) 27–40.
[35] P. Colonna, A. Buleon, C. Mercier, Physically modified starches, in: T. Galliard (Ed.), Starch: Properties and Potential, first ed., John Wiley & Sons, New York, 1987, pp. 79–115 (Chapter 4).
[36] A. Buléon, C. Gérard, C. Riekel, R. Vuong, H. Chanzy, Details of the crystalline ultrastructure of C-starch granules revealed by synchrotron microfocus mapping, Macromolecules 31 (1998) 6605–6610.
[37] R. Vermeylen, B. Goderis, H. Reynaers, J.A. Delcour, Amylopectin molecular structure reflected in macromolecular organization of granular starch, Biomacromolecules 5 (2004) 1775–1786.
[38] T.Y. Bogracheva, V.J. Morris, S.G. Ring, C.L. Hedley, The granular structure of C-type pea starch and its role in gelatinization, Biopolymers 45 (1998) 323–332.
[39] H.F. Zobel, Starch crystal transformation and their industrial importance, Starch/Stärke 40 (1988) 1–7.
[40] P. Cairns, T.Y. Bogracheva, S.G. Ring, C.L. Hedley, V.J. Morris, Determination of the polymorphic composition of smooth pea starch, Carbohydr. Polym. 32 (1997) 275–282.
[41] A.H. Young, Fractionation of starch, in: R.L. Whistler, J.N. Bemiller, E.F. Paschall (Eds.), Starch: Chemistry and Technology, second ed., Academic Press, San Diego, CA, 1984, pp. 249–283 (Chapter 8).
[42] N. Charoenkul, D. Uttapap, W. Pathipanawat, Y. Takeda, Molecular structure of starches form

cassava varieties having different cooked root textures, Starch/Stärke 58 (2006) 443–452.
[43] A. Buléon, F. Duprat, Single crystals of amylose with low degree of polymerization, Carbohydr. Polym. 4 (1984) 161–173.
[44] A. Becker, S.E. Hill, J.R. Mitchell, Relevance of amylose–lipid complexes to the behaviour of thermally processed starches, Starch/Stärke 53 (2001) 121–130.
[45] J. Karkalas, S. Ma, W.R. Morrison, R.A. Pethrick, Some factors determining the thermal properties of amylose inclusion complexes with fatty acids, Carbohydr. Res. 268 (1995) 233–247.
[46] K. Takeo, T. Kuge, Complexes of starchy materials with organic compounds .3. X-ray studies on amylose and cyclodextrin complexes, Agric. Biol. Chem. 33 (1969) 1174–1180.
[47] M.J. Gidley, S.M. Bociek, ^{13}C CP/MAS NMR studies of amylose inclusion complexes, cyclodextrins, and the amorphous phase of starch granules: relationships between glycosidic linkage conformation and solid-state ^{13}C chemical shifts, J. Am. Chem. Soc. 110 (1988) 3820–3829.
[48] M.C. Godet, B. Bouchet, P. Colonna, D.J. Gallant, A. Buléon, Crystalline amylose–fatty acid complexes: morphology and crystal thickness, J. Food Sci. (1996) 1196–2101.
[49] S. Simmons, E.L. Thomas, Structural characteristics of biodegradable thermoplastic starch/poly(ethylene vinyl alcohol) blends, J. Appl. Polym. Sci. 58 (1995) 2259–2285.
[50] C. Bastioli, Properties and applications of Mater-Bi starch-based materials, Polym. Degrad. Stab. 59 (1998) 263–272.
[51] N.K. Matheson, R.A. Caldwell, a(1–4) Glucan chain disposition in models of a(1–4)(1–6) glucans: comparison with structural data for mammalian glycogen and waxy amylopectin, Carbohydr. Polym. 40 (1999) 191–209.
[52] D.B. Thompson, On the non-random nature of amylopectin branching, Carbohydr. Polym. 43 (2000) 223–239.
[53] Z. Nikuni, Studies on starch granules, Starch/Stärke 30 (1978) 105–111.
[54] H. Liu, J. Lelievre, W.A. Ayoung-Chee, A study of starch gelatinization using differential scanning calorimetry, X-ray, and birefringence measurements, Carbohydr. Res. 210 (1991) 79–87.
[55] P.J. Jenkins, A.M. Donald, Gelatinization of starch: a combined SAXS/WAXS/DSC and SANS study, Carbohydr. Res. 308 (1998) 133–147.
[56] C.G. Biliaderis, C.M. Page, T.J. Maurice, B.O. Juliano, Thermal characterization of rice starches: a polymeric approach to phase transitions of granular starch, J. Agric. Food Chem. 34 (1986) 6–14.
[57] J.L. Kokini, L.S. Lai, L.L. Chedid, Effects of starch structure on starch rheological properties, Food Technol. 46 (1992) 124–139.
[58] G.H. Robertson, D.W.S. Wong, C.C. Lee, K.W. Wagschal, M.R. Smith, W.J. Orts, Native or raw starch digestion: a key step in energy efficient biorefining of grain, J. Agric. and Food Chem. 54 (2006) 353–365.
[59] A.A. Koutinas, R. Wang, C. Webb, Evaluation of wheat as generic feedstock for chemical production, Ind. Crops and Prod. 20 (2004) 75–88.
[60] J.F. Kennedy, J.M.S. Cabral, I. Sá-Correia, C.A. White, Starch biomass: a chemical feedstock for enzyme and fermentation processes, in: T. Galliard (Ed.), Starch: Properties and Potential, first ed., John Wiley & Sons, New York, 1987, pp. 115–148 (Chapter 5).
[61] F.H. Otey, W.M. Doane, Chemicals from starch, in: R.L. Whistler, J.N. Bemiller, E.F. Paschall (Eds.), Starch: Chemistry and Technology, second ed., Academic Press, San Diego, CA, 1984, pp. 389–416 (Chapter 11).
[62] D. Garlotta, A literature review of poly(lactic acid), J. Polym. Environ. 9 (2001) 63–84.
[63] G.J.L. Griffin, Gelatinized starch products, in: G.J.L. Griffin (Ed.), Chemistry and Technology of Biodegradable Polymers, first ed., Blackie Academic & Professional, London, 1984, pp. 135–150 (Chapter 7).
[64] V.F. Pfeifer, V.E. Sohns, H.F. Conway, E.B. Lancaster, S. Dabic, E.L. Griffin, 2-Stage process for dialdehyde starch using electrolytic regeneration of periodic acid, Ind. Eng. Chem. 52 (1960) 201–206.
[65] R.P. Westhoff, F.H. Otey, C.L. Mehltretter, C.R. Russell, Starch-filled polyvinyl chloride plastics – preparation and evaluation, Ind. Eng. Chem. Prod. Res. Dev. 13 (1974) 123–125.

[66] G.J.L. Griffin, Particulate starch based products, in: G.J.L. Griffin (Ed.), Chemistry and Technology of Biodegradable Polymers, first ed., Blackie Academic & Professional, London, 1984, pp. 18–47 (Chapter 3).
[67] R. Arnaud, P. Dabin, J. Lemaire, S. Al-Malaika, S. Chohan, M. Coke, G. Scott, A. Fauve, A. Maaroufi, Photooxidation and biodegradation of commercial photodegradable polyethylenes, Polym. Degrad. Stab. 46 (1994) 211–224.
[68] F.H. Otey, R.P. Westhoff, Biodegradable film compositions prepared from starch and copolymers of ethylene and acrylic acid, US Patent # US 4133784, January 9, 1979.
[69] F.H. Otey, R.P. Westhoff, Biodegradable starch-based blown films, US Patent # US 4337181, June 29, 1982.
[70] F.H. Otey, R.P. Westhoff, W.M. Doane, Title: starch-based blown films, Ind. Eng. Chem. Prod. Res. Dev. 19 (1980) 592–595.
[71] F.H. Otey, R.P. Westhoff, C.R. Russell, Biodegradable films from starch and ethylene–acrylic acid copolymer, Ind. Eng. Chem. Prod. Res. Dev. 16 (1977) 305–308.
[72] F.H. Otey, R.P. Westhoff, W.M. Doane, Starch-based blown films .2, Ind. Eng. Chem. Res. 26 (1987) 1659–1663.
[73] M.H. Gomez, J.M. Aguilera, A physicochemical model for extrusion of corn starch, J. Food. Sci. 49 (1984) 40–49.
[74] H. Röper, H. Koch, The role of starch in biodegradable thermoplastic materials, Starch/Stärke 42 (1990) 123–130.
[75] R.L. Shogren, G.F. Fanta, W.M. Doane, Development of starch based plastics – a reexamination of selected polymer systems in historical perspective, Starch/Stärke 45 (1993) 276–280.
[76] L. Gustav, J. Rehm, R.F. Stepto, R. Thoma, J-P. Sachetto, D.J. Lentz, J. Silbiger, Polymer composition containing destructurized starch, US Patent # US 5095054, 1992.
[77] C. Bastioli, R. Lombi, G. Deltredici, I. Guanella, De-structuring starch for use in biodegradable plastics articles – By heating non-dried, non-water-added starch with plasticizer and enzyme in an extruder, Eur. Pat. # EP400531–A1, 1991.
[78] N.L. Lacourse, P.A. Altieri, Biodegradable shaped products and method of preparation, US Patent # US 5035930, 1991.
[79] J. Lörcks, Properties and applications of compostable starch-based plastic materials, Polym. Degrad. Stab. 59 (1998) 245–249.
[80] W. Wiedmann, E. Strobel, Compounding of thermoplastic starch with twin-screw extruders, Starch/Stärke 43 (1991) 138–145.
[81] J.J.G. van Soest, D. de Wit, F.G. Vliegenthart, Mechanical properties of thermoplastic waxy maize starch, J. Appl. Polm. Sci. 61 (1996) 1927–1937.
[82] L. Avérous, Biodegradable multiphase systems based on plasticized starch: a review, J. Macromol. Sci. Part-C C44 (2004) 231–274.
[83] P.D. Orford, R. Parker, S.G. Ring, A.C. Smith, Effect of water as a diluent on the glass transition behavior of malto-oligosaccharides, amylase and amylopectin, Int. J. Biol. Macromol. 11 (1989) 91–96.
[84] R.L. Shogren, C.L. Swanson, A.R. Thompson, Extrudates of cornstarch with urea and glycols: structure/mechanical property relations, Starch/Stärke 44 (1992) 335–338.
[85] M.T. Kalichevsky, J.M.V. Blanshard, The effect of fructose and water on the glass transition of amylopectin, Carbohydr. Polym. 20 (1993) 107–113.
[86] K. Poutanen, P. Forsell, Modification of starch properties with plasticizers, Trends Polym. Sci. 4 (1996) 128–132.
[87] D. Lourdin, H. Bizot, P. Colonna, "Anti-plasticization" in starch–glycerol films? J. Appl. Polym. Sci. 64 (1997) 1047–1053.
[88] D. Lourdin, L. Coignard, H. Bizot, P. Colonna, Polymer 38 (1997) 5401–5406.
[89] S.H.D. Hulleman, F.H.P. Janssen, H. Feil, The role of water during plasticization of native starches, Polymer 39 (1998) 2043–2048.
[90] A.P. Mathew, A. Dufresne, Plasticized waxy maize starch: effect of polyols and relative humidity on material properties, Biomacromolecules 3 (2002) 1101–1108.
[91] X. Ma, J. Yu, Formamide as the plasticizer for thermoplastic starch, J. Appl. Polym. Sci. 93 (2004) 1769–1773.
[92] M. Huang, J. Yu, X. Ma, Ethanolamine as a novel plasticizer for thermoplastic starch, Polym. Degrad. Stab. 90 (2005) 501–507.

[93] E.D. Teixeira, A.L. Da Roz, A.J.F. de Carvalho, A.A.S. Curvelo, Preparation and characterisation of thermoplastic starches from cassava starch cassava root and cassava bagasse, Macromol. Symp. 229 (2005) 266–275.
[94] A.L. Da Roz, A.J.F. Carvalho, A. Gandini, A.A.S. Curvelo, The effect of plasticizers on thermoplastic starch compositions obtained by melt processing, Carbohydr. Polym. 63 (2006) 417–424.
[95] I. Tan, C.C. Wee, P.A. Sopade, P.J. Halley, Investigation of the starch gelatinization phenomena in water–glycerol system: application of modulate temperature differential scanning calorimetry, Carbohydr. Polym. 58 (2004) 191–204.
[96] P.A. Perry, A.M. Donald, The role of plasticization in starch granule assembly, Biomacromolecules 1 (2000) 424–432.
[97] R.I. Derby, B.S. Miller, B.H. Miller, H.B. Trimbo, Visual observation of wheat starch gelatinization in limited water systems, Cereal Chem. 52 (1975) 702–713.
[98] C. Mercier, R. Charbonniere, J. Grebaut, et al., Formation of amylose–lipid complexes by twin-screw extrusion cooking of manioc starch, Cereal Chem. 57 (1980) 4–9.
[99] J.J.G. van Soest, P. Essers, Influence of amylose–amylopectin ratio on the properties of extrude starch plastics shets, Pure Appl. Chem. A34 (1997) 1665–1689.
[100] S.H.D. Hulleman, M.G. Kalisvaart, F.H.P. Janssen, H. Feil, J.F.G. Vliegenthart, Origins of B-type crystallinity in glycerol-plasticized, compression moulded potato starches, Carbohydr. Polym. 39 (1999) 351–360.
[101] C. Mercier, P. Feillet, Modification of carbohydrate components by extrusion-cooking of cereal products, Cereal Chem. 52 (1975) 283–297.
[102] B.T. Lawton, G.A. Henderso, E.J. Derlatka, The effect of extruder variables on the gelatinization of corn starch, Can. J. Chem. Eng. 50 (1972) 168–172.
[103] J.L. Willett, M.M. Millard, B.K. Jasberg, Extrusion of waxy maize starch: melt rheology and molecular weight degradation of amylopectin, Polymer 38 (1997) 5983–5989.
[104] C. Mercier, Effect of extrusion-cooking on potato starch using a twin-screw French extruder, Starke 29 (1977) 48–52.
[105] J.J.G. van Soest, K. Benes, D. de Wit, J.F.G. Vliegenthart, The influence of starch molecular mass on the properties of extruded thermoplastic starch, Polymer 37 (1996) 3543–3552.
[106] C. Bastioli, V. Bellotti, A. Rallis, Microstructure and melt flow behavior of a starch-based polymer, Rheol. Acta 33 (1994) 307–316.
[107] O. Myllymäki, T. Eerikainen, T. Suortti, P. Forssele, P. Linko, K. Poutanen, Depolymerization of barley starch during extrusion in water glycerol mixtures, Food Sci. Technol. Lebensm. Wiss. Technol. 30 (1997) 351–358.
[108] A.J.F. Carvalho, M.D. Zambon, A.A.S. Curvelo, A. Gandini, Size exclusion chromatography characterization of thermoplastic starch composites 1. Influence of plasticizer and fibre content, Polym. Degrad. Stab. 79 (2003) 133–138.
[109] A.J.F. Carvalho, M.D. Zambon, A.A.S. Curvelo, A. Gandini, Thermoplastic starch modification during melting processing: hydrolysis catalyzed by carboxylic acids, Carbohydr. Polym. 62 (2005) 387–390.
[110] X.L. Wang, K.K. Yang, Y.Z. Wang, Properties of starch blends with biodegradable polymers, J. Macromol. Sci. Part C C43 (2003) 385–409.
[111] W. Amass, A. Amass, B. Tighe, A review of biodegradable polymers: uses, current development in the synthesis and characterization of biodegradable polyesters, blends of biodegradable polyesters and recent advances in biodegradable studies, Polym. Int. 47 (1998) 89–144.
[112] A.J.F. Carvalho, A.E. Job, N. Alves, A.A.S. Curvelo, A. Gandini, Thermoplastic starch/natural rubber blends, Carbohydr. Polym. 53 (2003) 95–99.
[113] H.W. Ritter, D.R. Bergner, K.W. Kempf, Starch-based materials and/or molded parts modified by synthetic polymer compounds and process for production the same. US Patent # US 5439953, 1995.
[114] N.M. Bortnick, R.K. Graham, E.E. LaFleur, W.J. Work, J.C. Wu, Melt-processed polymer blends, US Patent # US 5447669, 1995.

[115] L. Yu, K. Dean, L. Li, Polymer blends and composites from renewable resources, Prog. Polym. Sci. 31 (2006) 502−576.
[116] C. Bastioli, Global status of the production of biobased packaging materials, Starch/Stärke 53 (2001) 351−355.
[117] N. St-Pierre, B.D. Favis, B.A. Ramsay, J.A. Ramsay, H. Verhoogt, Processing and characterization of thermoplastic starch/polyethylene blends, Polymer 38 (1997) 647−655.
[118] F.J. Rodriguez-Gonzalez, B.A. Ramsay, B.D. Favis, High performance LDPE/thermoplastic starch blends: a sustainable alternative to pure polyethylene, Polymer 44 (2003) 1517−1526.
[119] N. Follain, C. Joly, P. Dole, B. Roge, M. Mathlouthi, Quaternary starch based blends: influence of a fourth component addition to the starch/water/glycerol system, Carbohydr. Polym. 63 (2006) 400−407.
[120] H. Wang, X.Z. Sun, P. Seib, Mechanical properties of poly(lactic acid) and wheat starch blends with methylenediphenyl diisocyanate, J. Appl. Polym. Sci. 84 (2002) 1257−1262.
[121] D. Demirgöz, C. Elvira, J.F. Mano, A.M. Cunha, E. Piskin, R.L. Reis, Chemical modification of starch based biodegradable polymeric blends: effects on water uptake, degradation behaviour and mechanical properties, Polym. Degrad. Stab. 70 (2000) 161−170.
[122] M. Avella, M.E. Errico, P. Laurienzo, E. Martuscelli, M. Raimo, R. Rimedio, Preparation and characterization of compatibilized polycaprolactone/starch composites, Polymer 41 (2000) 3875−3881.
[123] R. Mani, M. Bhattacharya, Properties of injection moulded starch/synthetic polymers blends-III. Effect of amylopectin to amylose ratio in starch, Eur. Polym. J. 34 (1998) 1467−1475.
[124] D. Bikiaris, J. Prinos, K. Koutsopoulos, N. Vouroutzis, E. Pavlidou, N. Frangis, C. Panayiotou, LDPE/plasticized starch blends containing PE−g−MA copolymer as compatibilizer, Polym. Degrad. Stab. 59 (1998) 287−291.
[125] A. Biswas, J.L. Willet, S.H. Gordon, V.L. Finkenstadt, H.N. Cheng, Complexation and blending of starch, poly(acrylic. acid), and poly(N-vinyl pyrrolidone), Carbohydr. Polym. 65 (2006) 397−403.
[126] I. Arvanitoyannis, I. Kolokuris, A. Nakayama, S. Aiba, Preparation and study of novel biodegradable blends based on gelatinized starch and 1,4-trans-polyisoprene (gutta percha) for food packaging or biomedical applications, Carbohydr. Polym. 34 (1997) 291−302.
[127] A. Rouilly, L. Rigal, R.G. Gilbert, Synthesis and properties of composites of starch and chemically modified natural rubber, Polymer 45 (2004) 7813−7820.
[128] H. Chanvrier, P. Colonna, G. Della Valle, D. Lourdin, Structure and mechanical behaviour of corn flour and starch−zein based materials in the glassy state, Carbohydr. Polym. 59 (2005) 109−119.
[129] E. Corradini, E.S. de Medeiros, A.J.F. Carvalho, A.A.S. Curvelo, L.H.C. Mattoso, Mechanical and morphological characterization of starch/zein blends plasticized with glycerol, J. Appl. Polym. Sci. 101 (2006) 4133−4139.
[130] A.J.F. Carvalho, A.A.S. Curvelo, J.A.M. Agnelli, Wood pulp reinforced thermoplastic starch composites, Int. J. Polym. Mater. 51 (2002) 647−660.
[131] A.A.S. Curvelo, A.J.F. Carvalho, J.A.M. Agnelli, Thermoplastic starch-cellulosic fibers composites: preliminary results, Carbohydr. Polym. 45 (2001) 183−188.
[132] A. Dufresne, M.R. Vignon, Improvement of starch film performance using cellulose microfibrils, Macromolecules 31 (1998) 2693−2696.
[133] A. Avérous, C. Frigant, L. Moro, Plasticized starch−cellulose interactions in polysaccharide composites, Polymer 42 (2001) 6565−6572.
[134] A.J.F. Carvalho, A.A.S. Curvelo, J.A.M. Agnelli, A first insight on composites of thermoplastic starch and kaolin, Carbohydr. Polym. 45 (2001) 189−194.
[135] M.N. Belgacem, A. Gandini, Partial or total oxypropylation of natural polymers and the use of the ensuing materials as composite or polyol macromonomers, in: Mohamed Naceur Belgacem, Alessandro Gandini (Eds.),

Monomers, Polymers and Composites from Renewable Resources, Elsevier, Oxford, UK, 2008, pp. 273−288 (Chapter 12).

[136] S. Kalambur, S. RizviS, An overview of starch-based blends from reactive extrusion, J. Plastic Film Sheeting 22 (2006) 39−58.

[137] J.M. Raquez, R. Narayan, P. Dubois, Macromol. Mater. Eng. 293 (2008) 447−470.

[138] H. Liu, F. Xie, L. Yu, L. Chen, L. Li, Thermal processing of starch-based polymers, Prog. Polym. Sci. 34 (2009) 1348−1368.

[139] G. Moad, Chemical modification of starch by reactive extrusion, Prog. Poym. Sci. 36 (2011) 218−237.

[140] A.J.F. Carvalho, Starch as source of polymeric materials, in: Susheel Kalia, Luc Avérous (Eds.), Biopolymers: biomedical and environmental applications, Scrivener Publishing/ Wiley, 2011 (Chapter. 4).

[141] R.L. Shogren, J.W. Lawton, K.F. Tiefenbacher, Baked starch foams: starch modifications and additives improve process parameters, structure and properties, Ind. Crops. Prod. 16 (2002) 69−79.

[142] L. Avérous, L. Moro, P. Dole, C. Fringant, Properties of thermoplastic blends: starch−polycaprolactone, Polymer 41 (2000) 4157−4167.

[143] G.J.L. Griffin, Starch polymer blends, Polym. Degrad. Stab. 45 (1994) 241−247.

[144] T. Seidenstucker, H.G. Fritz, Innovative biodegradable materials based upon starch and thermoplastic poly(ester-urethane) (TPU), Polym. Degrad. Stab. 59 (1998) 279−285.

[145] J.F. Zhang, X.Z. Sun, Mechanical properties of poly(lactic acid)/starch composites compatibilized by maleic anhydride, Biomacromolecules 5 (2004) 1446−1451.

[146] O. Martin, L. Avérous, Poly(lactic acid): plasticization and properties of biodegradable multiphase systems, Polymer 42 (2001) 6209−6219.

[147] M.A. Kotnis, G.S. Obrien, J.L. Willett, Processing and mechanical-properties of biodegradable poly(hydroxybutyrate-co-valerate)−starch compositions, J. Environ. Polym. Degrad. 3 (1995) 97−105.

[148] L. Avérous, C. Fringant, Association between plasticized starch and polyesters: processing and performances of injected biodegradable systems, Polym. Eng. Sci. 41 (2001) 727−734.

[149] E. Schwach, L. Avérous, Starch-based biodegradable blends: morphology and interface properties, Polym. Int. 53 (2004) 2115−2124.

[150] S. Baumberger, C. Lapierre, B. Monties, G. Della Valle, Use of kraft lignin as filler for starch films, Polym. Degrad. Stab. 59 (1998) 273−277.

[151] L.C. Morais, A.A.S. Curvelo, M.D. Zambon, Thermoplastic starch−lignosulfonate blends. 1. Factorial planning as a tool for elucidating new data from high performance size-exclusion chromatography and mechanical tests, Carbohydr. Polym. 62 (2005) 104−112.

[152] M. Wollerdorfer, H. Bader, Influence of natural fibres on the mechanical properties of biodegradable polymers, Ind. Crops. Prod. 8 (1998) 105−112.

[153] R. Mittenzwey, T. Seidenstücker, H. Fritz, R. Sübmuth, Prüfung der umweltverträglichkeit neu entwickelter polymer-werkstoffe auf der basis nachwachsender rohstoffe durch ein einfaches testsystem, Starch/Stärke 10 (1998) 438−443.

[154] A.S. Herrmann, J. Nickel, U. Riedel, Construction materials based upon biologically renewable resources − from components to finished parts, Polym. Degrad. Stab. 59 (1998) 251−261.

[155] D. Puglia, A. Tomassucci, J.M. Kenny, Processing, properties and stability of biodegradable composites based on Mater-Bi-(R) and cellulose fibres, Polym. Adv. Technol. 14 (2003) 749−756.

[156] U. Funke, W. Bergthaller, M.G. Lindhauer, Processing and characterization of biodegradable products based on starch, Polym. Degrad. Stab. 59 (1998) 293−296.

[157] X.F. Ma, J.G. Yu, J.F. Kennedy, Studies on the properties of natural fibers-reinforced thermoplastic starch composites, Carbohydr. Polym. 62 (2005) 19−24.

[158] M.N. Anglès, A. Dufresne, Plasticized starch/ tunicin whiskers nanocomposites materials. 2. Mechanical behavior, Macromolecules 34 (2001) 2921−2931.

[159] H. Angellier, S. Molina-Boisseau, P. Dole, A. Dufresne, Thermoplastic starch−waxy

maize starch nanocrystals nanocomposites, Biomacromolecules 7 (2006) 531−539.

[160] R.L. Reis, A.M. Cunha, P.S. Allan, M.J. Bevis, Structure development and control of injection-molded hydroxylapatite-reinforced starch/EVOH composites, Adv. Polym. Technol. 16 (1997) 263−277.

[161] H.M. Park, W.K. Lee, C.Y. Park, W.J. Cho, C.S. Ha, Environmentally friendly polymer hybrids, J. Mater. Sci. 38 (2003) 909−915.

[162] K. Bagdi, P. Müller, B. Pukánszky, Thermoplastic starch/layered silicate composites: structure, interactions, properties, Compos. Interfaces 13 (2006) 1−17.

8 Cellulose-Based Composites and Nanocomposites

Alain Dufresne

OUTLINE

8.1 Introduction

Composite materials (or composites for short) are engineered materials made from two or more constituents with significantly different mechanical properties, which remain separate and distinct within the finished structure. There are two categories of constituent materials: matrix and reinforcement. At least one portion of each type is required. The matrix surrounds and supports the reinforcements by maintaining their relative positions. The reinforcements impart special physical (mechanical and electrical) properties to enhance the matrix properties. A synergism produces material properties unavailable from naturally occurring materials. Due to the wide variety of matrixes and reinforcements available, the design potential for composite is huge.

The so-called natural composites like bones and woods are constructed by biological processes. The most primitive man-made composite materials comprised straw and mud in the form of bricks for building constructions. The most advanced examples are used on spacecrafts in highly demanding environments. The most visible applications are paved roadways in the form of either steel and Portland cement concrete or asphalt concrete. Engineered composite materials must be formed to shape. This involves strategically placing the reinforcements while manipulating the matrix properties. A variety of methods are used according to the end-item design requirements. These fabrication methods are commonly named molding or casting processes, as appropriate, and both have numerous variations. The principle factors impacting the methodology are the nature of the chosen matrix and reinforcement materials. Another important factor is the gross quantity of material to be produced. Large quantities can be used to justify high capital expenditures for rapid and automated manufacturing technology. Small production quantities are accommodated with lower capital expenditures, but higher labor costs at a correspondingly lower rate.

Many commercially produced composites use a polymer matrix often called a resin. The reinforcements are often fibers but also commonly ground minerals. Strong fibers such as fiberglass, quartz, Kevlar, or carbon fibers give the composite its tensile strength, while the matrix binds the fibers together, transferring the load from broken fibers to unbroken ones and between fibers that are not oriented along the tension lines. Also, unless the matrix chosen is especially flexible, it prevents the fibers from buckling in compression. In terms of stress, any fiber will provide resistance to tension, the

Ebnesajjad: Handbook of Biopolymers and Biodegradable Plastics. http://dx.doi.org/10.1016/B978-1-4557-2834-3.00008-2
© 2008 Elsevier Ltd. All rights reserved. Reproduced from a chapter in: Belgacem, *Monomers, Polymers and Composites from Renewable Resources* (2008).

matrix will resist shear, and all materials present will resist compression. Composite materials can be divided into two main categories, normally referred to as short-fiber reinforced materials and continuous-fiber reinforced materials, the latter often constituting a layered or laminated structure. Shocks, impacts, loadings, or repeated cyclic stresses can cause the laminate to separate at the interface between two layers, a condition known as delamination. Individual fibers can separate from the matrix through a mechanism called "fiber pull-out".

Nanoscience and nanotechnology correspond to science and technology that extend from about 100 nm down to atomic orders of magnitude around 0.2 nm, and to the physical phenomena and material properties observed when operating in this size range. Conceptually, nanocomposites refer to multiphase materials where at least one of the constituent phases has one dimension less than 100 nm. This field has attracted the attention, scrutiny, and imagination of both scientific and industrial communities in recent years, and has opened a large window of opportunity to overcome the limitations of traditional micrometer-scale composites. Research in this area is literally exploding, because of the intellectual appeal of building blocks on the nanometer scale and because the technical innovations permit the designing and creation of new materials and structures with unprecedented flexibility, improvements in physical properties, and significant industrial impact.

The large interest in the nanoscale range originates from outstanding properties. Enhanced properties can often be reached for low-filler volume fraction, without a detrimental effect on other properties such as impact resistance or plastic deformation capability. Though industrial exploitation of nanocomposites is still in its infancy, the rate of technology implementation is increasing.

8.2 Natural Fibers

Agro-based resources, also referred to as lignocellulosics, are resources that contain cellulose, hemicelluloses, and lignin. When considering lignocellulosics as possible engineering materials, there are several basic concepts that must be taken into account [1]. First, lignocellulosics are hygroscopic resources that were designed to perform, in nature, in a wet environment. Second, nature is programmed to recycle lignocellulosics in a timely way through biological, thermal, aqueous, photochemical, chemical, and mechanical degradation. In simple terms, nature builds a lignocellulosic structure from carbon dioxide and water and has all the tools to recycle it back to the starting chemicals.

There is a wide variety of agro-based or natural fibers to consider for utilization. They can be subdivided based on their origin, namely vegetable, animal, or mineral. Cellulose as a material is used by the natural world in the construction of plants and trees, and by man to make shipping sails, ropes, and clothes, to give but a few examples. It is also the major constituent of paper and further processing can be performed to make cellophane and rayon. Depending on the part of the plant from which they are taken, cellulose fibers can be classified as follows:

- *Grasses and reeds*: The fibers come from the stem of the plants, such as bamboo or sugar cane.
- *Leaf or hard fibers*: These fibers are most commonly used as reinforcing agents in polymers. They can be extracted for instance from sisal, henequen, abaca, or pineapple.
- *Bast or stem fibers*: These fibers come from the inner bark of the stem of the plants. Common examples are jute, flax, hemp, kenaf, and ramie.
- *Straw fibers*: Examples include rice, wheat, and corn straws.
- *Seed and fruit hairs*: These fibers come from seed-hairs and flosses and are primarily represented by cotton and coconut.
- *Wood fibers*: Examples include maple, yellow poplar, and spruce.

In any commercial development, there must be a long-term guaranteed supply of resources. The growing of natural fibers is spread across all the five continents. Quality and yield depend on the kind of plant, the grown variety, the soil, and the climatic conditions. Tanzania and Brazil are the two largest producers of sisal. Henequen is produced in Mexico, kenaf is grown commercially in the United States, and flax is a commodity crop grown in the European Union, as well as in many diverse agricultural systems and environments throughout the world, including Canada, Argentina, India, and Russia. Hemp originated in Central Asia, from where it spread to China, the Philippines, and many other countries. Ramie fibers, mostly available and used in

China, Japan, and Malaysia, are the longest and one of the strongest textile fibers. The largest producers of jute are India, China, and Bangladesh, and coir is produced in tropical countries. The price for natural fiber varies depending on the economy of the countries where such fibers are produced. Table 8.1 shows the inventory of some of the larger sources of agricultural bast fiber that could be utilized for fiber–polymer composites. However, only a small part of these fibers has been used for industrial applications up to now, which shows that the potential of the existing bast plants has not yet been exhausted and that huge natural resources are still available.

The traditional source of natural fibers has been wood and for many countries, this will continue to be the case. Other large sources come from recycling agro-fiber-based products such as paper, waste wood, and point-source agricultural residues. Recycling paper products back into paper requires a wet processing and the removal of inks, salts, and adhesives. Recycling the same products into composites can be done by using dry processing whereby all components are incorporated into the composite, eliminating the need of costly separation procedures. The major point source fibers are rice hulls from a rice processing plant, sunflower seed hulls from an oil processing unit, and bagasse from a sugar mill.

In order to maintain the high quality of the fibers, their separation from the original plant is best done by retting, scrapping, or pulping. Basically, there are two working principles to separate the bast fibers from the wood [3]. The conventional method uses breaking rollers, which alternatingly bend, buckle, and soften the stalks. This method requires an intensive retting before processing, which is induced by microorganisms that dissolve the lignin and pectins of the stalk. Modern technologies use swing hammer mills in most cases. The fiber decortication is provoked by the impact stress of the hammers directly on the surface of the stalks. This working principle ensures a complete separation of the fibers from the wood, even when processing freshly harvested, nonretted plants. The effective mechanical separation of the fibers and the wood inside the decorticator simplifies the subsequent fiber cleaning. The availability of large quantities of lignocellulosic fibers with well-defined mechanical properties is a general prerequisite for the successful subsequent use of these materials.

Table 8.1 Inventory of Major Potential Bast Fiber Sources for the Year 2000/2001 [2]

Fiber Source	World Production (metric tons)
Jute	2,900,000
Linseed	942,240
Kenaf	470,000
Flax	464,650
Sisal	380,000
Ramie	170,000
Hemp	157,800
Abaca	98,000

Plant fibers are bundles of elongated thick-walled dead plant cells. They are like microscopic tubes, that is cell walls surrounding the center lumen that contributes to their water uptake behavior (Fig. 8.1). Natural fibers display a multilevel organization and consist of several cells formed out of semicrystalline oriented cellulose microfibrils connected to a complete layer by lignin, hemicelluloses, and in some cases pectins. Climatic conditions, age, and digestion process influence not only the structure of the fibers but also their chemical composition. Table 8.2 reports the mean chemical composition of some natural fibers. With the exception of cotton, the components of natural fibers are cellulose, hemicelluloses, and lignin, which determine their physical properties. Several of such cellulose–lignin/hemicellulose layers in one primary and three secondary cell walls stick together to form a multilayer composite. Such microfibrils have typically a diameter of about 2–20 nm, are made up of 30–100 cellulose macromolecules in extended chain

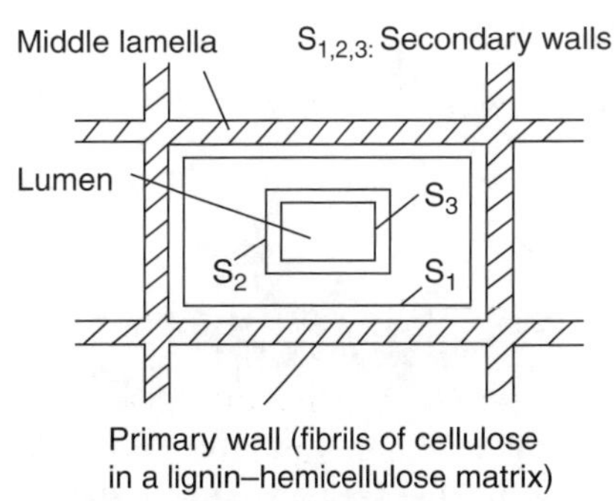

Figure 8.1 Schematic structure of a natural fiber cell. *Reproduced with permission from Reference [4].*

Table 8.2 Mean Chemical Composition of Some Natural Fibers [7]

	Cotton	Jute	Flax	Ramie	Sisal
Cellulose	82.7	64.4	64.1	68.6	65.8
Hemicelluloses	5.7	12.0	16.7	13.1	12.0
Pectin	5.7	0.2	1.8	1.9	0.8
Lignin	–	11.8	2.0	0.6	9.9
Water soluble	1.0	1.1	3.9	5.5	1.2
Wax	0.6	0.5	1.5	0.3	0.3
Water	10.0	10.0	10.0	10.0	10.0

conformation, and provide the mechanical strength to the fiber.

The cell walls differ among themselves in their composition and orientation of the cellulose microfibrils. In most plant fibers, these microfibrils are oriented at an angle to the normal axis called the microfibrillar angle (Fig. 8.2). The characteristic value for this structural parameter varies from one plant fiber to another.

The outer cell wall is porous and contains almost all of the noncellulose compounds, except proteins, inorganic salts, and coloring matters and it is this outer cell wall that creates poor absorbency, poor wettability, and other undesirable textile properties. In most applications, fiber bundles or strands are used rather than individual fibers. Within each bundle, the fiber cells overlap and are bonded together by pectins that give strength to the bundle as a whole. However, the strength of the bundle structure is significantly lower than that of the individual fiber cell and thus the potential of the individual fibers is not fully exploited.

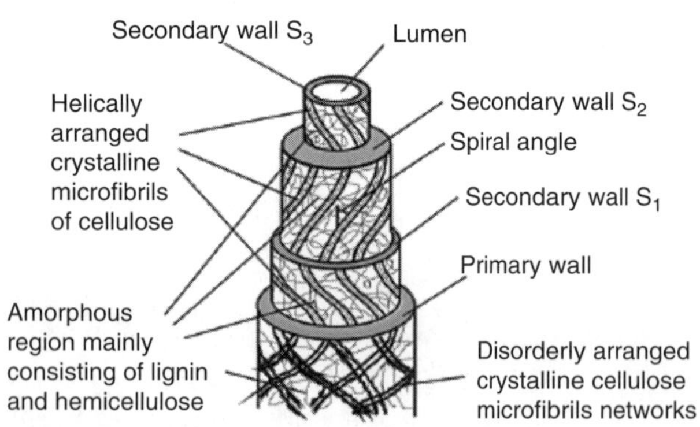

Figure 8.2 Schematic structure of an elementary plant fiber (cell). The secondary cell wall, S_2, makes up about 80% of the total thickness. *Reproduced with permission from Reference [5].*

The properties of natural fibers are strongly influenced by many factors, particularly chemical composition, internal fiber structure, microfibrillar angle, cell dimensions, and defects, which differ between different parts of a plant, as well as between different plants. A weak correlation between strength and cellulose content and microfibril or spiral angle is found for different plant fibers. In general, the fiber strength increases with increasing cellulose content and decreasing spiral angle with respect to the fiber axis. This means that the most efficient cellulose fibers are those that have a high-cellulose content, coupled with a low-microfibril angle. Other factors that may affect the fiber properties are the maturity, the separating process, the microscopic and molecular defects, such as pits and knots, the type of soil, and the weather conditions under which the vegetable was grown. Differences in fiber structure due to the environmental conditions during growth result in a broad range of characteristics. The mechanical properties of plant fibers are in general much lower when compared to those of the most widely used reinforcing glass fibers (Table 8.3). However, because of their low density, the specific properties which are property-to-density ratio dependent, namely strength and stiffness, are comparable to those of glass fibers. Thus, natural fibers are in general suitable to reinforce polymer matrices, both thermoplastics and thermosets.

8.3 Composites

The use of additives in polymers is likely to grow with the introduction of improved compounding technologies and new coupling agents that permit the use of high filler/reinforcement contents. Fillings up to

Table 8.3 Physical Properties of Some Natural Fibers (Properties of Some Synthetic Organic and Inorganic Fibers are Added for Comparison)

Fiber	Density (g cm^{23})	Young's Modulus (GPa)	Tensile Strength (MPa)	Elongation (%)	Microfibrillar Angle (°)
Cotton	1.5	5.5–27.6	300–1500	3–8	–
Jute	1.3–1.5	13–26.5	393–800	1.2–1.8	8
Flax	1.5	27.6	345–1500	2.7–3.2	5–10
Hemp	1.5	70	690	1.6	2–6.2
Ramie	1.55	61.4–128	400–938	1.2–3.8	7.5
Sisal	1.45	9.4–22	468–700	2–7	10–22
Coir	1.15–1.46	4–6	130–220	15–40	30–49
Viscose	–	11	593	11.4	–
Soft wood kraft	1.5	40	1000	–	–
E-glass	2.5	70	2000–3500	2.5	–
S-glass	2.5	86	4570	2.8	–
Aramide	1.4	63–67	3000–3150	3.3–3.7	–
Carbon	1.4	230–240	4000	1.4–1.8	–

75 pph could be common in the future and this would have a tremendous impact in lowering the use of petroleum-based polymers [6]. Since the price of plastics has risen sharply over the past few years, adding a natural powder or fiber to them provides a cost reduction to industry (and in some instances increases performance as well). To the agro-based industry, this represents an increased value for the agro-based component. Ideally, of course, a bio-based renewable polymer reinforced with agro-based fibers would be the most environment-friendly material.

Over the past decade there has been a growing interest in the use of lignocellulosic fibers as reinforcing elements in polymeric matrices [7,8]. A number of researchers have been involved in investigating the exploitation of cellulosic fibers as load-bearing constituents in composite materials. Prior work on lignocellulosic fibers in thermoplastics has concentrated on wood-based flour or fibers [9–13]. The majority of these studies has been on polyolefins, mainly polypropylene (PP). Compared to inorganic fillers, the main advantages of lignocellulosics are listed below:

1. Low density: Their density, around 1.5 g cm^{23}, is much lower than that of glass fibers, around 2.5 g cm^{23}.
2. Low cost and low-energy consumption.
3. High specific strength.
4. Renewability and biodegradability.
5. Abundant availability in a variety of forms throughout the world.
6. Flexibility: Unlike brittle fibers, lignocellulosic fibers will not be fractured during processing.
7. Nonabrasive nature to processing equipment, which allows high filling levels, resulting in significant cost savings and high stiffness properties.
8. Nontoxicity.
9. Ease of handling.
10. Reactive surface, facilitating its chemical modification.
11. Organic nature, resulting in the possibility to generate energy without residue after incineration at the end of their life-cycle.
12. Economic development opportunity for nonfood farm products in rural areas.

Despite these attractive aspects, lignocellulosic fibers are used only to a limited extent in industrial practice due to difficulties associated with surface interactions. It is important to keep these limitations in perspective when developing end-use applications. The primary drawback of agro-based fibers is associated with their inherent polar nature and the nonpolar characteristics of most thermoplastics, which causes difficulties in compounding the filler and the matrix and, therefore, in achieving acceptable dispersion levels, which in turn generates inefficient composites. Another drawback of lignocellulosic fillers is their hydrophilic character, which favors moisture absorption with a consequent swelling of the fibers and the decrease in their mechanical properties. Moisture absorption and the corresponding dimensional changes can be largely prevented if the hydrophilic filler is thoroughly encapsulated by the hydrophobic polymer matrix and there is a good adhesion between both components. However, if the adhesion level between the filler and the matrix is not good enough, diffusion pathways for moisture can preexist or can be created under mechanical solicitation. The existence of such pathways is also related to the filler connection and therefore to its percolation threshold.

Yet another limitation associated with the use of lignocellulosic fillers is the fact that the processing temperature of composites must be restricted to just above 200 °C (although higher temperatures can be used for short periods of time), because of their susceptibility to degradation and/or the possibility of volatile emissions that could affect the composite properties. This limits the types of thermoplastics that can be used to polymers like polyethylene, PP, poly-vinyl chloride, and polystyrene, which constitute, however, about 70% of all industrial thermoplastics. Nevertheless, technical thermoplastics like polyamides, polyesters, and polycarbonates, which are usually processed at temperatures higher than 250 °C, cannot be envisaged as matrices for these types of composite.

8.4 Composite Processing

Drying the fibers is an essential prerequisite that must be applied before processing, because water on the fiber surface acts as a separating agent at the fiber–matrix interface. In addition, because of the water evaporation during processing at temperatures higher than 100 °C, voids appear in the matrix. Both phenomena obviously lead to a decrease in the mechanical properties of the ensuing composites. Fiber drying can be done under different conditions, which results in different degrees of their residual moisture.

Extrusion and injection-molding are the economically most attractive processing methods of thermoplastic-based composites. The extrusion press processing (express-processing) has been developed for the production of flax fiber-reinforced PP at the research center of Daimler Benz [7]. In this process, flax fiber nonwovens and PP melt films are alternatively deposited and molded. A production process for PP semiproducts reinforced with lignocellulosic fibers in the form of mats has been developed by BASF AG [7]. Fiber mats are produced by stitching together layers of fibers that have previously been crushed.

Beginning with bakelite in the early 1900s, engineers and scientists have continued to work to improve the various attributes of thermosets through the addition of natural fibers. Unsaturated polyester, epoxy, and vinylester resins are commonly used for preparing such composites. Fabrication techniques suitable for manufacturing natural fiber reinforced thermoset composites include the hand lay-up technique for unidirectional fibers/mats/fabric, and filament winding and pultrusion for continuous fibers. Resin transfer molding (RTM) and prepregs can also be used. Semiproducts, such as sheet molding compounds (SMC) and bulk molding compounds (BMC), can be obtained with short and chopped fibers.

8.5 Composite Properties

The major factors that govern the properties of short-fiber thermoplastic composites are fiber volume fraction, fiber dispersion, fiber aspect ratio and length distribution, fiber orientation, and fiber–matrix adhesion. Each of these parameters is briefly discussed further.

8.5.1 Fiber Volume Fraction

Like other composite systems, the properties of short-fiber composites are strongly determined by the fiber concentration. The variation of the composite properties with fiber content can be predicted using the rule of mixtures, which involves the extrapolation of both matrix and fiber properties to a fiber volume fraction of 0 and 1. The following criteria must be taken into account:

1. The composite fracture has to be fiber-controlled.
2. The modulus of elasticity of the fiber should be greater than that of the matrix.
3. The strain to failure of the matrix must be greater than that of the fiber.

In the case of unidirectional (or longitudinal) fiber-reinforced composites, the stress is transferred from the matrix to the fiber by shear. When stressed in tension, both the fiber and the matrix elongate equally according to the principle of combined action [14]. Hence, the mechanical properties of the composite can be evaluated on the basis of the properties of the individual constituents. For a given elongation of the composite: both constituents, fiber and matrix, may be in elastic deformation; the fiber may be in elastic deformation, whereas the matrix may be in plastic deformation; or, both the fiber and the matrix may be in plastic deformation (Fig. 8.3).

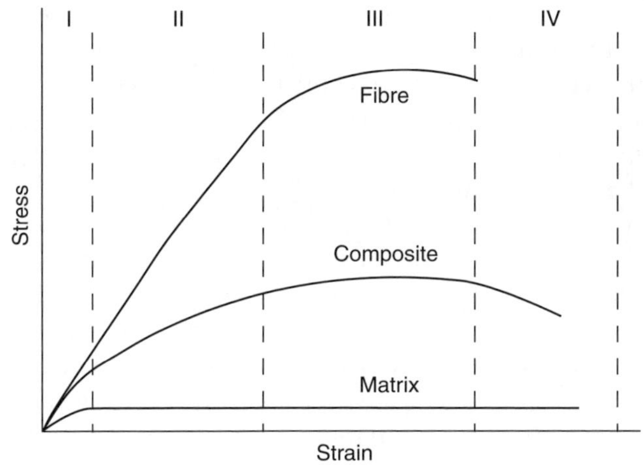

Figure 8.3 Illustration of four stages of deformation of fibers, matrix, and composite. Stage I: elastic deformation of both fibers and matrix; stage II: elastic deformation of fibers and plastic deformation of matrix; stage III: plastic deformation of both fibers and matrix; stage IV: failure of both fibers and matrix. *Reproduced with permission from Reference [14].*

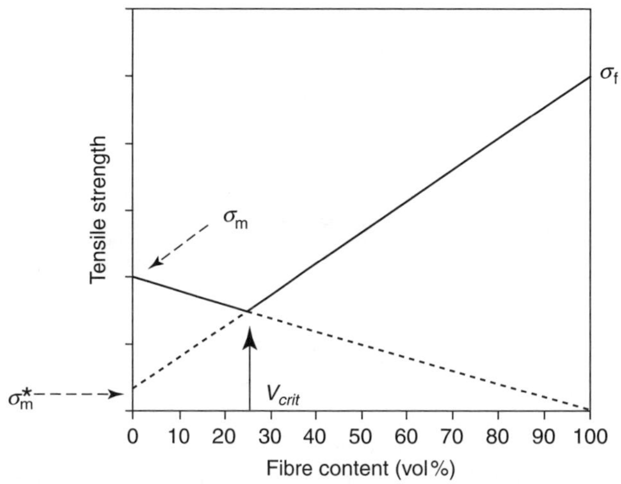

Figure 8.4 Model for the prediction of the ultimate tensile strength of unidirectional fiber-reinforced composites for which the fracture is fiber-controlled.

At low fiber volume fraction, a decrease in the tensile strength is usually observed (Fig. 8.4). This is ascribed to the dilution of the matrix and the introduction of flaws at the fiber ends where a high stress concentration occurs, causing the bond between fiber and matrix to break. At high volume fraction, the stress is more evenly distributed and a reinforcement effect is observed. For all values of strain, the stress value in the composite is given by a simple mixing rule balanced by the volume fraction of each constituent, namely:

$$p_c^9 = p_f^9 V_f + p_m^9 V_m \tag{8.1}$$

where p represents the stress value of each component at a particular strain value and V the volume fraction of each component of the composite. The subscripts c, f, and m correspond to the composite, the fiber, and the matrix, respectively.

The fiber volume fraction for which the strength ceases to decrease and begins to increase is called the critical fiber volume fraction, V_{crit}. Below this value, the behavior of the composite is only governed by the matrix:

$$\begin{array}{lll} \text{For } V_f \,, & V_{crit} & p_c 5 p_m V_m \\ \text{For } V_f \,. & V_{crit} & p_c 5 p_f V_f 1 p_m^* V_m \end{array} \tag{8.2}$$

where p_m is the ultimate tensile strength of the matrix, p_f is that of the ultimate tensile strength of the fiber, and σ_m^* is the stress on the matrix at a strain value where p_f is reached. V_{crit} is an important parameter because it corresponds to the volume fraction of the fibers above which they begin to strengthen, rather than weaken the matrix. It can be calculated from the following equation:

$$V_{\text{crit}} = \frac{p_{\text{m}}\text{-}p^*_{\text{m}}}{p_{\text{f}}\text{-}p^*_{\text{m}}} \qquad (8.3)$$

For a given matrix, the critical fiber volume fraction decreases with the increasing strength of the fibers. This means that for fibers which are much stiffer than the matrix, V_{crit} is very low.

The modulus of elasticity is also an important factor. Within strain limits for which both the fiber and the matrix are in elastic deformation, the modulus of the composite can be calculated using the rule of mixture:

$$E_{\text{c}} = E_{\text{f}}V_{\text{f}} + E_{\text{m}}V_{\text{m}} \qquad (8.4)$$

where (Fig. 8.4) E_{c}, E_{f}, and E_{m} are the modulus of elasticity of composite, fiber, and matrix, respectively. When the fiber is in elastic and the matrix is in plastic deformation, the equation becomes:

$$E_{\text{c}} = E_{\text{f}}V_{\text{f}} + \left(\frac{p^*_{\text{m}}}{\text{e}}\right)V_{\text{m}} \qquad (8.5)$$

The ratio $\sigma^*_{\text{m}}/\varepsilon$ is the slope of the stress–strain curve of the matrix at a given strain beyond the proportional limit of the matrix.

The length of some individual natural fibers can reach up to 4 m, and when bundled with other fibers, this maximum length will be even higher. However, lignocellulosic materials are mainly used as discontinuous short fibers and are ground into fine particles with relatively low aspect ratios. These fillers generally increase the stiffness of the composites, but the strength is generally lower than that of the pristine matrix [10]. For instance, residual softwood sawdust was used as a reinforcing material in PP [15] and it was found that the tensile strength decreased regularly from 35 MPa for the unfilled matrix down to 10 MPa for the 60 wt% filled system. Conversely, the addition of henequen fibers to a low-density polyethylene matrix increased the tensile strength by 50% (from 9.2 MPa to 14 MPa) at a fiber loading of 30 vol% [16]. At the same time, the modulus increased from 275 MPa to 860 MPa and the strain at break decreased from 42% to 5%. The increase in stiffness results from the fact that lignocellulosic fillers or fibers have a higher Young's modulus, as compared to commodity thermoplastics, thereby contributing to the higher stiffness of the composites. However, an anchoring effect of the lignocellulosic filler acting as nucleating agents for the polymeric chains has been reported [17], resulting in an increase in the degree of crystallinity of the matrix. This effect seems to be strongly influenced by the lignin content and the surface aspect of the fiber [18,19]. This transcrystallization phenomenon at the fiber–matrix interface participates in the reinforcing effect of the filler.

In order to use models to estimate composite properties, it is necessary to know the properties of the fibers, which vary widely depending on the source, age, separating techniques, moisture content, speed of testing, history of the fiber, etc. The properties of the individual fibers are therefore very difficult to measure. Moreover, in a natural fiber–polymer composite, the lignocellulosic phase is present in a wide range of diameters and lengths, some in the form of short filaments and others in forms that seem closer to the individual fiber.

Continuous regenerated cellulose fibers are extensively used as reinforcements in composites such as tyres. However, very few studies are available on their use as reinforcement for polymer composites. Because of the strong hydrogen bonds that occur between cellulose chains, cellulose does not melt or dissolve in common solvents. Thus, it is difficult to convert the short fibers from wood pulp into continuous filaments. Regenerated cellulose fibers are produced on a commercial scale under the generic name "Lyocell" by a spinning process from a cellulose *N*-methylmorpholine-*N*-oxide/water solution. The mechanical properties of these fibers were found to depend on the draw ratio [20,21]. The low mechanical properties reported for unidirectional composites composed of Lyocell fibers embedded in a poly(3-hydroxybutyrate-*co*-3-hydroxyvalerate) matrix were ascribed to weak interfacial adhesion due to both the smooth topography of the fibers and the hydrophobic properties of the matrix [22].

8.5.2 Fiber Dispersion

The primary requirement for obtaining good performances from short-fiber composites is a good dispersion level in the host polymer matrix, which is obtained if the fibers are separated from each other and each fiber is surrounded by the matrix. Clumping and agglomeration must therefore be avoided. Insufficient fiber dispersion results in an inhomogeneous mixture composed of matrix-rich and fiber-rich domains. Mixing the polar and hydrophilic fibers with a nonpolar and hydrophobic matrix can result in dispersion difficulties.

There are two major factors affecting the extent of fiber dispersion: fiber–fiber interaction, such as hydrogen bonding between the fibers, and fiber length, because of the possibility of entanglements. As mentioned above, one of the specificity of cellulose fibers as reinforcement is their poor dispersion characteristics in many thermoplastic melts, due to their hydrophilic nature. Several methods have been suggested and described in the literature to overcome this problem and improve the dispersion. The following are among them:

1. Fiber surface modification: The surface energy is closely related to the hydrophilicity of the lignocellulosic fibers.
2. Use of dispersing agents, such as stearic acid or a mineral oil: The dispersion of lignocellulosic fibers can be improved by pretreatment with lubricants or thermoplastic polymers. An addition of 1–3% stearic acid is sufficient to achieve a maximum reduction in size and number of aggregates in PP and polyethylene [7]. The use of stearic acid in high density polyethylene (HDPE)/wood fibers was reported to improve the fiber dispersion and the wetting between the fiber and the matrix [9].
3. Fiber pretreatments, such as acetylation, or use of a coupling agent.
4. Increased shear force and mixing time: The best processing method involves twin-screw extruder.

Some physical methods have also been suggested to improve the dispersion of short fibers within the matrix. Treatments such as stretching, calendering, thermal treatment, and the production of hybrid yarns do not change the chemical composition of the fiber, but modify their structural and surface properties and thus influence their mechanical bonding with polymers.

8.5.3 Fiber Aspect Ratio and Length Distribution

The efficiency of a composite also depends on the amount of stress transferred from the matrix to the fibers. This can be maximized by improving the interaction and adhesion between both phases and also by maximizing the length of the fibers retained in the final composite. However, long fibers sometimes increase the amount of clumping resulting in poor dispersion of the reinforcing phase within the host matrix. The ultimate fiber length present in the composite depends on the type of compounding and molding equipment used and the processing conditions. Several factors contribute to the fiber attrition, such as the shearing forces generated in the compounding equipment, the residence time, the temperature, and the viscosity of the compound. Using a polystyrene matrix, it was shown that the extent of breakage was most severe and rapid for glass fibers, less extensive for Kevlar fibers, and the least for cellulose fibers [23]. The effect of twin-screw blending of wood fibers and polyethylene was also reported [12] and it was shown that the level of fiber attrition depended on the configuration and the processing temperature.

The fiber aspect ratio, which is its length to diameter ratio, is a critical parameter in a composite. A relationship has been proposed by Cox to relate the critical fiber aspect ratio, l_c/d, to the interfacial shear stress, q_y, namely:

$$\frac{l_c}{d} = \frac{p_{fu}}{2q_y} \tag{8.6}$$

where, p_{fu} is the fiber ultimate strength in tension. At controlled fiber ultimate strength in tension, this equation shows an inverse relationship between the critical aspect ratio and the interfacial shear stress, where the former decreases as the latter increases, because of efficient transfer. This means that, for each short-fiber composite system, there is a critical fiber aspect ratio that corresponds to its minimum value for which the maximum allowable stress can be achieved for a given load. This parameter is determined by the fiber properties, the matrix properties, and the quality of the fiber–matrix interface.

The condition for maximum reinforcement, that is the condition ensuring maximum stress transfer to the fibers, before the composite fails, is to have a length higher than the critical length l_c (Fig. 8.5). If

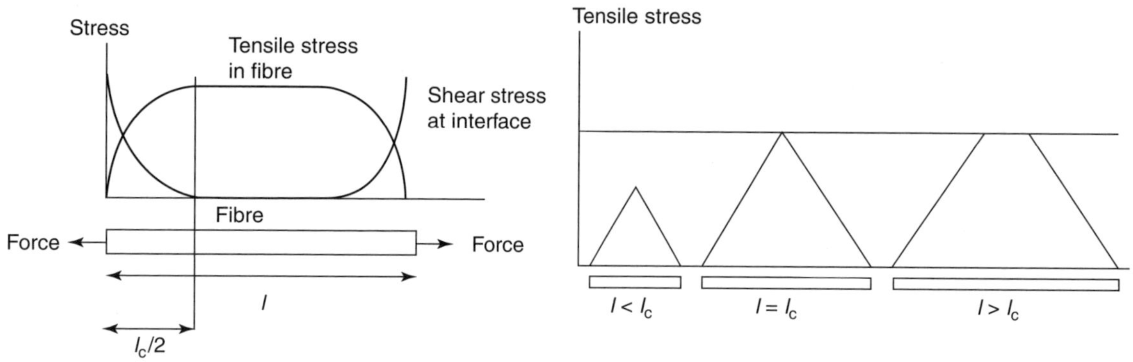

Figure 8.5 Variation of tensile stress in fiber and shear stress at interface occurring along the fiber length. If the fiber aspect ratio is lower than its critical value, l_c, the fibers are not loaded to their maximum stress value.

the fiber aspect ratio is lower than its critical value, the fibers are not loaded to their maximum stress value. A specificity of cellulose fibers is their flexibility compared to glass fibers which allows a desirable fiber aspect ratio to be maintained after processing – around 100 or 200 for high performance short-fiber composites.

8.5.4 Fiber Orientation

Fiber orientation is another important parameter that influences the mechanical behavior of short-fiber composites. This is because the fibers in such composites are rarely oriented in a single direction, which is necessary to obtain the maximum reinforcement effects. During the processing of short-fiber composites, a continuous and progressive orientation of individual fibers occurs (Fig. 8.6). This change is related to the geometrical properties of the fibers, the viscoelastic properties of the matrix, and the change in shape produced by the processing. In these operations, the polymer melt undergoes both elongational and shear flow.

8.5.5 Fiber–Matrix Adhesion

Fiber to matrix adhesion plays a very important role in the reinforcement of composites with short fibers. During loading, loads are not applied directly to the fibers but to the matrix. It is necessary to have an effective load transfer from the matrix to the fibers for the ensuing composites to have good mechanical properties. This requires good interaction as well as adhesion between the fibers and the matrix, that is strong and efficient fiber–matrix interface.

As already pointed out, strongly hydrophilic cellulose fibers are inherently incompatible with hydrophobic polymers. When two materials are incompatible, it is often possible to introduce a third material having intermediate properties capable of reducing their interfacial energy. One way of applying this concept to the present context is to impregnate the fibers with a polymer compatible with the matrix and, in general, this is achieved using low-viscosity polymer solutions or dispersion. For a number of interesting polymers, however, the lack of solvents limits the use of this method. The following example illustrates the less frequent approach, based on the use of a surface modifier that bears a structure very close to that of the matrix, but which has been appropriately modified so that its macromolecules can react at the fibers' surface. Figure 8.7 shows SEM micrographs from the fractured surface of PP reinforced with cellulose fibers

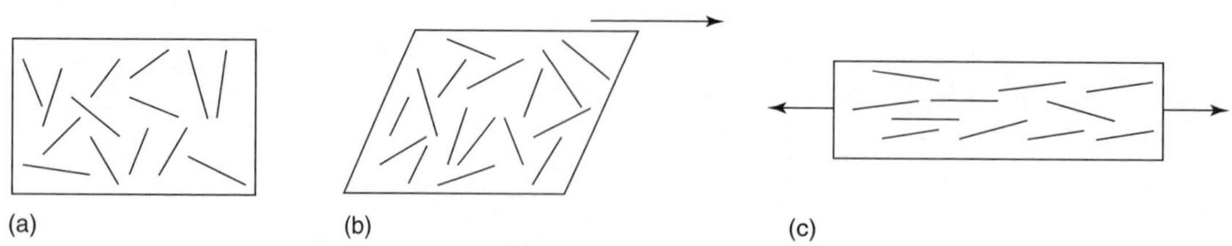

Figure 8.6 Orientation of individual fibers during processing: (a) initial random distribution, (b) rotation during shear flow, and (c) alignment during elongational flow.

[15]. With the untreated matrix (Fig. 8.7a), a poor interfacial adhesion is clearly observed because of the absence of any physical contact between the fiber and the matrix. The micrograph in Fig. 8.7b corresponds to fibers in contact with a maleic anhydride polypropylene (MAPP) graft copolymer (PP chains with pendant succinic acid moieties), which shows a good wetting, with absence of holes around the fibers. The mechanism of the reaction of MAPP with cellulose fibers can be divided into two steps (Fig. 8.8), the first being the activation of the copolymer by heat before the fiber treatment and the second, the esterification of cellulose. The fact of generating covalent bonds across the interface improved the adhesion between the matrix and the fibers, and both the Young's modulus and the tensile strength were found to be higher than those obtained with the untreated fibers [15].

It has also been found that moisture absorbance of the natural fiber–polymer composite can be prevented if the fiber–matrix adhesion is optimized [15,24]. Indeed, whereas composites based on standard PP and cellulosic fibers displayed high water content at the interface, due to the presence of microcavities, the encapsulation of the fibers with MAPP decreased the water sensitivity of the composites in terms of both the water uptake and its diffusion coefficient [25], as shown in Fig. 8.9.

8.6 Nanocomposites

As previously mentioned, natural fibers present a multilevel organization and consist of several cells formed out of semicrystalline oriented cellulose microfibrils. Each microfibril can be considered as a string of cellulose crystallites, linked along the chain axis by amorphous domains (Fig. 8.10) and having a modulus close to the theoretical limit for cellulose. They are biosynthesized by enzymes and deposited in a continuous fashion. Nanoscale dimensions and impressive mechanical properties make polysaccharide nanocrystals, particularly when occurring as high aspect ratio rod-like nanoparticles, ideal candidates to improve the mechanical properties of the host material. These properties are profitably exploited by Mother Nature.

The promise behind cellulose-derived nanocomposites lies in the fact that the axial Young's modulus of the basic cellulose crystalline nanocrystal, derived from theoretical chemistry, is potentially higher than that of steel and similar to that of Kevlar. It was first experimentally studied in 1962 from the crystal deformation of cellulose I, using highly oriented fibers of bleached ramie [26]. A value of 137 GPa was reported, which differed from the theoretical estimate of 167.5 GPa calculated by Tashiro and Kobayashi [27]. The latter value is thought to be higher because the calculations had been carried out for low temperature. Force deflection data from the compression of cubes of potato tissues were fed into a model containing two structural levels, the cell structure and the cell wall structure [28], giving a maximum modulus value of 130 GPa. Eichhorn and Young [29] observed a decrease of cellulose crystallites when their crystallinity decreased. Recently, Raman spectroscopy was used to measure the elastic modulus of native

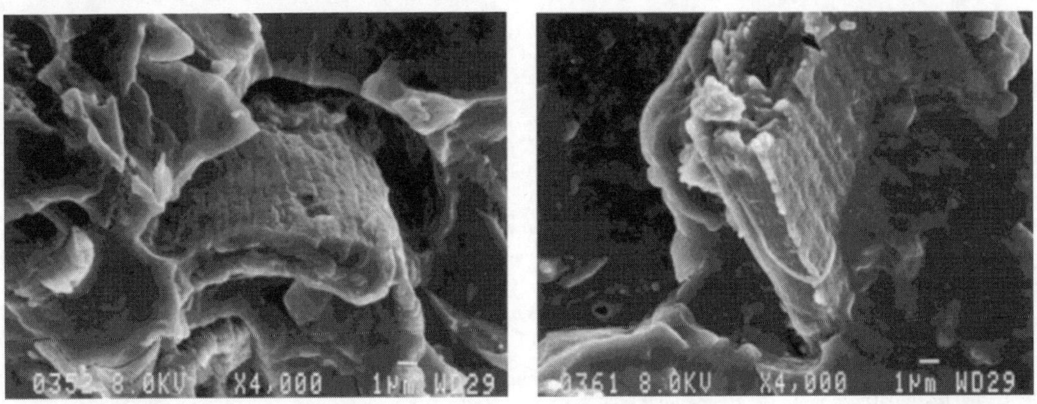

(a) (b)

Figure 8.7 Scanning electron micrographs of a freshly fractured surface of a PP film filled with 20 wt% of raw untreated softwood fibers (a), and MAPP-coated softwood fibers (b). *Reproduced with permission from Reference [24].*

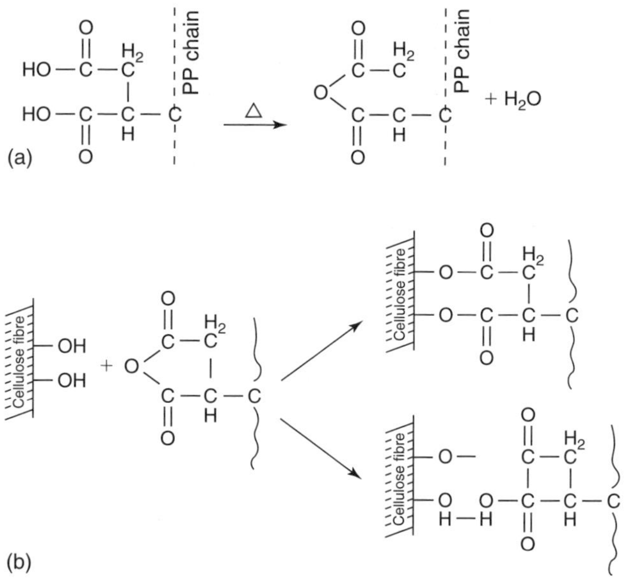

Figure 8.8 Reaction mechanism involved during the treatment of cellulose fibers with PP maleic anhydride copolymer (MAPP): (a) activation of MAPP (T 5/170 °C) before fiber treatment and (b) esterification of cellulose.

cellulose crystals [30] and a value around 143 GPa was reported. However, it is worth noting that these measurements were made on epoxy/tunicin whiskers composites.

Stable aqueous suspensions of polysaccharide nanocrystals can be prepared by the acid hydrolysis of vegetable biomass. Different descriptors of the resulting colloidal suspended particles are used, including whiskers, monocrystals, and nanocrystals. The designation "whiskers" is used to describe elongated rod-like nanoparticles. These crystallites have also often been referred in the literature as microfibrils, microcrystals, or microcrystallites, despite their nanoscale dimensions. Most of the studies reported in the literature refer to cellulose nanocrystals. A recent review described the properties and applications of cellulose whiskers in nanocomposites [31].

The procedure for the preparation of such colloidal aqueous suspensions is described in detail in the literature for cellulose and chitin [32,33]. The biomass is generally first submitted to a bleaching treatment with NaOH in order to purify cellulose or chitin by removing other constituents. The bleached material is then disintegrated in water, and the resulting suspension submitted to acid hydrolysis. The amorphous regions of cellulose or chitin act as structural defects and are responsible for the transverse cleavage of the microfibrils into short monocrystals by acid hydrolysis. Under controlled conditions, this transformation consists in the disruption of the amorphous regions surrounding and embedded within the cellulose or chitin microfibrils, while leaving the microcrystalline segments intact; because of the very large difference in the rate of hydrolysis between the amorphous and the crystalline domains, the latter is obviously much more resistant. The resulting suspension is subsequently diluted with water and washed by successive centrifugations. Dialysis against distilled water is then performed to remove the free acid in the dispersion. Complete dispersion of the whiskers is obtained by a sonication step. The dispersions are stored in a refrigerator after filtration to remove residual aggregates and addition of several drops of chloroform. This general procedure has to be adapted in terms of the acid hydrolysis conditions, such as time, temperature, and purity of materials depending on the nature of the substrate and the geometrical characteristics of the nanocrystals.

Figure 8.9 (a) Water uptake at equilibrium and (b) water diffusion coefficient of PP/*Opuntia ficus-indica* cladode flour composites conditioned at 98% RH versus filler loading: untreated filler (d) and MAPP-coated filler (s) (the solid line serves to guide the eye). *Reproduced with permission from Reference [25].*

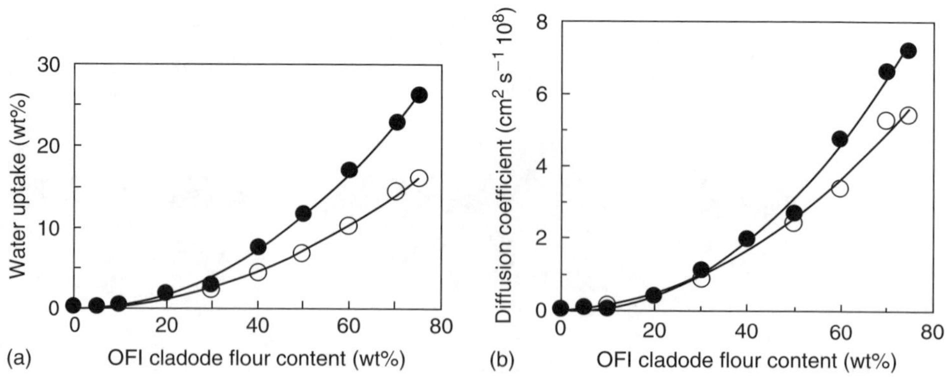

Figure 8.10 Schematic diagram showing the hierarchical structure of a semicrystalline cellulose fiber.

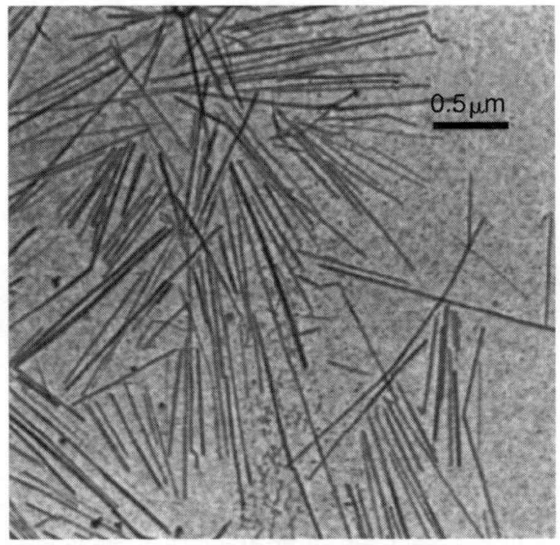

Figure 8.11 TEM of a dilute suspension of tunicin. *Reproduced with permission from Reference [34].*

The constitutive cellulose or chitin nanocrystals occur as elongated rod-like particles or whiskers. The length is generally of the order of few hundred nanometers and the width is of the order of a few nanometers. The aspect ratio of these whiskers is defined as the ratio of the length to the width. The high axial ratio of the rods is important for the determination of anisotropic phase formation and reinforcing properties. Figure 8.11 shows a transmission electron micrograph (TEM) obtained from a dilute suspension of tunicin whiskers, that is cellulose nanocrystals obtained from tunicate, a sea animal. Their average length and diameter are around 1 mm and 15 nm, respectively, and their aspect ratio was estimated to be around 67 [34].

Aqueous suspensions of starch nanocrystals can also be prepared by the acid hydrolysis of starch granules in aqueous medium using hydrochloric acid or sulfuric acid at 35 °C. Residues from the hydrolysis are called "lintners" and "nägeli" or amylodextrin. The degradation of native starch granules by acid hydrolysis depends on many parameters, which include the botanical origin of starch, namely crystalline type, granule morphology (shape, size, surface state), and the relative proportion of amylose and amylopectin. It also depends on the acid hydrolysis conditions, namely acid type, acid concentration, starch concentration, temperature, hydrolysis duration, and stirring. A response surface methodology was used by Angellier *et al.* [35] to investigate the effect of five chosen factors on the selective sulfuric acid hydrolysis of waxy maize starch granules in order to optimize the preparation of aqueous suspensions of starch nanocrystals. These predictors were temperature, acid concentration, starch concentration, hydrolysis duration, and stirring speed. The preparation of aqueous suspensions of starch nanocrystals with a yield of 15.7 wt%, was achieved after 5 days using 3.16 M H_2SO_4 at 40 °C, 100 rpm and with a starch concentration of 14.7 wt%.

Compared to cellulose or chitin, the morphology of constitutive nanocrystals obtained from starch is completely different. Figure 8.12 shows a TEM obtained from a dilute suspension of waxy maize starch nanocrystals. They consist of 5–7 nm thick platelet-like particles with a length in the range of 20–40 nm and a width in the range of 15–30 nm. The detailed investigation on the structure of these platelet-like nanoparticles was reported [36].

Because of the high stability of aqueous polysaccharide nanocrystals dispersions, water is the preferred processing medium. High level of dispersion of the filler within the host matrix in the resulting composite is expected when processing nanocomposites in an aqueous medium. Therefore, this restricts the choice of the matrix to hydrosoluble polymers. The use of aqueous dispersed polymers, i.e. latexes, is a first alternative, which makes it possible to employ hydrophobic polymers as matrices and ensure a good dispersion level of the filler, indispensable for homogenous composite processing. The possibility of dispersing polysaccharide

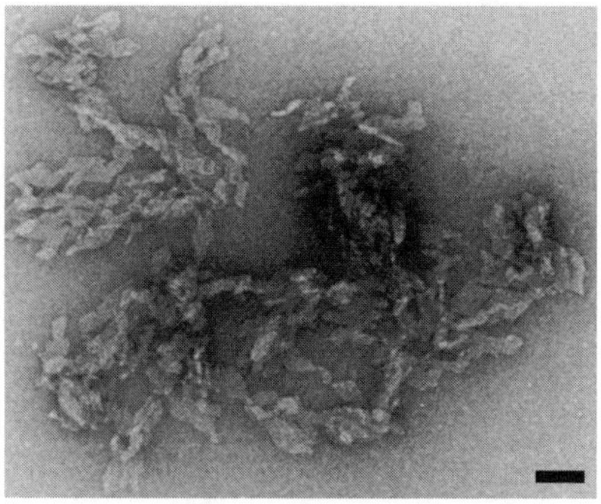

Figure 8.12 Transmission electron micrograph of a dilute suspension of hydrolyzed waxy maize starch (scale bar 50 nm). *Reproduced with permission from Reference [35].*

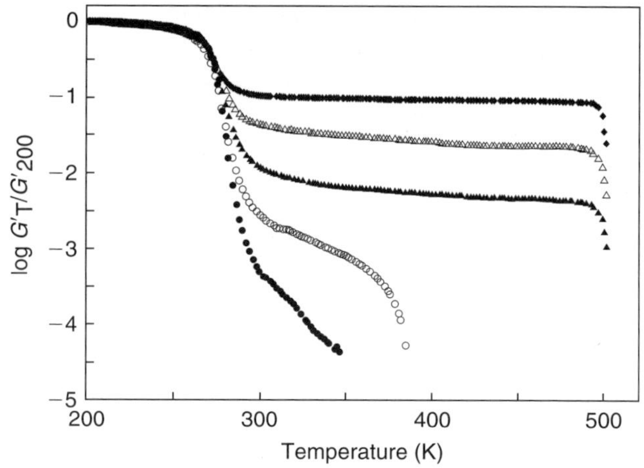

Figure 8.13 Logarithm of the normalized storage shear modulus (log G9T/G9200, where G9200 corresponds to the experimental value measured at 200 K) versus temperature at 1 Hz for tunicin whiskers reinforced poly(St-*co*-BuA) nanocomposite films, obtained by water evaporation and filled with 0 (d), 1 (s), 3 (m), 6 (n), and 14 wt% (r) of cellulose whiskers. *Reproduced with permission from Reference [31].*

nanocrystals in nonaqueous media is a second alternative, which opens other possibilities for nanocomposite processing.

The first demonstration of the reinforcing effect of cellulose whiskers in a poly(St-*co*-BuA) matrix was reported by Favier *et al.* [37]. The authors measured, using DMA in the shear mode, a spectacular improvement in the storage modulus after adding tunicin whiskers, even at a low content, into the host polymer. This increase was especially significant above the glass–rubber transition temperature of the thermoplastic matrix, because of its poor mechanical properties in this temperature range. Figure 8.13 shows the isochronal evolution of the logarithm of the relative storage shear modulus (log *G*9T/*G*9200, where *G*9200 corresponds to the experimental value measured at 200 K) at 1 Hz as a function of temperature for such composites prepared by water evaporation. In the rubbery state of the thermoplastic matrix, the modulus of the composite with a loading level as low as 6 wt% is more than two orders of magnitude higher than that of the unfilled matrix. Moreover, the introduction of 3 wt% or more cellulosic whiskers provides an outstanding thermal stability to the matrix modulus up to the temperature at which cellulose starts to degrade (500 K).

The macroscopic behavior of polysaccharide nanocrystals-based nanocomposites depends, as for any heterogeneous materials, on the specific behavior of each phase, the composition (volume fraction of each phase), the morphology (spatial arrangement of the phases), and the interfacial properties. The outstanding properties observed for these systems were ascribed to a mechanical percolation phenomenon [37]. A good agreement between experimental and predicted data was reported when using the series–parallel model of Takayanagi, modified to include a percolation approach. Therefore, the mechanical performance of these systems was not only the result of the high mechanical properties of the reinforcing nanoparticles. It was suspected that the stiffness of the material was due to infinite aggregates of cellulose whiskers. Above the percolation threshold, the cellulose nanoparticles can connect to form a three-dimensional continuous pathway through the nanocomposite film. For rod-like particles such as tunicin whiskers with an aspect ratio of 67, the percolation threshold is close to 1 vol%. The formation of this cellulose network was supposed to result from strong interactions, like hydrogen bonds, between whiskers. This phenomenon is similar to the high mechanical properties observed for a paper sheet, which result from the hydrogen-bonding forces that hold the percolating network of fibers. This mechanical percolation effect explains both the high reinforcing effect and the thermal stabilization of the composite modulus for evaporated composite films.

Any factor that affects the formation of the percolating whisker network, or interferes with it, changes the mechanical performances of the composite [38]. Three main parameters were reported to affect the mechanical properties of such materials, namely the morphology and dimensions of the nanoparticles, the processing method, and the microstructure of the matrix and matrix–filler interactions.

Apart from the mechanical performances, some other properties are interesting and can be improved by adding polysaccharide nanocrystals, for instance swelling properties. It was shown that the water uptake of tunicin whiskers/thermoplastic starch nanocomposites decreased as a function of the filler content [34]. For starch nanocrystals/natural rubber nanocomposites, it was shown that both the toluene uptake at equilibrium and its diffusion coefficient decreased when adding starch nanocrystals [39]. The evolution of the diffusion coefficient of toluene displayed a discontinuity around 10%, suggesting a possible percolation effect of the starch nanocrystals.

The barrier properties of starch nanocrystals/natural rubber nanocomposites were also investigated [39]. For these systems, the water vapor transmission rate, the diffusion coefficient of oxygen, the permeability coefficient of oxygen and its solubility, were measured. It was observed that the permeability to water vapor, as well as to oxygen, decreased when starch nanocrystals were added. These effects were ascribed to the platelet-like morphology of the nanocrystals.

8.7 Conclusions

There is a growing trend to use lignocellulosic fibers in applications for which synthetic fibers were traditionally employed, which is ascribed to their numerous well-known advantages. Present applications of natural fiber-filled composites are in the field of energy and impact absorption, such as car fenders and bicycle helmets. They also include markets that target cheaper, renewable and nonrecyclable, or biodegradable materials, such as packaging and structural elements. Other uses of natural fiber-based composites are deck surface boards, picnic tables, industrial flooring, etc. In cars, about 10–15 kg of these composites, typically made up of 50% natural fibers and 50% PP, along with other additives, are presently being used. Examples are door panels, roof headliners, seat backs, rear decks, and trunk liners.

Another interesting property of natural fibers is their hierarchical structure and the possibility to choose the scale linked to the application. Polysaccharide nanocrystals are building blocks biosynthesized to provide structural properties to living organisms. They can be isolated from cellulose-containing materials under strictly controlled conditions. Polysaccharide nanocrystals are inherently low-cost materials, available from a variety of natural sources in a wide range of aspect ratios. The corresponding polymer nanocomposites display outstanding mechanical properties and can be used to process high modulus thin films. Practical applications of such fillers and their transition into industrial technology require a favorable ratio between the expected performances of the composite material and its cost.

In conclusion, this area is moving fast toward novel outstanding composite materials based on renewable resources in the form of both traditional natural fibers and their nanomorphologies, but there are still significant scientific and technological challenges to be met.

References

[1] R.M. Rowell, Property enhanced natural fiber composite materials based on chemical modification (Proceedings of the Fourth International Conference on Frontiers of Polymers and Advanced Materials, Cairo, Egypt, 4–9 January 1997), in: Science and Technology of Polymers and Advanced Materials: Emerging Technologies and Business Opportunities, Plenum Press, New York N.Y, 1998, pp. 717–732.

[2] Anonym, Information Bulletin of the FAO European Cooperative Research Network on Flax and other Bast Plants. 2(16), Institute of Natural Fibres Coordination Centre, Poznan, Poland, 2001. Dec. 2001.

[3] F. Munder, C. Fürll, H. Hempel, Processing of bast fiber plants for industrial application, in: A.K. Mohanty, M. Misra, L.T. Drzal (Eds.), Natural Fibers, Biopolymers and Biocomposites, CRC Press Taylor & Francis Group, Boca Raton, 2005, pp. 109–140 (Chapter 3).

[4] A. Bismarck, I. Aranberri-Askargorta, J. Springer, T. Lampke, B. Wielage, A. Stamboulis, I. Shenderovich, H.-H. Limbach, Surface characterization of flax, hemp and cellulose fibers; surface properties and the water uptake behavior, Polym. Compos. 23 (5) (2002) 872–894.

[5] M.Z. Rong, M.Q. Zhang, Y. Liu, G.C. Yang, H.M. Zeng, The effect of fiber treatment on the

mechanical properties of unidirectional sisal-reinforced epoxy composites, Compos. Sci. Technol. 61 (10) (2001) 1437–1447.
[6] H.S. Katz, J.V. Milewski, Handbook of Fillers for Plastics, Van Nostrand Reinhold, New York, 1987, p. 512.
[7] A.K. Bledzki, J. Gassan, Composites reinforced with cellulose based fibres, Prog. Polym. Sci. 24 (1999) 221–274.
[8] S.J. Eichhorn, C.A. Baillie, N. Zafeiropoulos, L.Y. Mwaikambo, M.P. Ansell, A. Dufresne, K.M. Entwistle, P.J. Herrera-Franco, G.C. Escamilla, L. Groom, M. Hugues, C. Hill, T.G. Rials, P.M. Wild, Review: Current international research into cellulosic fibres and composites, J. Mater. Sci. 36 (2001) 2107–2131.
[9] R.T. Woodhams, G. Thomas, D.K. Rodges, Wood fibers as reinforcing fillers for polyolefins, Polym. Eng. Sci. 24 (15) (1984) 1166–1171.
[10] B.V. Kokta, R.G. Raj, C. Daneault, Use of wood flour as filler in polypropylene: Studies on mechanical properties, Polym. Plast. Technol. Eng. 28 (3) (1989) 247–259.
[11] P. Bataille, L. Ricard, S. Sapieha, Effect of cellulose in polypropylene composites, Polym. Compos. 10 (2) (1989) 103–108.
[12] K.L. Yam, B.K. Gogoi, C.C. Lai, S.E. Selke, Composites from compounding wood fibers with recycled high-density polyethylene, Polym. Eng. Sci. 30 (11) (1990) 693–699.
[13] A.R. Sanadi, R.A. Young, C. Clemons, R.M. Rowell, Recycled newspaper fibers as reinforcing fillers in thermoplastics: Part I. Analysis of tensile and impact properties in polypropylene, J. Reinf. Plast. Comp. 13 (1) (1994) 54–67.
[14] J.W. Weeton, D.M. Peters, K.L. Thomas, Engineers' Guide to Composite Materials, American Society for Metals, Metals Park, OH, 1987.
[15] M.N. Anglès, J. Salvadó, A. Dufresne, Mechanical behavior of steam exploded residual softwood filled polypropylene composites, J. Appl. Polym. Sci. 74 (1999) 1962–1977.
[16] P.J. Herrera-Franco, M.J. Aguilar-Vega, Effect of fiber treatment on mechanical properties of LDPE-henequen cellulosic fiber composites, J. Appl. Polym. Sci. 65 (1997) 197–207.
[17] A. Dufresne, D. Dupeyre, M. Paillet, Lignocellulosic flour reinforced poly(hydroxybutyrate-co-valerate) composites, J. Appl. Polym. Sci. 87 (8) (2003) 1302–1315.
[18] S. Luo, A.N. Netravali, Mechanical and thermal properties of environment-friendly 'green' composites made from pineapple leaf fibers and poly(hydroxybutyrate-co-valerate) resin, Polym. Compos. 20 (3) (1999) 367–378.
[19] V.E. Reinsch, S.S. Kelley, Crystallization of poly(hydroxybutyrate-co-hydroxyvalerate) in wood fiber-reinforced composites, J. Appl. Polym. Sci. 64 (9) (1997) 1785–1796.
[20] S.A. Mortimer, A.A. Peguy, R.C. Ball, Influence of the physical process parameters on the structure formation of Lyocell fibers, Cell. Chem. Technol. 30 (3–4) (1996) 251–266.
[21] S.A. Mortimer, A.A. Peguy, The formation of structure in the spinning and coagulation of Lyocell fibers, Cell. Chem. Technol. 30 (1–2) (1996) 117–132.
[22] C. Bourban, E. Karamuk, M.J. de Fondaumiere, K. Rufieux, J. Mayer, E. Wintermantel, Processing and characterization of a new biodegradable composite made of a PHB/V matrix and regenerated cellulosic fibers, J. Environ. Polym. Degr. 5 (3) (1997) 159–166.
[23] L. Czarnecki, J.L. White, Shear flow rheological properties, fiber damage and mastication characteristics of aramid-, glass-, and cellulose fiber-reinforced polystyrene melts, J. Appl. Polym. Sci. 25 (6) (1980) 1217–1244.
[24] H. Faria, N. Cordeiro, M.N. Belgacem, A. Dufresne, Dwarf cavendish as a source of natural fibers in polypropylene-based composites, Macromol. Mater. Eng. 291 (1) (2006) 16–26.
[25] M.E. Malainine, M. Mahrouz, A. Dufresne, Lignocellulosic flour from cladodes of *Opuntia ficus-indica* reinforced polypropylene composites, Macromol. Mater. Eng. 289 (10) (2004) 855–863.
[26] I. Sakurada, Y. Nukushina, T. Ito, Experimental determination of the elastic modulus of crystalline regions oriented polymers, J. Polym. Sci. 57 (165) (1962) 651–660.
[27] K. Tashiro, M. Kobayashi, Theoretical evaluation of three-dimensional elastic constants of native and regenerated celluloses: Role of hydrogen bonds, Polymer 32 (8) (1991) 1516–1526.
[28] D.G. Hepworth, D.M. Bruce, A method of calculating the mechanical properties of nanoscopic plant cell wall components from tissue properties, J. Mater. Sci. 35 (23) (2000) 5861–5865.

[29] S.J. Eichhorn, R.J. Young, The Young's modulus of a microcrystalline cellulose, Cellulose 8 (3) (2001) 197–207.
[30] A. Šturcová, G.R. Davies, S.J. Eichhorn, Elastic modulus and stress-transfer properties of tunicate cellulose whiskers, Biomacromolecules 6 (2) (2005) 1055–1061.
[31] M.A.S. Azizi Samir, F. Alloin, A. Dufresne, Review of recent research into cellulosic whiskers, their properties and their application in nanocomposite field, Biomacromolecules 6 (2) (2005) 612–626.
[32] L.E. Wise, M. Murphy, A.A. D'Addiecco, Chlorite holocellulose, its fractionation and bearing on summative wood analysis and on studies on hemicelluloses, Paper Trade J. 122 (1946) 35–43.
[33] R.H. Marchessault, F.F. Morehead, N.M. Walter, Liquid crystal systems from fibrillar polysaccharides, Nature 184 (1959) 632–633.
[34] M.N. Anglès, A. Dufresne, Plasticized starch/tunicin whiskers nanocomposites: 1. Structural analysis, Macromolecules 33 (22) (2000) 8344–8353.
[35] H. Angellier, L. Choisnard, S. Molina-Boisseau, P. Ozil, A. Dufresne, Optimization of the preparation of aqueous suspensions of waxy maize starch nanocrystals using a response surface methodology, Biomacromolecules 5 (4) (2004) 1545–1551.
[36] J.L. Putaux, S. Molina-Boisseau, T. Momaur, A. Dufresne, Platelet nanocrystals resulting from the disruption of waxy maize starch granules by acid hydrolysis, Biomacromolecules 4 (5) (2003) 1198–1202.
[37] V. Favier, G.R. Canova, J.Y. Cavaillé, H. Chanzy, A. Dufresne, C. Gauthier, Nanocomposites materials from latex and cellulose whiskers, Polym. Adv. Technol. 6 (1995) 351–355.
[38] A. Dufresne, Comparing the mechanical properties of high performances polymer nanocomposites from biological sources, J. Nanosci. Nanotechnol. 6 (2) (2006) 322–330.
[39] H. Angellier, S. Molina-Boisseau, L. Lebrun, A. Dufresne, Processing and structural properties of waxy maize starch nanocrystals reinforced natural rubber, Macromolecules 38 (9) (2005) 3783–3792.

9 Synthesis, Properties, Environmental and Biomedical Applications of Polylactic Acid

Luc Avérous

OUTLINE

9.1 Introduction

Tailoring new materials within a perspective of eco-design or sustainable development is a philosophy that is applied to more and more materials. It is the reason why material components such as biodegradable polymers can be considered as "interesting", environmentally safe, alternatives. Besides, ecological concerns have resulted in a resumed interest in renewable resources-based products.

Figure 9.1 shows an attempt to classify the biodegradable polymers into two groups and four different families. The main groups are (i) the agro-polymers (polysaccharides, proteins, etc.) and (ii) the bio-polyesters (biodegradable polyesters) such as polylactic acid (PLA), polyhydroxyalkanoate (PHA), and aromatic and aliphatic copolyesters [1]. Biodegradable polymers show a large range of properties and can now compete with nonbiodegradable thermoplastics in different fields (packaging, textile, biomedical, etc.). Among these biopolyesters, PLA is at present one of the most promising biodegradable polymers. PLA has been the subject of an abundant literature with several reviews and book chapters [2–9], mainly during the last decade. PLA can be

Ebnesajjad: Handbook of Biopolymers and Biodegradable Plastics. http://dx.doi.org/10.1016/B978-1-4557-2834-3.00009-4
© 2013 Elsevier Inc. All rights reserved. Adapted from a chapter in: Belgacem, *Monomers, Polymers and Composites from Renewable Resources* (2008).

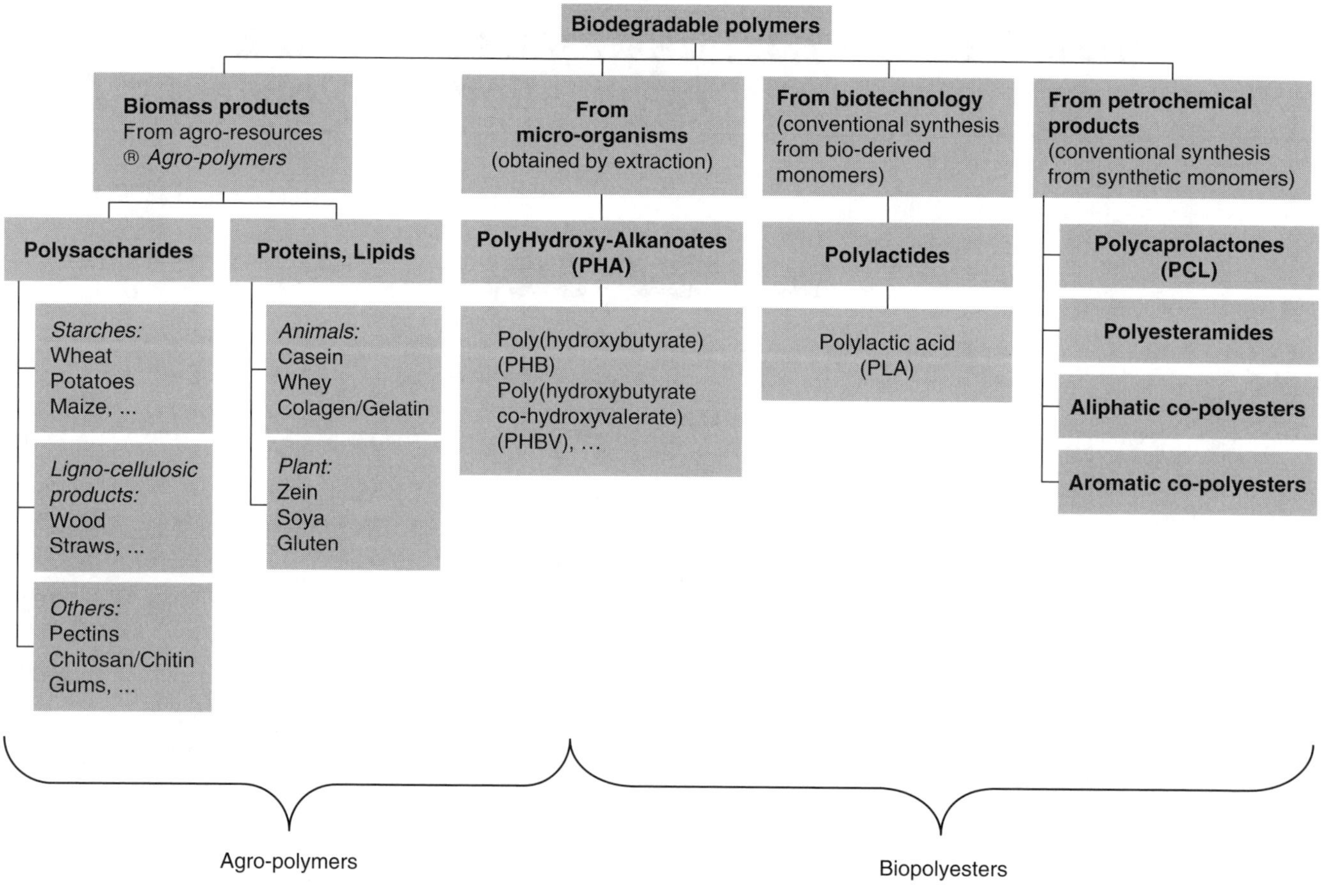

Figure 9.1 Classification of the biodegradable polymers. *Adapted from Reference [1]*

processed with a large number of techniques. PLA is commercially and largely available (large-scale production) in a wide range of grades. It has a reasonable price and some remarkable properties to fulfil different applications. For instance, the PLA production capacity of NatureWorks (Cargill—PTT, Thailand/US) in 2011 was around 150 kT per year, at about €2 per kg. Recently, some new productions have emerged such as Teijin (Japan), Zhejiang Hisun (China), Futerro (Galactic—Total, Belgium), and Purac (Netherland/Thailand). Some of them are mainly focused on the biomedical market like Boeringher Ingelheim (Germany) or Phusis (France), because the constraints of this market are very specific. However, according to different sources, PLA consumption in 2010 was only about 113,000 tons per year according to the European Bioplastics Association (http://en.european-bioplastics.org) and, only around one-third of lactic acid is used for PLA production. Thus, this polymer presents a high potential for development.

PLA belongs to the family of aliphatic polyesters commonly made from α-hydroxy acids, which also includes, e.g., polyglycolic acid (PGA). It is one of the few polymers in which the stereochemical structure can easily be modified by polymerizing a controlled mixture of L and D isomers (Fig. 9.2) to yield high molecular weight and amorphous or semicrystalline polymers. Properties can be modified through the variation of both the isomers (L/D ratio) and the homo and (D,L) copolymers relative contents. Besides, PLA can be tailored by formulation involving adding plasticizers, other polymers, fillers, etc.

PLA is considered both as biodegradable (e.g., adapted for short-term packaging) and as biocompatible in contact with living tissues (e.g., for biomedical applications such as implants, sutures, drug encapsulation, etc.). PLA can be degraded by abiotic degradation (i.e., simple hydrolysis of the ester bond without requiring the presence of enzymes to catalyze it). During the biodegradation process, and only in a second step, the enzymes degrade the

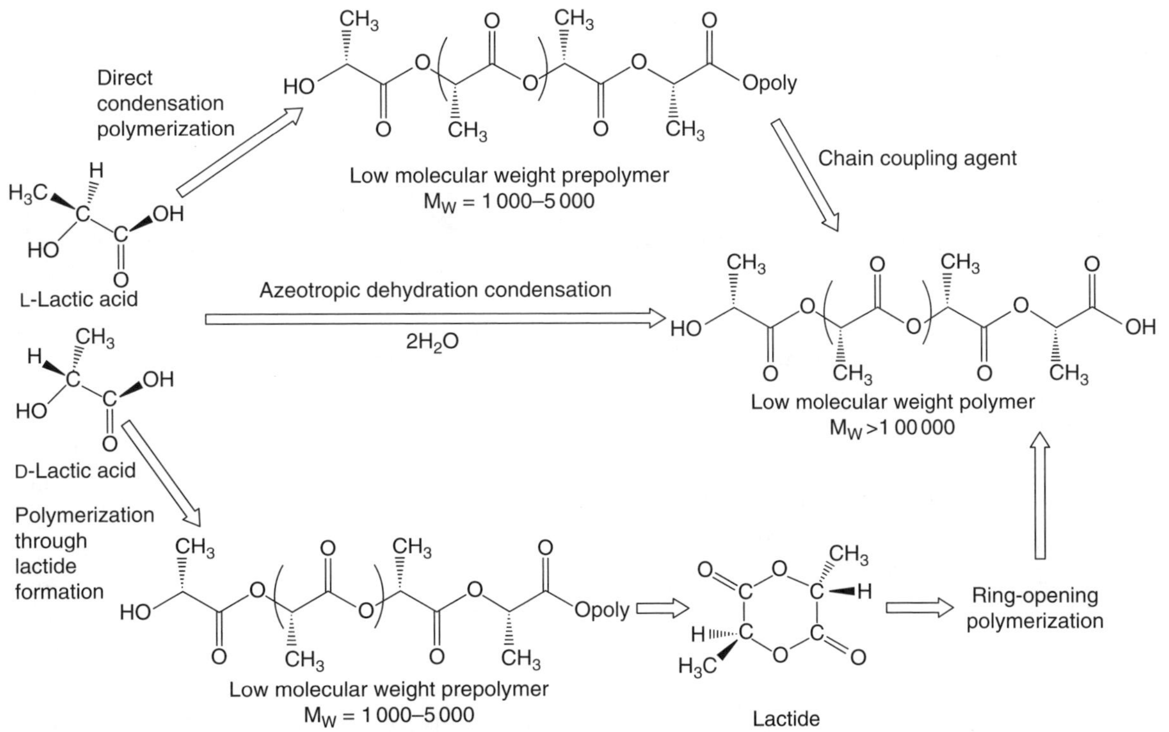

Figure 9.2 Synthesis methods for obtaining high molecular weight PLA. *Adapted from Reference [3]*

residual oligomers till final mineralization (biotic degradation).

As long as the basic monomers (lactic acid) are produced from renewable resources (carbohydrates) by fermentation, PLA complies with the rising worldwide concept of sustainable development and is classified as an environmentally friendly material.

9.2 Synthesis of PLA

The synthesis of PLA is a multistep process which starts from the production of lactic acid and ends with its polymerization [2–4,6–7]. An intermediate step is often the formation of the lactide. Figure 9.2 shows that the synthesis of PLA can follow three main routes. Lactic acid is condensation polymerized to yield a low-molecular-weight, brittle polymer, which, for the most part, is unusable, unless external coupling agents are employed to increase its chain length. The second route is the azeotropic dehydrative condensation of lactic acid. It can yield high-molecular-weight PLA without the use of chain extenders or special adjuvants [3]. The third and main process is ring-opening polymerization (ROP) of lactide to obtain high-molecular-weight PLA, patented by Cargill (US) in 1992 [8]. Finally, lactic acid units can be part of a more complex macromolecular architecture as in copolymers.

9.2.1 Precursors

9.2.1.1 Lactic Acid

Lactic acid is a compound that plays a key role in several biochemical processes. For instance, lactate is constantly produced and eliminated during normal metabolism and physical exercise. Lactic acid has been produced on an industrial scale since the end of the nineteenth century and is mainly used in the food industry to act, e.g., not only as an acidity regulator, but also in cosmetics, pharmaceuticals, and animal feed. It is, additionally, the monomeric precursor of PLA. It can be obtained either by carbohydrate fermentation or by common chemical synthesis. Also known as "milk acid," lactic acid is the simplest hydroxyl acid with an asymmetric carbon atom and two optically active configurations, namely the L and D isomers (Fig. 9.2), which can be produced in bacterial systems, whereas mammalian organisms only produce the L isomer, which is easily assimilated during metabolism.

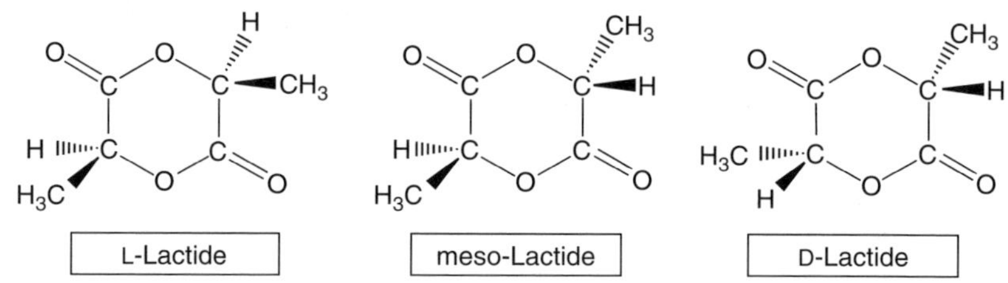

Figure 9.3 Chemical structures of L-, *meso*-, and D-lactides.

Lactic acid is mainly prepared in large quantities (estimated to more than 250 kT per year, in 2011) by the bacterial fermentation of carbohydrates. These fermentation processes can be classified according to the type of bacteria used: (i) the hetero-fermentative method, which produces less than 1.8 mol of lactic acid per mole of hexose, with other metabolites in significant quantities, such as acetic acid, ethanol, glycerol, mannitol, and carbon dioxide; (ii) the homo-fermentative method, which leads to greater yields of lactic acid and lower levels of by-products, and is mainly used in industrial processes [3]. The conversion yield from glucose to lactic acid is more than 90%.

The majority of the fermentation processes use species of *Lactobacilli* which give high yields of lactic acid. Some organisms predominantly produce the L isomer, such as *Lactobacilli amylophilus*, *L. bavaricus*, *L. casei*, and *L. maltaromicus*, whereas, *L. delbrueckii*, *L. jensenii*, and *L. acidophilus* produce the D isomer or a mixture of L and D [3,4]. These different bacteria are homofermentative. In general, the sources of basic sugars are glucose and maltose from corn or potato, and sucrose from cane or beet sugar, etc. In addition to carbohydrates, other products, such as B vitamins, amino acids, and different nucleotides, are formed. The processing conditions are an acid pH close to 6, a temperature around 40 °C, and a low oxygen concentration. The major method of separation consists in adding $CaCO_3$, $Ca(OH)_2$, $Mg(OH)_2$, NaOH, or NH_4OH to neutralize the fermentation acid and to give soluble lactate solutions, which are filtered to remove both the cells (biomass) and the insoluble products. The product is then evaporated, crystallized, and acidified with sulfuric acid to obtain the crude lactic acid. If the lactic acid is used in pharmaceutical and food applications, it is further purified to remove the residual by-products. If it is to be polymerized, it is purified by separation techniques including ultrafiltration, nanofiltration, electrodialysis, and ion-exchange processes.

9.2.1.2 Lactide

Figure 9.3 shows the different stereoforms of lactide. The cyclic dimer of lactic acid combines two of its molecules and gives rise to L-lactide or LL-lactide, D-lactide or DD-lactide, and *meso*-lactide or LD-lactide (a molecule of L-lactic acid associated with another one of D-lactic acid). A mixture of L- and D-lactides is a racemic lactide (*rac*-lactide). Lactide is usually obtained by the depolymerization of low-molecular-weight PLA under reduced pressure to give a mixture of L-, D-, and *meso*-lactides. The different percentages of the lactide isomers formed depend on the lactic acid isomer feedstock, temperature, and the catalyst's nature and content [3,4]. A key point in most of the processes is the separation between each stereoisomer to control the final PLA structure (e.g., by vacuum distillation), which is based on the boiling point differences between the *meso*- and the L- or D-lactide.

9.2.2 PLA Polymerization

9.2.2.1 Lactic Acid Condensation and Coupling

The condensation polymerization is the least expensive route, but it is difficult to obtain high molecular weights by this method. The use of coupling or esterification-promoting agents is required to increase chain length [3,4], but at the expense of an increase in both cost and complexity (multistep process). The role of chain coupling agents is to react with either the hydroxyl (OH) or the carboxyl end-groups of the PLA [3,4,7], thus giving telechelic polymers [10]. The nature of the chain end-groups should be fully controlled [2,3]. The use of chain-extending agents brings some advantages, because reactions involving small amounts of them are economical and can be carried out in the melt without the need of separating the different process steps. The tunability to design copolymers with various functional groups is also greatly expanded. The disadvantages are that the final polymer may contain unreacted chain-extending agents, oligomers, and residual metallic impurities from the catalyst. Moreover, some extending agents could be associated with a lack of biodegradability [2]. Examples of chain-extending agents are anhydrides, epoxides, and isocyanates [11]. Similar products are used to develop compatibilization for PLA-based blends. The disadvantages of using isocyanates as chain extenders are their (eco)toxicity [3].

The advantages of esterification-promoting adjuvants are that the final product is highly purified and free from residual catalysts and/or oligomers. The disadvantages are higher costs due to the number of steps involved and the additional purification of the residual by-products [3], since these additives produce by-products that must be neutralized or removed.

9.2.2.2 Azeotropic Dehydration and Condensation

The azeotropic condensation polymerization is a method used to obtain high chain lengths without the use of chain extenders or adjuvants and their associated drawbacks. Mitsui Chemicals (Japan) has commercialized a process wherein lactic acid and a catalyst are azeotropically dehydrated in a refluxing, high boiling, aprotic solvent under reduced pressures to obtain high-molecular-weight PLA (M_w 300,000) [2,3]. A general procedure consists in the reduced pressure distillation of lactic acid for 2–3 h at 130 °C to remove most of the condensation water. The catalyst and diphenyl ether are then added and a tube packed with molecular sieves is attached to the reaction vessel. The refluxing solvent is returned to the vessel by way of the molecular sieves during 30–40 h at 130 °C. Finally, the ensuing PLA is purified [12].

This polymerization gives considerable catalyst residues because of its high concentration needed to reach an adequate reaction rate. This can cause many drawbacks during processing, such as degradation and hydrolysis. For most biomedical applications, the catalyst toxicity is a highly sensitive issue. The catalyst can be deactivated by the addition of phosphoric acid or can be precipitated and filtered out by the addition of strong acids such as sulfuric acid. Thus, residual catalyst contents can be reduced to some ppm [3].

9.2.2.3 ROP of Lactide

The lactide method is the only method for producing pure high-molecular-weight PLA (M_w 100,000) [4,6,7,13]. The ROP of lactide was first demonstrated by Carothers in 1932 [14], but high molecular weights were not obtained until improved lactide purification techniques were developed by DuPont in 1954 [2]. This polymerization has been successfully carried out calling upon various methods, such as solution, bulk, melt, or suspension process. The mechanism involved in ROP can be ionic (anionic or cationic) or coordination–insertion, depending on the catalytic system [4,6,7,13]. The role of the racemization and the extent of transesterification in the homo or copolymerization are also decisive for the enantiomeric purity and chain architecture of the resulting macromolecules.

It has been found that trifluoromethane sulfonic acid and its methyl ester are the only cationic initiators known to polymerize lactide [15], and the mechanism of this process has been outlined in different papers [2,3,15].

Lactide anionic polymerizations proceed by the nucleophilic reaction of the anion with the carbonyl group and the subsequent acyl–oxygen bond cleavage, which produces an alkoxide end-group, which continues to propagate. The general mechanism for this anionic polymerization has been discussed in various publications [2,3,15,16]. Some authors [16] have shown that the use of alkoxides, such as potassium methoxide, can yield well-defined polymers with negligible racemization.

Both the anionic and cationic ROPs are usually carried out in highly purified solvents, and although they show a high reactivity, they are susceptible to give racemization, transesterification, and high impurity levels. For industrial and large commercial use, it is preferable to do bulk and melt polymerization with low levels of nontoxic catalysts. The use of less-reactive metal carboxylates, oxides, and alkoxides has been extensively studied in this context, and it has been found that high-molecular-weight PLA can be readily obtained in the presence of transition metal compounds of tin [6,7,13], zinc [17,18], iron [19], and aluminum [20], among others. A systematic investigation has led to the wide use of tin compounds, namely tin(II) bis-2-ethylhexanoic acid (stannous octoate) as a catalyst in PLA synthesis. This is mainly due to its high catalytic efficiency, low toxicity, food and drug contact approval, and ability to give high molecular weights with low racemization [15]. The mechanisms of the polymerization with stannous octoate have been studied in detail, and it is now widely accepted that this ROP is actually initiated from compounds containing hydroxyl groups, such as water and alcohols, which are either present in the lactide feed or can be added upon demand. Figure 9.4 shows that the global mechanism is of the "coordination−insertion" type [21], occurring in two steps. First, a complex between monomer and initiator is formed followed by a rearrangement of the covalent bonds. Second, the monomer is inserted within the oxygen−metal bond of the initiator, and its cyclic structure is thus opened through the cleavage of the acyl−oxygen link, thus, the metal is incorporated with an alkoxide bond into the propagating chain. It was found that the polymerization yield and the transesterification effect are affected by different parameters, such as the polymerization temperature and time, the monomer/catalyst ratio, and the type of catalyst. The interaction between the time and temperature is very significant in terms of limiting the degradation reactions, which affect the molecular weight and the reaction kinetics [22]. It has also been shown that the chain length is directly controlled by the amount of OH impurities [23].

To make an economically viable PLA, Jacobsen *et al.* [21] developed a continuous one-stage process based on reactive extrusion with a twin-screw extruder. This technique requires that the bulk polymerization be close to completion within a very short time (5−7 min), which is predetermined by the residence time in the extruder. These authors showed that the addition of an equimolar content of a Lewis base, particularly triphenyl-phosphine, to stannous octoate increased the lactide polymerization rate.

9.2.3 Copolymers Based on Lactic Acid Units

A large number of macromolecular architectures of copolymers based on lactic acid have been investigated [7,13]. Most of them are biodegradable or/and biocompatible. These copolymers can be prepared by using units containing a specific functionalized structure, thus giving rise to complex structure with unique properties. Examples of these materials are branched polyesters and graft copolymers (star, hyper-branched polymers) which involve different macromolecular architectures associated with novel materials properties and applications.

9.2.3.1 Ring-Opening Copolymerization

Several heterocyclic monomers can be used as comonomers with lactic acid in ring-opening copolymerizations—the most commonly used being glycolide (GA) for biomedical applications [24], caprolactone (CL), and valerolactone. The comonomer units can be inserted randomly or in block sequences.

9.2.3.2 Modification by High Energy Radiation and Peroxides

Radical reactions applied to PLA to modify its structure have been generated by peroxides or high energy radiation [7]. Branching has been suggested to be the dominant structural change in poly(L-lactide) (PLLA) with peroxide concentrations in the range of 0.1−0.25 wt% and crosslinking above 0.25 wt% [7]. The peroxide

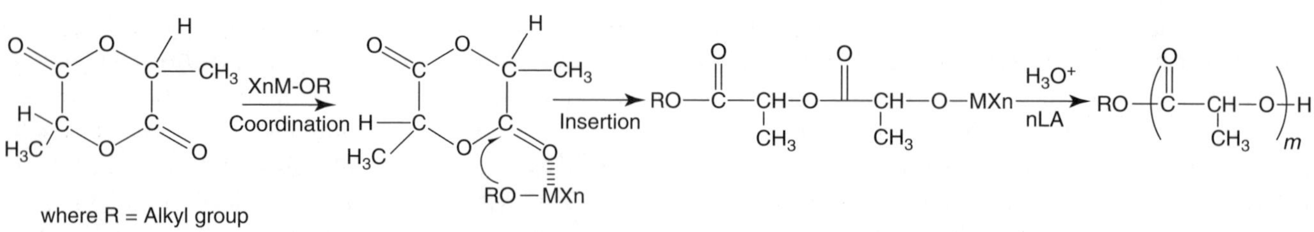

Figure 9.4 Coordination−insertion polymerization mechanism.

melt-reaction with PLA has been found to cause strong modifications of the original PLA properties. A similar approach was recently developed with starch-based blends without any major improvement in their mechanical properties [25]. Irradiation of PLA causes mainly chain-scissions or cross-linking reactions, depending on the radiation intensity [26].

9.2.3.3 Graft Copolymerization

Graft copolymers are often used as compatibilizers to improve the interfacial properties of blends or multiphase systems. Grafting reactions on a trunk polymer can be induced chemically, by plasma discharge, or by radiation (UV, X-rays or accelerated electrons)—the latter approach giving purer products at high conversions. Plasma-induced grafting is performed by introducing an organic vapor into the plasma of inorganic gases to modify the surface properties of a substrate. Depending on the penetration depth of the irradiation, grafting can be performed either at the surface, or both on the skin and in the bulk [7].

The chemical modification of lactic acid-based polymers by graft copolymerization has been reported for the homopolymer of L-lactide and for copolymers with different L-lactide/CL contents [7,13]. Carbohydrate polymers (e.g., amylose) can be modified by grafting lactic acid chains on their OH groups. A recent study [25] showed the interest of such a copolymer as a compatibilizer to improve the properties of starch/PLA blends to a better extent than the addition of peroxides or coupling agents (e.g., di-isocyanate) into the melt blend during the processing. Figure 9.5 shows the different steps involved in this grafting operation. After amylose purification to eliminate residual butanol and water, amylose-graft-PLA is obtained by the ROP of purified lactide with tin(II) bis(2-ethylhexanoate) in toluene at 100 °C for 20 h.

9.3 PLA Properties

9.3.1 Crystallinity and Thermal Properties

The properties of PLA, as indeed those of other polymers, depend on its molecular characteristics, as well as on the presence of ordered structures, such as crystalline thickness, crystallinity, spherulite size, morphology, and degree of chain orientation. The physical properties of polylactide are related to the enantiomeric purity of the lactic acid stereo-copolymers. Homo-PLA is a linear macromolecule with a molecular architecture that is determined by its stereochemical composition. PLA can be produced as totally amorphous or with up to 40% crystallinity. PLA resins containing more than around 93% of L-lactic acid are mainly semi-crystalline. Both *meso*- and D-lactides induce twists in the very regular PLLA architecture. Macromolecular imperfections are responsible for the decrease in both the rate and the extent of PLLA crystallization. In practice, most PLAs are made up of L-and D,L-lactide copolymers, since the reaction media often contain some *meso*-lactide impurities.

Table 9.1 gives the details of the different crystalline structures for neat PLA. Depending on the preparation conditions, PLLA crystallizes in different forms. The α-form exhibits a well-defined diffraction pattern [27]. This structure, with a melting temperature of 185 °C, is more stable than its β counterpart, which melts at 175 °C [27]. The latter form can be prepared at a high draw ratio and a high drawing temperature [28]. The γ-form is formed by epitaxial crystallization [29]. It has been observed that a blend with equivalent poly(L-lactide) PLLA and poly(D-lactide) PDLA contents gives stereo-complexation (racemic crystallite) of both polymers. This stereocomplex has higher mechanical properties than those of both PLAs and a higher melting temperature of 230 °C. The literature reports different density data [4] for PLA, with most values for the crystalline polymer around 1.29 compared with 1.25 for the amorphous material.

The crystallization kinetics of PLA have been extensively studied and found to be rather slow, as in the case of poly(ethylene terephthalate) (PET). The rate of crystallization increases with a decrease in the molecular weight and is strongly dependent on the (co)polymer composition [4]. PLLA can crystallize in the presence of D-lactide [30]; however, as the structure becomes more disordered, the rate of crystallization decreases. It has been reported that the crystallization rate is essentially determined by the decrease in the melting point of the different copolymers. PDLA/PLLA stereocomplexes are very efficient nucleating agents for PLLA, with increases in both crystallization rate and crystallinity, the latter of up to 60% [31]. Quenching decreases the time

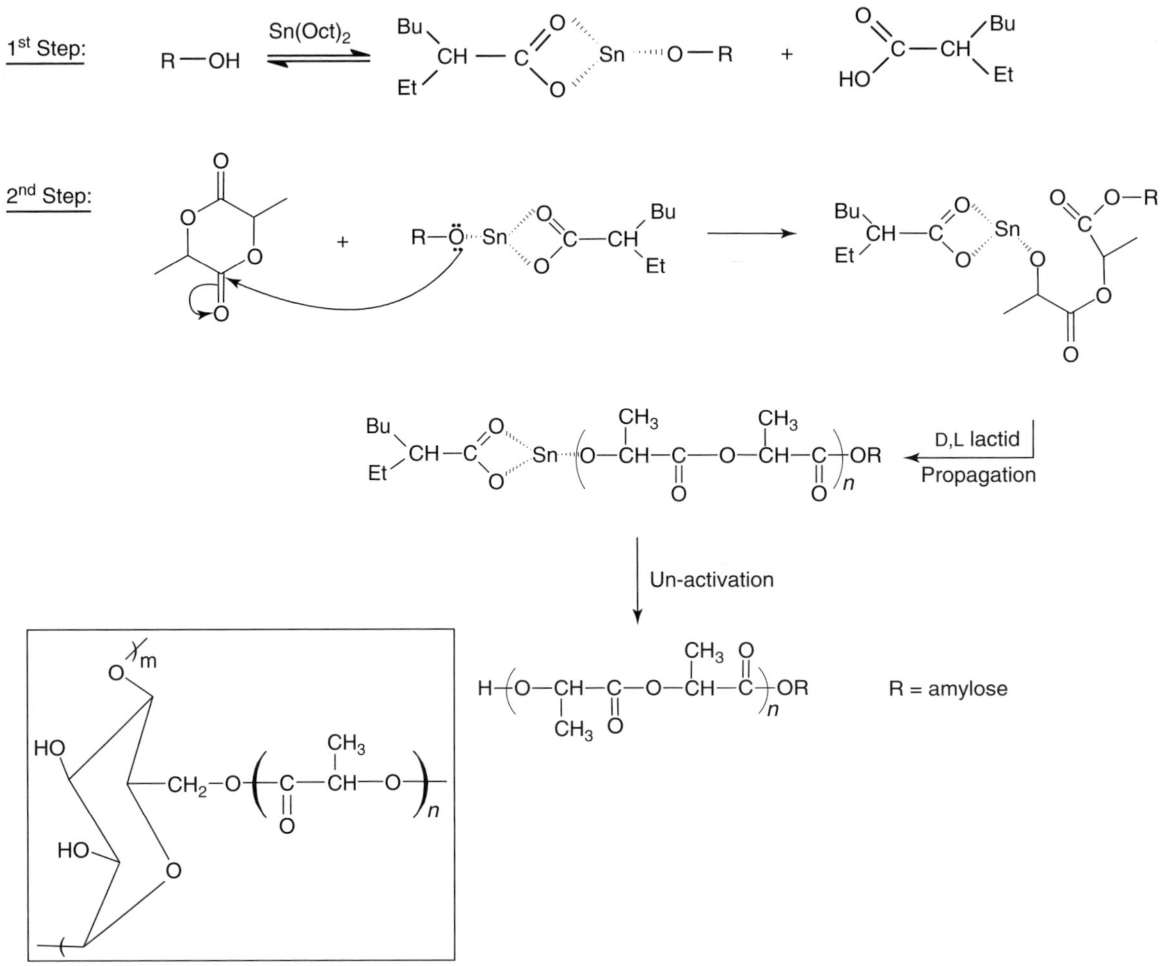

Figure 9.5 Mechanism of the ROP synthesis of amylose-graft PLA.

taken for crystallization [30]. As PET, PLA can be oriented by processing, and the chain orientation increases the mechanical strength of the polymer. If orientation is performed at low temperature, the resulting PLLA has a higher modulus without any significant increase in crystallinity. To determine the crystallinity levels by differential scanning calorimetry (DSC), the value most often referred to in the literature concerning the PLA melt enthalpy at 100% crystallinity is 93 J g^{-1} [7,8,9,32]. The crystallization of the thermally crystallizable, but amorphous, PLA can be initiated by annealing it at temperatures between 75 °C and the melting point. Annealing crystallizable PLA copolymers often produces two melting peaks [32] and different hypotheses have been put forward to explain this feature. Yasuniwa *et al.* [33] found a double melting point in PLLA polymers and attributed them to slow rates of crystallization and recrystallization.

The typical PLA glass transition temperature (T_g) ranges from 50 °C to 80 °C, whereas its melting temperature ranges from 130 °C to 180 °C. For instance, enantiomerically pure PLA is a semicrystalline polymer with a T_g of 55 °C and a T_m of 180 °C. For semicrystalline PLA, the T_m is a function of the different processing parameters and the initial PLA structure. According to Ikada and Tsuji [30], T_m increases with increasing molecular weight (M_w) to an asymptotic value, but the actual crystallinity decreases with increasing M_w. T_m, moreover, decreases with the presence of *meso*-lactide units in its structure [4]. Both, the degree of crystallinity and the melting temperature of PLA-based materials can be reduced by random copolymerization with different comonomers (e.g., GA, CL, or valerolactone).

The T_g of PLA is also determined by the proportion of the different types of lactide in its macromolecular chain.

Table 9.1 PLA Crystalline Structures. Unit Cell Parameters for Nonblended PLLA and Stereocomplex Crystals

		Space Group	Chain Orientation	Number Helices/ Unit Cell	Helical Conformation	*a* (nm)	*b* (nm)	*c* (nm)	α (degrees)	β (degrees)	γ (degrees)
PLLA form	α	Pseudo-orthorhombic	–	2	103	1.07	0.645	2.78	90	90	90
PLLA form	α	Pseudo-orthorhombic	–	2	103	1.07	0.62	2.88	90	90	90
PLLA form	α	Orthorhombic	Parallel	2	103	1.05	0.61	–	90	90	90
PLLA form	β	Orthorhombic	–	6	31	1.031	1.821	0.90	90	90	90
PLLA form	β	Trigonal	Random up-down	3	31	1.052	1.052	0.88	90	90	120
PLLA form	γ	Orthorombic	Antiparallel	2	31	0.995	0.625	0.88	90	90	90
Stereo complex		Triclinic	Parallel	2	31	0.916	0.916	0.870	109.2	109.2	109.8

Source: Adapted from Reference [8]

9.3.2 Surface Energy

Surface energy is critically important to many processes (printing, multilayering, etc.) and it influences the interfacial tension. The surface energy of a PLA made up of 92% L-lactide and 8% *meso*-lactide was found to be 49 mJ m^{-2}, with dispersive and polar components of 37 and 11 mJ m^{-2}, respectively [34], which suggests a relatively hydrophobic structure compared with that of other biopolyesters.

9.3.3 Solubility

A good solvent for PLA and for most of the corresponding copolymers is chloroform. Other solvents are chlorinated or fluorinated organic compounds, dioxane, dioxolane, and furan. Poly(*rac*-lactide) and poly(*meso*-lactide) are soluble in many other organic solvents like acetone, pyridine, ethyl lactate, tetrahydrofuran, xylene, ethyl acetate, dimethylformamide, and methyl ethyl ketone. Among nonsolvents, the most relative compounds are water, alcohols (e.g., methanol and ethanol), and alkanes (e.g., hexane and heptane) [7].

9.3.4 Barrier Properties

Because PLA finds a lot of applications in food packaging, its barrier properties (mainly to carbon dioxide, oxygen, and water vapor) have been largely investigated [4]. The CO_2 permeability coefficients for PLA polymers are lower than those reported for crystalline polystyrene at 25 °C and 0% relative humidity (RH) and higher than those for PET. Since diffusion takes place through the amorphous regions of a polymer, an increase in the extent of crystallization will inevitably result in a decrease in permeability. Figure 9.6 shows the oxygen permeability for poly(98% L-lactide) films as a function of the water activity. A significant increase in the oxygen permeability coefficient is shown as the temperature is increased, but its decrease with water activity at temperatures close to T_g and its stabilization at temperatures well below T_g are clearly visible. PET and PLA are both hydrophobic and the corresponding films absorb very low amounts of water, showing similar barrier properties, as indicated by the values of their water vapor permeability coefficient determined from 10 °C to 37.8 °C in the range of 40–90% RH. Auras *et al.* [4] have shown that the permeability

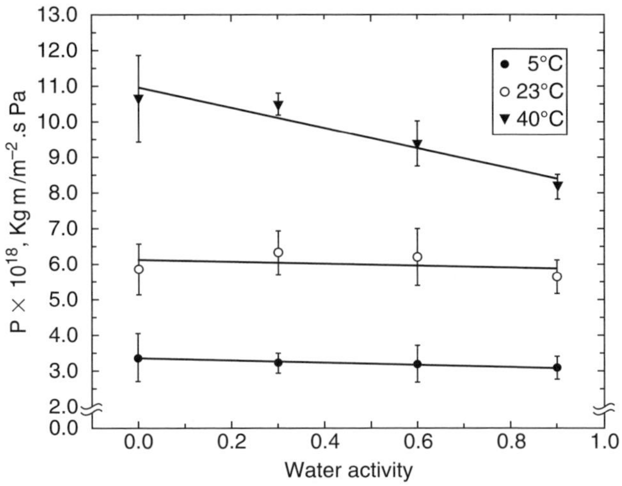

Figure 9.6 Oxygen permeability versus water activity at different temperatures, for poly(98% L-lactide) films. *Source: Reference [4].*

for 98% L-lactide polymers is almost constant over the range studied, despite PLA being a rather polar polymer [4].

9.3.5 Mechanical Properties

9.3.5.1 Solid State

The mechanical properties of PLA can vary to a large extent, ranging from soft and elastic materials to stiff and high strength materials, according to different parameters, such as crystallinity, polymer structure and molecular weight, material formulation (plasticizers, blend, composites, etc.), and processing (e.g., orientation). For instance, commercial PLA, such as poly(92% L-lactide, 8% *meso*-lactide), has a modulus of 2.1 GPa and an elongation at break of 9%. After plasticization, its Young's modulus decreases to 0.7 MPa and the elongation at break rises to 200%, with a corresponding T_g shift from 58 °C to 18 °C [32]. This example indicates that mechanical properties can be readily tuned to satisfy different applications.

The mechanical properties of PLA-related polymers were recently reviewed by Sodergard and Stolt [7], who showed, among other features, that the PLLA fiber modulus can be increased from 7–9 GPa to 10–16 GPa by going from melt to solution spinning. The mechanical behavior can also be modified by preparing suitable copolymers, as in the case of the use of CL, which, with its soft segments, induces

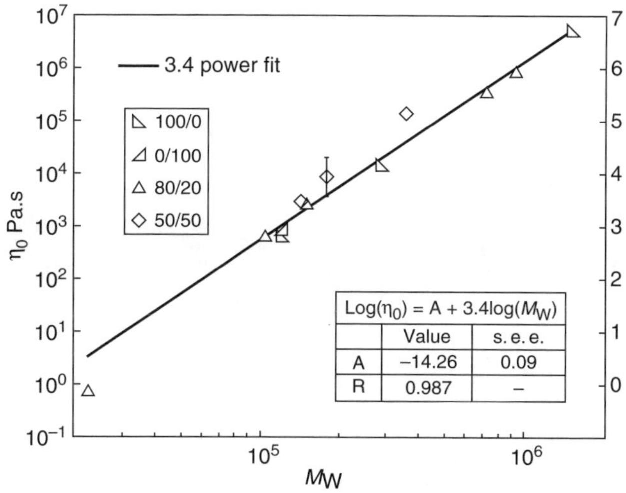

Figure 9.7 Zero-shear viscosity versus molecular weight for different L/D ratios (%). *Adapted from Reference [35].*

a decrease in modulus and an increase in the elongation at break, respectively.

9.3.5.2 Molten Behavior

For processing and for the corresponding applications, the knowledge of PLA melt rheology is of particular interest. A power law equation has been applied successfully by, e.g., Schwach and Averous [34]. The pseudoplastic index is in the range 0.2–0.3, depending on the PLA structure. For instance, poly(92% L-lactide, 8% *meso*-lactide) displays a pseudoplastic index of 0.23. Figure 9.7, based on data published by Dorgan *et al.* [35], shows the evolution of the zero-shear viscosity versus molecular weight (M_w) for a wide range of L/D ratios (%), the latter parameter having virtually no effect. Static and dynamic characterizations have shown that the molecular weight between entanglements is around 10^4. Some other studies suggested that chain branching and molecular weight distribution have a significant effect on the melt viscosity of PLA [5].

9.4 Degradation

9.4.1 Abiotic Degradation

The main abiotic phenomena involve thermal and hydrolysis degradations during the life cycle of the material.

9.4.1.1 Thermal Degradation

The thermal stability of biopolyesters is not significantly high—a fact that inevitably limits their range of applications. The PLA decomposition temperature lies between 230 °C and 260 °C. Gupta and Deshmukh [36] concluded that the carbonyl carbon–oxygen linkage is the most likely bond to split under isothermal heating, as suggested by the fact that a significantly larger amount of carboxylic acid end-groups were found compared with hydroxyl end-groups. The reactions involved in the thermal degradation of lactic acid-based polymers can follow different mechanisms [7], such as thermohydrolysis, zipper-like depolymerization [36] in the presence of catalyst residues, thermo-oxidative degradation [37,38], and transesterification reactions, which give simultaneous bond breaking and bond making.

9.4.1.2 Hydrolytic Degradation

PLA hydrolysis is an important phenomenon since it leads to chain fragmentation [4,7,39], and can be associated with thermal or biotic degradation. This process can be affected by various parameters such as the PLA structure, its molecular weight and distribution, its morphology (crystallinity), the shape of its samples and its thermal and mechanical history (including processing), as well as, of course, the hydrolysis conditions. Hydrolytic degradation is a phenomenon that can be both desirable (e.g., during the composting stage) and undesirable (e.g., during processing or storage). The hydrolysis of aliphatic polyesters starts with a water uptake phase, followed by hydrolytic splitting of the ester bonds in a random way. The amorphous parts of the polyesters have been known to undergo hydrolysis before their crystalline regions because of a higher rate of water uptake. The initial stage is therefore located at the amorphous regions, giving the remaining non-degraded chains more space and mobility, which leads to their reorganization and hence an increased crystallinity. In the second stage, the hydrolytic degradation of the crystalline regions of the polyester leads to an increased rate of mass loss and finally to complete resorbtion [40]. The PLA degradation in an aqueous medium has been reported by Li *et al.* [40] to proceed more rapidly in the core of the sample. The explanation for this specific behavior is an autocatalytic effect due to the increasing amount of compounds containing carboxylic end-groups. These

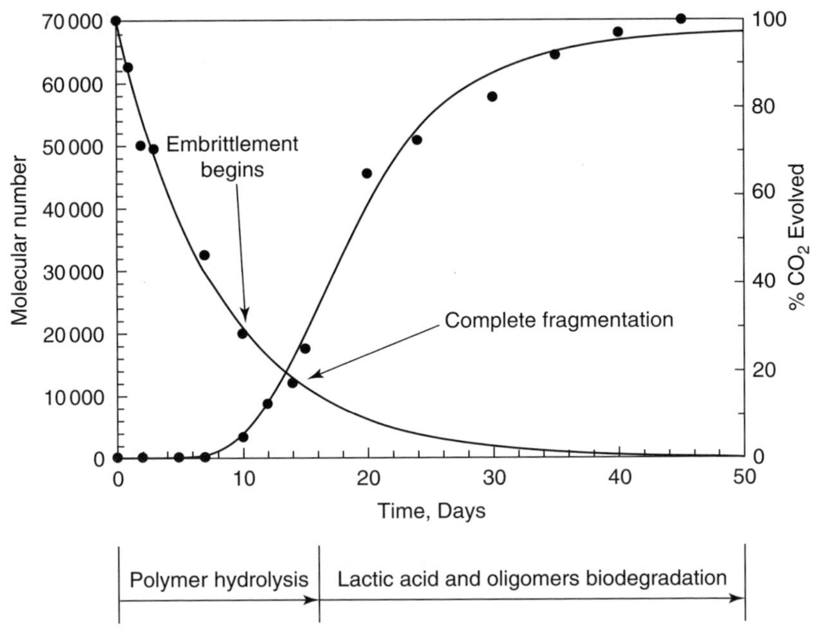

Figure 9.8 Abiotic and biotic degradations during composting stage. *Adapted from Reference [3].*

low-molar-mass compounds are not able to permeate the outer shell. The degradation products in the surface layer are instead continuously dissolved in the surrounding buffer solution [40]. As expected, temperature plays a significant role in accelerating this type of degradation.

9.4.2 Biotic Degradation

The biodegradation of aliphatic biopolyesters has been widely reported in the literature [5,7,39]. The biodegradation of lactic acid-based polymers for medical applications has been investigated in a number of studies *in vivo* [41] and some reports can also be found on their degradation in other biological systems [42]. The *in vivo* and *in vitro* degradations have been evaluated for PLA-based surgical implants [41]. *In vitro* studies have shown that the pH of the solution plays a key role in the degradation and that this analysis can be a useful predicting tool for *in vivo* PLA degradation [4]. Enzymes, such as proteinase K and pronase, have been used to bring about the *in vivo* PLA hydrolysis, although enzymes are unable to diffuse through the crystalline parts. As expected, little enzymatic degradation occurs at the beginning of the process, but pores and fragmentation are produced, widening the accessible area to the different enzymes.

Figure 9.8 shows that during the composting stage, PLA degrades in a multistep process with different mechanisms [39]. Primarily, after exposure to moisture by abiotic mechanisms, PLA degrades by hydrolysis. First, random nonenzymatic chain-scissions of the ester groups lead to a reduction in molecular weight, with the consequent embrittlement of the polymer. This step can be accelerated by acids or bases and is affected by both temperature and moisture levels [3]. Then, the ensuing PLA oligomers can diffuse out of the bulk polymer and be attacked by microorganisms. The biotic degradation of these residues produces carbon dioxide, water, and humus (mineralization).

Studies on PLA-based multiphase materials have been carried out. Gattin *et al.* [43] have found that the physical and morphological properties of the blend play an important role in its degradation behavior, as in the case of their comparative study of the degradation of PLA with and without plasticized starch materials [43]. These authors reported that the nature of the degradation strongly depends on the experimental biodegradation conditions. Sinha Ray *et al.* [44] prepared PLA nano-biocomposites filled with montmorillonite, and studied and characterized their biodegradability.

9.5 Processing

9.5.1 Multiphase Materials

The extrusion of PLA-based materials is generally linked with another processing step such as thermoforming, injection molding, fiber drawing, film blowing, bottle blowing, and extrusion coating. The properties of the polymer will therefore depend on the specific conditions during the processing steps (e.g., the thermomechanical input). The main parameters during the melt processing are temperature, residence time, moisture content, and atmosphere [1]. But the major problem in the manufacturing of PLA-based products is the limited thermal stability during the melt processing. To overcome such a drawback or to give PLA new properties, a large number of multiphase materials have been developed, mainly by mixing PLA with others products.

9.5.1.1 Plasticization

The brittleness and stiffness of PLA can be major drawbacks for some applications. According to Ljungberg *et al.* [45], any factor influencing PLA crystallinity, such as the isomer ratio, could disturb the distribution and compatibility of plasticizers with PLA and induce low efficiency and phase separation.

Lactide monomer is an effective plasticizer for PLA, but presents high migration due to its small molecular size. Oligomeric lactic acid (OLA) seems to be a better answer, since it shows low migration and high efficiency [32]. For instance, adding 20 wt% of OLA into poly(92% L-lactide, 8% *meso*-lactide) induces T_g and modulus decreases of 20 °C and 63%, respectively. A significant improvement of PLA (mainly PLLA) flexibility is accomplished by the incorporation of different types of citrates [45–48] or maleates [49] whose efficiency was evaluated in terms of T_g shift and mechanical properties improvement [32]. These plasticizers are miscible with PLA up to ~ 25 wt%, but increasing the plasticizer content can raise the PLA crystallinity by enhancing chain mobility [32]. Low-molecular-weight polyethylene glycol (PEG) [32], polypropylene glycol, and fatty acid are also compatible with PLA and can act as plasticizers [5].

9.5.1.2 Blends and Compatibilization

A great number of articles has been published during the last few decades on PLA-based blends [4,5,8,32], including starch/PLA blends, which allow reducing the material cost without sacrificing its biodegradability and maintaining certain mechanical and thermal properties. Native starch, which is composed of semicrystalline granules, can be physically blended with PLA, but remains in a separate conglomerate form in the PLA matrix [50]. Thus, starch is typically characterized as a solid filler with poor adhesion with PLA. Such biocomposites are used as a model to test (e.g., carbohydrate–PLA compatibilization [51]). Most of the studies that are focused on the production of starchy blends are based on plasticized starch, the so-called thermoplastic starch. Such a processable material is obtained by the disruption of the granular starch and the transformation of its semicrystalline granules into a homogeneous, rather amorphous material with the destruction of hydrogen bonds between the macromolecules. Disruption can be accomplished by casting (e.g., with dry drums) or by applying thermomechanical energy in a continuous process. The combination of thermal and mechanical inputs can be obtained by extrusion. After the processing, a homogeneous material is obtained [1,32]. A dependence of the PLA glass transition temperature on the blend composition was observed by DSC and DMA, indicating a small degree of compatibility between the blend components [32]. However, the mechanical characteristics of the blends were modest. The blend morphology (discontinuous versus co-continuous) has been investigated by Schwach and Averous [34] by microscopic observations. The full co-continuity is obtained in the domain of 60–80% in volume of PLA. Despite the interest in developing plasticized starch/PLA materials, some limitations, due to the lack of affinity between the respective constituents, seem difficult to overcome. This low compatibility is mainly due to the PLA hydrophobic character.

To improve the affinity between the phases, compatibilization strategies are generally developed. This implies the addition of a compound, the compatibilizer, which can be obtained by the modification of at least one of the polymers initially present in the blend. For PLA/starch compatibilization, the literature proposes different approaches, which can be classified in four groups [1,25]: (i) the functionalization of PLA with, e.g., maleic anhydride [51]; (ii) the functionalization of starch with, e.g., urethane functions [25]; (iii) the starch–polyester cross-linking with a coupling agent such as a peroxide [25]; and (iv) the use of copolymers, e.g., starch-graft PLA [25], following the mechanism discussed above and illustrated in Fig. 9.5, for which the length of the grafts can be controlled to obtain a comb structure [52].

It is known that PLA forms miscible blends with polymers such as PEG [53]. PLA and PEG are miscible with each other when the PLA fraction is below 50% [53]. The PLA/PEG blend consists of two semimiscible crystalline phases dispersed in an amorphous PLA matrix. PHB/PLA blends are miscible over the whole range of composition. The elastic modulus, stress at yield, and stress at break decrease, whereas the elongation at break increases, with increasing polyhydroxybutyrate (PHB) content [54]. Both PLA/PGA and PLA/PCL blends give immiscible components [55], the latter being susceptible to compatibilization with P(LA-*co*-CL) copolymers or other coupling agents.

9.5.1.3 Multilayers

Developing compostable and low cost multilayer materials based, for instance, on plasticized starch and PLA is interesting in more than one sense. Martin *et al.* [56] carried out several studies on such a system and showed that the basic requisites for the preparation of multilayered products are to obtain sufficient adhesion between the layers, good moisture barrier properties, and a uniform layer thickness distribution. Two different techniques were used to prepare the multilayers, namely coextrusion and compression molding. Peel strength was controlled by the compatibility between plasticized starch and PLA, which stayed low without compatibilizer. It was possible to increase the adhesion properties of the film by up to 50% (e.g., by blending low polyester contents into the starchy core layer). There exist some inherent problems due to the multilayer flow conditions encountered in coextrusion, such as encapsulation and interfacial instability phenomena [57]. Addressing these problems is a crucial issue, since they can be detrimental to the product, affecting its quality and functionality.

9.5.1.4 Biocomposites and Nano-Biocomposites

Different types of fillers have been tested with PLA, such as calcium phosphate or talc [58], which show an increase in its mechanical properties. Concerning inorganic fillers, the greatest reinforcing effect is obtained with whiskers of potassium titanate and aluminium borate with a high aspect ratio. Carbon or glass fibers [59] improve the mechanical properties, particularly with fiber surface treatments capable of inducing strong interactions with PLA matrix. Different organic fillers can be associated with PLA. Biocomposites with improved mechanical properties are obtained by the association of ligno-cellulose fillers, such as paper-waste fibers and wood flour, with PLA by extrusion and compression molding.

A significant and increasing number of papers have been published during the last 5 years on nano-biocomposites (i.e., nanocomposites based on a biodegradable matrix). Polylactide/layered silicate nanocomposites were largely investigated by Sinha Ray *et al.* [60,61] and other authors [62,63]. They successfully prepared a series of biodegradable PLA nano-biocomposites using mainly melt extrusion of PLA, principally with modified montmorillonites (O-MMT), targeting nanofillers exfoliation into the matrix. Because of the interactions between the organo-clay particles, which present large surface area (several hundreds m^2g^{-1}), and the PLA matrix, the nano-biocomposites dislayed improved properties, such as mechanical moduli, thermal stability, crystallization behavior, gas barrier, and biodegradability. The preparation of biodegradable nano-cellular polymeric foams via nanocomposites technology based on PLA and layered silicate has been reported by different authors [61,64] who used supercritical carbon dioxide as a foaming agent, with the silicate acting as nucleating site for cell formation. Cellular PLA structures can also be obtained by producing a co-continuous structure and extracting the coproducts [65].

9.6 Applications

At present, PLA-based materials are mainly referenced on different markets such as biomedical (initial market), textile, and packaging (mainly food, i.e., short-term applications). For instance, reported types of manufactured products are blow-molded bottles, injection-molded cups, spoons and forks, thermoformed cups and trays, paper coatings, fibers for the textile industry, and sutures, films, and various molded articles [8].

9.6.1 Biomedical Applications

PLA has been widely studied for use in medical applications because of its bioresorbability and biocompatible properties in the human body. The main reported examples on medical or biomedical products are fracture fixation devices like screws, sutures, delivery systems, and microtitration plates [8].

PLA-based materials are developed for the production of screws and plates. As the bone healing progresses, it is desirable that the bone is subjected to a gradual increase in stress, thus reducing the stress-shielding effect. This is possible only if the plate loses rigidity in *in vivo* environment. To meet this need, researchers introduced resorbable polymers for bone plate applications. PLA resorbs or degrades upon implantation into the body, but most of its mechanical properties are lost within a few weeks [41]. Tormala *et al.* [66] proposed fully resorbable composites by reinforcing matrices with resorbable PLLA fibers and calcium phosphate-based glass fibers. One of the advantages often quoted for resorbable composite prostheses is that they do not need to be removed with a second operative procedure, as with metallic or nonresorbable composite implants. To improve the mechanical properties, PLA is reinforced with variety of nonresorbable materials, including carbon and polyamide fibers. Carbon fiber/PLA composites possess very high mechanical properties before their implantation, but they lose them too rapidly *in vivo* because of delamination. The long-term effects of resorbed products and biostable or slowly eroding fibers in the living tissues are not fully known, and are concerns yet to be resolved [41].

Although PLA fibers are used in different textile applications as, e.g., nonwoven textile for clothes, they achieved their first commercial success as resorbable sutures. One of the first commercially available fiber-formed bioresorbable medical product is based on copolymers of GA in combination with L-lactide (Vicryl) [67]. Fibers can be produced both by solvent and by melt-spinning processes, and are drawn under different conditions to orient the macromolecules [7].

Micro- and nanoparticles are an important category of delivery systems used in medicine, and the use of PLA is interesting due to its hydrolytic degradability and low toxicity. The most important properties of the micro- and nanoparticles are the drug release rate and the matrix degradation rate, which are affected by the particle design and the material properties [7]. Copolymers of GA and *rac*-lactide [5] seem to be the most suitable combinations for use as drug delivery matrices.

Porous PLA scaffolds have been found to be potential reconstruction matrices for damaged tissues and organs. There are several techniques reported for the manufacturing of such materials [7].

9.6.2 Packaging Applications

Commercially available PLA packaging can provide better mechanical properties than polystyrene and have properties more or less comparable to those of PET [4,8,9]. Market studies show that PLA is an economically feasible material for packaging. With its current consumption, it is at the present the most important market in volume for biodegradable packaging [4,8,9]. Due to its high cost, the initial use of PLA as a packaging material has been in high value films, rigid thermoforms, food and beverage containers, and coated papers. One of the first companies to use PLA as a packaging material was Danone (France) in yoghurt cups for the German market at the end of the 1990s. But the production of these cups was rapidly stopped. In 2011, Danone launched new yoghurt cups for German market, 100% compostable, with a bigger success. During the last decade, the use of PLA as a packaging material has increased all across Europe, Japan, and the United States, mainly in the area of fresh products, where PLA is being used as a food packaging for short shelf-life products, such as fruit and vegetables. Package applications include containers, drinking cups, sundae and salad cups, wrappings for sweets, lamination films, blister packages, and water bottles [9]. Currently, PLA is used in compostable yard bags to promote national or regional composting programs. In addition, new applications such as cardboard or paper coatings are being pursued, e.g., for the fast-food market (cups, plates, and the like) [9]. However, to cater for a larger market, some PLA drawbacks must be overcome, such as its limited mechanical and barrier properties and heat resistance, and, in order to meet market expectations, the world production of PLA must be increased.

However, since 2009 the last trends for PLA applications have been in durable and biobased markets, such as automotive (seats, insulation, etc.) and building applications (acoustic and thermal insulation, etc.). For these last cases, the main advantages of PLA are the high bio-based content and the good fibrality conditions, and the corresponding physical and mechanical properties.

References

[1] L. Avérous, Biodegradable multiphase systems based on plasticized starch: A review, Polym. Rev. 4 (3) (2004) 231–274.

[2] D. Garlotta, A literature review of poly(lactic acid), J. Polym. Environ. 9 (2) (2002) 63–84.

[3] H. Hartmann, High molecular weight polylactic acid polymers, in: D.L. Kaplan (Ed.), Biopolymers from Renewable Resources, firstst ed., Springer-Verlag, Berlin, 1998, pp. 367–411.

[4] R. Auras, B. Harte, S. Selke, An overview of polylactides as packaging materials, Macromol. Biosci. 4 (2004) 835–864.

[5] J.F. Zhang, X. Sun, Poly(lactic acid)based bioplastics, in: R. Smith (Ed.), Biodegradable Polymers for Industrial Applications, CRC, Woodhead Publishing Limited, Cambridge -England, 2005, pp. 251–288. Chapter 10.

[6] R. Mehta, V. Kumar, H. Bhunia, S.N. Upahyay, Synthesis of poly(lactic acid): A review, J. Macromol. Sci., Polym. Rev. 45 (2005) 325–349.

[7] A. Sodergard, M. Stolt, Properties of lactic acid based polymers and their correlation with composition, Prog. Polym. Sci. 27 (2002) 1123–1163.

[8] Y. Doi, A. Steinbüchel, Biopolymers, Applications and Commercial Products – Polyesters III, Wiley-VCH, Weiheim – Germany, 2002. p. 410.

[9] S. Domenek, C. Courgneau, V. Ducruet, Characteristics and Applications of PLA, in: S. Kalia, L. Avérous (Eds.), Biopolymers: Biomedical and Environmental Applications, first ed. Wiley-Scrivener, New-York, 2011, pp. 183–224. Chapter 8.

[10] K. Hiltunen, M. Harkonen, J.V. Seppala, T. Vaananen, Synthesis and characterization of lactic acid based telechelic prepolymers, Macromolecules 29 (27) (1996) 8677–8682.

[11] K. Hiltunen, J.V. Seppala, M. Harkonen, Lactic acid based poly(ester-urethanes): Use of hydroxyl terminated prepolymer in urethane synthesis, J. Appl. Polym. Sci. 63 (1997) 1091–1100.

[12] M. Ajioka, K. Enomoto, K. Suzuki, A. Yamaguchi, The basic properties of poly lactic acid produced by the direct condensation polymerisation of lactic acid, J. Environ. Polym. Degrad. 3 (8) (1995) 225–234.

[13] K.M. Stridsberg, M. Ryner, A.C. Albertsson, Controlled ring-opening polymerization: Polymers with designed macro-molecular architecture, Adv. Polym. Sci. 157 (2001) 41–65.

[14] H. Carothers, G.L. Dorough, F.J. Van Natta, The reversible polymerization of six membered cyclic esters, J. Am. Chem. Soc. 54 (1932) 761–772.

[15] H.R. Kricheldorf, M. Sumbel, Polymerization of l, l-lactide with tin(II) and tin(IV) halogenides, Eur. Polym. J. 25 (6) (1989) 585–591.

[16] P. Kurcok, A. Matuszowicz, Z. Jedlinski, H.R. Kricheldorf, P. Dubois, R. Jerome, Substituent effect in anionic polymerization of b-lactones initiated by alkali metal alkoxides, Macromol. Rapid Commun. 16 (1995) 513–519.

[17] C.K. Williams, L.E. Breyfogie, S.K. Choi, W. Nam, V.G. Young, A highly active zinc catalyst for the controlled polymerization of lactide, J. Am. Chem. Soc. 125 (37) (2003) 11350–11359.

[18] F. Chabot, M. Vert, S. Chapelle, P. Granger, Configurational structures of lactic acid stereocopolymers as determined by 13C–1H N.M.R, Polymer 24 (1983) 53–59.

[19] M. Stolt, A. Sodergard, Use of monocarboxylic iron derivatives in the ring-opening polymerization of L-lactide, Macromolecules 32 (20) (1999) 6412–6417.

[20] P. Dubois, C. Jacobs, R. Jerome, P. Teyssie, Macromolecular engineering of polylactones and polylactides. 4. Mechanism and kinetics of lactide homopolymerization by aluminum isopropoxide, Macromolecules 24 (1991) 2266–2270.

[21] S. Jacobsen, H. Fritz, P. Degée, P. Dubois, R. Jérôme, New developments on the ring opening polymerisation of polylactide, Ind. Crop. Prod. 11 (2–3) (2000) 265–275.

[22] G. Schwach, J. Coudane, R. Engel, M. Vert, Stannous octoate-versus zinc-initiated polymerization of racemic lactide, Polym. Bull. 32 (1994) 617–623.

[23] Y.J. Du, P.J. Lemstra, A.J. Nijenhuis, H.A.M. Van Aert, C. Bastiaansen, ABA type copolymers of lactide with poly(ethylene glycol). Kinetic, mechanistic, and model studies, Macromolecules 28 (7) (1995) 2124–2132.

[24] D.K. Gilding, A.M. Reed, Biodegradable polymers for use in surgery — Polyglycolic/poly(-lactic acid) homo-and copolymers, Polymer 20 (1979) 1459—1464.

[25] E. Schwach, Etude de systèmes multiphasés biodegradables à base d'amidon de blé plastifié. Relations structure — propriétés, Approche de la compatibilisation (2004). Ph.D Thesis URCA, Reims- France.

[26] M.C. Gupta, V.G. Deshmukh, Radiation effects on poly(lactic acid), Polymer 24 (1983) 827—830.

[27] W. Hoogsten, A.R. Postema, A.J. Pennings, G. Brinke, P. Zugenmair, Crystal structure, conformation and morphology of solution-spun polyL-lactide) fibres, Macromolecules 23 (1990) 634—642.

[28] J. Puiggali, Y. Ikada, H. Tsuji, L. Cartier, T. Okihara, B. Lotz, The frustrated structure of poly(L-lactide), Polymer 41 (2000) 8921—8930.

[29] L. Cartier, T. Okihara, Y. Ikada, H. Tsuji, J. Puiggali, B. Lotz, Epitaxial crystallization and crystalline polymorphism of polylactides, Polymer 41 (2000) 8909—8919.

[30] H. Tsuji, Y. Ikada, Properties and morphologies of poly(L-lactide): 1. Annealing effects on properties and morphologies of poly(l-lactide), Polymer 36 (1995) 2709—2716.

[31] K.S. Anderson, M.A. Hillmyer, Melt preparation and nucleation efficiency of polylactide stereocomplex crystallites, Polymer 47 (2006) 2030—2035.

[32] O. Martin, L. Avérous, Poly(lactic acid): Plasticization and properties of biodegradable multiphase systems, Polymer 42 (14) (2001) 6237—6247.

[33] M. Yasuniwa, S. Tsubakihara, Y. Sugimoto, C. Nakafuku, Thermal analysis of the double-melting behavior of poly(L-lactic acid), J. Polym. Sci., Polym. Phys. 42 (2004) 25—32.

[34] E. Schwach, L. Averous, Starch-based biodegradable blends: Morphology and interface properties, Polym. Int. 53 (12) (2004) 2115—2124.

[35] J.R. Dorgan, J.S. Williams, D.N. Lewis, Melt rheology of poly(lactic cid): Entanglement and chain architecture effects, J. Rheol. 43 (5) (1999) 1141—1155.

[36] M.C. Gupta, V.G. Deshmukh, Thermal oxidative degradation of poly-lactic acid. Part II: Molecular weight and electronic spectra during isothermal heating, Colloid Polym. Sci. 260 (1982) 514—517.

[37] X. Zhang, U.P. Wyss, D. Pichora, M.F.A. Goosen, An investigation of the synthesis and thermal stability of poly(DL-lactide), Polym. Bull. 27 (1992) 623—629.

[38] I.C. McNeill, H.A. Leiper, Degradation studies of some polyesters and polycarbonates. 2. Polylactide: degradation under isothermal conditions, thermal degradation mechanism and photolysis of the polymer, Polym. Degrad. Stab. 11 (1985) 309—326.

[39] W. Amass, A. Amass, B. Tighe, A review of biodegradable polymers: Uses, current developments in the synthesis and characterization of biodegradable polyesters, blends of biodegradable polymers and recent advances in biodegradation studies, Polym. Int. 47 (1998) 89—144.

[40] S.M. Li, H. Garreau, M. Vert, Structure-property relationships in the case of the degradation of massive aliphatic poly-(a-hydroxy acids) in aqeous media. Part 1. Poly(D, L-lactic acid), J. Mater. Sci. Mater. Med. 1 (1990) 123—130.

[41] S. Ramakrishna, J. Mayer, E. Wintermantel, K.W. Leong, Biomedical applications of polymer-composite materials: A review, Compos. Sci. Technol. 61 (2001) 1189—1224.

[42] M. Hakkarainen, S. Karlsson, A.C. Albertsson, Rapid (bio)degradation of polylactide by mixed culture of compost microorganisms low molecular weight products and matrix changes, Polymer 41 (2000) 2331—2338.

[43] R. Gattin, A. Copinet, C. Bertrand, Y. Couturier, Biodegradation study of a coextruded starch and poly(lactic acid) material in various media, J. Appl. Polym. Sci. 88 (2003) 825—831.

[44] S. Sinha Ray, Y. Kazunobu, M. Okamoto, K. Ueda, Control of biodegradability of polylactide via nanocomposite technology, Macromol. Mater. Eng. 288 (3) (2003) 203—208.

[45] N. Ljungberg, T. Andersson, B. Wesslen, Film extrusion and film weldability of poly(lactic acid) plasticized with triacetine and tributyl citrate, J. Appl. Polym. Sci. 88 (2003) 3239—3247.

[46] L.V. Labrecque, R.A. Kumar, V. Dave, R.A. Gross, S.P. McCarthy, Citrate esters as plasticizer for poly (lactic acid), J. Appl. Polym. Sci. 66 (18) (1997) 1507—1513.

[47] J.F. Zhang, X. Sun, Physical characterization of coupled poly (lactic acid)/starch/maleic anhydride blends by triethyl citrate, Macromol. Biosci. 4 (2004) 1053–1060.
[48] N. Ljungberg, B. Wesslen, The effects of plasticizers on the dynamic mechanical and thermal properties of poly(lactic acid), J. Appl. Polym. Sci. 86 (5) (2002) 1227–1234.
[49] J.F. Zhang, X. Sun, Mechanical and thermal properties of poly (lactic acid)/starch blends with dioctyl maleate, J. Appl. Polym. Sci. 94 (2004) 1697–1704.
[50] R.A.D. Graaf, L.P.B.M. Janssen, Properties and manufacturing of a new starch plastic, Polym. Eng. Sci. 41 (3) (2001) 584–594.
[51] D. Carlson, L. Nie, R. Narayan, P. Dubois, Maleation of polylactide (PLA) by reactive extrusion maleation of polylactide (PLA) by reactive extrusion, J. Appl. Polym. Sci. 72 (4) (1999) 477–485.
[52] C. Nouvel, P. Dubois, E. Dellacherie, J.L. Six, Controlled synthesis of amphiphilic biodegradable polylactide-grafted dextran copolymers, J. Polym. Sci., Polym. Chem. 42 (11) (2004) 2577–2588.
[53] H. Tsuji, H. Muramatsu, Blends of aliphatic polyesters. IV. Morphology, swelling behavior, and surface and bulk properties of blends from hydrophobic poly(L-lactide) and hydrophilic poly(vinyl alcohol), J. Appl. Polym. Sci. 81 (2001) 2151–2160.
[54] M.L. Focarete, M. Scandola, P. Dobrzynski, M. Kowalczuk, Miscibility and mechanical properties of blends of (L)-lactide copolymers with atactic poly(3-hydroxybutyrate), Macromolecules 35 (2002) 8472–8477.
[55] M. Dell'Erba, G. Groeninckx, G. Maglio, M. Malinconico, A. Migliozzi, Immiscible polymer blends of semicrystalline biocompatible components: Thermal properties and phase morphology analysis of PLLA/PCL blends, Polymer 42 (18) (2001) 7831–7840.
[56] O. Martin, E. Schwach, L. Avérous, Y. Couturier, Properties of biodegradable multilayer films based on plasticized wheat starch, Starch/Starke 53 (8) (2001) 372–380.
[57] O. Martin, L. Avérous, Comprehensive experimental study of a starch/polyesteramide coextrusion, J. Appl. Polym. Sci. 86 (10) (2002) 2586–2600.
[58] C. Courgneau, S. Domenek, A. Guinault, L. Averous, L. Ducruet, Analysis of the structure-properties relationships of different multiphase systems based on plasticized PLA, J. Polym. Environ. 19 (2) (2011) 362–371.
[59] Y.Z. Wan, Y.L. Wang, Q.Y. Li, X.H. Dong, Influence of surface treatment of carbon fibers on interfacial adhesion strength and mechanical properties of PLA-based composites, J. Appl. Polym. Sci. 80 (2001) 367–376.
[60] S. Sinha Ray, M. Okamoto, Polymer/layered silicate nanocomposites: A review from preparation to processing, Prog. Polym. Sci. 28 (11) (2003) 1539–1641.
[61] S. Sinha Ray, M. Okamoto, New polylactide/layered silicate nanocomposites part 6, Macromol. Mater. Eng. 288 (2003) 936–944.
[62] P. Bordes, E. Pollet, L. Avérous, Nano-biocomposites: Biodegradable polyester/nanoclay systems, Prog. Polym. Sci. 34 (2009) 125–155.
[63] V.P. Martino, A. Jiménez, R.A. Ruseckaite, L. Averous, Structure and properties of clay nano-biocomposites based on Poly(lactic acid) plasticized with polyadipates, Polym. Advan. Technol. 22 (2011) 2206–2213.
[64] Y. Fujimoto, S. Sinha Ray, M. Okamoto, A. Ogami, K. Yamada, K. Ueda, Well-controlled biodegradable nanocomposite foams: From microcellular to nanocellular, Macromol. Rapid. Commun. 24 (2003) 457–461.
[65] P. Sarazin, X. Roy, B. Favis, Controlled preparation and properties of porous poly (image-lactide) obtained from a co-continuous blend of two biodegradable polymers, Biomaterials 25 (2004) 5965–5978.
[66] P. Tormala, J. Vasenius, S. Vainionpaa, J. Laiho, T. Pohjonen, P. Rokkanen, Ultra-high-strength absorbable self-reinforced polyglycolide (SR-PGA) composite rods for internal fixation of bone fractures: *In vitro* and *in vivo* study, J. Biomed. Mater. Res. 25 (1991) 1–22.
[67] A.C. Albertsson, I.K. Varma, Recent developments in ring opening polymerization of lactones for biomedical applications, Biomacromolecules 4 (2003) 1466–1486.

10 Compostable Polymer Materials: Definitions, Structures, and Methods of Preparation

Ewa Rudnik

OUTLINE

"Biodegradable polymers" or "compostable polymers" were first commercially introduced in the 1980s. These first-generation biodegradable products were made from a conventional polymer, usually polyolefin (e.g., polyethylene) mixed together with starch or some other organic substance. When starch was eaten by microorganisms, the products were broken down, leaving small fragments of polyolefins.

In 1994 Narayan *et al.* wrote: "The U.S. biodegradables industry fumbled at the beginning by introducing starch filled (6–15%) polyolefins as true biodegradable materials. These at best were only biodisintegradable and not completely biodegradable. Data showed that only the surface starch biodegraded, leaving behind a recalcitrant polyethylene material" [1].

The situation confused consumers and government regulators, and put into question the biodegradable plastics market for some years. Since then the confusion or misunderstanding appeared about what was and what was not biodegradable and/or compostable. Additionally, no scientifically based test methods or standards existed to support claims made by plastics manufacturers for the "biodegradability" or "compostability" of their products.

More recently, international and national standards bodies, i.e., International Organization for Standardization (ISO), American Society for Testing and Materials (ASTM), Japanese Standards Association (JIS), and European Organization for Standardization (EN), have developed definitions related to the degradation of plastics. Nowadays, ISO and ASTM standards exist describing in detail the purposes of "biodegradable" and "compostable."

The ASTM D6400 standard establishes the requirements for the labeling of materials and products, including packaging made from plastics, as "compostable in municipal and industrial composting facilities" (Table 10.1).

ISO 17088 specifies test methods and requirements to determine and label plastic products and products made from plastics that are designed to be recovered through aerobic composting. It particularly establishes the requirements for labeling of materials and products, including packaging made from plastics, as "compostable," "compostable in municipal and industrial composting facilities," and "biodegradable during composting."

Ebnesajjad: Handbook of Biopolymers and Biodegradable Plastics. http://dx.doi.org/10.1016/B978-1-4557-2834-3.00010-0
© 2013 Elsevier Inc. All rights reserved. Adapted from a chapter in: Rudnik, *Compostable Polymer Materials* (2008).

Table 10.1 Definitions of Compostability According to ASTM D6400 [2]

Compostable Plastic A plastic that undergoes degradation by biological processes during composting to yield carbon dioxide, water, inorganic compounds, and biomass at a rate consistent with other known compostable materials and leaves no visually distinguishable or toxic residues.
Composting A managed process that controls the biological decomposition and transformation of biodegradable materials into a humus-like substance called compost: the aerobic mesophilic and thermophilic degradation of organic matter to make compost, the transformation of biologically decomposable material through a controlled process of biooxidation that proceeds through mesophilic and thermophilic phases and results in the production of carbon dioxide, water, minerals, and stabilized organic matter (compost or humus). Composting uses a natural process to stabilize mixed decomposable organic material recovered from municipal solid waste, yard trimmings, biosolids (digested sewage sludge), certain industrial residues, and commercial residues.
Degradable Plastic A plastic designed to undergo a significant change in its chemical structure under specified environmental conditions, resulting in a loss of some properties that may be measured by standard test methods appropriate to the plastic and the application in a period of time that determines its classification.

Table 10.2 Definitions of Compostability According to ISO 17088 [3]

Compostable Plastics A plastic that undergoes degradation by biological processes during composting to yield CO_2, water, inorganic compounds, and biomass at a rate consistent with other known compostable materials and leaves no visible, distinguishable, or toxic residue.
Composting The autothermic and thermophilic biological decomposition of biowaste (organic waste) in the presence of oxygen and under controlled conditions by the action of micro- and macroorganisms in order to produce compost.
Compost Organic soil conditioner obtained by biodegradation of a mixture consisting principally of vegetable residues, occasionally with other organic material and having a limited mineral content.
Disintegration The physical breakdown of a material into very small fragments.

The definition of "compostable plastic" proposed in ISO 17088 is identical to that given in the ASTM D 6400 standard (Table 10.2).

In spite of its very large use (and abuse), the term "biodegradable" is not helpful because it is not informative. The term does not convey any information about the specific environment where the biodegradation is supposed to take place, the rate that will regulate the process (fast, slow), and the extent of biodegradation (partial or total conversion into CO_2).

The definition of "biodegradable" has been assessed during the past decade. Some examples of definitions of "biodegradable plastic" are given in the following.

ASTM definition [2]: "a degradable plastic in which the degradation results from the action of naturally occurring microorganisms such as bacteria, fungi, and algae."

ISO and CEN definition [4]: "degradable plastic in which degradation results in lower molecular weight fragments produced by the action of naturally occurring microorganisms such as bacteria, fungi and algae."

According to ISO definition [4] degradable plastic means "A plastic designed to undergo a significant change in its chemical structure under specific environmental conditions resulting in a loss of some properties that may vary as measured by standard test methods appropriate to the plastic and the application in a period of time that determines its classification."

Japanese Biodegradable Polymers Society (BPS) defines biodegradable plastics (called GreenPla) as plastics that can be used as conventional plastics, while on disposal they decompose to water and carbon dioxide by the action of microorganisms commonly existing in the natural environment [5].

Most of the definitions of biodegradation are based on the same concept: the action of microorganisms on the material and its conversion into carbon dioxide or methane and water.

A plastic can be degradable without being biodegradable, i.e., it might disintegrate into pieces or even an invisible powder, but not be assimilated by microorganisms. A plastic can be degradable and even biodegradable without being compostable, i.e., it might biodegrade at a rate that is too slow to be called compostable [6].

The difference between biodegradable and compostable polymers lies in additional requirements related to the latter. Besides biodegradation into carbon dioxide, water, inorganic compounds, and biomass, compostable polymers must fulfill other criteria such as compatibility with the composting process, no negative effect on quality of compost and a degradation rate consistent with other known composting materials.

It is noteworthy that compostable plastics are *a priori* designed for a given method of safe disposal, i.e., composting. This means that after their useful life they will biodegrade in a composting process. The idea of compostable polymers is in agreement with life-cycle thinking.

To summarize, the requirements a material must satisfy to be termed "compostable" include mineralization (i.e., biodegradation to carbon dioxide, water, and biomass), disintegration into a composting system, and completion of its biodegradation during the end-use of the compost, which, moreover, must meet relevant quality criteria, e.g., no ecotoxicity. The satisfaction of requirements should be proved by standardized test methods.

Compostable polymers can be divided according to source of origin or method of their preparation (Fig. 10.1).

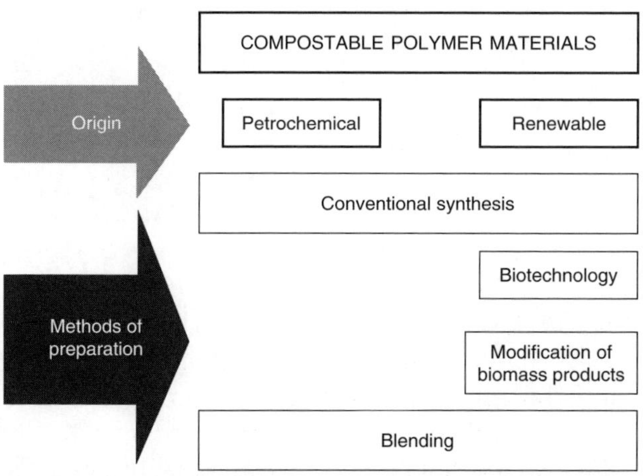

Figure 10.1 Classification of compostable polymers.

On the basis of origin, compostable polymers are derived from renewable and petrochemical resources.

Biodegradable polymers from renewable resources include the following:

1. polylactide (PLA)
2. polyhydroxyalkanoates: poly(3-hydroxybutyrate) (PHB)
3. thermoplastic starch (TPS)
4. cellulose
5. chitosan
6. proteins

Biodegradable polymers from petroleum sources comprise the following:

1. aliphatic polyesters and copolyesters (e.g., poly(butylene succinate)—PBS; poly(butylene succinate adipate)—PBSA)
2. aromatic copolyesters (e.g., poly(butylene adipate terephthalate)—PBAT)
3. poly(ε-caprolactone)—PCL
4. polyesteramides—PEAs
5. poly(vinyl alcohol)—PVA

There are three principal ways to produce polymers from renewable resources, i.e., bio-based polymers:

1. to make use of natural polymers that may be modified but remain intact to a large extent (e.g., starch polymers);
2. to produce bio-based monomers by fermentation which are then polymerized (e.g., polylactic acid, PLA);
3. to produce bio-based polymers directly in microorganisms or in genetically modified crops (polyhydroxyalkanoates).

In general, on the basis of methods of preparation, compostable polymer materials can be prepared via:

1. conventional synthesis
   - polymerization from nonrenewable monomer feedstocks, e.g., poly(ε-caprolactone)—PCL—copolyesters;
   - polymerization from renewable monomer feedstocks, e.g., PLA;

2. biotechnological route (extraction, fermentation), e.g., poly(hydroxybutyrate-*co*-hydroxyvalerate)—PHBV;
3. preparation directly from biomass, e.g., plants—starch;
4. blending, e.g., starch–PCL blends.

A method based on blending of biodegradable polymers is very often used in order to improve the properties of compostable polymer materials or to decrease their cost. The various polymers used are both renewable and of petrochemical origin. Novamont's Mater-Bi is an example of such a material.

10.1 Biodegradable Polymers from Renewable Resources

10.1.1 Poly(lactic acid)—PLA

The molecular structure of PLA is schematically presented in Fig. 10.2. PLA, linear aliphatic thermoplastic polyester, is prepared from lactic acid. Lactic acid (2-hydroxy propionic acid) is one of the simplest chiral molecules and exists as two stereoisomers, L- and D-lactic acid (Fig. 10.3).

Lactic acid is the most widely occurring carboxylic acid in nature [7]. It was discovered by the Swedish chemist Scheele in 1780 as a sour component of milk, and was first produced commercially by Charles E. Avery at Littleton, Massachusetts, USA, in 1881. Lactic acid can be manufactured by chemical synthesis or carbohydrate fermentation. First, lactic acid was petrochemically derived [8]. The commercial process for chemical synthesis is based on lactonitrile ($CH_3CHOHCN$) obtained from acetaldehyde (CH_3CHO) and hydrogen cyanide (HCN). After recovery and purification by distillation, lactonitrile is then hydrolyzed to lactic acid [7,8]. Lactic acid produced by the petrochemical route exists as a racemic (optically inactive) mixture of D and L forms. Though chemical synthesis produces a racemic mixture, stereospecific lactic acid can be made by carbohydrate fermentation depending on the strain being used.

Lactic acid-based polymers are prepared by polycondensation, ring-opening polymerization (ROP), and other methods (chain extension, grafting). High-molecular-weight PLA is generally produced by the ROP of the lactide monomer. The conversion of lactide to high-molecular-weight polylactide is achieved commercially by two routes. Recently, Cargill Dow used a solvent-free process and a novel distillation process to produce a range of PLA polymers. The process consists of three separate and distinct steps that lead to the production of lactic acid, lactide, and PLA high polymer [8] (Fig. 10.4).

Each of the process steps is free of organic solvent—water is used in fermentation while molten lactide and polymer serve as the reaction media in monomer and polymer production. The essential novelty of the process lies in the ability to go from lactic acid to a low-molecular-weight PLA, followed by controlled depolymerization to produce the cyclic dimer, commonly referred to as lactide. An organometallic catalyst, e.g., tin octanoate, is used to enhance the rate and selectivity of the intramolecular cyclization reaction [9]. This lactide is maintained in liquid form and purified by distillation. Catalytic ROP of the lactide intermediate results in the

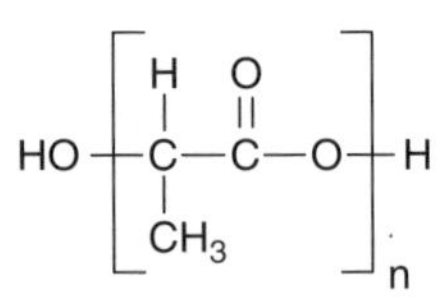

Figure 10.2 Structure of poly(lactic acid).

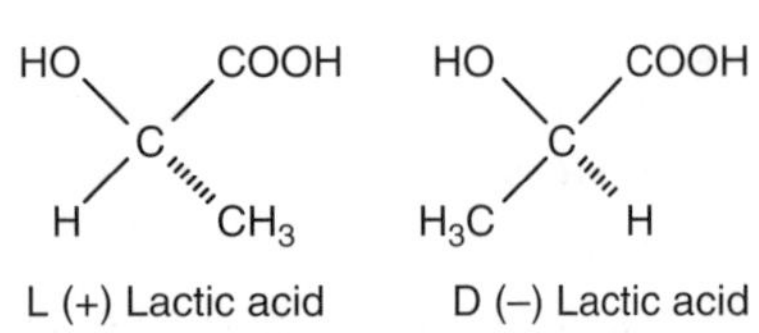

Figure 10.3 Stereoforms of lactic acid.

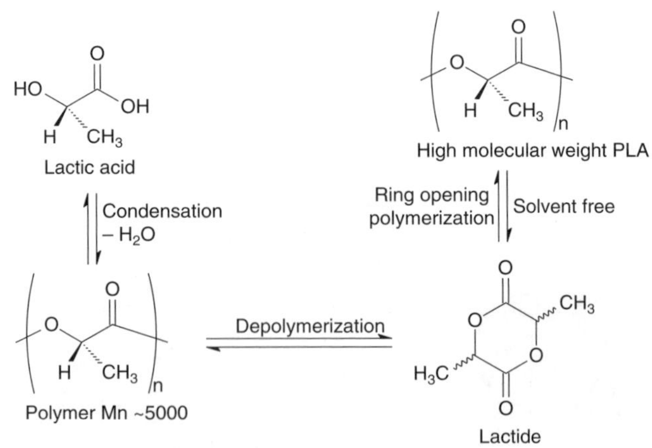

Figure 10.4 Manufacturing route to poly(lactic acid) according to the Cargill Dow process.

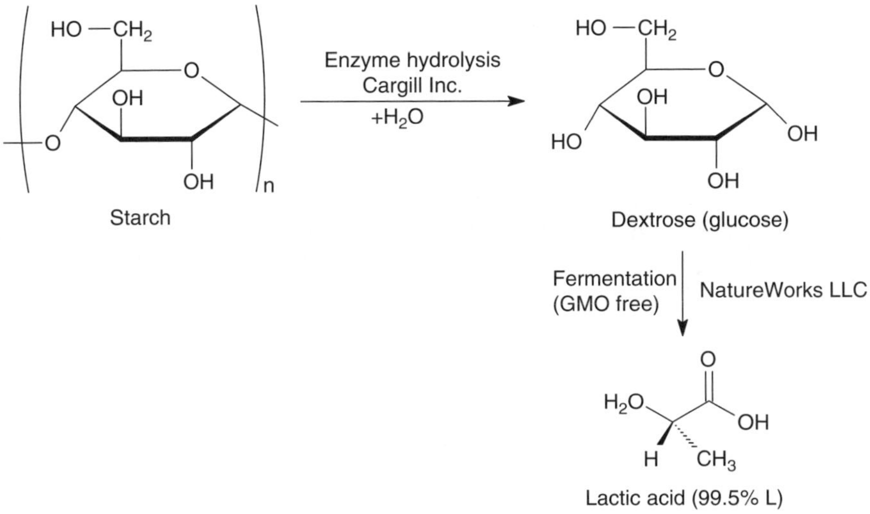

Figure 10.5 Cargill route to lactic acid.

production of PLA with controlled molecular weights. The process is continuous with no necessity to separate the intermediate lactide.

Lactic acid used in the preparation of PLA is derived from annually renewable resources. Cargill Dow uses sugar from maize as feedstock, due to its low cost and abundance, but it is envisaged to use local plant sources containing starch or sugar, such as wheat, sugar beets, or agricultural waste (Fig. 10.5).

The ROP of lactic acid monomers is catalyzed by compounds of transition metals: tin, aluminium, lead, zinc, bismuth, iron, and yttrium. A collection of more than 100 catalysts for PLA synthesis was reviewed [10,11]. The catalysts used mainly consist of metal powders, Lewis acids, Lewis bases, organometallic compounds and different salts of metals. However, organometallic compounds are very effective in the synthesis of high-molecular-weight PLA particularly alkali metals and metal halides, oxides, carboxylates, and alkoxides.

In contrast, Mitsui Toatsu (presently Mitsui Chemicals) utilizes a solvent-based process, in which a high-molecular-weight PLA is produced by direct condensation using azeotropic distillation to remove the water of condensation continuously (Fig. 10.6).

The synthesis of PLA through polycondensation of the lactic acid monomer gave an average molecular weight lower than 1.6×10^4, whereas ROP of lactides gave average molecular weights ranging from 2×10^4 to 6.8×10^4 [7].

Purac, producer of lactic acid, developed the technology of formation of stereocomplex PLA in a solid status by melt-blending PLLA and PDLA through a transesterification process using a catalyst [9]. The PLLA and PDLA polymers originate from separately polymerized L-lactide and D-lactide.

Copolymerization and blending of PLA has been extensively investigated as a useful route to obtain a product with a particular combination of desirable properties. Other ring formed monomers are also incorporated into the lactic acid-based polymer by ROP [7,12]. The most utilized comonomers are glycolide (1,4-dioxane-2,5-dione), ε-caprolactone (2-oxepanone), γ-valerolactone (2-pyranone), 1,5-dioxepane-2-one, and trimethylene carbonate (1,3-dioxan-2-one). Examples of repeating units of comonomers are given in Table 10.3.

Lactic acid-based polyesters could also be produced by enzymatic catalysis. For example, lipase-catalyzed ROP of cyclic lactides is applicable for the synthesis of PLA [12–14].

The polymers derived from lactic acid by the polycondensation route are generally referred to as poly(lactic acid) and the ones prepared from lactide by ROP as polylactide [15]. Both types are generally referred to as PLA.

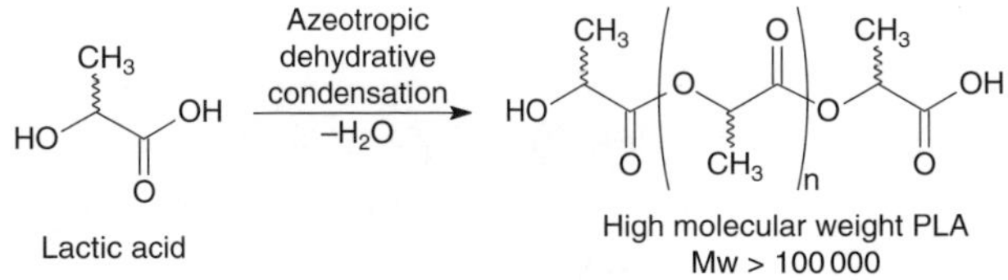

Figure 10.6 Manufacturing route to poly(lactic acid) according to the Mitsui process.

Table 10.3 Repeating Units of the Most Common Lactic Acid Comonomers

Name	Lactones	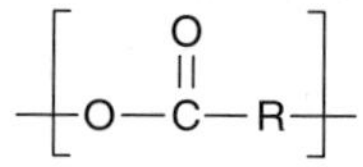 Structure where R
Poly(glycolide)	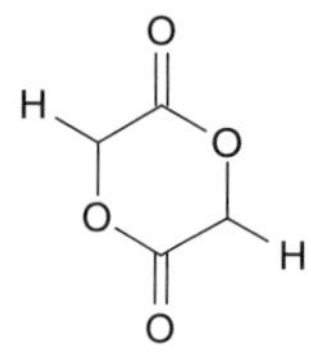	$-CH_2-$
Poly(lactide)	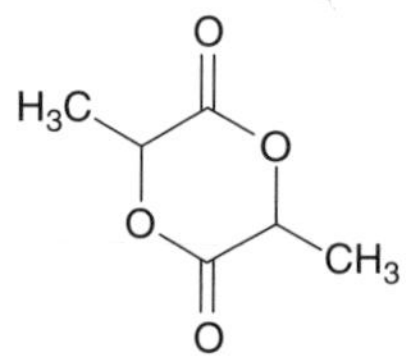	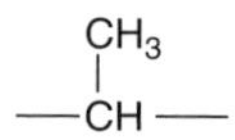
Poly(δ-valerolactone)	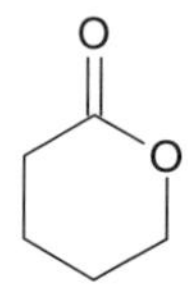	$-(CH_2)_4-$
Poly(ε-caprolactone)	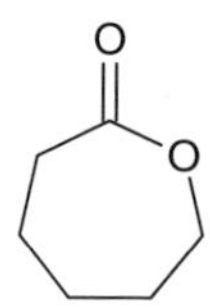	$-(CH_2)_5-$
Poly(β-hydroxybutyrate)	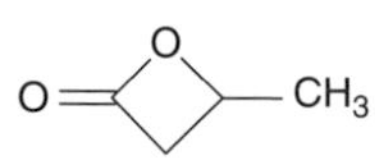	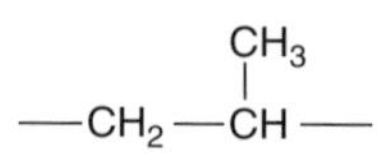
Poly(β-hydroxyvalerate)	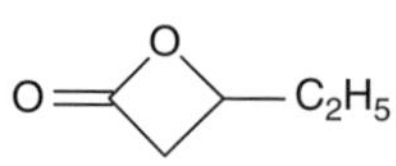	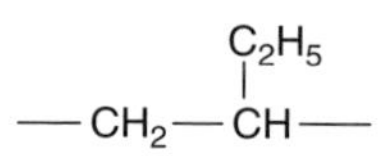
Poly(1,5-dioxepane-2-one)	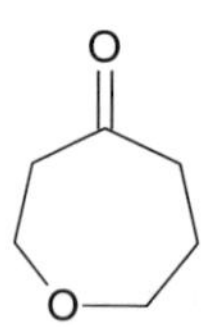	$-(CH_2)_2-O-(CH_2)_2-$
Poly(trimethylene carbonate)	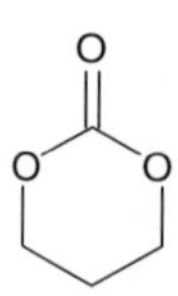	$-O-(CH_2)_3-$

Table 10.4 Commercially Available PLA Polymers

Trade Name	Supplier	Origin	Website
Lacea	Mitsui Chemicals	Japan	www.mitsui-chem.co.jp/e
Lacty	Shimadzu	Japan	www.shimadzu.co.jp
NatureWorks	Cargill Dow	USA	www.NatureWorksLLC.com
Hycail	Hycail b.v.	The Netherlands	www.hycail.com
Biofront	Teijin	Japan	http://www.teijin.co.jp/english
Futerro	Galactic/Total Petrochemical	Belgium	www.futerro.com
PLA	Zhejiang Hisun Biomaterials	China	http://hisunpla.en.gongchang.com/
PURAC[1]	Purac Biochem	The Netherlands	http://www.purac.com

[1]Partnership between PURAC, Sulzer, and Synbra. PURAC provides lactide and Synbra polymerizes the lactide into PLA, using PLA technology that was jointly developed by PURAC and Sulzer. Synbra processes the polymer into expanded PLA foam [9].

Table 10.4 lists the commercially available PLA polymers.

10.1.2 Polyhydroxyalkanoates—PHA

Figure 10.7 shows the generic formula for PHAs, where x is 1 for all commercially relevant polymers and R can be hydrogen or hydrocarbon chains of up to C15 in length.

Polyhydroxyalkanoates (PHA) are polyesters of various hydroxyalkanoates that are synthesized by many Gram-positive and Gram-negative bacteria from at least 75 different bacteria [16]. These polymers are accumulated intracellularly to levels as high as 90% of the cell dry weight under conditions of nutrient stress and act as a carbon and energy reserve.

In 1920s French bacteriologist Lemoigne discovered aliphatic polyester—poly(3-hydroxybutyrate) (PHB) as a granular component in bacterial cells [17]. PHB is the reserve polymer found in many types of bacteria, which can grow in a wide variety of natural environments and which have the ability to produce and polymerize the monomer [R]-3-hydroxybutyric acid. The repeating unit of PHB has a chiral center (Fig. 10.8) and the polymer is optically active.

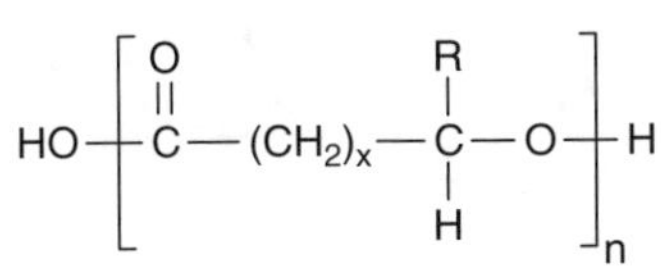

Figure 10.7 Structure of polyhydroxyalkanoates.

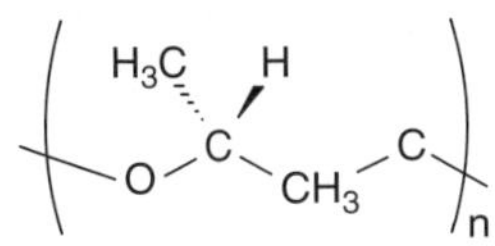

Figure 10.8 Repeating unit of PHB.

It was determined by Stanier, Wilkinson, and coworkers that PHB granules in bacteria serve as an intracellular food and energy reserve [17]. PHB polymer is produced by the cell in response to a nutrient limitation in the environment in order to prevent starvation if an essential element becomes unavailable [17]. It is consumed when no external carbon source is available.

Since the discovery of the simple PHB homopolymer by Lemoigne in the mid-1920s, a family of over 100 different aliphatic polyesters of the same general structure has been discovered. PHB is only the parent member of a family of natural polyesters having the same three-carbon backbone structure but differing in the type of alkyl group at the β or 3 position [17]. These polymers are referred to in general as polyhydroxyalkanoates (PHAs) and have the same configuration for the chiral center at the 3 position, which is very important both for their physical properties and for the activities of the enzymes involved in their biosynthesis and biodegradation. PHAs are also named bacterial polyesters since they are produced inside the cells of bacteria.

A wide range of PHA homopolymers, copolymers, and terpolymers have been produced, in most cases at the laboratory scale. Bacteria that are used for the production of PHAs can be divided into two groups based on the culture conditions required for PHA synthesis [18]. The first group of bacteria requires the limitation of an essential nutrient such as nitrogen, phosphorous, magnesium, or sulfur for the synthesis of PHA from an excess carbon source. The following bacteria are included in this group: *Alcaligenes eutrophus*, *Protomonas extorquens*, and *Protomonas oleovorans*. The second group of bacteria, which includes *Alcaligenes latus*, a mutant strain of *Azotobacter vinelandii*, and recombinant *Escherichia coli*, do not require nutrient limitation for PHA synthesis and can accumulate polymer during growth.

PHAs exist as discrete inclusions that are typically 0.2 $\pm$ 0.5 mm in diameter localized in the cell cytoplasm [18]. The molecular weight of PHAs ranges from 2×10^5 to 3×10^6, depending on the microorganism and the growth conditions.

Today, PHAs are separated into three classes: short chain-length PHA (scl-PHA, carbon numbers of monomers ranging from C3 to C5), medium chain-length PHA (mcl-PHA, C6–C14), and long chain-length PHA (lcl-PHA, >C14). The main members of the PHA family are the homopolymers PHB, which has the generic formula in Fig. 10.7 with R = 1(methyl), and poly(3-hydroxyvalerate) (PHV), with the generic formula with R = 2 (ethyl). PHAs containing 3-hydroxy acids have a chiral center and hence are optically active (Table 10.5).

Mcl-PHAs were first discovered in 1983 when *Pseudomonas oleovorans* was grown in octane [19]. Since then many fluorescent *Pseudomonas* species have been used for their production. To date more than 150 units of mcl-PHA monomers have been produced by culturing various *Pseudomonas* strains on different carbon substrates [*x*]. The versatility of *Pseudomonas* species in using a range of carbon sources and low substrate specificity of the mcl-PHA synthase, the key enzyme involved in the polymerization of medium chain-length hydroxyacyl coenzyme A (CoA) into mcl-PHA, is responsible for the diversity in mcl-PHA monomers. *Pseudomonas* species can be grown on both structurally related and unrelated carbon sources for producing PHAs. Structurally related carbon sources such as alkanes, alkenes, and aldehydes produce precursor substrates that exhibit structures related to the constituents of the mcl-PHAs. Mcl-PHAs are more structurally diverse than scl-PHAs and hence can be more readily tailored for specific applications. Studies on mcl-PHAs are still somewhat limited to P(3HO) and its copolymers and P(3HB-*co*-3HHx), which are available in large quantities.

Copolymers of PHAs vary in the type and proportion of monomers, and are typically random in sequence. PHBV is made up of a random arrangement of the monomers R 5 = 1 and R = 2. Poly(3-hydroxybutyrate-*co*-3-hydroxyhexanoate) (PHBH) consists of the monomers R = 1 (methyl) and R = 3 (propyl). The Nodax® family of copolymers are poly(3-hydroxybutyrate-*co*-3-hydroxyalkanoate)s with copolymer content varying from 3 to 15 mol% and chain length from C7 up to C19 [20].

Large-scale commercial production of PHAs uses fermentation technologies. A generic process for PHA produced by bacterial fermentation consists of three basic steps: fermentation, isolation and purification, and blending and palletizing [20]. Subsequent to inoculation and small-scale fermentation, a large fermentation vessel is filled with mineral medium and inoculated with seed ferment (containing the microbe or bacteria). The carbon source is fed at various rates until it is completely consumed and cell

Table 10.5 Polyhydroxyalkanoates Family

PHA	3-Hydroxy Acids With Side Chain R
P(3HB)	$—CH_3$
P(3HV)	$—CH_2 CH_3$
P(3HB-*co*-3HV) (Biopol®)[1]	$—CH_3$ and $—CH_2CH_3$
P(3HB-*co*-3HHx) (Kaneka)[2], (Nodax®)[3]	$—CH_3$ and $—CH_2 CH_2 CH_3$
P(3HB-*co*-3HO) (Nodax®)	$—CH_3$ and $—CH_2 CH_2 CH_2 CH_2CH_3$
P(3HB-*co*-3HOd) (Nodax®)	$—CH_3$ and $—(CH_2)_{14} CH_3$

[1]Patent held by Metabolix, Inc. [2]Kaneka holds the patent on chemical composition. [3]P&G holds processing and application patents.

growth and PHA accumulation is complete. Current carbon sources for producing PHA are carbohydrates (glucose, fructose, sucrose); alcohols (methanol, glycerol); alkanes (hexane to dodecane); and organic acids (butyrate upward). In the United States, the raw material source is chiefly corn steep liquor; in the Europe beet sugar predominates. The total fermentation step typically takes 38–48 h. To isolate and purify PHA, the cells are concentrated, dried, and extracted with hot solvent. The residual cell debris is removed from the solvent containing dissolved PHA by a solid–liquid separation process. The PHA is then precipitated by addition of a nonsolvent and recovered by the solid–liquid separation process. PHA is washed with solvent to enhance the quality and dried under vacuum and moderate temperatures (in certain cases where high purity product is not needed, solvent extraction may not be required). The solvents are distilled and recycled separately. The neat polymer is typically preformed into pellets with or without other polymer ingredients [20].

PHAs are produced from a wide variety of substrates such as renewable resources (sucrose, starch, cellulose, triacylglycerols), fossil resources (methane, mineral oil, lignite, hard coal), by-products (molasses, whey, glycerol), chemicals (propionic acid, 4-hydroxybutyric acid), and carbon dioxide [16].

As the major cost in the production of PHA is the medium, efforts are focused on finding cheap media. Extensive studies to select cheap sources for fermentation include media containing molasses, corn steep liquor, whey, wheat and rice bran, starch and starchy wastewaters, effluents from olive mill and palm olive mill, activated sludge, and swine waste [21,22].

The microorganisms of choice for the industrial production of PHA varies depending on factors that include the cell's ability to utilize an inexpensive carbon source, the cost of the medium, the growth rate, the polymers synthesis rate, the quality and quantity of PHAs, and the cost of downstream processes [22]. Although more than 300 different microorganisms synthesize PHAs, only a few-such as *Cupriavidus necator* (formerly known as *Ralstonia eutropha* or *Alcaligenes eutrophus*), *Alcaligenes latus*, *Azotobacter vinelandii*, *Pseudomonas oleovorans*, *Paracoccus denitrificans*, *Protomonas extorquens*, and recombinant *Escherichia coli*-are able to produce sufficient PHA for large-scale production [22].

There are different approaches and pathways for the synthesis of PHAs. Zimm *et al.* [23] distinguished four biosynthetic approaches to produce PHA: *in vitro* via PHA-polymerase catalyzed polymerization, and *in vivo* with batch, fed-batch, and continuous (chemostat) cultures.

The biosynthetic pathway of P(3HB) in *Alcaligenes eutrophus* (now renamed *Ralstonia eutropha*) consists of three enzymatic reactions catalyzed by three different enzymes [16,18] (Fig. 10.9).

The first reaction consists of the condensation of two acetyl-CoA molecules into acetoacetyl-CoA by β-ketoacyl-CoA thiolase. The second reaction is the reduction of acetoacetyl-CoA to (R)-3-hydroxybutyryl-CoA by an NADPH-dependent acetoacetyl-CoA dehydrogenase. Lastly, the (R)-3-hydroxybutyryl-*co*-A monomers are polymerized into PHB by P(3HB) polymerase.

Homopolymer PHB is a brittle, crystalline thermoplastic and undergoes thermal decomposition just at its melting point, thus making processing difficult and limiting its commercial usefulness. Therefore, extensive efforts have been directed toward synthesis of copolymers that have better properties than PHB. Zeneca (formerly Imperial Chemical Industries (ICI)) has developed the PHB copolymer PHBV also known as Biopol, which is less stiff and less brittle

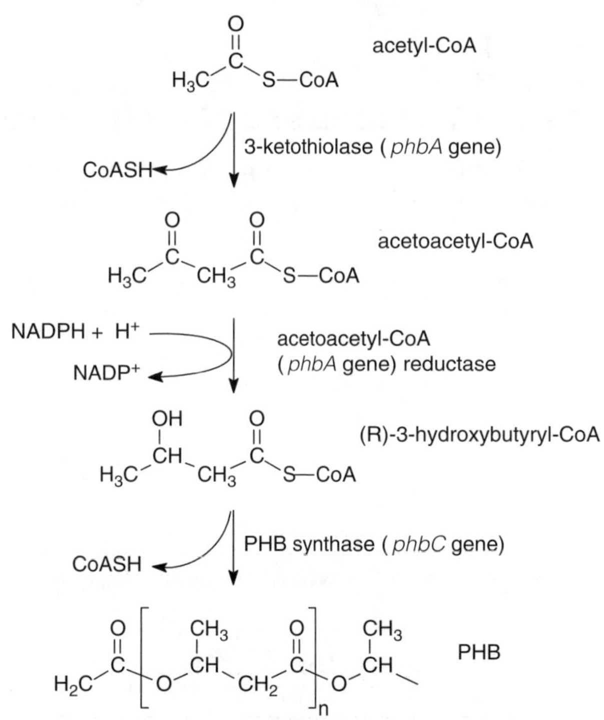

Figure 10.9 PHB synthesis in *Ralstonia eutropha.*

than homo-polymer PHB. The ratio of HB to HV monomer can be varied by changing the glucose to propionic acid ratio. By increasing the ratio of HV to HB, the melting temperatures are lower and mechanical properties are improved. In 1996 Zeneca sold its Biopol business to Monsanto, and then in 2001 Metabolix acquired Monsanto Biopol technology. Recently, Metabolix began work on a $15 million program, supported by the US Department of Energy, to produce PHAs in high yield from native American prairie grass. In 2006 Metabolix and ADM established a joint venture, Telles, to sell PHA-based bioplastics under trade name Mirel in the United States, Europe, and other countries. Mirel is a product of corn sugar fermentation with proprietary genetically engineered bacteria.

Another company, Procter & Gamble, has directed efforts into development and commercialization of a variety of PHA copolymers under the name Nodax. The Nodax® family of copolymers are poly(3-hydroxybutyrate-*co*-3-hydroxyalkanoate)s with a copolymer content varying from 3 to 15 mol% and chain length from C7 up to C19 [20]. In 2003 Procter & Gamble licensed recovery and processing routes for PHAs to the Japanese company Kaneka Corporation. The companies have a joint agreement to commercialize the Nodax family of PHAs, made from corn or sugar beet and vegetable oils.

Commercially available PHAs are given in Table 10.6.

10.1.3 Thermoplastic Starch—TPS

Starch, the storage polysaccharide of cereals, legumes and tubers, is a renewable and widely available raw material, being the end product of photosynthesis. Starch is composed of a mixture of two substances, an essentially linear polysaccharide, amylose, and a highly branched polysaccharide, amylopectin (Figs 10.10 and 10.11).

Both forms of starch are polymers of α-D-glucose. The ratio of both forms varies according to the botanical origin of the starch. Natural starches contain 15–30% amylose and 85–70% amylopectin [24]. Both amylose and amylopectin have a distribution of sizes with different average numbers (degree of polymerization) of glucose residues. The average number of glucose residues for amylose can vary from 250 to 5000, and the average number of glucose residues for amylopectin can vary from 10,000 to 100,000.

Amylose is a relatively long, linear α-glucan containing around 99% (1→4)-α- and 1% (1→6)-α-linkages [25]. Amylose has a molecular weight of approximately 1×10^5 to 1×10^6, a degree of polymerization (DP) by number (DP_n) of 324–4920 with around 9–20 branch points equivalent to 3–11 chains per molecule. Amylopectin is a much larger than amylose with a molecular weight of $1 \times 10^7$to 1×10^9 and a heavily branched structure built from about 95% (1→4)-α- and 5% (1→6)-α-linkages. The DP_n is typically within the range 9600–15,900.

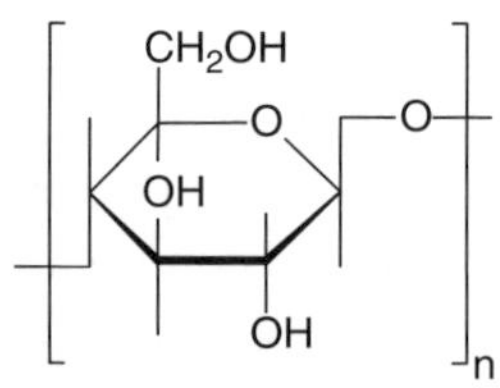

Figure 10.10 General structure of starch.

Table 10.6 Commercially Available PHA Polymers

Trade Name	Structure	Supplier	Origin	Website
Biopol®	poly(3-hydroxybutyrate-*co*-3-hydroxyvalerate)	Metabolix	USA	www.metabolix.com
Mirel	PHA	Telles (Metabolix)	USA	www.mirelplastics.com
Nodax®	poly(3-hydroxybutyrate-*co*-3-hydroxyalkanoate)s	Kaneka/P&G	Japan	www.nodax.com
Biogreen	poly(3-hydroxybutyrate)	Mitsubishi Gas Chemical	Japan	www.mgc.co.jp
Biomer	poly(3-hydroxybutyrate)	Biomer	Germany	www.biomer.de
Enmat™	PHB/PHBV	Tianan	China	www.tienan-enmat.com
Biocycle	PHB/PHBV	PHB Industrial S.A.	Brasil	www.biocycle.com.br

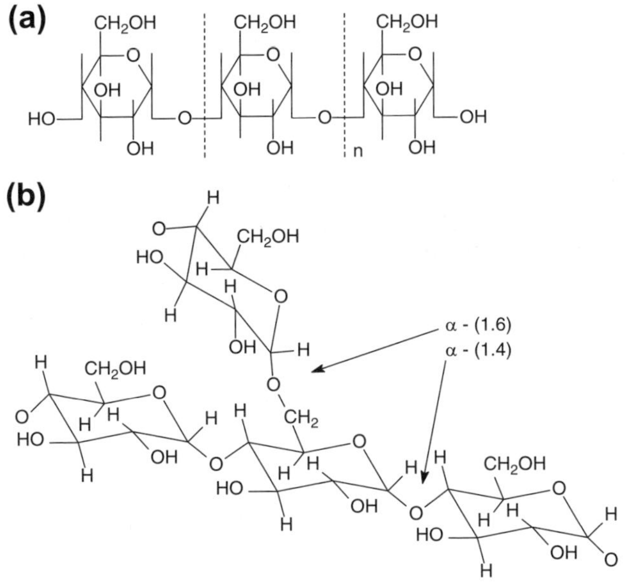

Figure 10.11 Schematic structure of (a) amylose and (b) amylopectin.

The size, shape, and morphology of the starch granules are characteristic of the particular botanical source (Fig. 10.12). Starch granules, typically ranging in size from 2 to 30 μm, depending on the plant origin, are partially crystalline and insoluble in cold water (Table 10.7).

The conventional processing of starch, including food processing and processing to produce pastes, thickeners and adhesives, is in the presence of heat and excess water [27].

In 1980s a breakthrough occurred by processing starch at approximately its natural water content (15%) in a closed volume at temperatures above 100 °C. Using conventional injection molding, glassy, amorphous, TPS polymers (T_g 60 °C) were obtained with moduli similar to those of polypropylene and high-density polyethylene.

TPS can be produced from native starch using a swelling or plasticizing agent while applying a dry starch in compound extruders without adding water. When starch with a water content higher than 5% is plastified or pasted under pressure and temperature, a destructured starch is always formed. In the production procedure of TPS, the mainly water-free raw material is homogenized and melted in an extrusion process with a plastifing material. Several plasticizers have been studied, including water, glycerol, sorbitol, glycol, poly(ethylene glycol), urea, glucose, maltose, as well as melt-flow accelerators,

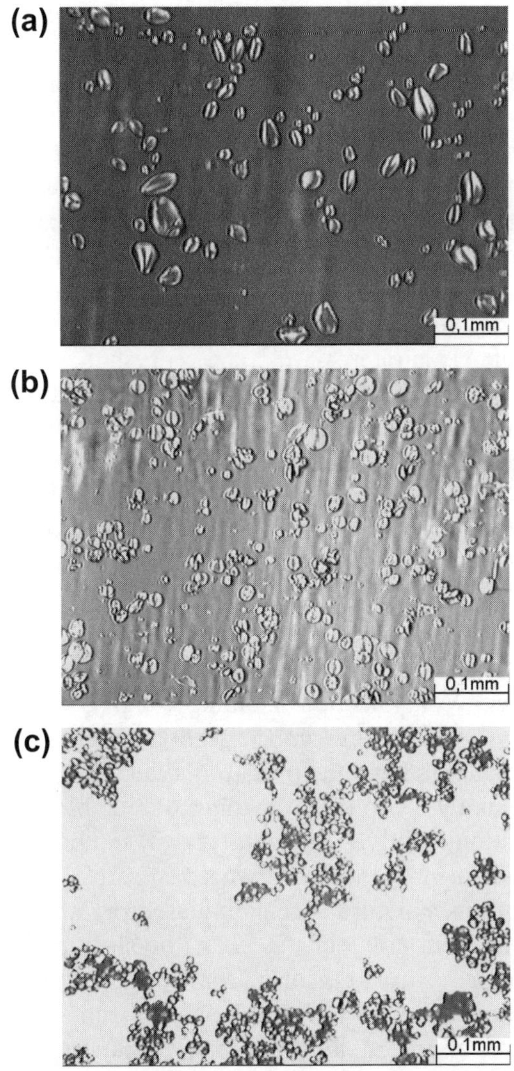

Figure 10.12 Optical micrographs of starch granules: (a) potato, (b) wheat, and (c) maize.

Table 10.7 Diameter and Gelatinization Temperature of Starch Granules [26]

Source	Mean Diameter, μm	Gelatinization Temperature, °C
Corn	15	62–71
Wheat	20–22	53–64
Rice	5	65–73
White potato	33	62–68
Sweet potato	25–50	82–83
Tapioca	20	59–70

such as lecithin, glycerol monostearate, and calcium stearate [28].

The glass temperature of starch-containing materials is a function of plasticizer content. Depending on the processing conditions and plasticizer content, thermomechanical processing of granular starch with the aid of plasticizers and melt-flow accelerators gives a complex starch plastic material. This is composed of residual swollen granular starch, partially melted, deformed and disrupted granules, completely molten starch, and recrystallized starch. The degree of disruption and melting of the various granular starches is regulated by the plasticizer content and by the processing parameters (shear stress, melt viscosity, and temperature).

The starch destructurization is defined as a partial fragmentation of the crystalline structure within the polysaccharides. By the transformation of native starch materials to highly amorphous thermoplastics, the compounded TPS formulation is remeltable and extrusion or injection molding is processable by renewing the energy input. Native starches can be destructurized within co-rotating twin screw extruder systems by a controlled feeding of suitable destructurization additives (water, glycerol) in combination with defined operating parameters [29].

Various mature technologies for processing conventional polymers, such as film/sheet extrusion, foaming extrusion, injection and compression molding, and casting, as well as new techniques like reactive extrusion, have been used to produce starch-based polymers [30]. However, as the processing of starch is much more difficult than conventional polymers, modifications to traditional processing techniques, carefully controlled processing conditions and the judicious use of additives have been used to overcome the various challenges presented in the processing of starch-based polymers. The achievements in this area are reviewed in Ref. [30]. The processing of starch is much more complicated and difficult to control than for conventional polymers, due to the unsatisfactory processing properties as a result of its unique phase transitions, high viscosity, water evaporation, fast retrogradation, etc. However, with proper formulation development and suitable processing conditions, many of these challenges can be overcome. Formulation developments include the following:

- Adding appropriate plasticizers;
- Adding appropriate lubricants;
- Using modified starch in which the hydroxyls have been replaced with ester and ether groups (e.g., carboxymethyl starch and hydroxypropylated starch);
- Blending starch with a hydrophobic polymer (e.g., PLA, PCL, or cellulose) in the presence of an appropriate compatibilizer (often a starch-graft-copolymer grafted with the hydrophobic polymer);
- Using copolymers of starch-graft-hydrophobic polymer, such as starch-graft-PLA, starch-graft-PCL, etc;
- Blending starch with a nanoclay to form starch nanocomposites.

Blends or composites materials have been produced by the processing of starch with biodegradable polymers such as PCL, PLA, PVA, PHBV, and PEA. The most common are Mater-Bi from Novamont and Ecostar from National Starch. Commercially available starch-based polymers are listed in Table 10.8.

Table 10.8 Commercially Available Starch-Based Polymers

Trade Name	Structure	Supplier	Origin	Website
Solanyl	Starch based	Rodenburg Biopolymers	The Netherlands	www.biopolymers.nl
Bioplast TPS	TPS	Biotem	Germany	www.biotec.de
EverCorn	Starch based	Japan Corn Starch	Japan	www.japan-cornstarch.com
Plantic	Starch based	Plantic Technologies	Australia	www.plantic.com.au
Biopar	Starch based	BIOP Biopolymer Technologies AG	Germany	www.biopag.de
Placorn	Starch based	Nihon Shokuhin Kako	Japan	www.nisshoku.co.jp

10.2 Other Compostable Polymers from Renewable Resources

10.2.1 Cellulose

Cellulose, the most abundant organic compound on earth, is the major structural component of the cell wall of higher plants [24]. It is a major component of cotton (95%), flax (80%), jute (60–70%), and wood (40–50%). Cellulose pulps can be obtained from many agricultural by-products such as sugarcane, sorghum bagasse, corn stalks, and straws of rye, wheat, oats, and rice.

Cellulose is a polydisperse linear polysaccharide consisting of β-1,4-glycosidic linked D-glucose units (so-called anhydroglucose unit) (Fig. 10.13).

The consequence of the supra-molecular structure of cellulose is its insolubility in water, as well as in common organic liquids [24,31]. Poor solubility in common solvents is one of the reasons why cellulose is converted to its cellulose esters. Another reason is that cellulose is not melt-processible, because it decomposes before it undergoes melt flow [32].

Cellulose esters have been commercially important polymers for nearly a century, and have found a variety of applications, including solvent-borne coatings, separation, medical and controlled release applications as well as composites and laminates, and plastics.

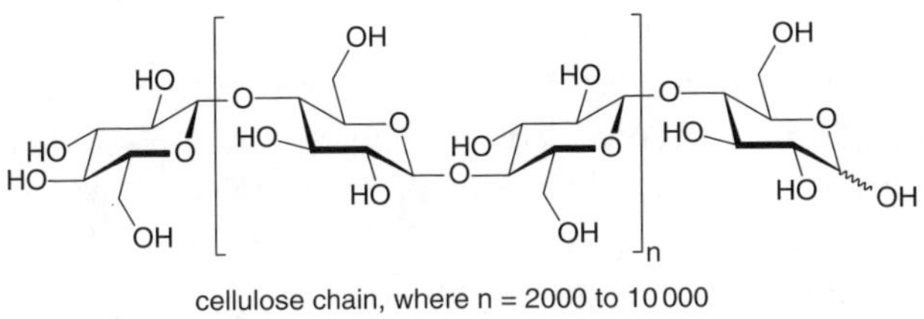

Figure 10.13 Schematic structure of cellulose.

The most common cellulose esters comprise cellulose acetate (CA), cellulose acetate propionate (CAP), and cellulose acetate butyrate (CAB). They are thermoplastic materials produced through esterification of cellulose. Different raw materials such as cotton, recycled paper, wood cellulose, and sugarcane are used to make the cellulose ester biopolymers in powder form. Bioceta, plasticized cellulose acetate, is prepared from cotton flakes and wood pulp through an esterification process with acetic anhydride. Cellulose acetate propionate (CAP) and cellulose acetate butyrate (CAB) are mixed esters produced by treating cellulose with appropriate acids and anhydrides in the presence of sulfuric acid.

Cellulose-based polymers are given in Table 10.9.

10.2.2 Chitosan

Chitin (poly(*N*-acetyl-D-glucosamine)) represents the second most abundant polysaccharide after cellulose. It is found in the exoskeleton of crustaceans and insects and in the cell wall of fungi and microorganisms [33]. Arthropod shells (exoskeletons), the most easily accessible sources of chitin, contain 20–50% of chitin on a dry basis. Wastes of seafood processing industries are used for the commercial production of chitin.

The structure of chitin is essentially the structure of cellulose, with the hydroxyl group at C-2 of the D-glucopyranose residue substituted with an *N*-acetylamino group [24] (Fig. 10.14).

Chitosan, poly-β(1,4)-2-amino-2-deoxy-D-glucopyranose, is the deacetylated product of chitin (Fig. 10.15).

Table 10.9 Cellulose-Based Polymers Commercially Available

Trade name	Structure	Supplier	Origin	Website
Natureflex	Cellulose based	Innovia Films (formerly Surface Specialties-UCB)	UK	www.innoviafilms.com
Tenite	Cellulose esters	Eastman	USA	www.eastman.com
Bioceta	Cellulose acetate	Mazzucchelli	Italy	www.mazzucchelli1849.it
Cellidor	Cellulose acetate propionate; cellulose acetate butyrate	Albis Plastics	Germany	www.albis.com

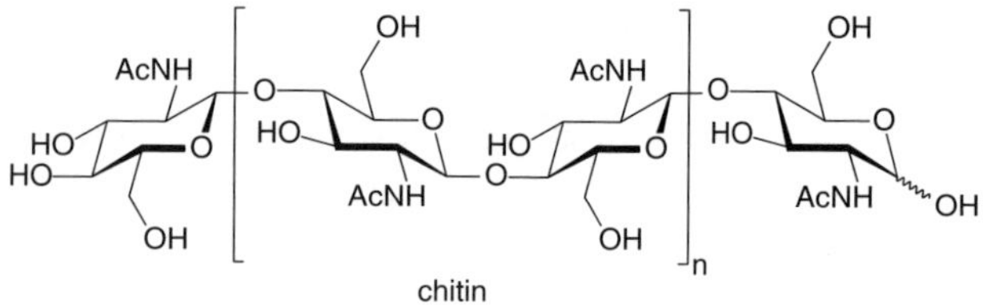

Figure 10.14 Schematic structure of chitin.

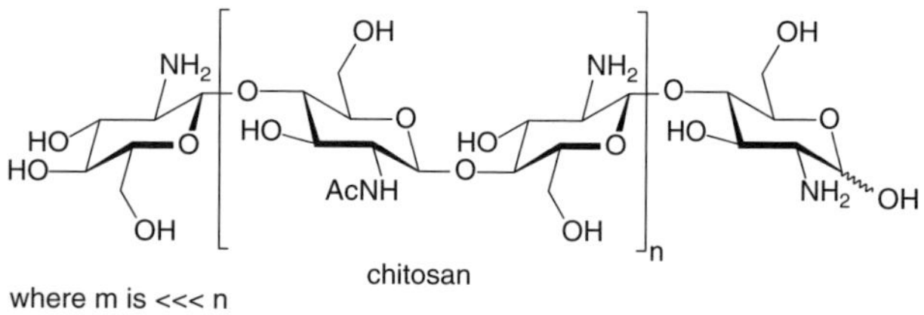

Figure 10.15 Schematic structure of chitosan.

Chitosan is composed of glucosamine (2-amino-2-deoxy-glucopyranose) and *N*-acetyl glucosamine (2-acetamido-2-deoxy-glucopyranose) linked in a β (1,4)-manner; the glucosamine to *N*-acetyl glucosamine ratio being referred to as the degree of deacetylation [34]. Depending on the source and preparation procedure, its molecular weight may range from 300 to over 1000 kD with degrees of deacetylation from 30 to 95%.

Chitosan is obtained on an industrial scale by the alkaline deacetylation of chitin [33,34]. The main commercial sources of chitin are shells of shellfish (mainly crabs, shrimps, lobsters, and krills) as wastes of the seafood processing industry. Basically, the process consists of deproteinization with a dilute NaOH solution, demineralization with a dilute HCl solution and decoloration of the raw shell material. Chitin is obtained as an almost colorless to off-white powdery material. Chitosan is produced by deacetylating chitin using 40–50% aqueous alkali at 100–160 °C for a few hours. The resultant chitosan has a degree of deacetylation up to 0.95.

Chitosan has been found to be nontoxic, biodegradable, biofunctional, biocompatible, and was reported by several researchers to have strong antimicrobial and antifungal activities. Thus, chitosan-based films have attracted serious attention in food preservation and packaging technology. The potential of chitosan as ingredient for active bio-based films production and the different methods used for chitosan-based films preparation and their perspectives in the modern food packaging technology are summarized in Ref. [35].

There are many producers of chitin and chitosan worldwide; Table 10.10 gives producers found in Europe.

10.2.3 Proteins

A protein is considered to be a random copolymer of amino acids. A generic protein monomeric unit is given in Fig. 10.16, where R represents the side chain of an amino acid. Proteins can be divided into proteins from plant origin (e.g., gluten, soy, pea, and potato) and proteins from animal origin (e.g., collagen (gelatin), casein, silk, keratin, whey).

Potential candidates for use in the fabrication of biodegradable films include soy proteins, wheat gluten, corn proteins, myofibrillar proteins from fish, and pea proteins [36,37]. Proteins are considered as structured heteropolymers [37]. Two classes of proteins can be distinguished, globular or pseudo-globular proteins such as globulins or gliadins and fibrous or "polymerized" proteins such as collagen or glutenins.

Gluten is a mixture of monomeric proteins (gliadins) and polymerized proteins (glutenins) linked through intermolecular disulfide bridges. Gluten is the main storage protein in wheat. In general, gluten-based plastics require the addition of plasticizer

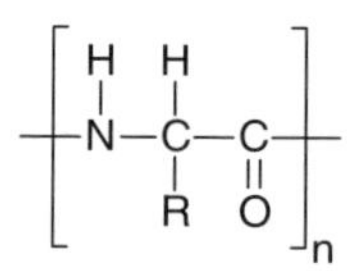

Figure 10.16 Schematic structure of proteins.

Table 10.10 Chitosan Producers in Europe

Structure	Supplier	Origin	Website
Chitosan	France Chitine	France	www.france-chitine.com
Chitosan	Nova Matrix	Norway	www.novamatrix.biz
Chitosan	Primex	Iceland	www.primex.is
Chitosan	Heppe GmbH	Germany	www.biolog-heppe.de

agents. Hydrophilic compounds (water, polyols, oligosaccharides) and lipidic compounds (waxes, oils, fatty acids, monoglycerols) are used as protein plasticizers—the most frequently used is glycerol [38,39]. Plasticizers decrease the protein interactions and increase polymer chain mobility and intermolecular spacing, decreasing also the glass transition temperature of proteins.

Soy protein-based plastics are another group of biodegradable, environmentally friendly polymer materials from an abundantly renewable resource [40–42]. There are several types of soybean products that can potentially be utilized for engineering structural applications [40].

Two processes are currently used to prepare protein-based films: the wet method ("casting"), which involves the solubilization of protein and a plasticizer in a solvent followed by the formation of a protein network on evaporation of the solvent; and the dry method, which is based on thermoplastic characteristics of proteins and combines the use of pressure and heat to plasticize protein chains [36,43]. Dehulled soybean, after solvent defatting and meal grinding, becomes a fat-free, low-fiber soy flour (48.5% protein). The soy flour, after leaching out of the water/alcohol soluble sugars, is termed soy protein concentrate (above 65% protein). The soy protein concentrate, if it is further extracted by alkali and reprecipitated by acidification, becomes the purest commercially available soy protein isolate (above 90% protein).

Vegetable and animal proteins have been used in many nonfood applications, but despite the potential, protein-based plastics have not yet made significant progress in commercialization at a large scale.

10.3 Biodegradable Polymers from Petrochemical Sources

Aliphatic polyesters are the representatives of synthetic biodegradable polymers.

Synthetic biodegradable polyesters are generally made by the polycondensation method and raw materials are obtained from petrochemical feed stocks. Aliphatic polyesters such as poly(butylene succinate) and poly(ε-caprolactone) are commercially produced. Besides these aliphatic polyesters, various types of synthetic biodegradable polymers have been designed [44]. They are, e.g., poly(ester amide)s, poly(ester carbonate)s, poly(ester urethane)s, etc. Most of them are still at a premature stage.

The traditional way of synthesizing polyesters has been by polycondensation using diols and a diacid (or an acid derivative), or from a hydroxy acid [44,45].

Polycondensation can be applicable for a variety of combinations of diols and diacids, but it requires, in general, higher temperature and longer reaction time to obtain high-molecular-weight polymers. In addition, this method suffers from such shortcomings as the need for removal of reaction by-products and a precise stoichiometric balance between reactive acid and hydroxy groups. The ROP of lactones, cyclic diesters (lactides and glycolides), is an alternative method, which can be carried out under milder conditions to produce high-molecular-weight polymers in a shorter time. Furthermore, recent progress in catalysts has enabled the production of polyesters of controlled chain lengths.

Recently, enzyme-catalyzed polymer synthesis has been established as another approach to biodegradable polymer preparation [46–48].

10.3.1 Aliphatic Polyesters and Copolyesters

One of the most promising polymers in this family is poly(butylene succinate) (PBS), which is chemically synthesized by the polycondensation of 1,4-butanediol with succinic acid (Fig. 10.17). High-molecular-weight PBS is generally prepared by a coupling reaction of relatively low-molecular-weight PBS in the presence of hexamethylene diisocyanate as a chain extender.

Bionolle is produced through the polycondensation reaction of glycols such as ethylene glycol and butanediol-1,4, and aliphatic dicarboxylic acids such as succinic and adipic acid used as principal raw materials [49]. Aliphatic polyesters, trademarked "Bionolle," such as polybutylene succinate (1000 series), polybutylene succinate adipate copolymer (3000 series), and polyethylene succinates (6000 series), with high molecular weights ranging from several tens of thousands to several hundreds of thousands, were invented in 1990 and produced through the polycondensation reaction of glycols with aliphatic dicarboxylic acids and others.

Commercially available aliphatic polyesters and copolyesters are given in Table 10.11.

Figure 10.17 Aliphatic polyesters and copolyesters.

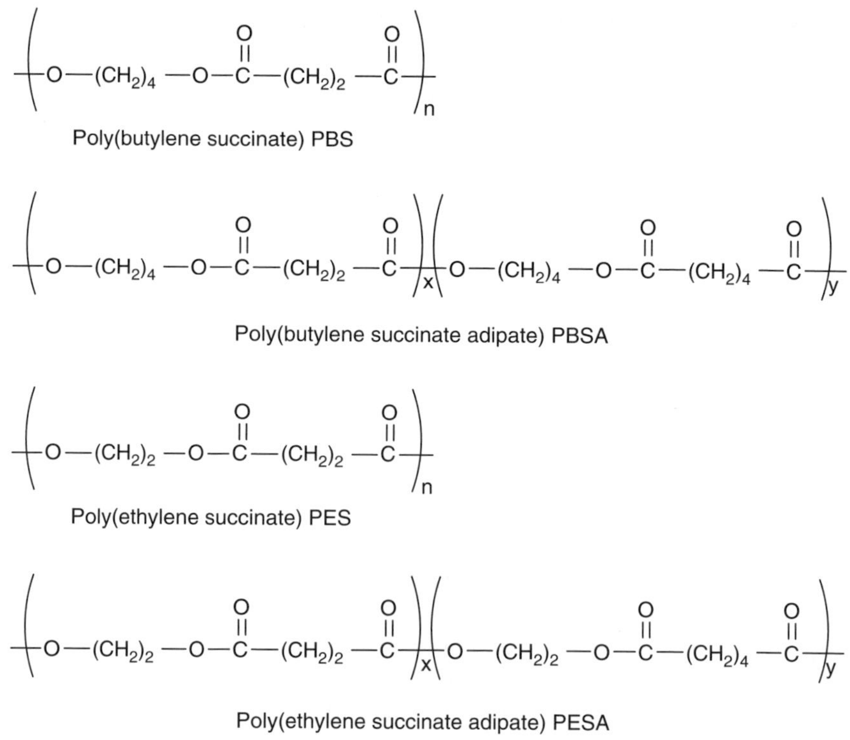

10.3.2 Aromatic Polyesters and Copolyesters

While the biological susceptibility of many aliphatic polyesters has been known for many years, aromatic polyesters such as polyethylene terephthalate (PET) or polybutylene terephthalate are regarded as nonbiodegradable [50]. To improve the use properties of aliphatic polyesters, an attempt was made to combine the biodegradability of aliphatic polyesters with the good material performance of aromatic polyesters in novel aliphatic–aromatic copolyesters (Fig. 10.18).

Using standard polycondensation techniques, copolyesters with molar masses in a range necessary for technical application were obtained [51,52]. The best results with regard to the use properties were achieved with a combination of 1,4-butanediol, adipic acid, and terephthalic acids.

Commercially available aromatic copolyesters are given in Table 10.12.

Poly(trimethylene terephthalate) (PTT) is a linear aromatic polyester produced by polycondensation of 1,3-propanediol (trimethylene glycol or PDO) with either purified terephthalic acid (PTA) or trimethylene terephthalate (Fig. 10.19).

While both these monomers—the diacid and the diol component—are conventionally derived from petrochemical feedstocks, DuPont, Tate & Lyle, and Genencor have recently succeeded in introducing PDO using an aerobic bioprocess with glucose from corn starch as the feedstock, opening the way for bulk production of PTT from a bio-based monomer.

The natural fermentation pathway to PDO involves two steps: yeast first ferments glucose to glycerol, then bacteria ferment this to PDO. In the bioprocess developed by DuPont, dextrose derived from wet-milled corn is metabolized by genetically engineered *Escherichia coli* bacteria and converted within the organism directly to PDO via an aerobic respiration pathway (Fig. 10.20). The PDO is then separated from the fermentation broth by filtration, and concentrated by evaporation, followed by purification and distillation. The PDO is then fed to the polymerization plant. PTT can be produced by transesterification of dimethyl terephthalate (DMT) with PDO, or by the esterification route, starting with PTA and PDO (Fig. 10.21) [20]. The polymerization can be a continuous process and is similar to the production of PET. In the first stage of polymerization, low-molecular-

Table 10.11 Commercially Available Aliphatic Polyesters and Copolyesters

Trade name	Supplier	Origin	Website
Bionolle® 1000 Poly(butylene succinate) PBS	Showa Highpolymer	Japan	www.showa-denko.com
Bionolle® 2000 Bionolle® 3000 Poly(butylene succinate adipate) PBSA	Showa Highpolymer	Japan	www.showa-denko.com
Bionolle 6000® Poly(ethylene succinate) PES	Showa Highpolymer	Japan	www.showa-denko.com
Bionolle 7000® Poly(ethylene succinate adipate) PESA	Showa Highpolymer	Japan	www.showa-denko.com
SkyGreen SG100 Poly(butylene succinate) PBS SG200 Poly(butylene succinate adipate) PBSA	SK Polymers	Korea	www.skchemicals.com/english
EnPol Poly(butylene succinate) PBS	Ire Chemicals	Korea	http://irechem.en.ecplaza.net
PBS	Anqing Hexing Chemical	China	http://hexingpbs.en.china.cn
GS-PLA Poly(butylene succinate) PBS	Mitsubishi Chemical	Japan	http://www.dia-chem.co.jp/en/products/gspla/index.html

weight polyester is produced in the presence of excess PDO, with water of esterification (in the case of PTA) or methanol (in the case of DMT) being removed. In the second stage, polycondensation, chain growth occurs by removal of PDO and remaining water/methanol. As chain termination can occur at any time (due to the presence of a monofunctional acid or hydroxyl compound), both monomers must be very pure. As the reaction proceeds, removal of traces of PDO becomes increasingly difficult. This is compensated for by having a series of reactors operating under progressively higher temperatures and lower pressures. In a final step, highly viscous molten polymer is blended with additives in a static mixer and then pelletized.

Table 10.13 summarizes commercially available PTT polymers.

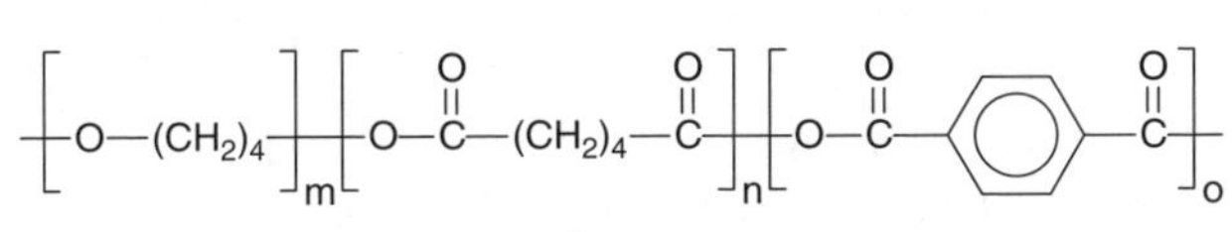

Figure 10.18 Aromatic copolyesters.

10.3.3 Poly(caprolactone)—PCL

PCL was one of the earliest polymers synthesized by the Carothers group in the early 1930s. Poly-(ε-caprolactone) is a linear polyester manufactured by ROP of a seven-membered lactone, ε-caprolactone (Figs 10.22 and 10.23). Catalysts such as stannous octoate are used to catalyze the polymerization and low-molecular-weight alcohols can be used to control the molecular weight of the polymer.

Anionic, cationic, coordination, or radical polymerization routes are all applicable [53,54]. Recently, enzymatic catalyzed polymerization of

Table 10.12 Commercially Available Aromatic Copolyesters

Trade Name	Supplier	Origin	Website
Biomax® Poly(butylene succinate terephthalate) PBST	DuPont	USA	www.dupont.com
Eastar Bio® Poly(butylene adipate terephthalate) PBAT	Eastman Chemicals1	Japan	www.eastman.com
Ecoflex® Poly(butylene adipate terephthalate) PBAT	BASF	Germany	www.bioplastics.basf.com
EnPol G8060 Poly(butylene adipate terephthalate) PBAT	Ire Chemicals	Korea	http://irechem.en.ecplaza.net
Origo-Bi	Novamont	Italy	www.novamont.com

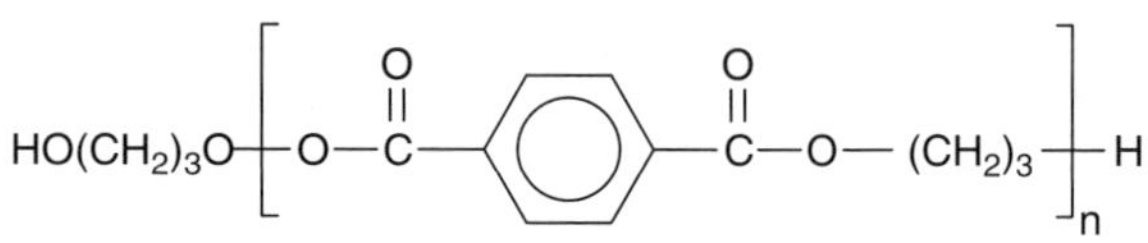

Figure 10.19 Schematic structure of PTT.

ε-caprolactone has been reported [47]. It is a semicrystalline polymer with a degree of crystallinity around 50%. It has a rather low glass transition temperature (−60 °C) and melting point (61 °C).

PCL was recognized as a biodegradable and nontoxic material, and a promising candidate for controlled release applications, especially for long-term drug delivery. The superior rheological and viscoelastic properties over many of its aliphatic polyester counterparts renders PCL easy to manufacture and manipulate into a large range of implants and devices. The application of PCL as a biomaterial over the last two decades focusing on medical devices, drug delivery, and tissue engineering was reviewed in Ref. [55]. PCL may be copolymerized with many other lactones, such as glycolide, lactide, δ-valerolactone, ε-decalactone, poly(ethylene oxide), and alkyl-substituted ε-caprolactone. Blends of PCL with other biodegradable polymers such as PHB, PLA, and starch have been prepared. Commercially available PCLs are listed in Table 10.14.

10.3.4 Poly(esteramide)—PEA

Polyesteramide BAK 1095 is based on caprolactam (Nylon 6), butanediol, and adipic acid; BAK 2195 is based on adipic acid and hexamethylenediamine (Nylon 6,6) and adipic acid with butanediol and diethylene glycol as ester components [56]. The

Figure 10.20 Biotechnological route to 1,3-propanediol.

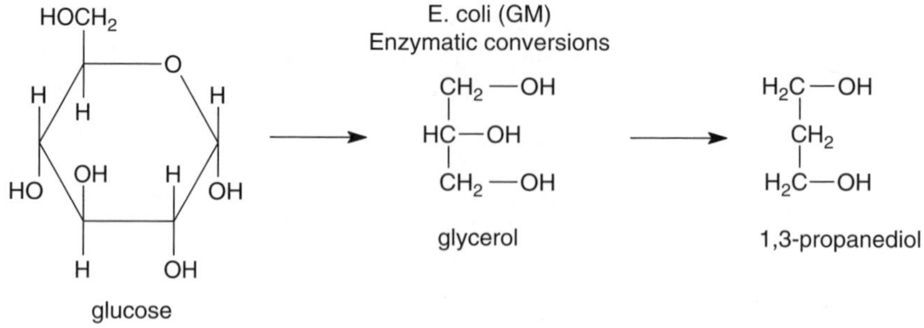

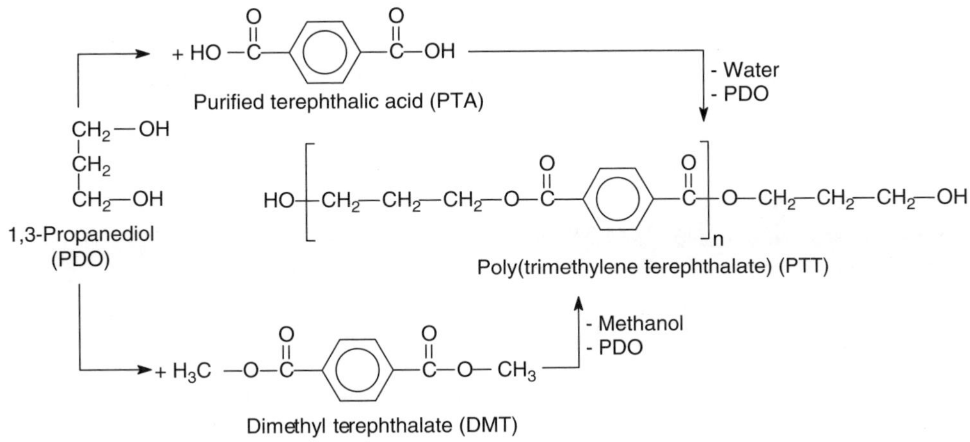

Figure 10.21 Manufacturing routes to PTT.

Table 10.13 PTT Polymers

Trade Name	Supplier	Origin	Website
Sorona™	DuPont	USA	www.dupont.com
Corterra®	Shell	Canada	www.shellchemicals.com
PermaStat	RTP	USA	www.rtpcompany.com

Table 10.14 Commercially Available PCL Polymers

Trade Name	Supplier	Origin	Website
Tone	Union Carbide	USA	www.unioncarbide.com
CAPA	Perstorp	UK	www.perstorp.com
Placcel	Daicel Chemical Indus.	Japan	www.daicel.co.jp/english/kinouhin/category/capro.html

production process is solvent and halogen free. Commercially available PEAs are detailed in Table 10.15 and the structure of PEAs is shown in Fig. 10.24.

10.3.5 Poly(vinyl alcohol)—PVA

Poly(vinyl alcohol) (PVA) (Fig. 10.25) is the largest volume water-soluble polymer produced today. PVA is not produced by direct polymerization of the corresponding monomer, since vinyl alcohol tends to convert spontaneously into the -enol form of acetaldehyde, driven by thermodynamic reasons and with extremely limited kinetic control [57]. PVA is attained instead from the parent homopolymer poly(vinyl acetate) (PVAc). The polymerization of vinyl acetate occurs via a free-radical mechanism, usually in an alcoholic solution (methanol, ethanol)

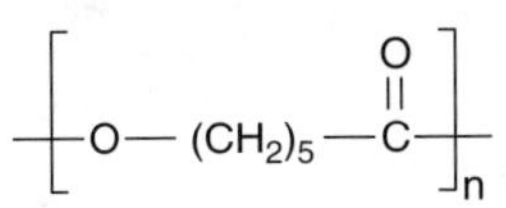

Figure 10.22 Structure of PCL.

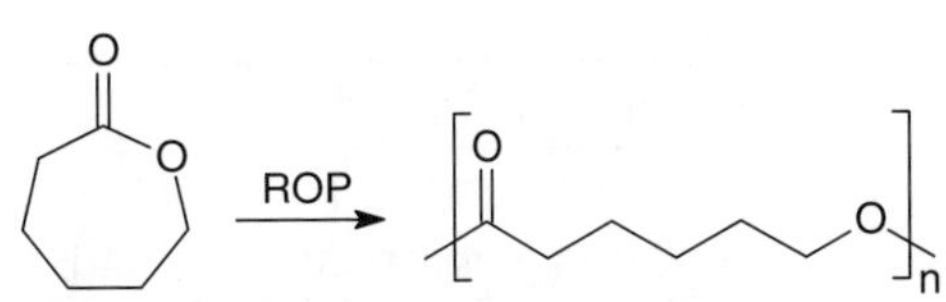

Figure 10.23 Schematic route to PCL.

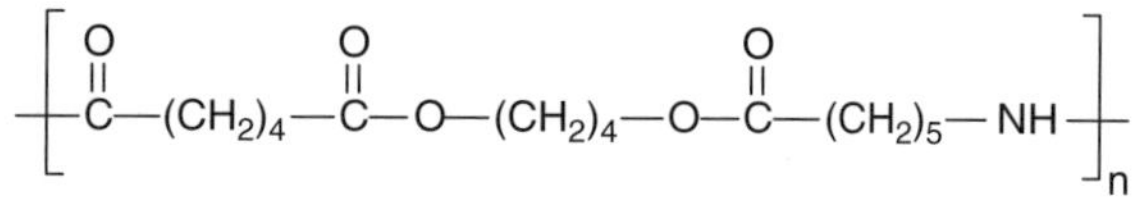

Figure 10.24 Schematic structure of PEAs.

Table 10.15 Commercially Available PEAs

Trade name	Supplier	Origin	Website
BAK[1]	Bayer AG	Germany	www.bayer.com

[1] *In 2002 the production was suspended.*

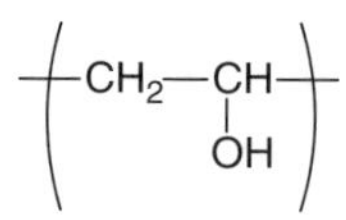

Figure 10.25 Schematic structure of PVA.

although for some specific applications a suspension polymerization can be used. The scheme for industrial production of PVA is given in Fig. 10.26.

PVA is produced on an industrial scale by hydrolysis (methanolysis) of PVAc, often in a one-pot reactor. Different grades of PVA are obtained depending upon the degree of hydrolysis (HD). Polymerization reactions can be carried out in batch

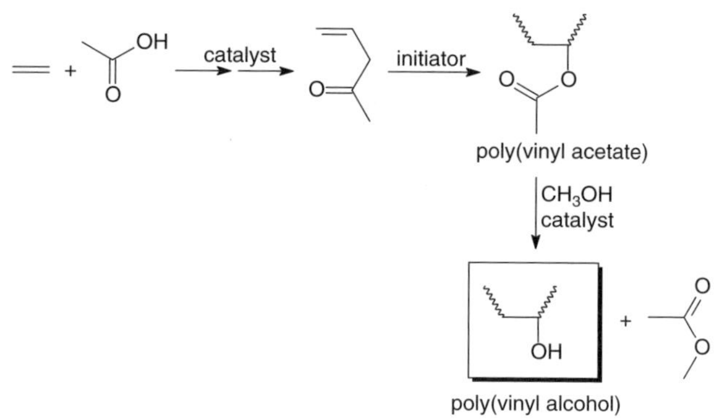

Figure 10.26 Manufacturing route to PVA.

or in continuous processes, the latter being used mostly for large-scale production. In the continuous industrial process, the free-radical polymerization of vinyl acetate is followed by alkaline alcoholysis of PVAc. The molecular weight of PVAc is usually controlled by establishing the appropriate residence time in the polymerization reactor, vinyl acetate feed rate, solvent (methanol) amount, radical initiator concentration, and polymerization temperature.

The main producers of PVA are given in Table 10.16.

10.3.6 Blends

One of the strategies adopted in producing compostable polymer materials is blending of biodegradable polymers. Blending is a common practice in

Table 10.16 PVA Producers

Trade Name	Supplier	Origin	Website
Mowiol	Clariant GmbH	Germany	www.cepd.clarinet.com
Erkol	Erkol SA	Spain	www.erkol.com
Sloviol	Novacky	Slovakia	www.nchz.sk
Polyvinol	Vinavil SpA	Italy	www.mpaei.it/it/vinavil/home.htm
Elvanol	DuPont	USA	www.dupont.com/industrial-polymers/elvanol/index.html
Cevol	Celanep	USA	www.celanesechemicals.com
Airvol	Air Products	USA	
Kuraray Poval	Kuraray Co. Ltd	Japan	www.kuraray.co.jp/en
Unitika Poval	Unitika Ltd	Japan	www.unitika.co.jp/e/home_e2.htm
Gohsenol	Nippon Gohsei—The Nippon Synthetic Chemical Industry Co. Ltd	Japan	www.nippongohsei.com/gohsenol/index.htm
Hapol	Hap Heng	China	

Table 10.17 Commercially Available Blends

Trade Name	Supplier	Origin	Website
Mater-Bi	Novamont	Italy	www.novamont.com
Ecostar	National Starch	USA	www.nationalstarch.com
Ecofoam	National Starch	USA	www.nationalstarch.com
Biograde (cellulose blends)	FKuR	Germany	www.fkur.com
Bioflex (PLA blends)	FKuR	Germany	www.fkur.com
Ecoflex (blends of Ecoflex and PLA)	BASF	Germany	www.bioplastics.basf.com
Fasal (cellulose based)	Austel 1 IFA	Austria	www.austel.at
Cereplast	Cereplast, Inc.	USA	www.cereplast.com

polymer science to improve unsatisfactory physical properties of the existing polymer or to decrease cost. By varying the composition and processing of blends, it is possible to manipulate properties. The leading compostable blends are starch-based materials. The aim is to combine the low cost of starch with higher cost polymers having better physical properties. An example of such material is Mater-Bi manufactured by Novamont [58]. Mater-Bi is prepared by blending starch with other biodegradable polymers in an extruder in the presence of water or plasticizer. The following three main classes of Mater-Bi are commercially available (see also Table 10.17):

- Class Z—TPS and PCL;
- Class Y—TPS and cellulose derivatives;
- Class V—TPS more than 85%.

References

[1] R. Narayan, Y. Doi, K. Fukada, Impact of Government Policies, Regulations, and Standards Activities on an Emerging Biodegradable Plastics Industry. Biodegradable Plastics and Polymers, Elsevier, New York, 1994, p. 261.

[2] ASTM D 6400−04 "Standard Specification for Compostable Plastics".

[3] ISO 17088: 2008 − Specifications for compostable Plastics.

[4] EN ISO 472:2001 − Plastics-Vocabulary.

[5] www.bpsweb.net/02_english

[6] E.S. Stevens, How green are green plastics? Biocycle 43 (2002) 42.

[7] N. Narayanan, P.K. Roychoudhury, A. Srivastava, L(1) lactic acid fermentation and its product polymerization, Electronic J. Biotechnol. 7 (2004) 167.

[8] J. Lunt, Large-scale production, properties and commercial applications of polylactic acid polymers, Polym. Degrad. Stab. 59 (1998) 145.

[9] L. Shen, J. Haufe, M.P. Patel, Product overview and market projection of emerging bio-based plastics. PRO-BIP 2009. Final Report. Utrecht University. June 2009. Revised in November 2009.

[10] A.P. Gupta, V. Kumar, New emerging trends in synthetic biodegradable polymers − Polylactide: A critique, Europ. Polym. J 43 (2007) 4053.

[11] K.M. Nampoothiri, N.R. Nair, R.P. John, An overview of the recent developments in polylactide research, Biores. Technol. 101 (2010) 8493.

[12] C. Gao, C. Ma, P. Xu, Biotechnological routes based on lactic acid production from biomass, Biotechnol. Adv. 29 (2011) 930.

[13] S. Kobayashi, Recent developments in lipase-catalyzed synthesis of polyesters, Macromol. Rap. Commun. 30 (2009) 237.

[14] R.A. Gross, M. Ganesh, W. Lu, Enzyme-catalysis breathes new life into polyester condensation polymerizations, Trends Biotechnol. 28 (2010) 435.

[15] A. Södergård, M. Stolt, Properties of lactic acid based polymers and their correlation with composition, Prog. Polym. Sci. 27 (2002) 1123.

[16] C.S.K. Reddy, R. Ghai, R. Rashmi, V.C. Kalia, Polyhydroxyalkanoates: An overview, Bioresour. Technol 87 (2003) 137.

[17] R.W. Lenz, R.H. Marchessault, Bacterial polyesters: Biosynthesis, biodegradable plastics and biotechnology, Biomacromolecules 6 (2005) 1.

[18] S. Khanna, A.K. Srivastava, Recent advances in microbial polyhydroxyalkanoates, Process Biochem. 40 (2005) 607.

[19] R. Rai, T. Keshavarz, J.A. Roether, A.R. Boccaccini, I. Roy, Medium chain length polyhydroxyalkanoates, promising new biomedical materials for the future, Mater. Sci. Eng. 72 (2011) 29.

[20] Techno-economic feasibility of large-scale production of bio-based polymers in Europe (PRO-BIP), Final Report, Utrecht/Karlsruhe, October 2004.

[21] T. Keshavarz, I. Roy, Polyhydroxyalkanoates: Bioplastics with a green agenda, Curr. Opin. Microbiol. 13 (2010) 321.

[22] S. Chanprateep, Current trends in biodegradable polyhydroxyalkanoates, J. Biosci. Bioeng. 110 (2010) 621.

[23] M. Zinn, B. Witholt, T. Egli, Occurrence, synthesis and medical application of bacterial polyhydroxyalkanoates, Adv. Drug Deliv. Rev. 53 (2001) 5.

[24] J.F. Robyt, Essentials of Carbohydrate Chemistry, Springer-Verlag, New York, 1998.

[25] R.F. Tester, J. Karkalas, X. Qi, Starch-composition, fine structure and architecture, J. Cer. Sci. 39 (2004) 151.

[26] J.R. Daniel, A.C.J. Voragen, W. Pilnik, Starch and other polysaccharides in: Ullmann's Encyclopedia of Industrial Chemistry, Vol. A25, VCH, Verlagsgesellschaft, 1994.

[27] S.B. Ross-Murphy, R.T. Stepto, Greening polymers for the 21st century: Real prospects and virtual realities, in: Emerging Themes in Polymers Science, Royal Society of Chemistry, London, 2001.

[28] J.J.G. Van Soest, J.F.G. Vliegenthart, Crystallinity in starch plastics: Consequences for material properties, TIBTECH 15 (1997) 208.

[29] W. Aichholzer, H.-G. Fritz, Rheological characterization of thermoplastic starch materials, Starch 50 (1998) 77.

[30] H. Liu, F. Xie, L. Yu, L. Chen, L. Li, Thermal processing of starch-based polymers, Prog. Polym. Sci. 34 (2009) 1348.

[31] T. Heinze, T. Liebert, Unconventional methods in cellulose functionalisation, Prog. Polym. Sci. 26 (2001) 1605.

[32] K.J. Edgar, C.M. Buchanan, J.S. Debenham, P.A. Rundquist, B.D. Seiler, M.C. Shelton, D. Tindall, Advances in cellulose ester performance and application, Prog. Polym. Sci. 26 (2001) 1605.

[33] K. Kurita, Controlled functionalisation of the polysaccharide chitin, Prog. Polym. Sci. 26 (2001) 1921.

[34] A. Di Martino, M. Sittinger, M.V. Risbud, Chitosan: A versatile biopolymer for orthopaedic tissue engineering, Biomaterials 26 (2005) 5983.

[35] M. Aider, Chitosan application for active bio-based films productions and potential in the food industry review, LWT-Food Sci.Technol. 43 (2010) 837.

[36] O. Orliac, A. Rouilly, F. Silvestre, L. Rigal, Effects of additives on the mechanical properties, hydrofobicity and water uptake of thermo-moulded films produced from sunflower isolate, Polymer 43 (2002) 5417.

[37] C. Larré, C. Desserme, J. Barbot, J. Guéguen, Properties of deamidated gluten films enzymatically cross-linked, J. Agric. Food Chem. 48 (2000) 5444.

[38] M. Pommet, A. Redl, M.-H. Morel, S. Guilbert, Study of wheat gluten plasticization with fatty acids, Polymer 44 (2003) 115.

[39] A.Ch Sánchez, Y. Popineau, C. Mangavel, C. Larré, J. Guéguen, Effect of different plasticizers on the mechanical and surface properties of wheat gliadin films, J. Agric. Food Chem. 46 (1998) 4539.

[40] H.-J. Sue, S. Weng, J.-L. Jane, Morphology and mechanical behaviour of engineering soy plastics, Polymer 38 (1997) 5036.

[41] J. Zhang, P. Mungara, J. Jane, Mechanical and thermal properties of extruded soy protein sheets, Polymer 42 (2001) 2569.

[42] P. Lodha, A.N. Netravali, Thermal and mechanical properties of environment-friendly "green" plastics from stearic acid modified-soy protein isolate, Ind. Crops Prod. 21 (2005) 49–64.

[43] V. Micard, M.-H. Morel, J. Bonicel, S. Guibert, Thermal properties of raw and processed wheat gluten in relation with protein aggregation, Polymer 42 (2001) 477.

[44] M. Okada, Chemical syntheses of biodegradable polymers, Prog. Polym. Sci. 27 (2002) 87.
[45] A.-Ch. Albertsson, I.K. Varma, Aliphatic polyesters: Synthesis, properties and applications in degradable aliphatic polyesters, Adv. Polym. Sci. 157 (2002) 1. Springer.
[46] S. Namekawa, S. Suda, H. Uyama, S. Kobayashi, Lipase-catalysed ring-opening polymerization of lactones to polyesters and its mechanistic aspects, Int. J. Biol. Macromol. 25 (1999) 145.
[47] H. Uyama, S. Kobayashi, Enzyme-catalysed polymerization to functional polymers, J. Mol. Cat. B 19–20 (2002) 117.
[48] R. Marcilla, M. de Geus, D. Mecerreyes, C.J. Duxbury, C.E. Koning, A. Heise, Enzymatic polyester synthesis in ionic liquids, Eur. Polym. J. 42 (2006) 1215.
[49] T. Fujimaki, Processability and properties of aliphatic polyesters, "BIONOLLE", synthesised by polycondensation reaction, Polym. Degrad. Stab. 59 (1998) 209–214.
[50] E. Marten, R.-J. Müller, W.-D. Deckwer, Studies on the enzymatic hydrolysis of polyesters, Polym. Degrad. Stab. 88 (2005) 371.
[51] U. Witt, R.-J. Müller, W.-D. Deckwer, Biodegradation behaviour and material properties of aliphatic/aromatic polyesters of commercial importance, J. Eviron. Polym. Degrad. 5 (1997) 81.
[52] R.-J. Müller, U. Witt, E. Rantze, W.-D. Deckwer, Architecture of biodegradable copolyesters containing aromatic constituents, Polym. Degrad. Stab. 59 (1998) 203.
[53] K.M. Stridsberg, M. Ryner, A.-Ch. Albertsson, Controlled ring-opening polymerization: polymers with designed architecture in degradable aliphatic polyesters, Adv. Polym. Sci. 157 (2002) 41. Springer.
[54] U. Edlund, A.-Ch. Albertsson, Degradable polymer microspheres for controlled drug delivery in degradable aliphatic polyesters, Adv. Polym. Sci. 157 (2002) 67. Springer.
[55] M.A. Woodruff, D.W. Hutmacher, The return of a forgotten polymer - polycaprolactone in the 21st century, *Prog. Polym. Sci.* 35 (2010) 1217.
[56] E. Grigat, R. Koch, R. Timmermann, BAK 1095 and BAK 2195: Completely biodegradable synthetic thermoplastics, Polym. Degrad. Stab. 59 (1998) 223–226.
[57] E. Chiellini, A. Corti, S. D'Antone, R. Solaro, Biodegradation of poly(vinyl alcohol) based materials, Prog. Polym. Sci. 28 (2003) 963.
[58] C. Bastioli, Properties and applications of Mater-Bi starch-based materials, Polym. Degrad. Stab. 59 (1998) 263.

11 Biodegradability Testing of Compostable Polymer Materials

Ewa Rudnik

OUTLINE

11.1 Definitions Related to Biodegradation Testing

Activated sludge (ISO 14851)

Biomass produced in the aerobic treatment of wastewater by the growth of bacteria and other microorganisms in the presence of dissolved oxygen.

Activated vermiculite (ISO 14855-1)

Vermiculite colonized by an active microbial population during a preliminary growth phase.

Biochemical oxygen demand (BOD) (ISO 14851)

The mass concentration of the dissolved oxygen consumed under specified conditions by the aerobic biological oxidation of a chemical compound or

Ebnesajjad: Handbook of Biopolymers and Biodegradable Plastics. http://dx.doi.org/10.1016/B978-1-4557-2834-3.00011-2
© 2013 Elsevier Inc. All rights reserved. Adapted from a chapter in: Rudnik, *Compostable Polymer Materials* (2008).

organic matter in water, expressed as milligrams of oxygen uptake per milligram or gram of test compound.

Biodegradation phase (ISO 14855 part 2)

Time, measured in days, from the end of the lag phase of a test until about 90% of the maximum level of biodegradation has been reached.

Digested sludge (ISO 14853)

Mixture of settled sewage and activated sludge which has been incubated in an anaerobic digester at about 35 °C to reduce the biomass and odor and to improve the dewaterability of the sludge. Digested sludge contains an association of anaerobic fermentation and methanogenic bacteria producing carbon dioxide and methane.

Dissolved inorganic carbon (DIC)(ISO 14852)

That part of inorganic carbon (IC) in water that cannot be removed by specific phase separation, e.g., by centrifugation at 40,000 m s^{-2} for 15 min or by membrane filtration using membranes with pores of 0.2–0.45 μm diameter.

Dissolved organic carbon (DOC) (ISO 14851)

That part of the organic carbon (OC) in water which cannot be removed by specified phase separation, e.g., by centrifugation at 40,000 m s^{-2} for 15 min or by membrane filtration using membranes with pores of 0.2–0.45 μm diameter.

Inorganic carbon (IC) (ISO 14853)

IC which is dissolved or dispersed in the aqueous phase of a liquid and is recoverable from the supernatant liquid after the sludge has been allowed to settle.

Lag phase (ISO 14855 part 2)

Time, measured in days, from the start of a test until adaptation and/or selection of the degradation microorganisms is achieved and the degree of biodegradation of a chemical compound or organic matter has increased to about 10% of the maximum level of biodegradation.

Maximum level of biodegradation (ISO 14855 part 2)

Degree of biodegradation, measured as a percentage, of a chemical compound or organic matter in a test, above which no further biodegradation takes place during the test.

Plateau phase (ISO 14855 part 2)

Time, measured in days, from the end of the biodegradation phase until the end of the test.

Primary anaerobic biodegradation (ISO 14853)

Structural change (transformation) of a chemical compound by microorganisms, resulting in the loss of a specific property.

Theoretical amount of evolved biogas (Thbiogas) (ISO 14853)

Maximum theoretical amount of biogas (CH_4 + CO_2) evolved after complete biodegradation of an organic material under anaerobic conditions, calculated from the molecular formula and expressed as milliliters of biogas evolved per milligram of test material under standard conditions.

Theoretical amount of evolved carbon dioxide ($ThCO_2$) (ISO 17088, ISO 14855 part 2)

Maximum theoretical amount of carbon dioxide evolved after completely oxidizing a chemical compound, calculated from the molecular formula and expressed as milligrams of carbon dioxide evolved per milligram or gram of test compound.

Theoretical amount of evolved methane ($ThCH_4$) (ISO 14853)

Maximum theoretical amount of methane evolved after complete reduction of an organic material, calculated from the molecular formula and expressed as milligrams of methane evolved per milligram of test material.

Theoretical oxygen demand (ThOD) (ISO 14851)

The theoretical maximum amount of oxygen required to oxidize a chemical compound completely, calculated from the molecular formula, expressed as milligrams of oxygen uptake per milligram or gram of test compound.

Total dry solids (ISO 17088; ISO 14855 part 2)

Amount of solids obtained by taking a known volume of test material or compost and drying at about 105 °C to constant mass.

Total organic carbon (TOC) (ISO 14851)

All the carbon present in the organic matter which is dissolved or suspended in water.

Ultimate aerobic biodegradation (ISO 14853)

Breakdown of an organic compound by microorganisms in the absence of oxygen to carbon dioxide, methane, water, and mineral salts of any other elements present (mineralization) plus new biomass.

Ultimate aerobic biodegradation (ISO 17088; ISO 14855 part 2)

Breakdown of an organic compound by microorganisms in the presence of oxygen into carbon dioxide, water, and mineral salts of any other elements present (mineralization) plus new biomass.

Volatile solids (ISO 17088)

Amount of solids obtained by subtracting the residue of a known volume of test material or compost after incineration at about 550 °C from the total dry solids of the same sample.

11.2 International Standards Related to Composting

Internationally recognized standardization bodies, such as the International Organization for Standardization (ISO), as well as regional standardization bodies, such as the American Society for Testing and Materials (ASTM) and the European Committee for Standardization (CEN), are actively involved in developing standards related to composting and biodegradation. In addition, national standardization bodies, such as the German Deutsches Institut für Normung (DIN) and the Biodegradable Plastics Society (BPS) of Japan, contribute to the development and issuing of standards on compostable polymers. Recently, interest in developing national standards related to compostability and biodegradation testing has appeared in other regions of the world, e.g., in China, Taiwan, and Australia.

Several ISO standards for determining the ultimate aerobic/anaerobic biodegradability of plastic materials have been published. In particular, ISO 14855-1 is a common test method that measures evolved carbon dioxide using such methods as continuous infrared analysis, gas chromatography, or titration.

11.3 Principles of Main Standards Related to Composting and Biodegradability Testing

ISO 14855-1:2005—Determination of the ultimate aerobic biodegradability of plastic materials under controlled composting conditions—Method by analysis of evolved carbon dioxide—Part 1: General method

Scope: This standard specifies a method for the determination of the ultimate aerobic biodegradability of plastics, based on organic compounds, under controlled composting conditions by measurement of the amount of carbon dioxide evolved and the degree of disintegration of the plastic at the end of the test. This method is designed to simulate typical aerobic composting conditions for the organic fraction of solid-mixed municipal waste. The test material is exposed to an inoculum which is derived from compost. The composting takes place in an environment wherein temperature, aeration, and humidity are closely monitored and controlled. The test method is designed to yield the percentage conversion of the carbon in the test material to evolved carbon dioxide as well as the rate of conversion.

It contains also a variant of the method, using a mineral bed (vermiculite) inoculated with the thermophilic microorganisms obtained from compost with a specific activation phase, instead of mature compost. This variant is designed to yield the percentage of carbon in the test substance converted to carbon dioxide and the rate of conversion.

Principle: The test method determines the ultimate biodegradability and the degree of disintegration of the test material under conditions simulating an intensive aerobic composting process. The inoculum used consists of stabilized, mature compost derived, if possible, from composting the organic fraction of solid municipal waste.

The test material is mixed with the inoculum and introduced into a static composting vessel where it is intensively composted under optimum oxygen, temperature, and moisture conditions for a test period not exceeding 6 months.

During the aerobic biodegradation of the test material, carbon dioxide, water, mineral salts, and new microbial cellular constituents (biomass) are the ultimate biodegradation products. The carbon dioxide produced is continuously monitored, or measured at regular intervals, in test and blank vessels to determine the cumulative carbon dioxide production. The percentage biodegradation is given by the ratio of the carbon dioxide produced from the test material to the maximum theoretical amount of carbon dioxide that can be produced from the test material. The maximum theoretical amount of carbon dioxide produced is calculated from the measured TOC content. The percentage biodegradation does not include that amount of carbon converted to new cell biomass, which is not metabolized in turn to carbon dioxide during the course of the test.

Additionally, the degree of disintegration of the test is determined at the end of the test, and the loss in mass of the test material may also be determined.

Table 11.1 ISO Standards Related to Composting

Standard	Title
ISO 17088:2008	Specifications for compostable plastics
ISO 14021:1999	Environmental labels and declarations – Self-declared environmental claims (Type II environmental labeling)
ISO 14851:1999 ISO 14851:1999/Cor 1:2005	Determination of the ultimate aerobic biodegradability of plastic materials in an aqueous medium – Method by measuring the oxygen demand in a closed respirometer
ISO 14852:1999 ISO 14852:1999/Cor 1:2005	Determination of the ultimate aerobic biodegradability of plastic materials in an aqueous medium – Method by analysis of evolved carbon dioxide
ISO 14853:2005 ISO 14853:2005/Cor 1:2009	Plastics – Determination of the ultimate anaerobic biodegradation of plastic materials in an aqueous system – Method by measurement of biogas production
ISO 14855-1:2005 ISO 14853-1:2005/Cor 1:2009	Determination of the ultimate aerobic biodegradability of plastic materials under controlled composting conditions – Method by analysis of evolved carbon dioxide – Part 1: General method
ISO 14855-2:2007 ISO 14855-2:2007/Cor 1:2009	Determination of the ultimate aerobic biodegradability of plastic materials under controlled composting conditions – Method by analysis of evolved carbon dioxide – Part 2: Gravimetric measurement of carbon dioxide evolved in a laboratory-scale test
ISO 15985:2004 ISO 15985:2004/Cor 1:2007	Plastics – Determination of the ultimate anaerobic biodegradation and disintegration under high-solids anaerobic-digestion conditions – Method by analysis of released biogas
ISO 16929:2002	Determination of the degree of disintegration of plastic materials under defined composting conditions in a pilot-scale test
ISO 17556:2003	Determination of the ultimate aerobic biodegradability in soil by measuring the oxygen demand in a respirometer or the amount of carbon dioxide evolved
ISO 20200:2004	Determination of the degree of disintegration of plastic materials under simulated composting conditions in a laboratory-scale test
ISO/FDIS 10210	Plastics – Methods for the preparation of samples for biodegradation testing of plastic materials

Table 11.2 EN Standards Related to Biodegradation and Composting

Standard	Title
EN ISO 14851:2004	Determination of the ultimate aerobic biodegradability of plastic materials in an aqueous medium – Method by measuring the oxygen demand in a closed respirometer
EN ISO 14852:2004	Determination of the ultimate aerobic biodegradability of plastic materials in an aqueous medium – Method by analysis of evolved carbon dioxide
EN ISO 14855-1:2007	Determination of the ultimate aerobic biodegradability of plastic materials under controlled composting conditions – Method by analysis of evolved carbon dioxide – Part 1: General method
EN ISO 14855-1:2007/AC:2009 EN ISO 14855-2:2009	Determination of the ultimate aerobic biodegradability of plastic materials under controlled composting conditions – Method by analysis of evolved carbon dioxide – Part 2: Gravimetric measurement of carbon dioxide evolved in a laboratory-scale test
EN ISO 17556:2004	Determination of the ultimate aerobic biodegradability in soil by measuring the oxygen demand in a respirometer or the amount of carbon dioxide evolved
EN ISO 20200:2005	Determination of the degree of disintegration of plastic materials under simulated composting conditions in a laboratory-scale test

Table 11.3 EN Standards Related to Packaging and Composting

Standard	Title
EN 14045:2003	Packaging – Evaluation of the disintegration of packaging materials in practical oriented tests under defined composting conditions
EN 14046:2003	Packaging – Evaluation of the ultimate aerobic biodegradability of packaging materials under controlled composting conditions – Method by analysis of released carbon dioxide
EN 14806:2005	Preliminary evaluation of the disintegration of packaging materials under simulated composting conditions in a laboratory-scale test

Vermiculite should be used instead of mature compost under the following conditions:

- Whenever the determination of the degree of biodegradation is affected by a priming effect induced by the test material, and/or
- When performing a final carbon balance with biomass determination and retrieval of the residual test material.

Priming effect: The organic matter present in large amounts in the mature compost can undergo polymer-induced degradation, known as the "priming effect," which affects the measurement of the biodegradability.

The inorganic vermiculite bed substantially reduces the priming effect, thus improving the reliability of the method. A further advantage of using vermiculite is the very small amount of carbon dioxide evolved in the blank vessel (nearly zero), because of the low level of microbial activity. This permits low levels of degradation activity to be evaluated precisely. The mineralization rates obtained with the activated vermiculite are identical,

Table 11.4 ASTM Standards Related to Composting and Biodegradation

Standard	**Title**
ASTM D6400-04	Standard specification for compostable plastics
ASTM D6868-11	Standard specification for labeling of end items that incorporate plastics and polymers as coatings or additives with paper and other substrates designed to be aerobically composted in municipal or industrial facilities
ASTM D 6340-98(2007)	Standard test methods for determining aerobic biodegradation of radiolabeled plastic materials in an aqueous or compost environment
ASTM D 6954-04	Standard guide for exposing and testing plastics that degrade in the environment by a combination of oxidation and biodegradation
ASTM D 7081-05	Standard specification for non-floating biodegradable plastics in the marine environment
ASTM D 5210-92(2007)	Standard test method for determining the anaerobic biodegradation of plastic materials in the presence of municipal sewage sludge
ASTM D 5929-96(2009)	Standard test method for determining biodegradability of materials exposed to municipal solid waste composting conditions by compost respirometry
ASTM D 5338-2011	Standard test method for determining aerobic biodegradation of plastic materials under controlled composting conditions. Incorporating thermophilic temperatures
ASTM D 5526-94(2011)e1	Standard test method for determining anaerobic biodegradation of plastic materials under accelerated landfill conditions
ASTM D 5988-03	Standard test method for determining aerobic biodegradation in soil of plastic materials or residual plastic materials after composting
ASTM D 6691:09	Test method for determining aerobic biodegradation of plastic in the marine environment by a defined microbial consortium or natural sea water inoculum
ASTM D 5511	Standard test method for determining anaerobic biodegradation of plastic materials under high-solids anaerobic-digestion conditions
Proposed new standards	
WK32805	New test method for disintegration of compostable plastics and products in a pilot scale aerobic composting system
WK34454	New test methods for standard method for determining the disintegration of compostable plastics and other materials in aerobic industrial composting environments
WK35342	New specification for home composting of biodegradable plastics

or very similar, to those obtained with mature compost, in terms of both the final degradation level and the degradation rate.

ISO 14855-2—Determination of the ultimate aerobic biodegradability of plastic materials under controlled composting conditions—Method by analysis of evolved carbon dioxide—Part 2: Gravimetric measurement of carbon dioxide evolved in a laboratory-scale test.

In order to ensure the activity of compost inoculum, inert material which works as soil texture is mixed into compost inoculum. The carbon dioxide evolved from the test vessel is determined by using gravimetric analysis of carbon dioxide absorbent. The method, which consists of a closed system to capture evolved carbon dioxide, is available to determine the ultimate aerobic biodegradability of plastic materials under controlled composting conditions in a laboratory-scale test. The valuable information of degradation on the molecular structure of copolymers can frequently be obtained by means of isotopic labeling studies based on this test method of a closed system.

Scope: This test method specifies a method for determining the ultimate aerobic biodegradability of plastic materials in controlled composting conditions by gravimetric measurement of the amount of evolved carbon dioxide.

Principle: The method is designed to yield an optimum degree of biodegradability by adjusting the humidity, aeration ratio, and temperature in a composting vessel. It also aims to determine the ultimate biodegradability of the test material by using a small-scale reactor. The degradation rate is periodically measured by increasing the weight of the evolved carbon dioxide using an absorption column charged with soda lime and soda talc on an electronic balance. The test material is mixed with the inoculum derived from mature compost and inert material such as sea sand. The sea sand takes an active part of the holding body for humidity and microorganism activity.

The amount of carbon dioxide evolved is measured at intervals on the electronic balance and the carbon dioxide content is determined. The level of biodegradation, expressed as a percentage, is determined by comparing the amount of carbon dioxide evolved with the theoretical amount ($ThCO_2$).

The test is terminated when the plateau phase of biodegradation has been attained; the standard time for termination is 45 days, but the test could continue for 6 months, at the latest.

ISO 20200:2004—Plastics—Determination of the degree of disintegration of plastic materials under simulated composting conditions in a laboratory-scale test

Scope: This standard specifies a method of determining the degree of disintegration of plastic materials when exposed to a laboratory composting environment. The method is not applicable to the determination of the biodegradability of plastic materials under composting conditions.

Principle: The method determines the degree of disintegration of test materials on a laboratory scale under conditions simulating an intensive aerobic composting process. The solid matrix used consists of a synthetic solid waste inoculated with mature compost taken from a commercial composting plant. Pieces of the plastic test material are composted with this prepared solid matrix. The degree of disintegration is determined after a composting cycle, by sieving the final matrix through a 2 mm sieve in order to recover the nondisintegrated material. The reduction in mass of the test sample is considered as disintegrated material and used to calculate the degree of disintegration.

EN ISO 14851:2004—Determination of the ultimate aerobic biodegradability of plastic materials in an aqueous medium—Method by measuring the oxygen demand in a closed respirometer (ISO 14851:1999)

ISO 14851:1999/Cor 1:2005

Scope: This standard specifies a method by measuring the oxygen demand in a closed respirometer, for the determination of the degree of aerobic biodegradability of plastic materials, including those containing formulation additives. The test material is exposed in an aqueous medium under laboratory conditions to an inoculum from activated sludge, compost, or soil.

If an unadapted sludge is used as the inoculum, the test simulates the biodegradation processes that occur in a natural aqueous environment; if a mixed or pre-exposed inoculum is used, the method can be used to investigate the potential biodegradability of a test material.

Principle: The biodegradability of a plastic material is determined using aerobic microorganisms in an aqueous system. The test mixture contains an inorganic medium, the organic test material (the sole source of carbon and energy) with a concentration between 100 mg/l and 2000 mg/l of OC, and activated sludge or a suspension of active soil or compost as the inoculum.

The mixture is stirred in closed flasks in a respirometer for a period not exceeding 6 months. The carbon dioxide evolved is absorbed in a suitable absorber in the headspace of the flasks. The consumption of oxygen (BOD) is determined, e.g., by measuring the amount of oxygen required to maintain a constant volume of gas in the respirometry flasks, or by measuring the change in volume or pressure (or a combination of the two) either automatically or manually.

The level of biodegradation is determined by comparing the BOD with the theoretical amount (ThOD) and expressed in percent. The influence of possible nitrification processes on the BOD has to be considered. The test result is the maximum level of biodegradation determined from the plateau phase of the biodegradation curve. There is the possibility of improving the evaluation of biodegradability by calculating a carbon balance.

Table 11.5 Summary of Biodegradability and Composting Methods

Standard	Medium	Duration	Temperature	Reference Material	Measurements
ISO 14855-1:2005	Mature compost, optionally vermiculite	Not exceeding 6 months	58 ± 2 °C	Thin-layer chromatography grade cellulose as positive reference	1. CO_2 evolution (by IR analysis, gas chromatography, titration method, etc.) 2. Disintegration (visual evaluation, relevant physical properties measurements)
ISO 14855-2	Mature compost, one inert material (sea sand)	Standard time (45 days); up to 6 months is allowed	58 ± 2 °C	Thin-layer chromatography grade cellulose as positive reference	CO_2 evolution (by gravimetric method)
EN ISO 14851: 2004	Aqueous	Not exceeding 6 months	Preferably between 20 and 25 °C	Aniline, microcrystalline cellulose powder, ashless cellulose filters or poly-β-hydroxybutyrate as positive reference	Oxygen consumption (by, e.g., respirometric method or measurements of changes in volume or pressure)
EN ISO 14852: 1999	Aqueous	Not exceeding 6 months	Preferably between 20 °C and 25 °C	Aniline, microcrystalline cellulose powder, ashless cellulose filters or poly-β-hydroxybutyrate as positive reference	CO_2 evolution (CO_2 or DIC analyser or apparatus for titrimetric determination after complete absorption in a basic solution)

ISO 14852:1999—Determination of the ultimate aerobic biodegradability of plastic materials in an aqueous medium—Method by analysis of evolved carbon dioxide

Scope: This standard specifies a method, by measuring the amount of carbon dioxide evolved, for the determination of the degree of aerobic biodegradability of plastic materials, including those containing formulation additives. The test material is exposed in a synthetic medium under laboratory conditions to an inoculum from activated sludge, compost, or soil. If an unadapted activated sludge is used as the inoculum, the test simulates the biodegradation processes that occur in a natural aqueous environment; if a mixed or pre-exposed inoculum is used, the method can be used to investigate the potential biodegradability of a test material. The standard is designed to determine the potential biodegradability of plastic materials or give an indication of their biodegradability in natural environments.

The method enables the assessment of the biodegradability to be improved by calculating a carbon balance.

Principle: The biodegradability of a plastic material is determined using aerobic microorganisms in an aqueous system. The test mixture contains an inorganic medium, the organic test material (the sole source of carbon and energy) with a concentration between 100 mg/l and 2000 mg/l of OC, and activated sludge or a suspension of active soil or compost as the inoculum. The mixture is agitated in test flasks and aerated with carbon dioxide-free air over a period of time depending on the biodegradation kinetics, but not exceeding 6 months. The carbon dioxide evolved during the microbial degradation is determined by a suitable analytical method. For example, the carbon dioxide evolved is absorbed in sodium hydroxide (NaOH) solution and determined as DIC using, e.g., a DOC analyzed without incineration. Another way is the titrimetric method using a barium hydroxide solution.

The level of biodegradation is determined by comparing the amount of carbon dioxide evolved with the theoretical amount ($ThCO_2$) and expressed in percent. The test result is the maximum level of biodegradation, determined from the plateau phase of the biodegradation curve. Optionally, a carbon balance may be calculated to give additional information on the biodegradation.

The standard is specially designed for the determination of the biodegradability of plastic materials. There is a possibility of improving the evaluation of the biodegradability by calculating a carbon balance.

ISO 14853:2005—Plastics—Determination of the ultimate anaerobic biodegradation of plastic materials in an aqueous system—Method by measurement of biogas production

Scope: This standard specifies a method for the determination of the ultimate anaerobic biodegradability of plastics by anaerobic microorganisms. The test calls for exposure of the test material to sludge for a period of up to 60 days, which is longer than the normal sludge retention time (25–30 days) in anaerobic digesters, though digesters at industrial sites can have much longer retention times.

Principle: The biodegradability of a plastic material is determined using anaerobic conditions in an aqueous system. Test material with a concentration of 20 to 200 mg/l OC is incubated at 35 °C ± 2 °C in sealed vessels together with digested sludge for a period normally not exceeding 60 days. Before use, the digested sludge is washed so that it contains very low amounts of IC and diluted to 1–3 g/l total solids concentration. The increase in headspace pressure or the volumetric increase (depending on the method used for measuring biogas evolution) in the test vessels resulting from the production of carbon dioxide and methane is measured. A considerable amount of carbon dioxide will be dissolved in water or transformed to bicarbonate or carbonate under the conditions of the test. The IC is measured at the end of the test. The amount of microbiologically produced biogas carbon is calculated from the net biogas production and the net IC formation in excess of blank values. The percentage biodegradation is calculated from the total amount of carbon transformed to biogas and IC and the measured or calculated amount added as test material. The course of biodegradation can be followed by making intermediate measurements of biogas production. As additional information, the primary biodegradability can be determined by specific analyses at the beginning and end of the test.

The test method is designed to determine the biodegradability of plastic materials under anaerobic conditions. Optionally, the assessment of the recovery rate may also be determined.

Reference material: Anaerobically biodegradable polymer, e.g., poly-β-hydoroxybutyrate, cellulose, or poly(ethylene glycol) 400.

ISO 15985:2004 Plastics—Determination of the ultimate anaerobic biodegradation and disintegration under high-solids anaerobic-digestion conditions—Method by analysis of released biogas

Scope: This standard specifies a method for the evaluation of the ultimate anaerobic biodegradability of plastics based on organic compounds under high-solids anaerobic-digestion conditions by measurement of evolved biogas and the degree of disintegration at the end of the test. This method is designed to simulate typical anaerobic digestion conditions for the organic fraction of mixed municipal solid waste. The test material is exposed in a laboratory test to a methanogenic inoculum derived from anaerobic digesters operating only on pretreated household waste. The anaerobic decomposition takes place under high-solids (more than 20% total solids) and static non-mixed conditions. The test method is designed to yield the percentage of carbon in the test material and its rate of conversion to evolved carbon dioxide and methane (biogas).

Principle: The test method is designed to be an optimized simulation of an intensive anaerobic digestion process and determines the ultimate biodegradability and the degree of disintegration of a test material under high-solids anaerobic conditions. The methanogenic inoculum is derived from anaerobic digesters operating on pretreated household waste, preferably only the organic fraction.

The test material is mixed with the inoculum and introduced into a static digestion vessel where it is intensively digested under optimum temperature and moisture conditions for a test period of 15 days or longer until a plateau in net biodegradation has been reached.

During the anaerobic biodegradation of the test material, methane, carbon dioxide, water, mineral salts, and new microbial cellular constituents (biomass) are produced as the ultimate biodegradation products. The biogas (methane and carbon dioxide) evolved is continuously monitored or measured at regular intervals in test and blank vessels to determine the cumulative biogas production. The percentage biodegradation is given by the ratio of the amount of biogas evolved from the test material to the maximum theoretical amount of biogas that can be produced from the test material. The maximum theoretical amount of biogas produced is calculated from the measured TOC. This percentage biodegradation does not include the amount of carbon converted to new cell biomass which is not metabolized in turn to biogas during the course of the test.

Additionally, the degree of disintegration of the test material is determined at the end of the test and the loss in mass of the test material may also be determined.

Reference material: Thin-layer chromatography (TLC) grade cellulose with a particle size of less than 20 μm is used as the positive reference material.

ISO 17556:2003—Determination of the ultimate aerobic biodegradability in soil by measuring the oxygen demand in a respirometer or the amount of carbon dioxide evolved

Scope: This standard specifies a method for determining the ultimate aerobic biodegradability of the plastic materials in soil by measuring the oxygen demand in a closed respirometer or the amount of carbon dioxide evolved. This method is designed to yield an optimum degree of biodegradation by adjusting the humidity of the test soil.

If a non-adapted soil is used as an inoculum, the test simulates the biodegradation processes that take place in a natural soil environment; if a pre-exposed soil is used, the method can be used to investigate the potential biodegradability of a test material.

Principle: This method is designed to yield the optimum rate of biodegradation of a plastic material in a test soil by controlling the humidity of the soil, and to determine the ultimate biodegradability of the test material.

The plastic material, which is the sole source of carbon and energy, is mixed with the soil. The mixture is allowed to stand in a flask over a period of time during which the amount of oxygen consumed (BOD) or the amount of carbon dioxide evolved is determined. The BOD is determined, e.g., by measuring the amount of oxygen required to maintain a constant gas volume in a respirometer flask, or by measuring either automatically or manually the change in volume or pressure (or a combination of the two). The amount of carbon dioxide evolved is measured at intervals dependent on the biodegradation kinetics of the test substance by passing carbon dioxide-free air over the soil and then determining the carbon dioxide content of the air by a suitable method.

The level of biodegradation, expressed in percent, is determined by comparing the BOD with the ThOD or by comparing the amount of carbon dioxide evolved with the theoretical amount ($ThCO_2$). The influence of possible nitrification processes on the BOD has to be considered. The test is terminated when a constant level of biodegradation has been attained or, at the latest, after 6 months.

11.4 Composting at Laboratory Scale

The composting test method based on activated vermiculite was proposed as a comprehensive system for the assessment of the environmental impact of compostable polymers [1,2]. Vermiculite, a clay mineral, can be activated (by an inoculation with an appropriate microbial population and fermentation) and used as a solid matrix in place of mature compost in the controlled composting test. The formula of vermiculite is $(Mg,Fe,Al)_3(Al,Si)_4O_{10}(OH)_2 \cdot 4H_2O$. The results obtained with two materials (cellulose and a starch-based blend) indicated that activated vermiculite affected neither the biodegradation rate nor the final biodegradation level. On the other hand, possible metabolic intermediates and polymeric residues left after biodegradation could be recovered more easily from activated vermiculite than from mature compost, a very complex organic matter. Therefore, at test termination it was possible to determine the carbon balance taking into account both the evolved CO_2 and a polymeric residue extracted from vermiculite, totaling 101% of the carbon present originally in the test material. To conclude, it allows, in a single test, (i) the measurement of the mineralization of the polymer under study, (ii) the retrieval of the final polymeric residues, (iii) the determination of the biomass (to make a final mass balance), and (iv) the detection of breakdown products of the original polymer. The vermiculite test method is also suitable to perform ecotoxicological studies [2].

Different vermiculite media were studied in order to determine the parameters of an inert solid medium which could simulate the degradation of a polymer in compost [3]. Five different vermiculite media have been tested according to the type of activation and the amount of inoculum used. The mineralization curves obtained for simulation tests have been compared with the mineralization curve of starch biodegradation in compost.

Glucose, starch, and cellulose can increase the biodegradation of the compost used as a solid matrix in the biodegradation test under composting conditions (priming effect). The enhanced evolution of carbon dioxide determines an overestimation of the biodegradation of the starch- and cellulose-based materials and, in some cases, values higher than 100% can be reached. Therefore, it was verified that by using activated vermiculite, an inorganic matrix, the priming effect can be reduced, improving the reliability of the test method [4]. Glucose, the most effective primer, causes the attainment of biodegradation values significantly higher than 100% in mature compost while this does not happen in activated vermiculite. Since all the initial carbon present in the activated vermiculite was converted into CO_2 within the test period, it was concluded that a substantial priming effect cannot occur for the lack of OC. Furthermore, by measuring in parallel both the consumption of glucose and the CO_2 evolution, the yield of CO_2 production (Y_{CO_2} 5 $C_{CO_2}/C_{glucose}$) was determined. In no case was a value higher than 1 found, a clear indication of the priming effect.

Variation of microbial population in the compost was examined at different stages of the composting [5]. Moisture content was controlled in the range $64 \pm 4\%$, and the thermophilic stage lasted about 2 weeks. The temperature during the composting was controlled not to exceed 58 °C. In the initial stage of the composting, mesophilic strains were more numerous than the thermophilic ones. As the thermophilic stage sets in, thermophilic bacteria and actinomycetes outnumbered mesophilic correspondents while fungi were not detected at all. In the cooling and maturing phases, a substantial number of actinomycetes were still found. However, bacteria decreased significantly in number, and only a small number of mesophilic fungi reappeared. When glucose was added to the compost, the so-called "priming effect" was observed, in that the amount of CO_2 evolved was larger than that predicted by assuming that all added glucose was mineralized into CO_2. However, the priming effect decreased as the quantity of the glucose in the compost increased. Addition of 5 wt% of glucose to the compost increased the number of microorganisms by 10–100 times.

Specimens in film shape as well as in powder shape were subjected to the biodegradation tests to investigate dependence of the test results on the shape of the specimens [6]. Biodegradation of plastics was tested in compost made with animal fodder. Polypropylene (PP) was chosen as a non-degradable plastic. Poly (L-lactic acid) (PLLA) and poly(butylene succinate) (PBS) were selected as slowly degrading plastics while polycaprolactone (PCL) and poly(butylene succinate-*co*-adipate) (PBSA) were chosen as easily degradable plastics. Biodegradability of PP in film shape as well as in powder shape was tested to investigate the possible change in the microbial aspiration, because the shape of the specimens may affect aeration behavior in the compost. Biodegradation results of PLLA and PBS depended on their shape all through the biodegradation test. In contrast, the shape of PCL and PBS exerted influences on their biodegradability only at the early stage of the biodegradation, while at the later stage, the biodegradation proceeded almost independently of their shape.

Some laboratory composting facilities were developed and described [7,8,9]. An automated multiunit composting facility for studying the biodegradation of polymers was developed in accordance with the guidelines included in standards ISO/DIS 14855 and ASTM D 5338-92 [7]. In the system, cellulose, newspaper, and two starch-based polymers were treated with compost in a series of 3 dm^3 vessels at 52 °C and under conditions of optimum moisture and pH. The degradation was followed over time by measuring the carbon dioxide evolved. Results showed that at 52 °C over 45 days, cellulose and starch-based blends degraded by 90, 87, and 72%, respectively. The cellulose and lignin-hemicellulose-based newspaper was degraded by approximately 50% under experimental conditions. A biological oxygen demand (BOD) measurement system was adapted to monitor biodegradation process in solid media [8]. BOD is widely used for the examination of sewage water, effluents, polluted water, and for the assessment of biodegradation of chemicals and biodegradable polymers, but exclusively in aquatic media.

After the optimization of sample concentration and test temperature, the measurement set-up possessing relatively small reaction vessels of 250 ml with 80 g of soil mix proved to supply reliable and reproducible results. The system was optimized with microcrystalline cellulose (MCE)—used as a reference material in aquatic and solid test as well—showing 89.3 ± 3.2% degree of degradation after 21 days. Two test systems for composting studies of different scales (up to 1500 ml; up to 100 l) were described [9]. The laboratory scale composting unit allows for the simulation of a composting process with all operating controls (aeration, moistening, turning) common to those in a composting facility. The developed set-up should simulate processes such as pressure-forced windrow and pile composting as well as tunnel, box, container, and channel systems.

The example of laboratory composting system and vessel is given in Figs 11.1 and 11.2, respectively [10]. The composting vessels were placed in the laboratory composting system. Humidified air was passed through flow meters and then into the composting vessel. External heat was applied to maintain a constant temperature of 52 °C. The exhaust air was directed through a two-way valve attached to a gas chromatograph to measure CO_2 concentration. Once per week, the compost in the vessels was stirred and compost samples removed to determine the moisture content, which ranged from 48% to 55% (calculated on wet weight basis).

The design characteristics of a laboratory aerobic biodegradation unit was discussed and a number of key design features with a particular focus on the effects of internal bioreactor design and aeration rate on the compost moisture content and overall sample biodegradation were compared [11].

The medium closest to the natural condition is a solid medium (soil, compost, inert solid media) [12]. The studies on solid-state biodegradation processes in field and laboratory conditions, and in various media such as compost, soil, or inert material, were reviewed [12]. The external parameters that influence biodegradation kinetics—the material concentration in the solid medium, the environmental conditions (temperature, pH, moisture, oxygen availability, composition, and concentration of inorganic nutrients of the solid medium), the microbial population (concentration, nature, and interactions), the presence or the absence of other degradable substances, and the conditions and properties of the test system (volume and shape of the vessels)—were presented. The most significant parameters would appear to be the substrate type, moisture content, and temperature.

Maximum temperature during the thermophilic phase and moisture content were controlled in the course of composting to examine the effects of these composting conditions on the quality of the compost used for the evaluation of the biodegradability of plastics [13]. The moisture content during composting was controlled at 65%, while keeping the maximum temperature below 46, 58, and 70 °C, respectively. In turn, the maximum temperature was controlled to be below 58 °C, while maintaining the moisture content at 45, 55, and 65%, respectively. Biodegradability tests for cellulose, PCL, and poly(butylene succinate-*co*-butylene adipate) were performed in the five compost samples. All the three samples were biodegraded faster in the compost prepared with a maximum temperature of 45 °C than in the composts prepared at 58 °C or 70 °C, due to a larger number of microbial cells in the former compost sample. The biodegradation proceeded faster in the compost prepared with a moisture content of 65% than in the compost prepared with a moisture content of 45 and 55%.

The critical review of norms and standards and corresponding tests to determine the compostability of biodegradable plastics, possibly applicable also to biodegradable agricultural plastics, was presented [14]. It was concluded that the media and conditions of testing cover mainly the conditions designed for industrial composting facilities, and only a few concern home composting conditions. Considering that the end of life management of biodegradable agricultural plastic products will be done at the farm to reduce the management of the waste and also its cost, only a few of these are considered to be suitable for adaptation to cover also biodegradable agricultural plastic products.

11.5 Biodegradability Testing Methods

An overview of the testing methods that have been used to evaluate biodegradability of polymers and packaging materials was given by Briassouliss *et al.* [15]. Two kinds of tests for biodegradability of polymers were proposed: screening tests and tests that simulate *in situ* conditions. Screening tests include enzymatic and aquatic test under anaerobic and aerobic (Sturm test) conditions. Real-life tests are based on three compost tests (compost environment, standard compost test, and CO_2 compost test elaborated at VTT). During the first test, compostability of the materials is determined as the weight loss of the sample. Evaluation of the compostability of the samples is performed visually at weekly intervals in connection with turning the biowaste, and weight loss is measured at the end of the test when the positive control sample has been completely degraded and the temperature decreased to the outdoor temperature. The other two tests are based on CO_2 evolution.

Different polymers (e.g., polyhydroxybutyrate–hydroxyvalerate, PCL, cellulose acetate) representing varied biodegradability levels were studied using an aerobic respirometric test in order to model degradation kinetics in a liquid medium [16]. The mathematical model was proposed that fitted as well as possible the CO_2 evolution curves. Three kinetic parameters were determined: one represents the maximal percentage of carbon converted into CO_2, the second the "half-life time" in days of the degrading part of the material, and the third one the curve radius.

Results of an international ring test of two laboratory methods were presented for investigating the

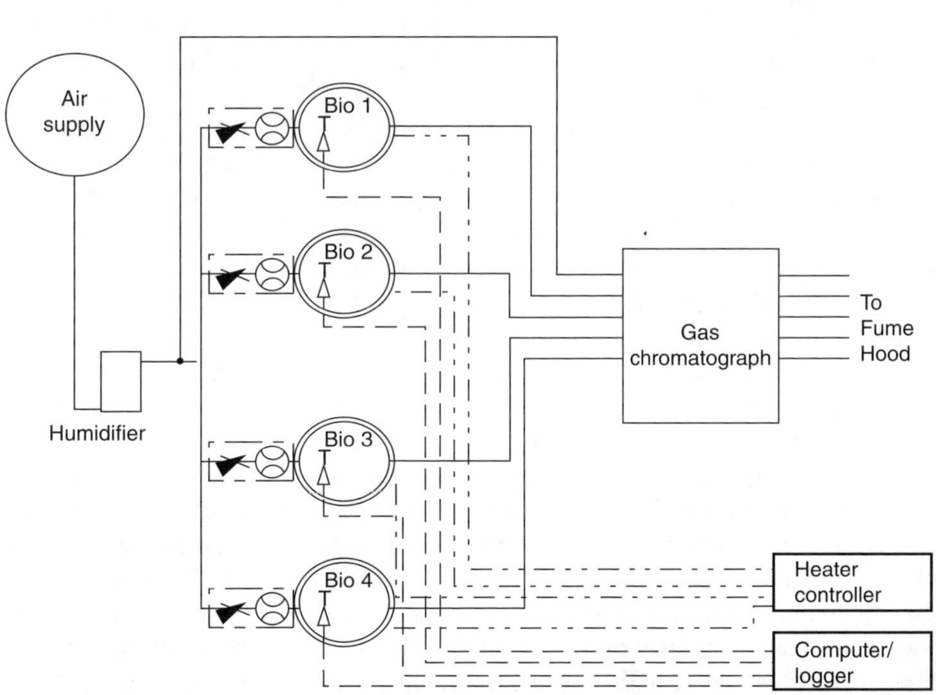

Figure 11.1 The laboratory composting system. *Reprinted with permission from Ref. [10].*

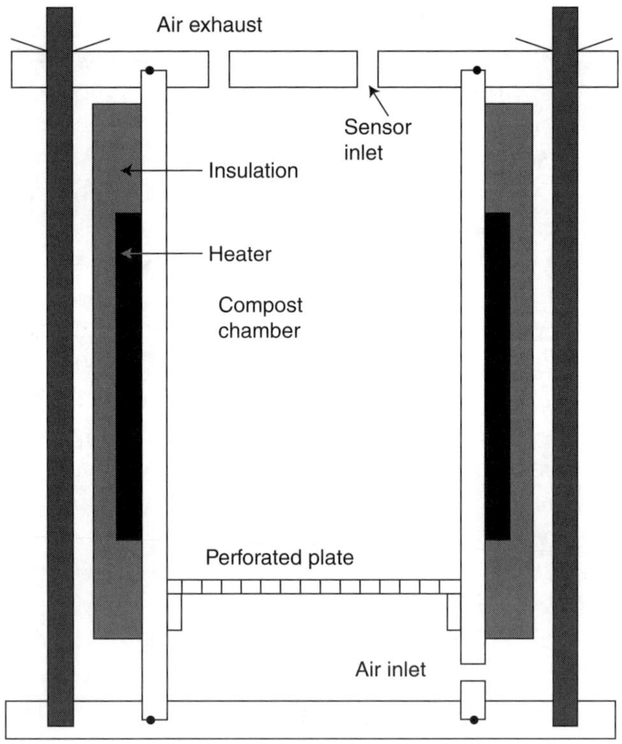

Figure 11.2 The laboratory composting vessel. *Reprinted with permission from Ref. [10].*

biodegradability of organic polymeric test materials in aquatic test systems based on respirometry and the evolution of carbon dioxide [17]. These methods were developed further from the well-known standardized biodegradation tests ISO 9408 (1999) and ISO 9439 (1999). A ring test was run using a PCL—starch blend and an aliphatic—aromatic copolyester as test materials and an MCE powder as a reference material. The most important improvements were the extension of the test period up to 6 months, the increase of the buffer capacity and nutrient supply of the inorganic medium, an optimization of the inoculum, and, optionally, the possibility of a carbon balance. Meanwhile, the test methods have been established as standards ISO 14851 (1999) and ISO 14852 (1999).

Test methods currently available for testing polymer degradability have been reviewed by Gu and Gu [18] and Eubeler *et al.* [19]. Table 11.6 presents a comparison of several methods available for testing degradability of different polymers and under a range of environmental and simulation techniques [18,19,20]. The gravimetric method is the most widely used technique with a long history of success. Requirements for the polymeric materials include that the polymer should be easily molded into some physical intact forms in sheet or strips and the specimens should not be sensitive to moisture to lose weight or easily hydrolyzed significantly upon exposure in a short period of time. Since the goal of this method is to obtain gravimetric information of exposed samples, specimens taken at different time intervals may also be used for chemical characterization including molecular weight and UV-visible spectra. When additional samples can be included initially, microbiological investigation including isolation of microorganisms from surfaces, characterization of the microorganisms, molecular analysis of pure species, mixed culture or the community can all be accomplished. The major advantage of this method is the simplicity and wide adaptability, while the drawback is that a large number of polymer samples are needed initially to carry out this kind of test.

The respirometric method measures either CO_2 produced or CO_2 consumed or both of them in an enclosed system with proper maintenance or regulation of air or oxygen supply. This technique is especially suitable for confirmation of the extent of mineralization. It can be used for measuring degradation of soluble powder from fragile polymeric materials. This method is easily adapted to a whole range of environmental conditions and/or specified or mixed culture microorganisms.

Examples of laboratory systems developed for biodegradation studies based on CO_2 evolution according to ISO 148551 and ISO/DIS 14855-2 standards are given in Figs 11.3 and 11.4, respectively [21]. The experimental set-up for biodegradation tests based on ISO 14855-1 shown in Fig. 11.3 is managed by Mitsui Chemical Analysis and Consulting Service, Inc., one of the research institutes that can determine the biodegradability of plastic products authorized by BPS for the GreenPla certification system in Japan. The CO_2 produced from the reaction vessels is trapped in alkaline solution bottles. The amounts of trapped CO_2 are determined by the titration of the acid solution to trap solutions.

Two hundred and sixteen strains of bacteria capable of degrading various biodegradable materials were isolated from several natural environments, including soil, river, and activated sludge [22]. Of the isolated strains, 20 strains with the strongest ability to degrade biodegradable materials were selected to construct a microbial community. The degradability of 14 kinds of biodegradable materials was investigated with both the original and improved ISO 14852 methods by substituting the natural inoculum with the constructed microbial community.

Table 11.6 Comparison of Testing Methods Available for Biodegradability Studies of Polymers [16,19,20]

Methods	Polymer Forms	Inoculum and Degradation Criteria Monitored	Comments
Gravimetry	Film or physical intact forms	A wide range of inocula can be used from soil, waters, sewage, or pure species of microorganisms from culture collections	This method is robust and also good for isolation of degradative microorganisms from environment of interest. Reproducibility is high. Disintegration of polymer cannot be differentiated from biodegradation
Respirometry	Film, powder, liquid, and virtually all forms and shapes	Either oxygen consumed or CO_2 produced under aerobic conditions. Under methanogenic conditions, produced methane can be monitored	This method is most adaptable to a wide range of materials. It may require a specialized instrument. When fermentation is the major mechanism of degradation, this method gives underestimation of the results
Surface hydrolysis	Films or others	Generally aerobic conditions, pure enzymes are used. Hydrogen ions (pH) released are monitored as incubation progresses	Prior information about the degradation of the polymer by microorganisms or particular enzymes is needed for the target specific test
Electrochemical impedance spectroscopy	Films or coatings resistant to water	The test polymers should be adhered to the surface of conductive materials and electrochemical conductance recorded	Polymer must be initially water impermeable for signal transduction. Degradation can proceed quickly and as soon as degradation is registered no further degradation processes can be distinguished
Radiolabeling	All kinds of materials	Marine, soil, sewage, compost sediment, etc.	Samples need to be 14C labeled
GPC/SEC	Virtually most polymers soluble in different solvents such as PEG, PVP, Ecoflex, Ecovio	Freshwater, salt water, CO_2 balance, DOC	Problems with environmental samples because extraction may be required
GC, GC/MS	Ecoflex and others, PHB, Xanthan, polysaccharide, Avicel. Requirement: small molecules, MWD low	Soil leachate, CO_2 balance; compost	Molecular weight can be a limiting factor for this type of analysis

(*Continued*)

Table 11.6 Comparison of Testing Methods Available for Biodegradability Studies of Polymers [16,19,20]—*Cont'd*

Methods	Polymer Forms	Inoculum and Degradation Criteria Monitored	Comments
MALDI-TOF	Ecoflex, Ecovio molecules with higher molecular weight	Freshwater, salt water, CO_2 balance, DOC	Parameters optimized, important for polymer analysis
AFM	Particles adhered or dispersed to a substrate		Surface analytical procedure
TEM	Thin and vacuum resistant, electron transparent samples	Surface water, sea water, activated sludge	Surface analytical procedure
NMR	Solid powder or liquid samples		
SEM	Gold sputtered solid samples	Bacterial degradation, surface area	Surface analytical procedure
FT-IR	PHB, Xanthan, polysaccharide, Avicel, solid or liquid samples		Fingerprinting technique
NIRS, MIRS	PLA, starch blends	Composting	

AFM, Atomic force microscopy; FT-IR, Fourier transform infrared spectroscopy; GC, gas chromatography; GC/MS, gas chromatography/mass spectrometry; GPC/SEC, gel permeation chromatography/size-exclusion chromatography; MALDI-TOF, matrix-assisted laser desorption/ionization-time-of flight; MIRS, mid-infrared spectroscopy; NIRS, near infrared spectroscopy; NMR, nuclear magnetic resonance; SEM, scanning electron microscopy; TEM, transmission electron microscopy.

The biodegradation test system with gravimetric measurement using the Microbial Oxidative Degradation Analyser (MODA) based on ISO/DIS 14855-2 uses the CO_2 trap system with CO_2 absorption column (Fig. 11.4). At first, room air is purged into a carbon dioxide trap to remove CO_2 in the air. Then, the air is moisturized and purged into the reaction vessel controlled at 58 and 70 °C using a thermosensor and ribbon heater. The air with the evolved CO_2 from biodegradation of the samples is poured into the ammonia trap to remove the produced ammonia from the compost for obtaining an accurate carbon balance using a gravimetric measurement. The air with its CO_2 is poured into dehumidifying traps to remove the moisture from the stream in air for an accurate carbon weight balance and then poured into an absorption column of carbon dioxide and an absorption column of water. In these two columns with soda lime (NaOH immobilized to slaked lime) and soda talc (NaOH immobilized to talc), the produced CO_2 is absorbed by the reactions indicated in Eqn (11.1):

$$CO_2 + 2NaOH \rightarrow Na_2CO_3 + H_2O \quad (11.1)$$

The produced H_2O is simultaneously trapped in these two columns. The weight of these two columns is increased to be the same as the weight of the produced CO_2, thus the produced CO_2 is easily obtained by a gravimetric method.

The enzymatic approach, based on the monitoring of pH changes in the degradation system and an increase of acidity, is a strong indication of surface hydrolysis of polymers after exposure to enzyme [18]. Because this kind of system may not be applicable for simulated environmental conditions involving microorganisms and the limitations of certain polymer chemistry, this method has a limited opportunity for wider applications. The advantage of this method is that a small quantity of material would be needed, especially for material in the development stage.

Electrochemical impedance spectroscopy (EIS) has been tested for monitoring the biodeterioration of high-strength materials and has very high sensitivity.

Table 11.7 Composting Studies of PLA Polymers

Polymer	Material Description	Method Used	Conditions/ Results	Remarks	References
PLLA	Poly-L-lactide; Neste Oy; In the form of non-woven fabrics and blown film	Bench-scale composting; carbon dioxide measurements	After 60 days final mineralization of PLLA films: 99%; PLLA fabrics: 73% and 48%	Newspaper as reference substance	[30]
PLA	PLA bottle, Biota	Composting, ISO 14855, ASTM D6400	At 63 days of exposure at 58 °C and 55% relative humidity: 64.2% mineralization	Corn starch as positive reference	[28]
PLA	PLA film	Composting; leaf compost rows, measurement of M_w	Temperature: 55–60 °C; humidity: 50–70%; PLA films required 2 weeks to disintegrate physically in the compost rows; degradation rate 109,173 and 68–532 M_w/week		[27]
PLA	PLLA (poly (L-lactide))—laboratory synthesized	Controlled composting test (prEN14046); CO_2 evolution measurement	Biodegradation: 92% (617%) for PLLA in 202 days (56% in 150 days)	Whatman chromatography paper as positive control	[29]
PLA	PLA (commercial; extruded 1.5 mm thickness sheets)	Composting; yard waste compost; CO_2 evolution measurement and molecular weight changes by GPC	Notable decrease in PLA molecular weight		[10]
PLA	Poly(lactic acid); commercial sample from Mitsui Chemicals	Composting (ISO 14855-1, ISO 14855-2, enzymatic degradation); CO_2 evolution measurement based on titration and gravimetric methods	Biodegradation of PLA powder was 91% for 31 days (ISO 14855-1 method) and 80% for 50 days at 58 °C (ISO 14855-2 method)	Cellulose powder was used as a reference material; PLA in the form of powders of different size was used	[21]

(*Continued*)

Table 11.7 Composting Studies of PLA Polymers—*Cont'd*

Polymer	Material Description	Method Used	Conditions/ Results	Remarks	References
PLA	Poly(lactic acid); commercial bottles and deli containers	Composting under real conditions (compost pile; temp. 65 °C; moisture 63%, pH 8.5); visual inspection; molecular weight changes (GPC method); glass transition and melting temperature (DSC method); decomposition temperature (TGA method)	Degradation of PLA containers, 30 days under composting conditions		[32]
PLA	Poly(lactic acid)	Composting at laboratory scale	Simulated aerobic composting facility (as *per se* ASTM D5338) more than 60% degradation before 100 days		[33]
Commercial Sample:					
Biomer L9000					

DSC, differential scanning calorimetry; GPC, gel permeation chromatography; TGA, thermogravimetry analysis.

Table 11.8 Biodegradation Results of Compostable Polymer Materials [183]

Polymer	Name	Company	Biodegradation Mineralization, %[1]
Polymers based on renewable resources			
PLA	NatureWorks	Cargill Dow	100
PHBV	Biopol D400G, HV 5 7%	Monsanto	100
Polymers based on petroleum resources			
PCL	CAPA 680	Solvay	100
PEA	BAK 1095	Bayer	100
PBSA	Bionolle 3000	Showa	90
PBAT	Eastar Bio 14766	Eastman	100

[1]At 60 days in controlled composting according to ASTM 5336.

Figure 11.3 Biodegradation evaluation method based on ISO 14855-1. *Reprinted with permission from Ref. [21].*

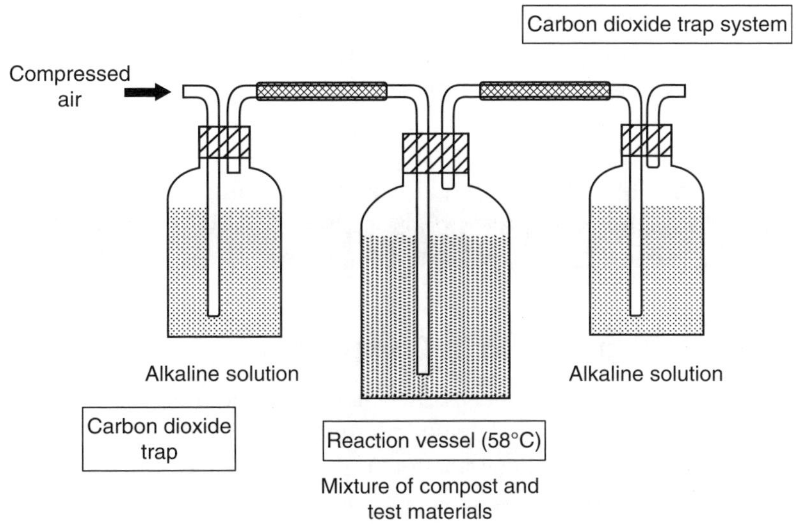

11.6 Biodegradation of Biodegradable Polymers from Renewable Resources

11.6.1 Biodegradation of Poly(lactic acid)

11.6.1.1 Degradation Mechanisms

Biodegradation of polylactic acid (PLA) proceeds via a two-stage mechanism [23]. In the first step, hydrolysis of ester linkage occurs. This step can be accelerated by acid or bases and is affected by both temperature and moisture levels [24]. In the primary degradation phase, no microorganisms are involved. As the average molecular weight diminishes, microorganisms present in the soil begin to digest the lower molecular weight lactic acid oligomers, producing carbon dioxide and water. This two-stage mechanism of degradation is a distinct advantage of PLA over other biodegradable polymers, which typically

Figure 11.4 Biodegradation evaluation method by gravimetric measurement of carbon dioxide evolved in laboratory-scale test using the Microbial Oxidative Degradation Analyser (MODA) instrument in controlled compost based on ISO/DIS 14855-2. *Reprinted with permission from Ref. [21].*

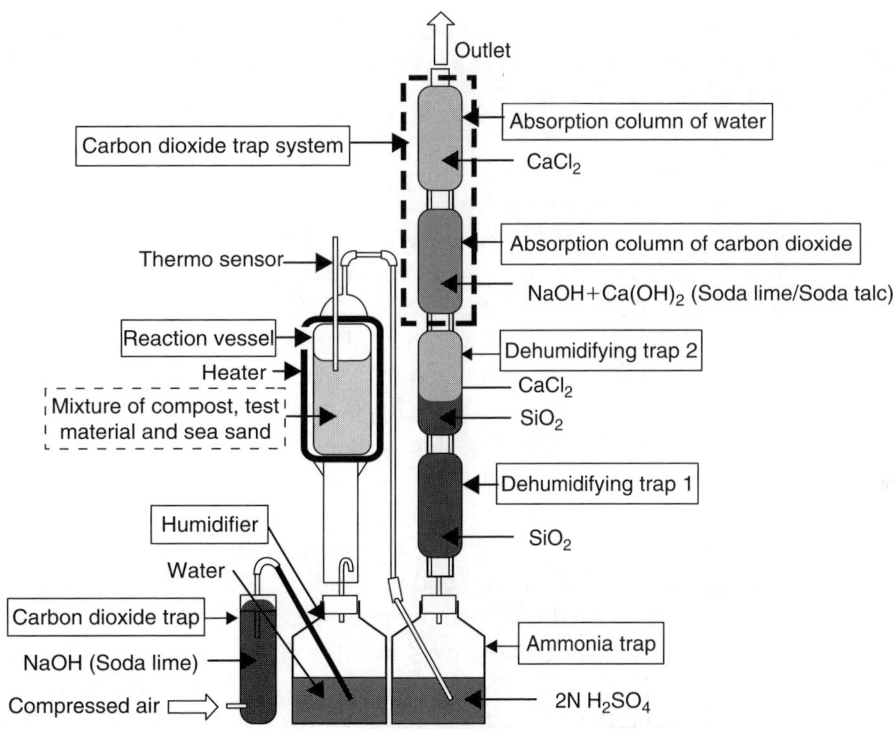

degrade by a single-step process involving bacterial attack on the polymer itself. This is a useful attribute, particularly for product storage and in applications requiring food contact. PLA degrades rapidly in the composting atmosphere of high humidity and temperature (55–70 °C). But, at lower temperatures and/or lower humidity, the storage stability of PLA products is considered to be acceptable.

11.6.1.2 Degradation in Compost

PLA is fully biodegradable when composted in a large-scale operation with temperatures of 60 °C and above. The first stage of degradation of PLA (2 weeks) is via hydrolysis to water-soluble compounds and lactic acid, then metabolization by microorganisms into carbon dioxide, water, and biomass proceeds [25].

PLA is largely resistant to attack by microorganisms in soil or sewage under ambient conditions. The polymer must first be hydrolyzed at elevated temperatures (>58 °C) to reduce the molecular weight before biodegradation can commence. Thus, PLA will not degrade in typical garden compost. Under typical use and storage conditions PLA is quite stable [26].

The degradation of PLA plastic films in Costa Rica soil and in a leaf composting environment was investigated [27]. The average soil temperature and moisture content in Costa Rica were 27 °C and 80%, respectively. The average degradation rate of PLA plastic films in the soil of the banana field was 7657 M_w/week. PLA films required 2 weeks to disintegrate physically in leaf compost rows.

PLA bottles were used as the test material to determine polymer biodegradation under simulated conditions using an automatic laboratory-scale respirometric system [28]. The results were compared with those for corn starch powder and poly(ethylene terephthalate) (PET) bottles. At 63 days of exposure at 58 °C and 55% relative humidity (RH), PLA, corn starch, and PET achieved 64.2, 72.4, and 2.7% mineralization, respectively. It was stated that, based on ASTM D 6400 and ISO 14855, PLA bottles qualified as biodegradable since mineralization was greater than 60%.

The biodegradability of lactic acid-based polymers was studied under controlled composting conditions (according to future CEN EN 14046), and the quality of the compost was evaluated [29]. All the polymers biodegraded to over 90% of the positive control in 6 months, which is the limit set by the CEN standard.

The biodegradation of polylactide (PLLA) was studied at different elevated temperatures in aerobic and anaerobic, aquatic and solid-state conditions. In the aerobic aquatic headspace test, the mineralization of PLLA was very slow at room temperature, but faster under thermophilic conditions [30]. The clear effect of temperature on the biodegradability of PLLA in the aquatic test indicates that its polymer structure has to be hydrolyzed before microorganisms can utilize it as a nutrient source. At similar elevated temperatures, the biodegradation of PLLA was much faster in anaerobic solid-state conditions than in aerobic aquatic conditions. The behavior of PLLA in the natural composting process was similar to that in the aquatic biodegradation tests, biodegradation starting only after the beginning of the thermophilic phase. These results indicate that PLLA can be considered as a compostable material, being stable during use at mesophilic temperatures, but degrading rapidly during waste disposal in compost or anaerobic treatment facilities.

It was demonstrated that PLA can be efficiently composted when added in small amounts (<30% by weight) to pre-composted yard waste (i.e., grass, wood mulch, and tree leaves in equal parts by weight) [10]. Garden waste and extruded PLA sheets were placed in laboratory composting vessels for 4 weeks. Evolved carbon dioxide concentration was measured by using gas chromatography to assess polymer degradation.

In all cases (0, 10, or 30% PLA), the amount of evolved CO_2 significantly increased as composting time increased (Fig. 11.5). Compost pH dropped (from 6.0 to 4.0) after 4 weeks of composting for 30% PLA, but remained unchanged (6.30 for 0 or 10% PLA). Most likely, in the case of 30% PLA,

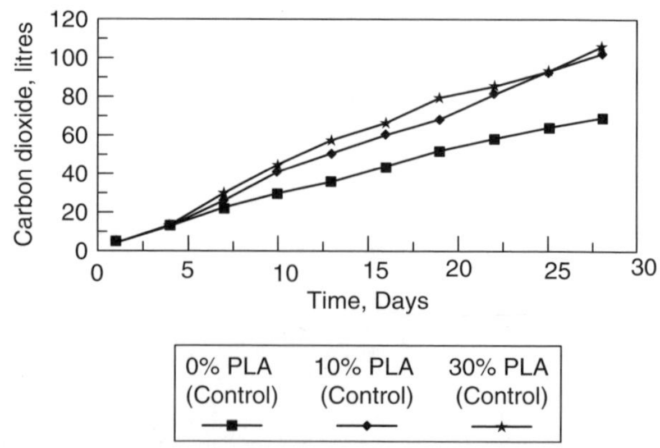

Figure 11.5 Generation of CO_2 during composting of yard waste compost/PLA mixtures (100%/0%, 90%/10%, or 70%/30% on dry weight basis). *Reprinted with permission from Ref. [10].*

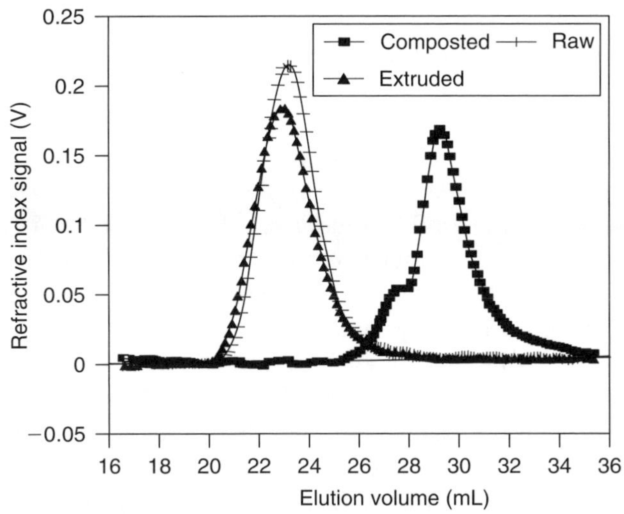

Figure 11.6 Gel permeation chromatograms of PLA resin, extruded PLA and extruded PLA composted for 4 weeks. *Reprinted with permission from Ref. [10].*

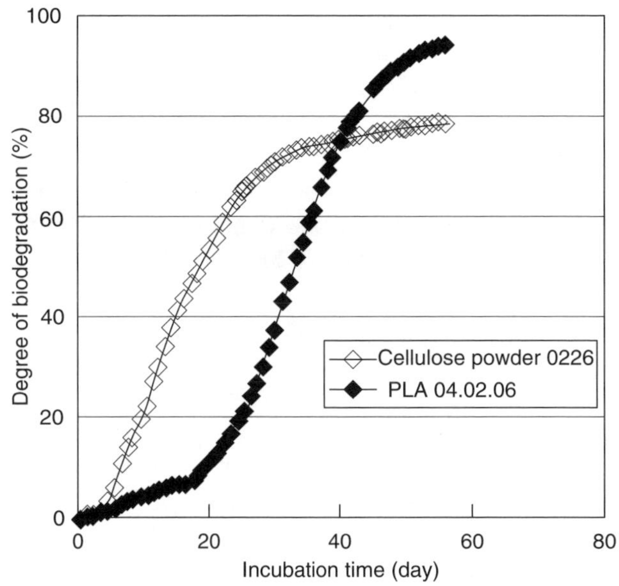

Figure 11.7 Biodegradation test of PLA and cellulose powders by ISO 14855-2 method using MODA instrument in controlled compost at 58 °C. *Reprinted with permission from Ref. [21].*

substantial chemical hydrolysis and lactic acid generation lowered the compost pH. The lowered pH likely suppressed microbial activity, thus explaining the lack of difference in carbon dioxide emissions between 10% and 30% PLA mixtures. The reduction in PLA molecular weight was observed after 4 weeks of composting (Fig. 11.6).

PLA crosslinked by using both triallyl isocyanurate and electron radiation or using dicumyl peroxide was studied with the aim of examining the behavior of the modified polymer under various environmental conditions including composting in an industrial pile [31]. It was found that neat PLA irradiated with high-energy electrons underwent degradation that increased during composting.

Recently, PLA powders were proposed as the reference test materials for the international standard of biodegradation evaluation methods [26]. Mechanical crushing at low temperature of polymer pellets using dry ice was applied as the method for producing polymer powder of PLA. After sieving, the average diameter of the PLA particles was 214.2 μm. The biodegradation speeds of these PLA polymer powders were evaluated by two methods based on the international standard and one *in vitro* method based on the enzymatic degradation. First, the degree of biodegradation for the PLA powder was 91% for 35 days in a controlled compost determined by a method based on ISO 14855-1 (JIS K6953) at 58 °C. Second, the polymer powders were measured for biodegradation by the MODA in a controlled compost at 58 and 70 °C based on ISO/DIS 14855-2 under many conditions. The degree of biodegradation for PLA powder was approximately 80% for 50 days (Fig. 11.7).

The degradation of two commercially available biodegradable packages made of PLA was investigated and compared under real compost conditions and under ambient exposure, using visual inspection, gel permeation chromatography (GPC), differential scanning calorimetry (DSC), and thermogravimetry analysis [32]. PLA bottles made of 96% L-lactide exhibited lower degradation than PLA delicatessen containers made of 94% L-lactide, mainly due to their highly ordered structure and therefore their higher crystallinity. Temperature, relative humidity, and pH of the compost pile played an important role in the rate of degradation of the packages. PLA deli containers degraded in $<$30 days under composting conditions (temperature $>$ 60 °C, RH $>$ 65%, pH $\approx$7.5).

Biodegradation studies of PLA and PCL and various components of green composites and their blends were studied at a laboratory scale simulated aerobic composting facility (as per ASTM D 5338) [33]. The individual polymers (PLA and PCL) showed inherent degradability of 60% of the OC to carbon dioxide within 180 days. Enhancement in degradation rate is observed in composites in comparison to their counterpart polymers PLA and PCL alone likely due to the presence of ready degradable natural biomass.

Pieces of PLA and PLA with 10% of corn (PLA-corn) of different thicknesses and shapes have been subjected to aerobic degradation at a constant temperature of 58 °C for 90 days, following EN 14806 Norm "Packaging—Preliminary evaluation of the disintegration of the packaging materials under simulated composting conditions in a laboratory-scale test" and ISO 20200:2004 Norm "Determination of the degree of disintegration of plastic materials under simulated composting conditions in a laboratory-scale test" [34]. It was found that the pieces made of PLA and PLA with foaming agent present an average biodisintegration degree of 63.6%. With regard to the pieces made of PLA-corn, an average biodisintegration degree of 79.7% was obtained.

The kinetics of C—CO_2 evolution during biodegradation of plastics materials including MCE and PLA [35]. The aerobic biodegradation under controlled composting conditions was monitored according to ISO 14855-1. The biodegradability of MCE and PLA were 94.34% and 85.75%, respectively. The work showed that MCE and PLA produced the high amounts of C—CO_2 evolution, which gave readily hydrolyzable carbon values of 55.49% and 40.17%, respectively, with readily hydrolysis rates of 0.338 day^{-1} and 0.025 day^{-1}, respectively. The mineralization rate of PLA was 0.500 day^{-1} as a lag phase was observed at the beginning of the biodegradability test. No lag phase was observed in the biodegradability testing of MCE.

11.6.1.3 Degradation in Other Environments

PLA undergoes enzymatic or non-enzymatic hydrolysis when it is exposed to an aqueous environment. Several factors, such as temperature, pH, additives, copolymerization, initial molar mass, specimen size, residual monomer, and the degree of crystallinity have been reported to affect the rate of hydrolysis of PLA. The biotic and abiotic degradation of poly(L-lactide) has been studied with pyrolysis gas chromatography mass spectrometry (Py-GC/MS) [36]. It was shown that degradation in the biotic medium proceeded mainly via a surface erosion mechanism, whereas bulk erosion was the predominant degradation mechanism in the abiotic medium. Based on the size-exclusion chromatography (SEC) and Py-GC/MS data, it was reported that degradation was faster in the biotic than in the abiotic sample.

Polyester-degrading ability of actinomycetes obtained from culture collections was investigated by the formation of clear zones on polyester-emulsified agar plates [37]. Using 41 genera (43 strains) of actinomycetes with phylogenetic affiliations, poly(L-lactide)-degraders were found to be limited to members of family Pseudonocardiaceae and related genera. On the other hand, poly(β-hydroxybutyrate) (PHB)-, PCL-, and PBS-degraders were widely distributed in many families.

Microbial and enzymatic degradation of PLA was reviewed by Tokiwa [38]. Most of the PLA-degrading microorganisms phylogenetically belong to the family of Pseudonocardiaceae and related genera such as *Amycolatopsis*, *Lentzea*, *Kibdelosporangium*, *Streptoalloteichus*, and *Saccharothrix*. Several proteinous materials such as silk fibroin, elastin, gelatin, and some peptides and amino acids were found to stimulate the production of enzymes from PLA-degrading microorganisms. In addition to proteinase K from *Tritirachium album*, subtilisin, a microbial serine protease and some mammalian serine proteases, such as α-chymotrypsin, trypsin, and elastase, could also degrade PLA.

The clear zone method using emulsified polyester agar plates was used to evaluate the population of polymer-degrading microorganisms in the environment. It was confirmed that the population of aliphatic polyester-degrading microorganisms at 30 and 50 °C decreased in the order of PHB = PCL > PBS > PLA [38,39,40]. Suyama *et al.* [41] reported that 39 bacterial strains of class Firmicutes and Proteobacteria isolated from soil were capable of degrading aliphatic polyesters such as PHB, PCL, and PBS, but no PLA-degrading bacteria were found. These results showed that PLA-degrading microorganisms are not widely distributed in the natural environment and thus PLA is less susceptible to microbial attack in the natural environment than other microbial and synthetic aliphatic polyesters. The biodegradability of PLA depends on the environment to which it is exposed. In human or animal bodies, it is believed that PLA is initially degraded by hydrolysis and the soluble oligomers formed are metabolized by cells. Soil burial tests show that the degradation of PLA in soil is slow and that it takes a long time for degradation to start. For instance, no degradation was observed on PLA sheets after 6 weeks in soil [42]. Urayama *et al.* [43] reported that the molecular weight of PLA films with different optical purity of the lactate units (100% L and 70%L) decreased by 20 and 75%, respectively, after 20 months in soil.

The degradation of PLA-based films by microorganisms extracted from compost was studied in a liquid medium [44]. The application of the ASTM standard (ASTM D 5209-92) did not produce biodegradation of pieces of PLA film. With the ISO/CEN standard method (ISO/CEN 14852-1998), the percentage biodegradation after 45 days was found to be 30%. The different temperature profile of medium used in two standards seemed to be the major factor in explaining the observed differences.

Commercial lipases were examined for their degradation efficiency of aliphatic polyester films in special emphasis on PLA [45]. Polyester films were immersed during 100 days in lipase solutions at 37 °C at pH 7.0. PBSA and poly(ε-caprolactone) (PCL) films were rapidly degraded during 4−17 days when either lipase Asahi derived from *Chromobacterium viscosum* or lipase F derived from *Rhizopus niveus* was used. Lipase Asahi could also degrade PBS film within 17 days. Lipase F-AP15 derived from *Rhizopus orizae* could degrade PBSA in 22 days. Lipase PL isolated from *Alcaligenes* sp. revealed its higher degradation activity of PLA film. PLA degraded completely at 55 °C, pH 8.5 with lipase PL during 20 days. Based on the results of GPC and HPLC analyses, it was concluded that complete degradation of PLA resulted from two processes. First, the chemical hydrolysis from PLA into oligomers at higher pH and/or under higher temperature conditions, because polyesters are generally not stable under such conditions. Second, the enzymatic hydrolysis from oligomers to the monomer.

Long-term degradation/disintegration behavior, indicative of the biodegradation in soil behavior of PLA films and fibers, was studied in natural Mediterranean soil environment during an 11-month trial in the experimental field [46]. In parallel, simulated soil burial experiments were carried out under controlled laboratory conditions. For comparison purposes, degradation/disintegration of PLA film was also studied under low temperature composting conditions (house composting). During long-term exposure under natural soil environment dominated by complex and uncontrolled biotic−abiotic conditions and Mediterranean climatic conditions and under house composting conditions, PLA film samples of different thickness were partially, to a rather low degree, degraded mechanically or slightly disintegrated. The results showed that degradation behavior of bio-based polymers like PLA in a real soil environment is a complex phenomenon, following different patterns regarding morphological changes.

11.6.2 Biodegradation of Polyhydroxyalkanoates

11.6.2.1 Degradation Mechanisms

The bacterially produced poly(hydroxyalkanoates) (PHAs) are fully biodegradable in both anaerobic and aerobic conditions, and also at a slower rate in marine environments.

PHAs are quite resistant to moisture, but they are rapidly biodegraded by a wide range of microorganisms [47]. The rate of enzymatic degradation of PHB and PHBV by PHA depolymerases was from 2 to 3 orders of magnitude faster than the rate of simple hydrolytic degradation. The enzymatic hydrolysis of PHB and PHV copolymers is a heterogeneous erosion process proceeding from the surface, where polymer chains are degraded initially by *endo*-scissions (randomly throughout the chain) and then by *exo*-scissions (from the chain ends) [47]. This results in subsequent surface erosion and weight loss. The average molecular weight and molecular weight distribution do not change during the enzymatic degradation because of selective degradation only at the surface, together with removal and dissolution of low molecular weight degradation products from the polymer matrix into the surrounding environment. It was reported that in the initial stages of degradation, only amorphous material was consumed. Later, however, both amorphous and crystalline regions were degraded without preference.

The biodegradable properties of Biopol, thermoplastic copolyester PHBV composed of HB units and between 0 and 30% HV units, incorporated randomly throughout the polymer chain, were discussed by Byrom [48]. Biopol biodegrades in microbially active environments. Biodegradation is initiated by the action of microorganisms growing on the surface of the polymer. Microorganisms that degrade Biopol include species of *Aspergillus*, *Streptomyces*, *Actinomyces*, and *Pseudomonas*. These microorganisms secrete extracellular enzymes, such as depolymerases and esterases, that solubilize the polymer in the immediate vicinity of the cell. The soluble degradation products are then absorbed through the cell wall and metabolized to CO_2 and H_2O under aerobic conditions. The rate of degradation is dependent on a number of factors. Particularly important are the level of microbial activity (determined by the moisture level, nutrient supply, temperature, and pH) and the surface area of the polymer. A series of tests was carried out in which Biopol was composted together with "biorefuse." A weight loss of 80% was observed after 15 weeks under these conditions when the stack was turned.

11.6.2.2 Degradation in Compost

Poly-β-hydroxybutyrate/valerate copolymer (Biopol) was used as a test material and cellulose powder as a reference material in a ring laboratory controlled composting test [49]. A laboratory method was presented for investigating the biodegradation of an organic test material in an aerobic composting system based on the evolution of carbon dioxide. The test becomes a basis of a European standard in connection with determining the compostability of packaging and packaging materials. The mean degree of Biopol biodegradation was 88% in comparison with 84% for MCE powder.

The compost activities of PHB and a copolymer of 20% β-hydroxyvalerate were studied in a simulated municipal solid waste compost test at a constant temperature of 55 °C and a constant moisture content of 54% [50]. Biodegradation was measured through weight loss and normalized for thickness. The compost activity was found to be divided into three stages with the maximum rate of polymer degradation occurring between the tenth and fifteenth day. The biodegradation rate of the valerate copolymer was seen to be much higher than that of the homopolymer.

The biodegradability of poly(hydroxybutyrate-*co*-hydroxyvalerate) containing 3 mol% of hydroxy-valerate (HV) was tested under composting conditions on both a pilot and a laboratory scale [51]. It was found that the biodegradability of PHBV in the pilot-scale composting conditions was similar to that in the laboratory scale. The PHBV film was completely disintegrated in the pilot-scale composting test, and the degree of biodegradation was 81% in the laboratory-scale control composting test.

To understand the influence of chemical structure on the biodegradability of PHA, the biodegradation behavior of PHB, poly(hydroxybutyrate-*co*-hydroxyvalerate) (PHBV, 40% mol HV), PHBV (20% mol HV), PHBV (3% mol HV), and poly(3-hydroxybutyrate-*co*-4-hydroxybutyrate) [P(3HB, 4HB), 10% mol 4HB] were investigated under the controlled composting conditions according to ISO 14855-1 [52]. It was found that the order of biodegradability was P(3HB, 4HB) (40% mol 4HB) ≈ PHBV (40% mol HV) > PHBV (20% mol HV) > PHBV (3% mol HV) > PHB.

The effect of abiotic factors such as water and air on the degradation of poly(3-hydroxybutyrate-*co*-3-hydroxyvalerate) (PHBV) in a compost was investigated using simulated and natural environments [53]. The results showed that during a period of 50 days, water and air have little or no effect on the degradation of PHBV in garden waste compost. It was suggested that the degradation was due to microbial action only.

Changes in physical and mechanical properties of poly(hydroxybutyrate-*co*-hydroxyvalerate) during degradation in a composting medium were studied by Luo and Netravali [54]. Fourier transform infrared spectroscopy - attenuated total reflectance (FTIR-ATR) spectra of the control and partly degraded PHBV specimens as a function of composting time are presented in Fig. 11.8. No detectable changes between the spectra of control and composted specimens were observed. Figure 11.9 presents typical stress vs. strain plots of control and composted PHBV specimens. The ultimate tensile strength and the strain at ultimate tensile strength decreased significantly as a function of composting time. The results from the analysis of weight loss, scanning electron microscopy (SEM), molecular weight, FTIR, DSC, and tensile testing suggested that the degradation of PHBV in compost medium was enzymatic rather than hydrolytic and occurred from surface and the degraded material leached out.

The biodegradation of poly-β-(hydroxybutyrate) (PHB) and poly-β-(hydroxybutyrate-*co*-β-valerate) (PHBV) was assessed by the loss of mass, tensile strength, and roughness of the polymer [55]. Both polymers showed similar biodegradation in soil composting medium at 46 °C and at room temperature (24 °C) and in a soil simulator. After aging in soil composting medium at 46 °C for 86 days, both polymers showed a decrease in the tensile strength at break (76% for PHB and 74% for PHBV). In agreement with this, the roughness of both polymers increased faster in soil composting medium at 46 °C. Surface damage can be assessed by measuring the surface roughness, a technique commonly used in mechanical engineering. It was suggested that roughness may be a useful parameter for evaluating the biodegradation of polymers.

The effect of temperature on the biodegradation of PHB, PHBV, and PCL was assessed based on the

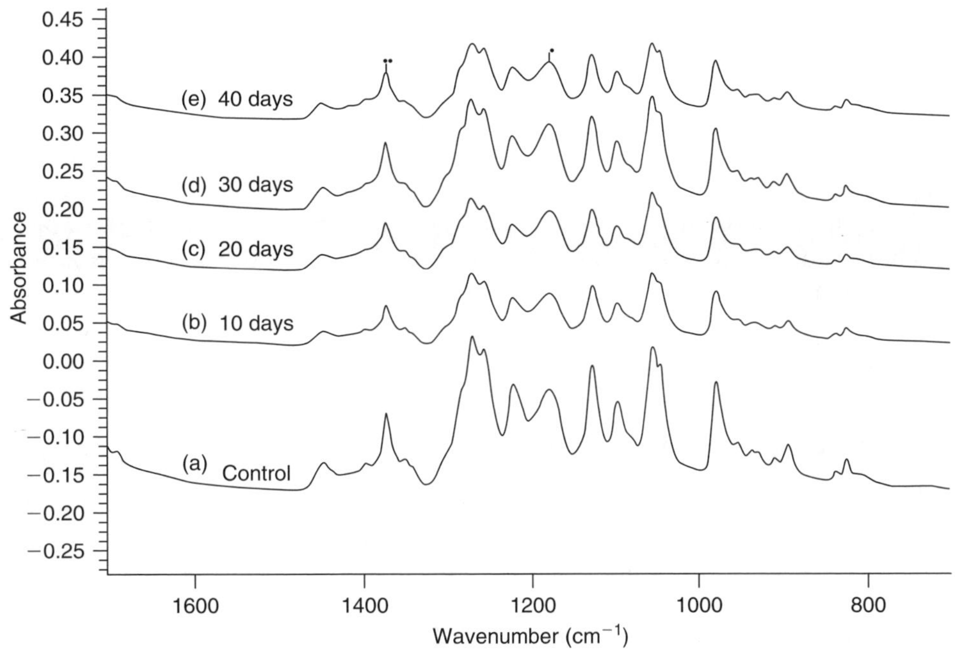

Figure 11.8 FTIR-ATR spectra of PHBV films exposed to the composting medium for different periods of time: (a) 0, (b) 10, (c) 20, (d) 30, and (e) 40 days. *Reprinted with permission from Ref. [54].*

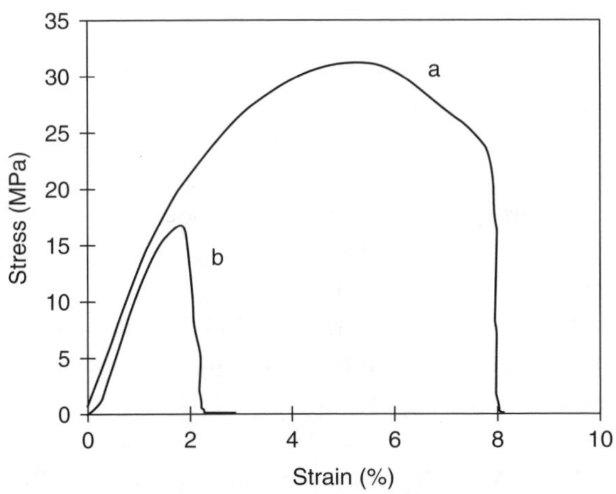

Figure 11.9 Typical stress vs. strain of control and partially degraded PHBV specimens: (a) control, (b) composted for 30 days. *Reprinted with permission from Ref. [54].*

mass retention when the polymers were incubated in soil compost at 46 and 24 °C [56]. Biodegradation was greatest at 46 °C for the three polymers studies. PHB and PHBV showed similar biodegradation at both temperatures. PHB and PHBV were totally degraded after 104 days of aging in soil compost at 46 °C and PCL degraded by 36% in 120 days. Degradation of the polymers at room temperature (24 °C) was relatively slow, with losses of 51 and 56% for PHB and PHBV, respectively, after 321 days of aging. In contrast, PCL showed no biodegradation at room temperature after almost 300 days.

The effect of thermal aging on the degradation of PHB, PHBV, and PCL in soil compostage was studied by Rosa *et al.* [57]. The biodegradability of PHB, PHBV, and PCL was examined following thermal aging in an oven for 192, 425, and 600 h. Different temperatures, 100, 120, and 140 °C for PHB and PHBV and 30, 40, and 50 °C for PCL, were used to assess the influence of this parameter on biodegradation. Thermal aging increased the biodegradability only for PHB at 120 and 140 °C.

Bacterial thermoplastic polyesters PHAs, produced by the fermentation of renewable materials, such as sugars or molasses, i.e., PHB and a copolymer of PHB(88%)/PHV(12%), were mixed with other biodegradable materials (additives) to improve their mechanical properties [58]. Plasticizers, glycerol, tributyrin, triacetin, acetyltriethylcitrate, acetyltributylcitrate, and a nucleation agent, saccharin, were used. Lubricants were glycerolmonostearate, glyceroltristearate, 12-hydroxystearate, and 12-hydroxystearic acid. The biodegradability of blends was investigated in the aerobic test, under compost conditions in soil and in river water. It was found that the blends were degraded more easily in the aerobic test, i.e., in the river water and compost, than in the soil.

Several types of biodegradable medium-chain-length polyhydroxyalkanoates (mcl-PHAs) were produced by *Pseudomonas putida* KT2442 at pilot and laboratory scales from renewable long-chain fatty acids and octanoic acid [59]. All purified polymers were subjected to *in vitro* aerobic biodegradation using a compost isolate. The extent of mineralization varied from 15% to 60% of the theoretical biochemical oxygen demand (ThBOD). The polymer weight loss after 32 days ranged from 40% to 90% for the different mcl-PHAs.

11.6.2.3 Degradation in Other Environments

PHAs are degraded upon exposure to soil, compost, or marine sediment [60]. Biodegradation is dependent on a number of factors such as microbial activity of the environment, and the exposed surface area, moisture, temperature, pH, and molecular weight. Biodegradation of PHA under aerobic conditions results in carbon dioxide and water, whereas in anaerobic conditions the degradation products are carbon dioxide and methane. PHAs are compostable over a wide range of temperatures, even at a maximum of around of 60 °C with moisture levels at 55%. Studies have shown that 85% of PHAs were degraded in 7 weeks. PHAs have been reported to degrade in aquatic environments (Lake Lugano, Switzerland) within 254 days even at temperatures not exceeding 6 °C.

Biodegradability patterns of two PHAs, a polymers of 3-hydroxybutyric acid (P-3HB) and a copolymer of 3-hydroxybutyric and 3-hydroxyvaleric acids (P-3HB/3HV) containing 11 mol% of HV, were studied in the tropical marine environment, in the South Sea [61]. No significant differences have been observed between the degradation rates of P-3HB and P-3HB/3-PHV polymers. PHA-degrading microorganisms were isolated. The PHA-degrading strains were identified as *Enterobacter* sp. (four strains), *Bacillus* sp., and *Gracilibacillus* sp.

Effective PHA destructors include various bacteria from widespread soil and water genera (*Pseudomonas, Alcaligenes, Comamonas, Streptomyces, Ilyobacter*), as well as fungi (*Ascomycetes, Basidiomycetes, Deuteromyces, Mastigiomycetes, Myxomycetes*) [62].

The degradation dynamics of PHAs of different compositions (a PHB homopolymer and a PHB/PHV copolymer with 14 mol% of HV) have been studied in a eutrophic storage reservoir for two seasons. It has been shown that the biodegradation of polymers under natural conditions depend not only on their structure and physicochemical properties but also, to a great extent, on a complex of weather—climatic conditions affecting the state of the reservoir ecosystem.

Comparative biodegradation of two types of aliphatic polyesters, based on renewable resources, PLA and PHA was studied in soil environment under real and simulated soil burial laboratory conditions [63]. For PHA, polymer biodegradation was shown to proceed much faster.

Results show that the microbes easily attack the surface of PHA film from the beginning of the soil burial exposure regardless of the temperature of environment. Visual inspection and FTIR and DSC analysis indicate a possible layer-by-layer degradation process for PHA.

11.6.2.4 Thermoplastic Starch

The suitability of an *in vitro* enzymatic method for assaying the biodegradability of starch-based materials was evaluated [64]. The materials studied included commercial starch-based materials and thermoplastic starch (TPS) films prepared by extrusion from glycerol and native potato starch, native barley starch, or crosslinked amylomaize starch.

In order to verify the response of the controlled composting test method (i.e., the ISO/DIS 14855:1997, the ASTM D 5338-92) to starch at different concentrations, the maximum amount prescribed by the test method (100 g) and lower amounts (60 and 30 g), as if starch were a coingredient in a blend, were tested [65]. After 44 days of incubation (at a constant temperature of 58 °C) the biodegradation curves were in a plateau phase, displaying the following final (referred to a nominal starch initial amount of 100 g: starch 100 g, 97.5%; starch 60 g, 63.7%; and starch 30 g, 32.5%. The data showed a CO_2 evolution roughly equal, in each case, to the theoretical maximum, indicating a complete starch mineralization. The average biodegradation of cellulose turned out to be 96.8% after 47 days.

The corn flour-based material biodegradation kinetics were assessed by an anaerobic biodegradation standard test (ISO 14853), an aerobic burial composting test (ISO/DIS 16929), and a test to assess the susceptibility of corn starch to hydrolysis by amylolytic enzymes [66]. It was found that degradation of the partially crystalline corn flour-based material under burial composting conditions proceeded in a selective manner: the amorphous regions being degraded prior to the crystalline ones.

The degradation of starch- and PLA-based plastic films by microorganisms extracted from compost was studied in a liquid medium [67]. The various degradation products produced (carbon dioxide, biomass formed by abstraction of some of the material's carbon, soluble organic compounds, and possibly nondegraded material) were measured throughout the duration of the experiment, and total carbon balances were estimated. The experiments were conducted according to ASTM and ISO/CEN standards and used two different physical states of the material, i.e., film and powder forms. The final mineralization percentage (Cg) of starch-based material was always greater than 60%, the minimum assigned value for a biodegradable material. Moreover, the percentage of biodegradation, defined as the sum of the mineralization (Cg) and bioassimilation (Cb), was between 82% and 90%. It was concluded that for an easily biodegradable material as starch, the evolution of the way carbon repartitioned between different degradation products was quite similar whatever the experimental condition or the type of substrate. On the other hand, for a resistant material (polylactic-based plastic) exposed to these microorganisms, the nature of the biodegradation depended strongly on the experimental conditions.

Biodegradation behavior of TPS and thermoplastic dialdehyde starch (TPDAS) under controlled composting conditions according to ISO 14855 was investigated [68]. It was found that chemical modification of starch could have a major impact on the biodegradation rate and final biodegradation percentage. The TPS degraded faster than TPDAS under controlled composting conditions. For the TPDAS, the degradation rate and final biodegradation percentage were closely related to the degree of oxidation of dialdehyde starch (DAS). Three kinds of actinomycete were isolated from compost and identified as *Micromonospora*, *Nocardia*, and *Streptomyces*, which were degrading microorganisms of the starch tested.

11.6.3 Biodegradation of Other Compostable Polymers from Renewable Resources

11.6.3.1 Biodegradation of Cellulose

TLC grade cellulose is used as the positive reference material during compostabilty studies according to international standards, e.g., ISO 14855. It was reported that the average biodegradation of cellulose during controlled composting method turned out to be 96.8 $\pm$ 6.7 (SD) after 47 $\pm$ 1 days [69].

Most of the cellulolytic microorganisms belong to eubacteria and fungi, even though some anaerobic protozoa and slime molds able to degrade cellulose have also been described [70]. Cellulolytic microorganisms can establish synergistic relationships with non-cellulolytic species in cellulosic wastes. The interactions between both the populations lead to complete degradation of cellulose, releasing carbon dioxide and water under aerobic conditions, and carbon dioxide, methane, and water under anaerobic conditions.

Microorganisms capable of degrading cellulose produce a battery of enzymes with different specificities, working together. Cellulases hydrolyze the β-1,4-glycosidic linkages of cellulose. Traditionally, they are divided into two classes referred to as endoglucanases (EGs) and cellobiohydrolases (CBHs). EGs (endo-1,4-β-glucanases) can hydrolyze internal bonds (preferably in cellulose amorphous regions) releasing new terminal ends. CBHs (exo-1,4-β-glucanases) act on the existing or EG-generated chain ends. Both enzymes can degrade amorphous cellulose but, with some exceptions, CBHs are the only enzymes that efficiently degrade crystalline cellulose. CBHs and EGs release cellobiose molecules. An effective hydrolysis of cellulose also requires β-glucosidases, which break cellobiose releasing two glucose molecules.

In 1999, there was considerable confusion regarding the biodegradation potential of cellulose esters [71]. There was a great deal of literature indicating that cellulose acetate (CA) above a degree of substitution (DS) of approximately 1.0 was not biodegradable while other reports suggested that CA might indeed be biodegradable. Since 1992, there have been several reports, which clearly demonstrate that CA having a DS of less than approximately 2.5 is inherently biodegradable. The general finding has been that as the DS of the CA decreases, the rate of biodegradation increases. Below a DS of ≈ 2.1, degradation rates of CA in composting environments approached or exceeded those of many other known biodegradable polymers. Regarding cellulose esters with longer side chains, it has been shown that cellulose propionates (CP) below a DS of ≈ 1.85 are also potentially useful as biodegradable polymers. In general, as the DS and the length of the acyl side group decreases, the rate of biodegradation increases.

A series of CA films, differing in DS, was evaluated in the bench-scale composting system [72]. Commercially available biodegradable polymers such as poly(hydroxybutyrate-*co*-valerate) (PHBV) and PCL were included as points of reference. Based on film disintegration and on film weight loss, CAs having a DS less than approximately 2.20 composted at rates comparable to that of PHB. Nuclear magnetic resonance (NMR) and GPC analyses of composted films indicated that low molecular weight fractions were removed preferentially from the more highly substituted and slower degrading CAs.

The biodegradability of CA films with DS values of 1.7 and 2.5 using laboratory-scale compost reactors maintained at a 60% moisture content and 53 °C [73]. It was found that the CA films (thickness values of 0.013 to 0.025 and 0.051 mm, respectively) had completely disappeared by the end of 7- and 18-day exposure periods, respectively. Moisture conditions in the laboratory-scale compost reactors were found to have a profound effect on the extent of CA film weight loss as a function of the exposure time. It was determined that for moisture contents of 60, 50, and 40% the time for complete CA DS-1.7 film disappearance was 6, 16, and 30 days, respectively.

The biodegradability of cellulose ester derivatives using a degradation assay based on commercially available cellulolytic enzyme preparations was found to depend on two factors: DS and substituent size [74]. The cellulose esters had acyl substituents ranging in size between propionyl and myristyl and DS values between 0.1 and nearly 3. The smaller the substituent, the higher is the DS that can be tolerated by cellulolytic enzymes.

Blends of CA having a DS of 2.49 with a CA having a DS of 2.06 were examined [75]. Bench-scale simulated municipal composting confirmed the biodestructurability of these blends and indicated that incorporation of a plasticizer (poly(ethylene glycol)) (PEG) accelerated the composting rates of the blends. *In vitro* aerobic biodegradation testing involving radiochemical labeling conclusively demonstrated that both the lower DS CA and the plasticizer significantly enhanced the biodegradation of the more highly substituted CA.

Several samples of CA polymers with varying DS between 0.7 and 1.7 have been prepared and tested for their biodegradation potential [76]. The DS of CA, i.e., the average number of acetyl groups per anhydroglucose unit, can range from 0 in the case of cellulose to 3 for the triacetate. It was found that the DS was a very significant factor in the biodegradation of these polymers. The lower the DS, the easier is the biodegradation. The higher DS polymers were amorphous, and the crystallinity increased with decreasing DS.

The biodegradation behavior of the chemically modified cellulose fibers from flax was investigated by using previously isolated cellulolytic bacterial strains [77]. The extent of biodegradation of acetylated fibers, evaluated from the weight percent remaining after 13 days of exposure to previously isolated cellulolytic bacteria *Cellvibrio* sp., decreased with increasing acetylation degree. After biodegradation the fibers showed a higher acetyl content than before the experiment, indicating that the bacteria preferentially biodegraded unsubstituted cellulose, though also acetylated chains were cleaved.

11.6.3.2 Biodegradation of Chitosan

Blends of PHB with chitin and chitosan biodegraded in an environmental medium [78]. PHB and all blends showed high biodegradability, over 60%. The PHB/α-chitin blend containing 25% PHB degraded much faster than the pure PHB or pure α-chitin. This acceleration of the biodegradation is supposed to have arisen from the lowered crystallinity of PHB. The pure chitosan film showed slower biodegradation compared to the other films. The biodegradability of the PHB/chitosan systems was found to be significantly improved.

11.6.3.3 Biodegradation of Proteins

Composting technique has been utilized to characterize the biodegradation of soy protein isolate (SPI)-based resin sheets with different additives [79]. Two different additives, i.e., Phytagel (the product of bacterial fermentation, composed of glucuronic acid, rhamnose, and glucose) and stearic acid were incorporated in order to improve mechanical properties of the SPI resin. The SPI resin containing stearic acid degraded at a slower rate than the SPI resin, whereas SPI containing Phyotogel degraded at the slowest rate. Based on the spectroscopic analysis and DSC studies, it was found that stearic acid and Phytagel were among the main residues in the modified SPI resins after composting. It was shown that the SPI resin degraded readily with 93.8% weight loss during the first 21 days of composting.

The effects of technological treatments of wheat gluten bioplastics on their biodegradation and on the formation of possible toxic products were studied [80]. To this end cast, hot-molded, and mixed gluten materials were investigated with a biodegradation test in liquid culture (modified Sturm test) and in farmland soil. All gluten materials were fully degraded after 36 days in aerobic fermentation and within 50 days in farmland soil. The tests of microbial inhibition experiments revealed no toxic effects of modified gluten or of its metabolites. Thus, it was concluded that the protein bulk of wheat gluten materials was non-toxic and fully biodegradable, whatever the technological process applied.

The chemiluminescence technique was used to study gelatine samples hydrolytically degraded under sterilization conditions and exposed to bacterial and fungal degradations [81]. It was found that the hydrolytic degradation mechanism was through a cleavage of the peptide bond of the protein without significant oxidation of the material. In contrast, biodegradation by bacteria and fungi at low temperatures decreased the molecular weight of the gelatine (viscosity) by the enzymatic activity but, also, produced an important oxidation in the material due to the reactive oxygen species generated in the microbial metabolism. This oxidation was detected by the drastic increase in the chemiluminescence emission of the materials. In general, much higher chemiluminescence emission intensities were observed for samples biodegraded by fungi with respect to those obtained for gelatine biodegraded by bacteria.

Proteic waste materials from pharmaceutical manufacturing, tanning, and agro industries have attracted increasing attention because their intrinsic agronomic values bound to the fairly high nitrogen (12−15%) [82]. The propensity to biodegradation behavior of casting films based on waste gelatin (WG) was investigated under incubation conditions aimed at simulating soil burial conditions. The results indicated the complete and very fast biodegradation of WG cast films. Pure WG films underwent about 60% biodegradation within 30 days of incubation. However, the negative effect of a crosslinker agent such as glutaraldehyde on the biodegradation extent and rate was observed for the films containing 1−5% crosslinking agent.

11.7 Biodegradation of Biodegradable Polymers from Petrochemical Sources

11.7.1 Biodegradation of Aliphatic Polyesters and Copolyesters

Aliphatic polyesters and copolyesters based on succinic acid and commercialized under the name Bionolle are biodegradable in compost, in moist soil, in fresh water with activated sludge, and in sea water [83].

A series of aliphatic homopolyesters and copolyesters was prepared from 1,4-butanediol and dimethyl esters of succinic and adipic acids through a two-step process of transesterification and polycondensation [84,85]. The biodegradation of the polymers was investigated by soil burial and enzymatic hydrolysis. It was suggested that the key factor affecting material degradation was its crystallinity.

The modified Sturm test showed that poly (ethylene adipate) (PEA) and PBS were assimilated to CO_2 at a similar rate [86]. As the degree of chain branching increased, the biodegradation rate of PEA increased to a greater extent than that of PBS due to the faster reduction in the crystallinity of PEA compared to the crystallinity of PBS. Poly(alkylene succinate)s were synthesized from succinic acid and aliphatic diols with 2 to 4 methylene groups by melt polycondensation [87]. A comparative biodegradability study of the three poly(alkyl succinate)s prepared, namely poly(ethylene succinate) (PESu), poly(propylene succinate) (PPSu), and poly(butylene succinate) (PBSu), was carried out using *Rhizopus delemar* lipase. Samples having the same average molecular weights were used. The biodegradation rates of the polymers decreased following the order PPSu $>$ PESu $\geq$ PBSu and it was attributed to the lower crystallinity of PPSu compared to other polyesters, rather than to differences in chemical structure.

The bio-catalyzed cleavage of ester bonds in low molecular mass model esters and aliphatic polyesters was studied [88]. The cleavage of ester bonds in liquid and solid low molecular mass model compounds by lipases exhibits substrate specificity, i.e., the cleavage rates are dependent on the chemical structure and on the molecular environment the ester bonds are embedded in. In contrast, when studying the degradation of polyesters by enzymatic hydrolysis, the substrate specificity plays only a minor role.

The most important quantity controlling the hydrolysis rate is the extent of mobility of the polyester chains in the crystallinity domains of the polymer. While the amorphous regions at the surface are easily degraded, the crystalline domains form a layer which protects the bulk material against enzymatic attack. Therefore, the low hydrolysis rate of the ester bonds in the crystallites is the limiting step of the overall degradation process. For aliphatic polyesters, the temperature difference between the melting point of the polymer and the temperature where degradation takes place turned out to be the primary controlling parameter for polyester degradation with the lipase. If this temperature difference is less than about 30 °C, the degradation rate increases significantly.

The biodegradation and hydrolytic degradation of the high molecular weight PBS homopolyester, poly (butylene adipate) homopolymer, and poly(butylene succinate-*co*-butylene adipate) copolyesters were investigated in the composting soil and NH_4Cl aqueous solutions at a pH level of 10.6 [89]. The biodegradability by microorganisms increased as the contents of butylene adipate increased, along with crystallinity and melting temperature, whereas the spherulite radius decreased. The biodegradability of poly(butylene succinate-*co*-butylene sebacate) P(BSu-*co*-BSe) and poly(butylene succinate-*co*-butylene adipate) P(BSU-co-BAd) samples, with different composition, was investigated under controlled soil burial conditions [90]. The influence of crystallinity, molar mass, chemical structure, and melting temperature upon biodegradation was studied. The weight loss of poly(3-hydroxybutyrate) (PHB), of PHBV 76/24, and of two commercial Bionolle samples was also investigated under soil burial conditions. PHB and PHBV 76/24 showed a higher biodegradation rate than Bionolle samples but lower than some P(BSu-*co*-BSe)s and P(BSU-*co*-BAd)s. Among the homopolyesters, P(BAd) appeared more susceptible to biodegradation. P(BAd) and P(BSe) had similar melting temperature and comparable crystallinity, but the former biodegraded twice as fast as the latter. It was suggested that adipate bonds were hydrolyzed faster than sebacate bonds.

The biodegradation behavior and mechanism of aliphatic copolyester PBSA by *Aspergillus versicolor* isolated from compost was studied by Zhao *et al.* [91]. Analysis of weight loss showed that more than 90% of PBSA film was assimilated within 25 days. The analyses of ^{1}H-NMR and DSC indicated that the preferred degradation took place in the adipate units and the succinate units are relatively recalcitrant to *A. versicolor.*

The biodegradation of homopolymer PBS was studied under controlled composting conditions [92]. Composting was performed according to ISO 14855 standard at 58 °C. After incubation for 90 days, the biodegradation percentage was 71.9, 60.7, and 14.1% for powder, film, and granule form sample, respectively. The ultimate biodegradation percentage revealed that the powder-formed sample showing the best biodegradability may be ascribed to the largest specific surface. The biodegradation process of PBS under controlled composting conditions exhibited three phases. The biodegradation in the first phase was slow (0−5 days), got accelerated in the second phase (6−66 days), and showed a leveling-off in the third phase (67−90 days). Four strains were isolated from compost and identified as *A. versicolor, Penicillum, Bacillus,* and *Thermopolyspora.* Among them, *A. versicolor* was the best PBS-degrading microorganism.

Ethylene glycol/adipic acid and 1,4-butanediol/succinic acid were copolymerized in the presence of 1,2-butanediol and 1,2-decanediol to produce ethyl and *n*-octyl branched PEA and PBS, respectively [93]. The modified Sturm test showed that the two polymers were assimilated to CO_2 at a similar rate. As the degree of chain branching increased, the biodegradation rate of PEA increased to a greater extent than that of PBS due to the faster reduction in the crystallinity of PEA compared to the crystallinity of PBS.

Unsaturated groups were introduced into the main chains of PBS by the condensation polymerization of 1,4-butanediol with succinic acid and maleic acid (MA) [94]. The resulting aliphatic polyesters were subjected to chain extension via the unsaturated groups with benzoyl peroxide (BPO), BPO/ethylene glycol dimethacrylate, or BPO/triallyl cyanurate. Chain extension increased the glass transition temperature, decreased the melting temperature and crystallinity, and improved mechanical properties such as elongation and tensile strength. The results of the modified Sturm tests showed that the biodegradability of the unsaturated aliphatic polyesters decreased greatly because of the chain extension.

PCL- and PHB-degrading microorganisms are distributed widely and they represent 0.2−11.4% and 0.8−11.0% of the total number of microorganisms in the environment, respectively [95]. The distribution of poly(tetramethylene succinate) (PTMS)-degrading microorganisms in soil environments was quite restricted compared with the distribution of microorganisms that degrade PCL. However, the ratios of the degrading microorganisms to the total microorganisms were almost the same for both PTMS and PCL. In soil samples in which the formation of a clear zone was observed, PTMS-degrading microorganisms constituted 0.2–6.0% of the total number of organisms, which was very close to the percentage (0.8–8.0%) observed for PCL-degrading microorganisms. Strain HT-6, an actinomycete, has good potential for the treatment of PTMS, since it can degrade and assimilate various forms of PTMS, including films. It assimilated about 60% of the ground PTMS powder after 8 days of cultivation.

PBSA-degrading bacterium was isolated from soil and identified as *Bacillus pumilus* [96]. It also degraded PBS and PCL. On the other hand, poly (butylene adipate terephthalate) (PBAT) and PLA were minimally degraded by strain. The NMR spectra of degradation products from PBSA indicated that the adipate units were more rapidly degraded than 1,4-butanediol and succinate units. It was proposed to be one of the reasons why *Bacillus pumilus* degraded more PBSA than PBS.

Fungal strain WF-6, belonging to *Fusarium solani*, that had not been reported was isolated from farmland as the PBS-degrading microorganism [97].

Polyesters, PBSA, PBS, PES, PBS/poly(caprolactone) blend and PBAT, were evaluated about their enzymatic degradation by lipases and chemical degradation in sodium hydroxide solution [98]. In enzymatic degradation, PBSA was the most degradable by lipase PS from *Pseudomonas* sp.; on the other hand, PBAT containing aromatic ring was little degraded by 11 kinds of lipases.

The extracellular depolymerase produced by the fungus *Aspergillus fumigatus* was found to have a broad hydrolytic activity toward bacterial and synthetic aliphatic polyesters [99]. The enzyme catalyzed the hydrolysis of the bacterial polyesters: poly(3-hydroxybutyrate-*co*-3-hydroxyvalerate) (PHB/HV) and poly (3-hydroxybutyrate-*co*-4-hydroxybutyrate) (P3HB/4HB), as well as synthetic polyesters: PEA, PES, poly(1,4-tetramethylene adipate) (PTMA), and commercial polyesters "Bionolle." By comparing the results of enzyme specificity experiments, degradation product analysis, and molecular modeling, it was suggested that polymer chain structure and conformation may strongly influence the activity of hydrolase toward specific polymers. Various thermophilic actinomycetes were screened for their ability to degrade a high melting point, aliphatic polyester, PTMS, at 50 °C [100]. By using the clear zone method, *Microbispora rosea*, *Excellospora japonica*, and *Excellospora viridilutea* were found to have PTMS-degrading activity. In a liquid culture with 100 mg PTMS film, *M. rosea* subsp. *aerate* IFO 14046 degraded about 50 mg film sample after 8 days.

A series of low-molecular-weight aliphatic biodegradable polyesters was synthesized from 1,3-propanediol and adipic acid and succinic acid and 1,4-cyclohexanedimethanediol by thermal polycondensation [101]. The biodegradability of the synthesized polyester films was tested by enzymatic degradation in phosphate buffer (pH 5 7.2) in the presence of *Rhizopus delemar* lipase incubated at 37 °C, and soil burial degradation at 30 °C. The biodegradability of the polyesters depended on the crystallinity of polymers. Synthesis of high molecular weight aliphatic polyesters by polycondensation of diester with diols with and without chain extension and the enzymatic degradation of those polyesters were investigated by Shirahama *et al.* [102]. Enzymatic degradation of the polyesters was performed using three different enzymes (cholesterol esterase, lipase B, and *Rhizopus delemar* lipase) before chain extension. The enzymatic degradability varied depending on both thermal properties (melting temperature and heat of fusion (crystallinity)) and the substrate specificity of enzymes. The enzymatic degradation of chain-extended polyesters was slightly smaller than that before chain extension, but proceeded steadily.

Eight polyester films derived from C_8 to C_{10} α, ω-aliphatic diols, and C_4 to C_{10} dicarboxylic acids were examined to determine differences in biodegradability [103]. Two test procedures were used to evaluate degradation: agar plate cultures with a mixture of *Aspergilli* and soil burial. In soil burial tests, weight loss of polymer from 3% to 40% was obtained after burial for 1 month. The order of polyester degradability in the agar culture test differed from that found in the soil burial test.

The effect of copolymer composition on the physical and thermal properties, as well as enzymatic degradation of a series of high molecular weight polyesters (butylene succinate-*co*-butylene adipate)s, was investigated [104]. The enzymatic degradation was performed in a buffer solution with *Candida cylindracea* lipase at 30 °C. The highest enzymatic degradation rate was observed for the copolyester containing 50 mol% butylene succinate units.

The filamentous fungus *Aspergillus oryzae* has been extensively used for traditional Japanese fermentation products, such as *sake* (rice wine), *shoyou* (soy sauce), and *miso* (soybean paste), for more than 1000 years [105]. This fungus could grow under culture conditions that contained emulsified PBS and PBSA as the sole carbon source, through the production of PBS-degrading enzyme in the medium, and could digest PBS and PBSA, as indicated by clearing of the culture supernatant.

11.7.2 Biodegradation of Aromatic Polyesters and Copolyesters

Within compostable polymer materials, polyesters play a predominant role, due to their potentially hydrolyzable ester bonds [106]. While aromatic polyesters such as PET exhibit excellent material properties, they proves to be almost resistant to microbial attack. Many aliphatic polyesters turn out to be biodegradable, but lack properties that are important for application. To combine good material properties with biodegradability, aliphatic–aromatic copolyesters have been developed. The review concerning the degradation behavior and the environmental safety of biodegradable polyesters containing aromatic constituents was given by Müller *et al.* [106].

Early investigations on the biologically induced degradation of aliphatic–aromatic copolyesters came to the conclusion that only at relatively low fractions of aromatic component can a significant degradation be observed. Later works reported that copolyesters of PET, poly(propylene terephthalate) (PPT), and PBT with adipic acid and sebacic acid, including statistical copolyesters, were degraded in a compost simulation test at 60 °C up to a content of terephthalic acid of about 50 mol% [107]. Based on the material properties concerns and price levels of raw materials, copolyesters of 1,4-butanediol, terephthalic acid, and adipic acid (BTA-copolyesters) are preferentially used for commercial biodegradable copolyesters [89]. The rate of biodegradation decreases significantly with an increasing fraction of terephthalic acid; the maximum content of terephthalic acid for BTA-materials intended to be around a maximum of 60 mol% (with regard to the acid component) [106].

The dependence of the degradation rate of BTA-copolyesters on the terephthalic acid content was investigated during degradation test on agar plates, where BTA-films were inoculated with a prescreening mixed microbial culture from compost at 60 °C [108]. Within a range of approximately 30–55 mol% terephthalic acid in the acid components, such copolymers are an acceptable compromise between use properties and degradation rate.

Model oligo esters of terephthalic acid with 1,2-ethanediol, 1,3-propanediol, and 1,4-butanediol were investigated with regard to their biodegradability in different biological environments (inoculated liquid medium, soil, and compost at 60 °C) [107]. SEC investigations showed a fast biological degradation of the oligomer fraction consisting of one or two repeating units, independent of the diol component used for polycondensation, while polyester oligomers with degrees of polymerization higher than two were stable against microbial attack at room temperature in a time frame of 2 months. At 60 °C in a compost environment, chemical hydrolysis also degraded chains longer than two repeating units.

Individual strains that are able to degrade aliphatic–aromatic copolyesters synthesized from 1,4-butanediol, adipic acid, and terephthalic acid were isolated by using compost as a microbial source [109]. Among these microorganisms, thermophilic actinomycetes dominate the initial degradation step. Two actinomycete strains identified as *Thermonospora fusca* exhibited high copolyester degradation rates.

The aerobic biological degradation of the synthetic aliphatic–aromatic copolyester of 1,4-butanediol, adipic acid, and terephthalic acid by 29 strains of enzyme-producing soil bacteria, fungi, and yeasts was investigated at moderate environmental conditions [110]. Results showed that the aliphatic–aromatic copolyester could be degraded by a number of different microorganisms. However, the biodegradation process was significantly slower at ambient temperatures than in a compost conditions. GPC results suggested exo-enzyme-type degradation, where the microbes hydrolyzed the ester bonds at the termini of the polymeric chains preferentially.

The degradation activities of bacteria that can degrade aliphatic polyesters on various aliphatic–aromatic copolyesters (poly butylene succinate/terephthalate/isophthalate)-*co*-(lactate) (PBSTIL), poly (butylene succinate/terephthalate)(PBST), Ecoflex, commercial name of copolyester synthesized from 1,4-butanediol, adipic acid, and terephthalic acid were investigated [111].

Poly(butylene adipate-*co*-succinate) (PBAS)/PET copolyesters prepared by the transesterification reaction of PBAS and PBT were characterized [112]. The biodegradability of copolyesters depended on the terephthalate unit in the composition and average block length of the aromatic unit.

The dependence of the enzymatic degradation of aliphatic–aromatic copolyesters on the polymer structure was investigated by Marten *et al.* [113]. A number of defined model copolyesters containing terephthalate units as aromatic component were synthesized. It was suggested that the mobility of the polymer chains (the ability of chain segments to temporarily escape for a certain distance from the embedding crystal) is the major and general controlling factor for the biodegradability of polyesters. The results showed that the lengths of aliphatic sequences in a copolymer were not correlated with the biodegradation rate. The major factor in controlling the biodegradation rate was how highly and tightly the polymer chains were fixed in the crystalline region of the material. The biodegradation rate of the copolyesters was mainly controlled by the chain mobility of the polymers, being correlated with the difference between the melting point of the polyester and the degradation temperature. The presence of longer aliphatic domains, e.g., in block copolyesters, does not facilitate the hydrolytic attack by the lipase, but longer aromatic sequences, which control the melting point of the crystalline regions, reduce the biodegradation rate. According to the authors, the concept of chain mobility seems to be a quite universal way to describe and predict the biodegradation rate of synthetic polyesters, independent on their composition or microstructure.

Generally, it seemed that many polyesters composed of aliphatic monomers were degradable by lipases, while most aromatic polyesters were characterized as biologically inert [114]. In aliphatic–aromatic copolyesters, the tendency was found that biodegradability decreases with the content of aromatic constituents. For copolyesters composed from adipic acid, terephthalic acid, and 1,4-butanediol a maximum content of about 50–60% terephthalic acid in the diacid component was reported to be the limit for biodegradability.

The model of chain mobility can generally describe the degradation behavior of a series of polyesters with lipases such as lipase from *Pseudomonas* sp. including the missing degradability of polyesters like PET or PBT, which exhibit very high melting points above 200 °C [114,115]. Recently, it was demonstrated that PET can be depolymerized by hydrolases from a new thermophilic hydrolase (TfH) *Thermobifida fusca* (former name *Thermonospora fusca*) [114,115]. Erosion rates of 8–17 μm per week were obtained upon incubation at 55 °C. This enzyme is especially active in degrading polyesters containing aromatic constituents and combines characteristics of lipases and esterases (activity optimum at 65 °C). It was suggested that the specific modification of the active site of enzymes like TfH may open the door for enzymatic PET recycling in the future [115].

Poly(ethylene terephthalate)/copoly(succinic anhydride/ethylene oxide) copolymers (PET/PES copolymers) were synthesized by the transreaction between PET and PES [116]. The enzymatic hydrolyzability by a lipase from *Rhizopus arrhizus* and biodegradability by activated sludge of the copolymers decreased with an increase in PET content. When the length of succinic acid unit in the copolymer was below 2, the hydrolyzability of the copolymers decreased considerably.

The biodegradation and hydrolysis rates of an aliphatic–aromatic copolyester (PBAT) were measured in manure-, food-, and yard-compost environments and in phosphate buffer solution (pH = 8.0) and vermiculite at 58 °C [117]. PBAT film was biodegraded at distinctive rates in manure, food, and yard compost environments having different microbial activities. Biodegradation of biodegradable polyesters such as PBAT was strongly influenced by the total microbial activity of the exposure environments, which was monitored by CO_2 emissions or C/N ratio. PBAT degraded more and faster in manure compost than in yard or food waste composts.

11.7.3 Biodegradation of PCL

PCL is fully biodegradable when composted. The low melting point (58–60 °C) of PCL makes the material suited for composting as a means of disposal, due to the temperatures obtained during composting routinely exceeding 60 °C [25].

PCL degradation proceeds through hydrolysis of backbone ester bonds as well as by enzymatic attack [118]. Hence, PCL degrades under a range of conditions, biotically in soil, lake waters, sewage sludge, *in vivo*, and in compost, and abiotically in phosphate buffer solution. Hydrolysis of PCL yields 6-hydroxycaproic acid, an intermediate of the ω-oxidation, which enters the citric acid cycle and is completely metabolized.

Generally, it has been shown that the biodegradation of PCL proceeds with rapid weight loss through surface erosion with minor reduction of the molecular weight [47]. In contrast, the abiotic hydrolysis of PCL proceeds with a reduction in molecular weight combined with minor weight loss.

PCL has been shown to biodegrade in many different environments, e.g., in pure fungal cultures, in compost, in active sludge, by enzymes, and in soil [47]. It was reported that degradation of PCL in a natural environment of compost and sea water is a result of enzymatic hydrolysis and of chemical hydrolysis of the ester bonds of PCL, the dominant role in this process being played by enzymatic hydrolysis [119].

During the biodegradation of film-blown PCL, both in compost and in thermophilic anaerobic sludge, regularly spaced grooves developed on the film surface [120]. Such grooves were not seen in the cases of samples degraded in an abiotic environment. The width of the grooves increased with increasing time of biodegradation. It was interpreted as indicating preferred degradation of the amorphous part of the material. The degree of crystallinity increased from 54% to 65% during composting. Figure 11.10 shows that a shoulder was detected on the low temperature side of the main melting point in the first heating after 10 days in compost. The appearance corresponds to the time of formation of the low molar mass fractions seen in the SEC chromatograms. The shoulder extended to lower temperatures with increasing degradation time. It was explained by the formation of lamellae thinner than the average thickness out of the low molar mass polymer chains formed by chain scission.

Compostability of the blends of semicrystalline and amorphous PCL having 20% branches, as checked by burying them in compost at 45 °C [121]. An increase in the rate of degradation was observed as compared to pure PCL with increased amount of amorphous PCL without affecting thermal stability.

A series of biodegradation tests was carried out according to the standard test method 14851 in order to compare the performance of different acitvated sludge inocula on different plastic materials (PCL and starch-based material (Mater-Bi NF01U)) [122]. Cellulose was used as a positive control. It was shown that the activated sludges, drawn from different wastewater treatment plants and used as

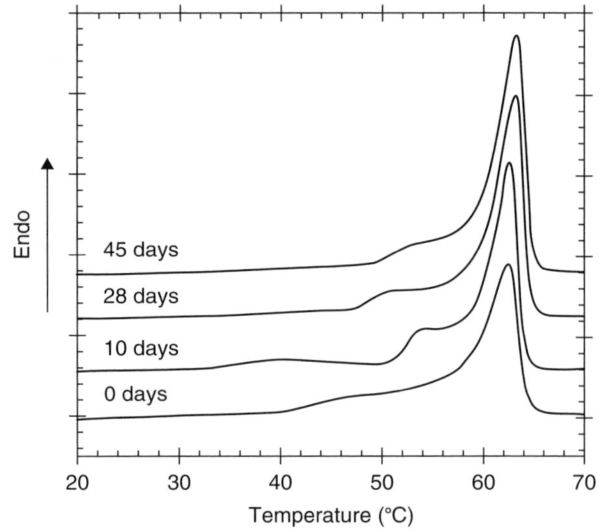

Figure 11.10 DSC curves from the first scan for the film-blown PCL degraded in compost for 0, 10, 28, and 45 days; first heating. *Reprinted with permission from Ref. [120].*

inocula, had different biodegradation activities. The starch-based material was degraded to similar or higher extents than PCL with municipal sludge. Industrial sludge gave good results with both materials (PCL = 100%; starch-based material = 89%), but was less active toward cellulose. Such results raise some questions about the opportunity of also using other reference materials besides cellulose for biodegradation tests. The use of mixtures of sludges from different origins seemed to be a successful strategy to increase biodiversity and therefore increase the overall activity of inoculum.

Ammonia is the greatest nuisance odor compound among the exhaust gases that evolve during the composting process, in which raw materials with high concentrations of nitrogen, such as wastewater sludge, are decomposed [123]. A reduction of NH_3 emission during composting of wastewater sludge was tried by mixing biodegradable plastic (i.e., PCL) into composting raw material. It was found that biodegradable plastic acted as "reserve acid," i.e., it was not acid itself but degraded and released acid intermediates during the composting progress. On the basis of the results obtained, it was concluded that PCL had the characteristic of being not only compostable, but also of being able to suppress NH_3 emission during composting.

The biodegradation of PCL was examined by measuring the release of CO_2 when the plastic was mixed not with maturated compost, as in the conventional method, but with the dog food used as a model fresh waste under controlled laboratory conditions [124]. From the composting in which the PCL was mixed with the dog food at various ratios, it was found that the quantity of CO_2 evolution in the presence and absence of PCL was in proportion to the PCL mixing level. The percentage of PCL decomposition, which was calculated as a ratio of the quantity of PCL decomposition to the mixing level of PCL, was 84% after 11 days in the composting using dog food, but was 59% after the same period using maturated compost.

The degradability of a biodegradable plastic depends not only on the specific kind of plastic, but also on the operational composting conditions such as the temperature and the type of incoculum used. The effects of temperature and the type of inoculum on the biodegradability of PCL were tested in a bench-scale composting reactor under controlled laboratory composting conditions [125]. The optimum composting temperature for the PCL was found to be approximately 50 °C, at which approximately 62% of the PCL was decomposed over 8 days. The degradability of PCL was significantly different for each of the two types of incocula used.

The lanthanide derivatives are known as very attractive catalysts in the ring-opening polymerization of cyclic esters [126]. The influence of the lanthanides on both the hydrolytic and enzymatic degradation of the PCL obtained by ring-opening polymerization of ε-caprolactone with different lanthanide-based catalysts such as lanthane chloride ($LaCl_3$), ytterbium chloride ($YbCl_3$), and samarium chloride ($SmCl_3$) was assessed. Samarium seemed to slightly accelerate the hydrolytic degradation of the polymer and to slow down or inhibit its enzymatic degradation, mainly when the molecular weight of the polymer was high. The behavior of PCL containing another lanthanide, lanthane, was dependent on the nature of the metallic ion. Complete degradation, by the lipase PS from *Pseudomonas cepacia*, was achieved only with ytterbium.

The biodegradation of electrospun nanofibers of PCL was investigated using pure-cultured soil filamentous fungi, *A. oryzae, Penicillium caseicolum, Penicillium citrinum, Mucor* sp., *Rhizopus* sp., *Curvularia* sp., and *Cladosporium* sp. [127]. Three kinds of nonwoven PCL fabrics with different mean fiber diameters (330, 360, and 510 nm) were prepared by changing the viscosities of the prespun PCL solutions. In the BOD test, the biodegradation of the 330 nm PCL nanofibers by *Rhizopus* sp. and *Mucor* sp. exceeded 20 and 30% carbon dioxide generation, respectively. The biodegradability of the PCL nonwoven fabrics decreased with the mean fiber diameter and the 330 nm PCL nanofiber exhibited the highest biodegradability.

PCL powders were prepared from PCL pellets using a rotation mechanical mixer [128]. PCL powders were separated by sieves with 60 and 120 meshes into four classes: 0–125, 125–250, 0–250, and 250–500 μm. Biodegradation tests of PCL powders and cellulose powders in an aqueous solution at 25 °C were performed using the coulometer according to ISO 14851. Biodegradation tests of PCL powders and cellulose powders in controlled compost at 58 °C were performed according to ISO 14855-1 and by using the MODA instrument according to ISO/DIS 14855-2. PCL powders were biodegraded more rapidly than cellulose powders. The reproducibility of biodegradation of PCL powders was excellent. Differences in the biodegradation of PCL powders with different classes were not observed by the ISO 14851 and ISO/DIS 14851-2. An enzymatic degradation test of PCL powders with different classes was studied using the enzyme Amano lipase PS. PCL with smaller particle size was degraded more rapidly by the enzyme. PCL powders with regulated sizes from 125 μm to 250 μm were proposed as a reference material for the biodegradation test.

11.7.4 Biodegradation of Poly(esteramide)s

Polyesteramides can be hydrolytically degraded through ester bond cleavages [129]. The degradation process is clearly accelerated at high temperatures, or in acid or basic pH media. In the same way, the polymer is susceptible to enzymatic attack with protease such as proteinase K.

Degradation of poly(esteramide)s differing in the amide–ester ratio under different media (water at 70 °C, acid or enzymatic catalysis at 37 °C) has been studied by evaluating the changes in intrinsic viscosity, in the NMR spectra, and in the surface texture of samples [130]. Results indicated that the amide–ester ratio had to be lower than certain values in order to obtain samples with a high susceptibility to enzymatic catalysis. Enzymes with a protease activity appeared more effective than those with only an esterase activity.

The influence of substitution of adipic acid by terephthalic acid units on degradability under different media of poly(esteramide)s were investigated by Lozano *et al.* [131]. The degradation rate decreased with the aromatic content in aqueous media as well as in those with acid or enzymatic (protease K) catalysis.

Two types of aliphatic poly(esteramide)s were subjected to microbial degradation in basal mineral salt broth, under the attack of an yeast, *Cryptococcus laurentii*, at 20 °C [132]. The first type of PEA was made by anionic ring-opening copolymerization of ε-caprolactone and ε-caprolactam, whereas the second one was synthesized by a two-step polycondensation reaction of hexanediol-1,6, hexanediamine-1,6 and adipoyl chloride. These copolymers were found to be readily degradable under biotic conditions, based on weight loss, GPC, NMR spectroscopy, and tensile property measurements. Furthermore, NMR spectroscopic analysis proved that the biodegradation of poly(esteramide)s involved the enzymatic hydrolysis of ester groups on the backbones of polymers into acid and hydroxyl groups. No breakdown of amide bonds was observed under the given biotic conditions.

Degradability of aliphatic poly(esteramide) derived from L-alanine has been studied in different media [133]. The poly(esteramide) showed a hydrolytic degradation that took place through the ester linkage and an enzymatic degradation that strongly depended on the type of enzyme. Thus, proteolytic enzymes such as papain and proteinase K were the most effective ones. Biodegradation by microorganisms from soil and activated sludges has also been evaluated.

BAK 1095, commercial polyesteramide based on caprolactam, butanediol, and adipic acid, was found to be completely biodegradable according to German compostability standard DIN 54900 [134]. Biodegradability of laboratory synthesized poly(esteramide) was studied in the controlled composting test according to EN 14046 standard [135]. It was found that poly(esteramide) meets the biodegradation criteria of the standard.

Polyesteramides (PEA) based on ε-caprolactam and ε-caprolactone differing in the content of esteramide structural units was subjected to biodegradation—composting in a big compost pile under controlled conditions [136]. The hydrolyzability of PEAs increases with the increasing fraction of ester bonds in macromolecules in accordance with the well-known higher sensitivity of ester bonds to hydrolysis compared to amide bonds. The degradation of PEAs was more effective in isothermal composting at 60 °C than in the standard composting process.

In order to establish the relationship between hydrophilicity and biodegradability of the aliphatic polyesters, the amide group was introduced to the biodegradable aliphatic polyester [137]. The effect of surface hydrophilicity was induced from the amide units in the polyesteramide. Biodegradability was evaluated from various methods including activated sludge test, enzyme hydrolysis, and soil burial test. It was found that the introduction of amide groups to the aliphatic polyester improved the biodegradability, although the increase of biodegradation rate was not directly proportional to the amide content. The biodegradability of aliphatic polyesters increased with the addition of amide functionality.

11.7.5 Biodegradation of Poly(vinyl alcohol)

Poly(vinyl alcohol) (PVA) has been considered to be a truly biodegradable synthetic polymer since the early 1930s [138,139,140]. Since 1936, it was observed that PVA was susceptible to ultimate biodegradation when submitted to the action of *Fusarium lini* [140]. Suzuki and Watanabe proposed two similar degradation pathways by using different *Pseudomonas* strains [138]. In both cases, the polymer is oxidized by oxidase-type enzymatic systems with the evolution of hydrogen peroxide and oxygen consumption; the result of this enzymatic attack is the production of carbonyl groups along the polymer chain. Activated β-diketones or α-keto groups are subsequently hydrolyzed with fission of the polymer carbon backbone [138].

The dependence of PVA biodegradation on several structural parameters, such as molecular weight, degree of saponification, and head-to-head junctions, was assessed in the presence of a selected PVA-degrading mixed culture and of the culture supernatant derived therefrom [139]. Respirometric tests carried out in the presence of selected microbial populations evidenced a limited but significant delay in the mineralization profile depending upon the degree of PVA hydrolysis, whereas no remarkable effect by molecular weight was detected. PVA is recognized as one of the very few vinyl polymers soluble in water that is also susceptible to ultimate biodegradation in the presence of suitably acclimated microorganisms. Nevertheless, the occurrence of specific PVA-degrading microorganisms in the environment appears to be uncommon and in most cases is strictly associated with PVA-contaminated environments [140]. Most PVA-degrading was attributed to aerobic bacteria belonging to *Pseudomonas*, *Alcaligenes*, and *Bacillus* genera [140].

Effect of minor chemical structures such as 1,2-diol content, ethylene content, tacticity, a degree of polymerization (DP), and a degree of saponification (DS) of the main chain on biodegradability of PVA was summarized [141]. The degradation of PVA is affected by chemical–structural characteristics, such as stereotacticity and 1,2-diol units, but is not significantly influenced by the DP and DS of PVA in the range of 0.5–100 kDa and over 80%, respectively.

In solution, the major biodegradation mechanism is represented by the random endocleavage of the polymer chains [140]. The initial step is the specific oxidation of 1,3-hydroxyl groups, mediated by oxidase and dehydrogenase-type enzymes, to give β-hydroxylketone as well as 1,3-diketone moities. The latter groups are susceptible to carbon–carbon bond cleavage promoted by specific β-diketone hydrolase, giving rise to the formation of carboxyl and methyl end groups.

The ultimate biological fate of PVA appears to be largely dependent upon the kind of environment it reaches [140]. Accordingly, high levels of biodegradation were observed in aqueous environments. On the other hand, moderate or negligible microbial attacks were repeatedly ascertained in soil and compost environments. Different hypotheses were tentatively suggested to account for these observations, such as the absence or scarce occurrence of PVA-degrading microorganisms in soil and compost matrices, the physical state of PVA-samples, and PVA's strong interactions with the organic and inorganic components of environmental solid matrices [140].

Biodegradation in an aqueous or soil environment very markedly depends on the microbe population present and the degradation conditions [140,142]. It proceeds quite slowly in an unadapted environment, e.g., inoculated municipal sludge gave 13% theoretical yield of CO_2 after 21 days, merely 8–9% after 74 days in soil, 7% after 48 days in compost, with a long initial lag phase of 22 days [138,142] (Figs 11.11 and 11.12). Very moderate PVA biodegradation was also detected when using compost extract as a microbial source [140,143].

In order to assess the effect of degree of hydrolysis (HD) on the biodegradation propensity of PVA, samples having a similar degree of polymerization (DPn) and noticeably different HD values were synthesized by controlled acetylation of commercial PVA (HD = 99%) and submitted to biodegradation tests in aqueous medium, mature compost, and soil by using respirometric procedures [144]. Reacetylated PVA samples characterized by HD of between

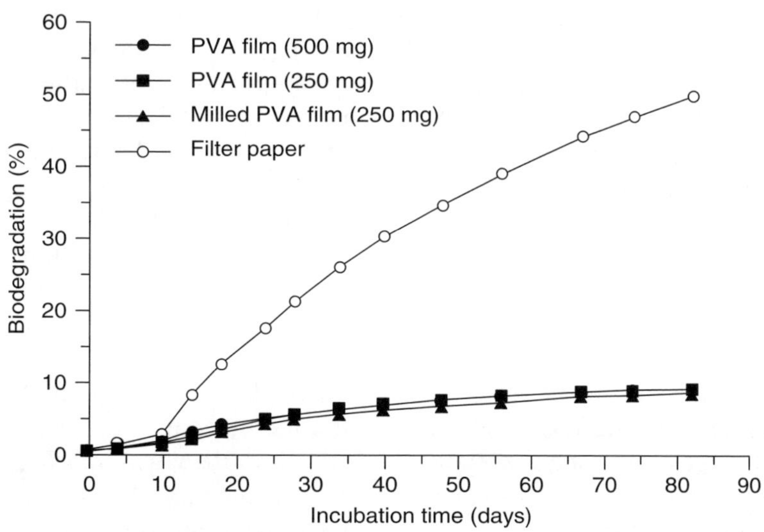

Figure 11.11 Biodegradation curves of a PVA-based film and filter paper recorded in simulated soil burial respirometric tests. *Reprinted with permission from Ref. [138].*

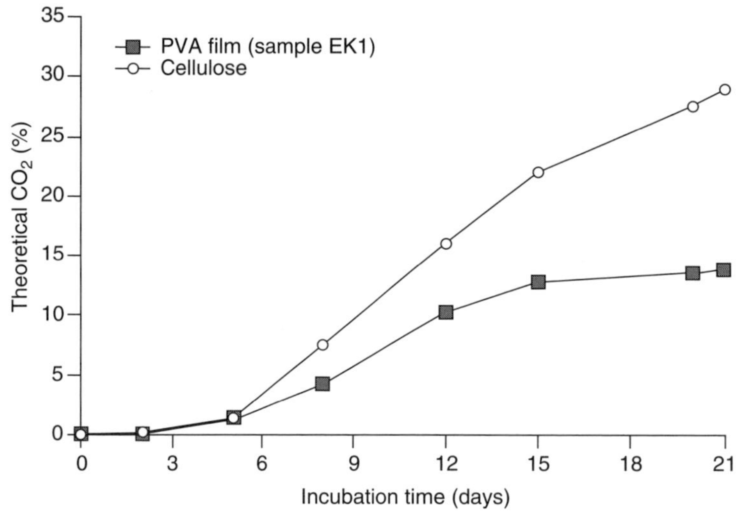

Figure 11.12 Biodegradation curves of a PVA-based film and cellulose recorded in the presence of municipal sewage sludge. *Reprinted with permission from Ref. [138].*

25 and 75% underwent extensive mineralization when buried in solid media, while PVA (HD = 99%) showed recalcitrance to biodegradation under those conditions. An opposite trend was observed in aqueous solution, in the presence of PVA-acclimated microorganisms. In these conditions, the driving parameter affecting the microbial assimilation of PVA appeared to be water solubility of the inspected samples; the higher the solubility, the faster is the biodegradation. It was suggested that biodegradation is not an absolute attribute directly related to structural features of the substrate under investigation; the conditions under which the tests are carried out have to be clearly defined.

The PVA degradation pathway by the enzyme from *Alcaligenes faecalis* KK314 was described by Matsumura *et al.* [145]. It was proposed that the hydroxy group of PVA was first dehydrogenated into the corresponding carbonyl group to form the β-hydroxy ketone moiety which was followed by the aldolase-type cleavage to produce the methyl ketone and the aldehyde terminals by the PVA-assimilating strain *Alcaligenes faecalis* KK314. Both the biodegradation steps of dehydrogenation and subsequent aldolase-type cleavage were catalyzed by the same protein.

A mathematical model that governs the temporal change of the weight distribution with respect to the molecular weight in order to determine the enzymatic degradation rate numerically was proposed by Watanabe and Kawai [146]. As an example, the GPC profiles of PVA were introduced into the numerical computation. PVA was degraded by random oxidation of hydroxyl groups and following cleavage of the carbon—carbon chain between two carbonyl groups/ a carbonyl group and an adjacent hydroxymethine group either by hydrolase or by aldolase.

The biodegradability of PVA was investigated under different conditions by respirometric determinations, iodometric analysis, and molecular weight evaluation [147]. Microbial inocula derived from the sewage sludge of municipal and paper mill wastewater treatment plants were used. A rather active PVA-degrading bacterial mixed culture was obtained from the paper mill sewage sludge. The influence of some polymer properties such as molecular weight and the degree of hydrolysis on the biodegradation rate and extent were investigated in the presence of either the acclimated mixed bacterial culture or its sterile filtrate. Kinetic data relevant to PVA mineralization and to the variation of PVA concentration, molecular weight, and molecular weight distribution revealed a moderate effect of the degree of hydrolysis.

The rates and the extents of absorption and desorption of PVA samples on different solid substrates comprising montmorillonite, quartz sand, and farm soil, as well as humic acid mixture were studied [148]. Biodegradation experiments carried out in liquid cultures of PVA adsorbed on montmorillonite showed that mineralization of the adsorbed PVA was much lower than that detected for the nonadsorbed PVA. It was suggested that irreversible adsorption of PVA on the clay component occurred in soil, thus substantially inhibiting PVA biodegradation.

11.8 Biodegradation of Blends

11.8.1 Blends of PLA

Biodegradability of PLA and PLA/corn starch composites with and without lysine diisocyanate (LDI) were evaluated by enzymatic degradation using proteinase K and burial tests [149]. The addition of corn starch resulted in a faster rate of enzymatic biodegradation and the composites with LDI were more difficult to degrade than those without it. In a burial test, pure PLA was little degraded but the composites gradually degraded. The degradation of the composite without LDI was faster than that of the composite with LDI.

Two different types of biodegradable polyester composites, PLLA fiber-reinforced PCL and PCL/PLLA blend films, were prepared with a PCL–PLLA ratio of 88/12 (w/w) and their enzymatic degradation was investigated by the use of *Rhizopus arrhizus* lipase and proteinase K as degradation enzymes for PCL and PLLA chains, respectively [150]. In the fiber-reinforced film, the presence of PLLA fibers accelerated the lipase-catalyzed enzymatic degradation of PCL matrix compared with that in the pure PCL film, whereas in the blend film, the presence of PLLA chains dissolved in the continuous PCL-rich domain retarded the lipase-catalyzed enzymatic degradation of PCL chains. In contrast, in the fiber-reinforced film, the proteinase K-catalyzed enzymatic degradation of PLLA fibers was disturbed compared with that of the pure PLLA film, whereas in the blend film, the proteinase K-catalyzed enzymatic degradation rate of particulate PLLA-rich domains was higher than that of pure PLLA film.

11.8.2 Blends of PHA

Blends of PHBV with corn starch were evaluated for their biodegradability in natural compost by measuring changes in physical and chemical properties over a period of 125 days [151]. The degradation of plastic material, as evidenced by weight loss and deterioration in tensile properties, correlated with the amount of starch present in the blends (neat PHBV, 30%, 50%). Incorporation of poly(ethylene oxide) (PEO) into starch–PHBV blends had little or no effect on the rate of weight loss. Starch in blends degraded faster than PHBV and it accelerated PHBV degradation. After 125 days of exposure to compost, neat PHBV lost 7% weight (0.056% weight loss/day), while the PHBV component of a 50% starch blend lost 41% of its weight (0.328% weight loss/day).

The degradation of atactic poly(R,S)-3-hydroxybutyrate (a synthetic amorphous analogue of natural PHB), binary blends with natural PHB and PLLA, respectively, has been investigated in soil [152]. In such a natural environment, a-PHB blend component was found to biodegrade. The degradation of a-PHB-containing blends proceeded faster than that of respective plain *n*-PHB and PLLA.

11.8.3 Blends of Starch

Commercially available biodegradable aliphatic polyesters, i.e., having high molecular weight PCL and PLA, were melt blended with polysaccharide/starch either as corn starch granules or as thermoplastic corn starch after plasticization with glycerol [153]. Interface compatibilization was achieved via two different strategies depending on the nature of the polyester chains. In the case of PLA/starch compositions, PLA chains were grafted with maleic anhydride (MAH) through a free-radical reaction conducted by reactive extrusion. As far as PCL/starch blends were concerned, the compatibilization was achieved via the interfacial localization of amphiphilic graft copolymers formed by grafting of PCL chains onto a polysaccharide backbone such as dextran. Finally, the biodegradability of the so-obtained PCL/starch blends has been investigated by composting. For doing so, thin films (~100 μm thick) were buried in an aerated composting bin for 120 days at 25–30 °C, then followed by 20 days more at a higher temperature of 35–40 °C. The film weight loss increased with the starch content. The degradation started first within the starch phase and then occurred within the polyester matrix. These compatibilized PCL/starch compositions displayed much more rapid biodegradation as measured by composting testing.

The biodegradability of native and compatibilized PCL–granular starch blends in composting and culture conditions was studied. The inherent biodegradability of the host polyester has been shown to increase with compatibilization within the PCL–starch compositions [154]. It was observed that the weight loss during composting increased with the decrease in interfacial tension between filler and polymer. In general, it was concluded that inherent biodegradability does not depend very significantly on the concentration of starch in the polyester matrix, but on the compatibilization efficiency.

Different proportions of starch were blended with poly(β-hydroxybutyrate)-*co*-poly(β-hydroxyvalerate) (PHB-V) or PCL by extrusion [155]. The biodegradability of the blends in soil compost was assessed after thermal aging for 192, 425, and 600 h at different temperatures. Two temperatures were chosen for each polymer: 100 and 140 °C for PHB-V and its blends and 30 and 50 °C for PCL and its blends. The samples of PHB-V degraded more than those of PCL, because after about 62 days of aging in soil compost, the first polymer had biodegraded almost 100%. The addition of starch to PCL slightly increased the loss of mass during biodegradation. For PHB-V the addition of 50% starch made the blend more susceptible to biodegradation, with PHB-V50 totally degraded in only 33 days. For the blends prepared, only the biodegradation of PHB-V25 was affected by thermal aging.

11.8.4 Blends of PCL

PCL was blended with PBS (PCL/PBS 5 30/70) to improve the heat stability of PCL [156]. The processability of the blended samples was improved by γ-ray irradiation. The soil degradation test showed that the blend film buried in the soil was almost degraded (97%) after 2 months and completely degraded after 2½ months. On the contrary, the samples placed on the surface of the soil degraded only 3.5% after 4 months. From these findings, it was confirmed that microorganisms contribute to degradation in soil. The blend sample used as garbage bags was well degraded (almost 50%) after a 2 month burial test.

The effects of replacing PCL with acrylic acid grafted PCL (PCL-*g*-AA) on the structure and properties of a PCL−chitosan composite were investigated [157]. Resistance to water was higher in the PCL-*g*-AA−chitosan blend, and consequently so was its resistance to biodegradation in soil and in an enzymatic environment. Nevertheless, weight loss of blends buried in soil or exposed to an enzymatic environment indicated that both blends were biodegradable, especially at high levels of chitosan content.

Biodegradation of blends of PCL with poly(vinyl butyral) blends was studied in the soil and by bacterial strains of *Bacillus subtilis* and *Escherichia coli* isolated from the soil [158]. Weight loss was observed in all the blends. PCL-rich blends showed more degradation, which was faster in the natural environment than in the laboratory. Blends in the *Bacillus subtilis* strain showed more degradation as compared to the *E. coli* strain.

PCL was blended with TPS prepared from regular corn starch [159]. PCL showed no significant reduction in mass after incubation with α-amylase, whereas blends containing corn starch were more susceptible to this enzyme. The biodegradation seen in simulated soil agreed with the findings for degradation by α-amylase.

PLLA and PCL, and their films blended with or without 50 wt% PEG, were prepared by solution casting [160]. Porous films were obtained by water extraction of PEG from solution-cast phase-separated PLLA-*blend*−PCL-*blend*−PEG films. Polymer blending as well as pore formation enhanced the enzymatic degradation of biodegradable polyester blends.

Modified PCL was synthesized by melt reaction of PCL and reactive monomers such as glycidyl methacrylate (GMA) and MAH in the presence of benzoyl perioxide in a Brabender mixer [161]. Reactive blends of the PCL-*g*-GMA and the gelatinized starch with glycerin were prepared and their mechanical properties and biodegradabilities were investigated. Reactive blends of PCL-*g*-GMA and starch showed a well-dispersed starch domain in the matrix and better mechanical strength than the unmodified PCL−starch blend. However, the reaction between PCL-*g*-GMA and starch induced a crosslinking during the reactive blending and this crosslinking in the blend lowered the biodegradation of the blend during the composting test.

Biodegradable polyester blends were prepared from PLLA and PCL (50/50) by melt-blending, and the effects of processing conditions (shear rate, time, and strain) of melt-blending on proteinase K- and lipase-catalyzed enzymatic degradability were investigated by gravimetry, DSC, and SEM [162]. The proteinase K-catalyzed degradation rate of the blend films increased and leveled off with increasing the shear rate, time, or strain for melt-blending, except for the shortest shear time of 60 s. It was revealed that the biodegradability of PLLA−PCL blend materials can be manipulated by altering the processing conditions of melt-blending (shear rate, time, or strain) or the sizes and morphology of PLLA-rich and PCL-rich domains.

The biodegradability properties of PCL and modified adipate starch blends, using EDENOL-3203 (an C_{18} alkyl epoxy stearate), were investigated in the laboratory by burial tests in agricultural soil [163]. The biodegradation process was carried out using the respirometric test according to ASTM D 5988-96, and the mineralization was followed by both variables such as carbon dioxide evolution and mass loss. It was found that the presence of modified adipate starch accelerated the biodegradation rate.

The biodegradability and biodegradation rate of PCL–starch blend and PBS were investigated under both aerobic and anaerobic conditions [164]. PCL–starch blend was easily degraded, with 88% biodegradability in 44 days under aerobic conditions, and showed a biodegradation rate of 0.07 day^{-1}, whereas the biodegradability of PBS was only 31% in 80 days under the same conditions, with a biodegradation rate of 0.01 day^{-1}.

11.8.5 Blends of Aliphatic–Aromatic Copolyesters

The blends of aliphatic–aromatic copolyesters synthesized from dimethyl succinate, dimethyl terephthalate, and butanediol with starch were studied by soil burial [165]. Blends of copolyesters with starch possessed higher degradation rate, but lower tensile strength as compared with unfilled copolyesters.

Biodegradation of natural and synthetic copolyesters in two different natural environments, i.e., in compost with activated sludge at a sewage farm and in the Baltic Sea, was studied by Rutkowska *et al.* [166]. The results revealed that the natural aliphatic copolyester PHBV and its blends with the synthetic aliphatic–aromatic copolyester of 1,4-butanediol with adipic and terephthalic acids degraded faster in compost than in seawater. In both natural environments, blends degraded faster than aliphatic–aromatic copolyester, but at a slower rate than natural component PHBV.

Biodegradability in soil of the PBSA–starch films prepared with starch contents of 5–30% by weight and processed by blown film extrusion was assessed [167]. The rate of biodegradation in soil, as measured by respirometry, increased significantly as the starch content was increased to 20% and then plateaued.

11.8.6 PVA Blends

Biodegradability in a typical environment medium of blend films composed of bacterial poly(3-hydroxybutyric acid) (PHB) and chemically synthesized PVA was investigated by BOD test [168]. Water from the River Tama (Tokyo, Japan) was used as an environmental medium. The degradation profile of the blend films was found to depend on their blend compositions. The blend films with PHB-rich composition showed higher degradation rate and higher final degradation ratio than the pure PHB film.

Hybrid blends based on PVA and collagen hydrolysate (CH), an abundant, added value waste product of the leather industry, have been processed by melt blow extrusion [169,170]. Biodegradation experiments performed under anaerobic conditions evidenced a positive effect of CH on the mineralization rate of PVA–CH blends. No differences in biodegradation under aerobic conditions of PVA and PVA–CH blends at 20 °C were observed when an adopted inoculum (i.e., obtained from a previous PVA biodegradation test) was used [169,170]. On the contrary, when at lower temperature (5 °C) the biodegradation level of CH-free PVA films was much lower than that detected for PVA–CH blend film.

Soil burial degradation behavior of miscible blend systems of PVA/partially deacetylated chitin, PVA/chitin-*graft*-poly(2-methyl-2-oxazoline), and PVA/chitin-*graft*-poly(2-ethyl-2-oxazoline) was investigated in comparison with the case of a pure PVA film [171]. The rate of weight decrease in these PVA–chitin derivative hybrids was higher than that of control PVA in the soil burial test. Fourier transform infrared spectra of the recovered samples of the blends showed an apparent increase of the absorption intensity due to β-diketone structure in PVA, which reflected the progress of biodegradation of PVA by PVA-oxidizing enzymes. The triad tacticity and number–average molecular weight of PVA in the hybrids after soil burial determined by ^{1}H-NMR and size exclusion chromatography, respectively, were almost the same as those before soil burial. It was suggested that enzymatic degradation of the hybrid films occurred mainly on the surface and that degradation of the PVA-based samples in the soil was accelerated by blending the chitin derivatives.

The effects of addition of the hydrophilic water-insoluble PVA on the non-enzymatic and enzymatic hydrolysis of hydrophobic PLLA were investigated [172]. The results of gravimetry, GPC, DSC, tensile testing, and SEM exhibited that the nonenzymatic and enzymatic hydrolysis of PLLA was accelerated by the presence of PVA and both the hydrolysis rates increased dramatically with a rise in PVA content in the blend films. The enhanced nonenzymatic hydrolysis of PLLA in the blend films was ascribed to the increased water concentration around PLLA molecules and water supply rate to them by the presence of hydrophilic PVA in both PLLA-rich and PVA-rich phases. However, the accelerated enzymatic hydrolysis of PLLA in the blend films was due to occurrence of enzymatic hydrolysis at the interfaces of PLLA-rich and PVA-rich phases inside the blend films as well as at the film surfaces.

The main shortcomings of biodegradable starch/PVA film are hydrophilicity and poor mechanical properties [173]. With an aim to overcome these advantages, corn starch was methylated and blend film was prepared by mixing methylated corn starch (MCS) with PVA. Enzymatic, microbiological, and soil burial biodegradation results indicated that the biodegradability of the MCS/PVA film strongly depended on the starch proportion in the film matrix.

The biodegradation of PVA blends with natural polymers, such as gelatin, lignocellulosic by-products (sugar cane bagasse), as well as poly(vinyl acetate), was investigated in respirometric tests aimed at reproducing soil burial conditions [174]. The collected data evidenced that the biodegradation of PVA and PVA-based materials was rather limited under soil conditions. Additionally, PVA depresses the biodegradation of some of the investigated blends, particularly when mixed with gelatin.

The biodegradation of chitosan modified PVA–starch blends by compost was reported and compared with unmodified film by Jayasekara *et al.* [175]. Within 45 days of composting, the starch and glycerol components were fully degraded, leaving the PVA component essentially intact for unmodified blends. The film characteristics were improved by surface modification with chitosan. There was slight evidence that PVA biodegradation had been initiated in composted, surface modified starch–PVA blends.

11.8.7 Miscellaneous

Biodegradation of plastics was tested in the compost stored at 220, 4, and 20 °C for different periods [176]. It was found that biodegradation of cellulose in the compost was almost independent of the storage time and temperature. In contrast, biodegradability of both PCL and PBS depended strongly on the storage conditions.

The degradation of poly(3-hydroxybutyrate) (PHB), a synthetic aliphatic polyester (Sky-Green) and a starch-based polymer material (Mater-Bi) was investigated in various soil types (i.e., forest soil, sandy soil, activated sludge soil, and farm soil), and the characteristics of fungi that degrade those polymers were examined [177]. Biodegradation of all the three polymers was most active in the activated sludge soil. In both the soil burial test and the modified Sturm test, the order of the biodegradation rate was PHB > Sky-Green > Mater-Bi.

The PCL and poly((R)-3-hydroxybutyrate) (R-PHB) films with a hydrophilic surface were prepared by the alkali treatment of their as-cast films in NaOH solutions of different concentrations [178]. The alkali-treated PCL and R-PHB films, as well as the as-cast PCL and R-PHB films, were biodegraded in soil controlled at 25 °C. The alkali treatment enhanced the hydrophilicities and biodegradabilities of the PCL and R-PHB films in the soil. The biodegradabilities of the as-cast aliphatic polyester films in controlled soil decreased in the following order: PCL > R-PHB > PLLA, in agreement with that in controlled static seawater.

Degradabilities of four kinds of commercial biodegradable plastics, copolyester of polyhydroxybutyrate (PHB, 92%) and valerate (8%) (PHBV), PCL, blends of starch and polyvinyl alcohol (SPVA) and CA, were investigated in waste landfill model reactors that were operated anaerobically and aerobically [179]. PCL showed film breakage under both conditions, which may have contributed to a reduction in the waste volume regardless of aerobic or anaerobic conditions. Effective degradation of PHBV plastic was observed in the aerobic conditions, though insufficient degradation was observed in the anaerobic condition. In contrast, aeration may not significantly enhance the volume reduction of SPVA and CA plastics.

Aerobic and anaerobic biodegradation of four different kinds of polymers, PLA, PCL, a starch–PCL blend (Mater-Bi), and poly(butylene adipate-*co*-terephthalate) (Eastar Bio), has been studied in the solid state under aerobic conditions and in the liquid phase under both aerobic and anaerobic conditions [180]. Several standard test methods (ISO 14851, ISO 14853, ASTM G 21-90, ASTM G-22-76, and NF X 41-514) were used to determine the biodegradability. To determine the efficiency of the biodegradation of polymers, quantitative (mass variations, oxygen uptake, pressure variations, biogas generation and composition, biodegradation percentages) and qualitative (variation of T_g and T_f, variation of molar mass by SEC, characterization by FTIR and NMR spectroscopy) analyses were made and materials were characterized before and after 28 days of degradation.

Melt-pressed films of PCL and PLA with processing additives, $CaCO_3$, SiO_2, and erucamide, were subjected to pure fungal cultures, *A. fumigatus* and *Penicillium simplicissimum*, and to composting [181]. The PCL films showed a rapid weight loss with a minor reduction in the molecular weight after 45 days in *A. fumigatus*. The addition of SiO_2 to PCL increased the rate of bio(erosion) in *A. fumigatus* and in compost. PLA without additives and PLA containing SiO_2 exhibited the fastest (bio)degradation, followed by PLA with $CaCO_3$. The degradation of the PLA films was initially governed by chemical hydrolysis, followed by acceleration of the weight change and of the molecular weight reduction.

Biodegradation of PCL, CA, and their blends using an aerobic biodegradation technique (the Sturm test) was compared [182]. The 40PCL–60CA blend showed faster biodegradation than the other blends. PCL was more susceptible to attack by a mixture of fungi on solid medium than was CA, but showed a lower loss of mass than the latter polymer; the 60 PCL–40CA blend showed the greatest loss of mass during the period of evaluation. In contrast, in liquid medium, PCL showed a greater loss of mass.

11.9 Summary of Composting

Composting is an alternative method of plastics waste management. To fulfill compostability criteria, the polymers should be tested using procedures established by relevant ISO, ASTM and EN standards. Some results of biodegradation testing under controlled composting conditions of main biodegradable polymers are given in Table 11.8.

References

[1] G. Bellia, M. Tosin, G. Floridi, F. Degli-Innocenti, Activated vermiculite: a solid bed for testing biodegradability under composting conditions, Polym. Deg. Stab. 66 (1999) 65.

[2] F. Degli-Innocenti, G. Bellia, M. Tosin, A. Kapanen, M. Itävaara, Detection of toxicity released by biodegradable plastics after composting in activated vermiculite, Polym. Deg. Stab. 73 (2001) 101.

[3] A. Longieras, A. Copinet, G. Bureau, L. Tighzert, An inert solid medium for simulation of material biodegradation in compost and achievement of carbon balance, Polym. Deg. Stab. 83 (2004) 187.

[4] G. Bellia, M. Tosin, F. Degli-Innocenti, The test method of composting in vermiculite is unaffected by the priming effect, Polym. Deg. Stab. 69 (2000) 113.

[5] J.-Ch Jang, P.-K. Shin, J.-S. Yoon, I.-M. Lee, H.-S. Lee, M.-N. Kim, Glucose effect on the biodegradation of plastics by compost from food garbage, Polym. Deg. Stab. 76 (2002) 155.

[6] H.-S. Yang, J.-S. Yoon, M.-N. Kim, Dependence of biodegradability of plastics in compost on the shape of specimens, Polym. Deg. Stab. 87 (2005) 131.

[7] R. Jayasekara, G.T. Lonergan, I. Harding, I. Bowater, P. Halley, G.B. Christie, An automated multi-unit composting facility for biodegradability evaluations, J. Chem. Technol. Biotechnol. 76 (2001) 411.

[8] L. Száraz, J. Beczner, G. Kayser, Investigation of the biodegradability of water-insoluble materials in a solid test based on the adaption of a biological oxygen demand measuring system, Polym. Deg. Stab. 81 (2003) 477.

[9] I. Körner, J. Braukmeier, J. Herrenklage, K. Leikam, M. Ritzkowski, M. Schlegelmilch, R. Stegmann, Investigation and optimization of composting processes – test systems and practical examples, Waste Manag. 23 (2003) 17.

[10] V.M. Ghorpade, A. Gennadios, M.A. Hanna, Laboratory composting of extruded poly(lactic acid) sheets, Biores. Technol. 76 (2001) 57.

[11] C. Way, D.Y. Wu, K. Dean, E. Palombo, Design considerations for high-temperature respirometric biodegradation of polymers in compost, Polym. Test. 29 (2010) 147.

[12] S. Grima, V. Bellon-Maurel, P. Feuilloley, F. Silvestre, Aerobic biodegradation of polymers in solid-state conditions: a review of environmental and physicochemical parameter settings in laboratory simulations, J. Polym. Environ. 8 (2000) 183.

[13] S.B. Joo, M.N. Kim, S.S. Im, J.S. Yoon, Biodegradation of plastics in compost prepared at different composting conditions, Macromol. Symp. 224 (2005) 355.

[14] M. Itävaara, M. Vikman, An overview of methods for biodegradability testing of

biopolymers and packaging materials, J. Environ. Polym. Degrad. 4 (1996) 29.
[15] D. Briassouliss, C. Dejean, P. Picuno, Critical review of norms and standards for biodegradable agricultural plastics Part II: composting, J. Polym. Environ 18 (2010) 364.
[16] A. Calmon, F. Silvestre, V. Bellon-Maurel, J.-M. Roger, P. Feuilloley, Modelling easily biodegradability of materials in liquid medium – relationship between structure and biodegradability, J. Environ. Polym. Deg. 7 (1999) 135.
[17] U. Pagga, A. Schäfer, R.-J. Müller, M. Pantke, Determination of the aerobic biodegradability of polymeric material in aquatic batch tests, Chemosphere 42 (2001) 319.
[18] J.-G. Gu, J.-D. Gu, Methods currently used in testing microbiological degradation and deterior-ation of a wide range of polymers with various degrees of degradability: a review, J. Polym. Environm 13 (2005) 65.
[19] J.P. Eubeler, M. Bernhard, S. Zok, T.P. Knepper, Environmental biodegradation of synthetic polymers I. Test methodologies and procedures, Trend. Anal. Chem. 28 (2009) 1057.
[20] W. Mulbry, J.B. Reeves, P. Millner, Use of mid- and near-infrared spectroscopy to track degradation of bio-based eating utensils during composting, Bioresour.Technol. 109 (2012) 93.
[21] M. Kunioka, F. Ninomiya, M. Funabashi, Biodegradation of poly(lactic acid) powders proposed as the reference test materials for the international standard of biodegradation evaluation methods, Polym. Deg. Stab. 91 (2006) 1919.
[22] W. Guo, J. Tao, C. Yang, Q. Zhao, C. Song, S. Wang, The rapid evaluation of material biodegradability using an improved ISO 14852 method with a microbial community, Polym. Test. 29 (2010) 832.
[23] J. Lunt, Large-scale production, properties and commercial applications of polylactic acid polymers, Polym. Deg. Stab. 59 (1998) 145.
[24] R. Auras, B. Harte, S. Selke, An overview of polylactides as packaging materials, Macromol. Biosci. 4 (2004) 835.
[25] Biodegradable Plastics – Development and environmental impacts, Nolan-ITU Pty Ltd, prepared in association with ExcelPLas, Australia, October 2002, www. environment. gov.au/settlements/ publications/waste/ degradables/biodegradable
[26] Techno-economic feasibility of large-scale production of bio-based polymers in Europe (PRO-BIP), Final Report, Utrecht/Karlsruhe, October 2004.
[27] K.-L.G. Ho, A.L. Pometto III, A. Gadea-Rivas, A. Briceño Rojas, Degradation of polylactic acid (PLA) plastic in Costa Rican soil and Iowa State University compost rows, J. Environ. Polym. Deg. 7 (1999) 173.
[28] T. Kijchavengkul, R. Auras, M. Rubino, M. Ngouajio, R.T. Fernandez, Development of an automatic laboratory-scale respirometric system to measure polymer biodegradability, Polym. Test. 25 (2006) 1006.
[29] J. Tuominen, J. Kylmä, A. Kapanen, O. Venelampi, M. Itävaara, J. Seppälä, Biodegradation of lactic acid based polymers under controlled composting conditions and evaluation of the ecotoxicological impact, Biomacromolecules 3 (2002) 445.
[30] M. Itävaara, S. Karjomaa, J.-F. Selin, Biodegradation of polylactide in aerobic and anaerobic thermophilic conditions, Chemosphere 46 (2002) 879.
[31] M. Żenkiewicz, R. Malinowski, P. Rytlewski, A. Richert, W. Sikorska, K. Krasowska, Some composting and biodegradation effects of physically or chemically crosslinked poly(lactic acid), Polym. Test. 31 (2012) 83.
[32] G. Kale, R. Aura, S.P. Singh, Comparison of the degradability of poly(lactide) packages in composting and ambient exposure conditions, Packag. Technol. Sci. 20 (2007) 49.
[33] R. Pradhan, M. Reddy, W. Diebel, L. Erickson, M. Misra, A. Mohanty, Comparative compostability and biodegradation studies of various components of green composites and their blends in simulated aerobic composting bioreactor, Int. J. Plast. Technol. 14 (Suppl. 1) (2010) S45.
[34] J. Sarasa, J.M. Gracia, C. Javierre, Study of the biodisintegration of a bioplastic material waste, Bioresour. Technol. 100 (2009) 3764.
[35] T. Leejarkapi, U. Suwanmanee, Y. Rudeekit, T. Mungcharoen, Biodegradable kinetics of plastics under controlled composting conditions, Waste Manag. 31 (2011) 1153.
[36] F. Khabbaz, S. Karlsson, Albertsson A.-Ch Py-GC/MS – An effective technique for characterizing the degradation mechanism of

poly(L-lactide) in different environments, J. Appl. Polym. Sci. 78 (2000) 2369.
[37] Y. Tokiwa, A. Jarerat, Microbial degradation of aliphatic polyesters, Macromol. Symp. 201 (2003) 283.
[38] Y. Tokiwa, B.P. Calabia, Biodegradability and biodegradation of poly(lactide), Appl. Microbiol. Biotechnol. 72 (2006) 244.
[39] H. Pranamuda, Y. Tokiwa, H. Tanaka, Polylactide degradation by an *Amycolatopsis* sp. Appl. Environ. Microbiol. 63 (1997) 1637.
[40] M.L. Tansengco, Y. Tokiwa, Comparative population study of aliphatic polyester-degrading microorganisms at 50 °C, Chem. Lett. 27 (1998) 1043.
[41] T. Suyama, Y. Tokiwa, P. Ouichanpagdee, T. Kanagawa, Y. Kamagata, Phylogenetic affiliation of soil bacteria that degrade aliphatic polyesters available commercially as biodegradable plastics, Appl. Environ. Microbiol. 64 (1998) 5008.
[42] T. Ohkita, S.H. Lee, Thermal degradation and biodegradability of poly(lactic acid)/corn starch biocomposites, J. Appl. Polym. Sci. 100 (2006) 3009.
[43] H. Urayama, T. Kanamori, Y. Kimura, Properties and biodegradability of polymer blends of poly(L-lactide) with different optical purity of the lactate units, Macromol. Mater. Eng. 287 (2002) 116.
[44] R. Gattin, A. Copinet, C. Bertrand, Y. Couturier, Comparative biodegradation study of starch- and polylactic acid-based films, J. Polym. Environ. 9 (2001) 11.
[45] A. Hoshino, Y. Isono, Degradation of aliphatic polyester films by commercially available lipases with special reference to rapid and complete degradation of poly (L-lactide) film by lipase PL derived from *Alcaligenes* sp. Biodegradation 13 (2002) 141.
[46] E. Rudnik, D. Briassoulis, Degradation behavior of poly(lactic acid) films and fibers in soil under Mediterranean field conditions and laboratory simulations testing, Ind. Crops Prod. 33 (2011) 648.
[47] M. Hakkarainen, Aliphatic polyesters: abiotic and biotic degradation and degradation products, Adv. Polym. Sci. 157 (2002) 115.
[48] D. Byrom, The synthesis and biodegradation of polyhydroxyalkanoates from bacteria, Int. Biodeter. Biodegrad. 31 (1993) 199.
[49] U. Pagga, D.B. Beimborn, J. Boelens, B. De Wilde, Determination of the aerobic biodegradability of polymeric material in laboratory controlled composting test, Chemosphere 31 (1995) 4475.
[50] C.L. Yue, R.A. Gross, S.P. McCarthy, Composting studies of poly(β-hydroxybutyrate-co-β-valerate), Polym. Deg. Stab. 51 (1996) 205.
[51] Y.-X. Weng, Y. Wang, X.-L. Wang, Y.-Z. Wang, Biodegradation behaviour of PHBV films in a pilot – Scale composting condition, Polym. Test. 29 (2010) 579.
[52] Y.-X. Weng, X.-L. Wang, Y.-Z. Wang, Biodegradation behaviour of PHAs with different chemical structures under controlled composting conditions, Polym. Test. 30 (2011) 372.
[53] C. Eldsätter, S. Karlsson, Albertsson A.-Ch effect of abiotic factors on the degradation of poly(3-hydroxybutyrate-*co*-3-hydroxyvalerate) in simulated and natural composting environments, Polym. Deg. Stab. 64 (1999) 177.
[54] S. Luo, A.N. Netravali, A study of physical and mechanical properties of poly(hydroxybutyrate-co-hydroxyvalerate) during composting, Polym. Deg. Stab. 80 (2003) 59.
[55] D.S. Rosa, N.T. Lotto, D.R. Lopes, C.G.F. Guedes, The use of roughness for evaluating the bio-degradation of poly-β-(hydroxybutyrate) and poly-β-(hydroxybutyrate-co-β-valerate), Polym. Test. 23 (2004) 3.
[56] N.T. Lotto, M.T. Calil, C.G.F. Guedes, D.S. Rosa, The effect of temperature on the biodegradation test, Mater. Sci. Eng. 24 (2004) 659.
[57] D. Dos Santos Rosa, M.R. Callil, C. Das Graças Fassina Guedes, T.C. Rodrigues, Biodegradability of thermally aged PHB, PHBV, and PCL in soil compostage, J. Polym. Environ. 12 (2004) 239.
[58] A. El-Hadi, R. Schnabel, E. Straube, G. Müller, S. Henning, Correlation between degree of crystallinity, morphology, glass temperature, mechanical properties and biodegradation of poly(3-hydroxyalkanoate) PHAs and their blends, Polym. Test. 21 (2002) 665.

[59] M.B. Kellerrhals, B. Kessler, B. Witholt, H. Tchouboukov Brandl, Renewable long-chain fatty acids for production of biodegradable medium-chain-length polyhydroxyalkanoates (mcl-PHAs) at laboratory and pilot plant scales, Macromolecules 33 (2000) 4690.
[60] C.S.K. Reddy, R. Ghai, Rashmi, V.C. Kalia, Polyhydroxyalkanoates: an overview, Bioresour. Technol. 87 (2003) 137.
[61] T.G. Volova, A.N. Boyandin, A.D. Vasiliev, V.A. Karpov, S.V. Prudnikova, O.V. Mishukova, U.A. Boyarskikh, M.L. Filipienko, V.P. Rudnev, B.B. Xuân, V.V. Dûng, I.I. Gitelson, Biodegradation of polyhydroxyalkanoates (PHAs) in tropical coastal waters and identification of PHA-degrading bacteria, Polym. Deg. Stab. 95 (2010) 2350.
[62] T.G. Volova, M.I. Gladyshev, M.Yu Trusova, N.O. Zhila, Degradation of polyhydroxyalkanoates and the composition of microbial destructors under natural conditions, Microbiology 75 (2006) 593.
[63] E. Rudnik, D. Briassoulis, Comparative biodegradation in soil behaviour of two biodegradable polymers based on renewable resources, J. Polym. Environ 19 (1) (2011) 18.
[64] M. Vikman, M. Itävaara, K. Poutanen, Measurement of the biodegradation of starch-based materials by enzymatic methods and composting, J. Envir. Polym. Deg. 3 (1995) 23.
[65] F. Degli-Innocenti, M. Tosin, C. Bastioli, Evaluation of the biodegradation of starch and cellulose under controlled composting conditions, J. Polym. Environ. 6 (1998) 197.
[66] F. Jbilou, S. Galland, F. Ayadi, L. Belard, P. Dole, V. Desjardin, R. Bayard, P. Degraeve, Biodegradation of corn flour-based materials assessed by enzymatic, aerobic, and anaerobic tests: influence of specific surface area, Polym. Test. 30 (2011) 131.
[67] R. Gattin, A. Copinet, C. Bertrand, Y. Couturier, Biodegradation study of a starch and poly(lactic acid) co-extruded material in liquid, composting and inert mineral media, Int. Biodeter. Biodegr. 50 (2002) 25.
[68] Y.-L. Du, Y. Cao, F. Lu, F. Li, Y. Cao, X.-L. Wang, Y.-Z. Wang, Biodegradation behaviours of thermoplastic starch (TPS) and thermoplastic dialdeyde starch (TPDAS) under controlled composting conditions, Polym. Test. 27 (2008) 924.
[69] F. Degli-Innocenti, M. Tosin, C. Bastioli, Evaluation of the biodegradation of starch and cellulose under controlled composting conditions, J. Polym. Environ. 6 (1998) 197.
[70] J. Pérez, J. Muñoz-Dorado, T. de la Rubia, J. Martinez, Biodegradation and biological treatments of cellulose, hemicellulose and lignin: an overview, Int. Microbiol. 5 (2002) 53.
[71] K.J. Edgar, C.M. Buchanan, J.S. Debenham, P.A. Rundquist, B.D. Seiler, M.C. Shelton, D. Tindall, Advances in cellulose ester performance and application, Prog. Polym. Sci. 26 (2001) 1605.
[72] R.M. Gardner, C.M. Buchanan, R. Komarek, D. Dorschel, C. Bogus, A.W. White, Compostability of cellulose acetate films, J. Appl. Polym. Sci. 52 (1994) 1477.
[73] J.-D. Gu, S. Yang, R. Welton, D. Eberiel, S.P. McCarthy, R.A. Gross, Effect of environmental parameters on the degradability of polymer films in laboratory-scale composting reactors, J. Environ. Polym. Degrad. 2 (1994) 129.
[74] W.G. Glasser, B. McCartney, G. Samaranayake, Cellulose derivatives with low degree of substitution. 3. The biodegradability of cellulose esters using a simple enzyme assay, Biotechnol. Prog. 10 (1994) 214.
[75] C.M. Buchanan, D. Dorschel, R.M. Gardner, R.J. Komarek, A.J. Matosky, A.W. White, M.D. Wood, The influence of degree of substitution on blend miscibility and biodegradation of cellulose acetate blends, J. Environ. Polym. Degrad. 4 (1996).
[76] E. Samios, R.K. Dart, J.V. Dawkins, Preparation, characterization and biodegradation studies on cellulose acetates with varying degrees of substitution, Polymer 38 (1997) 3045.
[77] G. Frisoni, M. Baiardo, M. Scandola, Natural cellulose fibers: heterogenous acetylation kinetics and biodegradation behaviour, Biomacromolecules 2 (2001) 476.
[78] T. Ikejima, Y. Inoue, Crystallization behaviour and environmental biodegradability of the blend films of poly(3-hydroxybutyric acid) with chitin and chitosan, Carbohydr. Polym. 41 (2000) 351.

[79] P. Lodha, A.N. Netravali, Effect of soy protein isolate resin modifications on their biodegradation in a compost medium, Polym. Deg. Stab. 87 (2005) 465.
[80] S. Domenek, P. Feuilloley, J. Gratraud, M.-H. Morel, S. Guilbert, Biodegradability of wheat gluten based plastics, Chemosphere 54 (2004) 551.
[81] C. Abrusci, D. Marquina, A. Santos, A. Del Amo, T. Corrales, F. Catalina, A chemiluminescence study on degradation of gelatine. Biodegradation by bacteria and fungi isolated from cinematographic films, J. Photochem. Photobiol. A: Chem. 185 (2007) 188.
[82] E. Chiellini, P. Cinelli, A. Corti, E.R. Kenawy, Composite films based on waste gelatin: thermal-mechanical properties and biodegradation testing, Polym. Deg. Stab. 73 (2001) 549.
[83] T. Fujimaki, Processability and properties of aliphatic polyesters, "BIONOLLE", synthesised by polycondensation reaction, Polym. Deg. Stab. 59 (1998) 209.
[84] V. Tserki, P. Matzinos, E. Pavlidou, D. Vachliotis, C. Panayiotou, Biodegradable aliphatic polyesters. Part I. Properties and biodegradation of poly(butylene succinate-*co*-butylene adipate), Polym. Deg. Stab. 91 (2006) 367.
[85] V. Tserki, P. Matzinos, E. Pavlidou, D. Vachliotis, C. Panayiotou, Biodegradable aliphatic polyesters. Part II. Synthesis and characterization of chain extended poly(-butylene succinate-*co*-butylene adipate), Polym. Deg. Stab. 91 (2006) 377.
[86] M.-N. Kim, K.-H. Kim, H.-J. Jin, J.-K. Park, J.-S. Yoon, Biodegradability of ethyl and n-octyl branched poly(ethylene adipate) and poly(butylene succinate), Eur. Polym. J. 37 (2001) 1843.
[87] D.N. Bikiaris, G.Z. Papageorgiou, D.S. Achilias, Synthesis and comparative biodegradability studies of three poly(alkylene succinate)s, Polym. Deg. Stab. 91 (2006) 31.
[88] E. Marten, R.-J. Müller, W.-D. Deckwer, Studies on the enzymatic hydrolysis of polyesters. I. Low molecular mass model esters and aliphatic polyesters, Polym. Deg. Stab. 80 (2003) 485.
[89] B.D. Ahn, S.H. Kim, Y.H. Kim, J.S. Yang, Synthesis and characterization of the biodegradable copolymers from succinic acid, and adipic acid with 1,4-butanediol, J. Appl. Polym. Sci. 82 (2001) 2808.
[90] P. Rizarrelli, C. Puglisi, G. Montaudo, Soil burial and enzymatic degradation in solution of aliphatic co-polyesters, Polym. Deg. Stab. 85 (2004) 855.
[91] J.-H. Zhao, X.-Q. Wang, J. Zeng, G. Yang, F.-H. Shi, Q. Yan, Biodegradation of poly(-butylene succinate-*co*-butylene adipate) by *Aspergillus versicolor*, Polym. Deg. Stab. 90 (2005) 173.
[92] J.-H. Zhao, X.-Q. Wang, J. Zeng, G. Yang, F.-H. Shi, Q. Yan, Biodegradation of poly (butylene succinate) in compost, J. Appl. Polym. Sci. 97 (2005) 2273.
[93] M.-N. Kim, K.-H. Kim, H.-J. Jin, J.-K. Park, J.-S. Yoon, Biodegradability of ethyl and n-octyl branched poly(ethylene adipate) and poly (butylene succinate), Eur. Polym. J. 37 (2001) 1843.
[94] H.-J. Jin, D.-S. Kim, B.-Y. Lee, M.-N. Kim, I.-M. Lee, H.-S. Lee, J.-S. Yoon, Chain extension and biodegradation of poly(butylene succinate) with maleic acid units, J. Polym. Sci. 38 (2000) 2240.
[95] H. Pranamuda, Y. Tokiwa, H. Tanaka, Microbial degradation of an aliphatic polyester with a high melting point, poly(tetramethylene succinate), Appl. Environ. Microbiol. 61 (1995) 1828.
[96] N. Hayase, H. Yano, E. Kudoh, Ch. Tsutsumi, K. Ushio, Y. Miyahara, S. Tanaka, K. Nakagawa, Isolation and characterization of poly(butylene succinate-*co*-butylene adipate)-degrading microorganism, J. Biosci. Bioeng. 97 (2004) 131.
[97] M. Abe, K. Kobayashi, N. Honma, K. Nakasaki, Microbial degradation of poly(-butylene succinate) by *Fusarium solani* in soil environments, Polym. Deg. Stab 95 (2010) 138.
[98] C. Tsutsumi, N. Hayase, K. Nakagawa, S. Tanaka, Y. Miyahara, The enzymatic degradation of commercial biodegradable polymers by some lipases and chemical degradation of them, Macromol. Symp. 197 (2003) 431.
[99] T.M. Scherer, R. Clinton Fuller, R.W. Lenz, S. Goodwin, Hydrolase activity of an extracellular depolymerase from *Aspergillus*

fumigatus with bacterial and synthethic polyesters, Polym. Deg. Stab. 64 (1999) 267.
[100] A. Jarerat, Y. Tokiwa, Degradation of poly(-tetramethylene succinate) by thermophilic actinomycetes, Biotechnol. Lett. 23 (2001) 647.
[101] A.S. Chandure, S.S. Umare, Synthesis, characterization and biodegradation of low molecular weight polyesters, Int. J. Polym. Mater. 56 (2007) 339.
[102] H. Shirahama, Y. Kawaguchi, M.S. Aludin, H. Yasuda, Synthesis and enzymatic degradation of high molecular weight aliphatic polyesters, J. Appl. Polym. Sci. 80 (2001) 340.
[103] M.J. Diamond, B. Freedman, J.A. Garibaldi, Biodegradable polyester films, Int. Biodeter. Biodegrad. 48 (2001) 219.
[104] M.S. Nikolic, J. Djonlagic, Synthesis and characterization of biodegradable poly (butylene succinate-*co*-butylene adipate), Polym. Deg. Stab. 74 (2001) 263.
[105] H. Maeda, Y. Yamagata, K. Abe, F. Hasegawa, M. Machida, R. Ishioka, K. Gomi, T. Nakajima, Purification and characterization of a biodegradable plastic-degrading enzyme from *Aspergillus oryzae*, Appl. Microbiol. Biotechnol. 67 (2005) 778.
[106] R.-J. Müller, I. Kleeberg, W.-D. Deckwer, Biodegradation of polyesters containing aromatic constituents, J. Biotechnol. 86 (2001) 87.
[107] U. Witt, R.-J. Müller, W.-D. Deckwer, Biodegradation of polyester copolymers containing aromatic compounds, J. Macromol. Sci. – Pure Appl. Chem. A32 (1995) 851.
[108] R.-J. Müller, U. Witt, E. Rantze, W.-D. Deckwer, Architecture of biodegradable copolyesters containing aromatic constituents, Polym. Deg. Stab. 59 (1998) 203.
[109] I. Kleeberg, C. Hetz, R.M. Kroppenstedt, R.-J. Müller, W.-D. Deckwer, Biodegradation of aliphatic-aromatic copolyesters by *Thermonospora fusca* and other thermophilic compost isolates, Appl. Environ. Microbiol. 64 (1998) 1731.
[110] F. Trinh Tan, D.G. Cooper, M. Maric, J.A. Nicell, Biodegradation of a synthetic copolyester by aerobic mesophilic microorganisms, Polym. Deg. Stab. 93 (2008) 1479.
[111] T. Nakajima-Kambe, F. Ichihashi, R. Matsuzoe, S. Kato, N. Shintani, Degradation of aliphatic-aromatic copolyesters by bacteria that can degade aliphatic polyesters, Polym. Deg. Stab. 94 (2009) 1901.
[112] H.J. Kang, S.S. Park, Characterization and biodegradability of poly(butylene adipate-*co*-succinate)/poly(butylene terephthalate) copolyester, J. Appl. Polym. Sci. 72 (1999) 593.
[113] E. Marten, R.-J. Müller, W.-D. Deckwer, Studies on the enzymatic hydrolysis of polyesters. II. Aliphatic-aromatic copolyesters, Polym. Deg. Stab. 88 (2005) 371.
[114] R.-J. Müller, H. Schrader, J. Profe, K. Dresler, W.-D. Deckwer, Enzymatic degradation of poly(ethylene terephthalate): rapid hydrolyse using a hydrolase from *T. fusca*, Macromol. Rapid Commun. 26 (2005) 1400.
[115] R.-J. Mueller, Biological degradation of synthetic polyesters – Enzymes as potential catalysts for polyester recycling, Process Biochem. 41 (2006) 2124.
[116] Y. Maeda, T. Maeda, K. Yamaguchi, S. Kubota, A. Nakayama, N. Kawasaki, N. Yamamoto, S. Aiba, Synthesis and characterization of novel biodegradable copolyesters by transreaction of poly(ethylene terephthalate) with copoly(succinic anhydride/ethylene oxide), J. Polym. Sci. Part A: Polym. Chem. 38 (2000) 4478.
[117] T. Kijchavengkul, R. Auras, M. Rubino, S. Selke, M. Ngouajio, R.T. Fernandez, Biodegradation and hydrolysis rate of aliphatic aromatic polyester, Poylm. Deg. Stab. 95 (2010) 2641.
[118] U. Edlund, A.-Ch. Albertsson, Degradable polymer microspheres for controlled drug delivery, Adv. Polym. Sci. 157 (2002) 67.
[119] K. Krasowska, A. Heimowska, M. Rutkowska, Enzymatic and hydrolytic degradation of poly (ε-caprolactone), Int. Polym. Sci. Technol. 33 (2006). T/57.
[120] C. Eldsäter, B. Erlandsson, R. Renstad, A.-C. Albertsson, S. Karlsson, The biodegradation of amorphous and crystalline regions in film-blown poly(ε-caprolactone), Polymer 41 (2000) 1297.
[121] S. Agarwal, C. Speyerer, Degradable blends of semi-crystalline and amorphous branched poly(caprolactone): effect of microstructure on blend properties, Polymer 51 (2010) 1024.

[122] V. Mezzanote, R. Bertani, F. Degli Innocenti, M. Tosin, Influence of inocula on the results of biodegradation tests, Polym. Deg. Stab. 87 (2005) 51.
[123] K. Nakasaki, Ohtaki A, H. Takano, Biodegradable plastic reduces ammonia emission during composting, Polym. Deg. Stab. 70 (2000) 185.
[124] A. Ohtaki, N. Sato, K. Nakasaki, Biodegradation of poly(ε-caprolactone) under controlled composting conditions, Polym. Deg. Stab. 61 (1998) 449.
[125] A. Ohtaki, N. Akakura, K. Nakasaki, Effects of temperature and inoculum on the degradability of poly(ε-caprolactone) during composting, Polym. Deg. Stab. 62 (1998) 279.
[126] R. Gattin, A. Cretu, D. Barbier-Baudry, Effect of the remaining lanthanide catalysts on the hydrolytic and enzymatic degradation of poly(ε-caprolactone), Macromol. Symp. 197 (2003) 455.
[127] K. Ohkawa, H. Kim, K. Lee, Biodegradation of electrospun poly(ε-caprolactone) nonwoven fabrics by pure-cultured soil filamentous fungi, J. Polym. Environ. 12 (2004) 211.
[128] M. Funabashi, F. Ninomiya, M. Kunioka, Biodegradation of polycaprolactone powders proposed as reference test materials for international standard of biodegradation evaluation method, J. Polym. Environ. 15 (2007) 7.
[129] E. Botines, A. Rodríguez-Galán, J. Puiggalí, Poly(ester amide)s derived from 1,4-butanediol, adipic acid and 1,6-aminohexanoic acid: characterization and degradation studies, Polymer 43 (2002) 6073.
[130] T. Ferré, L. Franco, A. Rodríguez-Galán, J. Puiggalí, Poly(ester amide)s derived from 1,4-butanediol, adipic acid and 1,6-aminohexanoic acid. Part II: composition changes and fillers, Polymer 44 (2003) 6139.
[131] M. Lozano, L. Franco, A. Rodríguez-Galán, J. Puiggalí, Poly(ester amide)s derived from 1,4-butanediol, adipic acid and 1,6-aminohexanoic acid. Part III: substitution of adipic acid units by terephthalic acid units, Polym. Deg. Stab. 85 (2004) 595.
[132] X. Chen, K.E. Gonsalves, J.A. Cameron, Further studies on biodegradation of aliphatic poly(ester-amides), J. Appl. Polym. Sci. 50 (1993) 1999.
[133] A. Rodríguez-Galán, L. Fuentes, J. Puiggalí, Studies on the degradability of a poly(ester amide) derived from L-alanine, 1,12-dodecanediol and 1,12-dodecanedioic acid, Polymer 41 (2000) 5967.
[134] E. Grigat, R. Koch, R. Timmermann, BAK 1095 and BAK 2195: completely biodegradable synthetic thermoplastics, Polym. Deg. Stab. 59 (1998) 223–226.
[135] R. Jayeskara, S. Sheridan, E. Lourbakos, H. Beh, G.B.Y. Christie, M. Jenkins, P.B. Halley, S. McGlashan, G.T. Lonergan, Biodegradation and ecotoxicity evaluation of a Bionolle and starch blend and its degradation products in compost, Int. Biodeter. Biodeg. 51 (2003) 77–81.
[136] J. Brožek, V. Šašek, I. Prokopová, D. Chromcová, J. Náhlik, P. Erbanová, Degradation of polyesteramides during composting under standard and isothermal conditions, Folia Microbiol. 54 (2009) 451.
[137] C. Park, E.Y. Kim, Y.T. Yoo, S.S. Im, Effect of hydrophilicity on the biodegradability of polyesteramides, J. Appl. Polym. Sci. 90 (2003) 2708.
[138] E. Chiellini, A. Corti, R. Solaro, Biodegradation of poly(vinyl alcohol) based blown films under different environmental conditions, Polym. Deg. Stab. 64 (1999) 305.
[139] A. Corti, R. Solaro, E. Chiellini, Biodegradation of poly(vinyl alcohol) in selected mixed culture and relevant culture filtrate, Polym. Deg. Stab. 75 (2002) 447.
[140] E. Chiellini, A. Corti, S. D'Antone, R. Solaro, Biodegradation of poly(vinyl alcohol) based materials, Prog. Polym. Sci. 28 (2003) 963.
[141] F. Kawai, X. Hu, Biochemistry of microbial polyvinyl alcohol degradation, Appl. Microbiol. Biotechnol. 84 (2009) 227.
[142] J. Hoffmann, I. Řezníčková, J. Kozáková, J. Růžička, P. Alexy, D. Bakoš, L. Precnerová, Assessing biodegradability of plastics based on poly(vinyl alcohol) and protein wastes, Polym. Deg. Stab. 79 (2003) 511.
[143] C. David, C. De Kesel, F. Lefebre, M. Weiland, The biodegradation of polymers: a recent studies, Angew. Makromol. Chem. 216 (1994) 21.
[144] E. Chiellini, A. Corti, G. Del Sarto, S. D'Antone, Oxo-biodegradable polymers –

effect of hydrolysis degree on biodegradation of poly(vinyl alcohol), Polym. Deg. Stab. 91 (2006) 3397.
[145] S. Matsumura, N. Tomizawa, A. Toki, K. Nishikawa, K. Toshima, Novel poly(vinyl alcohol)-degrading enzyme and the degradation mechanism, Macromolecules 32 (1999) 7753.
[146] M. Watanabe, F. Kawai, Numerical simulation for enzymatic degradation of poly(vinyl alcohol), Polym. Deg. Stab. 81 (2003) 393.
[147] R. Solaro, A. Corti, E. Chiellini, Biodegradation of poly(vinyl alcohol) with different molecular weights and degree of hydrolysis, Polym. Adv. Technol. 11 (2000) 873.
[148] E. Chiellini, A. Corti, B. Politi, R. Solaro, Adsorption/desorption of polyvinyl alcohol on solid substrates and relevant biodegradation, J. Polym. Environ. 8 (2000) 67.
[149] T. Ohkita, S.-H. Lee, Thermal degradation and biodegradability of poly(lactic acid)/corn starch biocomposites, J. Appl. Polym. Sci. 100 (2006) 3009.
[150] H. Tsuji, Y. Kidokoro, M. Mochizuki, Enzymatic degradation of biodegradable polyester composites of poly(L-lactic acid) and poly(ε-caprolactone), Macromol. Mater. Eng. 291 (2006) 1245.
[151] S.H. Imam, L. Chen, S.H. Gordon, R.L. Shogren, D. Weisleder, R.V. Greene, Biodegradation of injection molded starch-poly (3-hydroxybutyrate-*co*-3-hydroxyvalerate) blends in a natural compost environment, J. Envion. Polym. Deg. 6 (1998) 91.
[152] P. Rychter, R. Biczak, B. Herman, A. Smyłła, P. Kurcok, G. Adamus, M. Kowalczuk, Environmental degradation of polyester blends containing atactic poly(3-hydroxybutyrate). Biodegradation in soil and ecotoxicological impact, Biomacromolecules 7 (2006) 3125.
[153] P. Dubois, R. Narayan, Biodegradable compositions by reactive processing of aliphatic polyester/polysaccharide blends, Macromol. Symp. 198 (2003) 233.
[154] R.P. Singh, J.K. Pandey, D. Rutot, Ph. Degée, Ph. Dubois, : Biodegradation of poly(ε-caprolactone)/starch blends and composites in composting and culture environments: the effect of compatibilization on the inherent biodegradability of the host polymer, Carbohydr. Res. 338 (2003) 1759.
[155] D. Dos Santos Rosa, T.C. Rodrigues, C. das Graças Fassina Guedes, M.R. Calil, Effect of thermal aging on the biodegradation of PCL, PHB-V, and their blends with starch in soil compost, J. Appl. Polym. Sci. 89 (2003) 3539.
[156] P. Nugroho, H. Mitomo, F. Yoshi, T. Kume, K. Nishimura, Improvement of processability of PCL and PBS blend by irradiation and its biodegradability, Macromol. Mater. Eng. 286 (2001) 316.
[157] C.-S. Wu, A comparison of the structure, thermal properties, and biodegradability of polycaprolactone/chitosan and acrylic acid grafted polycaprolactone/chitosan, Polymer 46 (2005) 147.
[158] D. Rohindra, P. Sharma, J. Khurma, Soil and microbial degradation study of poly(ε-caprolactone)–poly(vinyl butyral) blends, Macromol. Symp. 224 (2005) 323.
[159] D. Dos Santos Rosa, J.E. Volponi, C. das Graças Fassina Guedes, Biodegradation and the dynamic mechanical properties of starch gelatinization in poly(ε-caprolactone)/corn starch blends, J. Appl. Polym. Sci. 102 (2006) 825.
[160] H. Tsuji, G. Horikawa, Porous biodegradable polyester blends of poly(L-lactic acid) and poly(ε-caprolactone): physical properties, morphology, and biodegradation, Polym. Int. 56 (2007) 258.
[161] C.-H. Kim, K.-M. Jung, J.-S. Kim, J.-K. Park, Modification of aliphatic polyesters and their reactive blends with starch, J. Polym. Environ. 12 (2004) 179.
[162] H. Tsuji, G. Horikawa, S. Itsuno, Melt-processed biodegradable polyester blends of poly(L-lactic acid) and poly(ε-caprolactone): effects of processing conditions on biodegradation, J. Appl. Polym. Sci. 104 (2007) 831.
[163] P.D.S.C. Mariani, A.P. Vinagre Neto, J.P. da Silva Jr., E.J.B.N. Cardoso, E. Esposito, L.H. Innocentini-Mei, Mineralization of poly(ε-caprolactone)/adipate modified starch blend in agricultural soil, J. Polym. Environ. 15 (2007) 19.
[164] H.S. Cho, H.S. Moon, M. Kim, K. Nam, J.Y. Kim, Biodegradability and biodegradation

rate of poly(caprolactone)-starch blend and poly(butylene succinate) biodegradable polymer under aerobic and anaerobic environment, Waste Manag. 31 (2011) 475.

[165] P. Zhang, F. Huang, B. Wang, Characterization of biodegradable aliphatic/aromatic copolyesters and their starch blends, Polym. Plast. Technol. Eng. 41 (2002) 273.

[166] M. Rutkowska, K. Krasowska, A. Heimowska, M. Kowalczuk, Degradation of the blends of natural and synthetic copolyesters in different natural environment, Macromol. Symp. 197 (2003) 421.

[167] J.A. Ratto, P.J. Stenhouse, M. Auerbach, J. Mitchell, R. Farrell, Processing, performance and biodegradability of a thermoplastic aliphatic polyester/starch system, Polymer 40 (1999) 6777.

[168] T. Ikejima, A. Cao, N. Yoshie, Y. Inoue, Surface composition and biodegradability of poly(3-hydroxybutyric acid)/poly(vinyl alcohol) blend films, Polym. Deg. Stab. 62 (1998) 463.

[169] P. Alexy, D. Bakoš, S. Hanzelová, L. Kukolíkowá, J. Kupec, K. Charvátová, E. Chiellini, P. Cinelli, Poly(vinyl alcohol)–collagen hydrolysate thermoplastic blends: I. Experimental design optimization and biodegradation behaviour, Polym. Test. 22 (2003) 801.

[170] P. Alexy, D. Bakoš, S. Hanzelová, G. Crkoňová, Z. Kramárová, J. Hoffman, M. Julinová, E. Chiellini, P. Cinelli, Poly(vinyl alcohol)–collagen hydrolysate thermoplastic blends: II. Water penetration and biodegradability of melt extruded films, Polym. Test. 22 (2003) 811.

[171] A. Takasu, K. Aoi, M. Tsuchiya, M. Okada, New chitin-based polymer hybrids. 4: soil burial degradation behaviour of poly(vinyl alcohol)/chitin derivative miscible blends, J. Appl. Polym. Sci. 73 (1999) 1171.

[172] H. Tsuji, H. Muramatsu, Blends of aliphatic polyesters: V. Non-enzymatic and enzymatic hydrolysis of blends from hydrophobic poly (L-lactide) and hydrophilic poly(vinyl alcohol), Polym. Deg. Stab. 71 (2001) 403.

[173] Z. Guohua, L. Ya, F. Cuilan, Z. Min, Z. Caiqiong, C. Zongdao, Water resistance, mechanical properties and biodegradability of methylated-corn starch/poly(vinyl alcohol) blend film, Polym. Deg. Stab. 91 (2006) 703.

[174] A. Corti, P. Cinelli, S. D'Antone, E.-R. Kenawy, R. Solaro, Biodegradation of poly(vinyl alcohol) in soil environment: influence of natural organic fillers and structural parameters, Macromol. Chem. Phys. 203 (2002) 1526.

[175] R. Jayasekara, I. Harding, I. Bowater, G.B.Y. Christie, G.T. Lonergan, Biodegradation by composting of surface modified starch and PVA blended films, J. Polym. Environ. 11 (2003) 49.

[176] H.-S. Yang, J.-S. Yoon, M.-N. Kim, Effects of storage of a mature compost on its potential for biodegradation of plastics, Polym. Deg. Stab. 84 (2004) 411.

[177] M.-N. Kim, A.-R. Lee, J.-S. Yoon, I.-J. Chin, Biodegradation of poly(3-hydroxybutyrate), Sky-Green and Mater-Bi by fungi isolated from soils, Eur. Polym. J. 36 (2000) 1677.

[178] H. Tsuji, K. Suzuyoshi, Y. Tezuka, T. Ishida, Environmental degradation of biodegradable polyesters: 3. Effects of alkali treatment on biodegradation of poly(ε-caprolactone) and poly((R)-3-hydroxybutyrate) films in controlled soil, J. Polym. Environ. 11 (2003) 57.

[179] T. Ishigaki, W. Sugano, A. Nakanishi, M. Tateda, M. Ike, M. Fujita, The degradability of biodegradable plastics in aerobic and anaerobic waste landfill model reactors, Chemosphere 54 (2004) 225.

[180] V. Massardier-Nageotte, C. Pestre, T. Cruard-Pradet, R. Bayard, Aerobic and anaerobic biodegradability of polymer films and physico-chemical characterization, Polym. Deg. Stab. 91 (2006) 620.

[181] R. Renstad, S. Karlsson, A. Sandgren, A.-Ch. Albertsson, Influence of processing additives on the degradation of melt-processing films of poly(ε-caprolactone) and poly(-lactic acid), J. Environ. Polym. Degrad. 6 (1998) 209.

[182] M.R. Calil, F. Gaboardi, C.G.F. Guedes, D.S. Rosa, Comparison of the biodegradation of poly(ε-caprolactone) (PCL), cellulose acetate (CA) and their blends by the Sturm test and selected fungi, Polym. Test. 25 (2006) 597.

[183] L. Avérous, Biodegradable multiphase systems based on plasticized starch: a review, Polym. Rev. 44 (2004) 231.

12 Pressure-Sensitive Adhesives, Elastomers, and Coatings from Plant Oil

Richard P. Wool

OUTLINE

In this chapter, we switch from highly cross-linked polymers used in composites to highly linear polymers made with the single fatty acids derived from the triglycerides. The rather simple linear chain architecture, so readily attainable with petroleum-based monomers with C = C functionality, such as polyethylene and polypropylene, presents challenges for triglyceride oils. The key to success is to obtain oils that are capable of providing mono-functionalized monomers, which are the fatty acids containing just one unsaturated C = C bond, such as the high-oleic oils. These oils can be optimally obtained through genetic engineering, crop selection, specialty high-oleic crops, and fatty acid separations. The fatty acid separation process, while being technically feasible at the laboratory scale, can be quite expensive to mass produce, especially if one is looking to keep prices for resin near \$2/kg. As with all bio-based materials, the least costly and greenest approach is to grow the monomers (or their precursor materials) in the field using free sunlight, water, oxygen, and carbon sources and, at the same time, remove global warming gases from the air.

In this chapter, we explore the development of pressure-sensitive adhesives (PSAs), elastomers, and coatings from high-oleic oils. The linear polymers with molecular weights of the order of 10^6 Da are made by emulsion polymerization using water as a solvent. The resulting waterborne latex particles are placed on a substrate such as paper or polymer, and as the water evaporates they coalesce by interdiffusion to form thin films. The resulting PSA products are the familiar Scotch® tapes, postage stamps, name labels, duct tape, masking tapes, packaging labels, etc. These products are typically disposable and amount to about 14 billion pounds

Ebnesajjad: Handbook of Biopolymers and Biodegradable Plastics. http://dx.doi.org/10.1016/B978-1-4557-2834-3.00012-4
© 2013 Elsevier Inc. All rights reserved. Adapted from a chapter in: Wool, *Bio-based Polymers and Composites* (2005).

per year. The U.S. Post Office alone uses about 11% of the total US market. The importance of the development of linear polymers from triglycerides is that it provides the technology platform to make other materials requiring linear architecture such as coatings, paints, elastomers, foams, Eco-Leather, and toughening agents.

12.1 Introduction to Pressure-Sensitive Adhesives

PSAs are almost indispensable in everyday life because they are used for labels, tapes, films, postage stamps, and many adhesive applications. Currently, the majority of PSAs are made from petroleum-based acrylate monomers, such as 2-EHA, *n*-butyl acrylate, and isooctyl acrylate [1]. To alleviate this dependency on petroleum, it is desirable to investigate the synthesis of these adhesives from a renewable resource, such as plant oil. Because most of the PSA applications are of a disposable nature, it would also be desirable to make these materials biodegradable. Plant oils are triglyceride esters of fatty acids, which vary in chain length and functionality. Their chemical versatility and abundance make them an ideal starting material [2]. The most common oils have a carbon—carbon double bond functionality. An example of a triglyceride molecule is shown in Fig. 12.1.

The C = C unsaturation of fatty acids has traditionally been used for oxidative coupling reactions leading to "air drying" of some plant oils. This is the chemistry of the well-known alkyd resins used for paint and varnish binders and the pre-vinyl, old-fashioned floor covering known as "Lino" (derived from cross-linked linseed oil), and now once again quite stylish, but expensive. Although there are many examples of the use of drying oils for surface-coating applications that date back hundreds of years to antiquity, the unsaturation on the fatty acid is not sufficiently reactive to allow homo- or copolymerization of the molecule directly to give resins with any degree of structural strength or stiffness. However, both triglycerides and individual fatty acids can be chemically modified in order to participate in free-radical polymerization reactions. The fatty acid molecule offers a number of reactive sites for functionalization. These include the double bond, the allylic carbons, the ester group, and the carbon alpha to the ester group, as shown in Fig. 12.1. Typical modifying reactants include maleic acid, maleic anhydride, methacrylic acid, and acrylic acid [3] Besides conventional bulk polymerization, these components can also be polymerized using emulsion polymerization, a common practice in the PSA industry. Solution polymerization should work as well. However, the PSA industry has moved toward eliminating solvent-based material for ecological and economical reasons. Therefore, this work focuses on optimizing a water-based emulsion system.

The polymer in a PSA is a viscoelastic material that is permanently, as well as aggressively, tacky and has enough cohesive strength and elasticity to be cleanly removed from a substrate surface [1]. These polymers are typically linear polymers with a slight degree of cross-linking. The degree of cross-linking is one of the key features controlling the balance between the cohesive and adhesive strengths of the polymer, in addition to the role of sticker and receptor groups. Monomers derived from plant oils possess an inherent degree of unsaturation that varies from plant to plant. The variation of unsaturation among the various plant oils, and hence the fatty acids, is an advantage. Depending on the property desired in the final product, various oils, or mixtures thereof, can be used in synthesizing the monomers. Functional groups to increase adhesive strength with particular substrates can also be placed onto the unsaturation sites (see Section 12.5).

Previous work in this area by S. P. Bunker and R. P. Wool [4] focused on synthesizing a monomer from

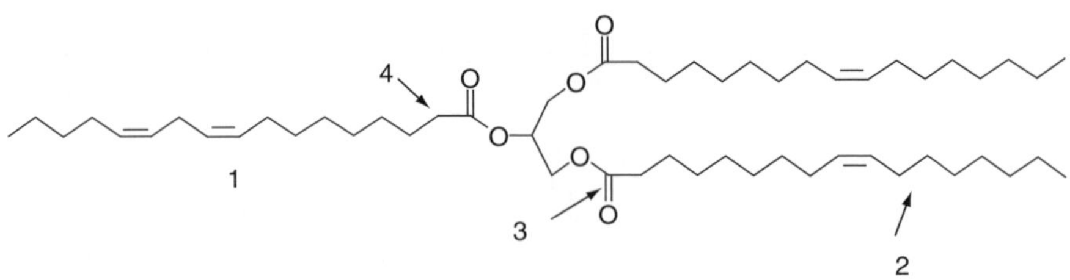

Figure 12.1 Diagram of a triglyceride molecule. A triglyceride is composed of three fatty acids connected at a glycerol center. The different functionalities are shown with the corresponding numbers: (1) double bond, (2) allylic carbons, (3) ester group, and (4) alpha carbon.

a fatty acid methyl ester that is capable of forming high-molecular-weight polymers using conventional (macro)emulsion polymerization. However, miniemulsion polymerization has several advantages over the normal emulsion technique. Miniemulsion is a good polymerization method for highly hydrophobic monomers because each droplet can be considered a minibatch reaction for the polymerization [5]. This is different from conventional emulsion polymerization, which has both monomer droplets and polymer particles. The conventional emulsion requires the transport of water-insoluble monomers from droplets to growing polymer particles, which can yield slower kinetics and, therefore, longer polymerization times [5,6].

In the next section, the mechanical properties of the renewable resource-based dispersions are compared to two petroleum-based dispersions. Specifically, the first one is a market standard for filmic label application, Acronal® A220 (www.basf.de/dispersions/), which is known for its high transparency, excellent water resistance, and outstanding adhesion to polyolefinic substrates. The second is a model dispersion of 2-ethylhexyl acrylate (2-EHA)-co-methyl methacrylate (MMA). This system was selected because the 2-EHA has a structure similar to that of the fatty acid methyl ester-based monomer. The side-by-side comparison of the properties of the petroleum-based PSA standards with the new bio-based polymers should be reassuring to most readers, especially if the economics are right and the bio-based PSA has additional benefits, such as biodegradability or being less energy intensive to produce. It may also be reassuring to some readers to know that the bio-based PSA could still be made in 2084 when global oil supplies may no longer support the existing petroleum-based materials.

12.2 Macroemulsion and Miniemulsion Polymerization

12.2.1 Macroemulsion Polymerization

Acrylated methyl oleate (AMO) was synthesized using methods reported by Bunker and Wool [4]. The monomer synthesis requires two steps. First, the unsaturated bond in oleic methyl ester (OME) must be epoxidized by a peroxy acid. The epoxidized fatty acid methyl ester is then acrylated using acrylic acid. The acrylate groups are able to participate in free-radical polymerization. A schematic of the monomer synthesis is shown in Fig. 12.2. The OME can also be

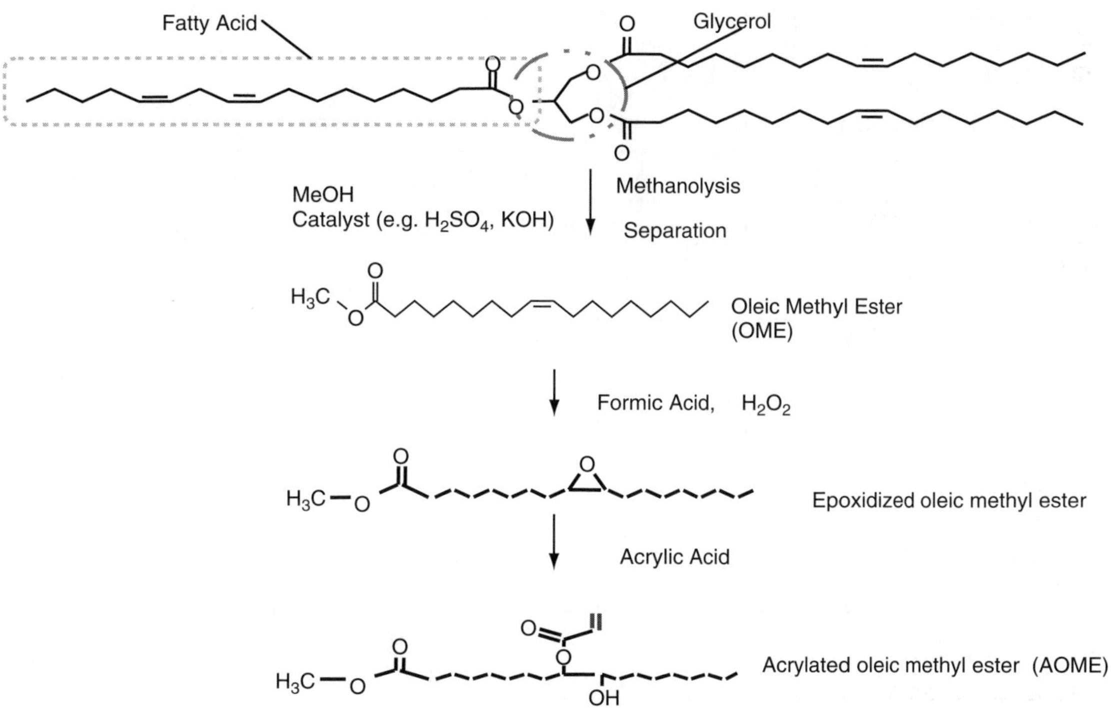

Figure 12.2 Schematic diagram of the monomer synthesis steps.

derived as a by-product from biodiesel, assuming that we have an efficient fatty acid separation process. The separation process was explored by Bunker and Wool and potentially can be done economically at large scale. This would circumvent the need for the development of specialty high-oleic oils and provide additional utilization of biodiesel plants currently being constructed in Delaware and elsewhere. From a green engineering perspective, the biodiesel is perhaps more valuable as a chemical feedstock rather than a combustible fuel feedstock and can attain this value when the current generation of internal combustion engines is replaced in the future by their fuel-cell equivalents.

The AMO monomer is polymerized using both macroemulsion (also referred to as conventional emulsion polymerization) and miniemulsion polymerization. The experimental conditions for the macroemulsion polymerization are outlined in detail by Bunker *et al.* [4]. The formulation of the macroemulsion is shown in Table 12.1 The reaction time was approximately 18 h at 70 °C.

12.2.2 Miniemulsion Polymerization

The specific formulations for each miniemulsion polymerization are listed in Table 12.1 (samples 1 to 5). Typically, the polymerizations were conducted in a 500 ml round-bottom flask equipped with a reflux condenser, nitrogen inlet, and a Teflon stirrer. First, the initiator was combined with the monomer using a magnetic stirrer, to ensure its complete dissolution in the monomer phase. After the initiator dissolution, the surfactant and water were mixed into the system using a magnetic stirrer for approximately 10 min. The miniemulsion was then prepared by continuous ultrasonification for 5 min. During sonification, the emulsion was submerged in an ice bath to maintain a temperature below 50 °C. This ensured that the initiator did not prematurely decompose. The glass reactor containing the monomer emulsion was then placed in an oil bath and heated to 85 °C for 1 h.

12.3 Polymer Characterization

The monomer conversion as a function of time is shown in Fig. 12.3. This plot tracks the intensity of the peak that corresponds to the carbon−carbon double bond of the monomer as well as the carbonyl group in the developing polymer. This chart indicates that the reactive monomer groups are completely depleted after 1 h of reaction time, which corresponds to the maximum intensity of the polymer carbonyl group. This is a significant improvement over the conventional emulsion polymerization. Figure 12.4 depicts the typical conversion of monomer to polymer in a conventional emulsion reaction as a function of time, as recorded using gravitational analysis. This study indicates that 18 h of reaction is required to achieve 90% monomer conversion.

Table 12.1 Emulsion Composition and the Properties of the Resulting Polymers

Component	Sample 1	Sample 2	Sample 3	Sample 4	Sample 5	Conventional
DDI H_2O (g)	40	40	40	40	40	30
SLS (15 wt%) (g)	1.33	1.33	1.33	1.33	1.33	2.25 g Aerosol® OT
AMO (g)	10	9	8.5	8.0	–	15
MMA (g)	–	1	1	1	1	0.5 g Acrylic Acid
BDDA (g)	–	–	0.5	1	–	–
2-EHA (g)	–	–	–	–	9	–
Vazo 67 (g)	0.1	0.05	0.05	0.05	0.05	0.05 g V-50
Emulsion droplet size (nm)	390	350	380	420	780	>1000
Particle diameter (nm)	350	350	380	400	800	>1000
K-value	26.8	42.7	*NA*	NA	NA	26.6
T_g (°C)	−49	−49	−50	−46	−58	−39

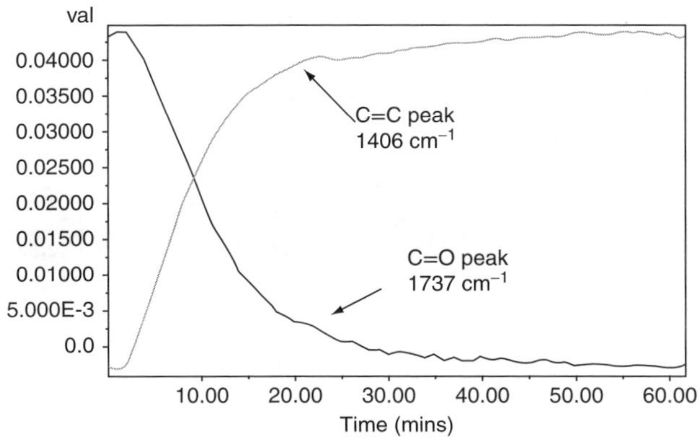

Figure 12.3 Conversion of monomer to polymer as a function of time for miniemulsion polymerization.

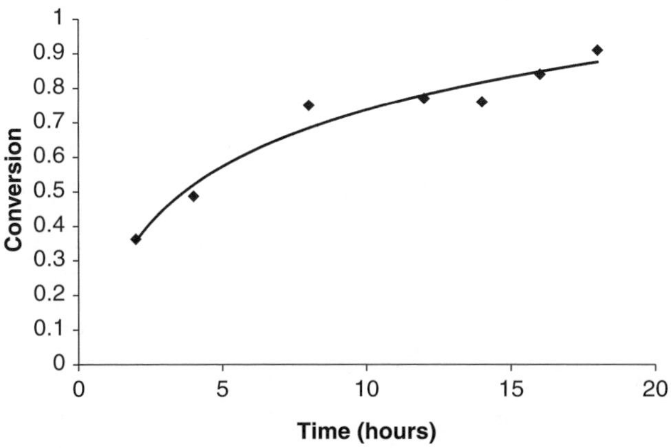

Figure 12.4 Conversion of monomer to polymer as a function of time for conventional emulsion polymerization.

Additional reaction time does not further increase this conversion.

12.4 Polymer Properties

Table 12.1 reports the particle size distribution (PSD) of the dispersion, the *K*-value, and the glass transition temperature (T_g) of the resulting polymers. Typical dispersions prepared by miniemulsion have particle sizes between 50 and 500 nm [5]. This corresponds to the PSD of samples 1–4, whereas sample 5, prepared with EHA, has a larger PSD. The effect of additional sonification time and additional surfactant on the mean particle size was examined. The additional sonification time (up to 10 min) seemed to have little effect on decreasing the particle size. An increase in surfactant levels from 2 to 5 wt% also resulted in no decrease in particle size. Surfactant concentrations above this amount were not used due to the well-known detrimental effect of excess surfactant on adhesive properties [7–9]. However, in conventional emulsion polymerization, 15 wt% of surfactant was required to form a stable dispersion [4].

The *K*-value of the AMO homopolymer (sample 1 in Table 12.1) from miniemulsion polymerization is similar to that of the polymer synthesized using conventional emulsion. Examination of the effect of comonomer (sample 2) on the *K*-value indicates that the addition of comonomer greatly increases the molecular weight of the polymer. Previous research in thermosetting polymers from acrylated-epoxidized soybean oil (AESO) found that a comonomer is required to increase the conversion of the AESO [10]. The comonomer behaves like a chain extender as well as a reactive diluent and reduces the mass transfer limitations associated with the reaction of the bulky AESO.

All of the glass transition temperatures presented here are significantly below room temperature. This is typical of PSA polymers because a majority of them are used at room temperature. The low T_g (typically 60–70 °C below room temperature) allows the polymer to flow and quickly form a bond to a substrate at room temperature. All of the T_g's are similar except for the polymer made using conventional emulsion polymerization, which has a higher T_g. This can be attributed to the large amount of surfactant used to stabilize the polymer particles. As previously stated, the polymers tested were not purified and therefore the surfactant remained in the polymer.

12.4.1 Dynamic Mechanical Analysis

The performance of a PSA is related to the viscoelastic response of the bulk adhesive. Storage and loss moduli for each polymer are shown in Fig. 12.5(a) and (b), respectively. The storage moduli of the AMO homopolymer and the AMO-co-MMA polymer are very similar. The lack of a rubbery plateau region indicates that the polymer is linear and has a molecular weight that is below or around the critical molecular weight M_c required for physical entanglements to form. Correspondingly, the polymers will have very little cohesive strength and therefore poor shear properties, but can easily wet

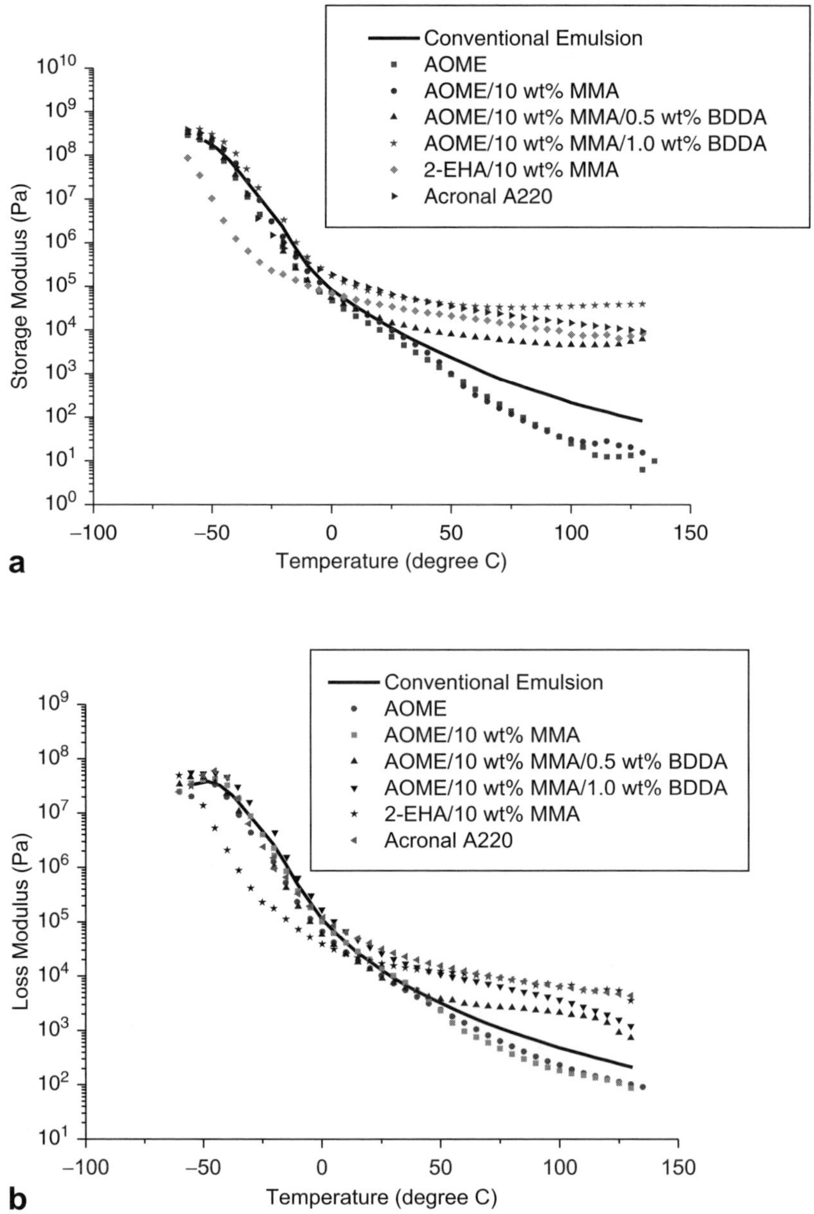

Figure 12.5 (a) The storage moduli of the polymers as a function of temperature. (b) The loss moduli of the polymers as a function of temperature.

rough surfaces, which is important for good contact and adhesion.

To improve these properties, the molecular weight needs to be greater than the critical entanglement molecular weight of about $M^* \approx 8M_c$. The most obvious method to accomplish this is to decrease the initiator concentration. However, on further analysis of the monomer, it was concluded that the monomer is the limiting factor of the molecular weight. Although ^{1}H nuclear magnetic resonance (NMR) and ^{13}C NMR of the AMO monomer indicated that the monomer contained 95% acrylate functionality, further analysis using gas chromatography showed that only 83% of the monomer had acrylate functionality. This discrepancy can be attributed to error in the NMR analysis.

Figure 12.6 shows the gas chromatography/Fourier transform infrared spectroscopy (FTIR) coupling results, which give a structure for the different side products. The monomer is composed of 83% of the AMO, 13% epoxidized methyl oleate (EMO), and 3% of the starting material, methyl oleate. These results were confirmed with the gas chromatography/mass spectrometry (GC/MS) analysis. As shown in Fig. 12.2, the EMO is the intermediate product in the monomer synthesis process. Therefore, the non-reacting part, approximately 17%, will behave as a plasticizer in the polymer and reduce the

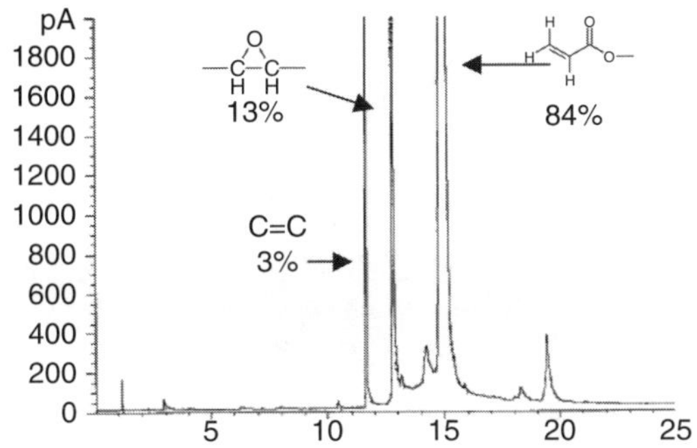

Figure 12.6 GC spectra of the monomer, AMO. The monomer is composed of 84% acrylate species, 16% epoxidized species, and 3% of the original OME.

mechanical and adhesive properties. Also, the non-reacting components limit the molecular weight of the polymer. Techniques to decrease the amount of EMO were explored. However, even with this limitation, the polymers synthesized with the current monomer show acceptable PSA properties, as demonstrated later.

Above the glass transition region, the storage modulus of 2-EHA-co-MMA (sample 5) exhibits a rubbery plateau with G' almost independent of temperature but at high enough temperatures viscous flow is dominating (G' falls below G''). This indicates that this is a linear but physically entangled polymer. From G' in the rubbery zone, the mean molecular weight between entanglements $M_e \approx M_c/2$ is calculated to be $M_e = 60$ kg/mol. This value is in good agreement with the literature data [11]. Such a high-molecular-weight physically entangled polymer is ideal for adhesive applications. Such polymers have long entangled chains that will impart cohesive strength to the system, but at the same time, the polymer chains are still mobile enough to form a good adhesive bond. These effects should show up in the application test results such as tack, shear, and peel.

The storage modulus of the polymer synthesized using conventional emulsion polymerization offers a slight improvement, indicated by the higher modulus values. Although the *K*-value of the conventional polymer and the miniemulsion AMO homopolymer are similar, the differing rheological properties may be explained by the different reaction times. It is well known that as polymerization reaction times increase, branching will increase due to chain transfer to polymer [12]. It was determined previously that the branching in the conventional emulsion is caused by the hydrogen abstraction of the tertiary backbone C–H bond [4]. Therefore, this chain transfer and resulting branched structure will have an effect on the dimethacrylate (DMA) properties as indicated by an increase in the storage modulus. Copolymerization of the stiff acrylic acid groups is another parameter expected to increase the storage modulus.

The addition of the cross-linking comonomer, 1,4-butanediol diacrylate (BDDA), to the miniemulsion system increased the modulus significantly. In fact, the resulting modulus of the AMO-MMA-co-BDDA (both 0.5 and 1 wt%) is comparable to the commercial Acronal® A220. Comparing the modulus profiles for these samples with the uncross-linked counterparts indicates that these polymers are indeed chemically cross-linked; G' is always higher than G'' even at high temperatures where the uncross-linked materials start to flow. Nevertheless, the plateau modulus of these slightly cross-linked polymers still fulfills the Dahlquist criterion, which states that polymers to be used for PSA applications should have a plateau modulus below 0.3 MPa [13].

12.4.2 Tack

The adhesion performance of PSAs is determined by three main properties: tack, peel strength, and shear resistance. Tack is a key property of PSAs and is defined as the ability of an adhesive to form a bond of measurable strength to another material under conditions of low contact pressure and short contact time [14]. Figures 12.7 and 12.8 are bar charts of the tack properties of the polymers tested with a polyethylene and stainless steel probe, respectively. In both cases, the conventional emulsion polymer displayed the lowest tack energy. This is attributed to the large amount of surfactant remaining in the polymer. It is well known that residual surfactant in the polymer migrates to the polymer–air surface, which decreases the tack and adhesive properties [14].

For both testing probes, the miniemulsion polymers display increasing tack values with the addition of comonomer. Against PE, the AMO-co-MMA copolymers show tack values almost as high as the petroleum-based polymers (2-EHA-co-MMA and Acronal® A220). Against stainless steel, the highest tack is observed for the sample with 0.5% BDDA. This may be due to the two carbonyl groups on the BDDA monomer that increase its attraction to a metal surface. However, there is a limit on the increase of tack with the increase in concentration of BDDA. The

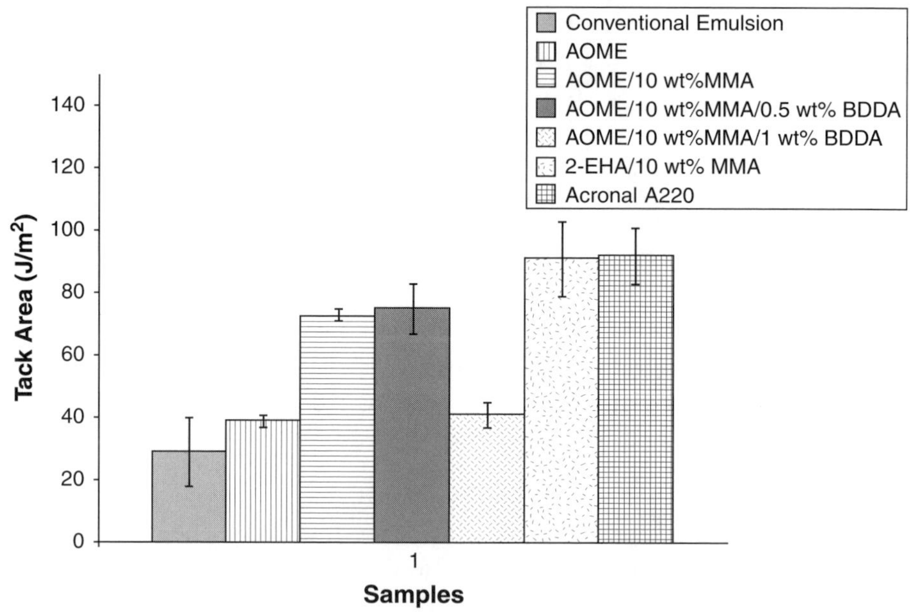

Figure 12.7 Tack results of polymers to a polyethylene probe.

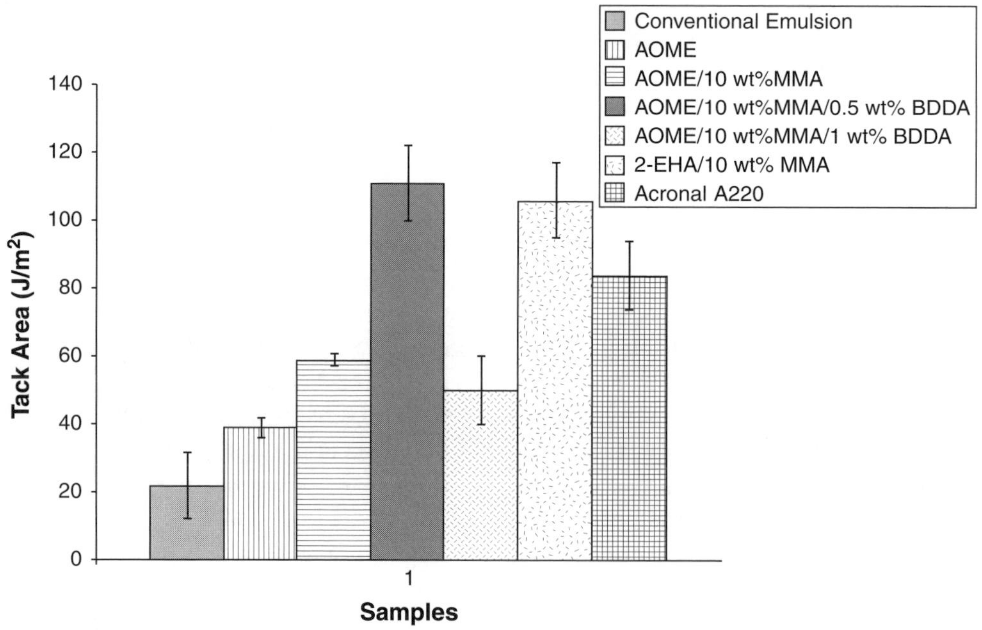

Figure 12.8 Tack results of polymers to a stainless steel probe.

polymer with 1 wt% BDDA showed a significant decrease in tack value. Again, this is attributed to the balance between a polymer that has cohesive strength while at the same time flows to make an adhesive bond quickly. The 1 wt% BDDA polymer has a tight network as indicated by the high storage modulus, providing good cohesive strength, but the flow is probably so strongly restricted that it cannot form a tight contact to the probe within the short dwell time. Overall, the AMO-co-0.5 wt% BDDA polymer shows tack results comparable to both the 2-EHA-co-MMA polymer and Acronal® A220.

12.4.3 Peel and Shear Tests

Figure 12.9(a) and (b) show the results of the peel and shear time-to-failure tests, respectively. The peel results of the conventional polymer, AMO

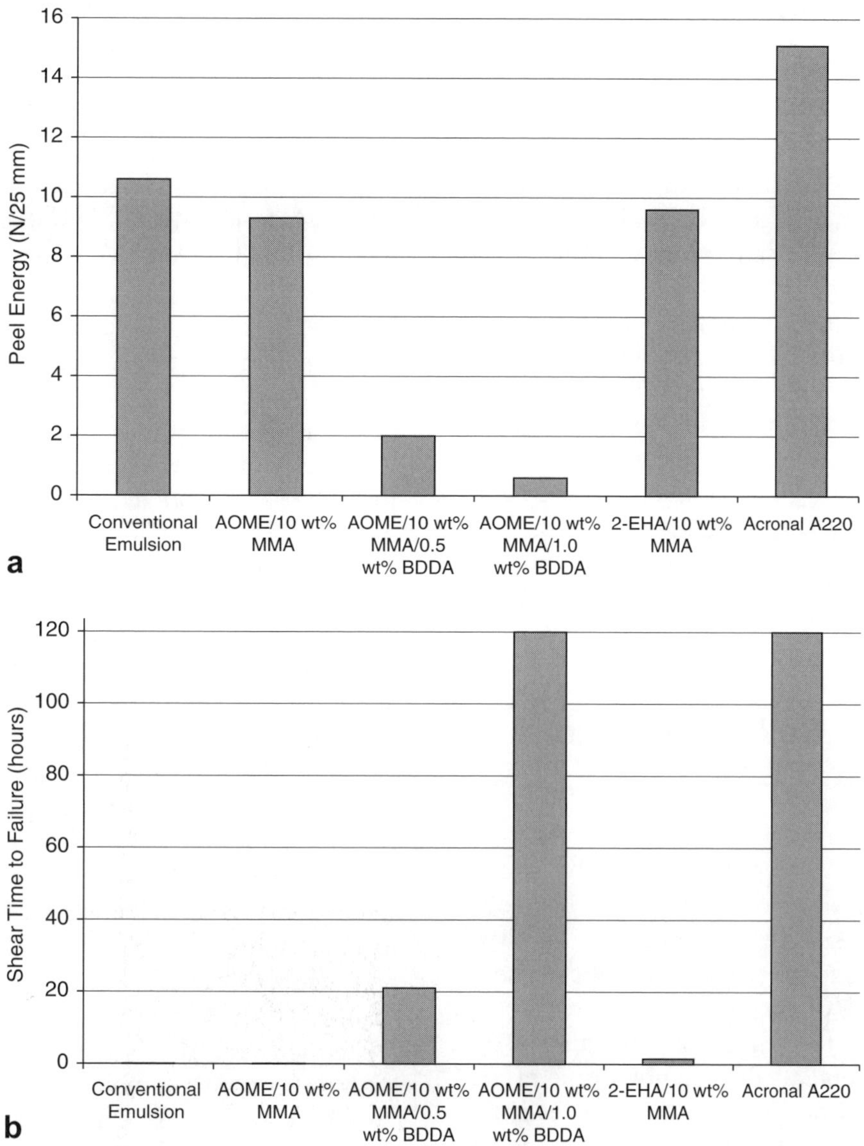

Figure 12.9 (a) 180° peel test results. The peel energy is in units of N/25 mm. (b) Shear time to failure, recorded in hours.

homopolymer, and AMO-co-MMA, and the 2-EHA-co-MMA are very comparable. The peel value for the conventional polymer is somewhat higher than for the other linear polymers, presumably due to the acrylic acid copolymerized in the conventional polymer. All of the linear polymers offer very little shear resistance, as observed with a time-to-failure on the order of minutes. These values are so low that they do not appear on Fig. 12.9(b). This is even true for the 2-EHA-co-MMA polymer, which clearly had a physically entangled network based on the storage modulus, although the time to failure is significantly longer than for the AMO linear polymers.

Introducing chemical cross-links is one strategy to improve the shear strength. Consequently, the chemically cross-linked AMO-copolymers did show a large increase in shear holding time, increasing from minutes to hours. However, with the increased amount of chemical cross-linking, the peel values decreased drastically. On the other hand, the 2-EHA-co-MMA and Acronal® A220 show a good balance between tack, peel, and shear properties, which is contributed to the high molecular weight of these systems. This clearly shows the future direction into which the development of the AMO polymers has to go in order to yield competitive PSA materials [15–17].

12.4.4 Conclusions on PSA Synthesis

Significant improvement of the polymerization of AMO latex was achieved by using miniemulsion polymerization. The reaction time was decreased from 18 h down to 1 h in addition to the surfactant concentration being drastically reduced from 15 to 2 wt%. The resulting PSA polymer has shown physical properties that are comparable to petroleum-based polymers. Most importantly, the polymers derived from a renewable resource display typical PSA properties suited to high-volume applications. Also, these materials, when exposed to soil, completely disappeared in a few months. They are also biocompatible, as determined from cytotoxicity tests and were found to support the growth of human tissue, which creates new opportunities for biomedical applications[1].

12.5 Polymer–Solid Adhesion Modification of PSAs

It is well known in the adhesion field that the addition of functional groups to a polymer backbone can enhance the adhesive potential of the polymer in contact with solid surfaces [1,18]. The adhesion potential was related to the surface energetics, where the number and type of chemical groups are important in predicting the adhesion potential [19]. However, a fundamental understanding of the effect of functional groups on adhesive behavior had not previously been achieved. Recent work on model polymer–substrate systems indicates that the fracture energy of polymer–solid interfaces is not a monotonic function of the surface energetics, as previously expected [4]. Lee and Wool modeled the behavior of the peel energy as a function of the number of functional groups (i.e., sticker groups) in the polymer and receptor groups on the substrate, as well as the interaction between these two groups [18].

Essentially, this model identifies the existence of an optimal concentration of functional sticker groups for reaching a maximum adhesive strength. The polymer–solid interface is a competition between cohesive failure of the polymer–polymer interface and adhesive failure of the adsorbed polymer–solid interfaces. The polymer–polymer interface involves the interpenetration of the adsorbed and free chains in a layer adjacent to the surface, whereas the adsorbed polymer–solid interface considers the adhesive strength of the adsorbed chains in contact with the solid surface. Thus, as the concentration of functional groups in a polymer chain increases and becomes strongly adsorbed on the surface, its conformation flattens out onto the surface, decreasing its ability to interpenetrate with the free chains in the bulk of the polymer.

Cohesive failure results when the adhesive forces required to debond the adhered chains exceed the cohesive forces required to break the poorly interpenetrated polymer–polymer interface. In contrast, adhesive failure results when there are few functional–receptor group interactions and the chain is highly interpenetrated in the polymer bulk. In PSA applications, the substrate usually cannot be controlled and therefore the effect of the receptor groups cannot be taken into account. However, the PSA polymer can be designed to incorporate functional groups that will result in optimal adhesive behavior, i.e., high peel energy and adhesive failure, for a given type of substrate. In this section, we examine the effect of acid functional groups on the adhesive performance on a metal substrate, i.e., acid–base interactions.

Figure 12.10 shows the increase in acid number with the increase in maleic acid added in the initial monomer mixture. This indicates that the maleic acid is being incorporated into the polymer backbone. However, the magnitude of the acid number is higher than expected for a given quantity of acid initially charged to the reaction vessel. Partial hydrolysis of the MMA and AOME monomers could account for this higher than expected value. However, the increasing trend confirms the incorporation of MA into the polymer.

12.5.1 Viscoelastic Properties

The storage modulus as a function of temperature at six different maleic acid concentrations is shown in Fig. 12.11. These are compared to the storage modulus of a miniemulsion polymer that contains no maleic acid. The storage moduli of the AOME-co-MMA-co-MA polymers are slightly higher than that of the AOME-co-MMA polymer. This is attributed

[1] C.M. Klapperich, C.L. Noack, J.D. Kaufman, *et al.* A novel biocompatible adhesive incorporating plant-derived monomers, J. Biomed. Mater. Res. A 91A (2) (Nov 2009) 378–384.

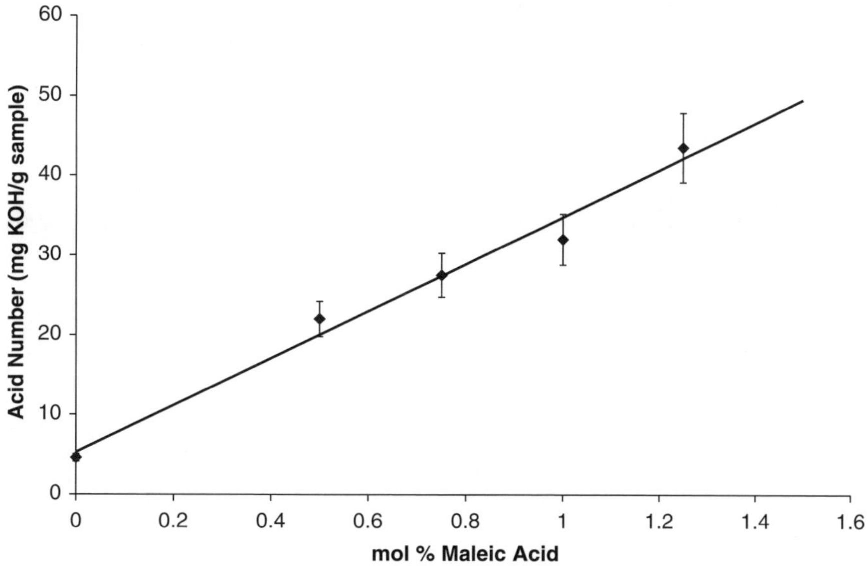

Figure 12.10 The experimentally recorded acid number as a function of maleic acid added to the monomer mixture.

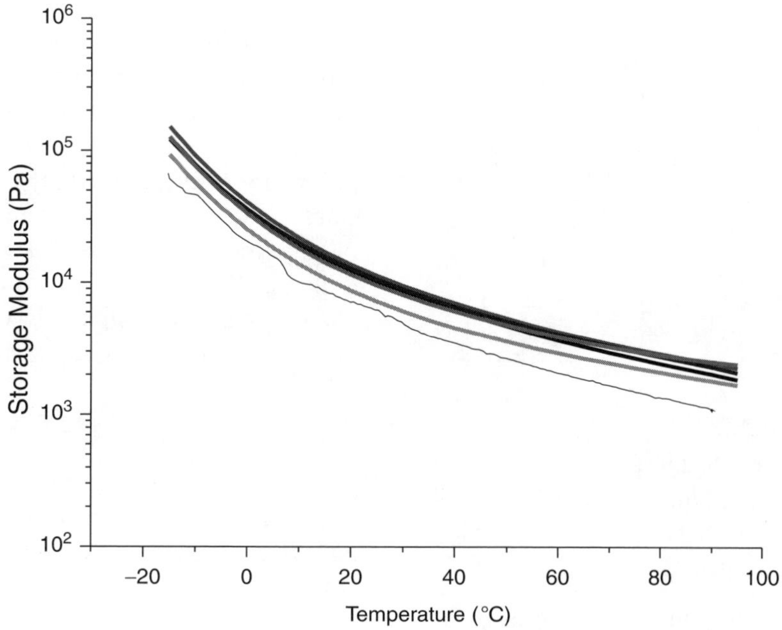

Figure 12.11 The storage modulus of the polymers that contain maleic acid as a function of temperature. The thin line represents a comparable polymer that contains no maleic acid.

to the stiff maleic acid group that is incorporated into the polymer chain. An example of the storage and loss modulus of an AOME-co-MMA-co-MA polymer is shown in Fig. 12.12. The G' is greater than the G'' at temperatures >0 °C, which indicates that the elastic behavior of the polymer dominates the properties and that physical entanglements are present. Similar behavior is observed for all of the polymers synthesized with maleic acid.

12.5.2 Adhesion Properties

The increase in peel energy with the increase in maleic acid content is shown in Fig. 12.13. The peel

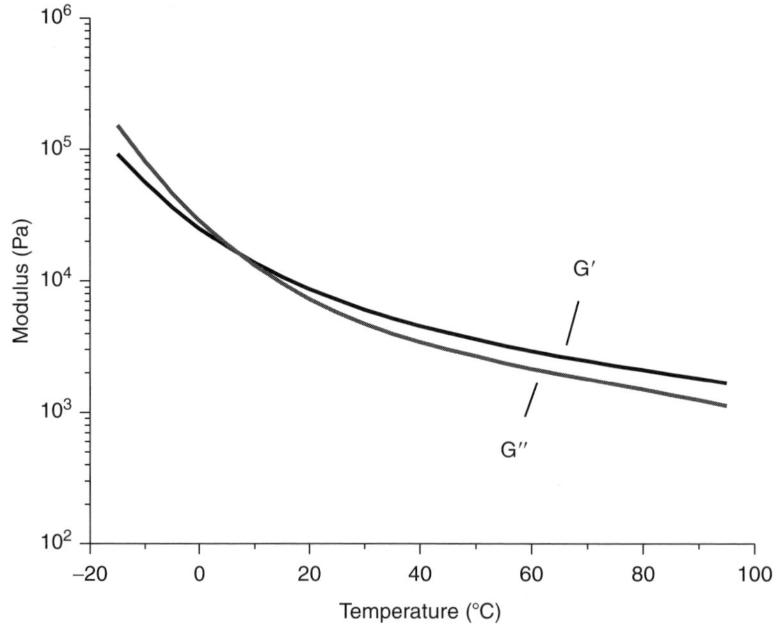

Figure 12.12 The storage and loss modulus of an AOME-co-MMA-co-MA polymer.

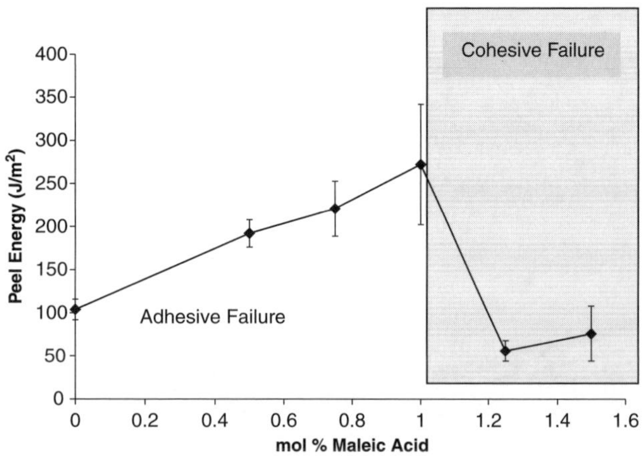

Figure 12.13 The peel energy as a function of maleic acid content. An optimum concentration of approximately 1 mol% maleic acid results in high peel energy and adhesive failure.

energy increases to an optimum concentration of maleic acid, approximately 1 mol%, with adhesive failure. An increase in maleic acid beyond this quantity results in lower peel energy as well as in cohesive failure. This behavior is similar to previously reported results using carboxylated poly(butadiene) [18]. At 1 mol%, small amounts of cohesive failure "patches" were observed. Specifically, randomly located small quantities of adhesive were observed on the stainless steel substrate. Therefore, the optimum level of maleic acid is slightly below 1 mol%.

An entanglement sink probability (ESP) model motivated by vector percolation explains the non-monotonic influences of functional group concentration, $\Phi_{x'}$, receptor concentration, $\Phi_{y'}$, and their interaction strength, χ, on the adhesion strength of the polymer—solid interface [18]. The ESP model quantifies the degree of interaction between adsorbed and neighboring chains based on the adsorbed chain domain. Specifically, the adsorbed chain domain changes thermodynamically with the energy of interaction, r, at the interface, as shown by

$$r = \chi\Phi_x\Phi_y \tag{12.1}$$

This parameter can be related to both the adhesive potential (G_A) and the cohesive potential (G_C). The adhesive strength potential behaves linearly as follows:

$$G_A \sim r = \chi\Phi_x\Phi_y \tag{12.2}$$

The cohesive strength between the adsorbed and neighboring chains is modeled as:

$$G_C \sim r^{-1} = (\chi\Phi_x\Phi_y)^{-1.0} \tag{12.3}$$

This model predicts maximum adhesion strength when the fracture stresses for the cohesive and adhesive failure are equal at an optimal value of $r^* = (\chi\Phi_x\Phi_y)^*$. Therefore, for a given χ value, there exist optimal values, Φ_x^*, and Φ_y^*, above or below which the fracture energy will not be optimized. In this study, the χ and the Φ_y parameters are considered to be constant. Therefore, the adhesive strength is solely linearly dependent on the concentration of functional groups, Φ_x, in the polymer chain, up to the optimal concentration of approximately 1 mol%. Beyond this, a sharp decrease in the peel energy is observed and the mode of failure becomes cohesive. In the cohesive region, the peel energy has an inverse relationship with the functional group concentration. However, the decrease in the peel energy is larger than that predicted by the preceding model.

The optimal sticker group value φ_x* can be deduced from the percolation model of entanglements [20] shown in Fig. 12.14, where the critical entanglement molecular weight M_c is determined when the chain is of sufficient length to cross an arbitrary plane three times. If the number of chains per unit area is Σ and the cross-section of a chain segment in the bulk is a, then the critical condition is met when $3a\Sigma = 1$. For the polymer—solid interface, we place a sticker group on the solid and begin a random walk and inquire using "Coin-Toss" statistics when that chain will return to the surface. In an entangled melt, a polymer chain segment of molecular weight M_c crosses any arbitrary plane three times, which is about half the number of times it returns to the plane (Fig. 12.14). Thus, the entanglement molecular weight M_e is about half that of the critical entanglement molecular weight

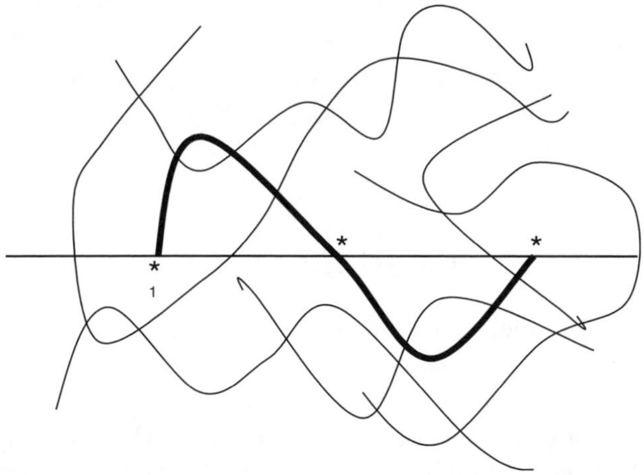

Figure 12.14 Entanglements in a polymer melt. The bold line represents the critical molecular weight (M_c) required to obtain a connected network. The asterisks (*) represent the points along the polymer chain that cross an arbitrary plane.

M_c. We find that this occurs after about $N_c \approx 30$ random walk steps and the entanglement molecular weight M_e is given by [20]:

$$M_e \approx 31 C_\infty M_j \quad (3/1 \text{ helices}) \tag{12.4}$$

$$M_e \approx 22 C_\infty M_j \quad (2/1 \text{ helices}) \tag{12.5}$$

in which C_∞ is the characteristic ratio of the random walk and M_j is the molecular weight per bond, e.g., $M_j = 14$ g/mol for polyethylene and 52 g/mol for polystyrene. The front factor of 31 is used for 3/1 helices (e.g., PS, PMMA, PVC, PVA, etc.), and is 22 for 2/1 helices (planar zig-zag structure, PE, polyamides, alkane-esters, etc.). This value depends on the local conformational details of the chain cross-section a [20]. Thus, to adhere an entanglement network to a solid, we place a sticker group at every "return touch point" of the random walk on the substrate, or on average, about two groups per M_e value, such that the optimal mole fraction of sticker groups is given by:

$$\phi_x^* = 2jM_j/M_e \tag{12.6}$$

in which $M_j = M_o/j$, where j is the number of backbone steps per monomer of molecular weight M_o. Substituting for M_c in the latter equation, and using $j = 2$, we obtain the optimal mole fraction of sticker groups as:

$$\phi_x^* \approx 0.129/C_\infty \quad (3/1 \text{ helices}) \tag{12.7}$$

$$\phi_x^* \approx 0.129/C_\infty \quad (2/1 \text{ helices}) \tag{12.8}$$

The characteristic ratio C_∞ is heavily dependent on the bond molecular weight M_j. It has been shown that for a homologous series of polymers of the 3/1 helical type $-CH_2-CHX-$, where X is the variable side group that C_∞ depends on M_j as:

$$C_\infty = 1.36\, M_j^{1/2} \tag{12.9}$$

Substituting the latter relation in Eqn (12.4), one obtains a very simple approximation for M_e and M_c for 3/1 type polymers as:

$$M_e \approx 42\, M_j^{3/2} \text{ g/mol} \tag{12.10}$$

$$M_e \approx 84\, M_j^{3/2} \text{ g/mol} \tag{12.11}$$

For example, with polystyrene, $M_j = 52$ g/mol and we predict that $M_c \approx 31{,}500$ g/mol, which is in accord with experimental values with $M_c \approx 30{,}000$ g/mol [20]. Since many polymers have C_∞ values in the range of 5−20 and $j = 2$ for vinyl polymers, it appears that there is an optimal sticker group number of $\varphi_x^* \approx$ 1−2%, which is a surprisingly small number to maximize adhesion.

At 1 mol% MA, a very small amount of cohesive failure "patches" were observed. Specifically, randomly located small quantities of adhesive were observed on the stainless steel substrate. Therefore, the optimum level of maleic acid is slightly below 1 mol%. This result is in excellent agreement with the predicted value of $\varphi_X^* \approx 1.0\%$ as given by Eqn (12.7). Because this polymer has three monomers, a weighted average monomer molecular weight M_o is calculated as follows[2]:

$$\overline{M_o} = x_1 M_{o1} + x_2 M_{o2} + x_3 M_{o3} \qquad (12.12)$$

where x_x is the weight fraction of the corresponding monomer. Based on the extreme cases of the terpolymer (0.5 mol% and 1.5 mol% maleic acid, the quantity of AOME and MMA remained fixed), then $\overline{M_o}$ is 356 g/mol and $M_j = 178$ g/mol. Using Eqn (12.9), we obtain $C_\infty \approx 18$. The M_e value can be calculated using the entanglement model (Eqn (12.7) for a 3/1 helix using $C_\infty = 18$) as $M_e = \approx 100{,}000$ g/mol. Alternatively, using $M_e = 42\ M_j^{3/2}$ (Eqn (12.10)), we also obtain $M_e \approx 100{,}000$ g/mol, which is in excellent agreement with the M_e value obtained from the dynamical mechanical analysis. Subsequently, we obtain $\varphi_x \approx 0.7$ % and this estimate is slightly less than the experimentally determined value of approximately 1%. An exact agreement with the 1% value would be obtained if $C_\infty = 13$ instead of $C_\infty =$ 18 used in the calculation. The characteristic ratio for this polymer was not determined independently but these values are reasonable when compared to similar polymers.

Thus in summary, adhesion at polymer−solid interfaces was explored for new bio-based PSAs in terms of sticker groups φ_X on the polymer phase, receptor groups φ_Y on the solid surface, and the bond strength of the sticker-receptor X−Y acid−base interaction, χ. The polymer−solid interface restructuring models of Gong and Lee *et al.* were extended with new percolation models of entanglements and interface strength to determine the optimal sticker group concentration φ_x^*. For the general case where φ_Y and χ are constant, it is predicted that when $\varphi_X < \varphi_x^*$, that the critical peel energy behaves as $G_{1c} \sim \varphi_X/\varphi^*_X$ and the locus of failure is adhesive between the polymer and the solid. However, when $\varphi_X > \varphi_x^*$, failure occurs cohesively in a polymer−polymer interface adjacent to the solid and the strength decreases as $G_{1c} \sim \varphi_X^*/\varphi_X$. The switch from adhesive to cohesive failure can be understood in terms of the changes in the chain conformations of the adhered chains and their decreasing interpenetration X_i with the bulk chains, via $X_i \sim 1/r$, where $r = \chi\varphi_X\varphi_Y$. The optimal value of φ_X which maximizes the adhesion and determines the mode of failure is given by $\varphi_X{}^* \approx 0.129/C_\infty$. For typical values of the characteristic ratio C_∞ in the range 7−20, $\varphi_x^* \approx 1\%$ mole fraction, corresponding to about two sticker groups per entanglement molecular weight M_e. This result was demonstrated for a bio-based PSA synthesized from an acrylated high-oleic fatty acid, which was copolymerized with 1% mol fraction maleic anhydride as the sticker group. The observed behavior is counterintuitive to the current wisdom for the effect of acid-based interactions on adhesion, where the strength is expected to increase with the number of X−Y contacts. The surprisingly low value of $\varphi_x^* \approx 1\%$ sticker groups, which maximizes the adhesion strength, can now be readily calculated using the percolation model of entanglements and fracture.

Using this fundamental understanding of the adhesion properties, the peel energy and mode of failure can be designed and controlled. This work could be extended by the incorporation of several different functional groups to enhance the adhesive properties of the polymer toward a variety of substrates.

12.6 Bio-Based Elastomers

Elastomers are widely used in automotive components, conveyor belting, transport, construction, footwear, and so on. The global market for elastomers and associated products is $40 billion [21]. A small amount of elastomers is derived from renewable resources such as rubber trees chiefly on plantations and small holdings in Malaysia, Indonesia, and other Asian countries [22]. Even though most of the elastomers used to be derived from the rubber tree, now most of the elastomers are

[2] R.P. Wool, S.P. Bunker, J. Adhes. 83 (10) (2007) 907−926.

synthesized from petroleum oil. Elastomers are soft (stiffness $E \sim 1$ MPa), highly extensible (~400%), and elastic. These unique properties come from the lightly cross-linked polymer network structure. The polymer chain in the network must be linear so that it is flexible enough to deform in any direction. In the relaxed state, the molecules between cross-links are coiled up in a random fashion. The cross-links limit the polymer chain extension but at the same time enable the network to recover to its original shape due to the entropic recovery force generated during the deformation. The properties of an elastomer are controlled by the nature of the cross-linked network. Hardness, modulus, tensile strength, elastic recovery, etc. are all influenced by the cross-link density v. In the most lightly cross-linked rubbers, v is in the range of 10^{-5} to 10^{-4} mol cm^{-3} [23]. Elastomers can be further reinforced by particulate fillers, such as carbon black and nanoclays, which causes an increase in ultimate properties, such as tear and tensile strength, abrasion resistance, and modulus of elasticity.

Some interpenetrating network (IPN) systems were developed for the application of elastomers from plant oils. Athawale and Kolekar [24,25] modified castor oil with linseed oil and tung oil and prepared urethanes and their IPNs with poly(methyl) methacrylate. In addition to this, Athawale and Raut developed new elastomers based on uralkyd resin blended with polystyrene [26], poly(methyl) methacrylate [27], or poly (butyl) methacrylate [28,29].

Triglycerides can be easily broken down into fatty acids such as OME, as shown earlier in Fig. 12.2. After epoxidation and acrylation, we can obtain acrylated oleic methyl ester (AOME). The double bond in the acrylate group is reactive and can easily undergo free-radical polymerization, as observed for the PSA in the previous section. The long fatty acid chain will give steric hindrance to the rotation of the main chain. However, at the same time, the high flexibility of the long fatty acid also allows them to work as plasticizers and makes the polymer more flexible. Therefore, AOME was considered by Zhu and Wool [30] to be a promising starting material for the development of new bio-based elastomers. Several molecular design strategies were developed using AOME as the starting monomer. The cross-link density was controlled by varying the cross-link ratio, cross-link agent, and reaction conditions. Various mechanical and thermal properties can be achieved by changing the network structure, and the material can be utilized in different applications.

12.6.1 Elastomer Molecular Design

For a rubber material in tri-axial tension with orthogonal deformation ratios λ_1, λ_2, and λ_3, the finite deformation strain energy density function $U(\lambda_1, \lambda_2, \lambda_3)$ is given by:

$$U(\lambda_1,\lambda_2,\lambda_3) - C_1(\lambda_1^2+\lambda_2^2+\lambda_3^2-3) + C_2(\lambda_1^{-2}+\lambda_2^{-2}+\lambda_3^{-2}-3) \quad (12.13)$$

in which C_1 and C_2 are constants. In the ideal Flory theory of rubber elasticity, $C_2 = 0$ and $C_1 = \mathrm{v_eRT}/2$, which is related to the tensile modulus for a rubber network with v_e cross-links per unit volume at temperature T and gas constant R. In this case, $U(\lambda_1, \lambda_2, \lambda_3) = C_1(\lambda_1^2 + \lambda_2^2 + \lambda_3^2 - 3)$, and for uniaxial deformation, one obtains the familiar expression:

$$U(\lambda) = C_1(\lambda^2 + 2^{\lambda-1} - 3) \quad \text{(Plane Stress)} \quad (12.14)$$

$$U(\lambda) \approx C_1(\lambda^2 + \lambda^{-2} - 2) \quad \text{(Plane Strain)} \quad (12.15)$$

In plane stress, $\lambda_2 = \lambda_3 = \lambda_1^{-1/2}$, while in plane strain, $\lambda_2 = \lambda_1^{-1/2}$ and $\lambda_3 = 1$ (no lateral contraction). In the ideal theory of rubber elasticity, the tensile modulus $E = 2C_1$. Consider an ideal rubber whose strain energy function $U(\lambda)$ is given by Eqn (12.14). Realistically, this function is defined for small λ values since strain hardening near $\lambda_{sh} \approx 4$ is not considered. However, we can inquire as to the requirements to fracture a rubber material at $\lambda < \lambda_{sh}$. The constant C_1 is related to the modulus E, via $C_1 = E/2$ and E is determined from Flory's entropy elasticity theory as:

$$E = v_e RT \quad (12.16)$$

where v_e is the cross-link density of the rubber. The strain energy density function for simple uniaxial deformation becomes:

$$U(\lambda) = 0.5\, v_e RT\, (\lambda^2 + 2\lambda^{-1} - 3) \quad (12.17)$$

The percolation dissipation fracture energy is:

$$U_f = 0.5\, v_e RT(\lambda^2 + 2\lambda^{-1} - 3) \quad (12.18)$$

Equating $U(\lambda)$ and U_f, we obtain the critical λ_c value as:

$$\lambda_c^2 + 2\lambda_c^{-1} = 3 + 2D_o[p - p_c]/RT \quad (12.19)$$

Notice that the entanglement density v_e canceled out, which is consistent with critical draw ratios being independent of M_e. An approximate solution for λ_c in Eqn (12.19) involves neglecting the $2/\lambda$ term compared to the λ^2 term such that:

$$\lambda_c \approx \{3 + 2D_o[p = p_c]/RT\}^{1/2} \quad (12.20)$$

For example, if we have a perfect net, $p = 1$, $p_c \approx ½$, $D_o = 80$ kcal/mol, $R = 1.986$ cal/mol, and $T = 300$ °K, which gives $2D_o\ [p - p_c]/RT = 134$, such that the critical value in Eqn (12.20) is $\lambda_c = 11.6$. Thus, since the critical value is much greater than the strain hardening value of 4, we can see that a typical ideal rubber material will not break without first strain hardening. Examining Eqns (12.19) and (12.20), it can be seen that to break an ideal rubber without strain hardening at $\lambda_c < 4$, one of the following events needs to occur:

(a) p decreases to p_c by either defect creation during synthesis, bond rupture during fatigue, biodegradation, or photodegradation. This is how a pencil eraser works with fragile rubber networks made with triglycerides wherein the network contains many defects, such that $p \sim p_c$. Also, the lifetime of a rubber subject to fatigue or oxidative aging with bond cleavage would be influenced. To achieve values of $\lambda_c \approx 4$, using Eqn (12.20), the term $[p - p_c]$ should decrease from its initial value of 0.5 to about $[p - p_c] = 0.05$.

(b) The bond energy D_o decreases, and this can occur with labile bonds, such as weak divalent cation linkages. This could be a method of designing recyclable elastomers, which would become thermoplastic at a certain temperature. At room temperature, the bond energy would have to decrease by a factor of about 10 to $D_o = 8$ kcal/mol in order to achieve the critical draw ratio of $\lambda_c \approx 4$.

(c) Increase the modulus E. The rubber modulus $E \sim T$ could be increased such that when $\lambda_c = 4$, Eqn (12.20) would be satisfied for the perfect rubber net with $T \approx 3000$ K, which corresponds to a 10-fold increase in E above that of room temperature. Of course, these temperatures are unrealistic, but the 10-fold increase in modulus is readily obtained during strain hardening, and that is the typical mechanism of fracture for rubber. We have all experienced this effect when attempting to break a rubber band. This phenomena is not unique to rubber, but occurs with all highly entangled polymers, and explains why crazes need to strain harden at local temperatures near their T_g during fracture of glassy polymers. The crazes typically do not fracture at their strain hardening draw ratio ($\lambda \approx 4$) but continue to bear increasing loads in the deformation zone at the crack tip before they fracture.

In evaluating the fracture of rubber, we have given an example for an ideal rubber using the strain energy term $U(\lambda)$. However, many other strain energy density functions can be used in a similar manner, and with different constraints on the λ values, as dictated by the sample geometry and tri-axial loading conditions, including compression ($\lambda < 1$). The stress resulting from fracture is determined from $U(\lambda)$ by the usual first derivative, $\sigma = dU(\lambda)/\delta\lambda$. Using Eqns (12.16) and (12.17) for the ideal rubber, we obtain the uniaxial fracture stress as [31]:

$$\sigma^* = E(\lambda_c - 1/\lambda_c) \quad (12.21)$$

in which the critical value of λ_c is determined by Eqn (12.20). The maximum fracture stress that one can obtain at $\lambda_c \approx 4$ is $\sigma^* = 3.75\, E$, before strain hardening occurs and the elasticity mechanism changes from entropic to enthalpic, where the internal energy of the molecules changes through bond and valence angle deformation. Because the density ρ of rubber is around 10^6 g/m^3, the cross-link density required to reach the strain-hardened area at $\lambda_c \approx 4$ is about 17 mol/m^3. At room temperature (300 K), this would provide a rubber modulus $G = 43$ kPa.

12.6.2 Elastomer Synthesis and Properties

To form an elastic network, cross-links were introduced into the system using ethylene glycol dimethacrylate (EGDMA) as the cross-linker [28, 29]. It has two double bonds and is miscible in AOME.

Table 12.2 Effects of Cross-Link Density on Mechanical Properties

Sample	Tensile Strength (MPa)	Elongation at Break (%)	Gel Fraction (%)	Cross-link Density (mol/m^3)
10%MMA1% EGDMA	0.26	85.51	86.11	199.71
40%MMA1% EGDMA	0.35	143.28	77.54	130.73
5%MMA1% EGDMA	0.13	84.85	79.12	112.82
5%MMA0.5% EGDMA	0.07	223.16	52.35	15.33

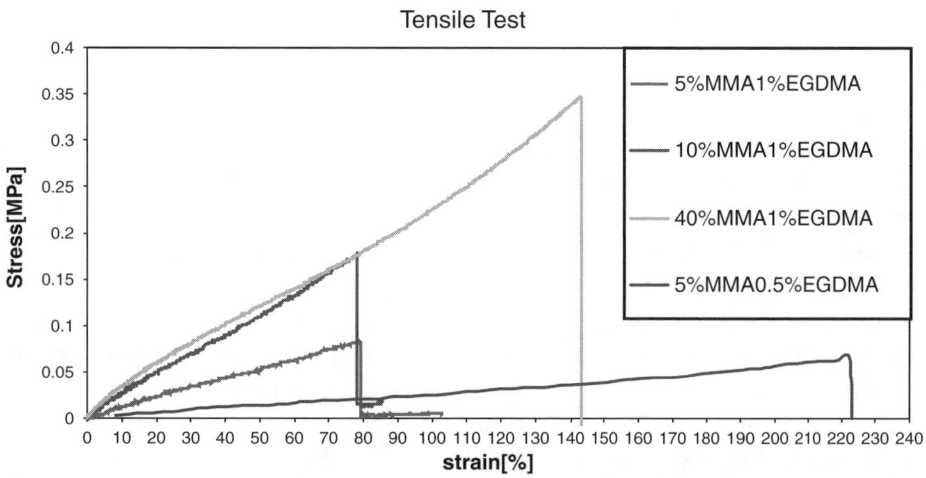

Figure 12.15 Stress–strain behavior of AOME elastomers with varying MMA and cross-linker.

By varying the amount of EGDMA, different cross-link densities can be achieved, as shown in Table 12.2 From the AOME structure, we can see that there are many polar groups in the chain, which contribute to making the final polymer tacky. In addition, although the long branches increase the flexibility of the chain, they also generate a weak network structure. To optimize the mechanical properties of the network and reduce its initial tack, we introduce an MMA comonomer. The methyl group increases the rigidity of the chain and MMA has a T_g of 102.8 °C, which can reduce the tackiness of the material by increasing the T_g of the copolymer[3]. At the same time, the polymer chain is extended due to the addition of MMA. The stress–strain relationship for different AOME elastomer samples is shown in Fig. 12.15.

Compared to the results of the bulk polymer sample, the introduction of MMA increases the mechanical properties significantly. With 5% MMA, the elongation at break increases to 85% although the tensile strength decreases slightly. Increasing the MMA ratio to 10% does not change the elongation much but the tensile strength increases. When the MMA ratio goes up to 40%, the tensile strength and elongation at break increases 169% and 68%, respectively, compared with the sample with 5% MMA. The cross-link density is mainly controlled by the concentration of the EGDMA in this method. The tensile tests shows that samples with 0.5% EGDMA have a 163% increase in elongation but 26% decrease in tensile strength. The elongation at break reaches 223%.

[3] R. P. Wool, J. Polym. Sci. B: Polym. Phys. 46 (24) (2008) 2765.

12.6.3 Elastomers Reinforced with Nanoclays

Clay is widely used as a nonblackening filler in the rubber industry. It is noted for its low cost and low-to-moderate reinforcement [31,33]. In 1987, the Toyota research group [34] replaced the inorganic exchange cations in the galleries of the native clay with alkylammonium surfactants and formed the organoclay. The surface chemistry was compatible with the hydrophobic polymer matrix and good dispersion was obtained. Nylon-6/clay nanocomposites were generated [35] and the research on organoclay-reinforced nanocomposites was extended to epoxy resins [36—44], polyamide [38,45—48], polystyrene [39—52], polyurethane [53, 54], polypropylene [55,56—58], etc. These nanocomposites demonstrate an increase in tensile properties, reduced gas permeability, thermal stability, and flame retardance [59].

Kojima *et al.* [60] studied the nanoclay-reinforced nitrile rubber and found that the permeability of hydrogen and water vapor was reduced by about one-third. Lopez-Manchado [61—63] prepared the organoclay nanocomposites based on natural rubber and noticed an increase in the cross-link density, degree of curing, structural order, and glass transition temperature. *cis*-1,4-Polyisoprene and epoxidized natural rubber were studied by Vu and coworkers [64]. Clays were incorporated into the elastomers by mixing the components in a standard internal blender or by mixing dispersions in toluene and methyl ethyl ketone. They found that the reinforcing effects depend on the degree of exfoliation. The morphology and mechanical properties of clay-reinforced styrene-butadiene rubber (SBR) were explored by Zhang *et al.* [65]. SBR latex was mixed with a clay/water dispersion to achieve the structure of layered bundles. The mechanical properties were increased compared with other fillers and regular rubber processing methods of mixing clay. Wang *et al.* [66] synthesized the silicone rubber/organomont-morillonite hybrid nanocomposites by a melt intercalation process. The properties of the resulting nanocomposites were quite close to the aerosilica-filled silicone rubber. Song *et al.* [67] prepared a high-performance nanocomposite consisting of a polyurethane elastomer and organoclay. An increase of 150% in tensile strength and strain was observed and the fatigue properties were improved. Pramanik *et al.* [68—70] used the solution method to obtain the thermoplastic elastomer/clay nanocomposites. The tensile strength was doubled by 4 wt% organophilic clay loading and the thermal stability was higher by about 34 °C.

Tsujimoto *et al.* [71] and Uyama *et al.* [72] developed green nanocomposites consisting of plant oils and clay. The epoxidized plant oil was cured in the presence of organophilic montmorillonite to produce the triglyceride-clay nanocomposites. A green nanocomposite coating was also developed by them. The hardness and mechanical properties were improved and good flexibility as well as high biodegradability were shown. Lu *et al.* [73] calculated the solubility parameter of the functionalized triglycerides and clay organic modifier and demonstrated their miscibility. Improvements were observed in the flexural modulus and thermal stability.

When polymers are associated with nanoclays, three possible structures can form, as shown in Fig. 12.16. The polymer infiltrates into the clay galleries in the intercalated and exfoliated structure. In the intercalated structure, the clay layers expand but still form an ordered structure. This structure is very similar to the thermoplastic polyurethane (TPU) structure in which the hard sections form physical cross-links. Therefore, in the intercalated structure, it is possible that the nanoclay generates a physical cross-linked network that has special properties.

Polymer—clay nanocomposites can be synthesized in four ways [74]: (1) exfoliation-adsorption, (2) *in situ* intercalation polymerization, (3) melt intercalation, and (4) template synthesis. Using the *in situ* polymerization method, different ratios of clay were mixed with AOME by mechanical stirring for 24 h. With the addition of 0.8 wt% cobalt naphthenate and 3 wt% Trigonox, the mixture was cured at room temperature. MMA and EGDMA were added before curing to modify the cross-link density of the final elastomer [34]. A Flory—Huggins solubility analysis indicated that Cloisite® 30B had a similar solubility parameter as poly(AOME) and was used in these experiments.

The morphology of the nanocomposite structure was measured by wide-angle X-ray diffraction (XRD) and transmission electron microscopy (TEM). Figure 12.17 shows how the degree of exfoliation depends on the clay ratio. The XRD peaks can be used to qualitatively identify the amount of the clay with certain spacing. With the increase of the

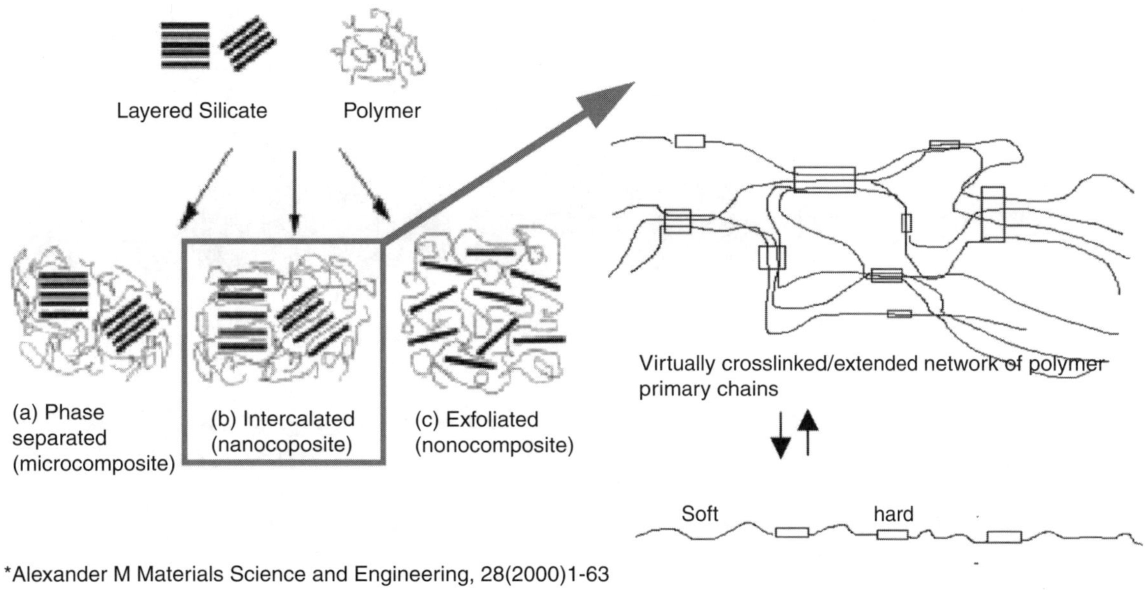

Figure 12.16 Structure of nanocomposites and polyurethane. *Source: M. Alexander, Materials Science and Engineering, 2000, 28, 1–63.*

clay ratios, the peak amplitude decreases at the low diffraction angles. It can be seen that Cloisite®30B remains well dispersed in AOME at loadings up to 10%. Figure 12.18(a) shows the TEM image of the distribution of clay in the polymer matrix. We can see that the clay is quite well dispersed. Some clay bundles are observed in Fig. 12.18(b). The polymer infiltrated into the clay layers, but the clay still keeps its regular layered structure. The exfoliated structure is shown in Fig. 12.18(c). In this state, the clay is not well aligned anymore.

Figure 12.19 shows the stress–strain curves of the 10% clay elastomers subjected to repeated loading–unloading strain cycles. It should be possible to design self-healing[4] elastomers using the nanoclay method. The intercalated layers should be able to debond and dissipate considerable energy, and upon removing the load, the stored strain energy in the clay nanobeams would allow them to heal again, thereby restoring the original structure. Zhu and Wool [30] noted that some healing did occur, but was comparable to typical healing in filled elastomers, as discussed and measured by Wool [20]. For the nanobeam healing effect to occur, the intercalation process could be first done

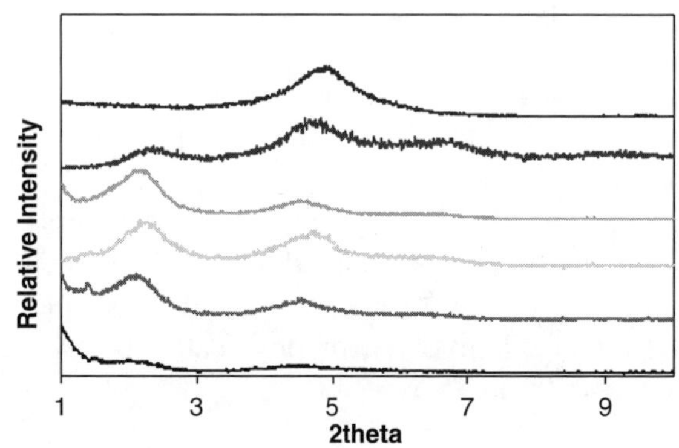

Figure 12.17 XRD for different clay ratios.

with a nonpolymerizing fluid and then inserted into the polymer matrix. However, the effect of the nanoclay on the mechanical properties was significant.

In Fig. 12.20, we see that the elongation at break and the tensile strength increase with clay content. The nanoclay significantly improves the mechanical properties of the elastomer: With only 3% clay loading, the tensile strength increases by 180% and the strain at break by 100%. When the clay loading goes up to 10%, the tensile strength increases to 0.58 MPa and the maximum strain to 190%. Increasing the amount of MMA increased the tensile strength, but the elongation remained the same. The thermal decomposition initiation temperature also increased

[4] R.P. Wool, Self-healing materials: a review, J. Soft Matter 4 (2008) 400–418.

Figure 12.18 TEM of 5% clay-filled elastomer: (a) low magnification, (b) intercalated structure, and (c) exfoliated structure.

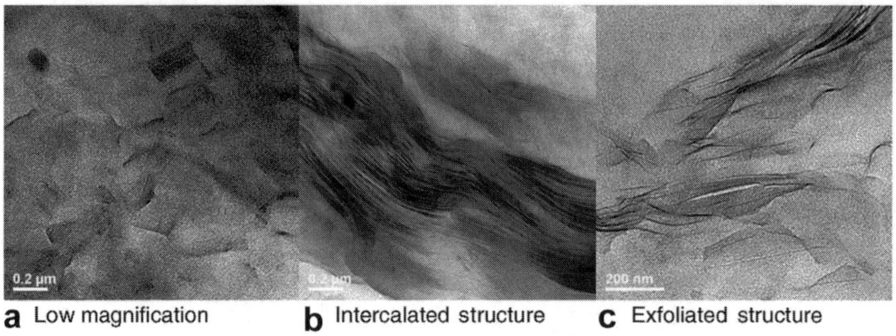

by about 12 °C from 138 °C to 144 °C with 3% nanoclay.

The fracture stress is related to the number of bonds in a percolation network. The critical stress σ required to break the network can be expressed as $\sigma = [2EvD_o(p - p_c)]^{1/2}$, where $[p - p_c]$ is the percolation fraction of bonds that must be broken to cause fracture in the network. Table 12.3 gives the calculated $[p - p_c]$ values. With increasing clay, $[p - p_c]$ increases linearly. Thus, the loading of the clay increases the perfection of the network by connecting the free chain end together and forming a more cross-linked structure. However, the network is still poor as indicated by the low $[p - p_c]$ value and that is the reason for the low mechanical properties. Percolation theory suggests that the optimal elastomer structure is obtained with high-molecular-weight linear polymers, which are subsequently cross-linked chemically, or physically, by intercalation with nanoclays.

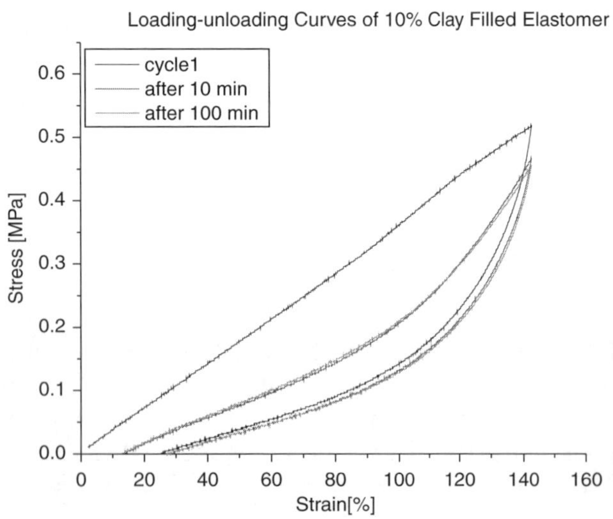

Figure 12.19 Loading–unloading curves of 10% clay-filled elastomers.

12.6.4 Biodegradability and Biocompatibility of Elastomers

12.6.4.1 Biodegradability

Weight loss measurements of samples buried in soil with different clay ratios were obtained as a function of time over a 100 day period. The weight loss was only of order 8 wt% after 100 days. The weight loss was larger in nanoclay-filled samples, but the degradation was not proportional to the clay content. The 3% clay loaded sample has a similar

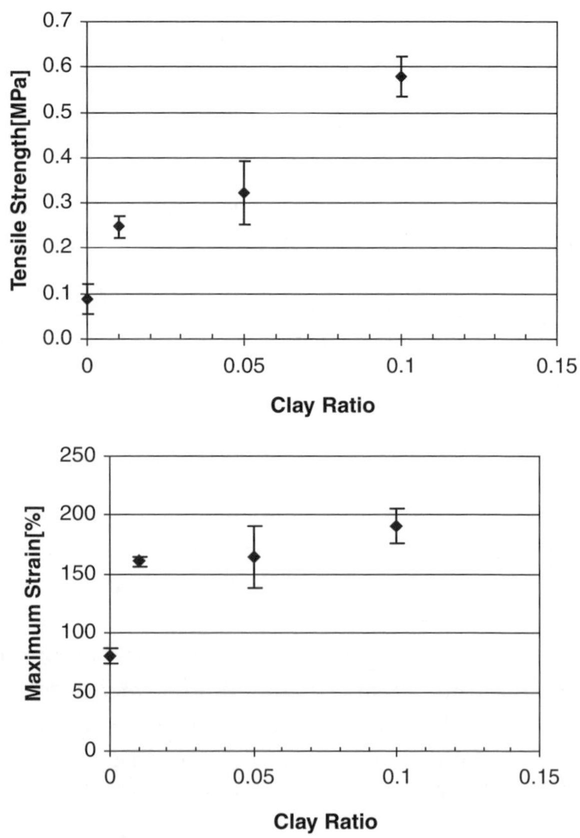

Figure 12.20 Fracture stress and elongation of elastomer versus clay ratio.

Table 12.3 Network Perfection

	σ (MPa)	E (MPa)	v (mol/m^3)	$p-p_c$
No clay	0.09	0.17	22.87	0.0033
3% Clay	0.25	0.19	25.56	0.0207
5% Clay	0.32	0.23	30.94	0.0239
10% Clay	0.58	0.28	37.67	0.0518

weight loss as the sample with 10% clay in the first 45 days and an even larger value afterward. All the samples have a large initial degradation rate and slow down later. High magnification optical images taken on the biodegraded surface of the poly(AOME) showed interesting structures: pearl-like and seashell-like structures were observed and rainbow colors were shown together with these interesting structures. The biodegradation mechanism needs further study, but the abnormal structures are believed to be related to degradation by different microorganisms. The increase in the rate of biodegradation has also been found in the study of poly (L-lactic acid) nanocomposites[5], polylactide-clay, and poly(butylenes succinate)-clay nanocomposites. The type of organoclay and loading level affect biodegradation kinetics. This suggests that the biodegradation can be fine tuned. The combination of improved mechanical properties and controlled biodegradability make these materials attractive for biomedical and packaging applications.

12.6.4.2 Biocompatibility

The biocompatibility experiments on the triglyceride-based materials (PSAs, elastomers, foams, composite resins) were done in collaboration with Professor Catherine Klapperich of Boston University. The goal of the tissue engineer is to generate a scaffold material that possesses the necessary microstructure and mechanical and surface chemical properties to stimulate and guide cells to regenerate diseased or destroyed tissues. Over the past two decades, a variety of natural and synthetic materials have been explored as potential scaffold materials. However, the number of materials that are used clinically is relatively small, and *de novo* design of materials for scaffold applications has been slow due to the stringent requirements for achieving approval for the implantation of new materials. This situation has led to a proliferation of studies looking in detail at a relatively narrow group of materials and little research into the design of scaffolds from scratch to satisfy performance requirements from the outset. Here we describe a flexible new class of materials suitable for a range of tissue engineering applications based on natural products that can be tailored at the molecular level to meet scaffold design objectives.

The materials described in Table 12.4 are based on monomers of modified triglycerides derived from plant sources. We hypothesized that the resulting polymers would be friendly to cells and possibly even encourage growth and extracellular matrix synthesis. Our goal in this study was to demonstrate that copolymers of AOME were supportive to cell attachment and proliferation. Once we determined that cells were able to attach and grow on these materials, we observed differential gene expression of genes involved in tissue remodeling between cells grown on the new materials and cells grown on control surfaces of tissue culture polystyrene (TCPS).

12.6.4.3 Cytotoxicity Assays

Cell viability on the AOME/MMA elastomers and PSAs was determined by performing an MTT-based *in vitro* toxicology assay (Sigma-Aldrich, St. Louis, MO). The MTT assay enables spectrophotometric measurement of mitochondrial dehydrogenase activity in living cells. WS-1 Human dermal fibroblasts (ATCC, Manassas, Virginia) were cultured in Dulbecco's Modified Eagle Medium (DMEM)

Table 12.4 Compositions of PSA Samples.

Sample	% AOME	% MMA	% EGDMA	% Styrene
A	100	0	0	0
B	91	0	0	9
C	95	5	0	0
D	79	20	1	0
E	59	40	1	0

[5] D. Schmidt, D. Shah, and E.P. Giannelis, Curr. Opin. Solid State Mater. Sci. 6 (3) (2002) 205–212.

(Invitrogen, Carlsbad, CA) containing fetal bovine serum (10%) and penicillin/streptomycin (1%). Cells were incubated at 37 °C in the presence of 5% CO_2 balance air at 100% humidity. Media was changed every 48 h, and cells were split as they approached 80% confluence. Population doubling numbers of cells used in these studies ranged from 25 to 40.

Test materials were placed in six well TCPS plates and seeded with approximately 800,000 cells in 2.5 ml of DMEM. Controls were created by seeding TCPS empty wells. The plates were incubated for 2 h to assess cell attachment via optical microscopy. Samples were then placed back in the incubator for a total of 24 h incubation with the materials. After optical images were recorded at 24 h, the MTT solution was added to the wells and incubated for an additional 4 h. The resulting formazan crystals were solubilized, and the absorbance of the solution was measured at a wavelength of 562 nm (A_{562}). The concentration of viable cells on each sample material was determined using a standard curve.

Pressure Sensitive Adhesive Cytocompatibility. After 2 h in culture, cells began to attach to all of the PSAs. Some materials were more cell adhesive than others, and this effect was seen both at the 2 h (images not shown) and 24 h time points. At 2 h, large numbers of rounded cells were observed beginning to attach to all of the PSA materials. After 24 h, the attached cells began to spread on all of the PSAs, with the exception of sample 100/0/0. On samples 95/5/0, 79/20/1, and 59/40/1, most of the attached cells elongated taking a typical fibroblast phenotype, which was still observed after 2 weeks in culture, while on sample 100/0/0, only single, rounded cells were observed.

The quantitative data from the Alamar Blue assay at the 24-h time point indicated that neither the blank media wells nor the material coated control wells showed a significant color change as expected, since no cells were present. The cells in the polystyrene positive control wells grew at a normal rate, doubling every 1–2 days, thus exhibiting more than the number of seeded cells after the 24 h incubation, while the cells in the glass positive control wells indicated that a portion of the initial 50,000 cells seeded did not attach (Fig. 12.4). Data for sample 100/0/0 showed a statistically significant decrease in the number of cells present from the initial seeding to the 24 h time point, also reflected by fluorescent imaging. Cells did not survive on sample 100/0/0 until imaging at the 2-week time point.

Quantitative data for sample 95/5/0 showed an increase in cell attachment in comparison to sample 100/0/0, but still had a statistically significantly lower number of attached cells in comparison to the positive control wells at the 24-h time point. Fluorescent imaging indicated the attached cells were beginning to elongate on the material at 24 h and continued to proliferate up to 2 weeks.

Sample 59/40/1 showed a high number of cells attached and proliferating at 24 h. The fluorescent images support the quantitative data, and the cells proliferated up to 2 weeks.

The sample 79/20/1 coating detached from all glass wells within 24 h of seeding with fibroblasts, so we were unable to run the Alamar Blue assay on these samples at 24 h. The duplicate samples prepared for fluorescent imaging did not fully detach from the flat, glass-bottomed plates and showed increased cell growth similar to that of sample 59/40/1, continuing to grow over a 2 week period. Samples 79/20/1 and 59/40/1 both had 1% EGDMA cross-linker added to the copolymers, while samples 100/0/0 and 95/5/0 did not. Of the two cross-linked samples, polymer 79/20/1 exhibited slightly less adhesive properties than sample 59/40/1, as evidenced by the detachment of the polymer from the glass well plates during the Alamar Blue proliferation assay and confirmed by the mechanical test data.

12.6.4.4 Elastomer Biocompatibility

The elastomers were the least biocompatible of the materials tested. All of the elastomers with nanoclay compositions greater than 1% resisted cell adhesion. Cells would attach to the tissue culture dish near the elastomers at early times, but did not attach to the elastomers in large numbers. After 24 h, a cell-free area surrounded almost all of the elastomer samples indicating that the materials were leaching something that was toxic to the cells. The quantitative analysis of the MTT assay for these samples was not carried out, since it was clear that the cell-killing effect was local, and that a large number of cells nearby were unaffected at 24 h. Due to the small size of the killed area, we felt that the quantitative data would not be helpful due to the resolution of the technique. Since the base materials of the elastomer are very similar to the base materials in the PSAs and in the biocompatible foams, and the fact that samples with 1% clay compositions were less toxic, we suspected that the nanoclay inclusions were the source of the cytotoxicity. We performed MTT assays on cells growing in a monolayer in the presence of the nanoclay, and the cultures were completely killed at 24 h. We have not yet tested elastomer samples without nanoparticle inclusions, but the result should be similar to the PSA since both are dominated by the AOME monomer.

12.6.4.5 Summary of Biocompatibility Studies

In summary, we have presented a class of materials derived from plant oils that have properties suitable for a wide range of biomedical applications. The copolymers are cytocompatible and encourage fibroblast attachment and growth as indicated by the Alamar Blue assay and fluorescent imaging. The homo-polymer sample 100/0/0 (polyAOME) did not support cell growth. This result could be due to the mechanical properties of the sample, since anchorage dependent fibroblasts have been shown to prefer stiffer substrates to attach to and migrate over. This polymer with a very low T_g is also outside the Twinkling Fractal range where the surface vibrations have disappeared. R. P. Wool has suggested an unusual idea that surfaces may be able to "communicate" with cells through their surface vibrational density of states determined by the Twinkling Fractal effect (Paper presented at the American Physical Society, Boston MA, March 2012). The surface vibrations exist in the approximate temperature range from $T_g - 20$ to $T_g + 50$ and thus, sample A (100/0/0) would be outside this range. It is also possible that the MMA monomers are driving the ability of cells to adhere and spread on the substrate, perhaps by reducing the mechanical tackiness of the PSA. Polymer 100/0/0 exhibited much stronger cohesive and adhesive properties than samples 95/5/0, 79/20/1, and 59/40/1. We speculate that the addition of more co-monomer MMA to sample 59/40/1 (40% by weight) as compared with the less adhesive sample 79/20/1 (20% by weight) appears to have affected the overall degree of cross-linking, with sample 59/40/1 remaining slightly more adhesive. If sample 79/20/1 has a higher degree of cross-linking, then it would follow that sample 79/20/1 would make a "worse" adhesive material than sample 59/40/1. The modified tack tests support this hypothesis.

Samples 95/5/0 and 59/40/1 also appeared to lose some adhesive properties following submersion in media over time and began to peel away from the glass wells after 48 h of incubation. The tackiest polymer sample 100/0/0 started to peel away after approximately 5 days. Submersion in media over time appeared to lower the attachment properties of the PSAs to glass. Data were taken from all samples except those that fully detached from the plate. Previous experiments using a spatula to thickly apply nonuniform layers (~ 0.5−1 mm) (data not shown) of each polymer did not detach after submersion in media for several days and did not require the initial solvation in $CHCl_3$.

The copolymer adhesives described here, in thin film form, may be appropriate for transdermal drug delivery applications, as they are both adhesive to skin and can be tailored to incorporate both water- and fat-soluble drugs. It is also possible that the addition of the fatty acid-based copolymer may enhance skin permeation of certain drugs. Again, it is critical to fully assess the degradation pathways and products of these materials before they are used as bioadhesives. We are currently synthesizing AOME-pHEMA copolymers that have more tunable hydrophilicity with composition for the transdermal delivery application. Now that the basic cytocompatibility has been demonstrated for a range of copolymers incorporating these plant-derived monomers, continuing work is focused on tissue specific applications and molecular level cell−biomaterial interactions.

12.7 Bio-Based Coatings

With increasing interest in "green chemistry" on the part of signatory nations to the Kyoto Accord on global warming, significant efforts among the international scientific communities were directed toward the area of renewable resources [75]. Plant oils were used in varnishes and alkyd resins used in the coatings industry. Varnishes are generally physical solutions of the natural or synthetic resin in plant oils. In alkyd resins, plant oils are chemically combined with polyester resins. The chemistry of plant oils allows paints based on these resins to "air dry" through oxidative coupling reactions [76]. However, the popularity of these resins in the coatings industry is waning with the increasing demand for waterborne emulsion polymers. In 2000, the world emulsion polymer demand was $15 billion, and the coatings industry supplied more than 50% of this demand [77]. Emulsion polymers used in latex paints are the highest volume coating resins in the industry [78] and are currently produced from petroleum derivatives. Incorporating renewable plant oils in latex technology will provide a renewable and sustainable alternative for the coatings industry as well as a new market for plant oils.

Organic coatings are complex mixtures of various substances. Components include polymers or resins, volatile organic compounds (VOCs), pigments, and additives. Polymers and resins, commonly called *binders* by the coatings industry, form the continuous film that adheres to the substrate, binds other substances in the film together, and imparts film strength and durability. VOCs are used to aid in film formation. However, due to ever-increasing environmental regulations on VOC emissions, the industry is focusing on low- to no-VOC paint formulations. Pigments impart color, opacity, and other visual effects to the coating film. Additives enhance the properties of the final product and include dispersants, colorants, and rheology modifiers.

The polymeric binder is the main vehicle of the coating. Many types of polymeric binders are used in coating formulations; primary examples include alkyd resins, polyester resins, isocyanates (polyurethanes), drying oils, and emulsion polymers (latexes). Latexes are the primary binders used in architectural coatings, particularly in the United States [79]. They offer superior durability, lower VOC emissions, and are much easier to use than their oil-based counterparts, making them more attractive to consumers.

Latex binders are formed via emulsion polymerization. Emulsion polymerization is a free-radical polymerization in which a monomer or mixture of monomers is polymerized in an aqueous surfactant solution to form a latex [80]. Emulsion polymers used in architectural coatings are typically linear, high-molecular-weight polymers that form films under ambient conditions by the evaporation of water and solvents, and the coalescence of latex particles. Common monomers in emulsion polymers include acrylic and vinyl esters, and their selection generally depends on specific requirements of the coating and cost.

Film formation of emulsion polymers occurs by coalescence of latex particles. A schematic of this process is shown in Fig. 12.21 [79,80]. Latex binders used in coatings typically have a high solids content ranging from 20 to 50%. Solid polymer particles are dispersed in an aqueous phase. Once applied to

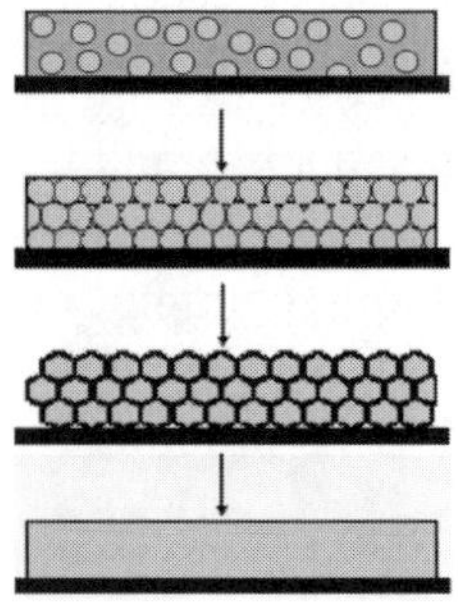

Figure 12.21 Film-formation process of latex polymers.

a substrate, the water and solvents within the emulsion begin to evaporate, leading to a close-packed layer of latex particles. This is the first and longest stage of film formation, and continues until the particles make up 60–70% volume fraction. The rate of evaporation is approximately equal to the rate of evaporation of water [80].

The second stage begins when the particles are concentrated and come into irreversible contact. A clear, continuous, but still weak film is formed due to particle deformation at temperatures greater than the minimum film-forming temperature (MFFT) [79,80]. The MFFT is the lowest temperature at which coalescence occurs sufficiently to form a continuous film.

The final stage of film formation occurs at temperatures above the T_g. Further coalescence transpires as polymer surface chains interdiffuse across interfaces of adjacent particles. Interdiffusion develops the mechanically coherent film[6], and the full strength of this film is reached when the surface chains diffuse a distance equal to their radius of gyration, R_g [20].

As mentioned earlier, plant oils are essential to oil-based coatings. These coatings form cross-linked films through autoxidation of fatty acids in the oils. The process is slow and the coatings often require high levels of solvent to aid in drying. The films also continue to oxidize and polymerize over long periods of time after application, causing eventual film degradation. Emulsion coatings are often preferable because they usually contain lower amounts of solvents and dry more rapidly. Efforts were made to extend the use of plant oils into the field of

[6] R.P. Wool, Adhesion at polymer-polymer interfaces: a rigidity percolation approach, C. R. Chim. 9 (1) (Jan 2006) 25–44; ERRATA, 9 (9) (Sep 2006) 1234–1234.

emulsion coatings in order to decrease the amount of petroleum used by the industry. Thames *et al.* [81] developed plant oil-based latex polymers for coatings that show good film-forming properties. The polymers were derived from ricinoleic acid, the primary fatty acid of castor oil. This acid comprises 90% of castor oil's triglycerides. It is a monounsaturated fatty acid with a hydroxyl functional group. The monomer is synthesized by converting ricinoleic acid to its methyl ester and acrylating the hydroxyl group. The monomer can then take part in free-radical polymerization of the acrylate group. The long carbon chain of the monomer also acts as a plasticizer in the films, decreasing the need for a solvent in the coating formulation. Additionally, the polymer undergoes cross-linking via oxidative polymerization of the residual double bond after surface application. This imparts additional strength to the film.

12.7.1 Design of Bio-Based Coatings

To obtain desired coating properties, most latex binders used in the field are actually copolymers. A critical decision in designing a latex emulsion is monomer and comonomer selection. A primary criterion for this selection is the T_g. It is crucial to select a monomer combination that produces a copolymer with the appropriate T_g; the T_g must be low enough to permit coalescence at the lowest application temperature, yet high enough to ensure coating durability [79]. Coalescence will not occur unless the temperature is at least slightly higher than the T_g. Typical T_g's of latex binders used in architectural coatings range from 0 °C to 25 °C. The T_g of the AOME polymer is approximately −60 °C [15]. This is too low to be used alone in formulating a latex binder for architectural coatings. A hard comonomer with a high T_g is essential to increase the T_g to an appropriate level. Alternatively, additional cross-linking reactions will also increase T_g. When using a comonomer, the amounts of different monomers necessary to produce a copolymer with the appropriate T_g can be estimated using several models. The comonomers considered in this work were MMA and styrene. The T_g of these monomers are 105 °C and 100 °C, respectively. MMA and styrene are typical comonomers used in acrylic binders. The Fox [82] equation predicts that approximately 40 wt% AOME and 60 wt% MMA or styrene is necessary to reach a common architectural coating with T_g of 15 °C. The petroleum content can be further reduced by cross-linking reactions.

The chemicals used in the miniemulsions include the AOME monomer, styrene, MMA, sodium dodecyl sulfate as the surfactant, and azodiisobutyrodinitrile as the initiator [83]. The miniemulsion was prepared by ultrasonification for 5 min and reaction at 80 °C for 1 h. The particle diameter, as determined by light scattering, ranged from approximately 90 to 170 nm, depending on the amount of comonomer used. This is consistent with the particle size for typical miniemulsions. Figure 12.22 shows the

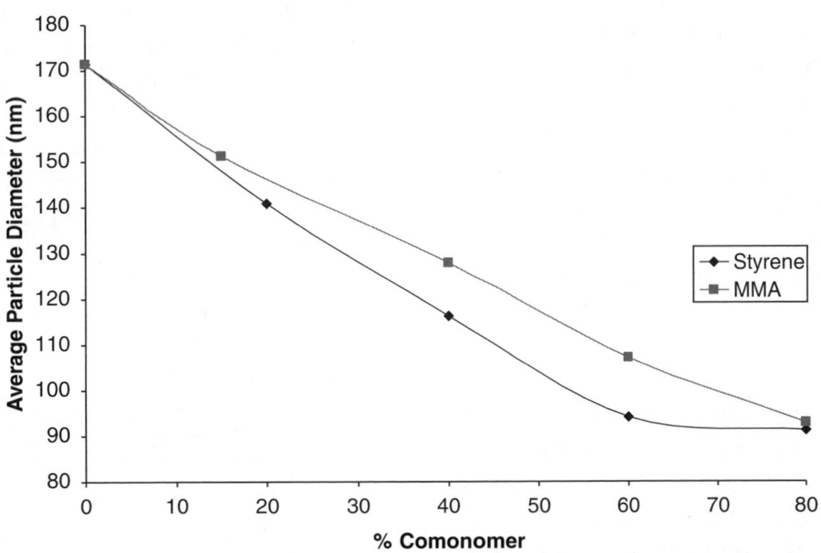

Figure 12.22 Average particle diameter of emulsions determined by dynamic light scattering.

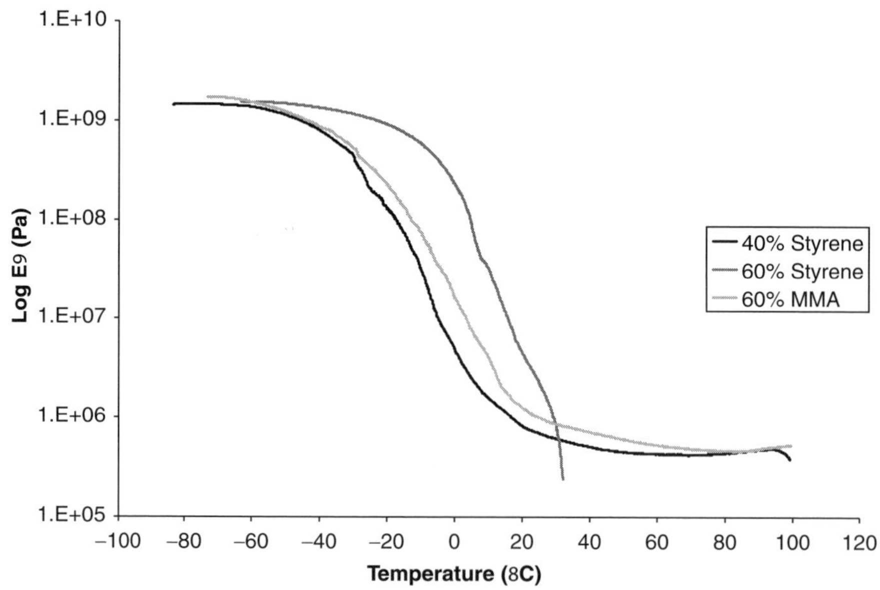

Figure 12.23 Storage modulus plot for films prepared from AOME copolymerized with styrene and MMA.

average particle diameter as a function of comonomer content. The particle size decreases with increasing styrene or MMA content.

12.7.2 Coating Properties

The storage modulus plot of the 40% styrene, 60% styrene, and 60% MMA films is shown in Fig. 12.23. The glassy regions are observed for each film sample at approximately 1.5 GPa. The modulus begins to decrease for the 40% styrene film and 60% MMA film at approximately −55 °C, whereas the modulus begins to decrease for the 60% styrene film at approximately −45 °C. The storage modulus remains highest for the 60% styrene film until the temperature reaches approximately 33 °C. Here, the modulus for the 60% styrene film continues to drop, whereas the other films begin to level into their rubbery plateau areas at approximately 46 kPa. This signifies that the 40% styrene and 60% MMA films are indeed cross-linked due to a higher concentration of technical-grade AOME polymer. The technical-grade OME used to produce the AOME monomer was only 70% pure. Levels of linoleic and linolenic methyl esters were present in the material. These have two and three double bonds, respectively, in the carbon chain, and these bonds have the ability to cross-link. The storage modulus of a polymer in the rubbery plateau region was used to determine the cross-link density. The cross-link density (Table 12.5) of the 40% styrene film sample at approximately 40 °C was 66.7 mol/m^3. The cross-link density of the 60% MMA film sample at approximately 50 °C was 77.1 mol/m^3.

12.7.3 Nano Coatings

The coatings discussed in the preceding sections were quite tacky and soft in comparison to commercial latex polymers, even with rather high comonomer content. One solution may reinforce the coatings with mineral clays, like montmorillonite. Montmorillonite clay naturally forms stacks of platelets. These platelets are <10 Å thick. A solution of 10 wt% Cloisite Na$^+$ clay to polymer in water was prepared. The films were allowed to dry under ambient conditions for approximately 16 h and then subsequently dried under vacuum to remove any excess moisture. The emulsion clay mixtures yielded hard, nontacky films. The extent of intercalation was studied through XRD. Pure Cloisite Na$^+$ has a peak at approximately 7°. This corresponds to a *d*-spacing of approximately 12.62 Å. The 5% and 10% clay film

Table 12.5 T_g and Cross-Link Density of the AOME and Styrene Films With 10 wt% Clay to Polymer.

% Styrene	T_g (°C)	v_e (mol/m^3)
20	−25	10,197
40	−15	6,797
60	35	4,405

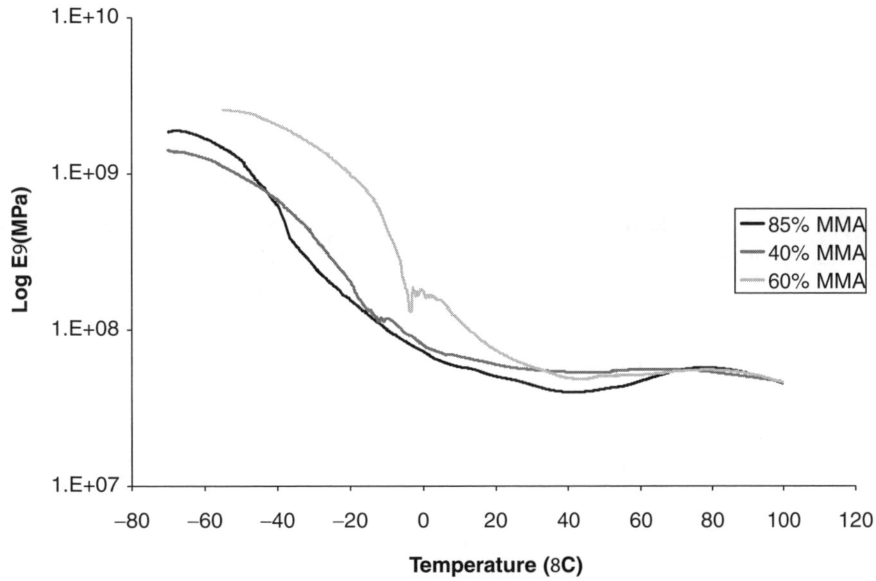

Figure 12.24 Storage modulus of films with 10 wt% clay to polymer, increasing MMA content.

samples showed peaks at approximately 2.4°, 4.5°, and 7° which correspond to *d*-spacing values of 36.78, 19.62, and 12.62 Å, respectively. The peak at 2.4° corresponds to clay intercalated by the polymer.

The storage moduli of the films as a function of temperature and 20, 40, and 60% comonomer content are shown in Fig. 12.24. All specimens show a large increase in storage modulus in both glassy and rubbery plateaus. The fact that a rubbery plateau is seen on the 60% styrene film sample with 10% clay content indicates that the clay imparts properties similar to that of a cross-linked film, as discussed in the last section. The rubbery plateau is increased by three orders of magnitude to approximately 55 MPa for the MMA samples and 40 MPa for the styrene samples. The T_g for each film was obtained from the tan δ peaks and is reported in Tables 12.4 and 12.5. Overall, the addition of clay seems to either decrease or have little effect on the T_g of the films. The T_g was expected to increase due to the decrease in tackiness of the film upon clay addition. The measured T_g only increased for the 60% styrene film, rising from approximately 20 °C to 35 °C. The cross-link densities for the samples and the temperatures at which they were calculated are listed in Tables 12.5 and 12.6. There is a significant increase in the cross-link density.

The gel fraction (in benzene) of the film sample without polymer was determined to be approximately 70%. The gel fraction of the sample with clay, determined by subtracting the amount of clay in the sample from the initial and final weights, was found to be approximately 97%. The gel fraction and therefore cross-link density were greatly increased by the addition of clay, which may account for the increase in the rubbery plateau modulus.

Film hardness is often described in the coatings industry by the ASTM D-3363 Film Hardness by Pencil Test. This test employs the use of a set of pencils each with a different grade of hardness. The grade is determined by the amount of baked graphite and clay in the composition of the pencil. The grades include 9H, 8H, 7H, 6H, 5H, 4H, 3H, 2H, H, F, HB, B, 2B, 3B, 4B, 5B, 6B, 7B, 8B, and 9B. The hardest is 9H, F is the middle of the scale, and 9B is the softest. The test is performed by flattening the lead of the pencil at a 90° angle using 400-grit sandpaper. Starting with the lowest grade pencil, the pencil is held at a 45° angle to the film and pushed forward ¼ inch using as much downward pressure as can be applied without breaking the lead. This is repeated with increasing grade pencils until the film is scratched. The hardness of the film is determined by

Table 12.6 T_g and Cross-Link Density of the AOME and MMA Films With 10 wt% Clay to Polymer.

% MMA	T_g (°C)	ν_e (mol/m³)
15	−36	9299
40	−23	8640

the grade of pencil that scratches the film. Typical coatings range in hardness from 3B to 9H [84,85]. Both emulsion films with and without nanoclay were cast on glass slides and allowed to dry overnight under ambient conditions and subsequently dried under vacuum. The pencil test was then performed as described. The nanoclay increased the film hardness of the emulsion polymers from 8B to 3B.

The addition of clay would allow films with low hard comonomer content to serve as coatings. In fact, the styrene copolymer nanocoatings showed increasing rubbery storage modulus with decreasing styrene content. This is very promising because the clay is both environmentally and economically friendly. Its use in combination with the plant oil-based AOME monomer could drastically reduce the amount of petroleum-based monomer needed in latex binders for architectural coatings. The current development of lignin-based monomers with T_gs of the order of 100−150 °C [86] offer considerable potential for the development of more bio-based coatings with higher properties for future generations.

References

[1] D. Satas (Ed.), Handbook of Pressure Sensitive Adhesive Technology, second ed., Van Nostrand Reinhold, New York, 1989.
[2] E.H. Pryde, Fatty Acids, American Oil Chemists' Society, Champaign, IL, 1979.
[3] R.P. Wool, S.H. Kusefoglu, G.R. Palmese et al., High Modulus Polymers and Composites from Plant Oils, U.S. Patent 6,121,398, (2000).
[4] S.P. Bunker, R.P. Wool, J. Polym. Sci. A: Polym. Chem. 40 (2001) 451−458.
[5] P.A. Lovell, M. El-Aasser, Emulsion Polymerization and Emulsion Polymers, John Wiley & Sons, New York, 1997.
[6] R.G. Gilbert, Emulsion Polymerization: A Mechanistic Approach, Academic Press, Inc., San Diego, 1995.
[7] J.Y. Charmeau, E. Kientz, Y. Holl, Prog. Org. Coat. 27 (1−4) (1996) 87−93.
[8] E. Kientz, Y. Holl, Coll. Surf. A: Physiochem. Eng. Aspects 78 (1993) 255−270.
[9] A. Zosel, B. Schuler, J. Adhes. 70 (1999) 179−195.
[10] S.N. Khot, Synthesis and application of triglyceride-based polymers, in: Chemical Engineering, University of Delaware, Newark, 2000.
[11] A. Zosel, Int. J. Adhes. Adhes. 18 (1998) 265.
[12] P.A. Lovell, T.H. Shah, Polym. Commun. 32 (4) (1991) 98−103.
[13] E. Chang, J. Adhes. 60 (1997) 233−248.
[14] A. Zosel, J. Adhes. 34 (1991) 201−209.
[15] S.P. Bunker, Ph.D. Thesis, University of Delaware 2002.
[16] S. Bunker, C. Staller, N. Willenbacher, et al., Int. J. Adhesion Adhesives 23 (2003) 29−38.
[17] R.P. Wool, S. P. Bunker, U.S. Patent 6,646,033, (2003).
[18] I. Lee, R.P. Wool, J. Polym. Sci. B: Polym. Phys. 40 (2002) 2343.
[19] F.M. Fowkes, M.A. Mostafa, Ind. Eng. Chem. Prod. Res. Dev. 17 (1978) 3.
[20] R.P. Wool, Polymer Interfaces: Structure and Strength, Hanser/Gardner Publications, Cincinnati, 1995.
[21] http://www.elastomersolutions.com.
[22] Synthetic Rubber—The Story of an Industry, International Institute of Synthetic Rubber Producers, International Institute of Synthetic Rubber Producers, Inc 1973.
[23] D.C. Blackley, Synthetic Rubbers: their Chemistry and Technology, Applied Science Publishers, 1983.
[24] V. Athawale, S. Kolekar, J. Macromolec. Sci. Pure Appl. Chem. 37 (2000) 65−79.
[25] V. Athawale, S. Kolekar, Polym. J. 30 (1998) 813−818.
[26] V. Athawale, S. Raut, Polym. J. 30 (1998) 963−967.
[27] V. Athawale, S. Raut, Phys. Chem. Chem. Phys. 2 (2000) 1249−1254.
[28] V. Athawale, S. Kolekar, J. Appl. Polym. Sci. 75 (2000) 825−832.
[29] S. Raut, V. Athawale, Eur. Polym. J. 36 (2000) 1379−1386.
[30] L. Zhu, R.P. Wool. Preprints, Amer. Chem. Soc. Philadelphia National Meeting, (August 2004).
[31] J.E. Mark, B. Erman, F.T. Eirich, Science and Technology of Rubber, Academic Press, 1994.
[32] R.P. Wool, J. Polym. Sci. Part B, Polym. Phys. 43 (2005) 168.
[33] M. Morton (Ed.), Rubber Technology, Van Nostrand Reinhold Company, New York, 1987.
[34] Y. Fukushima, S. Inagaki, J. Incl. Phenom. 5 (1987) 473−482.

[35] A. Usuki, Y. Kojima, M. Kawasumi, et al., J. Mater. Res. 8 (1993) 1179–1184.
[36] N.A. Salahuddin, Polym. Adv. Technol. 15 (2004) 251–259.
[37] T. Lan, T.J. Pinnavaia, Chem. Mater. 6 (1994) 2216–2219.
[38] D. Ratna, N.R. Manoj, R. Varley, et al., Polym. Int. 52 (2003) 1403–1407.
[39] T.J. Pinnavaia, T. Lan, Z. Wang, et al., In Nanotechnology (1996) 250–261.
[40] D. Ratna, O. Becker, R. Krishnamurthy, et al., J. Polym. 44 (2003) 7449–7457.
[41] J.H. Kang, S.G. Lyu, G.S. Sur, Polymer-Korea 24 (2000) 571–577.
[42] I.M. Daniel, H. Miyagawa, E.E. Gdoutos, et al., Exper. Mechan. 43 (2003) 348–354.
[43] C.R. Lee, K.J. Ihn, M.S. Gong, Polymer-Korea 27 (2003) 392–395.
[44] X. Kornmann, H. Lindberg, L.A. Berglund, Polymer 42 (2001) 1303–1310.
[45] T. Lan, P.D. Kaviratna, T. Pinnavaia, J. Chem. Mater. 6 (1994) 573–575.
[46] D.M. Delozier, R.A. Orwoll, J.F. Cahoon, et al., Polymer 43 (2002) 813–822.
[47] K. Yano, A. Usuki, A. Okada, et al., J. Polym. Sci. A: Polym. Chem. 31 (1993) 2493–2498.
[48] K. Yano, A. Usuki, A. Okada, J. Polym. Sci. A: Polym. Chem. 35 (1997) 2289–2294.
[49] K.Y. Kim, H.J. Lim, S.M. Park, et al., Polymer-Korea 27 (2003) 377–384.
[50] C.R. Tseng, J.Y. Wu, H.Y. Lee, et al., J. Appl. Polym. Sci. 85 (2002) 1370–1377.
[51] S. Qutubuddin, X.A. Fu, Y. Tajuddin, Polym. Bull. 48 (2002) 143–149.
[52] X. Fu, S. Qutubuddin, Mater. Lett. 42 (2000) 12–15.
[53] A.M. Chen, Y. Tian, B. Han, et al., Acta Polym. Sinica (2003) 591–594.
[54] J.S. Ma, Z.N. Qi, S.F. Zhang, Acta Polym. Sinica (2001) 325–328.
[55] Z. Wang, T.J. Pinnavaia, Chem. Mater. 10 (1998) 3769.
[56] Y.Q. Zhang, J.H. Lee, J.M. Rhee, et al., Compos. Sci. Technol. 64 (2004) 1383–1389.
[57] J.S. Ma, Z.N. Qi, Y.L. Hu, J. Appl. Polym. Sci. 82 (2001) 3611–3617.
[58] N. Hasegawa, H. Okamoto, M. Kawasumi, et al., J. Appl. Polym. Sci. 74 (1999) 3359–3364.
[59] M. Alexandre, P. Dubois, Mater. Sci. Eng. R: Rep. 28 (2000) 1–63.
[60] Y. Kojima, K. Fukumori, A. Usuki, et al., J. Mater. Sci. Lett. 12 (1993) 889–890.
[61] M.A. Lopez-Manchado, B. Herrero, M. Arroyo, Polym. Int. 52 (2003) 1070–1077.
[62] M. Arroyo, M.A. Lopez-Manchado, B. Herrero, Polymer 44 (2003) 2447–2453.
[63] M.A. Lopez-Manchado, M. Arroyo, B. Herrero, et al., J. Appl. Polym. Sci. 89 (2003) 1–15.
[64] Y.T. Vu, J.E. Mark, L.H. Pham, et al., J. Appl. Polym. Sci. 82 (2001) 1391–1403.
[65] L.Q. Zhang, Y.Z. Wang, Y.Q. Wang, et al., J. Appl. Polym. Sci. 78 (2000) 1873–1878.
[66] S.J. Wang, C.F. Long, X.Y. Wang, et al., J. Appl. Polym. Sci. 69 (1998) 1557–1561.
[67] M. Song, D.J. Hourston, K.J. Yao, et al., J. Appl. Polym. Sci. 90 (2003) 3239–3243.
[68] M. Pramanik, S.K. Srivastava, B.K. Samantaray, et al., Macromol. Res. 11 (2003) 260–266.
[69] M. Pramanik, S.K. Srivastava, B.K. Samantaray, et al., J. Appl. Polym. Sci. 87 (2003) 2216–2220.
[70] M. Pramanik, S.K. Srivastava, B.K. Samantaray, et al., J. Polym. Sci. B Polym. Phys. 40 (2002) 2065–2072.
[71] T. Tsujimoto, H. Uyama, S. Kobayashi, Macromol. Rapid Commun. 24 (2003) 711–714.
[72] H. Uyama, M. Kuwabara, T. Tsujimoto, et al., Chem. Mater. 15 (2003) 2492–2494.
[73] J. Lu, C.K. Hong, R.P. Wool, J. Polym. Sci. B Polym. Phys. 42 (2004) 1441–1450.
[74] R.A. Kalgaonkar, J.P. Jog, J. Polym. Sci. B Polym. Phys. 41 (2003) 3102–3113.
[75] P.T. Anastas, J.C. Warner, Green Chemistry, Theory and Practice, Oxford University Press, Oxford, 1998.
[76] D.H. Solomon, The Chemistry of Organic Film Formers, John Wiley and Sons, Inc., New York, 1967.
[77] The Freedonia Group, I. Adhesives and Sealants Industry; 2002 www.freedoniagroup.com/coatings.html
[78] Paint and Surface Coatings: Theory and Practice, Ellis Horwood Limited, New York, 1987.
[79] Z.W. Wicks, F.N. Jones, S.P. Pappas, Organic Coatings: Science and Technology, second ed., Wiley-Interscience, New York, 1999.
[80] P.A. Lovell, M.S. El-Aasser, Emulsion Polymerization and Emulsion Polymers, John Wiley and Sons, New York, 1997.
[81] S.F. Thames, H.B. Yu, R. Subramanian, J. Appl. Polym. Sci. 77 (2000) 8–13.

[82] T.G. Fox, Bull. Amer. Phys. Soc. 1 (1956) 123.
[83] A.L. Kulbick, M.S. Thesis, University of Delaware 2004.
[84] http://www.psrc.usm.edu/macrog/mpm/composit/nano/struct2_1.htm, 2004.
[85] http://www.rohmhaaspowdercoatings.com/tech/technical_briefs/techbriefs.hardpencil.jsp, 2004.
[86] J.F. Stanzione, III, R.P. Wool, PMSE ACS Preprints, Sandiego CA, March 27 (2012).

13 Biopolymer Films and Composite Coatings

Amos Nussinovitch

OUTLINE

13.1 Introduction

Water-soluble gums are valuable in many fields, including adhesives, agriculture, biotechnology, ceramics, cosmetics, explosives, food, paper, textiles and texturization, among many others [1]. The most recent developments and progress in the utilization of gums can be found in the field of edible coatings. Edible and biodegradable films have the potential to reduce packaging and limit moisture, aroma, and lipid migration between food components [2]. Such films can contain antioxidants, preservatives, or other additives to improve foods' mechanical integrity, handling and quality, and to change surface gloss [3]. Coatings are not limited to the food industry and thus gums are also used in coatings for fiberglass, fluorescent lamps, glass, metals, optical products, paper products, latex, and textiles [4,5]. The market for edible films has already shown impressive growth, from ~$1 million in 1999 [6] to more than $100 million in 2009 (http://www.ceepackaging.com/2006/08/289/film-going-down). Retail sales of edible film are expected to reach at least $2 billion in 2012 [7].

13.2 Mechanisms of Film Formation

Edible films can be produced from hydrocolloids, lipids, resins, and composites. There are many methods for forming films directly on food surfaces. For film-forming materials dispersed in aqueous solutions, solvent removal is required to achieve solid film formation and to control its properties [8]. For example, the temperature and rate of drying influence the mechanical properties and crystallinity of cellulosic films [9,10]. Proteins are heteropolymers that usually contain most of the 20 amino acids, allowing for an enormous number of sequential arrangements

Ebnesajjad: Handbook of Biopolymers and Biodegradable Plastics. http://dx.doi.org/10.1016/B978-1-4557-2834-3.00013-6
© 2013 Elsevier Inc. All rights reserved. Adapted from a chapter in: Kasapis, *Modern Biopolymer Science* (2009).

with a wide range of interactions and chemical reactions [11,12]. In contrast, polysaccharides contain only a few monomers, e.g., cellulose and starch contain only one monomer, glucose [8]. In polysaccharides, the hydroxyl is the only reactive group, while proteins present a large variety of possible interactions and chemical reactions [8]: they may participate in chemical reactions through covalent (peptide and disulfide) linkages and noncovalent interactions (ionic, hydrogen, and van der Waals bonding). In addition, hydrophobic interactions occur between nonpolar groups of amino acid chains [13]. Interlinkages between proteins participating in the formation of a film can lead to improved film properties [14]. Edible films can be formed via two main processes: a "wet process" in which biopolymers are dispersed or solubilized in a film-forming solution (solution casting) followed by evaporation of the solvent, and a "dry process," which relies on the thermoplastic behavior exhibited by some proteins and polysaccharides at low moisture levels in compression molding and extrusion [11,15,16].

A number of proteins have received particular attention for the production of edible coatings. The prolamin fraction of corn is known as zein [17]. Zein can form films when cast from appropriate solvent systems. The mechanisms that come into play upon solvent evaporation include hydrophobic and hydrogen bond development in the formed film matrix [18]. A limited number of disulfide bonds, due to low cystine content, may also be present. The produced films are tough, glossy, and scuff- and grease-resistant, and plasticizers should be included in their formulation to decrease brittleness [18]. Quantities higher than 0.5% of included plasticizer within an edible coating have been reported, in some cases, to result in reduced fruit protection [19]. Cross-linking agents may be involved in improving water resistance and tensile properties [20].

Wheat gluten films can be manufactured by the deposition and subsequent drying of wheat gluten dispersions. The solvent in many cases is aqueous ethanol under alkaline or acidic conditions. Homogeneous solutions are produced by heating and mixing [21,22]. The following mechanisms are involved in the film's formation: upon dispersing the gluten in alkaline environments, disulfide bonds in the gluten solution are reduced by the reducing agents [14]. Upon casting of the film-forming solution, disulfide bonds re-form, linking together polypeptide chains and producing a film structure. Mechanisms contributing to the re-formation of disulfide bonds include re-oxidation in air and sulfhydryl—disulfide interchange reactions [23,24]. Hydrogen and hydrophobic bonds also add to the film structure. Without plasticizer, such films are brittle; however, plasticizer addition decreases the rigidity of such films by mediating between polypeptide chains, disrupting some of the extensive intermolecular associations [25].

Edible films based on soybeans have long been produced in the Orient. Protein is the major component of these films, but lipids and carbohydrates are also incorporated. Consequently, soymilk films are in fact multicomponent films. The protein content in soymilk films was found to be higher than that in the initial soymilk [26]. Therefore, the film-formation mechanism was proposed to consist of the isolation and partial concentration of the proteins, probably via the mechanism of endothermic polymerization of heat-denatured protein together with surface dehydration [26]. Another description of this mechanism proposed formation of a protein matrix by heat-catalyzed protein—protein interactions, with hydrogen disulfide and hydrophobic bonds being the major associative forces in the film network [27]. Yet another suggested mechanism of polymerization involved intermolecular disulfide and hydrophobic bonds, whereby heating is required to alter the three-dimensional structure, thereby exposing sulfhydryl groups and hydrophobic side chains. Once dried, the unfolded protein macromolecules move toward each other and are linked through the hydrophobic and disulfide bonds [28].

Films can be produced from soy protein isolates by heating aqueous dispersions of same to form surface films or by the deposition and drying of soy protein solutions. Film production continues with polymerization and solvent evaporation at the film—air interface [14]. The mechanism of film formation is explained thus: when the protein is in solution, hydrophobic groups are oriented toward the interior of the protein molecule, away from water. At the air—water interface, these hydrophobic groups extend out of the water into the air, where they interact with each other, while hydrophilic groups remain submerged [29]. Another proposed mechanism involves interfacial coagulation, which may occur when the protein concentration in the interface monolayer exceeds a certain limit and the protein coagulates, forming a three-dimensional coagulum at the interface from the two-dimensional monolayer [8].

Proof that polymerization takes place through such bonds was provided by demonstrating its inhibition by cleavage of disulfide bonds [30]. Blockage of sulfhydryl with appropriate reagents also inhibited polymerization [31]. pH may also influence film formation. Near the isoelectric point of soy protein (~4.6), protein coagulation and, consequently, solution casting are impossible [32]. A method similar to film formation on soymilk surfaces was used to develop peanut protein–lipid films on the surface of heated peanut milk [33].

Cross-linking is responsible for the production of collagen films with desirable properties. In the presence of formaldehyde, this mechanism involves the combining of free amino groups of basic amino acids. Treatment with glyceraldehyde (nontoxic, in contrast to glutaraldehyde), which promotes cross-linking, increased the mechanical properties of the produced films and demonstrated, in part, the approach by which such films are produced [34]. Addition of lower alkyl diols with 4–8 carbon atoms also improved the mechanical properties of the collagen films (casings) by reducing internal hydrogen bonding while increasing intermolecular spacing [35]. Collagen films extruded from acid dispersions were subjected to irradiation in the presence of a photosensitive dye which catalyzed protein cross-linking and as a result, improved film properties [36]. The protein gelatin forms through partial hydrolysis of collagen. Gelatin forms a clear, flexible, strong and oxygen-impermeable film, whose properties are influenced by drying temperature. Above 35 °C, gelatin exists as a single molecule in a configuration that cannot form interchain hydrogen bonds, whereas at lower temperatures, gelatin has a "collagen fold" configuration capable of forming interchain hydrogen bonds [37]. The mechanism governing shellac and gelatin composite film formation can be described as hydrogen bonding between the carboxyl and hydroxyl groups of shellac and the amino and carboxyl groups of gelatin. The composite film exhibits higher wettability, surface free energy, and polarity with increasing gelatin content, indicating an increase in hydrophilicity. The formation of a composite film eliminates the drawbacks of shellac and could make a beneficial contribution to the application of film coating in the food and pharmaceutical industries [38].

Milk film is formed from milk proteins covering food surfaces, possibly via the effects of high relative humidity and elevated temperatures, which decrease lactose solubility and increase its granulation [39]. Milk film can also form at the air–water interface from small amounts of insoluble milk powder, the underlying mechanism thought to be the formation of sugar–protein complexes [40]. Casein is the major protein in milk, including αs1-, αs2-, β-, and κ-caseins and a small amount of γ-casein [41]. α-casein may form flexible films via increased intermolecular hydrogen bonding of its low hydrophobic and flexible random coil structure [42]. Electrostatic interactions can also play an important role. In general, edible coatings based on proteins may serve in many potential applications, such as for meat products, nuts, seafood, confectionary products, fruits and vegetables, and as ingredients for microencapsulation and controlled-release purposes [1].

13.3 Obtaining a Well-Matched Coating

The coating procedure involves wetting a surface with a coating gum solution, followed by the solution's possible penetration [43], and potential adhesion between the two commodities. The wetting stage is the shortest and most significant; if the solution used for spreading is suited to the food that is being coated, then spreadability is spontaneous [44]. Nevertheless, it is nearly impossible to find gum solutions (or their combinations) that are perfectly suited to the surface properties of a particular object, i.e., in terms of surface tension and polarity. As a result, the closest likely combination should be sought to obtain compatibility. Creating a successful coating that adheres to a food surface requires an estimation of the interfacial tension between the coating solution and the surface. This depends on the surface tension of the surface and the coating solution, and the contact angle between the two [45]. To estimate the surface tension of the solid, the critical surface tension needs to be calculated. Prior to this, the critical surface tension of the surface to be coated should be derived from Zisman plots followed by extrapolation. These plots are obtained by calculating the cosine of each measured contact angle of the prechosen liquid on this surface and plotting it against the already-known surface-tension values of the solvents being used [46,47].

Finding an appropriate coating solution can be quite a challenge. Coating solutions are water-based, i.e., they include mostly water with a surface tension of 72.8 dyne cm^{-1}. For many solid surfaces, lower surface-tension values must be taken into account. Hydrocolloids, as a general rule, have the potential to lower the surface tension of solutions designated for use as coating agents [48]. The lower the surface or interfacial tensions of a gum solution, the higher its surface or interfacial activity [48]. Bearing in mind that competent coating involves compatibility between liquid and solid surface tensions [49], a logical step is to reduce the surface tension of the coating solution to conform to the lower surface tension of the food surface, thereby lowering the interfacial tension and improving adhesion [50]. In previous studies [51], it was demonstrated that the addition of sterols effectively yields better adhesion between fruit and vegetable surfaces and coating films due to better compatibility with respect to the hydrophobicity of the two adhering surfaces [52]. Spreadability is an additional feature to be considered. It is enhanced by the surface roughness of the food for coating solutions having contact angles smaller than 90° [53], and inhibited by rough surfaces for solutions with contact angles greater than 90°. Surface roughness has been defined as the ratio of the true area of the solid to the apparent area. Wenzel was the first to propose a relationship between the contact angle of a liquid on a rough surface and its contact angle on an ideally smooth surface [54]. Surface roughness can be evaluated with a roughness tester [55] or by image-processing of atomic force microscopy micrographs [43]. Roughness decreases the interfacial tension as a result of improved spreadability [50]. The importance of the magnitude of the interfacial tension is well recognized by the polymer coating industry. Better compatibility between the coated object and the coating film can be achieved by incorporating surface-active agents within the coating gum solution. It can be concluded from the compatibility requirements that tailor-made hydrocolloid coatings for different food materials can be achieved only by further determination of the chemical and physical properties of the coating solutions and the objects to be coated.

13.4 Film-Application Stages and Methods for Testing Films

Films can be applied by dipping or spraying [56]. Brushes, falling-film enrobing technique, panning, or rollers can also be used to apply films to the surfaces of the coated objects [57]. Immersion of fresh produce in a gum solution takes between 15 and 120 s for a complete coating, where the duration depends on wettability, concentration, and viscosity of the hydrocolloid solution, surface roughness of the biological specimen, and possible penetration of the coating solution into the specimen [58]. In general, foods to be coated are dipped in a hydrocolloid solution followed by draining and drying. A coating solution can be dried to obtain a dry film adhering to the food surface; thus a film that never passes through a gel state is formed. An additional alternative is to use gum solutions which should be cross-linked before drying. In this case, a second immersion of the hydrocolloid-coated food in a cross-linking bath to induce gel formation (i.e., in alginate, low-methoxy pectin (LMP), κ- or λ-carrageenan, or gellan) takes between 30 s and 2 min, depending on the concentration and temperature of the cross-linking agent, the thickness of the coating gum solution, and the geometric complexity of the coated object [1]. The third stage of manufacturing a cross-linked coating is the continuous strengthening of the gel coating layer at high relative humidity. The fourth stage, namely drying, can result in different dry-film textures and structures, depending on the length of drying. The texture and structure of the dried film will vary according to the time required for the gelled film layer to dry. Properties of the dried films also depend on properties of the drying equipment and on thickness and composition of the coating film [1].

There are many methods to evaluate the properties of coatings. Sometimes films need to be produced by casting to obtain appropriate specimens for testing. Gas permeability of packaging films can be monitored in several ways [59–61]. Many devices for measuring film permeability to oxygen are commercially available [62,63]. Water-vapor transmission rates through dried coatings can be determined by ASTM E96-93 (standard testing methods for water-vapor transmission of materials used for gums destined for coatings and adhesives).

Peel testing, i.e., the force necessary to peel off the coating, is used to estimate the film's degree of adhesion to a surface (Fig. 13.1). The coating is peeled at 90° from the substrate, and the adhesion strength is estimated by the force per unit width necessary to peel the coating. It is important to study surface wetting and adhesion properties of coated commodities to obtain "good" coatings. The roughness of film surfaces can be

Figure 13.1 A coating is peeled at 90° from the substrate, and the adhesion strength is estimated by the force per unit width necessary to peel the coating.

studied using a roughness tester, and an electron and atomic force microscope (the latter being used for finer mapping of surface roughness). An important parameter is Ra, the arithmetic mean of the absolute values of the roughness profile's deviation from the center line within the length being evaluated. The surface tension of gelling and inducing solutions, and their contact angles on food and other objects' surfaces, can be studied with surface-tension instruments (maximum adhesion requires a contact angle of 0°). It is important to note that "if the coating does not spread spontaneously over the substrate surface, so that there is intermolecular contact between the substrate surface and the coating, there cannot be interactions and hence no contribution to adhesion" [64].

13.5 Selecting Biopolymers for Specific Applications

Edible films and coatings should be chosen based on their suitability to the task at hand. As a general rule, if the aim is to retard moisture migration, lipid-based or composite films are chosen, i.e., films consisting of a combination of lipid and hydrocolloid components present in a bilayer or conglomerate [65]. To retard oil and fat migration, hydrocolloid-type films are chosen. For all other applications, i.e., to retard gas or solute migration, to improve structural integrity or handling properties, to retain volatile flavor components or to convey food additives, hydrocolloid, lipid, or composite combinations are chosen in accordance with their suitability [65]. Hydrocolloid films can be used when control of water-vapor migration is not the objective. Such films are good barriers to oxygen, carbon dioxide, and lipids. Lipids, in addition to being a barrier against water vapor, might be added to increase the gloss of coated products [65]. Composite films combine the advantages of lipids and hydrocolloids while each compensates for the other's disadvantages [66].

In the storage and marketing of fruits and vegetables, one must bear in mind that after harvest, these products are not "static materials"—they consist of living tissue that needs to "breathe," or it will undergo certain anaerobic reactions and, as a consequence, "suffocate" [67]. Fruits and vegetables use up oxygen and release carbon dioxide as they respire, and lose water (transpiration), the amount depending on temperature, gaseous makeup, and humidity of the surrounding environment. To extend the shelf life of fruits and vegetables, hydrocolloids can be chosen to produce new types of coatings. Fruit and vegetable respiration is reduced by these films' selective permeabilities to oxygen and carbon dioxide, and thus the films serve as atmosphere modifiers [67]. The formulation of such films can include a "waxy" material to mimic the natural waxy coating of produce, and to give them a shiny appearance. Nevertheless, the emphasis should be on controlling gas exchange and creating a modified atmosphere inside the fruit that delays ripening and senescence, similar to the more costly practice of maintaining a controlled atmosphere (CA) [67]. Wax coatings were developed to mimic the natural coating of fruits and vegetables. However, these coatings inhibited respiratory gas exchange to such an extent that fermentation was induced. As a result, ethanol buildup was detected along with that of other volatiles, coinciding with fermented and bitter taste. At high levels, these volatiles are considered off-flavors and they reduce fruit quality [1]. To solve this problem, disturbances need to be created in the ordered, regular structure of the traditional wax coating, thereby improving fruit respiration [68,69].

After processing, i.e., washing, sorting, trimming, peeling, slicing, coring, etc., the freshness quality of the produce changes [70], because it remains biologically and physiologically active. Cutting a fruit or vegetable leads to tissue breakdown caused by enzymatic action, the formation of secondary metabolites, increased production of ethylene, increased respiration and changes in microbial flora. Approaches for the preservation of minimally processed products include storage at low temperatures, special preparation procedures, the use of additives, and atmospheric modification or control [56]. To minimize the undesirable changes in the processed product, coatings should be selected that are capable of forming an efficient barrier to moisture loss, exhibit selective permeability to gases, control migration of water-soluble solutes, and enable the incorporation of additives such as flavor, preservatives, or coloring [56]. In practice, achieving such an ideal coating is not a simple matter, and one needs to define which factors are undesirable or problematic. In general, dry films made up of layers may swell, dissolve, or disintegrate upon contact with fluids, and these are therefore not appropriate [57]. An emulsion coating might be appropriate, depending upon the stability of the preparation. Another factor to consider with cut surfaces is that because they are covered with fluid, the binding of lipid materials becomes problematic. In this case, biopolymers with functional groups for ionic cross-linking that include acetylated monoglycerides might be helpful. Combinations such as a caseinate/acetylated monoglyceride/alginate emulsion, or replacing the alginate with low methoxy pectin provide alternative options [57,71]. An example of the benefits of emulsion coatings is in the inclusion of ash gourd (*Benincasa hispida* Cogn.) peel wax in an edible coating (emulsion) for strawberries [72]. The emulsion that gave best results included 0.5% wax, 1 M sodium benzoate, and 3 min dip time. Fruits without wax coating spoiled completely in less than 2 days at 25 °C. Wax coating expanded their shelf life to 7 days at 25 °C and properties such as texture, color, weight loss, titratable acidity, and microbial counts were highly acceptable [72]. Meat and meat products may suffer from shrinkage, microbial contamination, and surface discoloration. For simply delaying moisture transport, a thin coating film (gel) produced from any of a variety of polysaccharides (i.e., alginate, carrageenan, pectin, and starch) can be used successfully due to evaporation of water within the gel [73]. If extended periods of storage are required, the hydrophobicity of the coating needs to be increased [71]. The same or different hydrocolloids can be used for such applications, but the formulation should include some lipids. These can be oils, mono-, di-, and triglycerides, waxes, or water-in-oil emulsions. If an antimicrobial agent needs to be incorporated into the coating, the water activity in the coating should not be high (i.e., $0.8 > a_w$), so as to avoid instability, but it should also not be lower than ~0.65, so that good permeability of the preservative can be achieved [74]. Many food products may contain high oil content, such as nuts or fried products. To eliminate oxidative off-flavors, the coating should have low oxygen permeability [75] and incorporation of an antioxidant in the coating is also recommended ([76]; see previous examples in this chapter). One general conclusion that can be drawn is that although edible films and coatings find uses in a variety of applications, the technical information on them is far from adequate, leaving the food scientist with the formidable task of developing a film for each food application [65].

13.6 Edible Protective Films

13.6.1 Packaging Materials Fit for Human Consumption

Solutions, highly viscous suspensions or gels can be produced from gums (hydrocolloids), which are high-molecular-weight molecules [77]. Due to the hydrophilic nature of many gums, the coatings that are produced from them have limited moisture-barrier abilities. However, in gel form, when no drying is applied, the films can retard moisture loss by serving as sacrificing agents, as found, for example, in soft white-brined cheeses and meats [78]. Reviews on food-packaging materials based on natural polymers can be found elsewhere [15,79–82]. The macromolecules found in edible films are mainly polysaccharides and proteins. Polysaccharide and protein films are fine gas barriers, but poor moisture barriers. Conversely, pure lipid films are good moisture barriers, but poor gas barriers. This is the main reason for the interest in developing composite edible coatings that include the virtues of each class of ingredients [81].

Manufacturers of novel edible films need to take many issues into consideration. These include barrier stability, mechanical properties, simplicity of application, biodegradability (environmentally friendly vs. plastic), nontoxicity, safety, and cost for the manufacturer and consumer, as well as organoleptic properties. Usually, coatings are tasteless; however, they can be designed with a unique taste that then influences the taste of the final product [79,83]. Biopolymers can be used as an alternative source in the production of edible coatings. However, these polymers are sensitive to temperature and relative humidity: solutions blending gelatin and polyvinyl alcohol (PVA) provide a way around this problem, at least with respect to humidity [84]. Biodegradable films obtained from chitosan and methylcellulose (MC) can also reduce environmental problems associated with synthetic packaging. However, the cost of biodegradable films is still prohibitive; the use of chitosan, a waste by-product of the fishing industry, may provide a good alternative [85]. Biodegradable films, which could serve as an environmentally friendly alternative to synthetic plastic packaging films, can be prepared from kafirin, the prolamin protein of sorghum. The resultant inferior functional properties can be improved by adding tannic acid or sorghum-condensed tannins as modifying agents during casting [86]. Vegetable starches, and fish-muscle protein prepared from blue marlin meat, and even low-quality fish meat can also serve as a good source for biodegradable films [87]. In the latter case, the bacterial population in the films can be dramatically reduced by the addition of polylysine [88]. Natural biopolymers are suitable for the construction of different types of wrappings and films. The produced biodegradable packaging can help in controlling physiological, microbiological, and physicochemical changes in food products. This is accomplished by controlling mass transfer between the food product and the ambient atmosphere or between components in heterogeneous food products, and by modifying and controlling the food's surface conditions (pH, level of specific functional agents, and slow release of flavor compounds) [89]. The material's characteristics (polysaccharide, protein, or lipid, plasticized or not, chemically modified or not, used alone, or in combination) and manufacturing procedures (casting of a film-forming solution, thermoforming) must be tailored to each specific food product and its surroundings (relative humidity, temperature). A few possible uses of these materials might be the wrapping of fabricated foods, protection of fruits and vegetables by controlling maturation, protection of meat and fish, control of internal moisture transfer in pizzas, all of which hinge on the organoleptic, mechanical, and gas- and solute-barrier properties of the films [89].

13.6.2 Inclusion of Food Additives within Edible Coatings

Edible films can be used as carriers for a range of food additives, including antibrowning and antimicrobial agents, colorants, flavors, nutrients, and spices [90–94]. Such additives could serve as a possible treatment for reducing the deleterious effects of minimal processing of fresh-cut fruit [92,94–97]. Different processing operations, including peeling, cutting, and shredding, induce enzymatic browning, which influences quality but can be successfully controlled by sulfites. However, ever since the ban on the use of sulfites for fresh fruits and vegetables [98], a replacement has been urgently sought. Several alternative chemical compounds have been suggested as enzymatic browning inhibitors [99]. Other common antibrowning agents include citric acid and oxyresveratrol. Ascorbic acid is extensively used to inhibit enzymatic browning of fruit: it reduces the *o*-quinones generated by the action of polyphenol oxidase enzymes back to their phenolic substrates [100]. Several thiol-containing compounds, such as cysteine, *N*-acetylcysteine, and reduced glutathione, have also been investigated as inhibitors of enzymatic browning [101]. These compounds react with quinones formed during the initial phase of the enzymatic browning reactions to yield colorless products or reduce *o*-quinones to *o*-diphenols [102]. Antibrowning agents incorporated into edible coatings are aimed at minimizing the browning of vegetative tissue on cut and exposed surfaces [103]. Consequently, the incorporation of antibrowning agents into edible films in minimally processed fruits was studied by several groups [104–106]. Most antibrowning agents are hydrophilic compounds that can enhance water-vapor permeability and water loss when incorporated into edible coatings [107]. In a study on fresh-cut "Fuji" apples, edible films based on alginate proved to be good carriers for antibrowning agents (*N*-acetylcysteine and glutathione) [108]. In this particular case, the edible coating was generally applied before the antibrowning agents such that the gel coating adhered to the fruit and the antibrowning agents were then incorporated in the dipping solution, which contained calcium for cross-linking and instant gelling of the coating [108]. The effect of coatings in combination with antibrowning agents on minimally processed apple slices was studied during storage. Chitosan-coating treatments effectively retarded enzymatic browning of minimally processed apples during storage and retarded or prevented tissue softening: apple slices underwent little loss of firmness. Chitosan coating alone did not perform very well as a water-vapor barrier in apple slices [109]. Fresh-cut mangoes are valued worldwide for their exotic flavor and nutritional composition. However, their shelf life is limited by changes in color, texture, appearance, and microbial growth. The quality parameters of "Tommy Atkins" mango slices pretreated with citric acid and cassava starch or sodium alginate edible coatings, with or without glycerol, were studied. The edible coatings acted as barriers to gas and water vapor, extending the storage time of the fresh-cut fruit. Thus, cassava starch and alginate are alternatives for preserving minimally processed mangoes, as they maintain the quality parameters of fresh fruit [110]. Similarly, the influence of treatment with ascorbic acid, citric acid, calcium lactate dipping, and cassava starch edible coatings on quality parameters and shelf life of fresh-cut pineapple slices was studied for 12 days at 5 °C. Edible coatings with and without calcium lactate were efficient in reducing weight loss and juice leakage, and in maintaining firmness during storage. However, these samples showed more browning and their ascorbic acid content was reduced. All treatments presented good sensory acceptance [111].

Peeling, cutting, or slicing of minimally processed fruits increases the product's functionality but induces wounding. Microorganisms present on the food surface may be involved in spoilage [112,113]. Therefore, incorporation of antimicrobial compounds into edible films provides a novel way of improving the safety and shelf life of ready-to-eat foods [114]. Frequently used antimicrobials include conventional preservatives such as benzoic and sorbic acids, bacteriocins (nisin and pediocin), and plant-derived secondary metabolites, such as essential oils and phytoalexins [115,116]. The effect of sodium alginate coating with preservatives on the quality of meat patties during refrigerated (4 ± 1 °C) storage was studied. A 2% sodium alginate coating significantly improved the overall appearance and color, juiciness, flavor, texture, and overall palatability of the product [117]. Essential oils have been evaluated for their ability to protect food against pathogenic bacteria in, for example, contaminated apple juice [115,118,119]. Essential oils can be added to edible films and coatings to modify flavor, aroma, and/or odor, as well as to introduce antimicrobial properties [114]. Cinnamon essential oil was employed in gelatin coatings to maintain the quality of refrigerated rainbow trout fillets over a period of 20 days. The coating enriched with cinnamon oil was suitable for the preservation of fresh fillets and efficiently maintained the quality attributes at an acceptable level during storage [120]. The effects of malic acid and essential oils of cinnamon, palmarosa, and lemon grass, and their main active compounds as natural antimicrobial substances incorporated into an alginate-based edible coating on the shelf life and safety of fresh-cut "Piel de Sapo" melon were investigated. Palmarosa oil incorporated at 0.3% into the coating appeared to be a promising preservation alternative for fresh-cut melon, since it had good acceptance by panelists, maintained the fruit's quality parameters, inhibited the growth of native flora, and reduced the population of *Salmonella enteritidis* [121]. Limonene and peppermint were also incorporated into modified chitosan (increasing its hydrophobicity and improving its stability and adhesion to the fruit product) to create bioactive edible coatings. These were tested for their ability to extend the shelf life of fresh strawberries during storage. Formulations based on modified chitosan containing limonene and Tween 80 were shown to perform better than other formulations [122].

Almost two decades ago, edible films made from fruit purees were developed and shown to be a promising tool for improving the quality and extending the shelf life of minimally processed fruit [123–125]. The inclusion of plant essential oils within coatings improved the shelf life of edible apple-puree films [126]. The effect of lemon grass, oregano oil, and vanillin incorporated into the edible apple puree–alginate coatings on the shelf life of fresh-cut "Fuji" apples was also investigated [108,127]. In general, all antimicrobial coatings significantly inhibited the growth of psychrophilic aerobes, yeasts, and molds. The antimicrobial effect of essential oils against *Listeria innocua* inoculated into apple pieces before coating was also proven [128]. When antimicrobial agents such as benzoic acid, sorbic acid, propionic acid, lactic acid, and nisin were incorporated into edible films, the coatings retarded surface growth of bacteria, yeasts, and molds on a wide range of products, including meats and cheeses [114]. Edible zein coatings containing nisin or nisin/ethylenediaminetetraacetic acid (EDTA) were used to preserve the quality of commercially manufactured fish balls. The coated fish balls underwent significantly less weight loss than their uncoated counterparts. In addition, formation of total basic volatile nitrogen was significantly reduced and the increase in microbial load on fish balls coated with the antimicrobial zein during 15 days of refrigerated storage was less than 1 log colony forming unit (cfu) g^{-1}, while the microbial load increased by about 3 log cfu g^{-1} for the control group without the coating treatment [129]. Benzoic acid ($C_7H_6O_2$) included in edible coatings inhibits the growth of molds, yeasts, and some bacteria. It was either added directly or created from reactions with its sodium, potassium, or calcium salt. Skinless tilapia fillet shelf life was prolonged under refrigeration after being coated with a gelatin coating containing benzoic acid as an antimicrobial agent [130]. Lipid materials such as wax, fatty acids, and neutral lipid resins and waxes are used to improve the water-barrier properties of biopolymer-based edible films. The water-vapor barrier properties of lipid-based biopolymer films are affected by the nature of the utilized lipid materials, the film structure, and factors such as temperature, vapor pressure, or physical state of the water contacting the films [131]. In laminated MC/corn zein-fatty acid films, water-vapor permeability was reported to decrease with increasing chain length and concentration of the fatty acids [132].

Another paper on composite films composed of fatty acids and soy protein isolate demonstrated that their physical properties are, to a great extent, subject to the type of fatty acid added and its concentration [133]. Many other studies reached the conclusion that lipid-based edible films and coatings can play an important role in the food industry by controlling the moisture-barrier properties of biopolymer-based edible films and coatings. Emulsifiers are extensively used in the food industry to improve texture and stability, among other features. Sucrose esters are manufactured by esterifying sucrose with edible fatty acids from palm oil. They are neutral in taste, odor, and color, stable at high temperatures for short times, soluble in cold water, kosher, non-GMO (genetically modified organisms), and vegetarian. When sucrose esters served as part of the composition of gelatin films, they decreased the films' water-vapor permeability and tensile strength [134]. Films containing fatty acids and their sucrose esters exhibited superior water-vapor permeability relative to those containing only the fatty acids (palmitic and stearic). Films that included fatty acids and their sucrose esters were generally more transparent than those with only fatty acids. The chain lengths of the fatty acids and their sucrose esters affected the properties of the films differently, depending upon the gelatin source [134]. Another study demonstrated that sucrose esters have a large influence on the stabilization of emulsified edible-film structures containing arabinoxylans and hydrogenated palm kernel oil [135]. Sucrose esters also improved the moisture-barrier properties of these coatings. Both lipophilic (90% di- and triester) and hydrophilic (70% monoester) sucrose esters can ensure the stability of the emulsion used to form the film, especially during preparation and drying [136].

The practice of adding probiotics to obtain functional edible films and coatings is in its infancy. Apparent health benefits and biological functions of bifidobacteria in humans include, among others, the intestinal manufacture of lactic and acetic acids, reduction of colon cancer risk, inhibition of pathogens, lessening of serum cholesterol, enhancement of calcium absorption, and activation of the immune system [137–139]. A viable bifidobacterial population of 10^5 cfu g^{-1} in the final product has been specified as the therapeutic minimum needed to gain the aforementioned health advantages [140]. A recent study described the formulation of alginate- and gellan-based edible films containing viable bifidobacteria for coating fresh-cut fruits. Values above 10^6 cfu g^{-1} *Bifidobacterium lactis* Bb-12 were maintained for 10 days during refrigerated storage of these fresh-cut fruits, demonstrating the ability of alginate- and gellan-based edible coatings to carry and support viable probiotics on these items [141].

13.6.3 Meat, Seafood and Fish Coatings

Meat is an animal tissue that is used as food. Most often it refers to the skeletal muscle and associated fat, but it may also include nonmuscle organs, such as the lungs, liver, skin, brain, bone marrow, and kidneys. The components of meat are muscle fiber, connective tissue, fat tissue, bone, and pigment. The proportions of these ingredients affect the eating quality of the meat. Commonly eaten animal meats include beef, veal, lamb, pork, fowl, and, less often, game animals [142]. Over several hundred different species of seafood are consumed in the United States. Seafood may be divided into two general groups, fish and shellfish. Fish, as a food, includes the edible parts of water-dwelling, cold-blooded vertebrates with gills. Other edible water-dwelling animals, such as mollusks and crustaceans, are often collectively referred to as shellfish. Crustaceans include crab, lobster, and shrimp. Mollusks include abalone, clam, oyster, and scallops. Crabs and lobsters are purchased fresh or cooked. Meat from cooked crustaceans is available chilled, frozen, and canned [142].

Alginate-based coatings have been used to coat meats [143−147] and are good oxygen barriers [148]. Calcium-alginate coating of lamb carcasses helped reduce surface microbial growth and achieve a faster chill rate [149]. Alginates, carrageenans, cellulose ethers, pectin, and starch derivatives have been used to improve stored meat quality. Such coatings serve as sacrificing agents, i.e., moisture loss is delayed until it evaporates from the film [71]. Meat, poultry, and seafood coated with calcium alginate exhibit reduced shrinkage, and decreased oxidative rancidity, moisture migration, and oil absorption during processing [73]. The effects of different concentrations of sodium alginate as an edible film on chemical changes of dressed kilka during frozen storage were studied. The coating prevented lipid oxidation, increased the fish's shelf life, and reduced its moisture loss [150]. Alginate coatings have been used to retard the development of oxidative off-flavors in precooked meat patties [71]. In a study of the physicochemical, microbiological, and sensorial qualities of cooked pork patty coated with pectin-based material containing green-tea leaf extract powder [151], the coated patties had higher moisture contents than the controls in both air- and vacuum-packaging. The numbers of total aerobic bacteria were significantly reduced by the coating treatments as well as by irradiation. No difference was detected in sensory characteristics due to gamma irradiation or coating treatments [151]. Combined treatments (irradation and coating) for extending the shelf life of meat products are an option. For example, moist beef biltong strips were inoculated with *Staphylococcus aureus* or sprayed with distilled water (noninoculated controls). Both the non-inoculated and inoculated biltong strips were coated with an edible casein-whey protein coating followed by irradiation to a target dose of 4 kGy. This level of irradiation effectively ensured the safety of moist beef biltong and, providing the initial fungal counts are not excessive, may extend the food's shelf life. The edible coating had no significant effect on microbial counts, possibly because the high moisture content of the biltong diminished the oxygen-barrier properties of the barrier [152]. Another combined treatment includes using high pressure plus an edible coating to inhibit microbial growth and prevent oxidation. Meat shelf life can be extended by the application of a surface coating, because it provides a barrier to water and oxygen. Collagen and gelatin coatings were used as barriers on meat products to reduce purge, color deterioration, aroma deterioration, and spoilage, improve sensory scores, and act as an antioxidant [153]. Coating muscle with gelatin-based films enriched with oregano or rosemary extracts increased the phenol content and the antioxidant power of muscle, particularly when used in association with high pressure, due to migration of antioxidant substances from the film [154]. The edible films with the included plant extracts lowered lipid-oxidation levels [154]. The extracts served as antioxidants, i.e., substances that terminate chain reactions by removing free radical intermediates and inhibit other oxidation reactions by being oxidized themselves [155], since they contained reducing agents (i.e., thiols and polyphenols). This finding was not surprising since plants maintain complex systems of multiple types of antioxidants, such as glutathione, vitamin C, and vitamin E, as well as enzymes, for example, various peroxidases [156,157]. In comparison, the gelatin−chitosan-based edible films reduced microbial growth (total count) more successfully, due to the antimicrobial activity of chitosan. This activity varies considerably with the type of chitosan, the target organism, and the environment in which it is applied [158]. The combination of high pressure and edible films yielded the best results in terms of both preventing oxidation and inhibiting microbial growth [154].

The shelf lives of frozen shrimp, fish, and sausages are extended by the use of alginate coatings [159,160]. Trout fillets were coated and stored at −18 °C for up to 7 months. The coated fillets were fried and analyzed for oil absorption and moisture content throughout the storage period [161]. It was found to be more advantageous to use gluten as the first coating, xanthan gum as the second coating, and wheat and corn flours at a ratio of 1:1 or 2:1 as the final coating. In the sensorial analysis, the coated samples were strongly preferred to those that were not coated. The coating layers also provided additional resistance against mass transfer during storage [161]. In Japan, an edible polysaccharide film (*Soafil*) has found wide use in the meat industry as a casing for processed smoked meats. Carrageenan coatings have also been used to reduce off-odor development in chicken carcasses [1]. Carrageenan coatings with soluble antibiotics have been found to be effective spoilage-retarding agents [77,162]. Agar coatings with water-soluble antibiotics were found to effectively extend the shelf life of poultry parts [163]. Agar coatings that included nisin (a bacteriocin) reduced contamination of *Salmonella typhimurium* in fresh poultry stored at 4 °C [164]. The effect of incorporating the additives trisodium phosphate into pea starch and acidified sodium chlorite into calcium alginate on their antimicrobial activity was studied against a three-strain cocktail of *Salmonella* inoculated on chicken skin. Coatings with 0.5% pea starch were absorbed quickly by the chicken skin and showed high skin adhesion, whereas those with >3.5% pea starch exhibited low skin adhesion and slow absorption [165]. Alginate coatings with or without acidified sodium chlorite were stable, and about 50% of the coating weight was retained at 120 h. The latter coatings appeared to have low absorptiveness because the skin gained ~1.0% of its weight within 60 min of application. These findings indicate that effects of the agents in coatings on skin pH, the extent of coating adhesion, and absorption may contribute to the overall antimicrobial behaviors [165].

Antimicrobial packaging materials could offer a potential alternative solution to preventing the spread of spoilage and pathogenic microorganisms via meat foodstuffs. Instead of mixing antimicrobial compounds directly with the food, incorporating them in films allows them to have their functional effect at the food surface—where most microbial growth is found [104,105,166]. Antimicrobial packaging includes systems such as adding a sachet to the package, dispersing bioactive agents within the packaging itself, coating bioactive agents on the surface of the packaging material, or utilizing antimicrobial macromolecules with film-forming properties or edible matrices [166]. Chitosan has film-forming abilities as well as a certain amount of antimicrobial activity, and therefore can serve as a potential and promising natural food-coating material. It was applied to improve the quality of various foods of agricultural, poultry, and seafood origin [167]. Chitosan films are clear, tough, flexible, and good oxygen barriers [168,169]. Chitosan has been used for the preparation of safely and stably coated mutton kebabs and streaky bacon [170]. The use of chitosan as an edible film was also evaluated for its antimicrobial activity against *Listeria monocytogenes* on the surface of ready-to-eat roast beef. An acetic acid–chitosan coating was more effective at reducing *L. monocytogenes* counts than a lactic acid–chitosan coating [171]. This is not surprising, since on an equimolar basis, acetic acid generally exhibits higher antimicrobial activity than other organic acids [172–174]. Coating of pink salmon fillets with chitosan was effective at reducing relative moisture loss compared to control, noncoated fillets. Chitosan postponed lipid oxidation, and there were no significant effects of coating on the color parameters or whiteness values of the cooked fillets after 3 months of frozen storage [175]. Chitosan film was also used for glazing skinless pink salmon fillets. Chitosan glazing delayed lipid oxidation in the salmon fillets after 8 months of frozen storage [176]. Chitosan treatment noticeably decreased bacterial counts and extended the shelf life of salted dried anchovy [177]. A chitosan edible coating for sardine was effective in reducing spoilage and considerably extending its shelf life [178]. The chitosan coating preserved good sensory properties for a longer time and reduced the formation of volatile bases and oxidation products [178]. Chitosan was also beneficial in the development of coatings for the shelf-life extension of fresh Atlantic cod and herring fillets. A significant reduction in relative moisture loss was also observed with cod samples coated with chitosan up to 12 days of storage. Chitosan coating also significantly reduced the chemical spoilage due to lipid oxidation and growth of microorganisms [179]. A chitosan–fish oil coating increased the total lipid and omega-3 fatty acid contents of fresh lingcod (*Ophiodon elongates*) fillets about threefold, reduced thiobarbituric acid-reactive substance values in both fresh and frozen samples, and also decreased drip loss of frozen samples by 14.1–27.6%. Such coatings may be used to extend shelf life and enhance omega-3 fatty acid content in lean fish [180]. In general, these observations can be explained by the unique properties of chitosan in comparison to other hydrocolloids. Chitosan has a positive charge in acidic solutions, due to the presence of primary amines on the molecule that bind protons [181]. Being cationic, chitosan has the potential to bind to many different food components, such as pectins, proteins, and inorganic polyelectrolytes (e.g., polyphosphate), allowing it to serve as a surface coating on meat products and fruits, or as an additive in acidic foods [181]. Another unique feature of chitosan is its natural antimicrobial activity against a wide range of target organisms. Yeasts and molds are the most sensitive group, followed by Gram-positive and then Gram-negative bacteria [182,183]. Therefore, it is not surprising that at the experimental level, there are numerous reports describing the use of chitosan as a food preservative [184–186]. Ferulic acid-incorporated films made from a starch–chitosan blend can find possible application as edible films or coatings. The barrier properties of the film blend improved considerably upon incorporation of oxidized ferulic acid. The amorphous nature of the ferulic acid-containing film blend supported good miscibility of the components, and Fourier transform infrared spectroscopy studies indicated intermolecular interactions between the different components. The ferulic acid-incorporated films were also found to reduce lipid-peroxide formation [187]. An edible coating composed of whey proteins and acetylated monoglycerides protected smoked salmon fillets against dehydration and lipid oxidation. A multicomponent coating comprised of collagen, casein, and cellulose derivatives reduced oil absorption during frying. Microbial proliferation, in particular *L. monocytogenes* growth, was reduced by coating fish products with a formulation of hydrocolloids, organic acids (lactic and acetic acids), and antimicrobial compounds [188]. In conclusion, edible films can have many potential applications in meat and fish processing, with their functional characteristics depending on their constituents.

13.6.4 Edible Coatings for Fruits and Vegetables

A seed is the product of the ripened ovule of gymnosperm and angiosperm plants which develops after fertilization and some growth in the mother plant. A major use of hydrocolloids in agriculture is as a seed coating. The seed surface is coated with a hydrophilic polymer, which can absorb water after planting, thereby increasing the rates, as well as the probabilities of germination. Since coating composition and manufacture can be controlled, coatings can be designed to delay germination, inhibit rot, control pests, fertilize, or bind the seed to the soil. Most of the reports on seed coatings deal with alginate coatings; reports on agar, various water-soluble cellulose ethers and hydrolyzed starch–polyacrylonitrile copolymers can also be found [189]. Soybean, cotton, corn, sorghum, sugar beet, and different vegetable seeds have been coated. As a result of coating, cotton and soybean yields increased by 20–30%. The poorer the growing conditions, the more pronounced the advantage of the coated seeds over controls [3,190].

A fruit is the ripened ovary—together with the seeds—of a flowering plant. The definition of a vegetable is traditional rather than scientific, and is somewhat arbitrary and subjective. Fresh fruits include 75–90% water. With the exception of a few fruits, their fat content is generally low. Fruits also include carbohydrates, minerals, acids, enzymes, pigments, aromatic compounds, and vitamins [142]. About 25% of freshly harvested fruits and vegetables are lost through spoilage [191]. Postharvest shelf-life extension can be achieved by applying edible coatings, which are semipermeable to water vapor and gases. Such coatings can enhance or replace other techniques used for the same purpose, such as modified-atmosphere or controlled-atmosphere storage [192,193]. Other achievements of coating applications include improvement of mechanical handling properties and retention of volatile flavor compounds [194,195]. As a result of coating, the permeability to oxygen and carbon dioxide changes and the coated commodity becomes an individual package with a modified atmosphere [1]. Respiration of the fruit or vegetable causes a reduction in oxygen and an increase in carbon dioxide. Therefore, care must be taken in designing the coating: if oxygen levels become too low, anaerobic reactions will occur, resulting in off-flavors and abnormal ripening [196]. Ethanol and acetaldehyde concentrations in the tissue can be used as monitors for final and next-to-final products of anaerobic respiration [196]. Therefore, if the coating can create moderate modified-atmosphere conditions, a climacteric fruit will exhibit decreased respiration, lower ethylene production, slower ripening, and extended shelf life [1].

The scientific literature is replete with examples of fruit and vegetable coatings. Therefore, the rest of this section is devoted to a few examples involving different gums and other ingredients, as well as a few commodities. Sodium alginate is the gum of choice for many experimentalists due to its easy dissolution and simple cross-linking reaction. Sodium-alginate coatings postponed the drying of mushroom tissue, thereby preventing changes in its texture during short periods of storage. As a result of the coating, fresh mushrooms had a better appearance and gloss, as did garlic [52,197,198]. Alginate or zein was used as edible coatings to delay postharvest ripening and to maintain tomato (*Solanum lycopersicon* Mill.) quality [199]. Coated tomatoes showed lower respiration rate and ethylene production than controls, with a twofold lower concentration of ethylene precursor. In addition, the evolution of parameters related to losses in tomato quality, such as softening, color evolution, and weight loss, was significantly delayed in coated tomatoes as compared to controls [199]. In peaches, sodium-alginate and MC coatings contributed to reductions in respiration rate and moisture loss, as well as to other changes relative to noncoated peaches. The sodium-alginate-coated and MC-coated samples maintained their acceptability up to 21 and 24 days, respectively, compared to 15 days for controls. These observations are the result of successful attempts to create modified atmospheres inside the coated fruit. The polysaccharide coatings retard gas migration, and delay ripening and senescence as effectively as the more costly controlled-atmosphere environments (in which specific levels of gases are maintained and regulated by external sources) [196]. In general, environmental levels of oxygen below 8% decrease ethylene production, and carbon dioxide levels above 5% prevent or delay many fruit-tissue responses to ethylene, including ripening [196]. Further discussion on postharvest physiology and effects of atmosphere and temperature can be found elsewhere [200], and the reader is also referred to Section 13.5.

The reduced respiration and transpiration rates as a result of the coatings were considered responsible for maintaining peach quality and increasing their shelf life [201]. Incorporation of ingredients found naturally in garlic skin (Fig. 13.2) in to the hydrocolloid (alginate- or gellan-based) gum solution before coating improved film adhesion to the surface of the coated commodity [51]. This was due to a reduction in the surface tension of the coating solution, and its improved wettability [202]. The effect of alginate- and gellan-based edible coatings on the shelf life of fresh-cut "Fuji" apples packed in trays with a plastic film of known permeability to oxygen was investigated by measuring changes in headspace atmosphere, color, firmness, and microbial growth during storage at 4 °C. Ethylene concentration in coated apples appeared to be delayed, while production of this gas was detected in uncoated apples from the very first day of storage. The coating included *N*-acetylcysteine (as an antibrowning agent), which helped maintain the firmness and color of the apple wedges throughout the storage period.

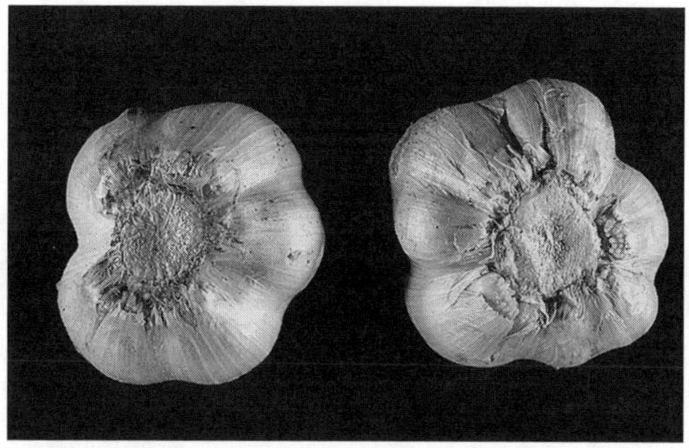

Figure 13.2 Alginate-coated garlic (left) vs. noncoated garlic (right). The thin coating (tens of microns) glued onto the garlic epidermis is strong and transparent. It eliminates the initial opening of the cloves seen in the noncoated commodity.

Application of the edible coatings also retarded the microbiological deterioration of the fresh-cut apples, and effectively prolonged their shelf life by 2 weeks in storage [203]. Pear wedges were immersed in a sodium chloride solution, followed by coating with chitosan or carboxymethyl chitosan solutions. The samples were packed in unsealed bags and stored at 4 °C for subsequent color, firmness, and weight loss measurements [204]. The sodium chloride treatment effectively inactivated *Escherichia coli* on the pear slices. Coating sodium chloride-treated samples with carboxymethyl chitosan significantly prevented the browning reaction and inhibited polyphenol oxidase activity. In addition, the coatings maintained tissue firmness and did not affect weight loss [204]. In parallel to the use of alginate and gellan or other familiar gums, the coating industry is trying to locate additional gums and/or raw materials, and less frequently, wastes or by-products from the food industry for use in future coatings. The effect of a whey protein isolate—pullulan—as a coating on fresh-roasted chestnut and roasted freeze-dried chestnut quality and shelf life was studied. Coatings were formed directly on the fruit surface, giving good adherence and coverage. The coating had a low, albeit significant effect on reducing moisture loss and decay incidence. The quality of the harvested chestnuts was improved and their shelf life increased [205]. The effects of cassava starch films and polyvinyl chloride (PVC) on the maintenance of postharvest quality of bell peppers stored at room temperature were evaluated. Fruit firmness and pH decreased, while titratable acidity and soluble solids increased toward the end of storage at room temperature. The treatments did not promote significant changes in total pectin content during the storage period, although lower soluble pectin content was observed in the fruits covered with PVC [206]. The application of gelatin—starch coatings delayed the ripening process of avocados, as indicated by better pulp firmness and retention of skin color, and lower weight loss of coated fruits in comparison with control avocados. The coatings also produced a delayed respiratory climacteric pattern [207]. Edible films based on candelilla wax with the potent antioxidant ellagic acid were able to significantly decrease the damage caused by *Colletotrichum gloeosporioides* to avocado fruits. The coating also significantly reduced the change in appearance and weight loss of the fruits [208].

Fruits and vegetables can be bought whole or in slices. The use of alginate- or gellan-based coating formulations on fresh-cut papaya pieces was studied to determine their ability to improve resistance to water vapor, to affect gas exchange, and to carry agents to help maintain the overall quality of the minimally processed fruit. Formulations containing glycerol plus ascorbic acid exhibited slightly improved water-barrier properties relative to the uncoated samples [209]. The incorporation of sunflower oil into the formulations resulted in increased water-vapor resistance of the coated samples. In general, coatings improved firmness of the fresh-cut product during the period studied [209]. Water loss and browning of cut apple slices were also inhibited by coatings of chitosan and lauric acid [210]. Application of surface coatings made up of shellac and aloe gel, singly or in combination, to apple slices resulted in a significant reduction in respiratory rates and a delay in peak ethylene synthesis rates during low-temperature storage. The coatings were found to minimize electrolyte leakage, changes in tristimulus color coordinates, and activity of the oxidizing enzymes, thereby enhancing the keeping quality of apple slices during storage [211]. Treatments to inhibit browning and decay and to extend the shelf life of fresh-cut "Keitt," "Kent," and "Ataulfo" mango cultivars were also investigated. Combinations of calcium chloride, the antioxidants ascorbic acid and citric acid, and two commercial film coatings resulted in reduced browning and deterioration of fresh-cut mangoes stored at 5 °C, especially for cv. Ataulfo [212]. A chitosan derivative used as a postharvest edible coating for fresh foods selectively forms permeable nontoxic films [184,213,214]. Synthetic polymers can modify the permeation response of a chitosan membrane to O_2 and CO_2 [215]. Extracellular microbial polysaccharides such as pullulan can be used to produce edible and biodegradable films that are clear, odorless, and tasteless and act as efficient oxygen barriers [168]. Transparent water-soluble films, with low permeability to oxygen, can be produced from hydroxypropylated amylomaize starch. The produced coatings exhibit increased bursting strength and elongation, and reduced tensile strength [216]. Other components have served to develop coatings of this nature for prunes, raisins, dates, figs, nuts, and beans. Other starch hydrolysate formulations were developed by the National Starch and Chemical Co. (West Bridgewater, NJ) to coat dried apricots, almonds, and apple slices [217]. A novel approach to developing coatings composed of fruit pulp is currently being developed [123]. One example of this is a mango film that provides a good oxygen barrier with sufficient mechanical properties to wrap whole and minimally processed mangoes. This film was found to reduce weight loss and extend the ripening period and shelf life of whole fresh mangoes [218].

Composite coatings of carboxymethylcellulose (CMC) with fatty-acid ester emulsifiers have been used for pears and bananas [193,219—223]. Delayed ripening and changes in internal gas levels were detected [219,220,224]. This type of marketable coating was first called Tal Pro-long (Courtaulds Group, London) and later Pro-long. It increases resistance to some types of fungal rot in apples, pears, and plums, but was not effective at decreasing respiration rate or water loss in tomato or sweet pepper [195,225]. Plums were treated with a coating material based on carbohydrate (Versa-Sheen) with sorbitol as a plasticizer and stored at 20 °C and 85% relative humidity. The coating treatment reduced the transmission rate of CO_2, O_2, and H_2O. The edible coating was effective at delaying the increase in pH and the loss of firmness, titratable acidity, hue angle, and malondialdehyde. The incorporation of sorbitol had beneficial effects by decreasing weight loss, CO_2, and ethylene exchange [226]. "Valencia" oranges coated with Tal Pro-long had improved essence and lower ethanol levels than controls [227]. Another coating with a similar composition, Semperfresh (United Agriproducts, Greely, CO), contains a higher proportion of short-chain unsaturated fatty-acid esters in its formulation [221]. These coatings retarded color advance and retained acids and firmness in apples relative to controls [193]. Semperfresh also extended the storage life of citrus [228]. The addition of waxes to Semperfresh produces glossier fruits with higher turgidity, less decay, and good flavor. On the other hand, Semperfresh is not effective at retarding water loss in melons [229]. A novel approach to extending the shelf life of fruits is the use of special ingredients within the coating (see Section 13.6.6). Aloe vera gel coating has been suggested as a means of preserving the quality and safety of cv. Crimson Seedless table grapes during cold storage and subsequent shelf life [230]. A similar approach for postharvest treatment of sweet cherry has been applied. Sweet cherry treated with aloe vera gel exhibited significant delays in the following: increases in respiration rate, rapid weight loss and color changes, accelerated softening and ripening, stem browning and increased microbial populations, without any detrimental effect on taste, aroma, or flavor [231].

13.6.5 Coatings for Fried Products

Frying is used to alter the edible quality of a food, to thermally destroy microorganisms and enzymes, and to reduce water activity on the food's surface [232]. Fats and oils are the cooking mediums. As an outcome of the high temperatures involved, fried foods cook rapidly, producing a unique flavor and texture. Oil uptake depends, among many other things, on frying temperature. It can increase up to 20% as the frying temperature is decreased [233], as also found in several other studies that looked at specific temperature ranges [234,235]. However, increasing the frying temperature is not always beneficial, as frying time is independent of oil temperatures in the range of 155–200 °C [236]. Thus many other factors affect oil uptake and these need to be considered on a case-by-case basis [237]. Since fried foods are high in calories relative to the same foods cooked in water or by other methods, the consistent demand for lower-calorie content of fried products can be satisfied, at least to a certain extent, by using coatings that retard oil and fat migration [142]. In addition to their properties as good film-formers, nonionic cellulose ethers are capable of yielding tough and flexible transparent films owing to the linear structure of the polymer backbone. Such coatings are soluble in water and resistant to fats and oils [238]. Hydrocolloids that serve in the construction of such films are gellan and LMP, among many others. Gellan-based coatings have been used for several years in Japan and other Asian countries with tempura-type fried foods [239]. An edible coating system called Fry Shield, developed and patented by Kerry Ingredients (Beloit, WI) and Hercules (Wilmington, DE), is based on calcium-reactive pectin and reduces fat intake during frying. French fries treated with pectin take up half the usual amount of oil [1].

MC and hydroxypropylmethylcellulose (HPMC) films are effective at reducing oil absorption by French fries, onion rings, and other fried, processed products [240] as a consequence of their resistance to fat and oil migration. The effect of an edible MC coating on reduced oil uptake during frying was analyzed on a dough system. The reduction in oil uptake was 30% for coated dough discs compared to noncoated ones; the coating did not modify either the water content of the samples or the quality attributes of the fried dough, such as color and texture. The coating reduced oil uptake, modifying the wetting properties relative to the interfacial tension and also becoming a mechanical barrier to lipids [241]. In addition to decreasing oil penetration and absorption by dry foods, the cellulose films can reduce weight loss and improve the adhesion of batter to products [1]. The effect of type, molecular weight, and concentration of cellulose ethers changed the microstructure of fried batter-coated potatoes [242]. For controlled-viscosity batters, the structure of fried batter containing MC of higher molecular weight and concentration with simultaneously higher moisture content showed greater hole sizes, which allowed a larger amount of oil penetration through the batter into the food substrate [242]. In contrast, the structure of batters with controlled initial moisture content with a higher molecular weight and concentration of MC was relatively more continuous; therefore, the batter helps in preventing oil penetration into the food substrate [242].

In addition to decreasing the oil content of a fried product, MC can be utilized to change its color. As an example, akara, which is a popular West African fried food, is prepared from a cowpea flour and soy flour blend. Application of an edible coating caused the product to absorb 26% less oil during frying than akara made from 100% cowpea flour. MC-coated akara was also found to be significantly different in total color from the control [243]: the 6% soy/MC-coated akara was significantly darker than the control. Akara formulated with 6% soy and coated with corn zein or MC was significantly more yellow and less red, respectively, than the control. These changes were due to the different oil absorption and reflectance of the coated surface of the fried product [243]. Frying in a gaseous environment influences the crispness of the product. When MC or whey protein isolate was incorporated into either the predust or batter before frying, fried chicken nuggets, using nitrogen gas, were crispier than those fried under steam [244]. Combinations of HPMC/MC with corn flour, seasoning, and salt were used to coat chicken pieces before frying at 180 °C for 10 min. The enrobing and frying time had a significant effect on the moisture and fat contents of the chicken pieces. The coating containing a 0.5% gum mixture was most acceptable from a sensory point of view. Enrobing increased the moisture content up to 42.6%, while fat uptake was reduced to 32% [245].

13.6.6 Miscellaneous Coatings

Studies of edible-film properties have progressed appreciably in the last century; the results are expected to be applied to a wide variety of foods, along with other applications [2,53]. Edible films can be utilized to coat nuts and peanuts, confectionery products and lightly processed agricultural products, to control the migration of preservatives and dough additives, in composite bilayers, blends and biopolymer films, and for many biotechnological uses [1]. More than 100 years ago (Fig. 13.3), a very interesting type of edible coating was developed in Japan. "Oblate" is the registered trade name for this edible paper which is prepared from starch and agar. It is 10−15 μm thick, and it is usually sold in drug stores in Japan for dosing medicines in powdered form. Oblate can also be employed as a convenient vehicle for wrapping condensed, sticky, sweet agar jelly: the nonsticky edible film coats the sticky confection and permits its consumption without having it stick to the fingers [4]. Gum arabic is used extensively in the coating industry due to the ease of preparation of solutions with over 50% gum, its compatibility with other plant hydrocolloids and its ability to produce stable emulsions with most oils over a wide range of pHs [77]. Gum arabic with or without gelatin has been used to produce protective films for chocolates, nuts, cheeses, and pharmaceutical tablets [246,247] and has also been reported to inhibit darkening of cooked potatoes [247]. A composite film of gum acacia and glycerol

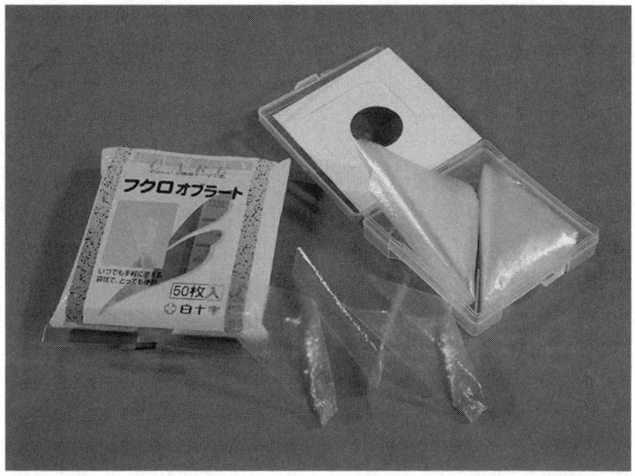

Figure 13.3 "Oblate" is the registered trade name for edible paper prepared from starch and agar. The product is generally sold in drug stores in Japan for dosing powdered medicines.

monostearate was reported to have good water-vapor barrier properties [248]. A composite coating based on gum arabic plus chitosan prolonged the storage life of banana fruits. The composite edible coating delayed color development and reduced the rate of respiration and ethylene evolution during storage as compared to uncoated controls. Sensory evaluation results also supported the coating's effectiveness at maintaining the overall quality of banana fruits [249]. Chitosan was also used to produce a novel emulsion coating for shelf-life extension of eggs in room temperature storage. The effects of mineral oil, chitosan solution, and their emulsions as coating materials in preserving the internal quality of eggs were evaluated during 5 weeks of storage at 25 °C [250]. Consumers evaluated surface properties and rated their purchase intent of freshly coated eggs. As storage time increased, Haugh unit and yolk index values decreased, whereas weight loss increased. Noncoated eggs rapidly changed from AA to B and C grades after 1 and 3 weeks, respectively. However, all emulsion-coated eggs maintained their A-grade quality for 4 weeks. The emulsion coatings preserved the eggs' internal quality and prolonged their shelf life [250].

A carrageenan-based coating, Soageena (edible polysaccharide), was developed by Mitsubishi International Corp. (Tokyo, Japan) for fresh produce [251,252]. Other carrageenan coatings have been used to retard moisture loss from coated foods [253]. A carrageenan-based coating applied to cut grapefruit halves successfully decreased shrinkage and taste deterioration [254]. Carrageenan, gellan, and alginate have been used to coat fresh, soft-brined cheeses (Fig. 13.4) [78]. Similar to coated meats, these hydrocolloid coatings offer only a weak barrier against moisture transport. However, the water contained within the different coatings serves as a sacrificing agent, and loss of product moisture is delayed until this liquid has been evaporated [78].

Chemically modified celluloses can be used to manufacture edible and biodegradable films. MC, HPMC, hydroxypropyl cellulose (HPC), and CMC are examples of raw materials for the production of coatings with moderate strength, resistance to oils and fats, elasticity and transparency, without odor or taste, and the ability to serve as a moderate barrier to moisture and oxygen [255,256]. Information on MC and HPMC can be found elsewhere [257]. MC films make fine barriers to oil- and fat-migration [258]. MC and HPMC can be used to manufacture composite films with solid lipids [132,259–261]. Bilayer films consisting of a solid lipid and a layer of MC or HPMC have been used to decrease the migration of water in model foods [262]. Application of edible HPMC–lipid composite coatings containing food preservatives was found to be an environmentally friendly method for reducing the losses caused by postharvest citrus diseases. These coatings could be used as an alternative to synthetic chemical fungicides for decay control in citrus packinghouses [263]. Nonionic cellulose ethers are capable of producing tough, transparent, and flexible films that are both water-soluble, and fat- and oil-resistant [238]. MC films prevent lipid migration [264]. HPMC and a bilayer film consisting of stearic–palmitic acid slowed moisture transfer from tomato paste to crackers [265]. Formulations involving MC, HPMC, and HPC delayed browning

Figure 13.4 Coated white-brined cheese.

and increased volatile flavor components [67]. HPC films retarded oxidative rancidity and moisture absorption in nut meats, coated nuts, and candies [238,266].

Whey is the watery, fluid part of the milk that remains after the curd, or casein precipitate, has been removed. α-Lactalbumin and β-lactoglobulin (whey protein fractions) and whey protein isolate are used to manufacture edible films due to their beneficial functional properties and being industrial surplus [267]. Whey proteins produce transparent, flexible, colorless, and odorless edible films which can serve as moderately good moisture barriers and excellent oxygen barriers [268,269]. Incorporation of plasticizers (e.g., polyols and mono-, di-, and oligosaccharides) reduces their brittleness. Whey-protein-isolate films have limited barrier properties, which can be enhanced by the inclusion of other film-forming polysaccharides [270]. Protein, sorbitol, beeswax, and potassium sorbate were found to influence the water-vapor permeability and water solubility of whey protein films. Beeswax was the most important factor influencing the stickiness and appearance of the films. The amount of protein had no effect on stickiness or appearance, while the amount of sorbitol (35–50%, w/w^{-1}) in the films had no influence on appearance [271]. Sorbitol and glycerol were used to plasticize a sodium caseinate-based edible coating. Berry cactus fruits were treated with this edible coating and their phytochemical contents were evaluated. Even though the edible coatings controlled moisture loss from the fruits, they did not provide an equally efficient barrier to the entry of oxygen. Thus it was determined that in the absence of any agents to prevent senescent breakdown, the decline in polyphenol content may not be preventable [272]. Scanning electron microscopy demonstrated these films' favorable structure and the addition of pullulan at low concentrations was sufficient to significantly modify their barrier properties and improve their potential characteristics for food applications [273].

13.7 Novel Products

Novel edible-coating products are ubiquitous in the marketplace. An example of these is Listerine PocketPaks®: small patches of edible film that melt instantly on the tongue, releasing breath freshener. PocketPaks contains at least four hydrocolloids: carrageenan, locust bean gum, pullulan, and xanthan gum. Listerine PocketPaks pushed the dental-care category into new territory by being the first item of this kind available to active consumers. Its success has paved the way for many other such products. Strips could prove appealing as a carrier for drug delivery. In 2004, the Swiss pharmaceutical giant Novartis introduced Triaminic® and Theraflu® Thin Strips™. Thin Strips were the first multisymptom cough and cold medicines to deliver an accurate dose in thin-film format and were selected from more than 100 applicants for their taste and quality ingredients, packaging, consumer acceptance, innovation, differentiation, and convenience (http://www.webpackaging.com; http://www.iqdurableink.com). Another report discussed novel polymeric film coatings that target the colon. The properties of these films, based on blends of ethylcellulose and Nutriose (a water-soluble, branched dextrin), were optimized, and shown to be highly promising for the site-specific delivery of drugs to the colon in patients suffering from inflammatory bowel diseases. Desired system properties could be easily adjusted by varying the polymer:polymer blend ratio, as well as plasticizer content [274].

Clarifoil, the producers of films for print and packaging uses, launched Clarisol, an innovative new range of water-soluble films: these are cast PVA/PVOH water-soluble films that are offered as consistent-quality products suitable for a wide range of functional uses with the following requirements: convenience, delivery of measured doses, and easy handling for consumers and industrial operatives. They could potentially also have uses in specialty products. Pets may be the next target for these "drugs-on-film," as edible films could serve as a convenient alternative to stuffing pills down pets' throats. Potentially, drugs could also be given at lower doses on films because they are better absorbed through the tongue.

The food industry has provided many new uses for edible-film technology. Meat producers are using films to cure and glaze hams. Athletes consume electrolyte strips in lieu of sports drinks to fight dehydration. Films may someday be used for separating the tomato sauce from the crust on a frozen pizza so that the crust stays crisp. Origami Foods, in cooperation with the USDA's Agricultural Research Service, has developed a new, unique food product in which almost any fruit, vegetable, or their combination can be used to create edible films. Such products are low-fat, low-calorie, tasty, and healthy, and were developed for people who have an aversion to seaweed, or are otherwise interested in an alternative (http://www.origami-foods.com; http://www.ceepackaging.com). Another development in the area of edible films relates to decorations. Today, it is possible to decorate cakes with computer-designed images. These images (sometimes clip-art) are printed with new high-quality food-grade inks on edible paper, using a standard inkjet printer. The attractive printouts can be placed on any cake or other baked goods. The same process can be used to produce designed candy wrappers, rolls of paper for pies or quiches, or decorative protective covers for condiments and tortes, and in hundreds of other applications where a cosmetic or personal touch is desired on a dessert product.

Novel products, developments, and research directions that could be important in the field of coatings or that use coatings as part of their achievement have been developed in recent years. These developments include (but are not limited to) a novel composite film for potential food-packaging applications that was prepared by plasticization of protein coatings on polypropylene film [275]. The optical and tensile properties of the films depend on the types of proteins and plasticizers used. High glossy surfaces were observed on films coated with whey protein isolate and corn zein, with the sucrose-plasticized whey protein isolate coating giving the highest gloss. Whey protein isolate-coated films also showed greater transparency and tensile strength than the other coated films. The additive nisin has an E number of 234 and is a polycyclic peptide antibacterial agent with 34 amino acid residues, used as a food preservative. It is a "broad-spectrum" bacteriocin used in processed cheese, meat, beverages, etc., at levels ranging from ~1–25 ppm, depending on the food type and regulatory approval during production, in order to extend shelf life by suppressing Gram-positive spoilage and pathogenic bacteria (http://en.wikipedia.org/wiki/Nisin). Nisin-incorporated whey-protein isolate coatings on polypropylene film exhibited significant growth inhibition of the bacterium *Lactobacillus plantarum* [275]. Another product in the packaging realm is a novel, quickly dissolvable, edible, and heat-sealable film consisting of a blend of konjac glucomannan and gelatin. This product was successfully prepared by using a solvent-casting technique with different blending ratios of the two polymers. Taking the degradability into account, this film blend might be a perfect material for edible inner packaging [276].

A novel research approach makes use of alginate coating to monitor the uptake of solids during osmotic dehydration of a model food system [277]. A combination of product coating with alternative settings of product/solution contact was investigated to monitor solids uptake during osmotic dehydration. Potato was used as a model plant material for short-term (i.e., 3 h) osmotic treatment in a series of solutions with decreasing or increasing concentrations of sucrose to simulate co-current or counter-current product/solution contact (flow) [277]. The alginate coating yielded significantly decreased solids uptake, without negatively affecting water removal. Moreover, the overall "dehydration efficiency" was drastically improved by combined coating and counter-current contact. The importance of this approach is that a mass-exchange prediction can be applied under alternative treatment scenarios regarding initial product solids [277].

13.8 Nonfood Gum Coatings

Paper is a thin material used mainly for writing, printing, or packaging. It is produced by pressing together moist fibers, typically cellulose pulp derived from wood or grasses, and drying them into flexible sheets. Many hydrocolloids can be utilized to obtain successful paper coatings. Agar has been found suitable for use in photographic papers when esterified with succinic or phthalic anhydride and after enzymic hydrolysis. Agar can also be used as an adhesive in the gloss-finishing of paper products. HPC is used for coating, due to its thermoplasticity and solvent solubility. It serves as an oil and fat barrier and is responsible for thermoplastic coating. Polyethylene oxide is also used for paper coating and sizing. As a processing additive, it is used as a fiber-formation aid [3]. Another article discusses the associative behavior of CMCs, hydroxy ethyl celluloses, and hydrophobically modified cellulosic thickeners in clay-based coatings and their effects on coating rheology and coated-paper properties [278]. PVA is used in paper sizings and coatings. Polyvinylpyrrolidone is used in all types of paper manufacture, mostly as an economical fluidizer and antiblocking agent in paper coating. The starches used in coating colors can be enzymatically converted, thermochemically converted, or oxidized to dextrins, hydroxyethyl starch ethers, and starch acetates [4].

Polyacrylic acid can be used as a film former. One important application is the use of the sodium salt of polyacrylic acid as the major component of nonglare coatings for headlights. Antifogging coatings for glass and transparent plastics for optical use have also been created by cross-linking polyacrylic acid with aminoplast resins to produce scratch- and water-resistant coatings [4]. Biological materials, which are generated from bio-sources, can significantly affect the chemical, mechanical, and visual properties of automotive coating systems. Such materials include mainly bird droppings, tree gum, and insect bodies. It was shown that depreciation of the surface free energy of additive loaded films could greatly improve their behavior against gum, apparently due to their enhanced nonstick characteristic [279]. In the manufacture of fluorescent lights, phosphor bonding to glass tubes is enhanced by using polyethylene oxide resin as a temporary binder in combination with barium nitrate as a bonding agent [77]. For the manufacture of coatings, paints, foams, or adhesives, xanthan gum (a polysaccharide produced by a process involving fermentation of glucose or sucrose by the *Xanthomonas campestris* bacterium) is compatible with the common types of latex emulsions, making it effective as a stabilizer, thickener, and modifier of rheological properties [77].

Recently, a novel method of producing chitosan-coated alginate–polyvinyl alcohol beads for the biodegradation of polycyclic aromatic hydrocarbons by microorganisms was reported [280]. Polycyclic aromatic hydrocarbons are potential hazards in the environment owing to their toxic, carcinogenic, and recalcitrant nature. An encapsulated form of the pollutant in chitosan-coated alginate–polyvinyl alcohol beads was developed. One advantage offered by this method seems to be the resultant negligible toxic effects of pyrene and solvents on the degrading microorganisms since these are encapsulated and therefore not in direct contact with the organism [280].

13.9 Next Generation of Edible Films

A nanometer equals one billionth of a meter, and can be written as 1×10^{-9} m [281]. Inclusion of nanoparticles in edible and nanocomposite films is expected to improve the mechanical and oxidative stability, barrier properties, and biodegradability of conventional polymeric matrices [282]. Four different types of chitosan-based nanocomposite films were prepared using a solvent-casting method by incorporation with four types of nanoparticles. A certain degree of intercalation occurred in the nanocomposite films, with the highest intercalation found in the unmodified montmorillonite-incorporated films, followed by films with organically modified montmorillonite and silver-zeolite [283]. In all nanocomposite films except those incorporating silver, nanoparticles were dispersed homogeneously throughout the chitosan polymer matrix. Consequently, mechanical and barrier properties of chitosan films were affected through intercalation of nanoparticles. In addition, chitosan-based nanocomposite films, especially silver-containing ones, showed a promising range of antimicrobial activities [283].

An example from a different product area is double-core tennis balls with a nanocomposite coating that keeps them bouncing twice as long as the old-style balls and also substantially extends the balls' shelf life [281]. To date, not many studies have been performed on the possibility of incorporating nanoparticles to improve the physical properties of biodegradable nanocomposites [284]. Adding clay montmorillonite to pectins could lower the diffusion of oxygen [284]. Similarly, a considerable improvement in physical properties was recorded for nanocomposites prepared with gelatin and montmorillonite [285,286]. Edible coatings and films constitute a viable means of incorporating food additives and other substances in order to enhance product color, flavor, and texture, and to control microbial growth [287]. To this aim, nanoparticles can be used as carriers of antimicrobials and additives. Nanotechnology provides a number of ways to create novel laminate films to be used in the food industry. A nanolaminate consists of two or more layers of material with nanometer dimensions that are physically or chemically bonded to each other. A deposition technique can be utilized to coat a charged surface with interfacial films consisting of multiple nanolayers of different materials [288]. Gum arabic has been regarded as a viable option for the surface coating of magnetic iron oxide nanoparticles. The potential for using coated magnetic nanoparticles as solid supports for the attachment of synthetic affinity ligands specific for particular antibodies was investigated. Success of the *in situ* triazine ligand synthesis was confirmed by fluorescence assays [289]. Success of drug delivery via microspheres is sometimes limited, owing to their short residence time at the site of absorption. It would, therefore, be advantageous to have means for providing intimate contact of the drug-delivery system with the absorbing membranes. This can be achieved by coupling bioadhesion characteristics with microspheres to produce mucoadhesive microspheres [290]. Galactomannan-coated glipizide microspheres were prepared and characterized in *in vitro* systems. In addition, these microspheres were evaluated *in vivo* for their sustained glucose-lowering effect. A single dose of galactomannan-coated glipizide microspheres reduced blood glucose levels for a longer period than an immediate-release formulation of glipizide [290].

Coating technology may also have a number of important applications in the food industry. Nanolaminates could be used as coatings that are attached to food surfaces, rather than as self-standing films [291,292]. To produce these layers, a variety of different adsorbing substances can be used. The functionality of the final film will depend on many factors, including type of materials, number of layers, their sequences, preparation conditions, etc. Thus research and development of bio-nanocomposite materials for food applications such as packaging and other food-contact surfaces is expected to grow in the next decade with the advent of new polymeric materials and composites with inorganic nanoparticles [282].

An appreciable increase in stability in chitosan-layered nanocomposites has also been reported [293]. Another publication described the inclusion of different-sized fillers, all less than 1.0 μm, in HPMC-based films, to improve the film's mechanical properties [294]. Such results will enable food scientists to envision a new generation of composite edible films and barriers; they will no longer be restricted to emulsified and bilayer films for current and novel applications [294]. The application of nanocomposites promises to expand the use of edible and biodegradable films [295,296] and to help reduce the packaging waste associated with processed foods [297,298]. In addition, inorganic particles may be used to add multiple functionalities, such as color and odor, but also to act as reservoirs for the controlled-release functions of drugs or fungicides [104,105,299]. In conclusion, future research directions will make use of nanotechnological tools to improve tensile, barrier, and delivery properties of edible films, to develop hydrophobic nanoparticles for improving water-barrier properties and to use nanoparticles to deliver health-promoting compounds, antimicrobials, sweetness, flavors, and other active agents. It is expected that future edible films will include antimicrobial edible films, added-value fruit and vegetable films, and invisible edible films, and that the tools of nanotechnology will be used to achieve developments in packaging and film-production technologies.

References

[1] A. Nussinovitch, Water-Soluble Polymer Applications in Foods, Blackwell Science Ltd, London, 2003.

[2] J.M. Krochta, C. De Mulder-Johnston, Edible and biodegradable polymer films, Food Technol. 51 (1997) 61–73.

[3] A. Nussinovitch, Hydrocolloid coating of foods: a review, Leatherhead Food Ind. J. 1 (1998) 174–188.

[4] R.L. Davidson, Handbook of Water-Soluble Gums and Resins, McGraw-Hill Book Company, New York, 1980.

[5] A. Nussinovitch, Hydrocolloids for coatings and adhesives, in: G.O. Phillips, P.A. Williams (Eds.), Handbook of Hydrocolloids, Woodhead Publishing Limited, Cambridge, UK, 2009.

[6] IMR, Food Hydrocolloids Conference Proceedings, Lisbon, Portugal, April 22–24, 2007.

[7] M.E. Embuscado, K.C. Huber, Edible Films and Coatings for Food Applications, Springer, London & New York, 2009.

[8] V.M. Hernandez-Izquierdo, J.M. Krochta, Thermoplastic processing of proteins for film formation—a review, J. Food Sci. 73 (2008) 30–39.

[9] I. Greener, Physical properties of edible films and their components, in: Ph.D. dissertation, University of Wisconsin, 1992.

[10] S. Reading, M. Spring, The effects of binder film characteristics on granule and tablet properties, J. Pharm. Pharmacol. 36 (1984) 421–426.

[11] M. Pommet, A. Redl, M.H. Morel, S. Domenek, S. Guilbert, Thermoplastic processing of protein-based bioplastics: Chemical engineering aspects of mixing, extrusion and hot molding, Macromol. Symp. 197 (2003) 207–217.

[12] M.P. Stevens, Polymer Chemistry. An Introduction, Oxford University Press, New York, 1999.

[13] J.L. Kokini, A.M. Cocero, H. Madeka, E. de Graaf, The development of state diagrams for cereal proteins, Trends Food Sci. Technol. 5 (1994) 281–288.

[14] S. Okamoto, Factors affecting protein film formation, Cereal Foods World 23 (1978) 256–262.

[15] B. Cuq, N. Gontard, S. Guilbert, Packaging materials based on natural polymers, Industries Alimentaires et Agricoles 114 (1997a) 110–116.

[16] L. Liu, J.F. Kerry, J.P. Kerry, Effect of food ingredients and selected lipids on the physical properties of extruded edible films/casings, Int. J. Food Sci. Technol. 41 (2006) 295–302.

[17] A. Esen, A proposed nomenclature for the alcohol-soluble proteins (zeins) of maize (Zea mays L.), J. Cereal Sci. 5 (1987) 117–128.

[18] R.A. Reiners, J.S. Wall, G.E. Inglett, Corn proteins, potential for their industrial use, in: Y. Pomerantz (Ed.), Industrial Uses of Cereals, American Association of Cereal Chemists, Inc, St. Paul, MN, 1973, pp. 285–302.

[19] J.A. Scramin, D. de Britto, L.A. Forato, R. Bernardes-Filho, L.A. Colnago, O.B.G. Assis, Characterization of zein–oleic acid films and applications in fruit coating, Int. J. Food Sci. Technol. 46 (2011) 2145–2152.

[20] R.T. Szyperski, J.P. Gibbons, Zein systems developed for heat cured coatings, Paint Varnish Prod. 53 (1963) 65–73.

[21] A. Gennadios, A.H. Brandenburg, C.L. Weller, R.F. Testin, Effect of pH on properties of wheat gluten and soy protein isolate films, J. Agric. Food Chem. 41 (1993a) 1835–1839.

[22] N. Gontard, S. Guilbert, J.L. Cuq, Edible wheat gluten films: Influence of the main process variables on film properties using response surface methodology, J. Food Sci. 57 (1992) 190–195. 199.

[23] E.E. McDermott, D.J. Stevens, J. Pace, Modification of flour proteins by disulfude interchange reactions, J. Sci. Food Agric. 20 (1969) 213–217.

[24] J.S. Wall, M. Friedman, L.H. Krull, J.F. Cavins, A.C. Beckwith, Chemical modification of wheat gluten proteins and related model systems, J. Polym. Sci. Part C 24 (1968) 147–161.

[25] J.S. Wall, A.C. Beckwith, Relationship between structure and rheological properties of gluten proteins, Cereal Sci. Today 14 (1969) 16–18.
[26] L.C. Wu, R.P. Bates, Soy protein lipid films. 1. Studies on the film formation phenomenon, J. Food Sci. 37 (1972) 36–39.
[27] C. Farnum, D.W. Stanley, J.L. Gray, Protein-lipid interactions in soy films, Can. Inst. Food Sci. Technol. J. 9 (1976) 201–206.
[28] D. Fukushima, J. Van Buren, Mechanisms of protein insolubilization during the drying of soy milk. Role of disulfide and hydrophobic bonds, Cereal Chem. 47 (1970) 687–696.
[29] D.F. Cheesman, J.T. Davies, Physiochemical and biological aspects of proteins at interfaces, Adv. Protein Chem. 9 (1954) 439–501.
[30] S.J. Circle, E.W. Meyer, R.W. Whitney, Rheology of soy protein dispersions. Effect of heat and other factors on gelation, Cereal Chem. 41 (1964) 157–172.
[31] W.J. Wolf, T. Tamura, Heat denaturation of soybean 11S protein, Cereal Chem. 46 (1969) 331–344.
[32] A. Gennadios, C.L. Weller, R.F. Testin, Temperature effect on oxygen permeability of edible protein-based films, J. Food Sci. 58 (1993b) 212–214.
[33] Y. Aboagye, D.W. Stanley, Texturization of peanut proteins by surface film formation. Influence of process parameters on film forming properties, Can. Inst. Food Sci. Technol. J. 18 (1985) 12–20.
[34] H.W. Jones, R.A. Whitmore, US Patent 3,694,234, (1972).
[35] K.A. Boni, Strengthened edible collagen casing and method of preparing same. US Patent 4,794,006, (1988).
[36] E. Kuntz, Preparation of collagenous materials. US Patent 3,152,976, (1964).
[37] C. Robinson, The hot and cold forms of gelatin, in: J.T. Randel (Ed.), Nature and Structure of Collagen, Academic Press, Inc, New York, 1953, pp. 96–105.
[38] S. Soradech, J. Nunthanid, S. Limmatvapirat, M. Luangtana-anan, An approach for the enhancement of the mechanical properties and film coating efficiency of shellac by the formation of composite films based on shellac and gelatin, J. Food Eng. 108 (2012) 94–102.
[39] R.C. Mabesa, R.T. Marshall, M.E. Anderson, Milk films exposed to high humidity: studies with electron microscopy and electrophoresis, J. Food Prot. 43 (1980) 29–35.
[40] B.E. Leach, N.H. Rainey, M.J. Pallansch, Physical and chemical analysis of film formed at air-water interface on reconstitution of whole milk powders, J. Dairy Sci. 49 (1966) 1465–1468.
[41] J.R. Brunner, in: J.R. Whitaker, S.R. Tannenbaum (Eds.), Milk Proteins in Food Proteins, Avi Publishers Inc, Westport, CT, 1977, pp. 175–208.
[42] D.G. Dalgleish, Milk proteins—chemistry and physics, in: J.E. Kinsella, W.G. Soucie (Eds.), Food Proteins Champagne, IL, American Oil Chemists Society, 1989, pp. 155–178.
[43] V. Hershko, D. Weisman, A. Nussinovitch, Methods for studying surface topography and roughness of onion and garlic skins for coating purposes, J. Food Sci. 63 (1998) 317–321.
[44] K.L. Mittal, The role of the interface in adhesion phenomena, Polym. Eng. Sci. 17 (1997) 467–473.
[45] A.W. Adamson, in: A.W. Adamson, A.P. Gast (Eds.), Physical Chemistry of Surfaces, third ed., John Wiley and Sons, New York, 1976, pp. 1–43.
[46] A.F.M. Barton, CRC Handbook of Solubility Parameters and Other Cohesion Parameters, 427, CRC Press, Boca Raton, FL, 1983. 441.
[47] J.R. Dann, Forces involved in the adhesive process. Critical surface tensions of polymeric solids as determined with polar liquids, J. Colloid Interface Sci. 32 (1970) 302–319.
[48] A.G. Gaonkar, Surface and interfacial activities and emulsion of some food hydrocolloids, Food Hydrocolloid 5 (1991) 329–337.
[49] M.T. Wu, D.K. Salunkhe, Control of chlorophyll and solanine synthesis and sprouting of potato tubers by hot paraffin wax, J. Food Sci. 37 (1972) 629–630.
[50] J.F. Oliver, S.G. Mason, Microspreading studies of rough surface by scanning electron microscopy, J. Colloid Interface Sci. 60 (1977) 480–487.
[51] A. Nussinovitch, V. Hershko, H.D. Rabinowitch, Israel Patent Application no. 111,495 PCT/US95/14252, (1994).

[52] A. Nussinovitch, V. Hershko, Gellan and alginate vegetable coatings, Carbohydr. Polym. 30 (1996) 185–192.

[53] V. Hershko, A. Nussinovitch, Relationships between hydrocolloid coating and mushroom structure, J. Agric. Food Chem. 46 (1998) 2988–2997.

[54] R.N. Wenzel, Surface roughness and contact angle, Ind. Eng. Chem. 28 (1936) 988–993.

[55] G. Ward, A. Nussinovitch, Gloss properties and surface morphology relationships of fruits, J. Food Sci. 61 (1996) 973–977.

[56] J.M. Krochta, E.A. Baldwin, M. Nisperos-Carriedo, in: Edible Coatings and Films to Improve Food Quality, Technomic Publishing Co, Basel, Switzerland, 1994.

[57] S. Guilbert, Technology and application of edible protective films, in: M. Mathlouthi (Ed.), Food Packaging and Preservation Theory and Practice, Elsevier Applied Science Publishing Co, London, 1986, p. 371.

[58] V. Hershko, E. Klein, A. Nussinovitch, Relationships between edible coatings and garlic skin, J. Food Sci. 61 (1996) 769–777.

[59] T.P. Aydt, C.L. Weller, R.F. Testin, Mechanical and barrier properties of edible corn and wheat protein films, Am. Soc. Agric. Eng. 34 (1991) 207–211.

[60] M. Karel, Protective packaging of foods, in: O. Fennema (Ed.), Principles of Food Science, Marcel Dekker, New York, 1975, pp. 399–464.

[61] S.A. Stern, T.P. Sinclair, T.P. Gaeis, An improved permeability apparatus of the variable-volume type, Mod. Plastics 10 (1964) 50–53.

[62] A.H. Landroac, B.E. Proctor, Gas permeability of films, Mod. Packag. 25 (10) (1952) 131–135. 199–201.

[63] D.G. Quast, M. Karel, Technique for determining oxygen concentration within packages, J. Food Sci. 37 (1972) 490–491.

[64] Z.W. Wicks, F.N. Jones, S. Peter-Pappas, Organic Coatings, Science and Technology, Wiley and Sons, Inc., New York, 1994.

[65] D. Greener, O. Fennema, Edible films and coatings: characteristics, formation, definitions and testing methods, in: J.M. Krochta, E.A. Baldwin, M.O. Nisperos-Carriedo (Eds.), Edible Coatings and Films to Improve Food Quality, Technomic Publishing Co. Inc, Lancaster, Basel, 1994, pp. 1–24.

[66] I. Greener, O. Fennema, Barrier properties and surface characteristics of edible bilayer films, J. Food Sci. 54 (1989) 1393–1399.

[67] M.O. Nisperos-Carriedo, E.A. Baldwin, P.E. Shaw, Development of edible coating for extending postharvest life of selected fruits and vegetables, Proc. Fla. State Hort. Soc. 104 (1992) 122–125.

[68] S. Chen, A. Nussinovitch, Galactomannans in disturbances of structured wax-hydrocolloid-based-coatings of citrus fruit (easy-peelers), Food Hydrocolloid 14 (2000) 561–568.

[69] S. Chen, A. Nussinovitch, Permeability and roughness of wax-hydrocolloid coatings and their limitations in determining citrus fruit overall quality, Food Hydrocolloid 15 (2001) 127–137.

[70] R.L. Shewfelt, Quality of minimally processed fruits and vegetables, J. Food Qual. 10 (1987) 143–156.

[71] J.J. Kester, O.R. Fennema, Edible films and coatings: a review, Food Technol. 42 (1988) 47–59.

[72] K.M. Sreenivas, K. Chaudhari, S.S. Lele, Ash gourd peel wax: extraction, characterization, and application as an edible coat for fruits, Food Sci. Biotechnol. 20 (2011) 383–387.

[73] R.A. Baker, E.A. Baldwin, M.O. Nisperos-Carriedo, Edible coatings and films for processed foods, in: J.M. Krochta, E.A. Baldwin, M.O. Nisperos-Carriedo (Eds.), Edible Coatings and Films to Improve Food Quality, Technomic Publishing Co, Basel, Switzerland, 1994, pp. 89–105.

[74] D.C. Rico-Pena, J.A. Torres, Sorbic acid and potassium sorbate permeability of an edible methylcellulose-palmitic acid film: water activity and pH effects, J. Food. Sci. 56 (1991) 497–499.

[75] H.A. Swenson, J.C. Miers, T.H. Schultz, H.S. Owens, Pectinate and pectate coatings. Applications to nuts and fruit products, Food Technol. 7 (1953) 232–235.

[76] H.B. Cosler, Methods of producing zein-coated confectionery. US Patent 2,791,509, (1957).

[77] A. Nussinovitch, Hydrocolloid Applications. Gum Technology in the Food and Other

Industries, Blackie Academic & Professional, London, 1997a.
[78] N. Kampf, A. Nussinovitch, Hydrocolloid coatings of cheeses, Food Hydrocolloid 14 (2000) 531–537.
[79] F. Debeaufort, J.A. Queada-Gallo, A. Voilley, Edible films and coatings: tomorrow's packaging: a review, Crit. Rev. Food Sci. 38 (1998) 299–313.
[80] D. Lin, Y.Y. Zaho, Innovations in the development and application of edible coatings for fresh and minimally processed fruits and vegetables, Compr. Rev. Food Sci. Food Saf. 6 (2007) 60–75.
[81] T.H. McHugh, Effects of macromolecular interactions on the permeability of composite edible films, in: N. Parris, A. Kato, L. Creamer, J. Pearce (Eds.), "Macromolecular Interactions in Food Technology", American Chemical Society, Washington, DC, 1996, pp. 134–144.
[82] A. Nussinovitch, Agricultural uses of hydrocolloids, in: A. Nussinovitch (Ed.), Hydrocolloid Applications. Gum Technology in the Food and Other Industries, Blackie Academic & Professional, London, 1997b, pp. 169–184.
[83] B. Ghanbarzadeh, A.R. Oromiehie, M. Musavi, P.M. Falcone, Z.E. D-Jomeh, E.R. Rad, Study of mechanical properties, oxygen permeability and AFM topography of zein films plasticized by polyols, Packaging Technol. Sci. 20 (2007) 155–163.
[84] P. Bergo, R.A. Carvalho, P.J.A. Sorbal, F.R.S. Bevilacqua, Microwave transmittance in gelatin-based films, Meas. Sci. Technol. 17 (2006) 3261–3264.
[85] A. Pinotti, M.A. Garcia, M.N. Martino, N.E. Zaritzky, Study on microstructure and physical properties of composite films based on chitosan and methylcellulose, Food Hydrocolloid 21 (2007) 66–72.
[86] M.N. Emmambux, M. Stading, J.R.N. Taylor, Sorghum kafirin film property modification with hydrolysable and condensed tannins, J. Cereal Sci. 40 (2004) 127–135.
[87] H.J. Bae, D.S. Cha, W.S. Whiteside, H.J. Park, Film and pharmaceutical hard capsule formation properties of mungbean, water chestnut, and sweet potato starches, Food Chem. 106 (2008) 96–105.
[88] P.Y. Hamaguchi, W. Weng, T. Kobayashi, J. Runglertkreingkrai, M. Tanaka, Effect of fish meat quality on the properties of biodegradable protein films, Food Sci. Technol. Res. 13 (2007) 200–204.
[89] S. Guilbert, B. Cuq, N. Gontard, Recent innovations in edible and/or biodegradable packaging materials, Food Addit. Contam. 14 (1997) 741–751.
[90] P. Li, M.M. Barth, Impact of edible coatings on nutritional and physiological changes in lightly processed carrots, Postharvest Biol. Technol. 14 (1998) 51–60.
[91] D.C.R. Pena, J.A. Torres, Sorbic acid and potassium sorbate permeability of an edible methylcellulose-palmitic acid films: water activity and pH effects, J. Food Sci. 56 (1991) 497–499.
[92] Y. Pranoto, V. Salokhe, K.S. Rakshit, Physical and antibacterial properties of alginate-based edible film incorporated with garlic oil, Food Res. Int. 38 (2005) 267–272.
[93] S.A. Valencia-Chamorro, L. Palou, M.A. del Río, M.B. Pérez-Gago, Antimicrobial edible films and coatings for fresh and minimally processed fruits and vegetables: a review, Crit. Rev. Food Sci. Nutr. 51 (2011) 872–900.
[94] D.W. Wong, K.S. Gregorski, J.S. Hudson, A.E. Pavlath, Calcium alginate films: thermal properties and permeability to sorbate and ascorbate, J. Food Sci. 61 (1996) 337–341.
[95] S.M. Alzamora, S. Guerrero, Plant antimicrobials combined with conventional preservatives for fruit products, in: S. Roller (Ed.), Natural Antimicrobials for the Minimal Processing Foods, CRC Press, Boca Raton, FL, 2003, pp. 235–249.
[96] E.A. Baldwin, M.O. Nisperos, X. Chen, R.D. Hagenmaier, Improving storage life of cut apple and potato with edible coating, Postharvest Biol. Technol. 9 (1996) 151–163.
[97] S.A. Burt, R.D. Reinders, Antibacterial activity of selected plant essential oils against *Escherichia coli* O157:H7, Lett. Appl. Microbiol. 36 (2003) 162–167.
[98] FDA, Chemical preservatives, in: Code of Federal Regulation, US Government Printing Office, Washington, DC, 1987, title 21, part 182.
[99] J.J. Nicolas, F.C. Richard-Forget, P.M. Goupy, M.J. Amiot, S.Y. Aubert, Enzymatic browning

reactions in apple and apple products, Crit. Rev. Food Sci. Nutr. 34 (1994) 109−157.
[100] A.J. McEvily, R. Iyengar, W.S. Otwell, Inhibition of enzymatic browning in foods and beverages, Crit. Rev. Food Sci. Nutr. 32 (1992) 253−273.
[101] G. Oms-Oliu, I. Aguiló-Aguayo, O. Martín-Belloso, Inhibition of browning on fresh-cut pear wedges by natural compounds, J. Food Sci. 71 (2006) 216−224.
[102] F.C. Richard, P.M. Goupy, J.J. Nicolas, J.M. Lacombe, A.A. Pavia, Cysteine as an inhibitor of enzymatic browning. 1. Isolation and characterization of addition compounds formed during oxidation of phenolics by apple polyphenol oxidase, J. Agri. Food Chem. 39 (1991) 841−847.
[103] R. Havenainen, New approaches in improving the shelf life of minimally processed fruit and vegetables, Trends Food Sci. Technol. 7 (1996) 179−187.
[104] C.H. Lee, D.S. An, H.J. Park, D.S. Lee, Wide spectrum antimicrobial packaging materials incorporating nisin and chitosan in the coating, Packag. Technol. Sci. 16 (2003a) 99−106.
[105] J.Y. Lee, H.J. Park, C.Y. Lee, W.Y. Choi, Extending shelf life of minimally processed apples with edible coatings and antibrowning agents, Leb. Wiss.Technol. 36 (2003b) 323−329.
[106] D.W.S. Wong, W.P. Camirand, A.E. Pavlath, Development of edible coatings for minimally processed fruit and vegetables, in: J.M. Krochta, E.A. Baldwin, M.O. Nisperos-Carriedo (Eds.), "Edible Coatings and Films to Improve Food Quality", Technomic Publishing Co, Switzerland, 1994, pp. 65−88.
[107] E. Ayranci, S. Tunc, The effect of edible coatings on water and vitamin C loss of apricots (*Armeniaca vulgaris* Lam.) and green peppers (*Capsicum annuum* L.), Food Chem. 87 (2004) 339−342.
[108] M.A. Rojas-Grau, R.M. Raybaudi-Massilia, R.C. Soliva-Fortuny, R.J. Avena-Bustillos, T.H. McHugh, O. Martin-Belloso, Apple puree-alginate edible coating as carrier of antimicrobial agents to prolong shelf-life of fresh-cut apples, Postharvest Biol. Technol. 45 (2007a) 254−264.
[109] H. Qi, W. Hu, A. Jiang, M. Tian, Y. Li, Extending shelf-life of fresh-cut 'Fuji' apples with chitosan-coatings, Innov. Food Sci. Emerg. Technol. 12 (2011) 62−66.
[110] M. Chiumarelli, C.C. Ferrari, I.G.L. Claire, M.D. Sarantopoulos Hubinger, Fresh cut 'Tommy Atkins' mango pre-treated with citric acid and coated with cassava (*Manihot esculenta* Crantz) starch or sodium alginate, Innov. Food Sci. Emerg. Technol. 12 (2011) 381−387.
[111] V.S. Bierhals, M. Chiumarelli, M.D. Hubinger, Effect of cassava starch coating on quality and shelf life of fresh-cut pineapple (*Ananas comosus* L. Merril cv "Perola"), J. Food Sci. 76 (2011) E62−E72.
[112] B.A. Del Rosario, L.R. Beuchat, Survival and growth of enterohemorrhagic *Escherichia coli* O157:H7 in cantaloupe and watermelon, J. Food Prot. 58 (1995) 105−107.
[113] R.L. Thunberg, T.T. Tran, R.W. Bennett, R.N. Matthews, N. Delay, Microbial evaluation of selected fresh produce obtained at retail markets, J. Food Prot. 65 (2002) 677−682.
[114] A. Cagri, Z. Uspunol, E. Ryser, Antimicrobial edible films and coating, J. Food Prot. 67 (2004) 833−848.
[115] S. Burt, Essential oils: their antibacterial properties and potential applications in foods—a review, Int. J. Food Microbiol. 94 (2004) 223−253.
[116] G. Fenaroli, Fenaroli's Handbook of Flavor Ingredients, CRC Press, Boca Raton, FL, 1995.
[117] C.R.C. Keshri, M.K. Sanyal, Effect of sodium alginate coating with preservatives on the quality of meat patties during refrigerated ($4 \pm 1°C$) storage, J. Muscle Foods 20 (2009) 275−292.
[118] M. Friedman, P.R. Henika, C.E. Levin, R.E. Mandrel, Antibacterial activities of plant essential oils and their components against *Escherichia coli* O157:H7 and *Salmonella enterica* in apple juice, J. Agric. Food Chem. 52 (2004) 6042−6048.
[119] R. Raybaudi-Massilia, J. Mosqueda-Melgar, O. Martin-Belloso, Antimicrobial activity of essential oils on *Salmonella enteritidis*, *Escherichia coli*, and *Listeria innocua* in fruit juices, J. Food Prot. 69 (2006) 1579−1586.
[120] G.T. Andevari, M. Rezaei, Effect of gelatin coating incorporated with cinnamon oil on the quality of fresh rainbow trout in cold storage, Int. J. Food Sci. Technol. 46 (2011) 2305−2311.

[121] R.M. Raybaudi-Massilia, J. Mosqueda-Melgar, O. Martin-Belloso, Edible alginate-based coating as carrier of antimicrobials to improve shelf-life and safety of fresh-cut melon, Int. J. Food Microbiol. 121 (2008) 313–327.
[122] K.D. Vu, R.G. Hollingsworth, E. Leroux, S. Salmieri, M. Lacroix, Development of edible bioactive coating based on modified chitosan for increasing the shelf life of strawberries, Food Res. Int. 44 (2011) 198–203.
[123] T.H. McHugh, C.C. Huxsoll, J.M. Krochta, Permeability properties of fruit puree edible films, J. Food Sci. 61 (1996) 88–91.
[124] T.H. McHugh, E. Senesi, Apple wraps: a novel method to improve the quality and extend the shelf life of fresh-cut apples, J. Food Sci. 65 (2000) 480–485.
[125] M.C. Salcini, R. Massantini, Minimally processed fruits: an update on browning control, Stewart Postharvest Rev. 3 (2005) 1–7.
[126] M.A. Rojas-Grau, R.J. Avena-Bustillos, M. Friedman, P.R. Henika, O. Martin-Belloso, T.H. McHugh, Mechanical, barrier and antimicrobial properties of apple puree edible films containing plant essential oils, J. Agric. Food Chem. 54 (2006) 9262–9267.
[127] H.P. Rupasinghe, J. Boulter-Bitzer, T. Ahn, J.A. Odumeru, Vanillin inhibits pathogenic and spoilage microorganisms in vitro and aerobic microbial growth in fresh-cut apples, Food Res. Int. 39 (2006) 575–580.
[128] M.A. Rojas-Grau, M.S. Tapia, F.J. Rodriguez, A.J. Carmona, O. Martin-Belloso, Alginate and gellan based edible coatings as support of antibrowning agents applied on fresh-cut Fuji apple, Food Hydrocolloid. 21 (2007b) 118–127.
[129] L.-S. Lin, Y.-M. Be-Jenwang, Quality preservation of commercial fish balls with antimicrobial zein coatings, J. Food Qual. 34 (2011) 81–87.
[130] C.-Y. Ou, T. Shwu-Fen, L. Chieh-Hsien, W. Yih-Ming, Using gelatin based antimicrobial edible coating to prolong shelf life of tilapia fillet, J. Food Qual. 25 (2002) 213–222.
[131] J.W. Rhim, Increase in water vapor barrier property of biopolymer-based edible films and coatings by compositing with lipid materials, Food Sci. Biotechnol. 13 (2004) 528–535.
[132] J.W. Park, R.F. Testin, H.J. Park, P.J. Vergano, C.L. Weller, Fatty acid concentration effect on tensile strength, elongation and water vapor permeability of laminated edible films, J. Food Sci. 59 (1994) 916–919.
[133] J.W. Rhim, Y. Wu, C.L. Weller, M. Schnepf, Physical characteristics of emulsified soy protein fatty acid composite films, Sci. Aliment. 19 (1999) 57–71.
[134] A. Jongjareonrak, S. Benjakul, W. Svisessanguan, M. Tanaka, Fatty acids and their sucrose esters affect the properties of fish skin gelatin-based film, Eur. Food Res. Technol. 222 (2006) 650–657.
[135] D.P. The, F. Debeaufort, C. Peroval, D. Despre, J.L. Courthaudon, A. Voilley, Arabinoxylan-lipid-based edible films and coatings. 3. Influence of drying temperature on film structure and functional properties, J. Agric. Food Chem. 50 (2002a) 2423–2428.
[136] D.P. The, C. Peroval, F. Debeaufort, D. Despre, J.L. Courthaudon, A. Voilley, Arabinoxylan-lipids-based edible films and coatings. 2. Influence of sucroester nature on the emulsion structure and film properties, J. Agric. Food Chem. 50 (2002b) 266–272.
[137] G.R. Gibson, M.B. Roberfroid, Dietary modulation of the human colonic microbiota: introducing the concept of probiotics, J. Nutr. 125 (1995) 1401–1412.
[138] W.S. Kim, T. Tanaka, H. Kumura, K. Shimazaki, Lactoferrin-binding proteins in *Bifidobacterium bifidum*, Biochem. Cell Biol. 80 (2002) 91–94.
[139] T. Mitsuoka, Bifidobacteria and their role in human health, J. Ind. Microbiol. 6 (1990) 263–267.
[140] A.S. Naidu, W.R. Bidlack, L.R.A. Clemens, Probiotic spectra of lactic acid bacteria (LAB), Crit. Rev. Food Sci. Nutr. 38 (1999) 13–126.
[141] M.S. Tapia, M.A. Rojas-Grau, F.J. Rodriguez, J. Ramirez, A. Carmona, O. Martin-Belloso, Alginate- and gellan-based edible films for probiotic coatings on fresh-cut fruits, J. Food Sci. 72 (2007) 190–196.
[142] J.C. Gates, Basic Foods, third ed., Holt, Rinehart and Winston, New York, 1987.
[143] L. Allen, A.I. Nelson, M.P. Steinberg, J.N. McGill, Edible corn-carbohydrate food coatings. Evaluation on fresh meat products, Food Technol. 17 (1963) 442–443.

[144] M. Glicksman, Red seaweed extracts, in: M. Glicksman (Ed.), Food Hydrocolloids, vol. 2, CRC Press Inc, Boca Raton, FL, 1983, p. 73.
[145] A.H. King, in: M. Glicksman (Ed.), Food Hydrocolloids, Brown seaweed extracts, vol. 2, CRC Press Inc, Boca Raton, FL, 1983, p. 115.
[146] K.G. Wanstedt, S.C. Seideman, L.S. Donnelly, N.M. Quenzer, Sensory attributes of precooked, calcium alginate-coated pork patties, J. Food Prot. 44 (1981) 732.
[147] S.K. Williams, J.L. Oblinger, R.L. West, Evaluation of calcium alginate film for use on beef cuts, J. Food Sci. 43 (1978) 292.
[148] K.R. Conca, T.C.S. Yang, Edible food barrier coatings, in: C. Ching, D. Kaplan, E. Thomas (Eds.), Biodegradable Polymers and Packaging, Technomic Publishing Co., Inc, Lancaster, Pennsylvania, 1993, pp. 357–369.
[149] C.R. Lazarus, R.L. West, J.L. Oblinger, A.Z. Palmer, Evaluation of a calcium alginate coating and a protective plastic wrapping for the control of lamb carcass shrinkage, J. Food Sci. 41 (1976) 639–641.
[150] N. Khanedan, A.A. Motalebi, A.A. Khanipour, A. Koochekian Sabour, M. Seifzadeh, A. Hasanzati Rostami, Effects of different concentrations of sodium alginate as an edible film on chemical changes of dressed Kilka during frozen storage, Iran. J. Fish. Sci. 10 (2011) 654–662.
[151] H.J. Kang, C. Jo, J.H. Kwon, J.H. Kim, H.J. Chung, M.W. Byun, Effect of a pectin-based edible coating containing green tea powder on the quality of irradiated pork patty, Food Control 18 (2007) 430–435.
[152] K. Nortje, E.M. Buys, A. Minnaar, Use of gamma-irradiation to reduce high levels of *Staphylococcus aureus* on casein-whey protein coated moist beef biltong, Food Microbiol. 23 (2006) 729–737.
[153] M.N. Antoniewski, S.A. Barringer, Meat shelf-life and extension using collagen/gelatin coatings: a review, Critical Rev. Food Sci. Nutr. 50 (2010) 644–653.
[154] J. Gomez-Estaca, P. Montero, B. Gimenez, M.C. Gomez-Guillen, Effect of functional edible films and high pressure processing on microbial and oxidative spoilage in cold-smoked sardine (*Sardina pilchardus*), Food Chem. 105 (2007) 511–520.
[155] J. German, Food processing and lipid oxidation, Adv. Exp. Med. Biol. 459 (1999) 23–50.
[156] G. Beecher, Overview of dietary flavonoids: nomenclature, occurrence and intake, J. Nutr. 133 (2003) 3248–3254.
[157] R.M. Ortega, Importance of functional foods in the Mediterranean diet, Public Health Nutr. 9 (2006) 1136–1140.
[158] C.R. Allan, L.A. Hadwiger, The fungicidal effect of chitosan on fungi of varying cell wall composition, Exp. Mycol. 3 (1979) 285–287.
[159] R. Daniels, Edible Coatings and Soluble Packaging, Noyes Data Corp, Park Ridge, NJ, 1973, pp. 360.
[160] R.D. Earle, C.E. Snyder, US Patent 3,255,021, (1996).
[161] O. Kilincceker, I.S. Dogan, E. Kucukoner, Effect of edible coatings on the quality of frozen fish fillets, LWT Food Sci. Technol. 42 (2009) 868–873.
[162] J.A. Pearce, C.G. Lavers, Frozen storage of poultry. V. Effects of some processing factors on quality, Can. J. Res. 27 (1949) 253–265.
[163] R.C. Meyer, A.R. Winter, H.H. Wiser, Edible protective coatings for extending the shelf life of poultry, Food Technol. 13 (1959) 146–148.
[164] N. Natrajan, B.W. Sheldon, Evaluation of bacteriocin-based packaging and edible film delivery systems to reduce Salmonella in fresh poultry, Poult. Sci. 74 (1) (1995) 31.
[165] G.F. Mehyar, J.H. Han, R.A. Holley, G. Blank, A. Hydamaka, Suitability of pea starch and calcium alginate as antimicrobial coatings on chicken skin, Poult. Sci. 86 (2007) 386–393.
[166] V. Coma, Bioactive packaging technologies for extended shelf life of meat-based products, Meat Sci. 78 (2008) 90–103.
[167] H.K. No, S.P. Meyers, W. Prinyawiwatkul, Z. Xu, Applications of chitosan for improvement of quality and shelf life of foods: a review, J. Food Sci. 72 (2007) 87–100.
[168] D.L. Kaplan, J.M. Mayer, D. Ball, J. McCassie, A.L. Allen, P. Stenhouse, Fundamentals of biodegradable polymers, in: C. Ching, D. Kaplan, E. Thomas (Eds.), Biodegradable Polymers and Packaging, Technomic Publishing Co., Inc, Lancaster, PA, 1993, pp. 1–42.
[169] P.A. Sandford, Commercial uses and potential applications, in: T. Skjak Braek, T. Anthonsen,

P. Sandford (Eds.), Chitin and Chitosan: Sources Chemistry, Biochemistry, Physical Properties and Applications, Elsevier Applied Science, New York, 1989, pp. 51–69.

[170] M.S. Rao, R. Chander, A. Sharma, Development of shelf-stable intermediate-moisture meat products using active edible chitosan coating and irradiation, J. Food Sci. 70 (2005) 325–331.

[171] R.L. Beverly, M.E. Janes, W. Prinyawlwatkula, H.K. No, Edible chitosan films on ready-to-eat roast beef for the control of *Listeria monocytogenes*, Food Microbiol. 5 (2008) 534–537.

[172] D.E. Conner, V.N. Scott, D.T. Bernard, Growth, inhibition, and survival *of Listena rnonocytogenes* as affected by acidic conditions, J. Food Prot. 53 (1990) 652–655.

[173] B.R. Ray, W.E. Sandine, Acetic, propionic, and lactic acids of starter culture bacteria as biopreservatives, in: B.R. Ray, M. Daeschel (Eds.), Food Biopreservatives of Microbial Origin, CRC Press, Boca Raton, FL, 1992, pp. 103–136.

[174] K.M. Sorrells, D.C. Enigel, J.R. Hatfield, Effect of pH, acidulant, time, and temperature on the growth and survival of *Listeria monocytogenes*, J. Food Prot. 52 (1989) 571–573.

[175] S. Sathivel, Chitosan and protein coatings affect yield, moisture loss, and lipid oxidation of pink salmon (*Oncorhynchus gorbuscha*) fillets during frozen storage, J. Food Sci. 70 (2005) 455–459.

[176] S. Sathivel, Q. Liu, J. Huang, W. Prinyawiwatkul, The influence of chitosan glazing on the quality of skinless pink salmon (*Oncorhynchus gorbuscha*) fillets during frozen storage, J. Food Eng. 83 (2007) 366–373.

[177] T.W. Augustini, S. Sedjati, The effect of chitosan concentration and storage time on the quality of salted-dried anchovy (*Stolephorus heterolobus*), J. Coastal Dev. 10 (2007) 63–71.

[178] C.O. Mohan, C.N. Ravishankar, K.V. Lalitha, T.K. Srinivasa Gopal, Effect of chitosan edible coating on the quality of double filleted Indian oil sardine (*Sardinella longiceps*) during chilled storage, Food Hydrocolloids 26 (2012) 167–174.

[179] Y.J. Jeon, J.Y.V.A. Kamil, F. Shahidi, Chitosan as an edible invisible film for quality preservation of herring and Atlantic cod, J. Agric. Food Chem. 50 (2002) 5167–5178.

[180] J. Duan, G. Cherian, Y. Zhaob, Quality enhancement in fresh and frozen lingcod (*Ophiodon elongates*) fillets by employment of fish oil incorporated chitosan coatings, Food Chem. 119 (2010) 524–532.

[181] N. Kubota, Y. Kikuchi, Macromolecular complexes of chitosan, in: S. Dumitriu (Ed.), Polysaccharides: Structural Diversity and Functional Versatility, Marcel Dekker Inc, New York, 1998, pp. 595–628.

[182] D.F. Kendra, L.A. Hadwiger, Characterisation of the smallest chitosan oligomer that is maximally antifungal to *Fusarium solani* and elicits pisatin formation in *Pisium satvium*, Exp. Mycol. 8 (1984) 276–281.

[183] G. Ralston, M. Tracey, P. Wrench, The inhibition of fermentation in baker's yeast by chitosan, Biochim. Biophys. Acta 93 (1964) 652–655.

[184] A. El Ghaouth, J. Arul, R. Ponnampalan, M. Boulet, Chitosan coating: effect on storability and quality of fresh strawberries, J. Food Sci. 56 (1991) 1618–1624.

[185] S. Sagoo, R. Board, S. Roller, Chitosan inhibits growth of spoilage micro-organisms in chilled pork products, Food Microbiol. 19 (2002) 175–182.

[186] D. Zhang, P.C. Quantick, Antifungal effects of chitosan coating on fresh strawberries and raspberries during storage, J. Hort. Sci. Biotechnol. 73 (1998) 763–767.

[187] S. Mathew, T.E. Abraham, Characterisation of ferulic acid incorporated starch-chitosan blend films, Food Hydrocolloids 22 (2008) 826–835.

[188] A. Sensidoni, D. Peressini, Edible films: potential innovation for fish products, Industrie Alimentari 36 (1997) 129–133.

[189] R. Ferraris, Gel seeding of sorghum into Mywybilla clay, in: M.A. Foale, B.W. Hare, R.G. Henzell (Eds.), Proceeding of the Australian. Sorghum Workshop, Toowoomba, Queensland, Australian Institute of Agricultural Science, Australia, Brisbane, 1989, pp. 66–73.

[190] H.F. Mark, D.F. Othmer, C.G. Overberger, G.T. Seaborg, Kirk-Othmer Encyclopedia of Chemical Technology, third ed., vol. 20, Wiley-Interscience, NY, 1985, pp. 219–220.

[191] R. Wills, T. Lee, D. Graham, B. McGlasson, E. Hall, Postharvest, an Introduction to the Physiology and Handling of Fruit and Vegetables, New South Wales University Press, Kensington, New South Wales, Australia, 1981, pp. 1−2.

[192] R. Barkai-Golan, Postharvest disease suppression by atmospheric modifications, in: M. Calderon, R. Barkai-Golan (Eds.), "Food Preservation by Modified Atmospheres", CRC Press, Boca Raton, FL, 1990, pp. 238−265.

[193] S.M. Smith, J.R. Stow, The potential of a sucrose ester coating material for improving the storage and shelf life qualities of Cox's Orange Pippin apples, Ann. Appl. Biol. 104 (1984) 383−391.

[194] W.M. Mellenthin, P.M. Chen, D.M. Borgic, In line application of porous wax coating materials to reduce friction discoloration of Bartlett and d'Anjou pears, Hort. Sci. 17 (1982) 215−217.

[195] M.O. Nisperos-Carriedo, E.A. Baldwin, Effect of two types of edible films on tomato fruit ripening, Proc. Fla. State Hort. Soc. 101 (1988) 217−220.

[196] A.A. Kader, Biochemical and physiological basis for effects of controlled and modified atmospheres on fruits and vegetables, Food Technol. 40 (1986) 99−104.

[197] A. Nussinovitch, N. Kampf, Shelf-life extension and texture of alginate coated mushrooms, in: IFTEC, The Hague, The Netherlands, 1992, p. 118.

[198] A. Nussinovitch, N. Kampf, Shelf-life extension and conserved texture of alginate coated mushrooms (*Agaricus bisporus*), J. Food Sci. Technol. 26 (1993) 469−475.

[199] P.J. Zapata, F. Guillen, D. Martinez-Romero, S. Castillo, D. Valero, M. Serrano, Use of alginate or zein as edible coatings to delay postharvest ripening process and to maintain tomato (*Solanum lycopersicon* Mill) quality, J. Sci. Food Agric. 88 (2008) 1287−1293.

[200] A.C. Hulme, Biochemistry of Fruits and their Products, vol. 1, Academic Press, London, 1970, p, 620,

[201] N. Maftoonazad, H.S. Ramaswamy, M. Marcotte, Shelf-life extension of peaches through sodium alginate and methyl cellulose edible coatings, Int. J. Food Sci. Technol. 43 (2008) 951−957.

[202] S. Wu, Polar and nonpolar interactions in adhesion, J. Adhes. 5 (1973) 39−55.

[203] M.A. Rojas-Grau, M.S. Tapia, O. Martin-Belloso, Using polysaccharide-based edible coatings to maintain quality of fresh-cut Fuji apples, LWT-Food Sci. Technol. 41 (2008) 139−147.

[204] Z. Xiao, Y. Luo, Y. Luo, Q. Wang, Combined effects of sodium chlorite dip treatment and chitosan coatings on the quality of fresh-cut d'Anjou pears, Postharvest Biol. Technol. 62 (2011) 319−326.

[205] M.E. Gounga, S.Y. Xu, Z. Wang, W.G. Yang, Effect of whey protein isolate-pullulan edible coatings on the quality and shelf life of freshly roasted and freeze-dried Chinese chestnut, J. Food Sci. 73 (2008) 155−161.

[206] E.T.D. Hojo, A.D. Cardoso, R.H. Hojo, E.V.D.V. Boas, M.A.R. Alvarenga, Use cassava starch films and PVC on post-harvest conservation of bell pepper, Ciencia E Agrotechnologia 31 (2007) 184−190.

[207] M.A. Aguilar-Mendez, E.S. Martin-Martinez, S.A. Tomas, A. Cruz-Orea, M.R. Jaime-Fonseca, Gelatine-starch films: physicochemical properties and their application in extending the post-harvest shelf life of avocado (*Persea americana*), J. Sci. Food Agric. 88 (2008) 185−193.

[208] S. Saucedo-Pompa, R. Rojas-Molina, A.F. Aguilera-Carbo, A. Saenz-Galindo, H. de La Garza, D. Jasso-Cantu, N. Cristobal, C.N. Aguilar, Edible film based on candelilla wax to improve the shelf life and quality of avocado, Food Res. Int. 42 (2009) 511−515.

[209] M.S. Tapia, M.A. Rojas-Graue, A. Carmona, F.J. Rodriguez, R. Soliva-Fortuny, O. Martin-Belloso, Use of alginate- and gellan-based coatings for improving barrier, texture and nutritional properties of fresh-cut papaya, Food Hydrocolloids 22 (2008) 1493−1503.

[210] E. Pennisi, Sealed in edible film, Science News 141 (1992) 12.

[211] O.P. Chauhan, P.S. Raju, A. Singh, A.S. Bawa, Shellac and aloe-gel-based surface coatings for maintaining keeping quality of apple slices, Food Chem. 126 (2011) 961−966.

[212] G.A. Gonzalez-Aguilar, J. Celis, R.R. Sotelo-Mundo, L.A. de la Rosa, J. Rodrigo-Garcia, E. Alvarez-Parrilla, Physiological and

biochemical changes of different fresh-cut mango cultivars stored at 5 degrees C, Int. J. Food Sci. Technol. 43 (2008) 91–101.
[213] M. Meheriuk, Skin color in Newton apples treated with calcium nitrate, urea, diphenylamine and a film coating, Hortscience 25 (1990) 775–776.
[214] M. Meheriuk, O.L. Lau, Effect of two polymeric coatings on fruit quality of Bartlett and d'Anjou pears, J. Am. Soc. Hort. Sci. 113 (1988) 222–226.
[215] R. Bai, M. Huang, Y. Jiang, Selective permeabilities of chitosan-acetic acid complex membrane and chitosan-polymer complex membranes for oxygen and carbon dioxide, Polym. Bull. 20 (1988) 83–88.
[216] W.B. Roth, C.L. Mehltretter, Some properties of hydroxypropylated amylomaize starch films, Food Technol. 21 (1967) 72–74.
[217] D.G. Murray, L.R. Luft, D.E. Low, Corn starch hydrolysates, Food Technol. 27 (1973) 32–40.
[218] R. Sothornvit, P. Rodsamran, Effect of a mango film on quality of whole and minimally processed mangoes, Postharvest Biol. Technol. 47 (2008) 407–415.
[219] N.H. Banks, Some effects of Tal Pro-long coating on ripening bananas, J. Exp. Bot. 35 (1984a) 127–137.
[220] N.H. Banks, Studies of the banana fruit surface in relation to the effects of Tal Pro-long coating on gaseous exchange, Sci. Hortic. 24 (1984b) 279–286.
[221] S.R. Drake, J.K. Fellman, J.W. Nelson, Postharvest use of sucrose polyesters for extending the shelf life of stored Golden Delicious apples, J. Food Sci. 52 (1987) 1283–1285.
[222] Mitsubishi-Kasei Ryoto, Sugar ester, in: Mitsubishi-Kasei Technical Information, RYOTO Sugar Ester (Food grade), 1989, pp. 1–20. http://www.mfc.co.jp/english/index.htm
[223] N.T. Ukai, T. Tsutsumi, K. Marakami, US Patent 3,997,674, (1975).
[224] N.H. Banks, Evaluation of methods for determining internal gases in banana fruit Musa acuminata x. Musa ballisiana, J. Exp. Bot. 34 (1983) 871–879.
[225] P.H. Lowings, D.G. Cutts, The preservation of fresh fruits and vegetables, in: "Proceedings of the International Institute of Food Science and Technology Annual Symposium, Nottingham, UK, 1982, p. 52.
[226] H.L. Eum, D.K. Hwang, M. Linke, S.K. Lee, M. Zude, Influence of edible coating on quality of plum (Prunus salicina Lindl. cv. 'Sapphire'), Eur. Food Res. Technol. 229 (2009) 427–434.
[227] M.O. Nisperos-Carriedo, P.E. Shaw, E.A. Baldwin, Changes in volatile flavor components of pineapple orange juice as influenced by the application of lipid and composite film, J. Agric. Food Chem. 38 (1990) 1382–1387.
[228] G.J. Curtis, Some experiments with edible coatings on the long-term storage of citrus fruits, in: R. Goren, K. Mendel (Eds.), Proceeding of 6th International Citrus Congress 3, Balaban Publishers, Margraf Scientific Book, Tel Aviv, Israel, 1988, pp. 1514–1520.
[229] M.E. Edwards, R.W. Blennerhassett, The use of postharvest treatments to extend storage life and to control postharvest wastage of Honey Dew melons (Cucumis melo L. var. inodorus Naud.) in cool storage, Aust. J. Exp. Agric. 30 (1990) 693–697.
[230] J.M. Valverde, D. Valero, D. Martinez-Romero, F. Guillen, S. Castillo, M. Serrano, Novel edible coating based on Aloe vera gel to maintain table grape quality and safety, J. Agric. Food Chem. 53 (2005) 7807–7813.
[231] D. Martinez-Romero, N. Alburquerque, J.M. Valverde, F. Guillen, S. Castillo, D. Valero, M. Serrano, Postharvest sweet cherry quality and safety maintenance by aloe vera treatment: a new edible coating, Poshharvest Biol. Technol. 39 (2006) 93–100.
[232] P.J. Fellows, Food Processing Technology, Ellis Harwood, New York, 1990.
[233] F. Pedreschi, P. Moyano, N. Santis, R. Pedreschi, Physical properties of pre-treated potato chips, J. Food Eng. 79 (2007) 1474–1482.
[234] L.L. Fan, J.A. Arce, Preparation of fried food products with oil containing emulsifiers. US Patent 4,608,264, (1986).
[235] H.V. Zeddelmann, P. Olendorf, Beeinflusung des fettvebrauches der herstellung von fettgeback, Influence of lipid consumption on donut production. Getreide, Mehl. und Brot. 33 (1979) 24–26.
[236] C.I. Pravisani, A. Calvelo, Minimum cooking time for potato strip frying, J. Food Sci. 51 (1986) 614–617.

[237] J.D. Selman, M. Hopkins, Factors Affecting Oil Uptake During the Production of Fried Potato Products. Technical Memorandum 475, Campden Food & Drink Research Association, Chipping Campden, Gloucestershire, UK, 1989.
[238] K.L. Krumel, T.A. Lindsay, Nonionic cellulose ethers, Food Technol. 30 (1976) 36–43.
[239] Kelco, Technical Bulletin: Rahway, Merck & Co, NJ, 1990. Gellan Gum Multifunctional Gelling Agent.
[240] G.R. Sanderson, Polysaccharides in foods, Food Technol. 35 (1981) 50–57.
[241] R.B. Suarez, L.A. Campanone, M.A. Garcia, N.E. Zaritzky, Comparison of the deep frying process in coated and uncoated dough systems, J. Food Eng. 84 (2008) 383–393.
[242] S. Naruenartwongsakul, M.S. Chinnan, S. Bhumiratana, T. Yoovidhya, Effect of cellulose ethers on the microstructure of fried wheat flour-based batters, LWT-Food Sci. Technol. 41 (2008) 109–118.
[243] H.L. Huse, Y.C. Hung, K.H. Mcwatters, Physical and sensory characteristics of fried cowpea (*Vigna unguiculata* L. Walp) paste formulated with soy flour and edible coatings, J. Food Qual. 29 (2006) 419–430.
[244] T.S. Ballard, P. Mallikarjunan, The effect of edible coatings and pressure frying using nitrogen gas on the quality of breaded fried chicken nuggets, J. Food Sci. 71 (2006) 259–264.
[245] S.G. Sudhakar, M.C. Pandey, M. Manral, K. Radhakrishna, A.S. Bawa, Effect of enrobing with carboxy methyl cellulose or hydroxy propyl methyl cellulose in corn or gram flour on moisture and fat content of chicken during frying, J. Food Sci. Technol. 43 (2006) 377–381.
[246] Colloides Naturels Inc, Sealgum: something new in films, in: Product Bulletin, Colloides Naturels Inc, Bridgewater, NJ, 1988.
[247] G. Mazza, H. Qi, Control of after-cooking darkening in potatoes with edible film-forming products and calcium chloride, J. Agric. Food Chem. 39 (1991) 2163–2166.
[248] M. Martin-Polo, A. Voilley, Comparative study of the water permeability of edible films composed of gum arabic and glycerolmonostearate, Sci. Aliment. 10 (1990) 473–483.
[249] M. Maqbool, A. Ali, P.G. Alderson, N. Zahid, Y. Siddiqui, Effect of a novel edible composite coating based on gum arabic and chitosan on biochemical and physiological responses of banana fruits during cold storage, J. Agric. Food Chem. 59 (2011) 5474–5482.
[250] D.D. Torrico, W. Jirangrat, H. Kyoon No, W. Prinyawiwatkul, B. Ge, D. Ingram, A novel emulsion coating and its effects on internal quality and shelf life of eggs during room temperature storage, Int. J. Food Sci. Technol. 45 (2010) 2241–2249.
[251] E.A. Baldwin, Edible coatings for fresh fruits and vegetables: past, present and future, in: J. Krochta, E. Baldwin, M. Nisperos-Carriedo (Eds.), "Edible Coatings and Films to Improve Food Quality", Technomic Publishing Co, Basel, Switzerland, 1994, pp. 25–65.
[252] IFT, New from Mitsubishi, in: "Annual Meeting and Food Expo Program and Exhibit Directory", Dallas Convention Center, Chicago, IL, 1991.
[253] J.A. Torres, M. Motoki, M. Karel, Microbial stabilization of intermediate moisture food surfaces. Control of surface preservative concentration, J. Food Process Preserv. 9 (1985) 75–92.
[254] D.S. Bryn, US Patent 3,707,383, (1972).
[255] R.D. Hagenmaier, P.E. Shaw, Moisture permeability of edible films made with fatty acid and (hydroxypropyl) methylcellulose, J. Agric. Food Chem. 38 (1990) 1799–1803.
[256] M.O. Nisperos-Carriedo, Edible coatings and films based on polysaccharides, in: J.M. Krochta, E.A. Baldwin, M.O. Nisperos-Carriedo (Eds.), Edible Coatings and Films to Improve Food Quality, Technomic Publishing Co, Basel, Switzerland, 1994, pp. 305–336.
[257] F. Vojdani, J.A. Torres, Potassium sorbate permeability of methylcellulose and hydroxypropyl methylcellulose coatings: effect of fatty acids, J. Food Sci. 55 (1990) 841.
[258] K.L. Nelson, O.R. Fennema, Methylcellulose films to prevent lipid migration in confectionery products, J. Food Sci. 56 (1991) 504–509.
[259] F. Debeaufort, M. Martin-Polo, A. Voilley, Polarity homogeneity and structure affect water vapor permeability of model edible films, J. Food Sci. 58 (1993) 426–434.

[260] C.M. Koelsch, T.P. Labuza, Packaging, waste disposal and food safety. II: incineration or degradation of plastics and a possible integrated approach, Cereal Foods World 36 (1991) 284–298.

[261] M. Martin-Polo, C. Mauguin, A. Voilley, Hydrophobic films and their efficiency against moisture transfer. Influence of the film preparation technique, J. Agric. Food Chem. 40 (1992) 407–412.

[262] D.C. Rico-Pena, J.A. Torres, Edible methylcellulose-based films as moisture impermeable barriers in sundae ice cream cones, J. Food Sci. 55 (1990) 1468–1469.

[263] S.A. Valencia-Chamorro, M.B. Perez-Gago, A. Miguel, M.A. Del Rio, L. Lluis Palou, Curative and preventive activity of hydroxypropyl methylcellulose-lipid edible composite coatings containing antifungal food additives to control citrus postharvest green and blue molds, J. Agric. Food Chem. 57 (2009) 2770–2777.

[264] A. Nussinovitch, Extending the shelf life of mushrooms by hydrocolloid coating, Hassadeh 74 (10) (1994) 1131–1132.

[265] S.L. Kamper, O.R. Fennema, Use of edible films to maintain water vapor gradients in foods, J. Food Sci. 50 (1985) 382–384.

[266] A.J. Ganz, CMC and hydroxypropylcellulose versatile gums for food use, Food Prod. Dev. 3 (1969) 65–71.

[267] H. Chen, Functional properties and applications of edible films made of milk proteins, J. Dairy Sci. 78 (1995) 2563–2583.

[268] P. Fairley, F.J. Monahan, J.B. German, J.M. Krochta, Mechanical properties and water vapor permeability of edible films from whey protein isolate and sodium dodecyl sulfate, J. Agric. Food Chem. 44 (1996) 438–443.

[269] J.I. Mate, J.M. Krochta, Comparison of oxygen and water vapor permeabilities of whey protein isolate and β-lactoglobulin edible films, J. Agric. Food Chem. 44 (1996) 3001–3004.

[270] K. Ciesla, S. Salmieri, M. Lacroix, Modification of the properties of milk protein films by gamma radiation and polysaccharide addition, J. Sci. Food Agric. 86 (2006) 908–914.

[271] M. Ozdemir, J.D. Floros, Optimization of edible whey protein films containing preservatives for water vapor permeability, water solubility and sensory characteristics, J. Food Eng. 86 (2008) 215–224.

[272] J. Correa-Betanzo, J.K. Jacob, C. Perez-Perez, G. Paliyath, Effect of a sodium caseinate edible coating on berry cactus fruit (*Myrtillocactus geometrizans*) phytochemicals, Food Res. Int. 44 (2011) 1897–1904.

[273] M.E. Gounga, S.Y. Xu, Z. Wang, Whey protein isolate-based edible films as affected by protein concentration, glycerol ratio and pullulan addition in film formation, J. Food Eng. 83 (2007) 521–530.

[274] Y. Karrout, C. Neut, D. Wils, F. Siepmann, L. Deremaux, P. Desreumaux, J. Siepmann, Novel polymeric film coatings for colon targeting: how to adjust desired membrane properties, Int. J. Pharm. 371 (2009) 64–70.

[275] J.W. Lee, S.M. Son, S.I. Hong, Characterization of protein-coated polypropylene films as a novel composite structure for active food packaging application, J. Food Eng. 86 (2008) 484–493.

[276] B. Li, J.F. Kennedy, Q.G. Jiang, B.J. Xie, Quick dissolvable, edible and heatsealable blend films based on konjac glucomannan-gelatin, Food Res. Int. 39 (2006) 544–549.

[277] H.N. Lazarides, G.E. Mitrakas, K.I. Matsos, Edible coating and counter-current product/solution contacting: a novel approach to monitoring solids uptake during osmotic dehydration of a model food system, J. Food Eng. 82 (2007) 171–177.

[278] T.S. Young, E. Fu, Associative behavior of cellulosic thickeners and its implications on coating structure and rheology, TAPPI J. 74 (1991) 197–207.

[279] B. Ramezanzadeh, M. Mohseni, A. Mohammad Rabea, H. Yari, Attributing the resistance against simulated tree gum of an acrylic/melamine film loaded with an active silicone additive to its surface free energy, Int. J. Adhes. Adhes. 31 (2011) 775–783.

[280] S.J. Sarma, K. Pakshirajan, B. Mahanty, Chitosan-coated alginate–polyvinyl alcohol beads for encapsulation of silicone oil containing pyrene: A novel method for biodegradation of polycyclic aromatic hydrocarbons, J. Chem. Technol. Biotechnol. 86 (2011) 266–272.

[281] J. Uldrich, The Next Big Thing Is Really Small: How Nanotechnology Will Change the Future

of Your Business, Crown Publishing Group, New York, 2003.
[282] A. Sorrentino, G. Gorrasi, V. Vittoria, Potential perspectives of bio-nanocomposites for food packaging applications, Trends Food Sci. Technol. 18 (2007) 84–95.
[283] J.W. Rhim, S.I. Hong, H.M. Park, P.K.W. Ng, Preparation and characterization of chitosan-based nanocomposite films with antimicrobial activity, J. Agric. Food Chem. 54 (2006) 5814–5822.
[284] P. Mangiacapra, G. Gorrasi, A. Sorrentino, V. Vittoria, Biodegradable nanocomposites obtained by ball milling of pectin and montmorillonites, Carbohydr. Polym. 64 (2005) 516–523.
[285] W.F. Lee, Y.T. Fu, Effect of montmorillonite on the swelling behavior and drug-release behavior of nanocomposite hydrogels, J. Appl. Polym. Sci. 89 (2003) 3652–3660.
[286] J.P. Zheng, P. Li, Y.L. Ma, K.D. Yao, Gelatine/montmorillonite hybrid nanocomposite. I. Preparation and properties, J. Appl. Polym. Sci. 86 (2002) 1189–1194.
[287] G.R. Siragusa, J.S. Dickson, Inhibition of *Listeria monocytogenes* on beef tissue by application of organic acids immobilized in a calcium alginate gel, J. Food Sci. 57 (1992) 293–298.
[288] G. Decher, J.B. Schlenoff, Multilayer Thin Films: Sequential Assembly of Nano Composite Materials, Wiley-VCH, Weinheim, Germany, 2003, p. 543.
[289] I.L. Batalha, A. Hussaina, A.C. Roquea, Gum Arabic coated magnetic nanoparticles with affinity ligands specific for antibodies, J. Mol. Recognit. 23 (2010) 462–471.
[290] P. Gaba, S. Singh, M. Gaba, G.D. Gupta, Galactomannan gum coated mucoadhesive microspheres of glipizide for treatment of type 2 diabetes mellitus: In vitro and in vivo evaluation, Saudi Pharm. J. 19 (2011) 143–152.
[291] N.A. Kotov, Layer-by-layer assembly of nanoparticles and nanocolloids: intermolecular interactions, structure and materials perspective, in: G. Decher, J.B. Schlenoff (Eds.), Multilayer Thin Films: Sequential Assembly of Nanocomposite Materials, Wiley-VCH, Weinheim, Germany, 2003, pp. 207–243.
[292] J. Weiss, P. Takhistov, D.J. McClements, Functional materials in food nanotechnology, J. Food Sci. 71 (2006) 107–116.
[293] M. Darder, M. Colilla, E. Ruiz-Hitzky, Biopolymer-clay nanocomposites based on chitosan intercalated in montmorillonite, Chem. Mat. 15 (2003) 3774–3780.
[294] N. Dogan, T.H. McHugh, Effects of microcrystalline cellulose on functional properties of hydroxy propyl methyl cellulose microcomposite films, J. Food Sci. 72 (2007) 16–22.
[295] J.M. Lagaron, L. Cabedo, D. Cava, J.L. Feijoo, R. Gavara, E. Gimenez, Improving packaged food quality and safety, Part 2: Nanocomposites. Food Addit. Contam. 22 (2005) 994–998.
[296] S. Sinharay, M. Bousmina, Biodegradable polymers and their layered silicate nanocomposites: in greening the 21st century materials world, Progr. Mat. Sci. 50 (2005) 962–1079.
[297] T.P. Labuza, W.M. Breene, Application of active packaging for improvement of shelf-life and nutritional quality of fresh and extended shelf-life foods, J. Food Process. Preserv. 13 (1988) 1–69.
[298] L. Vermeiren, F. Devlieghere, M. Van Beest, N. de Kruijf, J. Debevere, Developments in the active packaging of foods, Trends Food Sci. Technol. 10 (1999) 77–86.
[299] B. Li, J. He, D.G. Evans, X. Duan, Inorganic layered double hydroxides as a drug delivery system intercalation and in vitro release of fenbufen, Appl. Clay Sci. 27 (2004) 199–207.

14 Biopolymers in Controlled-Release Delivery Systems

Kunal Pal, Allan T. Paulson and Dérick Rousseau

OUTLINE

Ebnesajjad: Handbook of Biopolymers and Biodegradable Plastics. http://dx.doi.org/10.1016/B978-1-4557-2834-3.00014-8
© 2009 Elsevier Inc. All rights reserved. Reproduced from a chapter in: Kasapis, *Modern Biopolymer Science* (2009).

14.1 Introduction

In its ever-growing scope, polymer science includes various aspects of biology, chemistry, medicine, and materials science. The flexibility in polymer processing plays an important role in the development of controlled delivery devices for a range of applications related to drugs, food-related bioactive ingredients, and genes. The release of bioactive agents from a delivery system may be either diffusion-controlled (diffusion of drug through a rate-controlling barrier/matrix), degradation-controlled (chemical or physical breakdown of the matrix leads to bioactive agent release), or via an environmental trigger (change in pH, ionic strength, or pressure tailors the release of the bioactive agent).

Biopolymers are generally synthesized by an organism or plant and have a more complex chemical structure than that of synthetic polymers. They are usually biocompatible when compared to synthetic polymers [1] and can be easily tailored for the development of controlled-release matrices. A key disadvantage of biopolymers is their susceptibility to microbial contamination. Of late, the term biopolymer has been extended to polymers synthesized using naturally occurring monomers (e.g., lactic acid and glycolic acid) such as poly(lactic acid) (PLA), poly (glycolic acid) (PGA), and poly(lactic-co-glycolic acid) (PLGA) [1]. In general, these polymers are biocompatible, nontoxic, and biodegradable. Some examples of biopolymers used for biomedical or pharmaceutical applications are shown in Table 14.1.

Rather than developing new drugs or bioactive compounds (which can be expensive), many pharmaceutical companies now use controlled-/sustained-delivery technologies to improve existing therapeutic agents simply by controlling the rate at which they enter the bloodstream, thus reducing and ideally avoiding under- or overdosing [20].

Hydrogels represent a key means of controlled or sustained delivery. They consist of three-dimensional, insoluble (bio)polymeric networks capable of imbibing large amounts of water or biological fluids, and may be designed for stimuli-response release. Compared to other synthetic biomaterials, hydrogels closely mimic living tissues because of their similar chemical building blocks, thereby exhibiting a reduced incidence of toxicity and inflammation [1,30]. The hydrophilic nature of biopolymers imparts water-binding properties, whereas the presence of either physical or chemical cross-links results in the formation of a three-dimensional network that helps in retaining structural integrity when placed in an aqueous environment [1,30].

Parameters used to characterize the suitability of a hydrogel for a particular application include (i) the polymer volume fraction in the swollen state, (ii) the average molecular weight of the polymer chain between two neighboring cross-linking points, and (iii) the mesh/pore size between polymer chains. The polymer volume fraction in the swollen state is a measure of the amount of fluid imbibed and retained by the hydrogel [1,32]. The molecular weight between two consecutive cross-links can give a measure of the degree of cross-linking of the polymer. Finally, the mesh size, or the average distance between adjacent cross-links, can be described by a correlation length that provides

Table 14.1 Examples of Biopolymers Commonly Used for Controlled Delivery Systems

Material	Uses	References
Collagen and gelatin	Wound dressings, tissue engineering, drug delivery	[2,3,4,5,6,7]
Chitin and chitosan	Wound dressings, tissue engineering, drug delivery	[8,9,10]
Alginate	Drug delivery, cell encapsulation	[11,12]
Cellulose derivatives	Tissue engineering, drug delivery	[13,14]
Starch	Tissue engineering, drug delivery	[15,16]
Carrageenan	Tissue engineering, drug delivery	[17,18]
Hyaluronic acid	Wound dressings, tissue engineering, drug delivery	[19,20,21]
Pectin	Wound dressings, tissue engineering, drug delivery	[22,23,24]
Dextran	Wound dressings, tissue engineering, cell encapsulation, drug delivery	[25,26,27,28]

a measure of the space between biopolymer chains available for diffusion [32,33].

The preparation of hydrogels typically involves cross-linking functional groups (e.g., hydroxyl, amine, amide, ether, carboxylate, and sulfonate) along the polymer chains to increase network rigidity [30,33]. The design of a hydrogel's microstructure in part depends on the target route of administration as well as the properties of the incorporated compound. Hydrogels are often formed using specific molds to obtain the appropriate size and conformation to maximize delivery effectiveness, depending on the administration routes, e.g., orally (spherical beads, cylinders, and discs), in implants (drum-shaped, disc-shaped, and cylindrical preparations), and through the skin (films and gel slabs).

14.2 Drug Loading and Release

Release mechanisms can be chemical or physical in nature, but always involve some form of diffusion. Diffusion is defined as the movement of individual molecules of a substance through a semipermeable barrier from an area of higher concentration to an area of lower concentration [34]. It is fundamentally reliant on the properties of the polymeric network and the solvent—polymer interaction, and varies in magnitude depending on the phase; it is the fastest in gases (~10 cm min^{-1}), slower in liquids (~0.05 cm min^{-1}), and the slowest in solids (~0.00001 cm min^{-1}). Diffusion in hydrogels is more complex and the diffusivities of incorporated compounds will lie somewhere between those in liquids and solids.

For degradable matrices and systems loaded with the compound of interest, release will be controlled by the cleavage of the polymer bonds within the network, even though the diffusion of the liberated therapeutic compound may be rate-limiting [30,35]. For nonbiodegradable systems, release will be diffusion-controlled and will depend on the concentration gradient. However, release may also depend on osmotic pressure and/or matrix swelling [37,38].

Physically controlled release is classified into two types, depending on its mode of release: (i) reservoir-type diffusion or (ii) matrix-type diffusion [39,40].

Reservoir-type: In these hydrogels, therapeutic compounds (solid or liquid) are entrapped in a reservoir within a microporous or nonporous polymeric network. If the therapeutic agent is saturated, its transport will be constant (or follow zero-order release kinetics) since its chemical potential will remain unaffected [30,37,41,42]. To maintain zero-order release, the therapeutic compound must remain in a solid or suspension state. If the compound has high water solubility, saturation will be difficult to maintain and release will deviate from zero-order kinetics. Even under ideal conditions, release is generally not constant during the initial and final release phases [30,37]. When immersed in a releasing medium, the system will take time to attain a steady state. If there is little or no therapeutic compound present near the surface of the hydrogel, then an induction period will be required for saturation of the surface. In contrast, if the therapeutic compound is concentrated near the surface, the initial release rate will be higher than the equilibrium-state value, caused by a "bursting" effect. As release nears completion, the therapeutic compound concentration in the core will drop and the release rate will decrease. On the other hand, if boundary sink conditions are imperfect, i.e., if there is an appreciable concentration of the therapeutic compound in the medium at the interface, there will be a drop in the therapeutic compound chemical potential that will slow down release [30,37].

Matrix-type: These hydrogels, where the drug is dissolved or dispersed within a polymer network, tend to be the most common. A decreased release rate over time due to the increased diffusion distance is a characteristic of these systems [42]. The solubility of the therapeutic compound and its diffusivity in the polymer phase and polymer—compound interactions play an important role in the governing release. Along with these parameters, the shape and the path-length for diffusion also are critical [30,33,37]. In certain solvent-penetrating systems, release depends on polymer relaxation, i.e., the stress required to maintain the strain on the polymer decreases, which is a result of aqueous solution absorption [44,45]. When a hydrogel is immersed in an aqueous solution, it will relax into a state that has little resistance to the diffusion of the incorporated compound. The key factor dictating release kinetics becomes swelling, which is a result of polymer relaxation. Important parameters associated with release from a hydrogel include its bioactive compound hydrophilicity and network cross-linking density. Matrix-type hydrogels are commonly used given their ease in development and cost-effectiveness. They tend to follow Higuchi's model, where bioactive compound release is proportional to the square root of time [46].

14.3 Modeling Diffusion

Diffusion (D) can be determined in a double diffusion cell, where a concentrated solute solution (kept in the donor compartment) is separated from a pure solvent (kept in the receptor compartment) by a semipermeable membrane [47,48,49,50,51,52,53]. Two examples of such cells are the horizontal transport cell (Fig. 14.1) and the vertical transport cell (Fig. 14.2). During diffusion, the solute passes from the donor compartment (which is a region of high concentration) to the receptor compartment (which is a region of low concentration) through a semipermeable barrier (membrane). This leads to the decrease of the solute concentration in the donor compartment and corresponding increase in the concentration in the acceptor compartment. This process continues until equilibrium is reached, i.e., the concentration of the solute in both compartments is the same. Thus, a steady state is reached. In general, measurements of D are carried out using a kinetic method, by sampling from the donor and acceptor compartments at set time intervals. The measured solute concentration in the two chambers is then used to calculate the solute D value through the semipermeable matrix in the following [54,55]:

$$D = \frac{1}{\beta t} x \ln \frac{C_D(t) - C_R(t)}{C_D(0) - C_R(0)} \tag{14.1}$$

with

$$\beta = \frac{A_H}{W_H} x [\frac{1}{V_D} + \frac{1}{V_R}] \tag{14.2}$$

where $C_D(0)$ = initial concentration of drug in donor compartment; $C_R(0)$ = initial concentration of drug

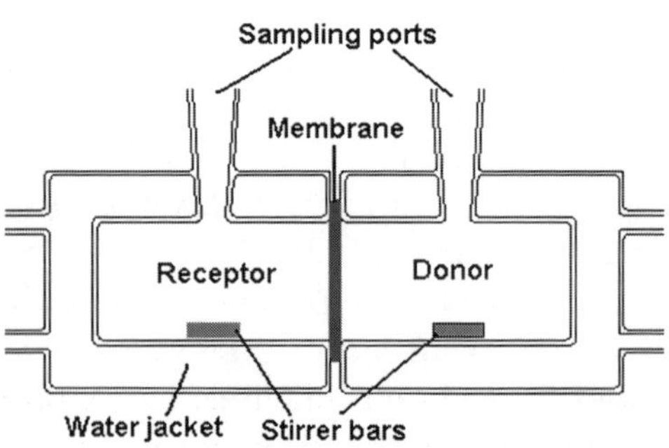

Figure 14.1 Schematic diagram of a horizontal diffusion cell.

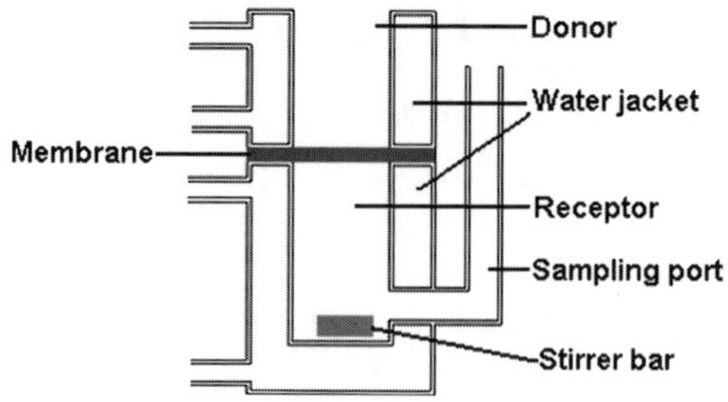

Figure 14.2 Schematic diagram of a vertical diffusion cell.

in receptor compartment; $C_D(t)$ = concentration of drug in donor compartment after time t; $C_R(t)$ = concentration of drug in receptor compartment after time t; A_H = effective cross-sectional area of diffusion in the hydrogel; W_H = width of the hydrogel sample; V_D = volume of drug solution in donor compartment; V_R = volume of receptor compartment fluid.

The slope of a plot of $-\ln \frac{C_D(T) - C_R(t)}{C_D(0) - C_R(0)}$ vs. time is used for calculating D.

Diffusion can be studied either under steady-state or under sink conditions [56,57]. In steady-state diffusion, the diffusion coefficient does not vary with time, i.e., dc/dt or dM/dt is constant and so diffusion is not allowed to attain equilibrium. During such experiments, the concentration of the solute in the donor and the acceptor compartments (i.e., hydrogel and surrounding medium, respectively) has to be maintained. This can be achieved by connecting the donor and acceptor to large reservoirs of solutions and recirculating these during the experimental period [34].

For experimental procedures under sink conditions, the concentration of the solute in the acceptor compartment is maintained at a lower level than that of the donor compartment [59,60]. The donor acts as a source of solute and the receptor acts as a sink, which is accomplished by connecting the receptor to a large reservoir from which the solution is recirculated to maintain a constant concentration gradient. For such experimental studies, sink conditions are preferred as maintaining a constant donor concentration can be difficult, and experiments with recirculation in one compartment are easier than recirculation in both compartments [34].

Diffusion coefficients can be described using Fick's first and second laws of diffusion [47]. These laws are applicable when a semipermeable barrier acts as a rate-controlling step for bioactive compound release.

14.3.1 Fick's First Law

The diffusion of molecules is denoted as dM/dt (rate of mass transfer) and is generally expressed as flux (J). J is defined as the rate of the mass transfer across a unit surface area of a barrier and is expressed as follows [34,56,61,62,63,64,66] (see also http://cnx.org/content/m1035/latest/):

$$J = \frac{\frac{1}{2} \times dM}{dt} \tag{14.3}$$

where dM = change in mass, g; S = barrier surface area, cm^2; dt = change in time, s.

The unit for flux is as follows:

$$J = \frac{\frac{1}{cm^2} \times g}{s} \tag{14.4}$$

which is often expressed as kg m^{-2} s^{-1}.

Fick's first law directly correlates flux with the concentration gradient and can be expressed as follows:

$$J = \frac{-D \times dC}{dx} \tag{14.5}$$

where dC = change in concentration of the solute, g/cm^3; D = diffusion coefficient of the solute, cm^2/s; dx = change in distance, cm.

The negative sign in Eqn (14.5) indicates a decrease in the concentration in the donor compartment. Since flux increases continuously during the process, it is always a positive entity. Combining Eqns (14.3) and (14.5) yields the rate of mass transfer:

$$\frac{dM}{dt} = -DS\frac{dC}{dx} \tag{14.6}$$

14.3.2 Fick's Second Law

Fick's second law deals with the change in the concentration with time at a definite location with respect to the x, y, and z axes. This law states that the change in concentration with time in a particular region is proportional to the change in the concentration gradient at that point of time. In general, this equation is not used in the pharmaceutical and nutraceutical industries. Briefly, it is expressed mathematically as follows [34,56,61,62,63,64,66] (see also http://cnx.org/content/m1036/latest/):

$$\frac{\partial C}{\partial t} = D\left[\frac{\partial^2 C}{\partial x^2} + \frac{\partial^2 C}{\partial y^2} + \frac{\partial^2 C}{\partial z^2}\right] \tag{14.7}$$

where x, y, and z are the coordinates.

D is affected by temperature, pressure, solvent properties, and chemical properties of the solute. Hence, it is regarded as a coefficient and not a constant. The effect of temperature on D can be expressed by the Arrhenius expression (see http://en.wikipedia.org/wiki/Fick's_law_of_diffusion):

$$D = D_0 \dot{c} e^{\frac{-E_a}{RT}} \tag{14.8}$$

where D = diffusion coefficient of the solute; D_0 = maximum diffusion coefficient (at infinite temperature); E_a = activation energy for diffusion; T = temperature, K; R = gas constant.

According to Hoffman [33], the structure of the polymer network, mesh/pore size, and polymer composition play an important role in the determination of D. Parameters such as average pore size, size distribution, and connectivity, which are often difficult to quantify, are usually included into a factor termed tortuosity. These factors are usually dependent on the properties of the polymer chains and the cross-link density of the polymer network. Tortuosity helps in the determination of the effective diffusion path length, which is defined as the product of film thickness and the ratio of porous fraction and tortuosity [33].

Release from a delivery system may either follow Fickian or non-Fickian diffusion. The three main release models used in the pharmaceutical industry include Higuchian kinetics, zero-order kinetics, and first-order kinetics, with the former taking Fickian diffusion into account.

14.4 Higuchian Model

This model is suitable when the active agent has been dispersed in an insoluble matrix that can swell. When a delivery system is exposed to an external medium, the exposed active agent layer dissolves and diffuses out of the matrix. This process continues as the boundary between the dissolution medium and the dispersed bioactive compound evolves with time [34]. The region within the polymer matrix devoid of any drug is referred to as the polymer ghost (Fig. 14.3).

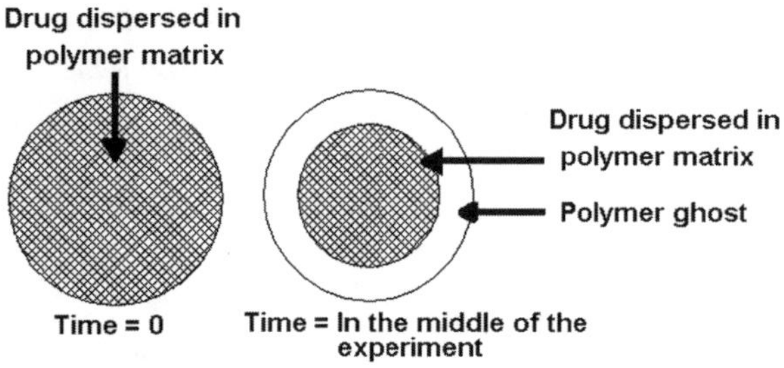

Figure 14.3 Change in the boundary between the dissolution medium and the drug for diffusion-controlled release systems. *Source: Modified from Ref. [33].*

Higuchi's model assumes that the rate of dissolution of the active agent is faster than the rate of diffusion of the active agent, which ensures the continuous release of the active agent to the surrounding medium. The rate equation for this kind of system is represented by [34]:

$$\frac{\mathrm{d}M}{\mathrm{d}x} = C_0 \cdot \mathrm{d}x - \frac{C_s}{2} \tag{14.9}$$

where $\mathrm{d}M$ = change in the amount of drug released per unit area; $\mathrm{d}x$ = change in polymer ghost thickness; C_0 = total amount of drug in unit volume of the matrix; C_s = saturated concentration of the drug in the matrix.

The diffusion theory expresses the rate of mass transfer as follows [34]:

$$k = 2C_sD_mC_0$$
$$\frac{\mathrm{d}M}{\mathrm{d}t} = \frac{D_mC_s}{x} \tag{14.10}$$

where D_m = diffusion coefficient in the matrix.

Equating, integrating, and solving Eqns (14.9) and (14.10) yields:

$$M = [C_sD_m(2C_0 - C_s)t]^{1/2} \tag{14.11}$$

When the amount of drug is in excess of the saturation concentration, i.e., $C_0 \gg C_s$:

$$M = [C_sD_m2C_0t]^{1/2} \tag{14.12}$$

or

$$M = k\,t^{1/2} \tag{14.13}$$

where $k = 2C_sD_mC_0$ is a constant as C_s, D_m, and C_0 are constant for a given delivery system [34]. If a plot of the amount of drug released vs. the square root of time is linear, then drug release from the matrix is said to be diffusion-controlled and Higuchian [4,67,69,70]. For example, the release of ibuprofen when dispersed in polystyrene microspheres can result in a delivery system following Higuchian kinetics. Depending on the composition of the ibuprofen–polystyrene microspheres, 30–80% of the drug is released into a dissolution medium in 24 h [71]. Similarly, the release of ketorolac tromethamine from albumin microspheres indicates diffusion-controlled release [72]. Pea protein has been used for preserving the biological activity of ascorbic acid, and the prepared microparticles released the ascorbic acid following Higuchian kinetics (Table 14.2) [73].

Zero-order release systems are structurally similar to the matrices following Higuchian kinetics, with the only exception being that release is constant over a period of time because the releasing surface remains constant [74]. The modeling of zero-order release systems is based on certain assumptions, namely a uniform distribution of the drug in the delivery system, and the average pore size in the delivery system being significantly smaller than the size of the bioactive compound. In practicality, however, zero-order release systems may often convert to first-order release systems [75].

14.5 Swelling

It is possible to design a delivery system that is incapable of releasing its therapeutic agent until it is placed in an appropriate environment that promotes the diffusion of the solvent into the delivery system. These swelling-controlled delivery systems may be initially dry and swell after absorbing dissolution media. The swelling increases the aqueous solvent content within the formulation with an increase in

Table 14.2 Summary of Equations for Release Kinetics

Model	Equation
Zero order	$F = k \cdot t$
First order	$\ln F = k \cdot t$
Higuchian	$F = k \cdot \sqrt{t}$

the drug dissolution thereby enabling the drug to diffuse out of the swollen network into the external environment [76]. As described later, there are numerous examples of food-related biopolymers used in the swelling-controlled-release of therapeutic compounds [77,78,79,80].

14.5.1 Stimuli-Responsive Delivery

An ideal delivery system should respond to physiological requirements, sense the changes and, accordingly, alter its release profile. The symptoms of most diseases follow a cyclic pattern and require drug delivery to mirror these cycles. If a drug possesses side effects, release, when not required, poses an extra burden on the body's metabolic system. Thus, delivery patterns need optimization for self-regulated mechanisms. Hydrogels can display a dramatic effect in their swelling behavior, network structure, permeability, or mechanical strength in response to different stimuli, such as temperature and pH [32,78,81,82,83].

14.6 Temperature-Sensitive Hydrogels

Temperature-sensitive hydrogels are probably the most commonly studied class of sensitive polymer systems in drug delivery research [84,85,86,87]. They have gained considerable attention in the pharmaceuticals field due to their ability to swell or shrink as a result of changes in temperature. Thermosensitive hydrogels can be classified as positive or negative temperature-sensitive systems. A positive temperature-sensitive hydrogel has an upper critical solution temperature (UCST). Such hydrogels contract upon cooling below the UCST. Negative temperature-sensitive hydrogels have a lower critical solution temperature (LCST) and contract upon heating above the LCST. Temperature can thus be utilized to modulate the volume of these hydrogels and consequently, their release profile can be altered [32].

14.6.1 pH-Sensitive Hydrogels

Hydrogels that exhibit pH-dependent swelling consist of networks comprised of ionic polymers. The polymers with a large number of ionizable groups are known as polyelectrolytes, and include food-grade materials (e.g., alginic acid and chitosan) [88,89]. In an aqueous medium of appropriate pH and ionic strength, these groups can ionize, resulting in changes to polymer conformation and behavior. Electrostatic repulsion between ionized polymer chains may increase the uptake of solvent in the network [18,19,90,93]. Anionic hydrogels are composed of polymer chains containing negatively charged functional groups. Ionization occurs when the environmental pH is above the pK_a of the ionizable moiety. As ionization increases, there is a resultant increase in electrostatic repulsions between the chains, which results in greater swelling. Conversely, cationic materials contain groups such as amines. These groups ionize in media that are at a pH below the pK_b of the ionizable species. Thus, in a low-pH environment, ionization increases. There are many advantages to using ionic over neutral networks in drug delivery. Their pH sensitivities can be easily exploited in a wide variety of biomedical applications, including corneal implants, controlled-release devices, and biocompatible materials [94,95,96]. Drug/bioactive-ingredient release is most often observed during the swelling of a hydrogel. However, a few instances have reported drug release during syneresis [97]. Another interesting characteristic of many responsive hydrogels is the reversibility of the induced changes in network microstructure. Such elastic deformability allows hydrogels to return to their original shape at the end of a triggering stimulus [30,98].

The majority of research efforts to date have focused on single stimulus response hydrogels for controlled delivery. The next few years should see developments in double or multiple stimuli response hydrogels. Already, progress is being made in this direction. An interpenetrating network of gelatin and dextran has been proposed as a dual stimuli-responsive biodegradable hydrogel, where lipid microspheres have been incorporated as reservoirs for bioactive compounds [99,100]. Hydrogels prepared below the sol–gel transition temperature were found to release the lipid microspheres in the presence of both α-chymotrypsin and dextranase; no release in the presence of either enzyme alone occurred.

14.7 Equilibrium Swelling and the Flory–Rehner Theory

Swelling is a characteristic of cross-linked polymer gels once they are immersed in a compatible fluid. If the network structure only swells and is not broken down in the presence of a solvent, then a state of equilibrium swelling may be attained, which may be explained by the Flory–Rehner theory [101,102]. This theory states that when a polymeric network swelling in a solvent is allowed to reach equilibrium, there are only two opposing forces at work—the force of thermodynamic mixing and the retractile force of the polymer. As more solvent penetrates the polymer network, the volume begins to increase and the network junction zones are forced to elongate and expand. The swelling effectively reduces the chain configurational entropy. Opposed to this is the increase in entropy due to the mixing of polymer and solvent during swelling. Once these opposing entropies or forces become equal in magnitude, equilibrium is reached. This situation is described in terms of the Gibbs free energy:

$$\Delta G_{\text{total}} = \Delta G_{\text{elastic}} + \Delta G_{\text{mixing}} \tag{14.14}$$

where $\Delta G_{\text{elastic}}$ is the change in energy due to the deformation of network chains between network junction zones and ΔG_{mixing} is the result of the solvent–polymer mixing. In the case of polyelectrolytes, ionic groups may complicate swelling behavior, where the ionic nature of the polymers (ΔG_{ion}) also contributes to the total change in Gibbs free energy:

$$\Delta G_{\text{total}} = \Delta G_{\text{elastic}} + \Delta G_{\text{mixing}} + \Delta G_{\text{ion}} \tag{14.15}$$

By taking the derivative of Eqn (14.15) with respect to the number of moles of solvent, an expression for the chemical potential can be derived:

$$\mu_1 - \mu_{1,0} = \Delta\mu_{\text{elastic}} + \Delta\mu_{\text{mixing}} + \Delta\mu_{\text{ion}} \tag{14.16}$$

where μ_1 is the chemical potential of the solvent in the polymer gel and $\mu_{1,0}$ is the chemical potential of the pure solvent. $\Delta\mu_{\text{elastic}}$, $\Delta\mu_{\text{mixing}}$, and $\Delta\mu_{\text{ion}}$ are the changes in chemical potential due to deformation of the polymer chain, solvent polymer mixing, and ionic character of the hydrogel, respectively. At equilibrium, the difference between the chemical potentials outside and inside the gel must be zero. A modified version of the original Flory–Rehner theory was proposed by Peppas and Merrill for hydrogels prepared in the presence of water [104]. Water effectively alters the change in chemical potential due to elastic forces and must be accounted for in the expression. In addition, the chemical potential has a strong dependency on the nature of the ions and the ionic strength of the surrounding medium. The following equation is derived for the swelling of an anionic hydrogel prepared in the presence of a solvent [18,19]:

$$\frac{V_1}{4I}\left(\frac{v_{2,s}}{\bar{v}}\right)^2\left(\frac{K_a}{10^{-\text{pH}} - K_a}\right)^2 = [\ln(1 - v_{2,s}) + v_{2,s} + \chi_1 v_{2,s}^2] + v_{2,r}\left(\frac{V_1}{\bar{v}\overline{M}_c}\right)\left(1 - \frac{2\overline{M}_c}{\overline{M}_n}\right)\left[\left(\frac{v_{2,s}}{v_{2,r}}\right)^{1/3} - \left(\frac{v_{2,s}}{2v_{2,r}}\right)\right] \tag{14.17}$$

and for the swelling of a cationic hydrogel:

$$\frac{V_1}{4I}\left(\frac{v_{2,s}}{\bar{v}}\right)^2\left(\frac{K_b}{10^{pOH} - K_a}\right)^2 = [\ln(1 - v_{2,s}) + v_{2,s} + \chi_1 v_{2,s}^2] + v_{2,r}\left(\frac{V_1}{\bar{v}\overline{M}_c}\right)\left(1 - \frac{2\overline{M}_c}{\overline{M}_n}\right) \times \left[\left(\frac{v_{2,s}}{v_{2,r}}\right)^{1/3} - \left(\frac{v_{2,s}}{2v_{2,r}}\right)\right] \tag{14.18}$$

where $\overline{M}_n$ is the molecular weight of the polymer without cross-linking; $\overline{M}_c$ is the number average polymer molecular weight between two adjacent cross-links determined from Eqn (14.17) or (14.18); $\bar{v}$ is the specific volume of the hydrogel prior to swelling; V_1 is the molar volume of the solvent, water (18 ml mol^{-1}); $v_{2,s}$ is the polymer volume fraction in the swollen state determined as approximately the inverse of the equilibrium swelling ratio; $v_{2,r}$ is the polymer volume fraction in the relaxed state (the state of the polymer immediately after cross-linking but before swelling); I is the ionic strength, K_a and K_b are the dissociation constants for the acidic and basic moieties on the polymer; and χ_1 is the Flory–Huggins parameter, which describes the polymer–solvent interaction [105].

Using the number average molecular weight between cross-links, $\overline{M}_c$, the cross-link density, q, can be determined from the following [106]:

$$q = \frac{\overline{M}_n}{\overline{M}_c} \tag{14.19}$$

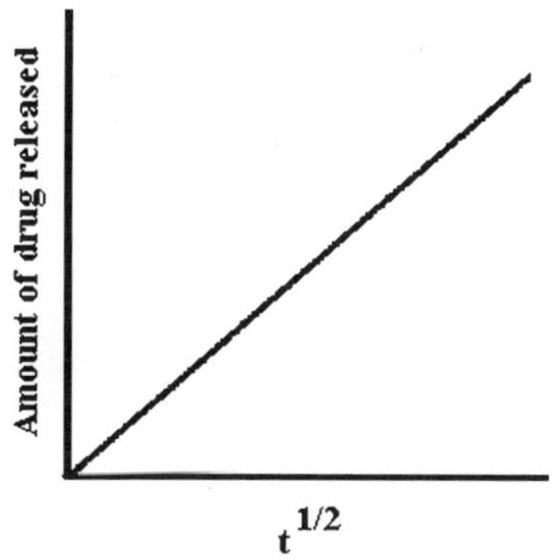

Figure 14.4 Schematic diagram of drug release.

The parameter $v_{2,s}$ is determined from the volume-swelling ratio, q_v[27]:

$$v_{2,s} = \frac{1}{q_v} \tag{14.20}$$

The volume-swelling ratio is calculated as:

$$q_v = 1 + \frac{(q_w - 1) \times \rho_2}{\rho_1} \tag{14.21}$$

where ρ_2 and ρ_1 are the densities of the polymer network and solvent, respectively. The weight-swelling ratio, q_w, is determined from the following:

$$q_w = \frac{m_s}{m_o} \tag{14.22}$$

where m_o and m_s are the mass of the unswollen gel and the mass of the swollen gel at equilibrium, respectively (Fig. 14.4).

14.8 Approaches to Cross-Linking

Cross-linking improves the thermal and mechanical stability of the matrix, and can be tailored to modify the release rate of the incorporated active agents [24,32,33,108,230]. Cross-linking methods can be broadly classified into two categories: physical (e.g., irradiation with UV and gamma radiation and dehydrothermal treatment) and chemical (use of cross-linkers such as genipin and glutaraldehyde) [109,111]. Desired release properties in a matrix modified with physical cross-linking methods are seldom achieved, as there is difficulty in controlling the cross-linking density. Chemical cross-linkers can be categorized into two types: non-zero-length cross-linkers and zero-length cross-linkers. Examples of the former include bi-/multifunctional molecules, which bridge free carboxylic acid groups, amino groups, and hydroxyl groups between adjacent polymer molecules (e.g., glutaraldehyde, polyepoxides, and isocyanates). With zero-length cross-linkers, reactive groups such as carboxylic acid and amine groups present in polymer network chains react with each other leading to formation of a covalent bond. Cross-linking with acyl azides and carbodiimides are classical examples of zero-length cross-linkers [112,113]. The main advantage of chemical cross-linking includes the easy control of the cross-linking density by tailoring the reaction conditions either by changing the cross-linking reaction period or by changing cross-linker concentration [76,114]. The pore size of the gelled matrix is governed by the cross-linking density, which in turn affects the diffusion of the solute particles through the gelled matrix. The pore size of a matrix is decreased with the increase in the cross-linking density, which will reduce the swelling ability of the (bio)polymeric network.

Though far from extensive, the following section describes the mechanisms and properties of some well-known and more recently examined chemical cross-linking agents.

14.9 Glutaraldehyde

Glutaraldehyde (GA) has seen extensive use as a cross-linking agent for biomedical applications such as enzyme and cell immobilization as well as hydrogel synthesis [115,117,118,119,120,230]. GA is a colorless liquid with a pungent odor at room temperature. It is a dialdehyde whose aldehydic groups are highly reactive and can form covalent bonds with functional groups such as amines, thiols, phenols, hydroxyl, and imidazoles. The schematic representation of the reaction of GA with hydroxyl and amino groups (which are very common in hydrogels) is shown in Fig. 14.5 [121,123].

The cross-linking of hydroxyl groups with GA must be carried out in acidic conditions. The presence of acid catalyzes acetalization among the hydroxyl and aldehydic groups [124]. Conversely, cross-linking of amines with GA in the presence of acidic conditions is slower compared to neutral and basic conditions. This phenomenon, i.e., slower reaction kinetics in acidic conditions, is used in developing hydrogels or scaffolds of various shapes and sizes [28,92,126].

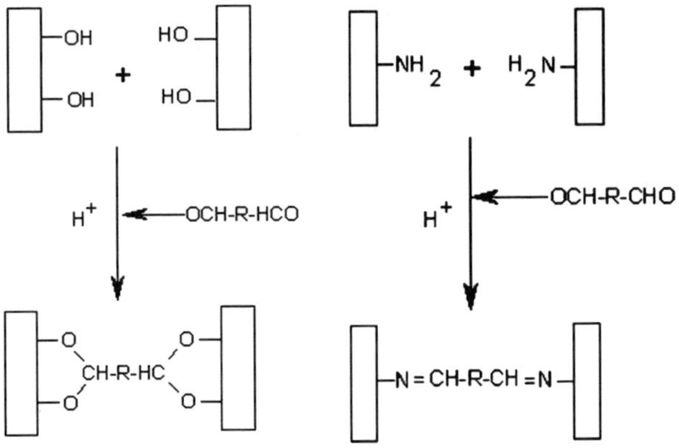

Figure 14.5 Schematic representation of the cross-linking reaction of GA with hydroxyl and amino groups.

GA is the most commonly used cross-linking agent due to its effectiveness in the stabilization of biomaterials, and it is easily accessible, economical, and its aqueous solutions can effectively cross-link collagenous tissues [11,127]. However, if released into the host due to degradation GA is toxic. There is also some local cytotoxicity if not neutralized properly, and calcification of long-term implants has been reported [118,128]. For this and other reasons, there is an increasing demand for cross-linkers that form stable, biocompatible cross-linked products and lack cytotoxicity.

14.10 Genipin

Genipin (Fig. 14.6) is an aglycone derived from an iridoid glycoside called geniposide present in the fruit of *Gardenia jasminoides*, a common flower in many parts of Asia. It is widely used in herbal medicine, and the dark blue pigments obtained by its spontaneous reaction with amino acids or proteins have been used in the fabrication of food dyes. Other

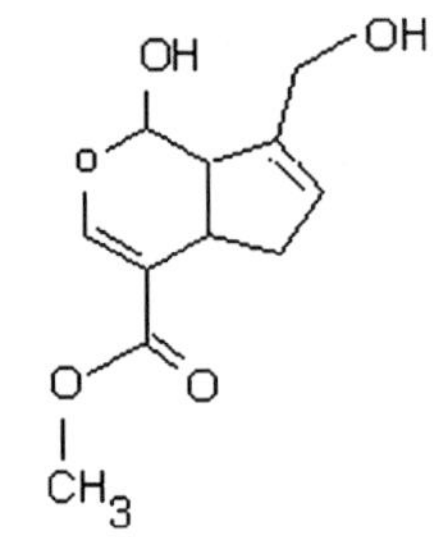

Figure 14.6 Molecular structure of genipin.

applications involve the preparation of gelatin capsules and the immobilization of enzymes [11].

It is a biodegradable molecule with low cytotoxicity [129] (see also http://www.wou.edu/las/physci/ch350/Projects_2006/Aaandering/Genipin.htm). The use of genipin as a biological glue has been evaluated extensively. It has been found that it results in lessened cytotoxicity and inflammatory responses as compared to aldehyde glues (e.g., formaldehyde and glutaraldehyde). In addition, wound healing has been shown to occur more rapidly when treated with genipin [79,130].

Genipin's cross-linking mechanism is complex and is not fully understood. It is known that it cross-links materials containing primary amine groups via two mechanisms [23,131,133]. The mechanism proposed by Zhu and Park [62,134] (Fig. 14.7) is based on the ring-opening reaction of genipin, which can be initiated by an amino group via a nucleophilic attack on the olefinic carbon atom of genipin. This is subsequently followed by a covalent grafting of the genipin onto the polymer with the amino group by a two-step reaction. An unstable intermediate formed during the reaction collapses to form a tautomeric aldehyde. The aldehyde group thus formed is again attacked by another amine group from another polymer forming another covalent bond resulting in the formation of the cross-link [134].

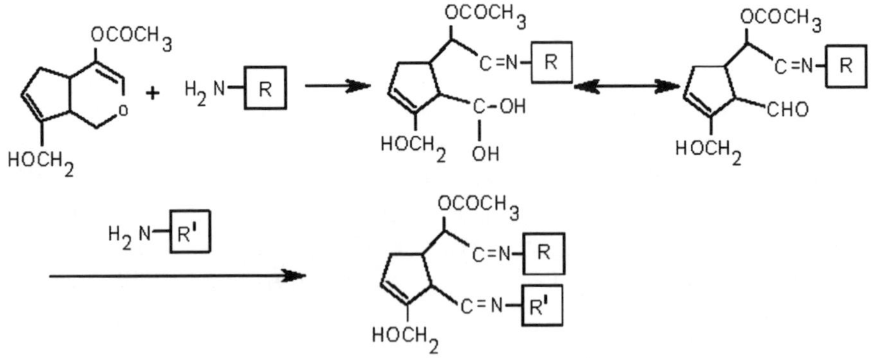

Figure 14.7 Schematic representation of mechanism of genipin cross-linking proposed by Zhu and Park. *Source: Modified from Ref. [134].*

14.11 Quinones and Phenols

The tanning of biological materials with quinones (Fig. 14.8) alters their mechanical and solubility properties, making them insoluble in water, detergents, organic solvents, and in strong acids and alkalies. Furthermore, there are changes to texture, with the cross-linked materials evolving from soft and pliable to hard and tough. This phenomenon has been exploited to develop protein-based biomaterials with desirable textural attributes [135]. Polymeric phenolics (e.g., catechol) are abundant in the plant kingdom. These polyphenols have the capacity to form networks with biopolymers (Fig. 14.9) in the presence of suitable enzymes, such as catechol oxidase. The enzymes result in the formation of reactive intermediates such as quinones and free radicals [136,139]. Lignification in plants is a classic example of this reaction [7,140,141,142]. Another example of this includes the curing of the mussel's adhesive proteins, which leads to the formation of a cross-linked network [22,136,138,139,142]. This phenomenon has been explored in the fabrication of hydrogels for biomedical and food applications [144]. These cross-linkers are natural, nontoxic (in general), and easily available.

14.12 Polyelectrolyte Cross-Linking and Complexes

The use of ionic cross-linking has recently gained importance in medical and pharmaceutical applications. The main advantages of this approach are that it is simple and does not require the use of catalysts.

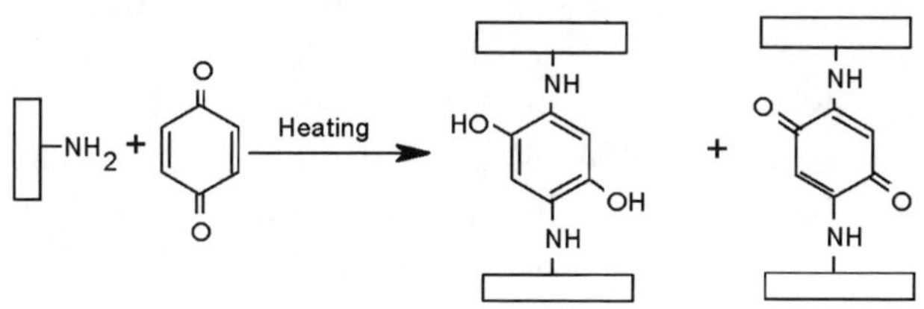

Figure 14.8 Schematic representation of mechanism of cross-linking of amino groups with quinones. *Source: Modified from Ref. [135].*

In order to prepare an ionically cross-linked polymeric network, an ionic polymer (e.g., sodium alginate or chitosan) and a charged counterion (also known as ionic cross-linker) (e.g., calcium chloride) or ionic polymer having an opposite charge are dispersed in a solvent, after which ionic interactions occur (Fig. 14.10) [9,34].

Anionic polymers such as alginates can be cross-linked and gelled using multivalent cationic cross-linkers in the presence of divalent cations such as Ca^{2+}, Sr^{2+}, or Ba^{2+}. Divalent cations bind with the guluronic acid residues of the alginates and form junction zones with other chains resulting in the formation of a cross-linked networked structure, which has the ability to retain its structure. Depending upon the concentration of the divalent cation used, alginates can give rise to either highly viscous thixotropic solutions (at low levels of Ca^{2+} ions) or a permanent gelled network (at high levels of Ca^{2+} ions) [3,145,146].

Metallic anions [e.g., Mo(VI), Pt(II) and phosphate-bearing groups (e.g., tripolyphosphates)] may be used for inducing ionic cross-links in cationic polymers. The pH of the polymer solution also plays an important role in cross-linking. For example, with a chitosan solution, if the pH is higher than the pK_a of the amino groups, the polymer solution will undergo coacervation-phase inversion without any cross-link formation, which is attributed to the precipitation of the cationic polymers [9].

The cross-links formed between two oppositely charged polymers lead to the formation of polyelectrolyte complexes (Fig. 14.11) [68,147,149,150]. Gels stabilized with this approach have been successfully used for the development of biomedical products, such as matrices for drug/gene delivery systems and tissue engineering. For example, calcium alginate cores encapsulated with a polyelectrolyte complex of alginate and poly(L-lysine) as well as alginate–chitosan polyelectrolyte systems have been used in various biomedical applications [149,151,153]. Ionic cross-linking can be easily controlled and hence is gaining importance in the development of hydrogels for biomedical and food-related applications.

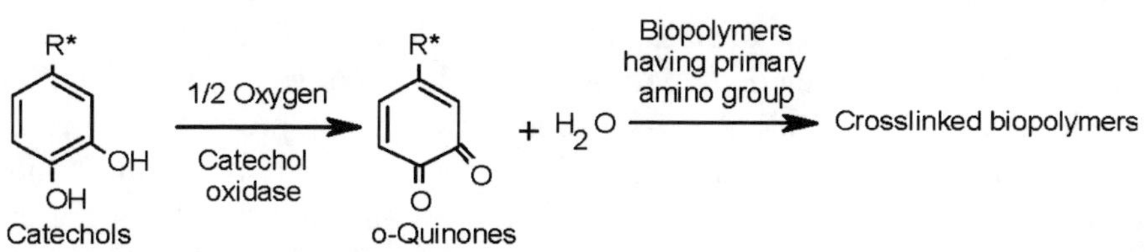

Figure 14.9 Schematic representation of the conversion of catechol into quinones, which subsequently takes part in the cross-linking of the biopolymers. *Source: Modified from Refs. [136,139].*

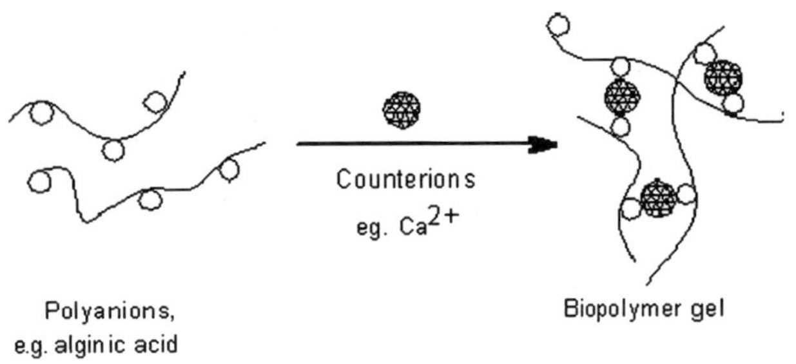

Figure 14.10 Schematic representation of hydrogel formation using a polyelectrolyte and an ionic cross-linker.

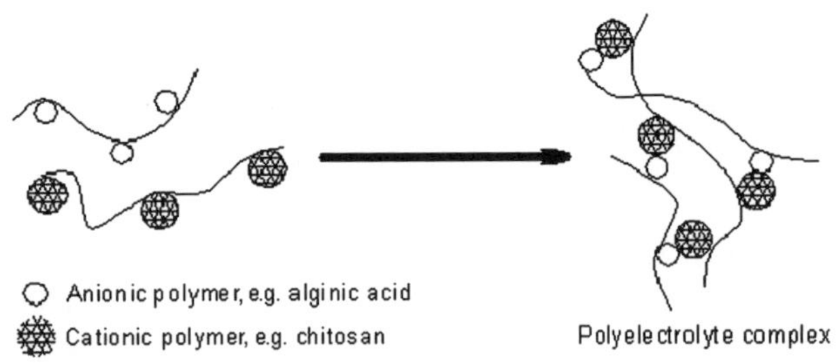

Figure 14.11 Schematic representation of hydrogel formation using polyelectrolytes with opposite charges.

Overall, the rate of cross-linking can be influenced by the size of the cross-linker. Smaller cross-linkers have the ability to diffuse more readily into the polymer solution, resulting in a faster rate of reaction. The overall charge on the polymer may affect the properties of the cross-linked gels. Larger polymers will also affect the final rheological properties of the resulting polymer network. Lastly, if the polymer solution is not fully cross-linked, the hydrogel may be pH-sensitive, given the presence of free charged groups in the polymer network [9]. This can be achieved either by shortening the reaction time or by using a stoichiometrically lower quantity/amount of the cross-linker.

14.13 Polymer–Drug Interactions

The interactions between drugs and polymers can be tailored to modify the release profile of a bioactive agent. Such interactions may be either chemical or physical in nature. When an ionic polymer comes in contact with an ionic drug having an opposite charge, the polymer interacts with the drug, resulting in complexation. Drug release from this type of matrix is dependent on the exchange of drug molecules with counter-ions in the dissolution medium, if present. As the drug is released, the polymer undergoes dissolution, exposing the inner drug–polymer complex and ensuring complete drug release. This approach has been used to develop various delivery systems [16,17]. Bonferoni *et al.* [16,17] prepared microparticles of chondroitin-6-sulfate (an anionic polymer) loaded with ciprofloxacin HCl (a cationic drug) via ionic interactions. These particles were subsequently entrapped in a Carbopol® matrix (a cross-linked acrylic acid matrix). The resulting complex was used in ocular delivery systems and showed a prolonged precorneal residence time and increased drug bioavailability in the anterior segment of the eye [16,17]. In another study, a complex of chitosan (which is cationic) and salicylic acid (an anionic drug) was tested as a transdermal delivery system [154].

In the following sections of this chapter, the physicochemical properties and applications of collagen, gelatin, chitosan, and alginates in pharmaceutical and biomedical applications are discussed.

14.14 Collagen

Collagen is connective tissue found in the hides of mammals. It is proteinaceous, occurs as fibers, and constitutes ~25% of total protein in mammals. It is mainly present in the extracellular matrix, which helps in supporting most tissues and cells [64,155,156] (see also http://en.wikipedia.org/wiki/Collagen). The length of a typical collagen fiber is ~300 nm with a diameter of ~1.5 nm. Structurally, it consists of a right-handed coiled triple helix, which is made up of three left-handed polypeptide helix strands stabilized by hydrogen bonds (see http://en.wikipedia.org/wiki/Collagen). Collagen molecules form bundles, which in turn are organized into fibrils that constitute the basic unit of collagen fibers. The component amino acids are arranged in a regular pattern within the chains, with the three main residues being glycine (every third amino acid), proline, and hydroxyproline. The carboxyl and secondary amino groups of proline and hydroxyproline are responsible for the spontaneous formation of left-handed helices (see http://en.wikipedia.org/wiki/Collagen). Glycine residues are present along the interior axis of the helix, whereas proline and hydroxyproline residues point outward. These latter residues are responsible for the thermal stability of the triple helix. Collagen has high tensile strength (50–100 MPa) and Young's modulus (1–2 GPa).

Collagen-based matrices have been widely used in delivery systems, wound healing, and various other biomedical applications given the biocompatibility and availability of the collagen. In the following section, applications of collagen in various delivery systems are discussed.

14.14.1 Cross-Linked Collagen in Controlled Drug Delivery

Koob and Hernandez [103] reported a novel method of stabilizing collagen-based materials with catechol-containing monomers. The authors oxidized *o*-catechols with *o*-quinone and polymerized bovine collagen with di-catechol nordihydroguaiaretic acid (NDGA), and reported an increase in the tensile strength and stiffness of the collagen fibers once treated with NDGA. The mechanical properties increased further with a second NDGA treatment. The group further reported that NDGA cross-linked fibers had better mechanical properties than the collagen fibers treated with other cross-linking agents, namely GA or carbodiimide [103]. The group recently reported the potential use of the NDGA cross-linked fibers as a matrix for drug delivery. In the study, the authors loaded the cross-linked fibers with dexamethasone and dexamethasone-21-phosphate, and determined the diffusion coefficients of these drugs into the fibers from the respective aqueous solutions. The diffusion coefficients of dexamethasone and dexamethasone-21-phosphate into the fibers were 1.86×10^{-14} and 2.36×10^{-13} $m^2\ s^{-1}$, respectively. The use of these kinds of fibers has also been proposed for the delivery of other active agents for the treatment of the human diseases in lieu of GA-treated fibers [158].

pH-responsive collagen gels have been prepared using alkali-treated collagen reacted with naturally derived cross-linkers based on citric acid (CA) or malic acid (MA). In this study, the authors reported that as the concentration of the CA and MA was increased, there was a corresponding decrease in the amino groups. But with increased CA in the cross-linked gels, there was an increase in the residual carboxylic groups, which accounted for the pH-sensitive nature of the treated gels. The authors concluded that the CA-treated samples have potential in sustained oral drug delivery and tissue engineering applications [159]. Collagen and algal sulfated polysaccharide films cross-linked with GA have been used in applications such as in the coating of cardiovascular prostheses, support for cellular growth, and in systems for controlled drug delivery [58]. Duan and Sheardown [161] developed a highly cross-linked collagen, reacted with poly-propyleneimine octaamine dendrimers. The cross-linking of collagen with multifunctional dendrimers was carried out by activating the carboxylic acid groups of glutamic and aspartic acid residues in collagen. The group subsequently reported the use of dendrimer-cross-linked collagen matrices for applications in cartilage tissue repair and in the development of corneal tissue engineering [164,165].

14.14.2 Collagen in Gene and Hormone/Growth Factor Delivery Systems

Biopolymers, because of their ability to serve as gene carriers and tissue engineering scaffolds, are poised to play an important role in the field of regenerative medicine [164,165,166]. Sarojini *et al.* [167] developed silica colloidal particles for localized recombinant DNA release that were coated with a collagen-containing viral vector, and transferred to human lung fibroblast cultures. The authors reported that only cells in contact with the particles were infected due to the gradual release of the viral particles from the collagen matrix, resulting in apoptotic cell death. This result indicated that the particles could be used as a delivery system to deliver genes of choice to localized subgroups of specific cells of interest [167]. In the year 2007, Takezawa *et al.* [168] reported the use of collagen vitrigel membranes (which are rigid glassy membranes obtained by drying collagen gel membranes) for the subcutaneous delivery of vascular endothelial growth factor (VGF). *In vivo* release studies of VGF in rats indicated angiogenesis.

14.14.3 Collagen in Ophthalmic Drug Delivery

Ophthalmic drug delivery systems are normally based on aqueous drops of drugs, water-insoluble drug suspensions in ointments or oil drops containing drugs. With such approaches, most of the drug is lost due to reflex blinking and lacrimation. Improvements in the ocular bioavailability of drugs have been achieved with the use of collagen shields [21]. Such shields consist of a collagen matrix loaded with a pharmacologically active agent [166,169]. Studies on delivery of fourth-generation fluoroquinolones (broad-spectrum antibiotics such as gatifloxacin, ofloxacin, and moxifloxacin) indicated that the shields could help in maintaining the desired drug concentration for a prolonged period of time in the aqueous and vitreous humor [166,169,170].

14.14.4 Collagen-Based Formulations in Oral Delivery Systems

The delivery of active agents to the stomach can be ineffective given the short residence time of the ingested delivery systems. Collagen sponges have been used successfully to increase gastric residence time. Tablets containing collagen sponges expand quickly after coming in contact with gastric fluids. The increase in the gastric residence time of the delivery system is attributed to an increase in the physical size of the matrix. For example, studies on riboflavin delivery from the collagen tablets have indicated sustained release for over 12 h [171]. Collagen treated with CA has also shown potential as an oral delivery matrix, though little work has been reported [159].

14.14.5 Collagen as a Matrix/Scaffold for Drug Delivery

Collagen matrices have been used to deliver antimicrobials to wound surfaces to inhibit microbes capable of hindering wound healing. For example, sustained delivery of ciprofloxacin (a fourth-generation fluoroquinolone) from succinylated type-I collagen matrix has been shown to improve healing by eliminating bacteria present in the wound [172].

Collagen may also be used in association with other polymers, e.g., poly(caprolactone) (PCL) [173]. PCL-modified collagen matrices, which are transparent, can help in the easy monitoring of wound surfaces during healing. These have been used for the sustained release of amikacin and gentamycin (classical examples of antibiotics) for 48 h.

The network properties of collagen blends can be altered by sterilization, which may affect the release profile of incorporated bioactive agents. For example, the glass transition temperature of collagen/PLGA blends is reduced by ethylene oxide sterilization, while there is no change in properties when these are subjected to beta- and gamma-irradiation [174].

14.15 Gelatin

Gelatin is a translucent, colorless, and brittle powder that is nearly tasteless. It has long been used as a gelling agent in the food, pharmaceutical, and cosmetic industries due to its ease of use and availability. The major source of gelatin is animal skin and bones and fish scales. It is prepared by the thermal degeneration of collagen present in these sources (see www.lsbu.ac.uk/water/hygel.html). The method of extraction of gelatin from collagen can be modified to yield either acidic or basic gelatin. Gelatin can undergo polyion complexation with either positively or negatively charged therapeutic agents, depending on the type of the gelatin. Acidic gelatin is generally used to deliver basic bioactive agents, whereas basic gelatin is used to deliver acidic bioactive agents [175]. In general, gelatin is extracted from type I collagen (with a triple helix structure), which contains two $\alpha 1$ (I) and one $\alpha 2$ (I) chains. Each of the α-strands has a molecular weight of ~95 kDa and is present in the gelatin along with several polypeptides (see www.lsbu.ac.uk/water/hygel.html) [176]. Like collagen, gelatin is also composed of mainly three amino acids, namely glycine, proline, and 4-hydroxyproline (see www.lsbu.ac.uk/water/hygel.html). Gelatin having higher levels of pyrrolidines forms stronger gels due to lower water absorption. This is usually related to the presence of higher triple helix content [176]. The gels formed by gelatin are thermo-reversible in nature. Its gel-to-sol transition takes place at ~35 °C, i.e., gelatin forms a gel at temperatures below 35 °C and has a sol-like consistency at temperatures above 35 °C (see www.lsbu.ac.uk/water/hygel.html) [14].

As described earlier, gelatin's gelling properties can be altered with chemical cross-links, an approach that has been used by various researchers for the development of controlled-release drug delivery systems [177,178,179,180]. A typical structural unit of gelatin is given in Fig. 14.12.

14.15.1 Gelatin in Peptide Delivery

Recombinant gelatin (e.g., HU4 gelatin), modified with acrylates has been used as a matrix for the delivery of proteins (e.g., lysozyme and trypsin inhibitor). The release of proteins loaded in gelatin

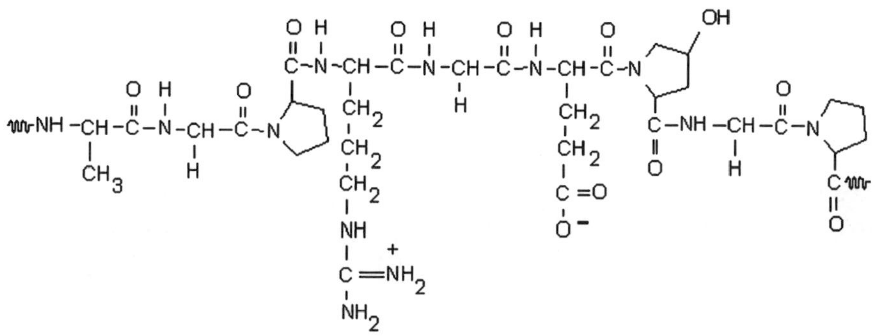

Figure 14.12 A typical structural unit of gelatin. *Source: Reproduced with permission from www.lsbu.ac.uk/water/hygel.html.*

matrices prepared in this manner has been found to be diffusion-controlled, with the complete release of the incorporated active agents taking place over 120 h. Furthermore, matrices prepared with this method have been found to be degradable in the presence of metalloproteinase 1, indicating biodegradability under *in vivo* conditions [177]. Gelatin has also been used for the development of a release system for basic fibroblast growth factor (bFGF). The matrix system, so obtained, was used for the topical administration of bFGF in rabbit models to study the effect of bFGF on angiogenesis and tissue blood perfusion of hind limb ischemia. The results indicated sustained release of bFGF from the hydrogel, which augmented angiogenesis and improved tissue blood flow [43]. Hayashi *et al.* [177] used the above-mentioned concept of bFGF delivery and developed a bFGF—gelatin hydrogel complex to coat implants for bone augmentation. *In vivo* results indicated new bone growth around the implant. The group found that an optimal amount of bFGF was needed for the development of new bone. Seki *et al.* [182] studied the effect of sperminated gelatin (SG), prepared by the addition of spermine to gelatin, on the nasal absorption of insulin in rats. They found a 5.3-fold increase in insulin absorption in the presence of 0.2% SG. The plasma glucose levels of the rats fell in a manner dependent on insulin levels, indicating the probable use of the SG for the intra-nasal delivery of insulin. These findings are promising examples for insulin-dependent diabetes treatment as insulin intake is mainly by subcutaneous injections and hence is one of the main reasons for noncompliance of patients.

Yamamoto [180] compared the efficacy of gelatin and chitosan capsules as colon-specific drug delivery systems. He incorporated prednisolone into capsules and studied the bioavailability of the steroid in rats by measuring its plasma concentration. He reported that, though the intestinal concentration of the steroid was greater in the case of chitosan capsules, the plasma concentration of the drug was higher for the gelatin capsules.

The use of aminated gelatin microspheres (AGMS) was investigated as a nasal drug delivery system for peptide drugs by Wang *et al.* [183]. Fluorescein-labeled insulin and FITC—dextran (MW = 4.4 kDa) were used as a model drug for *in vitro* release studies. The investigators found that the release of labeled insulin from AGMS was significantly slower than from native gelatin microspheres (GMS). However, there was no significant difference in the release profile of the dextran from the AGMS and GMS microspheres. The absorption-enhancing effect of insulin was determined by measuring the plasma glucose concentrations of healthy rats following intranasal administration of insulin-incorporated microspheres. AGMS microspheres indicated a higher hypoglycemic effect when delivered as an insufflation powder formulation rather than as a suspension. The group indicated that AGMS might be a new candidate carrier for the nasal delivery of peptide drugs.

14.15.2 Gelatin in Wound Healing and Implantable Delivery Systems

Mukherjee and Banthia [178] developed a PVA—gelatin hydrogel containing adrenochrome (a blood coagulant) for wound-healing applications. The hydrogel was prepared by mixing gelatin and PVA solutions in water followed by heating at 40 °C and subsequent addition of the drug. They suggested that the gel would help in instant coagulation of blood at the wound surface and thus help in wound healing.

The presence of proteases in wound surfaces results in the degradation of granulation tissue and endogenous biologically active proteins thereby hindering the wound-healing process. The inhibition of these enzymes facilitates wound healing. Apart from its antimicrobial property, doxycycline has been found to inhibit proteases on wounds, when applied topically, and promote wound healing. Adhirajan *et al.* [2] encapsulated doxycyline in microspheres of gelatin cross-linked with 1-ethyl-3-(3-dimethylaminopropyl) carbodiimide. It was found that the drug-loaded microspheres released the drug in a controlled manner. DeFail *et al.* [184] encapsulated doxorubicin into PLGA microspheres (using a double emulsion/solvent extraction method) and incorporated these into gelatin scaffolds. The release profile of doxorubicin was determined in phosphate-buffered solution for 30 days. The microspheres were added to aqueous gelatin during cross-linking/gelation with GA. The microspheres and the scaffolds were suspended in murine mammary mouse tumor cell line 4T1 for 48 h. Results indicated that the release of the drug was controlled by the incorporation of PLGA microspheres into gelatin scaffolds. Liu *et al.* [122] developed PLGA−placitaxel microspheres incorporated into cross-linked gelatin sponges, and then cross-linked with carbodiimide. The release of placitaxel (an anticancer drug) was governed by the degradation of PLGA. *In vivo* lymphatic delivery was performed in rats with orthotopic lung cancer. The implantation of the sponge intraperitoneally and intrapleurally resulted in spontaneous absorption of the particles in the lymphatic system, which indicated the potential use of these sponges for targeted delivery of drugs to the lymphatic system.

14.15.3 Gelatin in Stimuli-Responsive Delivery Systems

Hu *et al.* [188] developed ferroscaffolds of different compositions using biodegradable gelatin and iron oxide nanoparticles, which were produced *in situ*. The pore sizes of the scaffolds ranged from 50 to 200 mm depending on gelatin concentration. The yield of the iron oxide nanoparticles decreased with increasing gelatin content. The release rate of incorporated vitamin B_{12} was reduced when in the presence of a magnetic field. The authors suggested that the ferroscaffolds could be used as stimuli-responsive drug carriers.

14.15.4 Gelatin Micro- and Nanoparticles as Delivery Systems

Gelatin-based micro- and nanoparticles have been extensively studied for the controlled delivery of drugs and genetic material [189,190]. Such particles may be used to encapsulate and subsequently release genetic material when degraded or digested by enzymes. In addition to targeting tissues or cells in the body through surface functionalization, encapsulation helps in the protection of nucleic acids from harsh *in vivo* conditions, namely enzymatic degradation and phagocytosis [31]. Sha and Kaiming [192] encapsulated plasmid DNA in poly-ethylene-glycolated gelatin nanoparticles through a mild water−ethanol solvent displacement method under controlled pH and temperature. The developed particles delivered plasmid DNA into NIH 3T3 murine fibroblast cells and targeted solid tumors. Gelatin nanoparticles have been also used successfully to target lung cancer cells. Tseng *et al.* [193] grafted NeutrAvidin (FITC) (deglycosylated avidin having an affinity for biotin) on the surface of gelatin nanoparticles followed by a conjugation of biotinylated epithelial growth factor (bEGF) with NeutrAvidin (FITC), forming a core-shell structure (gelatin−avidin−bEGF). The conjugation with bEGF resulted in improved targeting efficiency for the detection of lung adenocarcinoma.

Pica *et al.* [194] suggested that intratumor administration of gelatin−methotrexate microspheres may minimize the systemic toxicity of methotrexate and may also help in overcoming drug resistance. A gelatin−methotrexate conjugate was prepared by blocking amino groups in gelatin with citraconic anhydride followed by a conjugation of amino group of methotrexate with the carboxylate group of gelatin. This was subsequently followed by cross-linking with GA to form microspheres. The *in vitro* evaluation of the drug-loaded microspheres was performed by simulating tumor conditions at pH 6.5 and 37 °C. The release of gelatin fragments in the presence of cathepsin B (a protease) and the drug was evaluated with HPLC. Results indicated that the release of fragments with molecular weights <10 kDa was low, whereas release of free methotrexate was negligible. The authors suggested that this method may be used for the development of other proteinaceous microspheres.

Rokhade *et al.* [195] successfully encapsulated ketorolac tromethamine, a nonsteroidal anti-inflammatory drug, in semi-interpenetrating polymer network (IPN) microspheres of gelatin and sodium carboxymethylcellulose (CMC) using GA as a cross-linker, resulting in particles 247−535 μm in diameter. *In vitro* release studies indicated non-Fickian behavior, with release depending on the extent of cross-linking and the amount of sodium CMC used to produce the microspheres. Liang *et al.* [79] made gelatin microspheres cross-linked with genipin, with GA-cross-linked gelatin microspheres used as a control. The biocompatibility of the microspheres was determined by injecting them intramuscularly in rats. Results indicated that the inflammatory reaction for genipin-cross-linked microspheres was significantly less than with GA-cross-linked microspheres. In addition, the degradation of the genipin-cross-linked microspheres was slower than their GA counterparts.

Gelatin alone or in combination with mucoadhesive polymers such as mucin may be used to encapsulate various acid-labile therapeutic substances in microspheres for rectal delivery. *In vitro* release studies of ceftrioxone (third-generation cephalosporin) encapsulated in gelatin and gelatin–mucin microspheres indicated diffusion-controlled non-Fickian release and sustained delivery [196]. Lu *et al.* [197] developed a sustained-release drug delivery system for indometacin, where the drug was encapsulated in a polymer shell consisting of gelatin and cellulose acetate phthalate (CAP). *In vitro* release studies of the drug prepared by complex coacervation indicated Higuchian kinetics.

14.15.5 Gelatin as a Matrix for Biologically Active Agents

Hiroyuki *et al.* (2006) examined the improved efficacy of erythropoietin–gelatin hydrogel sheets in the treatment of myocardial infarctions. Using rabbits, they conducted *in vivo* studies where the control group was infused with saline in a first experimental group, with erythropoietin subcutaneously injected in a second test group, and with erythropoietin–gelatin hydrogel sheets applied to the affected area on the heart in a third group. The results indicated that the hydrogel sheet decreased myocardial infarctus size and improved the function of the left ventricle compared to erythropoietin systemic injection [198].

Films of alginate and gelatin cross-linked with Ca^{2+} were developed by [199] using a solvent casting method. They incorporated ciprofloxacin hydrochloride as a model drug and found that its release from the gel films was dependent on the pH and ionic strength of the release solution. The drug release rate was faster when the pH of the dissolution medium was pH 7.4 compared to pH 3.6 and was accelerated by higher ionic strengths. The other factors playing an important role in drug release were the component ratio of alginate and gelatin, the amount of ciprofloxacin hydrochloride loaded in the gel films, the thickness of the drug-loaded films and the cross-linking time with Ca^{2+}.

14.16 Chitin and Chitosan

Chitin is a polysaccharide found in the outer skeleton of insects, crabs, shrimps, and lobsters, whereas chitosan is deacetylated chitin. Chitin resembles cellulose in structure, but it has an acetamido group instead of a hydroxyl group at the C-2 position of the backbone polymer chain. It is composed of 2-acetamido-2-deoxy-β-D-glucose attached with β (1→4) linkages (Fig. 14.13) and is degraded by chitinase [200,201,203]. To improve solubility, it is necessary to deacetylate chitin by 80–85% or higher, thereby yielding chitosan. With further deacetylation, enhanced solubility is achieved. Both chitin and chitosan are biocompatible, biodegradable, and nontoxic, and have good adsorption properties, which make them suitable for various drug delivery applications. Chitosan is one of the few cationic polyelectrolytes found in nature [204].

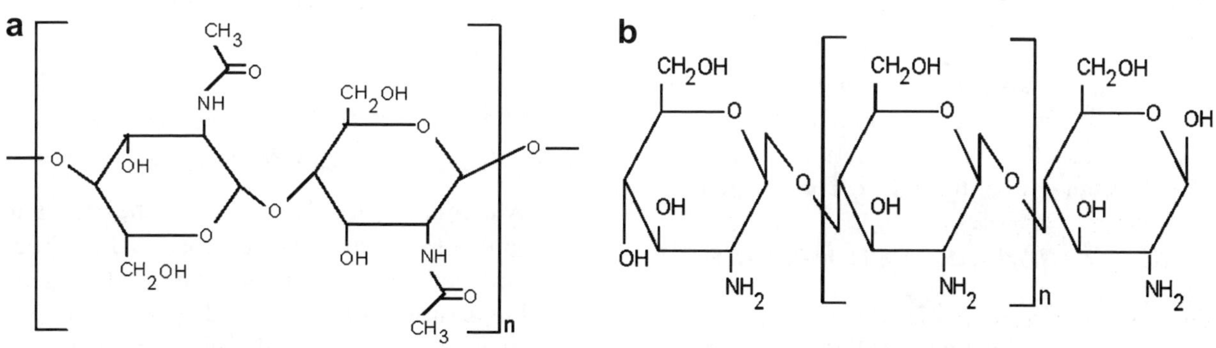

Figure 14.13 Chemical structure of (a) chitin and (b) chitosan. *Source: Modified from Ref. [249].*

Chitin and chitosan vary in composition depending on the origin and manufacturing process. Their use has been limited due to the relatively laborious isolation process which can increase production costs. With a better understanding of their inherent biological and physicochemical characteristics, however, they have seen much use in various biomedical applications (e.g., as a wound-healing agent and a delivery vehicle for pharmaceuticals and genes) [205].

14.16.1 Chitin/Chitosan as a Matrix for Biologically Active Agents

Saito *et al.* [206] developed adriamycin-containing chitosan sheets by mixing adriamycin with a chitosan suspension followed by freeze-drying. They inserted the chitosan sheet into the peritoneal cavity of mice and found that adriamycin was stable even after 2 months, indicating that the sheet could be used to improve therapeutic efficacy in the treatment of topical lesions.

14.16.2 Chitin/Chitosan in Stimuli-Responsive Delivery Systems

Sun *et al.* [207] developed chitosan-based hydrogel films having both thermal and pH sensitivity. They blended chitosan with poly(*N*-isopropyl-acrylamide) (PNIPAAm), a temperature-sensitive polymer, and polyethylene glycol (PEG). X-ray diffraction studies revealed higher crystallinity in the blended films as compared to chitosan or PNIPAAm films alone. Scanning electron microscopy showed that the films could be made more porous upon heating, indicating their temperature sensitivity. The films were found to be pH sensitive because of the pendant amino groups present in the chitosan. Guo *et al.* [208] also developed thermo- and pH-responsive hydrogels. They prepared a semi-IPN polyampholyte of carboxymethyl chitosan and poly(2-(dimethylamino)ethyl methacrylate) using *N*-*N*′-methylenebisacrylamide as cross-linker. Swelling and de-swelling of the hydrogel was reversible and consistent. The release profile of incorporated coenzyme A increased with temperature, but decreased as the proportion of carboxymethyl chitosan within the hydrogel was increased.

14.16.3 Chitin/Chitosan Particles as Delivery Systems

Reverchon and Antonacci [209] successfully micronized chitosan in a 1% acetic acid aqueous solution using supercritical fluid-assisted atomization. Characterization of the microparticles revealed a size range of 0.1−1.5 μm when the chitosan was precipitated from solutions having concentrations ranging between 1 and 10 mg ml^{-1}. Higher precipitation temperatures resulted in a decrease in particle crystallinity. The authors considered these microparticles as promising vehicles for drug delivery. Yuan *et al.* [210] studied the effects of genipin cross-linking on drug/protein release from chitosan microspheres. The release of incorporated albumin revealed that the amount of cross-linker, and the cross-linking duration played a significant role in release rate. Jayakumar *et al.* [211] synthesized water-soluble thiol-containing chitosan using a graft copolymerization technique and made beads of the modified chitosan, loading them with indometacin. *In vitro* release studies indicated pH dependency, with slower release at pH 1.4 than 7.4. This was attributed to the ionization of the thiol groups and the high solubility of indometacin in an alkaline medium. Peng *et al.* [163] prepared microspheres of *N*-methylated chitosan (NMC) cross-linked with GA to encapsulate ofloxacin, a fourth-generation fluoroquinolone. The ofloxacin was electrostatically cross-linked with the polymer, and its release depended on the molecular weight of the *N*-methylated chitosan, with higher NMC molecular weights slowing the release. Release profiles indicated non-Fickan diffusion through the swollen microspheres. Shi *et al.* [213] prepared beads via the polyelectrolyte complexation of chitosan and alginate solutions. Infrared spectroscopy revealed electrostatic interactions between the amino groups of chitosan and carboxyl groups of alginate. The beads were loaded with bovine serum albumin (BSA) and the *in vitro* release profile was studied at different pHs. Bead composition influenced the encapsulation and release of BSA. The release was also dependent on pH, with higher pH values leading to increased release.

14.16.4 Chemically Modified Chitin/Chitosan for Drug Delivery

Jayakumar *et al.* [211] discussed the generation of new bifunctional materials by chemical modification of chitin and chitosan with sulfates, which have been used in a variety of biomedical applications (e.g., as anticoagulants, in drug-delivery matrices and as antimicrobial polymers). *N*-alkyl-*O*-sulfated chitosan developed by Zhang *et al.* [214] was used to form micelles (100–400 nm diameter) in water. The group entrapped Taxol®, an anticancer drug, into the polymeric micelles by physical entrapment and suggested that the modified polymer could be used as a potential drug carrier. Another variant of sulfated chitosan, *N*-carboxymethylchitosan-*N*-*O*-sulfate, has been found to inhibit HIV-1 replication and viral binding with the CD4 cells, thus inhibiting the progression of the virus [215]. Lee *et al.* [216] developed nanoparticles of thiolated chitosan for intranasal delivery of theophylline to alleviate allergic asthma. They studied the efficacy of the delivery system to suppress inflammatory allergic response on ovalbumin (OVA)-sensitized BALB/c mice. The results revealed that the beneficial effects of theophylline were increased when it was complexed with the nanoparticles. Satoh *et al.* [217] successfully modified chitosan in a highly regioselective manner. They synthesized water-soluble (at neutral pH) 6-amino-6-deoxy-chitosan using *N*-phthaloyl-chitosan as a starting material. It was evaluated for its capability as a gene carrier. The transfection results for COS-1 cells revealed that the synthesized product was superior to standard chitosan. Jayakumar *et al.* [218] synthesized phosphorous-containing chitosan by graft copolymerization and prepared beads encapsulating indometacin by using tripolyphosphate at pH 4.0. *In vitro* drug release patterns were pH-dependent, with the release rate increasing as the pH of the dissolution media was increased, indicating the potential use of the polymer in developing a delayed drug delivery system for oral administration.

14.16.5 Chitin/Chitosan in Cardiovascular Delivery Systems

A novel biodegradable and rapidly expanding stent of chitosan cross-linked with an epoxy compound having a shape-memory property was developed by Chen *et al.* [219]. The stent prepared using cross-linked chitosan was able to withstand >30% deformation as compared to 10% in commercial metallic stents before loss of elasticity. Glycerol and poly(ethylene oxide) were used to reduce the crystallinity of the chitosan films. Preliminary *in vivo* studies in animals showed no thrombus formation and adverse reactions. The authors suggested the use of the degradable stent as a vehicle for local drug delivery.

14.16.6 Chitin/Chitosan Derivatives as In Situ Gelling Agents

Yu *et al.* [220] discussed the development of a novel composite hydrogel using dialdehyde konjac glucomannan as a cross-linking agent for chitosan. Cross-linking with this compound led to rapid gelling (in minutes). Drug release studies using ofloxacin indicated sustained release of the active agent, which may be due to the association of the drug with the polymer.

14.16.7 Chitin/Chitosan in Ocular Delivery Systems

Verestiuc *et al.* [221] prepared a series of hybrid polymeric hydrogels by reacting acrylic acid-functionalized chitosan with either *N*-isopropylacrylamide or 2-hydroxyethyl methacrylate monomers. The hybrid polymers prepared with this approach were pressed into minitablets and evaluated as a drug delivery system for ocular delivery. The matrices were loaded with chloramphenicol, atropine, norfloxacin, or pilocarpine. *In vitro* release patterns revealed drug-specific carrier compositions for the controlled delivery of these compounds.

14.17 Celluloses

Cellulose is a structural polysaccharide with the chemical formula $(C_6H_{10}O_5)_n$ (Fig. 14.14). β-Glucose molecules condense through β(1→4)-glycosidic bonds to give rise to cellulose. Cellulose is a linear polymer and is highly crystalline, which is attributed to extensive intramolecular hydrogen bonding (see http://www.lsbu.ac.uk/water/hycel.html). Acetic acid bacteria, some forms of algae and oomycetes, also have the capability to synthesize cellulose.

14.17.1 Cellulose as a Thermo-Sensitive Polymer

Aqueous solutions of most natural polymers (e.g., gelatin and carrageenan) form gels below a critical temperature usually known as the sol-to-gel transition temperature. Some cellulose derivates [e.g., methylcellulose (MC) and hydroxypropyl methylcellulose (HPMC)] form a gel upon heating, with gelation temperatures of 40–50 °C and 75–90 °C for MC and HPMC, respectively. The transition temperature of HPMC can be lowered to ~40 °C by reducing the

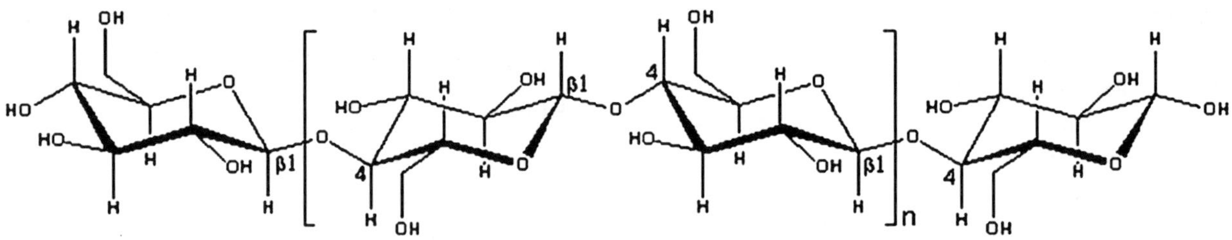

Figure 14.14 Chemical structure of cellulose. *Source: Reproduced with permission from http://www.lsbu.ac.uk/water/hycel.html.*

hydroxylpropyl molar substitution. The polymer chains are hydrated at lower temperatures while they start dehydrating as the temperature increases. Partial dehydration results in polymer–polymer association, thereby resulting in a network structure [222]. Tate *et al.* [223] developed MC-based scaffolds for the repair of brain defects. Their gels were biocompatible in both *in vitro* and *in vivo* conditions. Carlsson *et al.* [25,26] reported a change in the thermal behavior of aqueous solutions (1–4 wt%) of ethyl(hydroxyethyl) cellulose with the addition of ionic surfactants such as sodium dodecyl sulfate or cetyl triammonium bromide. The systems underwent sol-to-gel phase transitions at 30–40 °C, resulting in the formation of gels, with micelle-like surfactant clusters influencing gel properties. Scherlund *et al.* [224] used this system for the sustained delivery of lidocaine and prilocaine in periodontal pockets.

14.17.2 Cellulose Esters

Cellulose esters have been used in drug delivery, given their negligible cytotoxicity, stability, high water permeability, film-forming capability, and compatibility with most bioactive agents [225]. CAP is widely used and was first described by Malm and Fordycee in 1940. In 1951, Malm's group reported that CAP could be used for the development of enteric coated delivery systems [226]. Other cellulose esters of importance in this regard are cellulose acetate trimellitate (CAT) and cellulose acetate succinate. Levine *et al.* [227] coated capsules containing beclomethasone dipropionate with CAP and administered it to patients who underwent colostomies. The authors were able to recover more than triple the amount of the steroid and its metabolites from the patients administered with the CAP-coated capsules when compared to patients treated with capsules without coating. Liu *et al.* [228] developed microcapsules of CAP and cellulose acetate with encapsulated nitrofurantoin. The microcapsules slowed the release of the drug, with lower pHs further slowing the release. *In vivo* results indicated a decrease in the occurrence of stomach ulcers, which was attributed to the enteric nature of the formulation.

There are numerous examples of cellulose derivatives converted into microparticles for drug delivery. Pongpaibul and Whitworth [229] made microparticles of cellulose acetate butyrate (CAB) with propranolol entrapped in the matrix. *In vitro* studies indicated a decrease in the release rate of the drug with an increase in polymer content. Sprockel *et al.* [231] developed microparticles of CAB, which were used to entrap water-soluble drugs complexed with a sulfonated anion exchange resin. The release of the drug from the microparticles, when suspended in water, indicated sustained release. Palmieri *et al.* [232] developed microparticles of CAP, CAT, and hydroxypropyl methylcellulose phthalate entrapping ketoprofen. The microparticles showed good protection at gastric pH and rapid release at small intestinal pH, demonstrating a potential method for the development of pH-dependent release systems without the need to coat them with enteric polymers.

Cellulose diacetate (CA) is highly permeable to water, but is impermeable to salts and many organic compounds. This property has been used to develop osmotic delivery systems. For example, Theeuwes [233] reported that the release rate from such systems was not affected by agitation of the dissolution medium and the pH of the surrounding environment. Makhija and Vavia [234] reported an osmotic drug delivery system for pseudoephidrine. The group used PEG and diethylphthalate as a dopant to create pores in the CA semipermeable membranes. The drug's release rate depended on CA film thickness and PEG content.

14.18 Alginates

Alginates are isolated from brown seaweed using dilute alkaline extraction. The resulting solutions are treated with mineral acids and are subsequently converted to sodium alginate. Alginic acid is a linear polymer consisting of D-mannuronic acid and L-guluronic acid residues (Fig. 14.15) [235] (see also http://www.lsbu.ac.uk/water/hyalg.html). Alginic acid forms a high-viscosity acid gel in the presence of water, which is attributed to the hydration of the polymer chain and intermolecular hydrogen bonding. Alginate polymers form gels in the presence of divalent and multivalent cations (except Mg^{2+}) by cross-linking of the carboxylate groups on the polymer backbone [146].

Particles of alginates, cross-linked with calcium ions, have been studied extensively. In general, there are three different methods for the preparation of the particles. In the first method, the aqueous alginate solution is added dropwise into an aqueous calcium chloride solution. The second method involves the *in situ* release of cross-linking calcium ions from calcium carbonate by an acidification process, whereas the third method involves the addition of a dispersed calcium chloride solution to an alginate solution.

Under acidic conditions, the swelling of calcium alginate particles is negligible, as alginic acid is an anionic polymer and the carboxyl groups are not ionized at pHs lower than the pK_a of the polymer. Hence, any drug release will be mainly associated with the diffusion of the active agent through the insoluble matrix. Under neutral and basic conditions, however, the anionic polymer becomes ionized and swells. Drug release under these conditions will depend on the swelling of the beads and/or the erosion of the polymer matrix as well as the inherent properties of the drug. Because of the above properties, alginates have been used for the development of a multiple-unit, controlled-release drug delivery system. Higher concentrations of alginate in the beads and alginates rich in guluronic acid have been found to decrease the release rate of active agents. Low-molecular-weight alginates have been studied for their capacity to enhance the dissolution rate of acidic, basic, and neutral drugs [146].

14.18.1 Alginates in Diffusion-Controlled Delivery Systems

Alginates have been used to encapsulate drugs in microcapsules or for matrix-type drug delivery systems for a large number of proteins [145, 146,236]. Positively charged proteins (e.g., TGF β1) can react with the carboxylic acid groups of the alginates, which may result in protein denaturation. To prevent loss of activity, additives such as polyacrylic acid are often used. The protection of the protein has been attributed to the shielding effect of the polyacrylic acid from the low molecular fragments of the alginates [237]. Amsden and Turner [5] reported that protein diffusion was highest in alginate gels prepared from alginates with a low guluronic acid fraction, which was associated with greater flexibility of the alginate backbone. In general, higher backbone flexibility results in higher solute diffusion. Tomida *et al.* [238] reported that theophylline-loaded alginate gels showed zero-order release kinetics, stating that release could be controlled by changing coating thickness. Iannuccelli *et al.* [239] found that gentamycin sulfate (GS) could selectively interact with the mannuronic residues of the alginates and did not compete with the calcium ions involved in gelation. They observed that calcium ions preferentially reacted with the polygluronic sequences, though the polymannuronic sequences could also play a small role during cross-linking. Thus, alginates rich in mannuronic acid residues were preferred for GS delivery systems given their higher GS binding capacity. Chan and Heng [29] reported that

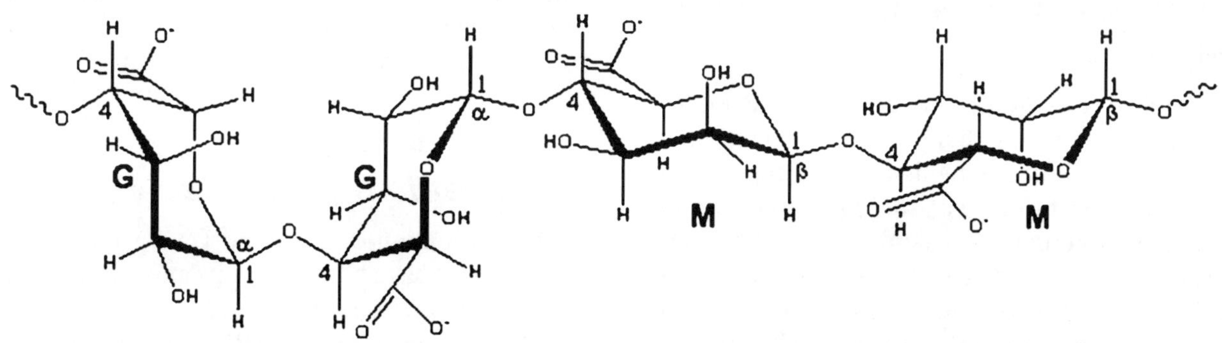

Figure 14.15 Chemical structure of alginic acid. *Source: Reproduced with permission from http://www.lsbu.ac.uk/water/hycel.html.*

the release of drugs from alginate microspheres was affected by the presence of various additives, such as poly(vinylpyrrolidone) and ethylcellulose. The group reported that high-viscosity ethylcellulose reduced the drug release rate while poly(vinylpyrrolidone) was able to increase the flowability of the alginate microspheres. Takka and Acartürk [240] reported that the alginate's molecular weight did not affect the release pattern of nicardipine HCl, a neutral molecule, but that the release rate was increased with the addition of various polymers such as Carbopol® 941, HPMC, and Eudragit® RS 30. They also reported that when Ca^{2+} was used as a cross-linker, drug release was prolonged when compared with particles cross-linked with Ba^{2+} and Sr^{2+}. These findings could be used to modulate the release of drugs from the alginate microspheres, in which other polymers could be deliberately incorporated or the cross-linker could be changed. Imai *et al.* [241] reported that the release of pindolol, a basic drug, was slow for gel beads prepared with the low-molecular-weight alginates. They proposed that drug release could be modulated by using alginates with different molecular weights. Kulkarni *et al.* [110] developed polymeric sodium alginate microparticles by precipitating sodium alginate in methanol and subsequently cross-linking the particles with GA. The group reported that with an increase in cross-linking, there was a decrease in microsphere swelling, though the loading efficiency of nimesulide, a water-soluble drug, increased. Hodsdon *et al.* [242] made tablets from blends of various drugs and alginate. They reported that water-soluble drugs were released at a faster rate in simulated gastric fluid than in the simulated intestinal fluid. However, drugs having poor water solubility showed the opposite effect.

14.18.2 Alginates as *In Situ* Gelling Agents

Miyazaki *et al.* [243] reported a novel method for the *in situ* gelation of sodium alginate for controlled oral delivery. The oral administration of the sodium alginate solution to rats was subsequently followed by a solution containing calcium ions. When the solutions reached the acidic stomach, the release of free Ca^{2+} promoted gelation. The group reported that there was an increase in the bioavailability of theophylline as compared to oral sustained-release formulations.

14.18.3 Alginates in Oral Delivery Systems

Dennis *et al.* (2002) filled hard gelatin capsules with a mixture of drug, alginate, and a pH-independent polymer (HPMC). Upon ingestion, the capsules absorbed gastric fluid thereby initiating surface hydration of the polymer. The formation of a surface gel layer led to air entrapment and the capsules began to float. Over time, the gel layer eroded resulting in the movement of the gel–dissolution interface toward the core of the capsules. Eventually, the device lost its floatability and passed into the intestinal tract where the alginates dissolved (due to the basic environment in the intestine), thereby making the delivery system more porous leading to drug release (see http://patents1.ic.gc.ca/details?patent_number=2081070).

Alginates have good mucoadhesive properties and have been used in combination with chitosan in various biomedical applications because of this property. Miyazaki *et al.* [244] developed alginate–chitosan tablets for the sublingual delivery of ketoprofen. Tonnesen and Karlsen [146] reported that alginate microparticles showed strong mucoadhesion toward the stomach mucosa. Coating of the particles with chitosan did not change their mucoadhesion property. Such examples demonstate that alginate-based systems can be developed for drug delivery in the stomach.

Mandel *et al.* [132] reported that alginate-based formulations containing antacids and H_2-receptor blockers can be used in the treatment of heartburn and esophagitis by acting as a barrier against acid reflux. Katayama *et al.* [245] developed a liquid preparation consisting of sodium alginate and ampicillin (an antibiotic) for the eradication of *Helicobacter pylori*. Once ingested, this preparation apparently spreads on the stomach wall releasing the incorporated drug on the mucosa. Finally, sodium alginate has been used to mask the bitter taste of drugs such as amiprilose hydrochloride.

14.18.4 Alginates as Encapsulating Agents

Encapsulation of cells and DNA within the alginate matrix has shown great promise in biomedical applications. Esquisabel *et al.* [246] encapsulated bacillus Calmette-Guérin (BCG) (a vaccine against tuberculosis) within a calcium alginate matrix using an emulsification method for microencapsulation.

They reported that the size of the particles depended on the viscosity of the oil used for emulsification. Smith [247] developed an enteric delivery system for DNA by encapsulating DNA into a calcium alginate matrix. Similarly, Alexakis *et al.* [3] encapsulated DNA in alginate microspheres, which were further coated with chitosan. The delivery system was orally administered to rats. The microparticles were recovered from the rat feces and were analyzed for the activity of the entrapped DNA molecules. The results indicated that the encapsulated DNA was biologically active and could be substantially recovered.

14.18.5 Alginates as Wound-Healing Materials

Tonnesen and Karlsen [146] also reported that alginates could be used as wound-dressing materials in the form of powder, films, or fibers. The calcium alginate matrix promoted the exchange of calcium ions with sodium ions on the wound surface, which enhanced the blood-coagulation cascade, imparting a hemostatic property to the matrix.

14.18.6 Alginates in Ophthalmic Delivery Systems

Cohen *et al.* [36] reported the *in situ* formation of an ophthalmic drug delivery system from alginates, which underwent gelation in the ocular sack without any external cross-linking agent. The extent of alginate gelation and release of an incorporated drug (pilocarpine) was dependent on the percent of glucoronic acid residues in the polymer backbone. The group also reported that alginates having a glucoronic acid content of >65% instantaneously gelled. *In vitro* results in rabbits indicated that the intra-ocular pressure-reducing effect of pilocarpine was increased to 10 h as compared to 3 h when the drug was administered in solution. In addition, the dissolution of the matrix in the dissolution medium was negligible for the first 12 h at 37 °C.

14.19 Summary

Although synthetic polymers are used more extensively in the field of drug delivery, biopolymers and their derivatives are rapidly gaining in importance. This is mainly due to their intrinsic properties that render them appealing. In general, they are non-carcinogenic, mucoadhesive, biocompatible and biodegradable. For example, chitosan has been used in oral and nasal delivery systems, where the mucoadhesive property of the polymer plays an important role. Furthermore, the properties of the biopolymers can be tailored via chemical modification to further expand their functionality. For example, sodium CMC has been used as a viscosity-building agent and as a binder in pharmaceutical formulations but does not show pH-dependent swelling behavior. However, when it is esterified with acryloyl or methacryloyl chloride, it shows pH-dependent swelling and becomes insoluble in water [55]. Along with being biocompatible and non-carcinogenic, biopolymers are also ideal candidates for the development of matrices for tissue engineering and wound dressings. For example, Pal *et al.* [248] developed transparent starch hydrogels for use as a wound dressing. In tissue engineering, researchers are physically entrapping growth factors within biopolymer matrices for the development of specific cells and tissues.

In closing, the importance of biopolymers in the development of matrices for the controlled release of drugs and other bioactive compounds will continue to increase. Other than developments in the biomedical and pharmaceutical industries, it is likely that biopolymer usage will see rapid growth in the areas of cosmetology and nutraceutical delivery.

Acknowledgments

The authors gratefully acknowledge funding from the Advanced Foods and Materials Network of Centres of Excellence (AFMNet) and the Natural Sciences and Engineering Research Council of Canada (NSERC) during the completion of this chapter.

References

[1] J. Enderle, S. Blanchard, J. Bronzino, Biomaterials: Properties, Types and Applications. Introduction to Biomedical Engineering, Academic Press, San Diego, CA, 2005.

[2] N. Adhirajan, N. Shanmugasundaram, M. Babu, Gelatin microspheres cross-linked with EDC as a drug delivery system for doxycyline: development and characterization, J. Microencapsul. 24 (7) (2007) 659–671.

[3] T. Alexakis, D.K. Boadi, D. Quong, A. Groboillot, I. Oneill, D. Poncelet, R.J. Neufeld, Microencapsulation of DNA within alginate microspheres and cross-linked chitosan membranes for in-vivo application, Appl. Biochem. Biotechnol. 50 (1) (1995) 93–106.
[4] J. Ali, S. Arora, A. Ahuja, A.K. Babbar, R.K. Sharma, R.K. Khar, S. Baboota, Formulation and development of hydrodynamically balanced system for metformin: in vitro and in vivo evaluation, Eur. J. Pharm. Biopharm. 67 (1) (2007) 196–201.
[5] B. Amsden, N. Turner, Diffusion characteristics of calcium alginate gels, Biotechnol. Bioeng. 65 (5) (1999) 605–610.
[6] C. Anchisi, M.C. Meloni, A.M. Maccioni, Chitosan beads loaded with essential oils in cosmetic formulations, J. Cosmet. Sci. 57 (3) (2006) 205–214.
[7] S.O. Andersen, M.G. Peter, P. Roepstorff, Cuticular sclerotization in insects, Comp. Biochem. Physiol. B–Biochem. Mol. Biol. 113 (4) (1996) 689–705.
[8] E.M. Bachelder, T.T. Beaudette, K.E. Broaders, J. Dashe, J.M.J. Frechet, Acetal-derivatized dextran: an acid-responsive biodegradable material for therapeutic applications, J. Am. Chem. Soc. 130 (32) (2008) 10494.
[9] J. Berger, M. Reist, J.M. Mayer, O. Felt, N.A. Peppas, R. Gurny, Structure and interactions in covalently and ionically cross-linked chitosan hydrogels for biomedical applications, Eur. J. Pharm. Biopharm. 57 (1) (2004) 19–34.
[10] M. Bertone, V. Dini, P. Romanelli, F. Rizzello, M. Romanelli, Objective analysis of heterologous collagen efficacy in hard-to-heal venous leg ulcers, Wounds 20 (9) (2008) 245–249.
[11] A. Bigi, G. Cojazzi, S. Panzavolta, N. Roveri, K. Rubini, Stabilization of gelatin films by cross-linking with genipin, Biomaterials 23 (24) (2002) 4827–4832.
[12] Y. Bin Choy, F. Cheng, H. Choi, K. Kim, Monodisperse gelatin microspheres as a drug delivery vehicle: release profile and effect of cross-linking density, Macromol. Biosci. 8 (8) (2008) 758–765.
[13] J.S. Boateng, K.H. Matthews, H.N.E. Stevens, G.M. Eccleston, Wound healing dressings and drug delivery systems: a review, J. Pharm. Sci. 97 (8) (2008) 2892–2923.
[14] H.B. Bohidar, S.S. Jena, Kinetics of sol–gel transition in thermoreversible gelation of gelatin, J. Chem. Phys. 98 (11) (1993) 8970–8977.
[15] C. Bolliet, M.C. Bohn, M. Spector, Non-viral delivery of the gene for glial cell line-derived neurotrophic factor to mesenchymal stem cells in vitro via a collagen scaffold, Tissue Eng. Part C: Methods 14 (3) (2008) 207–219.
[16] M.C. Bonferoni, G. Sandri, E. Gavini, S. Rossi, F. Ferrari, C. Caramella, Microparticle systems based on polymer-drug interaction for ocular delivery of ciprofloxacin – I. In vitro characterization, J. Drug Deliv. Sci. Technol. 17 (1) (2007a) 57–62.
[17] M.C. Bonferoni, G. Sandri, E. Gavini, S. Rossi, F. Ferrari, C. Caramella, Microparticle systems based on polymer-drug interaction for ocular delivery of ciprofloxacin – II. Precorneal residence times, J. Drug Deliv. Sci. Technol. 17 (1) (2007b) 63–68.
[18] L. Brannon-Peppas, N.A. Peppas, Equilibrium swelling behavior of dilute ionic hydrogels in electrolytic solutions, J. Control Release 16 (3) (1991a) 319–329.
[19] L. Brannon-Peppas, N.A. Peppas, Equilibrium swelling behavior of pH-sensitive hydrogels, Chem. Eng. Sci. 46 (3) (1991b) 715–722.
[20] L. Brannon-Peppas, Polymers in controlled drug delivery, Medical Plastics and Biomaterials Magazine (1997). Nov. 34.
[21] H.Z. Bu, H.J. Gukasyan, L. Goulet, X.J. Lou, C. Xiang, T. Koudriakova, Ocular disposition, pharmacokinetics, efficacy and safety of nanoparticle-formulated ophthalmic drugs, Curr. Drug Metab. 8 (2) (2007) 91–107.
[22] L.A. Burzio, J.H. Waite, Cross-linking in adhesive quinoproteins: studies with model decapeptides, Biochemistry 39 (36) (2000) 11147–11153.
[23] M.F. Butler, Y. Ng, P.D.A. Pudney, Mechanism and kinetics of the cross-linking reaction between biopolymers containing primary amine groups and genipin, J. Polym. Sci. A1 41 (24) (2003) 3941–3953.
[24] D. Campoccia, P. Doherty, M. Radice, P. Brun, G. Abatangelo, D.F. Williams, Semisynthetic resorbable materials from hyaluronan esterification, Biomaterials 19 (1998) 2101–2127.

[25] A. Carlsson, G. Karlström, B. Lindman, O. Stenberg, Interaction between ethyl (hydroxyethyl)cellulose and sodium dodecyl sulphate in aqueous solution, Colloid Polym. Sci. 266 (1988) 1031–1036.

[26] A. Carlsson, G. Karlström, B. Lindman, Thermal gelation of nonionic cellulose ethers and ionic surfactants in water, Colloids Surf. 47 (1990) 147–165.

[27] T. Caykara, Ö Kantoğlu, Thermal behavior and network structure of poly(*N*-vinyl-2-pyrrolidone-crotonic acid) hydrogels prepared by radiation-induced polymerization, Polym. Adv. Technol. 15 (3) (2004) 134.

[28] H. Chajra, C.F. Rousseau, D. Cortial, M.C. Ronziere, D. Herbage, F. Mallein-Gerin, A.M. Freyria, Collagen-based biomaterials and cartilage engineering. Application to osteochondral defects, Biomed. Mater. Eng. 18 (2008) S33–S45.

[29] L.W. Chan, P.W.S. Heng, Effects of poly (vinylpyrrolidone) and ethylcellulose on alginate microspheres prepared by emulsification, J. Microencapsul. 15 (1998) 409–420.

[30] P. Gupta, K. Vermani, S. Garg, Hydrogels: from controlled release to pH-responsive drug delivery, Drug Discov. Today 7 (10) (2002) 569–579.

[31] M.D. Chavanpatil, A. Khdair, J. Panyam, Nanoparticles for cellular drug delivery: mechanisms and factors influencing delivery, J. Nanosci. Nanotechnol. 6 (9–10) (2006) 2651–2663.

[32] N.A. Peppas, P. Bures, W. Leobandung, H. Ichikawa, Hydrogels in pharmaceutical formulations, Eur. J. Pharm. Biopharm. 50 (1) (2000) 27–46.

[33] A.S. Hoffman, Hydrogels for biomedical applications, Adv. Drug Deliv. Rev. 54 (1) (2002) 3–12.

[34] C.V.S. Subrahmanyam, Diffusion. Textbook of Physical Pharmaceutics, New Delhi, 2006.

[35] E. Lih, Y.K. Joung, J.W. Bae, K.D. Park, An in situ gel-forming heparin-conjugated PLGA-PEG-PLGA copolymer, J. Bioact. Compat. Polym. 23 (5) (2008) 444–457.

[36] S. Cohen, E. Lobel, A. Trevgoda, Y. Peled, A novel in situ-forming ophthalmic drug delivery system from alginates undergoing gelation in the eye, J. Control Release 44 (2-3) (1997) 201–208.

[37] K.W. Leong, R. Langer, Polymeric controlled drug delivery, Adv. Drug Deliv. Rev. 1 (3) (1988) 199–233.

[38] P. Markland, Y.H. Zhang, G.L. Amidon, V.C. Yang, A pH- and ionic strength-responsive polypeptide hydrogel: synthesis, characterization, and preliminary protein release studies, J. Biomed. Mater. Res. 47 (4) (1999) 595–602.

[39] H.Y. He, X. Cao, L.J. Lee, Design of a novel hydrogel-based intelligent system for controlled drug release, J. Control. Release 95 (3) (2004) 391–402.

[40] K. Juntanon, S. Niamlang, R. Rujiravanit, A. Sirivat, Electrically controlled release of sulfosalicylic acid from cross-linked, poly (vinyl alcohol) hydrogel, Int. J. Pharm. 356 (1–2) (2008) 1–11.

[41] C.M. Dorski, F.J. Doyle, N.A. Peppas, Preparation and characterization of glucose-sensitive P(MAA-g-EG) hydrogels, Polym. Mater. Sci. Eng. Proc. 76 (1997) 281–282.

[42] N. Isiklan, M. Inal, M. Yigitoglu, Synthesis and characterization of poly(*N*-Vinyl-2-pyrrolidone) grafted sodium alginate hydrogel beads for the controlled release of indomethacin, J. Appl. Polym. Sci. 110 (1) (2008) 481–493.

[43] K. Doi, T. Ikeda, A. Marui, T. Kushibiki, Y. Arai, K. Hirose, Y. Soga, A. Iwakura, K. Ueyama, K. Yamahara, H. Itoh, K. Nishimura, Y. Tabata, M. Komeda, Enhanced angiogenesis by gelatin hydrogels incorporating basic fibroblast growth factor in rabbit model of hind limb ischemia, Heart Vessels 22 (2) (2007) 104–108.

[44] R.W. Korsmever, N.A. Peppas, Macromolecular and modeling aspects of swelling-controlled systems, in: T.J. Roseman, S.Z. Mansdorf (Eds.), Controlled Release Delivery Systems, Marcel Dekker, New York, 1983, p. 77.

[45] C.S. Satish, K.P. Satish, H.G. Shivakumar, Hydrogels as controlled drug delivery systems: synthesis, cross-linking, water and drug transport mechanism, Indian J. Pharm. Sci. 68 (2) (2006) 133–140.

[46] T. Higuchi, Mechanisms of sustained action mediation. Theoretical analysis of rate of release of solid drugs dispersed in solid matrices, J. Pharm. Sci. 52 (1963) 1145–1149.

[47] M. Foldvari, C. Oguejiofor, S. Afridi, T. Kudel, T. Wilson, Liposome encapsulated prostaglandin E1 in erectile dysfunction: correlation between in vitro delivery through foreskin and efficacy in patients, Urology 52 (5) (1998) 838–843.
[48] S. Liang, L. Zhang, J. Xu, Morphology and permeability of cellulose/chitin blend membranes, J. Membr. Sci. 287 (1) (2007) 19–28.
[49] I.Z. Nagy, M. Ohta, K. Kitani, Effect of centrophenoxine and BCE-001 treatment on lateral diffusion of proteins in the hepatocyte plasma membrane as revealed by fluorescence recovery after photobleaching in rat liver smears, Exp. Gerontol. 24 (4) (1989) 317–330.
[50] K. Pal, A.K. Banthia, D.K. Majumdar, Biomedical evaluation of polyvinyl alcohol–gelatin esterified hydrogel for wound dressing, J. Mater. Sci. Mater. Med. 18 (9) (2007a) 1889–1894.
[51] K. Pal, A.K. Banthia, D.K. Majumdar, Preparation and characterization of polyvinyl alcohol-gelatin hydrogel membranes for biomedical applications, AAPS Pharm. Sci. Tech. 8 (1) (2007b) 21–24.
[52] K.B. Sloan, Prodrugs for dermal delivery, Adv. Drug Deliv. Rev. 3 (1) (1989) 67–101.
[53] S.N. Tenjarla, R. Kasina, P. Puranajoti, M.S. Omar, W.T. Harris, Synthesis and evaluation of *N*-acetylprolinate esters – Novel skin penetration enhancers, Int. J. Pharm. 192 (2) (1999) 147–158.
[54] B. Falk, S. Garramone, S. Shivkumar, Diffusion coefficient of paracetamol in a chitosan hydrogel, Mater. Lett. 58 (26) (2004) 3261–3265.
[55] K. Pal, A.K. Banthia, D.K. Majumdar, Characterizations of the prepared corn-starch based hydrogel membranes, J. Appl. Biomater. Biomech. 4 (2006a) 38–44.
[56] A. Gazzaniga, L. Palugan, A. Foppoli, M.E. Sangalli, Oral pulsatile delivery systems based on swellable hydrophilic polymers, Eur. J. Pharm. Biopharm. 68 (1) (2008) 11–18.
[57] D.M. Saylor, C. Kim, D.V. Patwardhan, J.A. Warren, Diffuse-interface theory for structure formation and release behavior in controlled drug release systems, Acta Biomater. 3 (6) (2007) 851–864.
[58] S.D. Figueiro, A.A.M. Macedo, M.R.S. Melo, A.L.P. Freitas, R.A. Moreira, R.S. de Oliveira, J.C. Goes, A.S.B. Sombra, On the dielectric behaviour of collagen-algal sulfated polysaccharide blends: effect of glutaraldehyde cross-linking, Biophys. Chem. 120 (2) (2006) 154–159.
[59] F.N. Christensen, F.Y. Hansen, H. Bechgaard, Mathematical model for in vitro drug release from controlled release dosage forms applied to propoxyphene hydrochloride pellets, J. Pharm. Sci. 71 (1982) 694–699.
[60] X. Tongwen, H. Binglin, A mechanism on the drug release into a perfect sink from a coated planar matrix with a super-saturation loading in the core, Int. J. Pharm. 197 (1–2) (2000) 23–24.
[61] K.A. Dill, S. Bromberg, Molecular Driving Forces. Garland Science, London, 2003.
[62] R.L. Fournier, Basic Transport Phenomena in Biomedical Engineering, Taylor and Francis, 1998.
[63] A.L. Gennaro, Remington's Pharmaceutical Sciences, Mack Publishing Company, Pennsylvania, 1990.
[64] L.L. Hench, J.R. Jones, Biomaterials, Artificial Organs and Tissue Engineering, CRC Press, New York, 2005.
[65] A. Gazzaniga, L. Palugan, A. Foppoli, M.E. Sangalli, Oral pulsatile delivery systems based on swellable hydrophilic polymers, Eur. J. Pharm. Biopharm. 68 (1) (2008) 11–18.
[66] A. Martin, J. Swarbrick, A. Cammarata, Physical Pharmacy, Varghese Publishing House, Bombay, 1991.
[67] G.S. Oladiran, H.K. Batchelor, Determination of ibuprofen solubility in wax: a comparison of microscopic, thermal and release rate techniques, Eur. J. Pharm. Biopharm. 67 (1) (2007) 106–111.
[68] S. Girod, M. Boissière, K. Longchambon, S. Begu, C.T. Pétheil, J.M. Devoisselle, Polyelectrolyte complex formation between iota-carrageenan and poly(-lysine) in dilute aqueous solutions: a spectroscopic and conformational study, Carbohydr. Polym. 55 (1) (2004) 37–45.
[69] P. Sriamornsak, J. Nunthanid, M. Luangtana-anan, Y. Weerapol, S. Puttipipatkhachorn, Alginate-based pellets prepared by extrusion/spheronization: effect of the amount and type of sodium alginate and calcium salts, Eur. J. Pharm. Biopharm. 69 (1) (2008) 274–284.

[70] D. Tebbe, R. Thull, U. Gbureck, Correlation between heparin release and polymerization degree of organically modified silica xerogels from 3-methacryloxypropylpolysilsesquioxane, Acta Biomater. 3 (6) (2007) 829–837.

[71] M. Saravanan, M.D. Dhanaraju, S.K. Sridhar, S. Ramachandran, S.K.G. Sam, P. Anand, K. Bhaskar, G.S. Rao, Preparation, characterization and in vitro release kinetics of ibuprofen polystyrene microspheres, Ind. J. Pharm. Sci. 66 (3) (2004) 287–292.

[72] S.T. Mathew, S.G. Devi, K.V. Sandhya, Formulation and evaluation of ketorolac tromethamine-loaded albumin microspheres for potential intramuscular administration. doi: 10.1208/pt0801014.AAPS, Pharm. Sci. Tech. 8 (1) (2007) 14.

[73] A.P.T.R. Pierucci, L.R. Andrade, E.B. Baptista, N.M. Volpato, M.H.M. Rocha-Leao, New microencapsulation system for ascorbic acid using pea protein concentrate as coat protector, J. Microencapsul. 23 (6) (2006) 654–662.

[74] J.E. Mockel, B.C. Lippold, Zero-order drug release from hydrocolloid matrices, Pharm. Res. 10 (7) (1993) 1066–1070.

[75] T. Ehtezazi, C. Washington, C.D. Melia, First order release rate from porous PLA microspheres with limited exit holes on the exterior surface, J. Control. Release 66 (1) (2000) 27–38.

[76] N. Kashyap, N. Kumar, M.N.V.R. Kumar, Hydrogels for pharmaceutical and biomedical applications, Crit. Rev. Ther. Drug Carrier Syst. 22 (2) (2005) 107–149.

[77] S. Gunasekaran, S. Ko, L. Xiao, Use of whey proteins for encapsulation and controlled delivery applications, J. Food Eng. 83 (1) (2007) 31–40.

[78] T. Hasegawa, A. Kawazome, G. Yanagimoto, T. Hayashi, T. Seki, M. Akimoto, H. Todo, K. Sugibayashi, Analysis of skin disposition of flurbiprofen after topical application using dual agar gel discs-inserted rats, Biol. Pharm. Bull. 30 (11) (2007) 2135–2140.

[79] H.C. Liang, W.H. Chang, K.J. Lin, H.W. Sung, Genipin-cross-linked gelatin microspheres as a drug carrier for intramuscular administration: in vitro and in vivo studies, J. Biomed. Mater. Res. 65a (2) (2003) 271–282.

[80] M.J. Montisci, G. Giovannuci, D. Duchene, G. Ponchel, Covalent coupling of asparagus pea and tomato lectins to poly(lactide) microspheres, Int. J. Pharm. 215 (1-2) (2001) 153–161.

[81] L.D. Deng, Y.L. Zhai, X.N. Lin, F.M. Jin, X.H. He, A.J. Dong, Investigation on properties of re-dispersible cationic hydrogel nanoparticles, Eur. Polym. J. 44 (4) (2008) 978–986.

[82] C.A. Kavanagh, Y.A. Rochev, W.A. Gallagher, K.A. Dawson, A.K. Keenan, Local drug delivery in restenosis injury: thermoresponsive co-polymers as potential drug delivery systems, Pharmacol. Ther. 102 (1) (2004) 1–15.

[83] J. Shang, X. Chen, Z.Z. Shao, The electric-field-sensitive hydrogels, Hua Xue Jin Zhan 19 (9) (2007) 1393–1399.

[84] J. Li, B.C. Wang, P. Liu, Possibility of active targeting to tumor by local hyperthermia with temperature-sensitive nanoparticles, Med. Hypotheses 71 (2) (2008a) 249–251.

[85] J. Li, B.C. Wang, Y.Z. Wang, P. Liu, W.L. Qiao, Preparation and characterization of thermosensitive nanoparticles for targeted drug delivery, J. Macromol. Sci. Part A Pure Appl. Chem. 45 (10) (2008b) 833–838.

[86] M.A.M.E. Vertommen, H.J.L. Cornelissen, C.H.J.T. Dietz, R. Hoogenboom, M.F. Kemmere, J.T.F. Keurentjes, Pore-covered thermoresponsive membranes for repeated on-demand drug release, J. Membr. Sci. 322 (1) (2008) 243–248.

[87] B. Wang, X.D. Xu, Z.C. Wang, S.X. Cheng, X.Z. Zhang, R.X. Zhu, Synthesis and properties of pH and temperature sensitive P(NIPAAm-co-DMAEMA) hydrogels, Colloids Surf. 64 (1) (2008) 34–41.

[88] Y.N. Dai, P. Li, J.P. Zhang, A.O. Wang, Q. Wei, Swelling characteristics, drug delivery properties of nifedipine-loaded pH sensitive alginate-chitosan hydrogel beads, J. Biomed. Mater. Res. 86b (2) (2008a) 493–500.

[89] Y.N. Dai, P. Li, J.P. Zhang, A.Q. Wang, Q. Wei, A novel pH sensitive N-succinyl chitosan/alginate hydrogel bead for nifedipine delivery, Biopharm. Drug. Dispos. 29 (3) (2008b) 173–184.

[90] A. Katchalsky, I. Michaeli, Polyelectrolyte gels in salt solutions, J. Polym. Sci. 15 (79) (1955) 69–86.

[91] S.I. Jeong, Y.M. Lee, H. Shin, Tissue engineering using a cyclic strain bioreactor and gelatin/PLCL scaffolds, Macromol. Res. 16 (6) (2008) 567–569.

[92] Y.M. Ju, B.Z. Yu, T.J. Koob, Y. Moussy, F. Moussy, A novel porous collagen scaffold around an implantable biosensor for improving biocompatibility. I. In vitro/in vivo stability of the scaffold and in vitro sensitivity of the glucose sensor with scaffold, J. Biomed. Mater. Res. A 87a (1) (2008) 136–146.

[93] S. Tomita, K. Sato, J.I. Anzai, pH-sensitive thin films composed of poly(methacrylic acid) and carboxyl-terminated dendrimer, Sens. Lett. 6 (1) (2008) 250–252.

[94] N. Farooqui, D. Myung, W. Koh, M. Masek, R. Dalal, M.R. Carrasco, J. Noolandi, C.W. Frank, C.N. Ta, Histological processing of pH-sensitive hydrogels used in corneal implant applications, J. Histotechnol. 30 (3) (2007) 157–163.

[95] N.A. Peppas, J.Z. Hilt, A. Khademhosseini, R. Langer, Hydrogels in biology and medicine: from molecular principles to bionanotechnology, Adv. Mater. 18 (11) (2006) 1345–1360.

[96] Y.S. Zhou, D.Z. Yang, G.P. Ma, H.L. Tan, Y. Jin, J. Nie, A pH-sensitive water-soluble *N*-carboxyethyl chitosan/poly(hydroxyethyl methacrylate) hydrogel as a potential drug sustained release matrix prepared by photopolymerization technique, Polym. Adv. Technol. 19 (8) (2008) 1133–1141.

[97] A. Gutowska, J. Seok Bark, I. Chan Kwon, Y. Han Bae, Y. Cha, S. Wan Kim, Squeezing hydrogels for controlled oral drug delivery, J. Control Release 48 (2–3) (1997) 141–148.

[98] Y.Y. Lang, S.M. Li, W.S. Pan, L.Y. Zheng, Thermo- and pH-sensitive drug delivery from hydrogels constructed using block copolymers of poly(*N*-isopropylacrylamide) and Guar gum, J. Drug Deliv. Sci. Technol. 16 (1) (2006) 65–69.

[99] M. Kurisawa, N. Yui, Dual-stimuli-responsive drug release from interpenetrating polymer network-structured hydrogels of gelatin and dextran, J. Control Release 54 (2) (1998a) 191–200.

[100] M. Kurisawa, N. Yui, Gelatin/dextran intelligent hydrogels for drug delivery: dual-stimuli-responsive degradation in relation to miscibility in interpenetrating polymer networks, Macromol. Chem. Phys. 199 (8) (1998b) 1547–1554.

[101] P.J. Flory, J.J. Rehner, Statistical mechanics of cross-linked polymer networks II. Swelling, J. Chem. Phys. 11 (11) (1943a) 521–526.

[102] P.J. Flory, J.J. Rehner, Statistical mechanics of cross-linked polymer networks I. Rubberlike elasticity, J. Chem. Phys. 11 (11) (1943b) 512–520.

[103] T.J. Koob, D.J. Hernandez, Material properties of polymerized NDGA–collagen composite fibers: development of biologically based tendon constructs, Biomaterials 23 (1) (2002) 203–212.

[104] N.A. Peppas, E.W. Merrill, Cross-linked poly(vinyl alcohol) hydrogels as swollen elastic networks, J. Appl. Polym. Sci. 21 (7) (1977) 1763–1770.

[105] C.M. Ofner III, W.A. Bubnis, Chemical swelling evaluations of amino group cross-linking in gelatin and modified gelatin matrices, Pharm. Res. 13 (12) (1996) 1821–1827.

[106] S. Rajvaidya, R. Bajpai, A.K. Bajpai, Effect of gamma irradiation on the interpenetrating networks of gelatin and polyacrylonitrile: aspect of cross-linking using microhardness and cross-link density measurements, J. Appl. Polym. Sci. 101 (4) (2006) 2581–2586.

[107] I. Krucinska, A. Komisarczyk, M. Chrzanowski, D. Paluch, Producing wound dressing materials from chitin derivatives by forming nonwovens directly from polymer solution, Fibres Text. East. Eur. 15 (5-6) (2007) 73–76.

[108] G.D. Prestwich, D.M. Marecak, J.F. Marecak, K.P. Vercruysse, M.R. Ziebell, Controlled chemical modification of hyaluronic acid, J. Control. Release 53 (1998) 93–103.

[109] K.L. Moffat, K.G. Marra, Biodegradable poly(ethylene glycol) hydrogels cross-linked with genipin for tissue engineering applications, J. Biomed. Mater. Res. 71b (1) (2004) 181–187.

[110] A.R. Kulkarni, K.S. Soppimath, M.I. Aralaguooi, T.M. Aminabhavi, W.E. Rudzinski, Preparation of cross-linked sodium alginate microparticles using glutaraldehyde in methanol, Drug Dev. Ind. Pharm. 26 (10) (2000) 1121–1124. http://0-www.informaworld.com.innopac.lib.ryerson.ca/smpp/title~content=t713597245~db=all~tab=issueslist~branches=26-v26, 2000.

[111] M.F.A. Taleb, Radiation synthesis of polyampholytic and reversible pH-Responsive hydrogel and its application as drug delivery system, Polym. Bull. 61 (3) (2008) 341–351.

[112] M.C. Kao, A. Matsuno-Yagi, T. Yagi, Subunit proximity in the H+-translocating NADH-quinone oxidoreductase probed by zero-length cross-linking, Biochemistry 43 (12) (2004) 3750–3755.

[113] H. Petite, V. Frei, A. Huc, D. Herbage, Use of diphenylphosphorylazide for cross-linking collagen-based biomaterials, J. Biomed. Mater. Res. 28 (2) (1994) 159–165.

[114] N.A. Peppas, Hydrogels in Medicine and Pharmacy, vol. I–III, CRC Press, Boca Raton, FL, 1987.

[115] Z. Long, J. Xu, J. Pan, Immobilization of *Serratia marcescens* lipase and catalytic resolution of trans-3-(4′-methoxyphenyl)glycidic acid methyl ester, Chin. J. Catal. 28 (2) (2007) 175–179.

[116] K.W. Leong, R. Langer, Polymeric controlled drug delivery, Adv. Drug Deliv. Rev. 1 (3) (1988) 199–233.

[117] A.P. Rokhade, N.B. Shelke, S.A. Patil, T.M. Aminabhavi, Novel interpenetrating polymer network microspheres of chitosan and methylcellulose for controlled release of theophylline, Carbohydr. Polym. 69 (4) (2007) 678–687.

[118] M.S. Sacks, H. Hamamoto, J.M. Connolly, R.C. Gorman, J.H. Gorman, R.J. Levy, In vivo biomechanical assessment of triglycidylamine cross-linked pericardium, Biomaterials 28 (35) (2007) 5390–5398.

[119] C. Silva, C.J. Silva, A. Zille, G.M. Guebitz, A.C. Paulo, Laccase immobilization on enzymatically functionalized polyamide 6,6 fibres, Enzyme Microb. Technol. 41 (6-7) (2007) 867–875.

[120] L. Wu, C.S. Brazel, Modifying the release of proxyphylline from PVA hydrogels using surface cross-linking, Int. J. Pharm. 349 (1–2) (2008) 144–151.

[121] K.J. Kim, S.B. Lee, N.W. Han, Kinetics of cross-linking reaction of PVA membrane with glutaraldehyde, Korean J. Chem. Eng. 11 (1) (1994) 41–47.

[122] J. Liu, D. Meisner, E. Kwong, X.Y. Wu, M.R. Johnston, A novel trans-lymphatic drug delivery system: implantable gelatin sponge impregnated with PLGA-paclitaxel microspheres, Biomaterials 28 (21) (2007) 3236–3244.

[123] D.R. Walt, V.I. Agayn, The chemistry of enzyme and protein immobilization with glutaraldehyde, Trends Anal. Chem. 13 (70) (1994) 425–430.

[124] K. Tomihata, Y. Ikada, Cross-linking of hyaluronic acid with glutaraldehyde, J. Polym. Sci. Part A Polym. Chem. 35 (1997) 3553–3559.

[125] Y.M. Ju, B.Z. Yu, T.J. Koob, Y. Moussy, F. Moussy, A novel porous collagen scaffold around an implantable biosensor for improving biocompatibility. I. In vitro/in vivo stability of the scaffold and in vitro sensitivity of the glucose sensor with scaffold, J. Biomed. Mater. Res. A 87a (1) (2008) 136–146.

[126] S.H. Yu, Y.B. Wu, F.L. Mi, S.S. Shyu, Polysaccharide-based artificial extracellular matrix: preparation and characterization of three-dimensional macroporous chitosan, and heparin composite scaffold, J. Appl. Polym. Sci. 109 (6) (2008) 3639–3644.

[127] E. Khor, Methods for the treatment of collagenous tissues for bioprostheses, Biomaterials 18 (2) (1997) 95–105.

[128] E. Pettenazzo, G. Thiene, A.M. Gatti, E. Pasquino, E. Talenti, G. Noera, T. Bottio, M. Valente, Is the tricuspid position suitable for testing replacement bioprosthetic valves in the sheep model? J. Heart Valve Dis. 10 (4) (2001) 513–519.

[129] M.P. Linnes, B.D. Ratner, C.M. Giachelli, A fibrinogen-based precision microporous scaffold for tissue engineering, Biomaterials 28 (35) (2007) 5298–5306.

[130] H.W. Sung, D.M. Huang, W.H. Chang, L.L.H. Huang, C.C. Tsai, I.L. Liang, Gelatin-derived bioadhesives for closing skin wounds: an in vivo study, J. Biomater. Sci. Polym. Ed. 10 (7) (1999) 751–771.

[131] F. Mi, H. Sung, S. Shyu, Characterization of ring-opening polymerization of genipin and pH-dependent cross-linking reactions between chitosan and genipin, J. Polym. Sci. 43 (10) (2005) 1985–2000. A1.

[132] K.G. Mandel, B.P. Daggy, D.A. Brodie, H.I. Jacoby, Review article: alginate-raft formulations in the treatment of heartburn and acid reflux, Aliment. Pharmacol. Ther. 14 (6) (2000) 669–690.

[133] F. Mi, H. Sung, S. Shyu, Synthesis and characterization of a novel chitosan-based network prepared using naturally occurring cross-linker, J. Polym. Sci. 38 (15) (2000) 2804−2814. A1.
[134] Y. Zhu, M.B. Chan-Park, Density quantification of collagen grafted on biodegradable polyester: its application to esophageal smooth muscle cell, Anal. Biochem. 363 (1) (2007) 119−127.
[135] K. Suyama, M. Tsunooka, Effects of cross-linking by quinones on dyeing of irradiated films of copolymers bearing photobase generating groups, J. Appl. Polym. Sci. 68 (1998) 1177−1184.
[136] J.H. Waite, Marine adhesive proteins − natural composite thermosets, Int. J. Biol. Macromol. 12 (2) (1990a) 139−144.
[137] M. Mayo-Pedrosa, N. Cachafeiro-Andrade, C. Alvarez-Lorenzo, R. Martinez-Pacheco, A. Concheiro, In situ photopolymerization-coated pellets for pH-dependent drug delivery, Eur. Polym. J. 44 (8) (2008) 2629−2638.
[138] L.M. McDowell, L.A. Burzio, J.H. Waite, J. Schaefer, Rotational echo double resonance detection of cross-links formed in mussel byssus under high-flow stress, J. Biol. Chem. 274 (29) (1999) 20293−20295.
[139] J.H. Waite, The phylogeny and phemical diversity of quinone-tanned glues and varnishes, Comp. Biochem. Physiol. 97B (I) (1990b) 19−29.
[140] M.R. Chase, K. Raina, J. Bruno, M. Sugumaran, Purification, characterization and molecular cloning of prophenoloxidases from *Sarcophaga bullata*, Insect. Biochem. Mol. Biol. 30 (10) (2000) 953−967.
[141] K. Kramer, M. Kanost, T. Hopkins, H. Jiang, Y. Zhu, R. Xu, J. Kerwin, F. Turecek, Oxidative conjugation of catechols with proteins in insect skeletal systems, Tetrahedron 57 (2) (2001) 385−392.
[142] L.Q. Wu, M.K. McDermott, C. Zhu, R. Ghodssi, G.F. Payne, Mimicking biological phenol reaction cascades to confer mechanical function, Adv. Funct. Mater. 16 (2006) 1967−1974.
[143] L.M. McDowell, L.A. Burzio, J.H. Waite, J. Schaefer, Rotational echo double resonance detection of cross-links formed in mussel byssus under high-flow stress, J. Biol. Chem. 274 (29) (1999) 20293−20295.
[144] G. Strauss, S.A. Gibson, Plant phenolics as cross-linkers of gelatin gels and gelatin-based coacervates for use as food ingredients, Food Hydrocol. 18 (1) (2004) 81−89.
[145] W.R. Gombotz, S.F. Wee, Protein release from alginate matrices, Adv. Drug Deliv. Rev. 31 (3) (1998) 267−285.
[146] H.H. Tonnesen, J. Karlsen, Alginate in drug delivery systems, Drug Dev. Ind. Pharm. 28 (6) (2002) 621−630.http://0-www.informaworld.com.innopac.lib.ryerson.ca/smpp/title~content=t713597245~db=all~tab=issueslist~branches=28-v28.
[147] S. Dumitriu, E. Chornet, Inclusion and release of proteins from polysaccharide based polyion complexes, Adv. Drug Deliv. Rev. 31 (3) (1998) 223−246.
[148] S. Girod, M. Boissière, K. Longchambon, S. Begu, C.T. Pétheil, J.M. Devoisselle, Polyelectrolyte complex formation between iota-carrageenan and poly(-lysine) in dilute aqueous solutions: a spectroscopic and conformational study, Carbohydr. Polym. 55 (1) (2004) 37−45.
[149] B. Thu, P. Bruheim, T. Espevik, O. Smidsrod, P.S. Shiong, G.S. Braek, Alginate polycation microcapsules I. Interaction between alginate and polycation, Biomaterials 17 (1996) 1031−1040.
[150] T.L. Whateley, Microencapsulation of Drugs. Harwood Academic Publishers, Amsterdam, 1992.
[151] M.G. Sankalia, R.C. Mashru, J.M. Sankalia, V.B. Sutariya, Reversed chitosan-alginate polyelectrolyte complex for stability improvement of alpha-amylase: optimization and physicochemical characterization, Eur. J. Pharm. Biopharm. 65 (2) (2007) 215−232.
[152] C.M. Ofner III, W.A. Bubnis, Chemical swelling evaluations of amino group cross-linking in gelatin and modified gelatin matrices, Pharm. Res. 13 (12) (1996) 1821−1827.
[153] B. Sarmento, A.J. Ribeiro, F. Veiga, D.C. Ferreira, R.J. Neufeld, Insulin-loaded nanoparticles are prepared by alginate ionotropic pre-gelation followed by chitosan polyelectrolyte complexation, J. Nanosci. Nanotechnol. 7 (8) (2007) 2833−2841.

[154] S. Puttipipatkhachorn, J. Nunthanid, K. Yamamoto, G.E. Peck, Drug physical state and drug-polymer interaction on drug release from chitosan matrix films, J. Control Release 75 (1−2) (2001) 143−153.
[155] J.B. Park, R.S. Lake, Biomaterials − An Introduction. Plenum, New York, 1990.
[156] R.J. Young, P.A. Lovell, An Introduction to Polymers. Chapman and Hall, London, 1991.
[157] T.J. Koob, D.J. Hernandez, Material properties of polymerized NDGA−collagen composite fibers: development of biologically based tendon constructs, Biomaterials 23 (1) (2002) 203−212.
[158] Y. Moussy, E. Guegan, T. Davis, T.J. Koob, Transport characteristics of a novel local drug delivery system using nordihydroguaiaretic acid (NDGA)-polymerized collagen fibers, Biotechnol. Prog. 23 (2007) 990−994.
[159] H. Saito, T. Taguchi, H. Aoki, S. Murabayashi, Y. Mitamura, J. Tanaka, T. Tateishi, pH-responsive cross-linkers swelling behavior of collagen gels prepared by novel based on naturally derived di- or tricarboxylic acids, Acta Biomater. 3 (1) (2007) 89−94.
[160] S.D. Figueiro, A.A.M. Macedo, M.R.S. Melo, A.L.P. Freitas, R.A. Moreira, R.S. de Oliveira, J.C. Goes, A.S.B. Sombra, On the dielectric behaviour of collagen-algal sulfated polysaccharide blends: effect of glutaraldehyde cross-linking, Biophys. Chem. 120 (2) (2006) 154−159.
[161] X. Duan, H. Sheardown, Cross-linking of collagen with dendrimers, J. Biomed. Mater. Res. A 75A (3) (2005) 510−518.
[162] S.M. Pawde, K. Deshmukh, Characterization of polyvinyl alcohol/gelatin blend hydrogel films for biomedical applications, J. Appl. Polym. Sci. 109 (5) (2008) 3431−3437.
[163] X.H. Peng, L.N. Zhang, J.F. Kennedy, Release behavior of microspheres from cross-linked *N*-methylated chitosan encapsulated ofloxacin, Carbohydr. Polym. 65 (3) (2006) 288−295.
[164] X.D. Duan, C. McLaughlin, M. Griffith, H. Sheardown, Biofunctionalization of collagen for improved biological response: scaffolds for corneal tissue engineering, Biomaterials 28 (1) (2007) 78−88.
[165] C.R. Vinas, A. Breen, G. Damodaran, T. Ritter, T. O'Brien, A. Pandit, A cross-linked collagen scaffold as a vehicle for gene delivery, Tissue Eng. 13 (7) (2007) 1635−1636.
[166] G. Kleinmann, S. Larson, B. Hunter, S. Stevens, N. Mamalis, R.J. Olson, Collagen shields as a drug delivery system for the fourth-generation fluoroquinolones, Ophthalmologica 221 (1) (2007) 51−56.
[167] H. Sarojini, K. Medepalli, D.A. Terry, B.W. Alphenaar, E. Wang, Localized delivery of DNA to the cells by viral collagen-loaded silica colloidal crystals, Biotechniques 43 (2) (2007) 213−221.
[168] T. Takezawa, T. Takeuchi, A. Nitani, Y. Takayama, M. Kino-oka, M. Taya, S. Enosawa, Collagen vitrigel membrane useful for paracrine assays in vitro and drug delivery systems in vivo, J. Biotechnol. 131 (1) (2007) 76−83.
[169] M.J. Taravella, J. Balentine, D.A. Young, P. Stepp, Collagen shield delivery of ofloxacin to the human eye, J. Cataract Refract. Surg. 25 (4) (1999) 562−565.
[170] S.M. Hariprasad, G.K. Shah, J. Chi, R.A. Prince, Determination of aqueous and vitreous concentration of moxifloxacin 0.5% after delivery via a dissolvable corneal collagen shield device, J. Cataract Refract. Surg. 31 (11) (2005) 2142−2146.
[171] R. Groening, C. Cloer, M. Georgarakis, R.S. Mueller, Compressed collagen sponges as gastroretentive dosage forms: in vitro and in vivo studies, Eur. J. Pharm. Sci. 30 (1) (2007) 1−6.
[172] R. Sripriya, M.S. Kumar, M.R. Ahmed, P.K. Sehgal, Collagen bilayer dressing with ciprofloxacin, an effective system for infected wound healing, J. Biomater. Sci. Polym. Ed. 18 (3) (2007) 335−358.
[173] P. Prabu, N. Dharmaraj, S. Aryal, B.M. Lee, V. Ramesh, H.Y. Kim, Preparation and drug release activity of scaffolds containing collagen and poly(caprolactone), J. Biomed. Mater. Res. A 79A (1) (2006) 153−158.
[174] W. Friess, M. Schlapp, Sterilization of gentamicin containing collagen/PLGA microparticle composites, Eur. J. Pharm. Biopharm. 63 (2) (2006) 176−187.
[175] S. Young, M. Wong, Y. Tabata, A.G. Mikos, Gelatin as a delivery vehicle for the controlled

release of bioactive molecules, J. Control Release 109 (1–3) (2005) 256–274.

[176] Y. Ikada, Tissue Engineering: Fundamentals and Applications, first ed. Academic Press, 2006.

[177] K. Hayashi, T. Kubo, K. Doi, Y. Tabata, Y. Akagawa, Development of new drug delivery system for implant bone augmentation using a basic fibroblast growth factor-gelatin hydrogel complex, Dent. Mater. J. 26 (2) (2007) 170–177.

[178] D. Mukherjee, A.K. Banthia, Preparation of adrenochrome hydrogel patch, gel, ointment, and the comparison of their blood coagulating and wound healing capability, Mater. Manuf. Processes 21 (3) (2006) 297–301.

[179] M. Sutter, J. Siepmann, W.E. Hennink, W. Jiskoot, Recombinant gelatin hydrogels for the sustained release of proteins, J. Control Release 119 (3) (2007) 301–312.

[180] A. Yamamoto, Study on the colon specific delivery of prednisolone using chitosan capsules, Yakugaku Zasshi J. Pharm. Soc. Japan 127 (4) (2007) 621–630.

[181] K. Doi, T. Ikeda, A. Marui, T. Kushibiki, Y. Arai, K. Hirose, Y. Soga, A. Iwakura, K. Ueyama, K. Yamahara, H. Itoh, K. Nishimura, Y. Tabata, M. Komeda, Enhanced angiogenesis by gelatin hydrogels incorporating basic fibroblast growth factor in rabbit model of hind limb ischemia, Heart Vessels 22 (2) (2007) 104–108.

[182] T. Seki, H. Kanbayashi, S. Chono, Y. Tabata, K. Morimoto, Effects of a sperminated gelatin on the nasal absorption of insulin, Int. J. Pharm. 338 (1-2) (2007) 213–218.

[183] J. Wang, Y. Tabata, K. Morimoto, Aminated gelatin microspheres as a nasal delivery system for peptide drugs: evaluation of in vitro release and in vivo insulin absorption in rats, J. Control Release 113 (1) (2006) 31–37.

[184] A.J. DeFail, H.D. Edington, S. Matthews, W.C.C. Lee, K.G. Marra, Controlled release of bioactive doxorubicin from microspheres embedded within gelatin scaffolds, J. Biomed. Mater. Res. A 79A (4) (2006) 954–962.

[185] M. Sadeghi, H. Hosseinzadeh, Synthesis of starch-poly(sodium acrylate-co-acrylamide) superabsorbent hydrogel with salt and pH-responsiveness properties as a drug delivery system, J. Bioact. Compat. Polym. 23 (4) (2008b) 381–404.

[186] H. Saito, T. Taguchi, H. Aoki, S. Murabayashi, Y. Mitamura, J. Tanaka, T. Tateishi, pH-responsive cross-linkers swelling behavior of collagen gels prepared by novel based on naturally derived di- or tricarboxylic acids, Acta Biomater. 3 (1) (2007) 89–94.

[187] J. Liu, D. Meisner, E. Kwong, X.Y. Wu, M.R. Johnston, A novel trans-lymphatic drug delivery system: implantable gelatin sponge impregnated with PLGA-paclitaxel microspheres, Biomaterials 28 (21) (2007) 3236–3244.

[188] S.H. Hu, T.Y. Liu, C.H. Tsai, S.Y. Chen, Preparation and characterization of magnetic ferroscaffolds for tissue engineering, J. Magn. Magn. Mater. 310 (2) (2007) 2871–2873.

[189] N. Sivadas, D. O'Rourke, A. Tobin, V. Buckley, Z. Ramtoola, J.G. Kelly, A.J. Hickey, S.A. Cryan, A comparative study of a range of polymeric microspheres as potential carriers for the inhalation of proteins, Int. J. Pharm. 358 (1-2) (2008) 159–167.

[190] C.L. Tseng, F.H. Lin, Preparation of gelatin nanoparticles with EGFR selection ability via biotinylated-EGF conjugation for lung cancer targeting, Biomed. Eng. Appl. Basis Commun. 20 (3) (2008) 161–169.

[191] M.D. Chavanpatil, A. Khdair, J. Panyam, Nanoparticles for cellular drug delivery: mechanisms and factors influencing delivery, J. Nanosci. Nanotechnol. 6 (9–10) (2006) 2651–2663.

[192] J. Sha, Y. Kaiming, Nanoparticle-mediated drug delivery and gene therapy, Biotechnol. Prog. 23 (1) (2007) 32–41.

[193] C.L. Tseng, T.W. Wang, C.C. Dong, S.Y.H. Wu, T.H. Young, M.J. Shieh, P.J. Lou, F.H. Lin, Development of gelatin nanoparticles with biotinylated EGF conjugation for lung cancer targeting, Biomaterials 28 (27) (2007) 3996–4005.

[194] K. Pica, R. Tchao, C.M. Ofner, Gelatin-methotrexate conjugate microspheres as a potential drug delivery system, J. Pharm. Sci. 95 (9) (2006) 1896–1908.

[195] A.P. Rokhade, S.A. Agnihotri, S.A. Patil, N.N. Mallikarjuna, P.V. Kulkarni, T.M. Aminabhavi, Semi-interpenetrating polymer network microspheres of gelatin and sodium carboxymethyl cellulose for controlled

release of ketorolac tromethamine, Carbohydr. Polym. 65 (3) (2006) 243−252.
[196] K.C. Ofokansi, M.U. Adikwu, V.C. Okore, Preparation and evaluation of mucin-gelatin mucoadhesive microspheres for rectal delivery of ceftriaxone sodium, Drug Dev. Ind. Pharm. 33 (6) (2007) 691−700.
[197] B. Lu, R. Wen, H. Yang, Y.J. He, Sustained-release tablets of indomethacin-loaded micro-capsules: preparation, in vitro and in vivo characterization, Int. J. Pharm. 333 (1-2) (2007) 87−94.
[198] H. Kobayashi, S. Minatoguchi, N. Bao, S. Yasuda, Y. Misao, Y. Uno, G. Takemura, T. Fujiwara, Y. Tabata, H. Fujiwara, New drug delivery system using an erythropoietin gelatin hydrogel sheet selectively protects the heart against myocardial infarction through angio-genesis, J. Card. Fail. 12 (8) (2006) S167. Suppl. 1.
[199] Z.F. Dong, Q. Wang, Y.M. Du, Alginate/gelatin blend films and their properties for drug controlled release, J. Membr. Sci. 280 (1−2) (2006) 37−44.
[200] R.A.A. Muzzarelli, Natural Chelating Poly-mers, Pergamon Press, New York, 1973.
[201] M.N.V. Ravi Kumar, A review of chitin and chitosan applications, React. Funct. Polym. 46 (1) (2000) 1−27.
[202] N. Sivadas, D. O'Rourke, A. Tobin, V. Buckley, Z. Ramtoola, J.G. Kelly, A.J. Hickey, S.A. Cryan, A comparative study of a range of polymeric microspheres as potential carriers for the inhalation of proteins, Int. J. Pharm. 358 (1-2) (2008) 159−167.
[203] J.P. Zikakis, Chitin, Chitosan and Related Enzymes, Academic Press, Orlando, 1984.
[204] T. Takayanagi, S. Motomizu, Chitosan as cationic polyelectrolyte for the modification of electroosmotic flow and its utilization for the separation of inorganic anions by capillary zone electrophoresis, Anal. Sci. 22 (9) (2006) 1241−1244.
[205] G. Cravotto, S. Tagliapietra, B. Robaldo, M. Trotta, Chemical modification of chitosan under high-intensity ultrasound, Ultrason. Sonochem. 12 (1-2) (2005) 95−98.
[206] K. Saito, T. Fujieda, H. Yoshioka, Feasibility of simple chitosan sheet as drug delivery carrier, Eur. J. Pharm. Biopharm. 64 (2) (2006) 161−166.
[207] G.M. Sun, X.Z. Zhang, C.C. Chu, Formulation and characterization of chitosan-based hydro-gel films having both temperature and pH sensitivity, J. Mater. Sci. Mater. Med. 18 (8) (2007) 1563−1577.
[208] B.L. Guo, J.F. Yuan, L. Yao, Q.Y. Gao, Preparation and release profiles of pH/temper-ature-responsive carboxymethyl chitosan/P (2-(dimethylamino) ethyl methacrylate) semi-IPN amphoteric hydrogel, Colloid Polym. Sci. 285 (6) (2007) 665−671.
[209] E. Reverchon, A. Antonacci, Chitosan micro-particles production by supercritical fluid pro-cessing, Ind. Eng. Chem. Res. 45 (16) (2006) 5722−5728.
[210] Y. Yuan, B.M. Chesnutt, G. Utturkar, W.O. Haggard, Y. Yang, J.L. Ong, J.D. Bumgardner, The effect of cross-linking of chitosan microspheres with genipin on protein release, Carbohydr. Polym. 68 (3) (2007) 561−567.
[211] R. Jayakumar, R.L. Reis, J.F. Mano, Synthesis and characterization of pH-sensitive thiol-containing chitosan beads for controlled drug delivery applications, Drug Deliv. 14 (1) (2007) 9−17.
[212] X.H. Peng, L.N. Zhang, J.F. Kennedy, Release behavior of microspheres from cross-linked *N*-methylated chitosan encapsulated ofloxacin, Carbohydr. Polym. 65 (3) (2006) 288−295.
[213] X.W. Shi, Y.M. Du, L.P. Sun, B.Z. Zhang, A. Dou, Polyelectrolyte complex beads composed of water-soluble chitosan/alginate: characterization and their protein release behavior, J. Appl. Polym. Sci. 100 (6) (2006) 4614−4622.
[214] C. Zhang, Q. Ping, H. Zhang, J. Shen, Prepa-ration of *N*-alkyl-*O*-sulfate chitosan derivatives and micellar solubilization of taxol, Carbohydr. Polym. 54 (2) (2003) 137−141.
[215] M.A.G. Sosa, F. Fazely, J.A. Koch, S.V. Vercellotti, R.M. Ruprecht, N-Carbox-ymethylchitosan-N, O-sulfate as an anti-HIV-1 agent, Biochem. Biophys. Res. Commun. 174 (2) (1991) 489−496.
[216] D.W. Lee, S.A. Shirley, R.F. Lockey, S.S. Mohapatra, Thiolated chitosan nano-particles enhance anti-inflammatory effects of intranasally delivered theophylline, Respir. Res. 7 (2006). http://respiratory-research.com/

content/pdf/1465-9921-7-112.pdf, 2006. Art. No. 112.
[217] T. Satoh, S. Kakimoto, H. Kano, M. Nakatani, S. Shinkai, T. Nagasaki, In vitro gene delivery to HepG2 cells using galactosylated 6-amino-6-deoxychitosan as a DNA carrier, Carbohydr. Res. 342 (11) (2007) 1427–1433.
[218] R. Jayakumar, R.L. Reis, J.F. Mano, Phosphorous containing chitosan beads for controlled oral drug delivery, J. Bioact. Compat. Polym. 21 (4) (2006) 327–340.
[219] M.C. Chen, H.W. Tsai, Y. Chang, W.Y. Lai, F.L. Mi, C.T. Liu, H.S. Wong, H.W. Sung, Rapidly self-expandable polymeric stents with a shape-memory property, Biomacromolecules 8 (9) (2007) 2774–2780.
[220] H.Q. Yu, J. Lu, C.B. Xiao, Preparation and properties of novel hydrogels from oxidized konjac glucomannan cross-linked chitosan for in vitro drug delivery, Macromol. Biosci. 7 (9-10) (2007) 1100–1111.
[221] L. Verestiuc, O. Nastasescu, E. Barbu, I. Sarvaiya, K.L. Green, J. Tsibouklis, Functionalized chitosan/NIPAM (HEMA) hybrid polymer networks as inserts for ocular drug delivery: synthesis, in vitro assessment, and in vivo evaluation, J. Biomed. Mater. Res. A 77A (4) (2006) 726–735.
[222] E. Ruel-Gariepy, J.C. Leroux, In situ-forming hydrogels – Review of temperature-sensitive systems, Eur. J. Pharm. Biopharm. 58 (2) (2004) 409–426.
[223] M.C. Tate, D.A. Shear, S.W. Hoffman, D.G. Stein, M.C. LaPlaca, Biocompatibility of methylcellulose-based constructs designed for intracerebral gelation following experimental traumatic brain injury, Biomaterials 22 (2001) 1113–1123.
[224] M. Scherlund, A. Brodin, M. Malmsten, Nonionic cellulose ethers as potential drug delivery systems for periodontal anesthesia, J. Colloid Interface Sci. 229 (2000) 365–374.
[225] K.J. Edgar, Cellulose esters in drug delivery, Cellulose 14 (2007) 49–64.
[226] C.J. Malm, J. Emerson, G.D. Hiatt, Cellulose acetate phthalate as an enteric coating material, J. Am. Pharm. Assoc. 40 (1951) 520–525.
[227] D.S. Levine, V.A. Raisys, V. Ainardi, Coating of oral beclomethasone dipropionate capsules with cellulose acetate phthalate enhances delivery of topically active anti-inflammatory drug to the terminal ileum, Gastroenterology 92 (1987) 1037–1044.
[228] J. Liu, S.Y. Chan, P.C. Ho, Polymer-coated microparticles for the sustained release of nitrofurantoin, J. Pharm. Pharmacol. 54 (2002) 1205–1212.
[229] Y. Pongpaibul, C.W. Whitworth, Preparation and in vitro dissolution characteristics of propranolol microcapsules, Int. J. Pharm. 33 (1986) 243–248.
[230] Trimukhe, K.D., Varma, A.J. (2008). Complexation of heavy metals by cross-linked chitin and its deacetylated derivatives. Carbohydr. Polym. 71(1), 66–73.
[231] O.L. Sprockel, J.C. Price, Evaluation of sustained release aqueous suspensions containing microencapsulated drug-resin complexes, Drug Dev. Ind. Pharm. 15 (1989) 1275–1287.
[232] G.F. Palmieri, G. Bonacucina, P.D. Martino, S. Martelli, Gastro-resistant microspheres containing ketoprofen, J. Microencapsul. 19 (2002) 111–119.
[233] F. Theeuwes, Elementary osmotic pump, J. Pharm. Sci. 64 (1975) 1987–1991.
[234] S.N. Makhija, P.R. Vavia, Controlled porosity osmotic pump-based controlled release systems of pseudoephedrine. I. Cellulose acetate as a semipermeable membrane, J. Control Release 89 (2003) 5–18.
[235] K.I. Draget, G. SkjakBraek, B.E. Christensen, O. Gaserod, O. Smidsrod, Swelling and partial solubilization of alginic acid gel beads in acidic buffer, Carbohydr. Polym. 29 (3) (1996) 209–215.
[236] A. Polk, B. Amsden, K. Deyao, T. Peng, M.F.A. Goosen, Controlled-release of albumin from chitosan-alginate microcapsules, J. Pharm. Sci. 83 (2) (1994) 178–185.
[237] R.J. Mumper, A.S. Hoffman, P.A. Puolakkainen, L.S. Bouchard, W.R. Gombotz, Calcium-alginate beads for the oral delivery of transforming growth factor-beta (1) (TGF-Beta (1)) – Stabilization of TGF-Beta (1) by the addition of polyacrylic-acid within acid-treated beads, J. Control Release 30 (3) (1994) 241–251.
[238] H. Tomida, C. Nakamura, H. Yoshitomi, S. Kiryu, Preparation of theophylline-loaded

calcium alginate gel capsules and evaluation of their drug-release characteristics, Chem. Pharm. Bull. 41 (1993) 2161–2165.

[239] V. Iannuccelli, G. Coppi, R. Cameroni, Biodegradable intraoperative system for bone infection treatment .1. The drug/polymer interaction, Int. J. Pharm. 143 (1996) 195–201.

[240] S. Takka, F. Acartürk, Calcium alginate microparticles for oral administration: III. The effect of cross-link agents and various additive polymers on drug release and drug entrapment efficiency, Pharmazie 54 (2) (1999) 137–139.

[241] T. Imai, C. Kawasaki, T. Nishiyama, M. Otagiri, Comparison of the pharmaceutical properties of sustained-release gel beads prepared by alginate having different molecular size with commercial sustained-release tablet, Pharmazie 55 (3) (2000) 218–222.

[242] A.C. Hodsdon, J.R. Mitchell, M.C. Davies, C.D. Melia, Structure and behaviour in hydrophilic matrix sustained release dosage forms: 3. The influence of pH on the sustained-release performance and internal gel structure of sodium alginate matrices, J. Control Release 33 (1) (1995) 143–152.

[243] S. Miyazaki, W. Kubo, D. Attwood, Oral sustained delivery of theophylline using in-situ gelation of sodium alginate, J. Control Release 67 (2-3) (2000) 275–280.

[244] S. Miyazaki, A. Nakayama, M. Oda, M. Takada, D. Attwood, Chitosan and sodium alginate based bioadhesive tablets for intraoral drug delivery, Biol. Pharm. Bull. 17 (1994) 745–747.

[245] H. Katayama, T. Nishimura, S. Ochi, Y. Tsuruta, Y. Yamazaka, K. Shibata, H. Yoshitomi, Sustained release liquid preparation using sodium alginate for eradication of *Helicobacter pyroli*, Biol. Pharm. Bull. 22 (1999) 55–60.

[246] A. Esquisabel, R.M. Hernandez, M. Igartua, A.R. Gascon, B. Calvo, J.L. Pedraz, Production of BCG alginate-PLL microcapsules by emulsification internal gelation, J. Microencapsul. 14 (5) (1997) 627–638.

[247] T.J. Smith, Calcium alginate hydrogel as a matrix for enteric delivery of nucleic acids, Biopharm. 4 (1994) 54–55.

[248] K. Pal, A.K. Banthia, D.K. Majumdar, Preparation of novel pH sensitive hydrogels of carboxymethyl cellulose acrylates: a comparative study, Mater. Manuf. Processes 21 (2006b) 877–882.

[249] E. Khor, L.Y. Lim, Implantable applications of chitin and chitosan, Biomaterials 24 (13) (2003) 2339–2349.

15 Hydrocolloids and Medicinal Chemistry Applications

Liam M. Grover and Alan M. Smith

OUTLINE

15.1 Drug Delivery

The techniques for delivering substances into the human body have been widely researched by mankind for literally thousands of years from both a medicinal and recreational perspective. Generally speaking, the approach is determined by the nature of the formulation that is to be delivered and therefore rational design and forethought is required to produce effective medicines that are fit for purpose. Biopolymers in particular are widely used within pharmaceutical products, traditionally as excipients such as binders, fillers, thickeners, and disintegrants. Indeed, glance through the *Pharmaceutical Handbook of Excipients* and you will find a large quantity of biopolymers listed for the pharmaceutical use. The description of excipients, however, is gradually shifting toward functionally active materials rather than the traditional definition of inactive or inert ingredients. Biopolymers are utilized in the majority of drug dosage forms including oral, ocular, nasal sprays, topical formulations (gels and ointments), pulmonary delivery, and even parenteral delivery. These traditional dosage forms are the subject of much research to improve stability and drug efficiency, to reduce costs, to increase patient compliance, and to improve therapeutic performance. The wide variety of biopolymers available with diverse compositions and properties, and natural origin along with their ease of production make them a popular research focus within pharmaceutics. Apart from numerous applications in traditional dosage forms, biopolymers have recently been used for the controlled delivery of biological material such as proteins, peptides, and vaccines. To discuss all of the biopolymers that have been investigated for use as drug delivery systems would require considerably more space than this chapter allows; therefore, attention is focused on biopolymers that have been investigated to provide modified-release properties

Ebnesajjad: Handbook of Biopolymers and Biodegradable Plastics. http://dx.doi.org/10.1016/B978-1-4557-2834-3.00015-X
© 2009 Elsevier Inc. All rights reserved. Reproduced from a chapter in: Kasapis, *Modern Biopolymer Science* (2009).

and the use of biopolymers in biopharmaceutical formulations.

15.1.1 Oral Delivery

The major challenge in relation to optimal delivery of drugs lies in achieving the required dose at the required time in the required place. This may relate to a clinical need for accuracy of dosing in the case of drugs that have potentially dangerous side effects if taken in excess, or to situations in which a consistent level of the drug in the body is required (such as with hormone replacement therapy) or simply to improve patient compliance when the treatment regimen is inconvenient. To achieve the uptake of the drug via the oral route, disintegration of the dosage form is required followed by the dissolution of the released drug. In the case of drug formulations intended for oral delivery, it is usually the case that an immediate-release formulation is delivered to reach or possibly exceed the therapeutic window. The problem with such an approach is that once the drug has been fully metabolized the therapeutic effect is lost and a repeat dose becomes necessary, a situation that is avoided with sustained delivery as illustrated in Fig. 15.1. Moreover, certain drugs are unsuitable for immediate release in the stomach and require delayed-release formulations to target the drug further down the gastrointestinal tract (GIT). The physiological variability in the GIT therefore needs to be considered when designing oral drug delivery systems, especially when formulating drugs with a narrow therapeutic index. Fluctuations in pH and enzyme concentrations can vary in the fed and fasted state, which can cause changes in the solubility of certain drugs which then alters the rate of absorption. The naturally fluctuating environment of the GIT not only has effects on drug molecules but also on the excipients used to formulate the dosage form.

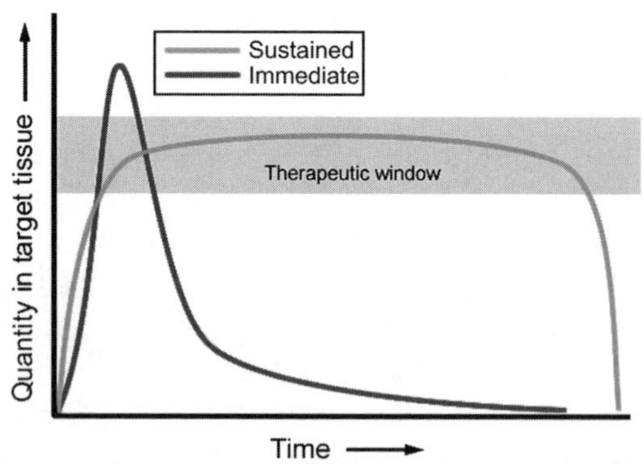

Figure 15.1 Simplified illustration of drug release profiles in relation to dosage form highlighting the difference in therapeutic activity of an immediate-release and a sustained-release system.

15.1.1.1 Tablets

Standard uncoated immediate-release tablet formulations that are prepared using methods such as direct compression of powder blends or compression following wet or dry granulation tend to include biopolymers such as starch or microcrystalline cellulose (MCC) as disintegrants or binders. Additionally, super swelling materials can be incorporated at low levels (2–4% w/w) as superdisintegrants. Materials such as sodium starch glycolate (trademarked Explotab® or Primojel®) and cross-caramelose sodium (cross-linked sodium carboxymethyl cellulose), which can swell to between 8 and 12 times their original size on contact with water, disrupt the integrity of tablet matrices resulting in disintegration. These simple tablets are designed to dissolve in the stomach to deliver drugs locally in the GIT or for systemic effects. To achieve modified-release from tablet matrices, there are several approaches that can be used to either sustain or delay the release of the drug depending on what drug release profile is required to achieve the optimum therapeutic effect. One method of controlling release from tablet matrices is by the addition of polysaccharides such as hydroxypropyl methylcellulose (HPMC). When a tablet (containing HPMC) encounters the aqueous environment of the stomach, hydration of the HPMC occurs, forming a gel layer surrounding the dosage form which creates a barrier to diffusion. Release of the drug then occurs by diffusion through the gel layer and as the gel layer gradually erodes within the aqueous dissolution medium. The rate of drug release is therefore controlled by the rate of drug diffusion through the swollen HPMC gel layer. Increasing the proportion of HPMC in the formulation reduces diffusion of the drug and delays the erosion of the tablet matrix due to an increased swelling volume and a resulting larger gel layer. The rate of polymer swelling and dissolution can be increased by using lower viscosity HPMC, which subsequently increases drug release rate. Xanthan is another polysaccharide that has shown to retard drug release; however, the rate of release is dependent on the ionic strength [1] of the release media. It has also been reported that when galactomannans [2], and more recently konjac glucomannan, are incorporated with xanthan in tablet matrices release rate is reduced further. Incorporation of gluco- or galactomannans within xanthan tablet matrices also provides the potential for colon-targeted delivery. The swollen gel layer formed by synergistic interactions of xanthan and gluco- or galactomannans, which retards release in the upper GIT, is degraded in the colon as the mannans are hydrolyzed by β-manannases secreted by colonic micro-flora increasing drug release rate [3,4].

15.1.1.2 Pharmaceutical Capsules

The majority of two-piece hard capsules used for pharmaceuticals are made from gelatin and were first produced in 1834 by Francois Mothes who dipped a mold into a solution of gelatin and allowed it to dry before removing the resultant capsule. The dip molding principle of manufacture has remained relatively unchanged since. Today manufacture of the capsules is carried out using automated machinery, where stainless steel pins (for the body and the cap) are dipped into an aqueous gelatin solution (25–30% w/w), which is maintained at about 50–55 °C. As the pins are withdrawn, they are rotated to distribute the gelatin evenly and cooled below the gelation temperature. The capsules are then allowed to dry at controlled humidity and then stripped from the pins, trimmed, and assembled. Usually capsules are prepared with a self-locking mechanism or can be hermetically sealed once filled with the pharmaceutical contents.

Gelatin is still the material of choice for capsule production due to excellent film-forming properties and rapid dissolution in gastric fluid; pharmaceutical companies, however, have been forced to develop capsules prepared from nonanimal sources for numerous reasons including a rise in vegetarianism, religious objections to animal-derived products, and instability with drugs that are hygroscopic or contain reactive aldehyde groups. Several alternatives to the gelatin capsule have been formulated [5–7] that are on the market, most notably Shionogi Quali-V® and Capsugel V-Caps®, both these capsules are based on low-viscosity HPMC as the film-forming material and a gelling agent (Quali-V uses κ-carrageenan, V-caps uses gellan gum) to provide structure to the HPMC during production. Both gelatin and HPMC capsules are designed to dissolve and release their contents in the stomach; however, for certain drugs that are unstable or reactive at gastric pH, dissolution of the capsule is required further down the GIT. The most successful method currently employed to achieve this is spray coating with synthetic-acid-insoluble methacrylate polymers either on the exterior surface of the capsule shell or the loaded drug itself; however, recently, research groups have focused on developing capsules with bulk enteric properties built into the capsule shell. Indeed, this has already been achieved in one-piece soft gel capsules used for oil-based preparations. Enteric Softgel capsules developed by Banner under the trade name Entericare™ claim enteric properties by blending solutions of polymers used for enteric coating (polymethacrylate, cellulose acetate phthalate, or shellac) with gelatin solutions; however, no two-piece hard capsule with enteric properties has been developed to date. Several attempts to form gelatin capsules by cross-linking with aldehydes have resulted in mixed success [8–10]. More recently, capsule films prepared from tertiary mixtures of HPMC/gellan/alginate have shown good film-forming properties and acid-delayed rupture at gastric pH due to the presence of alginate and gellan [11].

15.1.1.3 Oral Liquids

A wide variety of biopolymers are used in the formulation of oral pharmaceutical liquids, whether to thicken solutions, suspend dispersions, or stabilize emulsions. Functionality, however, is not only restricted to providing structure to oral liquids. Responsive characteristics desirable in directed drug delivery and applications of biopolymers in which triggered or environmentally induced changes in state utilized in addressing a clinical problem are frequently incorporated into oral liquid dosage forms. For instance, gel-forming biopolymer systems such as alginate and gellan have a rapid sol–gel transition on exposure to acid, which is a particularly an attractive property for oral liquids due to the low pH of the stomach. The most well-known example of a product that utilizes this form of physiological responsive sol–gel transition therapeutically is Gaviscon®, which has been commercially available for almost 40 years. Here, alginate is used to prevent gastric reflux in combination with bicarbonates as an oral liquid buffered at neutral pH. When swallowed and contact is made with H^+ ions in the gastric fluid, the alginate is rapidly cross-linked forming a gel "raft" on the surface of the gastric fluid, preventing acid reflux. This has been an extremely successful product with several formulations available in the product range, the majority of which are based on alginate. A more recent example of acid gelation in the stomach utilizes gellan gum to increase the gastric residence time of the antibiotic clathromysin, resulting in improved eradication of the pathogen *H. pylori* [12]. In this study, the authors describe the two main prerequisites for the *in situ* gelling system: optimum viscosity for ease of swallowing, and gelling capacity and rapid sol–gel transition due to ionic interaction. These prerequisites can relate to several polysaccharide systems and therefore provide many options for formulation. Another attractive property of solutions of polysaccharides is their ability to adhere to physiological tissue which has been investigated in oral liquids for targeted delivery of drugs to poorly accessible target sites such as the esophagus, where transit time is less than 16 s [13]. This bioadhesion or more specifically mucoadhesion can significantly increase the bioavailability of drugs formulated as an oral liquid or enable formulations to act as a protective barrier, coating physiological lesions. Interactions of polysaccharide with the mucus layer and other mechanisms of mucoadhesion will be discussed later in this chapter.

15.1.1.4 Alternative Oral Delivery Systems

Demand for alternative oral delivery systems is particularly strong in the pediatric and geriatric markets, where noncompliance is high due to patient difficulty in swallowing conventional tablets or capsules. To address this problem, there have been many developments utilizing the physical properties of hydrocolloids. One such dosage form is orally disintegrating tablets (ODTs) of which there are several commercially available patented formulations which include among others: Zydis (gelatin), OraSolve (Starch, HPMC), DuraSolve (MCC), FLASHDOSE® (polymaltodextrins, polydextrose), and WOWTAB (ethylcellulose). These dosage forms are prepared by different technologies but all incorporate biopolymers within the matrix to provide different functional properties. Of these ODT formulations, the Zydis® dosage form was the first to gain FDA approval and has the most preparations that are currently on the market. The Zydis® system consists of drug entrapped in a lyophilized preparation of gelatin and mannitol and dissolves almost instantly (2–3 s) on contact with saliva removing the need for water. The dispersed or dissolved drug and excipients are then swallowed with saliva. It has been claimed that bioavailability can be increased as a consequence of dissolution and subsequent absorption through the buccal and esophageal mucosa [14]. These systems are particularly useful for hydrophobic drugs accommodating up to 400 mg; however, they are limited to ~60 mg of hydrophilic drugs. Other major drawbacks of lyophilized preparations include the high cost of processing and poor mechanical properties, which cannot be accommodated in standard blister packs. OraSolve, DuraSolve, and WOWTAB are manufactured using conventional tableting techniques, such as direct compression of powder blends (at a reduced compression force) or compression following wet or dry granulation, which dramatically reduce production costs compared with lyophilization. These directly compressed tablets have the advantage of obtaining a high dosage of both hydrophobic and hydrophilic drugs and have good mechanical properties, which however significantly increase disintegration time. The fast disintegrating action of directly compressed ODT is facilitated by combinations of superdisintergants, effervescing agents, and highly water-soluble excipients, and mechanical properties are controlled by the addition of polysaccharides such as MCC, starch, or ethylcellulose. Successful formulation of these products is therefore a delicate balance of adequate structural properties suitable for processing and rapid disintegration on exposure to saliva. Other variants of ODTs have also been developed, using technology platforms such as molding and flossing [15]. Demand for inventive formulations that utilize melt-in-the-mouth properties continues to grow due to their convenience in administration. This is one area where there appears to be potential to apply technology developed for food applications to pharmaceutical products. A recent example by Agoub *et al.* [16] revealed melt-in-the-mouth behavior in synergistic xanthan–konjac gels prepared at pH 3.5 and below. The suggested application in this particular study relates to where melt-in-the-mouth characteristics are important for product quality, and where moderate acidity is acceptable or necessary (e.g., fruit jellies). Therefore, it seems reasonable to argue that technologies such as this could potentially be adapted for pharmaceutical orally disintegrating dosage forms. Additionally, dosage forms designed to disintegrate and dissolve in the mouth require considerations not generally given to standard oral dosage forms such as mouth feel and taste-masking. Mouth feel is improved in ODTs by the addition of effervescing excipients such as sodium bicarbonate and taste-masking achieved by incorporation of flavors, acidity regulators, artificial sweeteners such as aspartame allied with techniques such as polymer coating applied to bitter-tasting pharmaceutical actives prior to tablet formulation. For a detailed review on taste-masking oral pharmaceuticals refer to Ref. [17].

There are already numerous oral formulations on the market that are alternatives to classic tablets and capsules which include lozenges, chewable tablets, chewing gums, and buccal films that either use or have the potential to use the multifunctional properties of biopolymers. One particularly inventive dosage form is Clarosip® (Grunenthal Ltd.), which contains the antibiotic clarithromycin. This product is in the form of a drinking straw that is loaded with granules formulated using carrageenan, HPMC, and methacrylic acid ethylacrylate copolymer to provide enhanced mouth feel, taste-masking, and enteric properties. The drinking straw contains one dose which is administered using the straw to sip the patient's favorite drink (carbonated drinks are preferred as this helps taste-masking and improves mouth feel). The granules are then swallowed along with the drink. Each straw contains a gauge to inform the patient when the full dose has been taken. It is simple but innovative technologies such as this that have the potential to significantly improve patient compliance and therefore therapeutic efficacy.

15.1.2 Ocular Delivery

The normal response of the eye from the insertion of foreign material is the blink reflex and tear production, both of which serve to eliminate low-viscosity solutions more rapidly than higher viscosity solutions. As a consequence of this, common eye drops incorporate viscosity-enhancing polymers to increase contact time and bioavailability; however, due to high rates of shear produced when blinking, rheological properties of the formulations also need to be considered. The application of shear-thinning polymers in particular provides reduced resistance to blinking, resulting in greater patient acceptance. Although lacrimal clearance of viscous eye drops occurs at a slower rate than Newtonian solutions, a further increase in residence time can be achieved by utilizing biopolymers with a rapid sol–gel transition *in situ*. This approach was first developed by Pramoda and Lin [18], who exploited the pH-dependent sol–gel transitions of xanthan–locust bean gum (LBG) mixtures, which exist as a liquid formulation below pH 3.5, but once applied to physiological pH of the eye (~pH 7) gelation is induced achieving increased residence time. Application of acidic liquids to the eye may be uncomfortable to the patient and perhaps this technology would be better applied to a less-sensitive target site such as the nasal cavity where the average baseline pH is ~6.3 [19], which would still facilitate the formation of a synergistic xanthan–LBG gel.

A more practical approach to *in situ* gelation within the eye utilizes the composition of tear fluid. Tears are composed of a complex mixture of enzymes (lysozyme and lactoferrin), antibodies, organic acids, vitamins, glucose, cholesterol, and electrolytes including Na^+, K^+, and Ca^{2+} [20]. The presence of cations in tear fluid has provided pharmaceutical scientists with the opportunity to utilize the ionotropic gel-forming properties of polysaccharides, such as deacetylated gellan gum, for an increased residence time and subsequently an increased duration of therapy. Indeed, ophthalmic formulations are the most frequently encountered examples of deacetylated gellan gum in current pharmaceutical use. One such example is Timoptic®, an ophthalmic solution (that contains the active ingredient timolol malate, to treat glaucoma), which utilizes deacetylated gellan gum to provide the capacity for the formulation to thicken on application to the surface of the eye. This thickening is a result of a suppression of negatively charged carboxyl groups of gellan gum polymer chains by cations present in the tear fluid (Table 15.1) prior to which the solution viscosity is sufficiently low to enable simple dispensing. The initial application of the formulation

Table 15.1 Ionic Content of Tear Fluid

Electrolyte	Concentration mMol/l
Calcium	0.57
Sodium	~140
Potassium	15–29
Chloride	120–135
Bicarbonate	26

Source: Adapted from Ref. [20]

provokes the tear production providing the cations for gelation, and as more tears are produced (because of the increase in viscosity of foreign material in the eye) the gel is strengthened. The therapeutic effect of the inclusion of gellan gum has been shown to the extent that once-daily administration of a formulation containing gellan gum is equivalent to a twice-daily administration of a standard timolol ophthalmic solution [21]. This mechanism of *in situ* gelation by interaction with tear fluid has also been demonstrated in xanthan-based formulations, where sol—gel transition is induced by the presence of lysozyme [22]. Interactions between ionic polysaccharides and ocular mucins have been studied with increasing interest recently with the aim of developing mucoadhesive systems incorporating a range of ophthalmic drugs [23—25]. To appreciate the potential use of mucoadhesive polymers in drug delivery, it is important to understand the nature of mucoadhesion and how the physical and chemical properties of biopolymers are ideally suited for this purpose.

15.1.3 Mucoadhesion

Mucoadhesive biopolymers are of great interest to pharmaceutical science, as they can increase the residence time of the drug at the target site providing more time for absorption, ultimately increasing bioavailability. This can be vitally important where there is a constant flow of fluid or particulate matter as in the GIT. In a recent review by Zhang *et al.* [26], the authors state that in pharmaceutical terms the object of mucoadhesion can be defined as attachment of drug delivery applications to mucus or mucous membranes, and mucoadhesives described as natural or synthetic materials used in drug delivery systems that lead to mucoadhesion. Furthermore, the authors also describe the physicochemical properties a polymer should possess to be mucoadhesive. These properties include hydrophilicity, numerous hydrogen-bonding functional groups, viscoelastic properties when hydrated, and for pharmaceutical applications they should be easily combined with drugs to provide sufficient control over drug release.

Several mechanisms of mucoadhesion have been proposed which describe a number of polymer interactions with mucins, which are the main constituent of mucus. Mucins are glycoproteins that are highly glycosylated consisting of ~80% carbohydrates as oligosaccharide chains based on 5—15 monomers, primarily *N*-acetylgalactosamine, *N*-acetylglucosamine, fucose, galactose, and sialic acid, with few traces of mannose and sulfonic acid [27]. Due to a high prevalence of sialic acid and sulfonic acid terminating the oligosaccharide side chains, the mucin has a net negative charge [28], which provides the opportunity for electrostatic attraction. Moreover there are regions of the mucin that are free from oligosaccharide side chains providing domains for hydrophobic interactions also the size of the mucins (up to 50 MDa) and provide the possibility of physical entanglement. In depth analysis of these mechanisms is beyond the scope of this chapter, and for a more detailed account of these mechanisms refer to Refs. [29] and [30].

Chitosan in particular has well-known mucoadhesive properties and has been investigated widely for use in drug delivery systems. The mechanism of mucoadhesion in chitosan is thought to be due to electrostatic interactions between the positively charged amino groups of chitosan and the negatively charged sialic groups of the mucins. These interactions are strong at acidic pH, where the charge density of the chitosan is high (depending on the degree of deacetylation) [31]. Formulation of mucoadhesive chitosan systems has been developed for a wide range of target delivery sites, e.g., chitosan-based mucoadhesive tablets have been developed for oral, sublingual, and buccal delivery. Liquid formulations in the form of eye drops for mucoadhesive ocular delivery are another example [32]. The use of alginate in a mucoadhesive system has been shown to be particularly useful when targeting poorly accessible target sites such as the esophagus. In another study by Batchelor *et al.* [33], 2% solutions of sodium alginate (high *G*, medium molecular weight) were shown to retard transit time through the esophagus for up to 30 min when measured *in vitro* using porcine esophageal tissue. The incorporation of drugs into such a system would enable localized therapy on the esophageal epithelium. Measurements of esophageal retention using *in vivo* techniques such as gamma scintigraphy have not always correlated with predicted *in vitro* measurements as demonstrated by McCargar *et al.* [34]; this is probably due to the design of *in vitro* apparatus that had not taken into account peristalsis, posture position, or saliva flow. *In vitro* measurements have since been improved by using whole *ex vivo* esophagus tubes, periodic washing with artificial saliva, and simulated periodic peristaltic waves performed by roller along the length of the esophageal tube. This method was used by Richardson *et al.* [35] in demonstrating esophageal mucosa retention time of sodium alginate, which was prolonged for up to ~60 min when dispersed in glycerol.

Mucoadhesion of polysaccharides relies solely on noncovalent interactions; therefore, only weak levels of adhesion are achieved which can be problematic for certain drugs where sustained release is required. To overcome this problem, mucoadhesive polysaccharides such as alginate and chitosan can be synthetically thiolated to produce thiolomers, which enable the formation of disulfide bonds between the mucoadhesive thiolomer and cysteine-rich subdomains of mucin. Thiolated polymers can improve mucoadhesion to vastly different extents. For example, alginate–cysteine thiolomer has been shown to achieve a four-fold increase in mucoadhesion, whereas for chitosan–iminothiolane, thiolomer mucoadhesion is increased 250-fold in comparison with polymers prior to thiolation. The array of biopolymers with potential thiolation sites will surely lead to a greater number of biopolymer-based thiolomers for applications in mucoadhesive drug delivery systems.

This section has provided a brief overview of how biopolymers are used in a variety of traditional dosage forms and how developments within pharmaceutics are requiring more functional excipients for targeted drug delivery to improve efficiency and patient compliance. The remainder of this chapter will focus on utilization of biopolymers in the development of drug delivery systems for biopharmaceuticals and as tissue engineering scaffolds.

15.1.4 Medicine of the Future

The desire for personalized medicine is at the forefront of medical and pharmaceutical research. Advances in pharmacogenomics have released the potential to design-directed therapeutics formulated and prescribed based on the knowledge of patient's genetics, environment including genetic predispositions or treatment-related facts, such as a patient's particular drug response. In principle, this has the potential to revolutionize treatments, reducing long-term costs due to individually tailored treatments and improved patient compliance. Technical aspects of pharmacogenomics are beyond the scope of this chapter; however, biopolymers can play an important role in delivering personalized medicines and are already being exploited, most notably in tissue engineering. It is in this area of medical research where the use of biopolymers can have a great impact in the future.

15.1.4.1 The Development of Biopharmaceuticals

Originally, biopharmaceuticals were restricted to growth hormones and insulin. However, since the elucidation of the human genome, genetic disorders have become potentially treatable using peptides, proteins, and nucleic acids driven by pharmacogenomics and pharmacoproteomics, which in turn have provided pharmaceutical scientists with a great challenge to formulate these delicate molecules into effective drugs.

Proteins, peptides, genes, and live vaccines are limited to parenteral delivery. To highlight the difficulty of developing a viable nonparenteral biopharmaceutical production, in 2006 Pfizer produced the first commercially available form of inhaled insulin trademarked Exubera® and in early 2008 the product was removed from the market as it did not meet patient needs and financial expectations.

Oral delivery has historically been the preferred route of drug administration for patients, as it offers a convenient pain-free treatment. To deliver biopharmaceuticals orally, however, is an extremely tough challenge due to first-pass metabolism effects, poor bioavailability and susceptibility to extreme fluctuations in pH, and abundance of digestive enzymes in the GIT. However, the variety of physiological conditions within the GIT provides potential to utilize the properties of biopolymers to create stable formulations that are physiologically responsive. The development of suitable biopharmaceutical formulations using polysaccharide delivery systems is an attractive proposition due to mild processing conditions and low toxicity (formulation using synthetic materials can often involve organic solvents, high shear mixing, or high temperatures). Additionally, due to the high development costs of biotech drugs, there is great interest in reformulation as patents begin to expire on some of the early biopharmaceutical products in an attempt to keep one step ahead of a growing number of biogeneric companies creating their own version of off-patent products.

Physiological responsive hydrogels such as chitosan and alginate have attracted increasing attention for the delivery of biopharmaceuticals especially with respect to oral drug delivery due to favorable properties such as mucoadhesion and pH sensitivity, along with the ability to form microspheres and nanospheres. Indeed, a large amount of research published in the area of oral protein delivery using polysaccharides has focused on alginate microspheres as the vehicle of choice due to the mild cross-linking conditions having a minimal effect on protein denaturation. Microspheres can be prepared using several techniques to achieve the desired size. Simple extrusion of polymer into a solution of cross-linker is probably the simplest technique used. Other frequently used techniques involve single or double emulsion systems to form liquid polymer droplets in the aqueous dispersed phase that can be cross-linked by addition of a cross-linker with the particle size governed by the size of the polymer droplets formed in the emulsion which can be reduced by using high shear mixing.

The immobilization of proteins by Ca^{2+} cross-linking of continuous phase of alginate can provide a stabilizing effect for entrapped proteins, and the pH responsive sol–gel transition has also been investigated to transit proteins through the harsh acidic environment of the stomach, releasing the protein in the intestine. This method of delivery has been investigated for the entrapment and oral delivery of live bacteria as probiotics [36] or as vaccines [37]. Although alginates have been shown as promising vehicles for oral delivery, there are issues such as drug leaching, imbibing of acid, and competitive inhibition of carboxyl groups by some positively charged proteins affecting stability [38]. To overcome these problems, alginate formulations have been tailored further by coating with positively charged polymers such as poly-L-lysine and chitosan to create multilayer microspheres [39], also blending with other biopolymers such as pectin [40] and chitosan to produce polyelecrolyte complexes (PECs), which affects gel network complexity and pore size, parameters that have the potential to be manipulated to achieve the desired effect. Alginate–chitosan PECs, e.g., can be produced with a range of drug release profiles by varying parameters such as polymer ratio, molecular weight, the degree of acetylation of the chitosan, and the G:M ratio of the alginate.

It is known that chitosan can enhance drug absorption by opening the intercellular tight junctions of the GIT, which combined with mucoadhesive properties and pH sensitivity make it a popular choice as a coating material for particulate drug delivery systems. For example, chitosan has been investigated as a coating material for drug delivery systems such as liposomes for oral delivery of the peptide calcitonin and was found to possess excellent retention and penetrative property into the intestinal mucosa of rats [41]. Chitosan has also been used as a mucoadhesive coat on peptide-loaded DL-lactide/glycolide copolymer nanospheres for pulmonary delivery, which resulted in prolonged mucoadhesion for sustained drug release at the absorption site [42].

The versatility and intrinsic properties of biopolymers provide ideal vehicles for directed drug delivery and will continue to assist with the progress of future medical practices not only in the controlled release of therapeutic compounds but also in developing regenerative medicine. Indeed, the controlled release of bioactive proteins in modern medicine is not confined to delivery *in vivo* but also to the rapidly expanding area of tissue engineering *in vitro*. An important aspect of tissue engineering is to provide suitable delivery systems for growth factors that can stimulate desired cell responses both *in vitro* and *in vivo*, to accelerate tissue regeneration. Polysaccharide microspheres have been explored for the controlled release of growth factors within three-dimensional tissue engineering scaffolds, to augment cellular proliferation and extracellular matrix (ECM) formation. Lee *et al.* [43] presented results suggesting that transforming growth factor (TGF-β1), which controls cell proliferation, and differentiation loaded into chitosan microspheres then incorporated into cell scaffolds seeded with chondrocytes have the potential to enhance cartilage formation. In another study, vascular endothelial growth factor (VEGF), essential for angiogenesis, has shown high encapsulation efficiency in alginate microspheres followed by sustained zero-order release by diffusion over 3 weeks [44]. The *in vitro* delivery of growth factors is destined to provoke even greater interest in the future as the discipline of tissue engineering develops. The remainder of this chapter will focus on the functional role of biopolymers in recent tissue engineering research and potential future directions.

15.2 Tissue Engineering

The mean life expectancy of the developed world rose significantly in the twentieth century due to major advances in medical technology. As a consequence, the average age of the population has increased, which has resulted in a greater and growing demand for novel technologies that are aimed at the replacement of diseased and damaged tissues. Although the gold standard for the replacement of the majority of tissues is autograft (harvested from the patient) [45], there exist significant problems with donor site morbidity [46] and lack of availability [47]. Tissues derived from cadavers (allograft) and animals (xenograft) address both the harvesting and availability issues; however, there is a significant risk of implant rejection following implantation as a result of an immunogenic reaction or disease transmission [48]. One potential solution to avoid immunogenic response and availability issues would be to harvest the patient's cells and culture them *ex vivo* to produce tissues for eventual implantation. This approach to tissue replacement was originally termed tissue engineering in 1993 by Langer and Vacanti [49] and has since been the focus of a significant and high-profile global research effort.

Cells, to some extent, may organize themselves into crude tissue-like structures in a simple culture system [50,51]. It is now widely recognized, however, that in order to reproduce functional tissues *in vitro* it is necessary to culture the cell populations on substrates that provide relevant mechanical, chemical, and biological cues to direct tissue formation [52]. Such structures are typically referred to as scaffolds and have been formed from a range of ceramic [53], metallic [54], and polymeric materials (synthetic and biologically derived) [55]. Hydrogels are becoming widely used as scaffolds for the replacement of both hard and soft tissues due to their biological compatibility in a wide range of applications, which is attributed to their hydrophilic nature and the capability for a relatively high rate of molecular diffusion throughout their structures [56].

Such is the widespread application of hydrogels in the field of tissue engineering that an entire book could be dedicated to their use in a plethora of contrasting tissue types. As a consequence, this section focuses on the current state of the art in the area, with an emphasis on how cells interact with and can modify hydrogel matrices. This is of particular importance, since one of the principal advantages of hydrogel-based materials in this application is that they provide an environment more akin to human tissue than other synthetic materials traditionally used to support cell growth [57]. A range of processing technologies will be discussed that can be used to microstructurally modify hydrogel-based materials to counteract mass transport limitations. Finally, a range of currently available engineered tissue structures will be discussed in addition to an important emerging area—the development of complex tissue interfaces using hydrocolloid-derived structures.

15.2.1 Cell Adhesion

The adhesion of cells to polymeric matrices is an important consideration when selecting a hydrogel for use as a tissue engineering scaffold. Hydrogel materials derived from mammalian ECM components such as collagen, fibrin, and chitosan typically allow for cell adhesion [58–60]. Materials derived from other sources, such as alginate, however, do not readily interact with mammalian cells [61]. Hydrogels that are used in tissue engineering can therefore be defined either as cell adherent or noncell adherent. Each class of material has its advantages. Cell-adherent hydrogels can typically be restructured *in situ* by a cell population, which may secrete enzymes that are capable of resorbing the matrix and can concurrently produce their own ECM [60]. In contrast, alginate has been shown to preserve the phenotype of cells and maintain dimensional stability with time. In cartilage replacement, e.g., it is imperative that implanted cells maintain a chondrocytic phenotype in order that they synthesize a cartilage-like matrix. Bonaventure *et al.* [62] demonstrated that dedifferentiated chondrocytes encapsulated in an alginate matrix expressed cartilage-specific genes and allowed the formation of a cartilaginous tissue within 15 days.

15.2.2 Cell-Adhesive Hydrogels

The cell-adhesive hydrogels that are most frequently used in the fabrication of tissue engineering scaffolds include fibrin, collagen, and chitosan. The adhesion of cells to the surface of these hydrogels is mediated through the attachment of specific cell-adhesion molecules present on the surface of the cell membrane to cell-adhesion proteins adsorbed to the material surfaces. These membrane-bound proteins known as integrins are heterodimers, with an α and β group. There have been 18 α-units and 8 β-units described in the literature, which allow for a certain amount of selectivity to the cell adhesion process. Depending on the α–β subunit combination, the resulting integrins

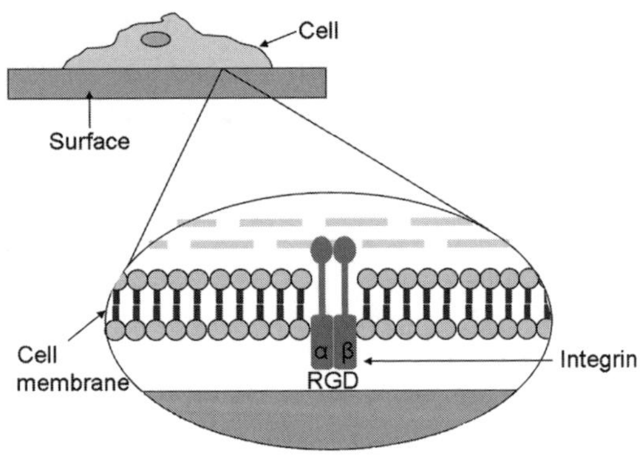

Figure 15.2 A schematic diagram of an adhesion between a 2D surface and a single cell. Integrins present within the membrane of the cells interact with proteins or ligands, such as the RGD sequence, that are readily absorbed onto the surface of the material. These initial adhesions may subsequently develop into focal adhesions, fibrillar adhesions, or 3D matrix adhesions. The attachment of the heterodimeric integrin to the actin cytoskeleton of the cell is the mechanism by which stresses and strains are detected by the cell and can result in significant changes in cell phenotype.

(24 different combinations are currently known) can bind to specific proteins, including fibronectin, laminin, vitronectin and collagen, and ligands such as the arginine–glycine–aspartic acid (RGD) sequence, which may be absorbed or attached onto a range of different substrates.

Integrins do not simply mediate adhesion, but are thought to be largely responsible for transmitting stresses to the cell cytoskeleton, which in turn enables the cell populations to respond to a range of mechanical stimuli. Integrins are transmembrane proteins and attach not only to ligands external to the cell, but also to the actin filaments present on the interior of the cell (Fig. 15.2). It is the attachment of cells in this manner that determines the morphology, function, and shape of cells and allows them to both migrate and reorder their local microenvironment. After cell adhesion mediated by binding of ligands, closely clustered integrins form dot-like complexes known as focal complexes. Focal complexes can be broken down rapidly to enable migration through an adhesive substrate in a "hand-over-hand" manner. Eventually, the focal complexes evolve into focal adhesions, which bind bundles of actin fibers and enable a very strong adhesion to the substrate. Focal adhesions may then further evolve to form fibrillar adhesions, which are thought to be involved in the organization of the pericellular matrix.

The attachment of cells to the surface of materials enables the migration of cells through the hydrogel, which is essential in the production of a structure that mimics the geometry and organization of the replaced tissue. Furthermore, the adhesion allows the cell to monitor its local geometry to some extent and reorganize the fibrous structure accordingly. A major drawback with both fibrin and collagen gels is that the cells compact and contract the gel via strong adhesions resulting in significant shrinkage following cell attachment and this must be taken into account when using these materials as scaffolds.

15.2.2.1 Collagen

There are 20 different forms of collagen found in the human body. The most abundant form of collagen found in many different tissues throughout the body is type I collagen. It is typically well accepted on implantation in the body and therefore has been used in a wide range of different medical applications, including dermal replacement [63], localized drug delivery [64], and bone graft replacement [65]. Collagen is typically harvested from rat tail tendons or from bovine cartilage by a process of acid digestion. The collagen is then stored in acidic conditions until required and may be gelled by returning the pH value of the colloid to physiological pH. The collagen monomers self-assemble to form collagen fibers, which subsequently become physically entangled to form a weak gel. By controlling either collagen concentration or pH value during the gelling process, it is possible to manipulate both the fiber diameter and the pore size of the resulting gels [66]. The weakness of the gel is attributed to the fact that gelling occurs only as a consequence of physical chain entanglement. It is possible to enhance strength by cross-linking the gel with glutaraldehyde, but this process is toxic to the cell population encapsulated within the gel structure. The mechanical properties exhibited by the gel improve on contraction by the cell population; however, this process can take many weeks, which makes clinical application unfeasible. Recent work has focused on uniaxially unconfined compaction of the forming collagen gels. Excess water is removed from the gel, resulting in densification and therefore a rapid improvement in mechanical properties and less dramatic shrinkage following cell seeding. Numerous recent advances in the use of the technique have seen the incorporation of aligned pores within the collagen structure [67], embossing of the surface with defined textures [68] and the fabrication of three-dimensional tissues by manipulating the collagen gels following multiple compactions [69]. Importantly collagen gels can support the growth of a range of different cell types, including kidney cells [70], smooth muscle cells [71], fibroblasts [72,73], endothelial cells [74], and cardiomyocytes [73]. Collagenase enzymes secreted by most cell types enable the modification of the gel structure by cell populations to provide favorable niches for the maintenance of their phenotype, and facilitating migration through the gel structure.

Collagen has been used widely in the *in vitro* production of skin replacements. Skin was first tissue engineered in the mid-1970s by the co-culture of fibroblasts and keratinocyte cells [75], with the intention of subsequently applying the co-culture to the surface of the damaged skin as a replacement for autograft tissue. Although this approach was later successful in the treatment of relatively superficial wounds, where only an epidermal replacement was required, in some cases it is necessary to also provide a replacement for the dermis that can be incorporated into the body with no significant deleterious side effects. Hydrogels of collagen and hyaluronic acid have been widely investigated for this application, although clinical success has been limited due to the extensive contraction of the engineered skin-like material and also localized skin blistering. Numerous workers have attempted to reduce contraction and thereby enhance clinical success by chemically or physically modifying the hydrogel structure. Although some success has been reported, to date there are only three products commercially available which comprise bovine collagen (Apligraf, Orcel, and Permaderm) [76].

15.2.2.2 Fibrin

Fibrin gels are used widely as supports to study the contractile ability of a range of cells [77] and to determine the propensity of both chemicals and materials to encourage angiogenesis (the formation of new blood vessels) [78]. In the human body, fibrin is essential to enable the creation of a seal following significant tissue damage. In order to form a gelled fibrin matrix, fibrinogen is combined with thrombin and $CaCl_2$. The use of three different precursors to the formation of the gel means that there is considerable scope to modify the mechanical and chemical properties exhibited by the gel [66], which typically exhibits shear and storage moduli of at least three orders of magnitude greater than those exhibited by collagen-based gels (Table 15.2). The resulting matrix is a densely gelled fibrin structure through which cells capable of secreting matrix metalloproteinases can migrate. Cells are typically seeded directly onto the surface of the gel and can then migrate through and modify its structure. This approach has been used to tissue engineer a range of soft tissues including ligaments [79], tendons [80], and muscles [81].

Table 15.2 Summary of Literature Values for a Selection of Mechanical Properties Exhibited by Collagen, and Fibrin Hydrogels

Gel → Mechanical Property ↓	Collagen	Fibrin
Shear storage modulus (Pa)	0.15–50	150–520
Shear loss modulus (Pa)	0.02–8	30
Tensile modulus (kPa)	1–33	31–112

Source: Values from Ref. [66]

15.2.2.3 Chitosan

Chitosan can be used in hydrogel form for the replacement of a range of tissues and as a tissue engineering scaffold. Similarly to collagen, chitosan is soluble in a weak acid and can be induced to gel by adjusting the pH value of the hydrocolloid to neutral or physiological pH, though with chemical modification it is possible to gel chitosan in response to the application of UV radiation [82] or a temperature change [83]. The main use for chitosan-based polysaccharides in tissue engineering is as a scaffold for the formation of new cartilage [84]. Due to its cationic nature, chitosan can form complexes with glycosaminoglycans and can immobilize chondroitin sulphate, thus mimicking the structure of a cartilaginous ECM. It has been shown that chitosan-based hydrogels can therefore effectively support cultures of articular chondrocytes maintaining their chondrocytic phenotype.

15.2.3 Noncell Adhesive Hydrogels

Alginate is the most often investigated hydrogel derived from natural sources that is used in tissue engineering that is nonadherent to cells [85]. Alginate was first used in cell-based therapies as an encapsulation medium for pancreatic islets in order to facilitate immunoisolation [86]. More recently, alginate has been investigated for use in a wide range of other medical applications, including the delivery of a range of pharmaceuticals and biopharmaceuticals [87], as a wound dressing [88] and as a carrier of cells for direct implantation or for engineering tissues *ex vivo*. Alginate has a number of advantageous properties, which make it favorable for use as a tissue engineering scaffold. Its mild gelation, with a range of different cations, means that cells can be encapsulated with little risk to their viability and they can remain vital in encapsulated form for a considerable period of time (at the time of writing the author has maintained fibroblast cultures in an encapsulated state for more than 150 days; Fig. 15.3). The biggest risk to cell viability is during the encapsulation process, when in alginate hydrocolloids of $\geq$2 wt% the shear forces are sufficient to cause cell death (Fig. 15.4).

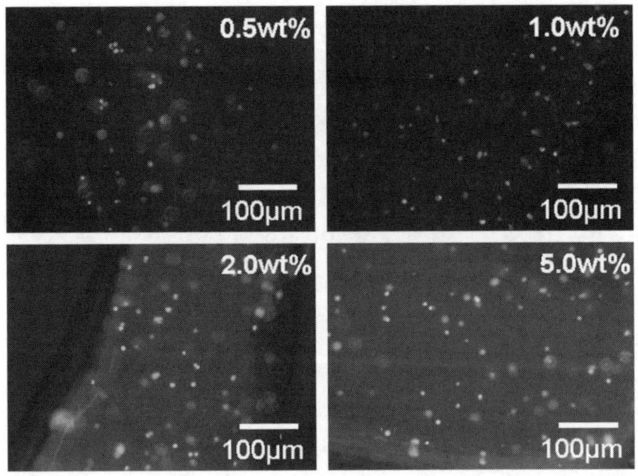

Figure 15.4 LIVE/DEAD assay of 3T3 fibroblast cells encapsulated in 0.5, 1.0, 2.0, and 5.0 wt% alginate and are cultured for 1 day. The cells stained in light gray are considered to be alive and those in dark gray are considered dead. An increase in gel concentration resulted in a concomitant increase in the incidence of cell death; this effect has previously been attributed to an increase in the shear stress required to homogenize the cell dispersion with an increase in alginate concentration.

A significant advantage that is associated with the use of alginate is that its mechanical properties can be relatively easily tailored to a given application. The mechanical properties that are exhibited by alginate can be influenced by changes in molecular weight distribution (*M*:*G* ratio), cross-linking cation density, gelation temperature, and alginate concentration. The high degree of control that can be exercised over the mechanical properties that are exhibited by alginate-based hydrogels is a significant advantage in the control of cell differentiation. It is now well established that cellular phenotype and to some extent function are heavily influenced by local fluctuations in elastic modulus. Relatively high moduli, e.g., are thought to favor the differentiation of mesenchymal stem cells to differentiate to cells exhibiting an osteoblastic phenotype, and gels of comparatively low modulus have been shown to favor the differentiation of marrow stromal cells to a fibroblastic phenotype.

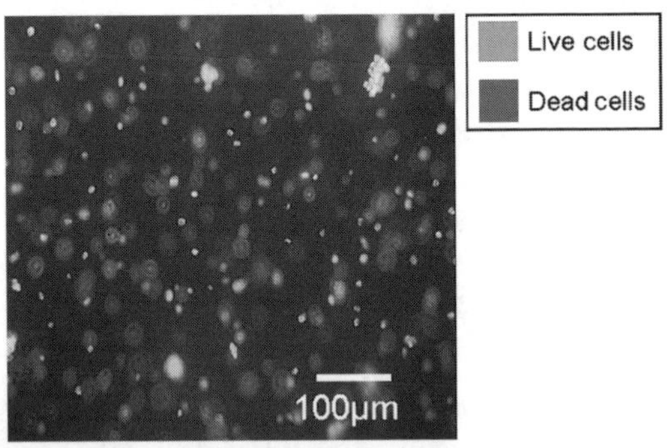

Figure 15.3 3T3 fibroblast cells seeded in an alginate bead (diameter 6 mm) and stained with a LIVE/DEAD assay. The cells were cultured for periods of up to 150 days following seeding. Live cells are stained in light gray and dead cells in dark gray. Even after 150 days in culture the majority of the cells were alive.

Interestingly, when cell–cell adhesions are stronger than cell–matrix adhesions, as is the case with alginate, cell aggregation tends to occur, which many researchers have exploited in order to form the precursors to many tissues or to form functional components of tissues. Hepatocytes and fibroblasts, e.g., have been aggregated following 3D culture in an agarose gel [89]. A significant problem with the formation of spheroids within a hydrogel matrix is poor mass transport to the center of the spheroid, which can cause cell cytotoxicity at its center.

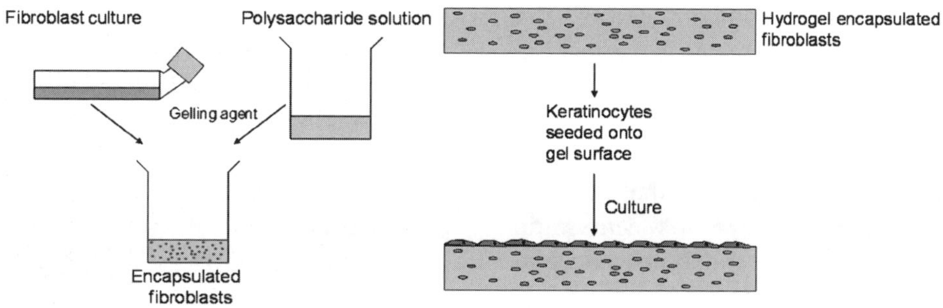

Figure 15.5 A schematic showing a method for the fabrication of a full-thickness dermal substitute comprising fibroblasts encapsulated in an alginate matrix and a keratinocyte layer.

While the relative dimensional stability of alginate is attractive for use and many applications and certain cell types (e.g., chondrocytes) have been shown to maintain their phenotype for prolonged periods of time in culture [62], alginate hydrogels have been widely modified in an attempt to induce cell attachment. As previously outlined, cell adhesion takes place as a result of interactions between cell adhesion molecules such as fibronectin, the biopolymer and integrins present in the membranes of adherent cells (Fig. 15.2). Workers have attempted to enhance cell adhesion by the incorporation of fibronectin, but have found that the cells tend to interact with the hydrogel in a nonspecific and nonuniform manner. The incorporation of RGD sequences, fibronectin-derived cell-adhesion complexes, yielded a significantly more homogeneous cell distribution within the hydrogel matrix [85] and this modification is now widely used in the production of tissues using alginate as a scaffold.

Alginate has also been used widely in the delivery of chondrocytes [90], osteoblasts [91], and bone marrow stromal cells with the intention of regenerating diseased or damaged bone and cartilage. It has been demonstrated that the delivery of osteoblasts derived from calvaria and chondrocytes can significantly enhance the rate of bone formation *in vivo*, although it has also been shown that in order to achieve good bone regeneration it is essential that the alginate gel can to some extent degrade within the body. The likely reason for this is that cells encapsulated in alginate matrices are known to be metabolically inhibited while encapsulated, which would minimize or even prevent the formation of new ECM by the cell population. Once released from encapsulation, the cells encapsulated within the gels were shown to return to their normal metabolic state. Alginate has also been used in the production of scaffolds that seek to augment osteochondral defects, which are complex interfaces formed between bone and cartilage [92]. The incorporation of a calcium phosphate component into the scaffold to form a layered biphasic structure has been shown to successfully support the cultivation of chondrocytes and osteoblasts *in vitro*; however, to date no data exist to prove that this approach would succeed *in vivo*.

Recent work in the author's group has sought to produce a bilayered skin replacement using an alginate hydrogel seeded with fibroblasts as the dermal component and a population of keratinocytes cultured on the surface of the hydrogel to form the epidermal component (Fig. 15.5). The encapsulation of the fibroblast cells within the alginate matrix inhibited cell proliferation, meaning that there was no need for mitotic inhibition of the fibroblast population and furthermore a stratified layer of keratinocytes has been shown to be formed on the upper surface (Fig. 15.6). The weak adhesion of the

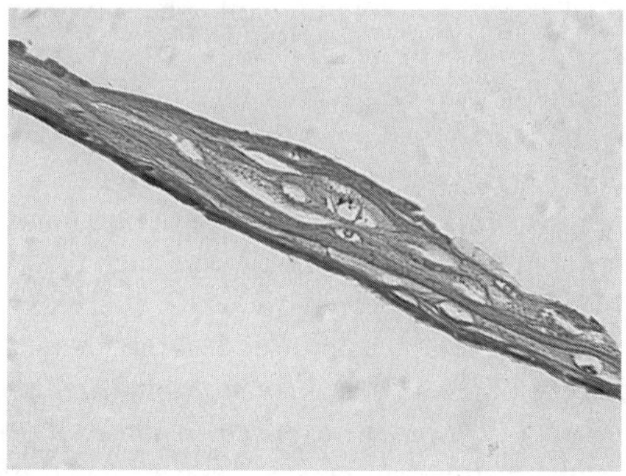

Figure 15.6 A histological section through a layer of stratified keratinocytes grown on the surface of an alginate hydrogel. The section was stained using hemotoxylin and eosin.

keratinocyte population to the surface of the unmodified alginate means that it is possible that the stratified layer could be transferred onto the surface of superficial epidermal injuries. The concurrent application of the alginate structure may also effectively "seal" the wound, lowering the chances of future infection. Indeed, due to its hemostatic nature alginate is already used widely as a wound dressing [88].

15.2.4 Mechanical Conditioning

Through their adhesion to materials and the ECM, cells are constantly monitoring and responding to external mechanical stimuli by migrating, differentiating, or laying down ECM. Consequently, bulk tissues are able to respond with time to the application of varying levels of load, remodeling their structures accordingly. Two of the most obvious and extreme examples of mechanical-conditioning-induced structural changes are the significant loss of bone mass in astronauts in microgravity or the significant gain in bone mass in the playing arm of a tennis player. One of the major problems associated with the *in vitro* production of tissues is that in the absence of mechanical conditioning the tissues will bear little structural or mechanical similarity to the tissue that they are designed to replace. The structures of tissues are directly influenced by cells, which modify the ECM to create their own niches. To further complicate matters, cellular phenotype is significantly affected by the mechanical properties that are exhibited by the ECM. To generate tissues that bear a resemblance to those *in vivo*, therefore, it is essential to provide the correct mechanical and chemical cues to control cell phenotype while allowing the cells to modify their own environments to generate new tissues representative of those found in the body. To generate tissues that are able to function *in vivo* researchers have developed a range of complex bioreactors in order to define loads and thereby condition the resulting tissue such that it will be able to resist the load to which it is exposed *in vivo*. The type of reactor used is largely dependent on the type of tissue that is to be produced. Tendons and ligaments, e.g., are exposed to uniaxial tensile loads, which stimulate the cells encapsulated within the scaffold to align in-line with the principal stress axis and produce a highly conserved matrix [80,93]. In contrast, blood vessels are exposed to pulsatile flows and are therefore stimulated in a reactor that simulates the flow of blood [94]. It is very important to consider, however, that upon initial seeding the matrix will bear little resemblance to that *in vivo* and therefore the way in which the load is transmitted to the seeded cells will be very different to the situation in mature tissues. When cells are seeded at relatively low densities in the hydrogel matrix, the mechanical properties exhibited by the hydrogel will dominate and it will therefore support the largest part of the load and shield the cells from both stress and strain. The situation will be particularly complex in a non-affine gel (such as collagen) where stress and strain distributions are highly complex and difficult to model computationally [66].

15.2.4.1 Cell Morphology

The morphologies of cells that are cultured in a 3D cell-adherent hydrogel are very different to those cultured on a 2D surface. Fibroblast cells that are cultured on surfaces, e.g., spread and exhibit very prominent cellular extensions. In comparison, when cultured in a 3D cell-adherent gel the cells tend to take on a spindle-like, stellate or dendritic morphology. Cell seeding density and then application of loads to cell-seeded gels are also well documented to have a marked influence on cell morphology and the organization of cells within the structure, which tend to align with the principal axis of stress. In addition to the stress and strain concentrations in tissues, the morphology of the cell population can result in the formation of different classes of adhesions between the cells and the ECM. It is thought that the chemical complexity of explanted tissue, e.g., is essential to the formation of 3D adhesions rather than simple focal or fibrillar cell adhesions. Matrix stiffness has also been implicated as having a significant influence on the rate of proliferation of cells, with stiffer materials allowing more rapid cell proliferation than softer materials.

15.2.4.2 Cell Migration

It has also been widely reported that many cell types will migrate toward stiffer regions of a substrate. One group has previously immobilized fibronectin-coated beads in an optical trap and demonstrated that cells tend to strengthen their integrin-mediated adhesion proportionally to the force required to restrain the bead. This property has been employed by one group in order to guide the formation of neuronal cells through a hydrogel-based tube [95]. Other authors have reported that although cells tend to move along stiffness gradients toward the stiffest part of a material, the rate at which the cells move is considerably faster on a soft substrate than on a hard substrate.

As well as influencing cell morphology, the mechanical environment to which a cell is exposed has an influence on cell phenotype. This is of obvious importance in tissue engineering, particularly since the cells responsible for the formation of bone, cartilage, and other commonly engineered tissues are each derived from the same pluripotent cell source, the mesenchymal stem cell. The differentiation of mesenchymal stem cells to form bone has been shown to occur more readily on substrates of relatively high modulus (GPa) such as ceramics and more compliant substrates have been shown to support the formation of fibroblasts and chondrocytes. Although it is a complex set of environmental stimuli, both chemical and mechanical, that are known to stimulate the formation of these different cell types and not mechanotransduction alone, providing a suitable mechanical environment to encourage the formation of the desired cell type is obviously an important consideration in the design of a suitable scaffold material.

15.2.5 Microengineering of Hydrogels

In the body, tissues are permeated by a network of blood vessels which serve to supply cells encapsulated within the ECM with nutrients and oxygen and also to remove their metabolic waste products. The absence of blood vessels within the majority of tissue engineering scaffolds means that the size of a tissue that can be engineered *in vitro* is hindered by mass transport limitations. There are two ways in which these mass transport problems can be solved: (1) "microgels" may be formed, which when laden with cells can be self-assembled to form tissues and (2) hydrogel monoliths can be precisely structured to form a network of capillary-like channels, which enable mass transport or the precise delivery of small quantities of nutrients or active molecules to the cell population. The production of both kinds of structure has been made possible by the increased availability of microfabrication facilities, which enable the production of precisely defined polymeric or silicon based molds.

15.2.5.1 Microgels

The fabrication of microgels of controlled morphology formed by a process of soft-lithography from alginate has previously been reported by Qui *et al.* [96]. They demonstrated that it was possible to generate alginate particles of defined morphology down to the order of 10 μm and suggested that the increased surface area to volume ratio of the gels would negate mass-transport problems. Such processing methodologies have enabled researchers to take a new "bottom-up" approach to the development of tissue-engineered structures, allowing more control over tissue structure than ever before. By directing the assembly of controlled morphology particles containing cells capable of forming different tissues, it is possible to effectively design complex tissue interfaces at the micro-level. Recent work by the Khademhosseini group [97], e.g., has demonstrated that it is possible to direct the assembly of similar gel-based structures into structurally complex multi-tissue constructs by exploiting the tendency of liquid–liquid systems to minimize their surface energy. The rapid progress in this area is extremely exciting as it will enable the production of ever more complex structures, opening up the possibility of forming extremely complex tissues with precisely defined biological and mechanical properties *in vitro*.

15.2.5.2 Microfluidic Scaffolds

Microchanneled hydrogel structures have also recently been reported in the literature. Choi *et al.* [98] formed a precisely microchanneled alginate structure using soft-lithography. Using a micromachined silicon surface, it is possible to form channeled structures with very precisely defined structures with feature sizes down to 30 μm by gelling the alginate on the surface of a silicon wafer (Fig. 15.7). The presence of the aligned pore structure throughout the hydrogel monolith would have helped to address mass transport issues, but importantly such channels allow the chemical environment of the encapsulated cells to be precisely controlled by means of the delivery of precise quantities of growth factors and other stimuli. In common with the manufacture of microgels, this approach to some extent would enable the production of complex tissue interfaces *in vitro*. In addition, it is possible that this technology could allow us to develop high-throughput assays to evaluate cell and tissue response

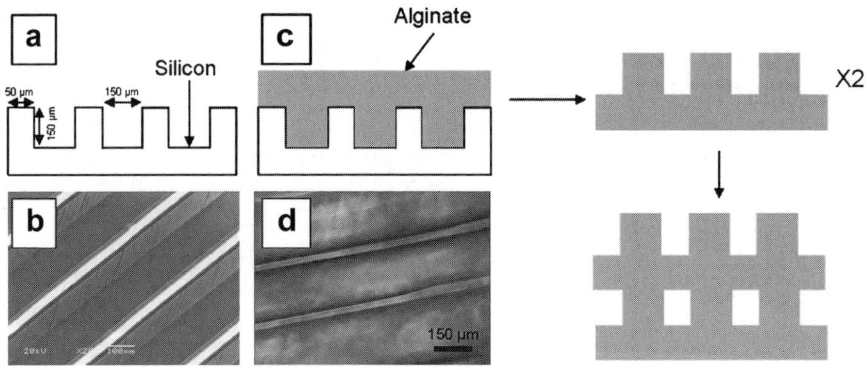

Figure 15.7 The fabrication of a microchanneled structure using a micromachined silicon surface as a mold (a, b, and c). The alginate structures cast onto the surface showed good feature reproduction (d).

to a range of chemical stimuli. Developing such technologies is becoming more important in a range of different industrial sectors as the use of animal testing to determine the biological response to a range of both chemicals and drugs is becoming increasingly taboo.

15.3 Future Horizons

Biopolymer-derived hydrogels are now used widely in the delivery of drugs and are finding increased use in the fabrication of tissue-engineered structures. Their biological compatibility in a range of applications and exciting recent developments in microscale processing technologies means that we can exercise unprecedented control over their microstructures and regional variations in chemical and physical environments. As the requirement for tissue replacements increases and with the development of new biopharmaceutical products, their widespread application in regenerative medicine in the next decade or so is set to increase significantly.

Acknowledgments

Dr. L M Grover would like to acknowledge Miss Nicola Hunt and Miss Pauliena Bohari for contributing their data to the chapter.

References

[1] M.M. Talukdar, R. Kinget, Swelling and drug-release behavior of xanthan gum matrix tablets, Int. J. Pharm. 120 (1995) 63–72.
[2] V.R. Sinha, B.R. Mittal, K.K. Bhutani, R. Kumria, Colonic drug delivery of 5-fluorouracil: an in vitro evaluation, Int. J. Pharm. 269 (2004) 101–108.
[3] H.Y. Fan, K. Wang, M.M. Liu, Z.M. He, In vitro evaluations of konjac glucomannan and xanthan gum mixture as the sustained release material of matrix tablet, Carbohydr. Polym. 73 (2008) 241–247.
[4] M.M. Liu, J.Y. Fan, K. Wang, Z. He, Synthesis, characterization, and evaluation of phosphated cross-linked konjac glucomannan hydrogels for colon-targeted drug delivery, Drug Deliv. 15 (2007) 397–402.
[5] D. Cade, R. Scott, X. He Polymer film compositions for capsules. US Patent 6,517,865, (2003).
[6] T. Ogura, et al., HPMC capsules: an alternative to gelatin, Pharm. Technol. Europe 10 (1998) 32–42.
[7] T. Yamamoto, S. Matsuura, K. Aka, Capsule shell. US Patent 5,431,917, (1995).
[8] J. Brown, N. Madit, E.T. Cole, I.R. Wilding, D. Cade, The effect of cross-linking on the in vivo disintegration of hard gelatin capsules, Pharm. Res. 15 (1998) 1026–1030.
[9] H. Marchais, G. Cayzeele, J.Y. Legendre, M. Skiba, P. Arnaud, Cross-linking of hard gelatin carbamazepine capsules: effect of dissolution conditions on in vitro drug release, Eur. J. Pharm. Sci. 19 (2003) 129–132.
[10] M.E. Pina, A.T. Sousa, Application of hydroalcoholic solutions of formaldehyde in preparation of acetylsalicylic acid gastro-resistant capsules, Drug Dev. Ind. Pharm. 443 (2002).
[11] A.M. Smith, Y. Perrie, Polymer film formulations for the preparation of enteric pharmaceutical capsules. Filed 29/03/07 GP 0706178.1 (Patent pending), 2007.

[12] P.S. Rajinikanth, B. Mishra, Floating in situ gelling system for stomach site-specific delivery of clarithromycin to eradicate *H-pylori*, J. Control. Release 125 (2008) 33–41.
[13] H. Batchelor, Bioadhesive dosage forms for esophageal drug delivery, Pharm. Res. 22 (2005) 175–181.
[14] H. Seager, Drug-delivery products and the Zydis fast-dissolving dosage form, J. Pharm. Pharmacol. 50 (1998) 375–382.
[15] L. Dobetti, Fast-melting tablets: developments and technologies, Pharm. Technol. (2001) 44–50.
[16] A.A. Agoub, A.M. Smith, P. Giannouli, R.K. Richardson, E.R. Morris, "Melt-in-the-mouth" gels from mixtures of xanthan and konjac glucomannan under acidic conditions: a rheological and calorimetric study of the mechanism of synergistic gelation, Carbohydr. Polym. 69 (2007) 713–724.
[17] H. Sohi, Y. Sultana, R.K. Khar, Taste masking technologies in oral pharmaceuticals: recent developments and approaches, Drug Dev. Ind. Pharm. 30 (2004) 429–448.
[18] M.K. Pramoda, S. Lin, Mixed xanthan gum and locust beam gum. US Patent 4136173, (1979).
[19] N. Washington, R.J. Steele, S.J. Jackson, D. Bush, J. Mason, D.A. Gill, K. Pitt, D.A. Rawlins, Determination of baseline human nasal pH and the effect of intranasally administered buffers, Int. J. Pharm. 5 (2000) 139–146.
[20] P. Edman, Biopharmaceuticals of Ocular Drug Delivery (Pharmacology & Toxicology) (2002).
[21] A.H. Shedden, J. Laurence, A. Barrish, T.V. Olah, Plasma timolol concentrations of timolol maleate: timolol gel-forming solution (TIMOPTIC-XE) once daily versus timolol maleate ophthalmic solution twice daily, Doc. Ophthalmol. 103 (2001).
[22] R. Bawa, R.E. Hall, B.P. Kabra, J.E. Teague, G.D. Cagle, K.L. Markwardt, M.V. Shah, Gelling ophthalmic compositions containing. US Patent 6261547, (1999).
[23] J. Ceulemans, I. Vinckier, A. Ludwig, The use of xanthan gum in an ophthalmic liquid dosage form: rheological characterization of the interaction with mucin, J. Pharm. Sci. 91 (2002) 1117–1127.
[24] A. Maltese, A. Borzacchiello, L. Mayol, C. Bucolo, F. Maugeri, L. Nicolais, L. Ambrosio, Novel polysaccharides-based viscoelastic formulations for ophthalmic surgery: rheological characterization, Biomaterials 27 (2006) 5134–5142.
[25] G. Sandri, M.C. Bonferoni, P. Chetoni, S. Rossi, F. Ferrari, C. Ronchi, C. Caramella, Ophthalmic delivery systems based on drug-polymer-polymer ionic ternary interaction: in vitro and in vivo characterization, Eur. J. Pharm. Biopharm. 62 (2006) 59–69.
[26] L. Zhang, D. Russell, B.R. Conway, H. Batchelor, Strategies and therapeutic opportunities for the delivery of drugs to the esophagus, Crit. Rev. Ther. Drug Carrier Syst. 25 (2008) 259–304.
[27] R. Bansil, B.S. Turner, Mucin structure, aggregation, physiological functions and biomedical applications, Curr. Opin. Colloid Interface Sci. 11 (2006) 164–170.
[28] R.B. Gandhi, J.R. Robinson, Oral cavity as a site for bioadhesive drug-delivery, Adv. Drug Deliv. Rev. 13 (1994) 43–74.
[29] J.D. Smart, The basics and underlying mechanisms of mucoadhesion, Adv. Drug Deliv. Rev. 57 (2005) 1556–1568.
[30] Y. Sudhakar, K. Kuotsu, A.K. Bandyopadhyay, Buccal bioadhesive drug delivery—a promising option for orally less efficient drugs, J. Control. Release 114 (2006) 15–40.
[31] S.E. Harding, Trends in mucoadhesive analysis, Trends Food Sci. Technol. 17 (2006) 255–262.
[32] O. Felt, P. Furrer, J.M. Mayer, B. Plazonnet, P. Buri, R. Gurny, Topical use of chitosan in ophthalmology: tolerance assessment and evaluation of precorneal retention, Int. J. Pharm. 180 (1999) 185–193.
[33] H.K. Batchelor, D. Banning, P.W. Dettmar, F.C. Hampson, I.G. Jolliffe, D.Q.M. Craig, An in vitro mucosal model for prediction of the bioadhesion of alginate solutions to the oesophagus, Int. J. Pharm. 238 (2002) 123–132.
[34] L. McCargar, D. Crail, R. Dansereau, W. Myers, M. Lane, The in-vitro porcine adhesion model is not predictive of the esophageal transit of risedronate tablets in humans, Int. J. Pharm. 222 (2001) 191–197.
[35] J.C. Richardson, P.W. Dettmar, F.C. Hampson, C.D. Melia, Oesophageal bioadhesion of sodium alginate suspensions 2. Suspension behaviour on oesophageal mucosa, Eur. J. Pharm. Sci. 24 (2005) 107–114.

[36] K.Y. Lee, T.R. Heo, Survival of *Bifidobacterium longum* immobilized in calcium alginate beads in simulated gastric juices and bile salt solution, App. Environ. Microbiol. 66 (2000) 869−873.
[37] F. Dobakhti, F. Rahimi, A.R. Dehpour, M. Taghikhani, S. Ajdary, S. Rafiei, M. Rafiee-Tehrani, Stabilizing effects of calcium alginate microspheres on mycobacterium bovis BCG intended for oral vaccination, J. Microencapsul. 23 (2006) 844−854.
[38] T. Espevik, M. Otterlei, G. Skjakbraek, L. Ryan, S.D. Wright, A. Sundan, The involvement of Cd14 in stimulation of cytokine production by uronic-acid polymers, Eur. J. Immunol. 23 (1993) 255−261.
[39] A.K. Anal, D. Bhopatkar, S. Tokura, H. Tamura, W.F. Stevens, Chitosan-alginate multilayer beads for gastric passage and controlled intestinal release of protein, Drug Dev. Ind. Pharm. 29 (2003) 713−724.
[40] P. Liu, T.R. Krishnan, Alginate-pectin-poly-L-lysine particulate as a potential controlled release formulation, J. Pharm. Pharmacol. 51 (1999) 141−149.
[41] H. Takeuchi, Y. Matsui, H. Sugihara, H. Yamamoto, Y. Kawashima, Effectiveness of submicron-sized, chitosan-coated liposomes in oral administration of peptide drugs, Int. J. Pharm. 303 (2005) 160−170.
[42] H. Yamamoto, Y. Kuno, S. Sugimoto, H. Takeuchi, Y. Kawashima, Surface-modified PLGA nanosphere with chitosan improved pulmonary delivery of calcitonin by mucoadhesion and opening of the intercellular tight junctions, J. Control. Release 102 (2005) 373−381.
[43] J.E. Lee, S.E. Kim, I.C. Kwon, H.J. Ahn, H. Cho, S.H. Lee, H.J. Kim, S.C. Seong, M.C. Lee, Effects of a chitosan scaffold containing TGF-beta 1 encapsulated chitosan microspheres on in vitro chondrocyte culture, Artif. Organs 28 (2004a) 829−839.
[44] Y.M. Elcin, V. Dixit, T. Gitnick, Extensive in vivo angiogenesis following controlled release of human vascular endothelial cell growth factor: implications for tissue engineering and wound healing, Artif. Organs 25 (2001) 558−565.
[45] S.C. Gamradt, J.R. Lieberman, Bone graft for revision hip arthroplasty, Clin. Orthop. Relat. Res. 417 (2003) 183−194.
[46] E.D. Arrington, W.J. Smith, H.G. Chambers, A.L. Bucknell, N.A. Davino, Complications of iliac crest bone graft harvesting, Clin. Orthop. Relat. Res. (1996) 300−309.
[47] G.K.B. Sandor, B.N. Rittenberg, C.M.L. Clokie, M.F. Caminiti, Clinical success in harvesting autogenous bone using a minimally invasive trephine, J. Oral Maxillofac. Surg. 61 (2003) 164−168.
[48] C. Patience, Y. Takeuchi, R.A. Weiss, Infection of human cells by an endogenous retrovirus of pigs, Nat. Med. 3 (1997) 282−286.
[49] R. Langer, J.P. Vacanti, Tissue engineering, Science 260 (1993) 920−926.
[50] N. L'Heureux, N. Dusserre, G. Konig, B. Victor, P. Keire, T.N. Wight, N.A.F. Chronos, A.E. Kyles, C.R. Gregory, G. Hoyt, R.C. Robbins, T.N. McAllister, Human tissue-engineered blood vessels for adult arterial revascularization, Nat. Med. 12 (2006) 361−365.
[51] A. Neagu, I. Kosztin, K. Jakab, B. Barz, M. Neagu, R. Jamison, G. Forgacs, Computational modeling of tissue self-assembly, Mod. Phy. Lett. B 20 (2006) 1217−1231.
[52] M.M. Stevens, Biomaterials for bone tissue engineering, Mater. Today 11 (2008) 18−25.
[53] W.J.E.M. Habraken, J.G.C. Wolke, J.A. Jansen, Ceramic composites as matrices and scaffolds for drug delivery in tissue engineering, Adv. Drug. Deliv. Rev. 59 (2007) 234−248.
[54] F. Witte, T. Calliess, H. Windhagen, Biodegradable synthetic implant materials. Clinical applications and immunological aspects, Orthopade 37 (2008) 125−130.
[55] J. Jagur-Grodzinski, Polymers for tissue engineering, medical devices, and regenerative medicine. Concise general review of recent studies, Polym. Adv. Technol. 17 (2006) 395−418.
[56] B. Baroli, Hydrogels for tissue engineering and delivery of tissue-inducing substances, J. Pharm. Sci. 96 (2007).
[57] A. Abbott, Cell culture: biology's new dimension, Nature 424 (2003) 870−872.
[58] C.D. Hoemann, A. Chenite, J. Sun, M. Hurtig, A. Serreqi, Z. Lu, E. Rossomacha, M.D. Buschmann, Cytocompatible gel formation of chitosan-glycerol phosphate solutions supplemented with hydroxyl ethyl cellulose is

due to the presence of glyoxal, J. Biomed. Mater. Res. Part A 83A (2007) 521–529.
[59] K. Imai, T. Sato, H. Senoo, Adhesion between cells and extracellular matrix with special reference to hepatic stellate cell adhesion to three-dimensional collagen fibers, Cell Struct. Funct. 25 (2000) 329–336.
[60] L. Urech, A.G. Bittermann, J.A. Hubbell, H. Hall, Mechanical properties, proteolytic degradability and biological modifications affect angiogenic process extension into native and modified fibrin matrices in vitro, Biomaterials 26 (2005) 1369–1379.
[61] A.B. Lansdown, M.J. Payne, An evaluation of the local reaction and biodegradation of calcium sodium alginate (Kaltostat) following subcutaneous implantation in the rat, J. R. Coll. Surg. Edinb. 39 (1994).
[62] J. Bonaventure, N. Kadhom, L. Cohensolal, K.H. Ng, J. Bourguignon, C. Lasselin, P. Freisinger, Reexpression of cartilage-specific genes by dedifferentiated human articular chondrocytes cultured in alginate beads, Exp. Cell Res. 212 (1994) 97–104.
[63] W.R. Otto, J. Nanchahal, Q.L. Lu, N. Boddy, R. Dover, Survival of allogeneic cells in cultured organotypic skin-grafts, Plast. Reconstr. Surg. 96 (1995) 166–176.
[64] O.C.M. Chan, K.F. So, B.P. Chan, Fabrication of nano-fibrous collagen microspheres for protein delivery and effects of photochemical cross-linking on release kinetics, J. Control Release 129 (2008).
[65] J.A. Leupold, W.R. Barfield, Y.H. An, L.A. Hartsock, A comparison of ProOsteon, DBX, and Collagraft in a rabbit model, J. Biomed. Mater. Res. Part B Appl. Biomater. 79B (2006) 292–297.
[66] J.A. Pedersen, M.A. Swartz, Mechanobiology in the third dimension, Ann. Biomed. Eng. 33 (2005) 1469–1490.
[67] S.N. Nazhat, E.A. bou Neel, A. Kidane, I. Ahmed, C. Hope, M. Kershaw, P.D. Lee, E. Stride, N. Saffari, J.C. Knowles, R.A. Brown, Controlled microchannelling in dense collagen scaffolds by soluble phosphate glass fibers, Biomacromolecules 8 (2007) 543–551.
[68] R.A. Brown, M. Wiseman, C.B. Chuo, U. Cheema, S.N. Nazhat, Ultrarapid engineering of biomimetic materials and tissues: fabrication of nano- and microstructures by plastic compression, Adv. Funct. Mater. 15 (2005) 1762–1770.
[69] E.A. Abou Neel, U. Cheema, J.C. Knowles, R.A. Brown, S.N. Nazhat, Use of multiple unconfined compression for control of collagen gel scaffold density and mechanical properties, Soft Matter 2 (2006) 986–992.
[70] G.G. Reid, S.D. Gorham, J.M. Lackie, The attachment, spreading and growth of baby hamster-kidney cells on collagen, chemically modified collagen and collagen-composite substrata, J. Mater. Sci. Mater. Med. 4 (1993) 201–209.
[71] J. Song, B.E. Rolfe, I.P. Hayward, G.R. Campbell, J.H. Campbell, Effects of collagen gel configuration on behavior of vascular smooth muscle cells in vitro: association with vascular morphogenesis, In Vitro Cell. Dev. Biol. Anim. 36 (2000) 600–610.
[72] Z. Feng, M. Yamato, T. Akutsu, T. Nakamura, T. Okano, M. Umezu, Investigation on the mechanical properties of contracted collagen gels as a scaffold for tissue engineering, Artif. Organs 27 (2003a) 84–91.
[73] Z. Feng, T. Matsumoto, T. Nakamura, Measurements of the mechanical properties of contracted collagen gels populated with rat fibroblasts or cardiomyocytes, J. Artif. Organs 6 (2003b).
[74] M. Nakamura, M. Mie, H. Mihara, M. Nakamura, E. Kobatake, Construction of multi-functional extracellular matrix proteins that promote tube formation of endothelial cells, Biomaterials 29 (2008) 2977–2986.
[75] J.G. Rheinwald, H. Green, Formation of a keratinizing epithelium in culture by a cloned cell line derived from a teratoma, Cell 6 (1975) 317–330.
[76] S. MacNeil, Progress and opportunities for tissue-engineered skin, Nature 445 (2007) 874–880.
[77] M.E. Carr, Development of platelet contractile force as a research and clinical measure of platelet function, Cell Biochem. Biophys. 38 (2003) 55–78.
[78] A.K.B. Maier, N. Kociok, G. Zahn, D. Vossmeyer, R. Stragies, P.S. Muether, A.M. Joussen, Modulation of hypoxia-induced

neovascularization by JSM6427, an integrin alpha 5 beta 1 inhibiting molecule, Curr. Eye Res. 32 (2007) 801–812.
[79] S. Hankemeier, M. van Griensven, M. Ezechieli, T. Barkhausen, M. Austin, M. Jagodzinski, R. Meller, U. Bosch, C. Krettek, J. Zeichen, Tissue engineering of tendons and ligaments by human bone marrow stromal cells in a liquid fibrin matrix in immunodeficient rats: results of a histologic study, Arch. Orthop. Trauma Surg. 127 (2007) 815–821.
[80] K. Baar, Uniaxial stretch increases collagen synthesis in fibrin-based engineered tendons, Tissue Eng. 13 (2007) 1773.
[81] Y.C. Huang, R.G. Dennis, L. Larkin, K. Baar, Rapid formation of functional muscle in vitro using fibrin gels, J. Appl. Physiol. 98 (2005) 706–713.
[82] B.G. Amsden, A. Sukarto, D.K. Knight, S.N. Shapka, Methacrylated glycol chitosan as a photopolymerizable biomaterial, Biomacromolecules 8 (2007) 3758–3766.
[83] A. Chenite, C. Chaput, D. Wang, C. Combes, M.D. Buschmann, C.D. Hoemann, J.C. Leroux, B.L. Atkinson, F. Binette, A. Selmani, Novel injectable neutral solutions of chitosan form biodegradable gels in situ, Biomaterials 21 (2000) 2155–2161.
[84] J.K.F. Suh, H.W.T. Matthew, Application of chitosan-based polysaccharide biomaterials in cartilage tissue engineering: a review, Biomaterials 21 (2000) 2589–2598.
[85] A.D. Augst, H.J. Kong, D.J. Mooney, Alginate hydrogels as biomaterials, Macromol. Biosci. 6 (2006) 623–633.
[86] F. Lim, R.D. Moss, Microencapsulation of living cells and tissues, J. Pharm. Sci. 70 (1981) 351–354.
[87] M.C. Peters, B.C. Isenberg, J.A. Rowley, D.J. Mooney, Release from alginate enhances the biological activity of vascular endothelial growth factor, J. Biomater. Sci.Polym. Ed. 9 (1998) 1267–1278.
[88] I.R. Matthew, R.M. Browne, J.W. Frame, B.G. Millar, Subperiosteal behavior of alginate and cellulose wound dressing materials, Biomaterials 16 (1995) 275–278.
[89] D.M. Dean, A.P. Napolitano, J. Youssef, J.R. Morgan, Rods, tori, and honeycombs: the directed self-assembly of microtissues with prescribed microscale geometries, FASEB J. 21 (2007) 4005–4012.
[90] J.F. Grimmer, C.B. Gunnlaugsson, E. Alsberg, H.S. Murphy, H.J. Kong, D.J. Mooney, R.A. Weatherly, Tracheal reconstruction using tissue-engineered cartilage, Arch. Otolaryngol. Head. Neck. Surg. 130 (2004) 1191–1196.
[91] K.Y. Lee, E. Alsberg, S. Hsiong, W. Comisar, J. Linderman, R. Ziff, D. Mooney, Nanoscale adhesion ligand organization regulates osteoblast proliferation and differentiation, Nano Lett. 4 (2004b) 1501–1506.
[92] M. Gelinsky, M. Eckert, F. Despang, Biphasic, but monolithic, scaffolds for the therapy of osteochondral defects, Int. J. Mater. Res. 98 (2007) 749–755.
[93] C.C. Berry, J.C. Shelton, D.L. Bader, D.A. Lee, Influence of external uniaxial cyclic strain on oriented fibroblast-seeded collagen gels, Tissue Eng. 9 (2003) 613–624.
[94] A. Solan, V. Prabhakar, L. Niklason, Engineered vessels: importance of the extracellular matrix, Transplant. Proc. 33 (2001) 66–68.
[95] Y. Luo, P.D. Dalton, M.S. Shoichet, Investigating the properties of novel poly(2-hydroxyethyl methacrylate-co-methyl methacrylate) hydrogel hollow fiber membranes, Chem. Mater. 13 (2001) 4087–4093.
[96] C. Qiu, M. Chen, H. Yan, H.K. Wu, Generation of uniformly sized alginate microparticles for cell encapsulation by using a soft-lithography approach, Adv. Mater. 19 (2007) 1603.
[97] Y. Du, E. Lo, S. Ali, A. Khademhosseini, Directed assembly of cell-laden microgels for fabrication of 3D tissue constructs, Proc. Natl. Acad. Sci. USA 105 (2008).
[98] N.W. Choi, M. Cabodi, B. Held, J.P. Gleghorn, L.J. Bonassar, A.D. Stroock, Microfluidic scaffolds for tissue engineering, Nat. Mater. 6 (2007) 908–915.

16 Natural Polymers in Tissue Engineering Applications

Manuela Gomes, Helena Azevedo, Patrícia Malafaya, Simone Silva, Joaquim Oliveira, Gabriela Silva, Rui Sousa João Mano and Rui Reis

OUTLINE

Objectives

- To understand the origin, structure, and properties of natural polymers used in tissue engineering (TE) applications
- To identify the characteristics that make natural polymers interesting for TE applications
- To understand the possible factors that may affect cells/tissue response to natural polymer-based scaffolds
- To understand the possible specific applications of each natural polymer in the context of tissue engineering
- To understand the processing possibilities of the different natural origin polymers for TE applications
- To recognize the most important achievements in this research field attained by different scientists
- To understand the versatility obtained by combining natural origin polymers with other materials in TE applications

Perhaps appropriately designed biodegradable templates can be used to regenerate segments of other tissues or organs which have become lost or dysfunctional due to disease or trauma [313].

Yannas et al (1982)

16.1 Introduction

Life as we know it could not exist without natural polymers. Just think of deoxyribonucleic acid (DNA) and ribonucleic acid (RNA). They are natural polymers essential in many life processes. In fact, long before there were plastics and synthetic polymers, nature was using natural polymers to make life possible. In the early 1900s, scientists began to understand the chemical makeup of natural polymers and how to make synthetic polymers with properties that complement those of natural materials. Nevertheless, for many purposes, we still do not think of natural polymers in the same way as we think about synthetic polymers. However, that does not make natural polymers less important; indeed, it turns out that they are more important in many ways. In fact,

Ebnesajjad: Handbook of Biopolymers and Biodegradable Plastics. http://dx.doi.org/10.1016/B978-1-4557-2834-3.00016-1
© 2008 Elsevier Inc. All rights reserved. Reproduced from a chapter in: van Blitterswijk, *Tissue Engineering* (2008).

after a century of developing synthetic polymers for use as materials, polymer science is turning back toward its roots, as natural polymers show promise in a wide range of biomedical uses, such as scaffolds for growing artificial human tissues, that is, for making life better after injury or disease.

Tissue engineering offers the possibility to help in the regeneration of tissues damaged by disease or trauma and, in some cases, to create new tissues and replace failing or malfunctioning organs. Typically, this is achieved through the use of degradable biomaterials to either induce surrounding tissue and cell ingrowth or to serve as temporary scaffolds for transplanted cells to attach, grow, and maintain differentiated functions. In any case, the role of the biomaterial scaffold is temporary, but still crucial to the success of the strategy. Therefore, the design and production of an appropriate scaffold material is the first, and one of the most important stages, in tissue engineering strategies. In this critical stage, the selection of the most adequate raw material is a primary consideration. Natural polymers were the first to be used as scaffold materials for tissue regeneration. They have frequently been used in tissue engineering applications because they are either components of, or have properties similar to, the natural extracellular matrix (ECM).

This chapter provides an overview of the natural origin polymers that are commercially available or currently being studied in different labs for tissue engineering applications, with some emphasis on the most widely studied systems. It describes their chemical structure, main properties, and potential applications within the field. Several aspects regarding the development and research status toward their final application are addressed. The main advantages and disadvantages of the use of natural origin polymers as compared to other materials used in tissue engineering scaffolding are also discussed.

16.2 Natural Polymers

Natural polymers are derived from renewable resources such as plants, animals, and microorganisms, and are, therefore, widely distributed in nature. These materials exhibit a large diversity of unique (and in most cases) rather complex structures, and different physiological functions, and may offer a variety of potential applications in the field of tissue engineering due to their various properties, such as pseudoplastic behavior, gelation ability, water-binding capacity, and biodegradability, among many others. In addition, they possess many functional groups (amino, carboxylic, and hydroxyl groups) available for chemical (hydrolysis, oxidation, reduction, esterification, etherification, cross-linking reactions, etc.) [1,2] and enzymatic [3,4] modification and/or conjugation with other molecules, which allows an overwhelming variety of products with tailorable chemistries and properties to be obtained. Protein materials may offer an additional advantage as they are able to interact favorably with cells through specific recognition domains present in their structure. On the other hand, the creation of hybrid materials—by means of combining the advantages of different natural polymers—may constitute a useful approach to mimicking the natural environment of the ECM and to obtaining scaffolding materials with superior mechanical and biological properties.

An intrinsic characteristic of natural origin polymers is their ability to be degraded by naturally occurring enzymes, which may indicate the greater propensity of these materials to be metabolized by the physiological mechanisms. Another important aspect to consider when using natural polymers, is that they can induce an undesirable immune response due to the presence of impurities and endotoxins (depending on their source), and their properties may differ from batch to batch during large-scale isolation procedures due to the inability to accurately control the processing techniques. Nevertheless, as knowledge about these natural polymers increases, new approaches (including methods for production, purification, controlling material properties, and enhancing material biocompatibility) are likely to be developed for designing better scaffolding materials to support the development of more natural and functional tissues.

In summary, both natural and synthetic polymers present important characteristics and, therefore, one must recognize that the best biodegradable polymer for biomedical applications might be found by taking steps toward the development of new biomaterials that combine the most favorable properties of synthetic and natural polymers. Several examples of the combination of natural and synthetic polymers will be described in further sections. Another approach that will be presented consists of the reinforcement of polymeric matrices with bioactive ceramic materials, such as hydroxyapatite and other calcium phosphates. These fillers have the ability, in most cases, to improve the mechanical properties and the biological behavior simultaneously.

It is well known that living organisms are able to synthesize a vast variety of polymers that can be divided into eight major classes according to their chemical structure: (1) polysaccharides, (2) proteins and other polyamides, (3) polyoxoesters (polyhydroxyalkanoic acids), (4) polythioesters, (5) polyanhydrides (polyphosphate), (6) polyisoprenoids, (7) lignin, and (8) nucleic acids [5]. However, only polymers belonging to the three first classes will be described in more detail in this chapter. This is due to their importance as raw materials in tissue engineering scaffolding. Although most of these natural polymers are obtained from plant [6–8] and animal [9,10] sources or from algae [11], there are a large number of microorganisms capable of synthesizing many biopolymers. In fact, with advances in biotechnology, there is an increasing interest in using microorganisms to produce polymers by fermentation (enabling large-scale production, avoiding complex and time-consuming isolation procedures, and the risk of animal-derived pathogens) [12,13] or *in vitro* enzymatic processes [14]. This makes it possible to control polymer molecular weight, branching patterns and branch chain lengths, and cross-linking between chains, altering the fine structure, and functional properties of polymers.

16.2.1 Classical Experiment

While the term *tissue engineering* was still to be "coined" (this happened only in 1987), researchers were already studying an approach to regenerate skin wounds, which resulted in a paper published in *Science* in 1982 [15]. This paper described the prompt and long-term closure of full-thickness skin wounds in guinea pigs and humans, achieved by applying a bilayer polymeric membrane, comprising of a top silicone layer and a bottom layer of a porous cross-linked network of collagen and glycosaminoglycan (GAG), seeded with a small number of autologous basal cells before grafting [15]. This study was conducted following three main stages, today recognized by many researchers as the three main phases of tissue engineering approach: The first stage corresponds to the development of the material matrix membranes; in the second stage, the membranes are seeded with cells; and finally, in the third stage, the cell seeded membranes are grafted onto the tissue defect. The first stage, i.e., the development of the appropriate membranes based on collagen and GAG, was in fact, extensively explored, as the authors have analyzed a range of different chemical compositions and several methods were compared for preparing membranes with different porosities and pore sizes [16–18]. Extensive characterization of the materials led to the selection of the most suitable formulation/structure to be used in the further stages of the development of these skin tissue equivalents. Neodermal tissue synthesis occurred in these membranes seeded with autologous cells and there was no evidence of conventional scar formation. New and apparently normal functional skin was generated in less than 4 weeks. It was demonstrated that, although the acellular membranes can also be used to regenerate skin defects, the cell-seeded membranes provide a means for closing the largest full-thickness skin wounds in a shorter period of time [15]. This system, mainly based on a 3D polymeric matrix obtained from two natural origin polymers (collagen and GAG), also resulted in the first tissue-engineered product to be approved by FDA (1996), and clearly opened the way to the concept of tissue engineering, i.e., to the regeneration of tissues using cells seeded onto a 3D matrix made from the natural-origin polymers.

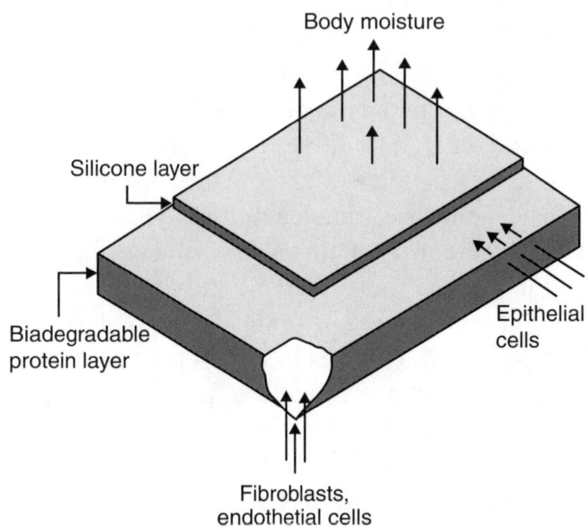

Schematic representation of the polymeric membrane developed by Yannas et al. [15] (*Wound tissue can utilize a polymeric template to synthesize a functional extension of skin.* Science, *215*(*4529*): 174–176) *to be used as a template to obtain skin substitutes. The top layer, made of medical grade silicone, is designed to be spontaneously ejected following formation of a confluent neoepidermal layer under it. The bottom layer, a cross-linked network of collagen and chondroitin 6-sulfate, was designed to undergo biodegradation at a controlled rate while it is replaced by neodermal tissue.*

16.2.2 State of the Art Experiment

16.2.2.1 Natural Polymers in Gene Delivery and Tissue Engineering

Research on natural polymers for the development of matrix-based gene delivery systems has opened the way to new and exciting possibilities to be explored within the field of regenerative medicine [19]. In fact, the combination of gene therapy and tissue engineering exploits the potential of genetic cell engineering to provide biochemical signals that direct cell proliferation and differentiation, and simultaneously, the ability of natural polymers to serve as gene carriers and tissue engineering scaffolds. It is true that synthetic polymers and viral carriers have been preferentially used in gene delivery applications, but natural polymers have unique and intrinsic properties that can make them more suitable candidates for this type of application. Such properties include their general biocompatibility, mucoadhesive character, and biodegradability. The biocompatibility of natural polymers, e.g., allows for cells infiltration into the matrix and transfection can occur as these cells come into contact with the imbedded DNA. The biodegradability of the matrices obtained from natural polymers may also assist the release of gene transfer agents into the surrounding environment and thus affect nearby cells. The current research suggests therefore, that natural polymeric carriers have a different mechanism for intracellular escape and transfection than synthetic polymers. However, a very limited number of studies have focused on the development of matrices based on natural polymers for gene delivery and for cell support. An interesting example is provided by research work from Lim *et al.* [20] in which the authors have investigated a 3D fiber-mesh scaffold, based on chitin and alginate, as a way to obtain a better spatial control of plasmid localization, in opposition to other available systems that are based on simple mixture to bond the matrix and the gene delivery elements. In this study, chitin and alginate fibers were formed by polyelectrolyte complexation of the water-soluble polymers, and PEI–DNA nanoparticles containing green fluorescent protein (GFP)-encoding plasmid were loaded during the fiber drawing process. These fibers were then processed into a nonwoven fiber-mesh scaffold, using a method based on the needle-punching technique. This system was then studied to analyze the tranfectability of human epithelial kidney (HEK293) cells and human dermal fibroblasts (HDFs) seeded on the scaffolds. In summary, the results obtained showed that nanoparticles released from the fibers over time retained their bioactivity and successfully transfected cells seeded on the scaffold in a sustained manner. Transgene expression in HEK293 cells and HDFs seeded on the transfecting scaffolds was significant even after 2 weeks of culture compared to 3-day expression in 2D controls. Fibroblasts seeded on scaffolds containing DNA encoding basic fibroblast growth factor (bFGF) demonstrated prolonged secretion of bFGF at levels significantly higher than baseline.

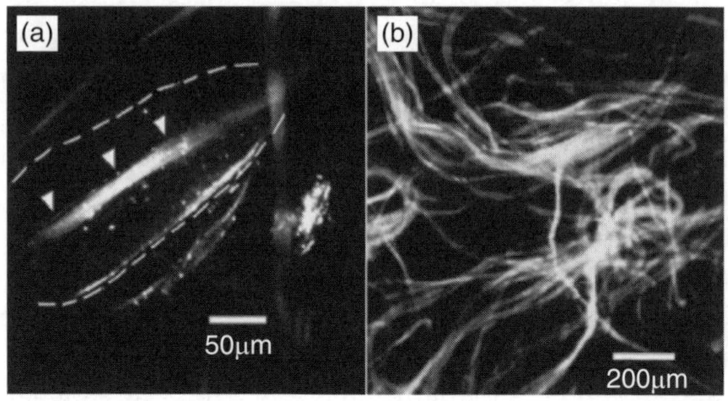

Confocal microscopy images of the fibrous scaffolds and the PEI–DNA nanoparticles (fluorescently stained) encapsulated within the fibers developed by Lim et al. *(Nonviral gene delivery from nonwoven fibrous scaffolds fabricated by interfacial complexation of polyelectrolytes.* Mol Ther, *13(6): 1163–1172, 2006). (a) Phase microscopy showing the bead region of a single fiber, depicting the nanoparticles dispersed within the bead (dotted lines) and at a higher density in the core fiber segment (arrows); (b) fibers containing nanoparticles.*

16.3 Polysaccharides

Polysaccharides, also known as glycans, consist of monosaccharides (aldoses or ketoses) linked together by *O*-glycosidic linkages. Each monosaccharide is classified according to the number of carbons in the monosaccharide chain (usually 3–9), into trioses (C3), tetroses (C4), pentoses (C5), hexoses (C6), heptoses (C7), octoses (C8), and nonoses (C9). Polysaccharides can be classified as homopolysaccharides or heteropolysaccharides if they consist of one type or more than one type of monosaccharide. Because glycosidic linkages can be made to any of the hydroxyl groups of a monosaccharide, polysaccharides form linear as well

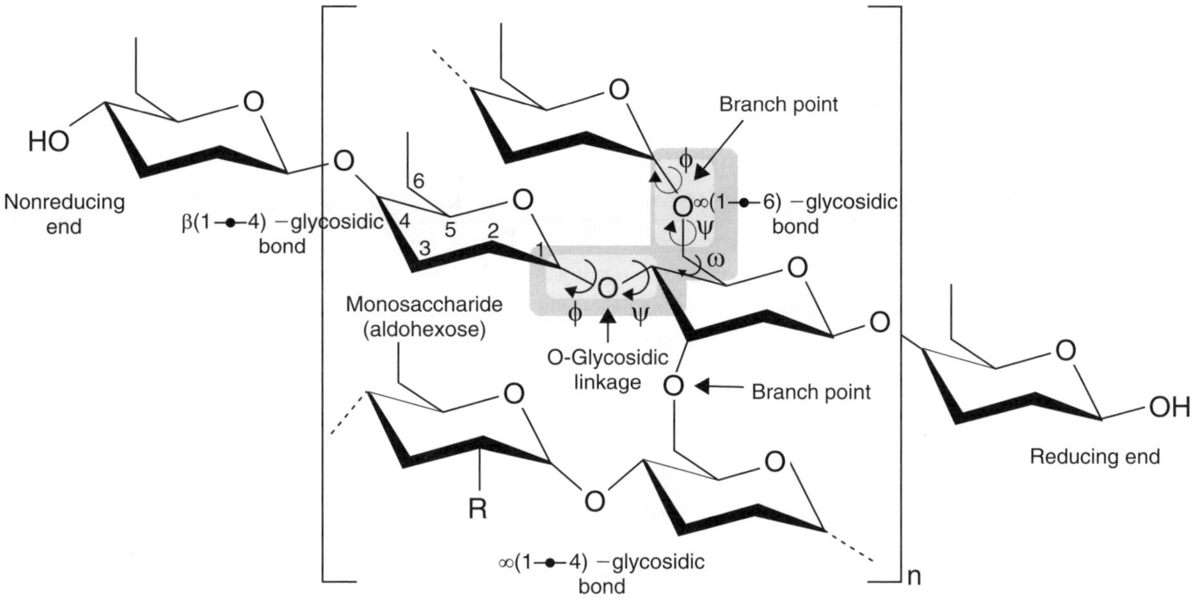

Figure 16.1 General structure of polysaccharides showing their diversity in terms of monosaccharide composition (nature and molar ratios of the monosaccharide building blocks), linkage patterns (linkage positions between the glycosidic linkages and branches), anomeric configuration (α- or β-configuration of the glycosidic linkage), substitutions (position and nature of OH_2 modifications), degree of freedom in (1→4) and (1→6)-glycosidic bonds. *Source: Izydorczyk, M. (2005).* Understand the Chemistry of Food Carbohydrates. Food Carbohydrates: Chemistry, Physical Properties, and Application *(Cui, S.W. ed.), Boca Raton, CRC Press, Taylor & Francis Group: 1–65.*

as branched polymers (Fig. 16.1). Differences in the monosaccharide composition, linkage types and patterns, chain shapes, and molecular weight, dictates their physical properties, including solubility, flow behavior, gelling potential, and/or surface and interfacial properties [9,21].

In the living organisms, polysaccharides perform a range of biological functions, such as maintenance and structural integrity (e.g., cellulose, chitin), energy reserve storage (e.g., starch, glycogen), and biological protection and adhesion (e.g., gum exudates, extracellular microbial polysaccharides). These functions can be found in the compilation presented in Table 16.1.

In this chapter, the authors have chosen to focus only on the polymers that have been proposed, by different researchers, for application within the tissue engineering field; namely alginate, dextran, chitosan, cellulose, starch, and hyaluronic acid (HA) polysaccharides. All of these polymers have been used as scaffold materials and will be described in more detail in the following sections.

16.3.1 Alginate and Dextran

Alginate is a biological material derived from sea algae, composed of linear block copolymers of 1–4 linked β-D-mannuronic acid (*M*) and α-L-guluronic acid (*G*) (Fig. 16.2). Divalent ions form cross-links in alginate by binding the guluronic residues, inducing a sol–gel transition in the material.

Dextran is a bacterial-derived polysaccharide, consisting essentially of α-1,6 linked D-glucopyranose residues with a few percentage of α-1,2-, α-1,3-, or α-1,4-linked side chains (Fig. 16.3) synthesized from sucrose by *Leuconostoc mesenteroides* and *Streptococcus* sp.

Because of their biocompatibility, abundance in source, and low prices, they have been widely used in the food industry as thickeners and emulsifying agents. Alginate can be ionically cross-linked by the addition of divalent cations (like Ca^{2+}) in aqueous solution. The gelation and cross-linking of the polymers are mainly achieved by the exchange of sodium ions from the guluronic acids with the divalent cations, and the stacking of these guluronic groups to form the characteristic egg-box structure shown in Fig. 16.4.

The cross-links are believed to create a stiff egg-box structure and they impart viscoelastic solid behavior to the material [22,23]. The properties of alginate derive from this behavior, and include [24–26] a relatively inert aqueous environment within the matrix; a high gel porosity that allows for

Table 16.1 Classification of Polysaccharides According with Their Origin, Function, Linkage Patterns, Sequence and Composition of Sugar Units in Polysaccharide Chains, and Presence of Ionizing Groups

Origin	Polysaccharide	Occurrence/Function	Glycosidic Linkage/ Repeating Unit	Nature and Distribution of the Monosaccharide Units
Plant	Starch	Starch is synthesized in amyloplasts of green in plants and deposited in the major depots of seeds, tubers, and roots in the form of granules. Energy storage material in almost higher plants (corn, rice, potato, wheat, tapioca, etc).	Amylose: α-(1→4)-D-Glc Amylopectin: α-(1→4, 1→6)-D-Glc	Homopolysaccharide: neutral Amylose: linear Amylopectin: branched
	Cellulose	Structural polysaccharide in the cell walls of higher plants (cotton, wood). Besides the mechanical strength of the plant cell, cellulose is a protective component against external attack by mechanical forces or microorganisms.	β-(1→4)-D-Glc	Homopolysaccharide: neutral, linear
	Arabinogalactan	Larch arabinogalactan is extracted from the heartwood of the western larch *Larix occidentalis*. It is an exudate gum polysaccharide that is produced on the exterior surfaces of the plant usually as a result of trauma or stress (physical injury and/or fungal attack).	Main chain: β-(1→3)-D-Gal Side chains: disaccharides β-D-Gal-(1→6)-β-D-Gal and β-L-Ara-(1→3)-α,-L-Ara	Heteropolysaccharide: neutral, branched
Algal	Alginate	It occurs combined with calcium and other bases in the cell walls and intracellular matrix of brown seaweeds (*Phaeophyceae*), being the main structural component. Contributes to ionic interactions and physical protection.	β-(1→4)-D-ManA-α(1→4)-L-GulA	Heteropolysaccharide: anionic, linear
	Agarose	Red algae (*Rhodophyceae*). Biological function in algae is antidessication at low tide and to provide mechanical support so that cells do not collapse.	-(1→3)-β-D-Gal-(1→4)-3,6-anhydro-α-L-Gal	Heteropolysaccharide: neutral, linear
	Carrageenans	Carrageenans are structural polysaccharides of the marine red algae (*Rhodophyceae*).	j-carrageenan: -(1→3)-β-D-Gal-4-sulfate-(1→]4)-3,6-anhydro-α-D-Gal-(1→3)-	Heteropolysaccharide: anionic, linear

(*Continued*)

Table 16.1 Classification of Polysaccharides According with Their Origin, Function, Linkage Patterns, Sequence and Composition of Sugar Units in Polysaccharide Chains, and Presence of Ionizing Groups—Cont'd

Origin	Polysaccharide	Occurrence/Function	Glycosidic Linkage/ Repeating Unit	Nature and Distribution of the Monosaccharide Units
			κ-carrageenan(1→3)-β-D-Gal-4-sulfate-(1→4)-3,6-anhydro-a-D-Gal-2-sulfate-(1→3)- k-carrageenan: (1→3)-β-D-Gal-2-sulfate-(1→4)–α-D-Gal-2,6-disulfate-(1→3)-	
Animal	Chitin/chitosan	Chitin is the main component of the exoskeleton of insects and shells of crustaceans (crab, shrimp, lobster, etc.). Structural/ supporting polysaccharide. Chitosan is a chitin derivative obtained by a deacetylation reaction.	Chitin: 1-(1→4)-D-GlcNAc Chitosan: β-(1→4)-D-GlcN-α-(1→4)-D-GlcNAc, distributed in a random way depending on the degree of acetylation.	Chitin: homopolysaccharide, neutral, linear Chitosan: heteropolysaccharide, cationic, linear
	Hyaluronic acid	Hyaluronan is an important glycosaminoglycan component of connective tissue (cartilage, tendon, skin, and blood vessel walls), synovial fluid (the fluid that lubricates joints) and the vitreous humor of the eye. It plays a significant role in wound healing.	-β(1→4)-D-GlcUA-β(1→3)-D-GlcNAc-	Heteropolysaccharide: anionic, linear
Microbial	Dextran	Extracellular polysaccharide produced by the bacterium *Leuconostoc mesenteroides.*	α-(1→2, 1→3, 1→4, 1→6)-Glc	Homopolysaccharide: neutral, branched
	Gellan gum	Extracellular polysaccharide produced by the bacterium *Sphingomonas elodea.*	→3)-β-D-Glc-(1→4)-b-D-GlcUA- (1→4)-β-D-Glc-(1→4)-α-L-Rha-(1→	Heteropolysaccharide, anionic, linear
	Pullulan	Extracellular polysaccharide produced by the fungus *Aureobasidium pullulans.*	α-(1→6)-maltotriose	Homopolysaccharide: neutral, branched

Glc, glucose; Ara, arabinose; GulA, guluronic acid; ManA, mannuronic acid; Gal, galactose; GlcNAc, N*-acetylglucosamine; GlcN,* N*-glucosamine; GlcUA, glucuronic acid, Rha, rhamnose*

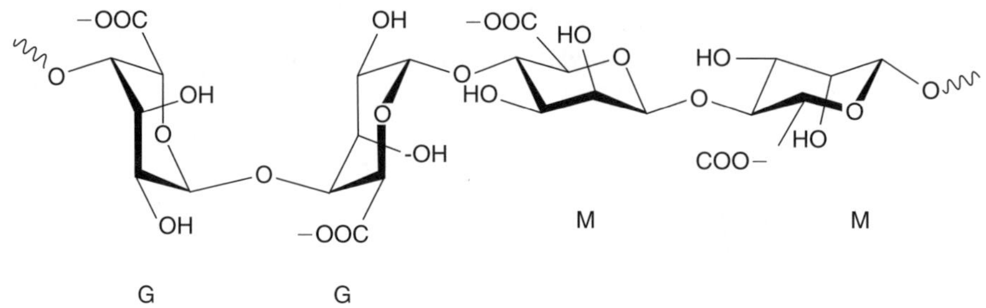

Figure 16.2 Alginate structure. M – mannuronic acid; G – guluronic acid.

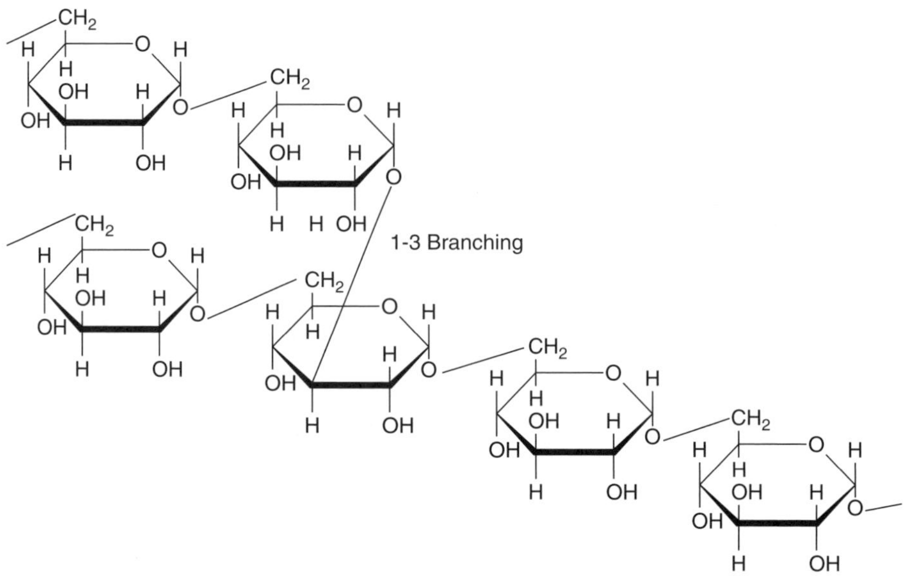

Figure 16.3 Dextran structure evidencing the 1–3 branching.

high diffusion rates of macromolecules; the ability to control this porosity with simple coating procedures and dissolution and biodegradation of the systems under normal physiological conditions and, at room temperature, a mild encapsulation process free of organic solvents. An attractive class of physically cross-linked gels are those where gel formation is not instantaneous, but occurs a certain time after mixing the hydrogel components or after a certain trigger (such as pH or temperature). These systems can be administered by injection as liquid formulation and gellify *in situ* [27–29].

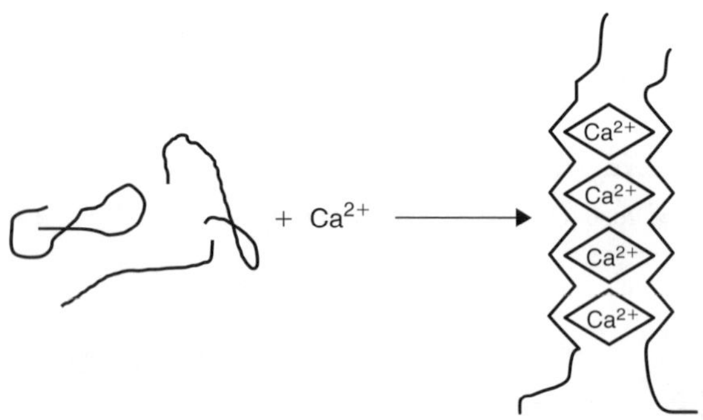

Figure 16.4 Egg-box model for alginate gel formation. It is shown the conversion of random coils to buckled ribbon-like structures containing arrays of Ca^{2+} ions (e).

It is the latter characteristic that drew the attention of researchers to use alginate for encapsulation of cells as well as bioactive agents. The material to be encapsulated is usually mixed with an alginate solution, and the mixture dripped into a solution containing Ca^{2+} ions, resulting in the instantaneous formation of microparticles that entrap cells or drugs within a 3D lattice [30]. Dextran hydrogels can be created by either physical or chemical cross-linking, taking advantage of the hydroxyl groups present on the α-1,6-linked D-glucose residues. Dextran particles have been widely used as separation matrices, such as Sephadex, as cell microcarriers, such as Cytodex, and as drug delivery vehicles [31]. There has been considerable interest on dextran scaffolds for tissue engineering applications [31,32].

Similarly, alginate cross-linked with Ca^{2+} has been popularized for *in vitro* cell culture [33,34] and tissue engineering applications [27,35–39] primarily because of the ability to immobilize and later recover cells from the culture matrix [23]. Alginate has also been used as a bioartificial matrix for cartilage generation and fundamental studies on entrapped chondrocytes [33,39]. The suspension of cells in a bioartificial matrix, such as alginate, is associated with significant changes in the local physical and mechanical environment of the cells compared to their native ECM [23]. ECM plays an important role in tissue engineering because cellular growth and differentiation, in the 2D cell culture as well as in the 3D space of the developing organism, require ECM with which the cells can interact [40]. In the artificial culture system, the physical properties of the artificial matrix will govern the deformations and tractions applied to the cells, altering important cell–matrix interactions present in the native system that appear to regulate cell activity in response to mechanical stress [33,41]. On the other hand, alginate is well known for forming strong complexes with polycations including, but not limited to, synthetic polymers, proteins, and polypeptides. This feature is particularly important when attempting to use alginate as a scaffold for tissue engineering applications such as cartilage and bone, since mechanical constraints limit its applications. Combining alginate with other polymers and ceramic materials has been shown to be able to obviate this feature [42–46].

Alginate has also been widely studied for engineering liver tissue [47–49]. The bioartificial liver-assist device or regeneration of the liver-tissue substitutes for liver tissue engineering requires a suitable ECM for hepatocyte culture because hepatocytes are anchorage-dependent cells and are highly sensitive to the ECM milieu for the maintenance of their viability and differentiated functions [40]. A potential approach to facilitate the performance of implanted hepatocytes is to enable their aggregation and reexpression of their differentiated function prior to implantation [49] and alginate has been shown to allow hepatocyte culture and function.

The differentiation and growth of adult stem cells within engineered tissue constructs are believed to be under the influence of cell-biomaterial interactions. Gimble et al. [50] have shown that alginate-based materials can have an enhancing effect over the differentiation of human adipose-derived adult stem (hADAS) cells, and manipulating the composition of these tissue engineered constructs may have significant effects on their mechanical properties. Additionally, the major role of alginate in tissue engineering has been defined as a vehicle for cell encapsulation and delivery to the site, and attachment of RGD sequences has shown to potentate bone cell attachment and upregulation of specific bone markers [51].

In summary, the possibility of having an injectable *in situ*-gellifying material that can serve as a filler and template for the regeneration/repair of tissues such as cartilage, is a very attractive one. Alginate and dextran have shown and continue to show excellent properties for this purpose. Allied to this, the potential of being tailored for several applications—hence a multitasking ability for these materials—renders interest from the scientific community.

16.3.2 Chitosan

During the past 30 years, a substantial amount of work has been published on chitosan and its potential use in various pharmaceutical applications [52], including tissue engineering. This is due to its similar structure to naturally occurring GAG and its degradability by enzymes in humans [53]. Figure 16.5 shows a chitosan structure. It is a linear polysaccharide of (1→4)-linked D-glucosamine and *N*-acetyl-D-glucosamine residues derived from chitin, a high molecular weight, the second most abundant natural biopolymer commonly found in arthropod exoskeletons such as shells of marine crustaceans and cell walls of fungi [54]. Chitosan has been proven to be biologically renewable, biodegradable [55–57], bioadhesive [58,59], and biocompatible [52,55,60–63], and used in wound dressing and healing [55,64,65], drug delivery systems [66–69], and various tissue engineering applications. We focus on these issues in this chapter.

Depending on the source and preparation procedure, chitosan's average molecular weight may range from 50 to 1000 kDa [70]. The degree of *N*-deacetylation usually varies from 50 to 90% [52]. Chitosan is a semicrystalline polymer and the degree

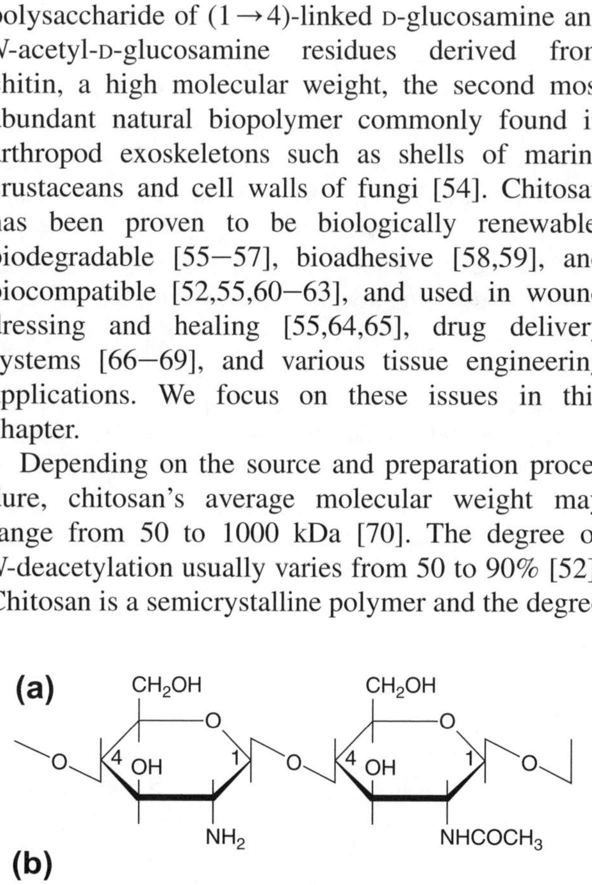

Figure 16.5 Structure of chitosan.

of crystallinity is a function of the degree of deacetylation. Crystallinity is maximum for both chitin (i.e., 0% deacetylated) and fully deacetylated (i.e., 100%) chitosan. Minimum crystallinity is achieved at intermediate degrees of deacetylation. Chitosan is degraded by lysozyme [18,57,71]; the kinetics of degradation is inversely related to the degree of crystallinity. Because of the stable crystalline structure, chitosan is normally insoluble in aqueous solutions above pH 7. However, in dilute acids, the free amino groups are protonated and the molecule becomes fully soluble below pH 5. The pH-dependent solubility of chitosan provides a convenient mechanism for processing under mild conditions. Viscous solutions can be extruded and gelled in high pH solutions [72] or baths of nonsolvents such as methanol. Such gel forms (particles, fibers, or blocks) can be subsequently drawn and dried to form high-strength materials. Much of the potential of chitosan as a biomaterial stems from its cationic nature and high charge density in solution. The charge density allows chitosan to form insoluble ionic complexes or complex coacervates with a wide variety of water-soluble anionic polymers. Chitosan derivatives and blends have also been gelled via glutaraldehyde cross-linking [73] and other cross-linking agents such as genipin [74], UV irradiation [53], and thermal variations [54]. Besides the referred gelling-based processing method, freeze-drying is undoubtedly the most widely processing technology used to process chitosan shapes. Furthermore, the cationic nature of chitosan is primarily responsible for electrostatic interactions with anionic GAGs, proteoglycans, and other negatively charged molecules. This property is of great interest because a large number of cytokines/growth factors are linked to GAG, and a scaffold incorporating a chitosan−GAG complex may retain and concentrate growth factors secreted by colonizing cells [54]. Several researchers have examined the host tissue response to chitosan-based materials. In general, these materials evoke a minimal foreign body reaction, with little or no fibrous encapsulation [75]. Formation of normal granulation tissue associated with accelerated angiogenesis, appears to be the typical course of the healing response. This immunomodulatory effect has been suggested to stimulate the integration of the implanted material by the host [76].

Due to its promising properties, chitosan has been applied in tissue engineering applications targeting several tissues and these are summarized in Table 16.2. The most commonly aimed tissues are bone, cartilage, and skin, but others such as liver or trachea have applied chitosan as scaffolds to support the temporary cell functions. Due to its easy processability, chitosan has been molded in a range of shapes including porous scaffolds, injectable gels, membranes, tubular systems, and particles as described in Table 16.2. Chitosan scaffolds for bone tissue engineering have been widely investigated and shown to enhance bone formation both *in vitro* and *in vivo*, mainly in the presence of other polymers such as gelatin [73] and alginates [44]. When one considers cartilage tissue engineering applications, in particular, chitosan seems to be a good candidate given the importance of GAGs in stimulating the chondrogenesis [77], the use of GAGs or GAG analogs such as chitosan as components of a cartilage tissue scaffold appears to be a logical approach for enhancing chondrogenesis as shown by several papers [76−79]. It thus shares some characteristics with various GAGs and HA present in articular cartilage [70].

At present, chitosan is one of the most promising natural origin polymers for tissue engineering. In particular, its chemical versatility and the possibility to generate structures with predictable pore sizes and degradation rates make chitosan a promising candidate scaffold for these applications. In fact, the combination of good biocompatibility, intrinsic antibacterial activity, and ability to bind to growth factors renders this material as a good potential for several tissue engineering applications.

16.3.3 Cellulose

Cellulose is the main component of plant cell walls. It also constitutes the most abundant, renewable polymer resource available today, existing mainly in lignocellulosic material in forests, with wood being the most important source. The primary structure of this linear polymer consists of up to 15,000 D-glucose residues linked by $\beta(1\rightarrow4)$-glycosidic bond [80] (Fig. 16.6). The fully equatorial conformation of β-linked glucopyranose residues stabilizes the chain structure, minimizing its flexibility. It is the ability of these chains to hydrogen-bond together into fibers (microfibrils) that give cellulose its unique properties of mechanical strength and chemical stability, leading also to insoluble materials with small degradability *in vivo* [81].

Table 16.2 Chitosan-Based Scaffolds for Different Tissue Engineering Applications

Material	Scaffold Structure	Processing Methodology	Cell type (Source)	TE Application	References
Chitosan	3D fiber meshes	Wet spinning	Osteoblast-like SAOS-2 (human osteosarcoma cell line)	Bone	[271]
Chitosan	3D porous blocks	Freeze-drying	Osteoblast-like ROS (rat osteosarcoma cell line)	Bone	[272,273]
Chitosan/ polyester	3D fiber meshes	Fiber extrusion	MSCs (human bone marrow, primary culture)	Bone	[274–276]
Chitosan/ alginate	3D porous cylinders	Freeze-drying	Osteoblast-like MG63 (human osteosarcoma cell line)	Bone	[44]
Chitosan/ alginate	Injectable gel	Gelation by sonication	MSCs (rat bone marrow, primary culture)	Bone	[277]
Chitosan/ (HA)	3D porous cylinders	Particle aggregation	ADAS cells (human adipose tissue, primary culture)	Bone	[72]
Chitosan/ (HA)	3D porous cuboids	3D-printing	Osteoblasts (human calvaria, primary culture)	Bone	[87]
Chitosan/ nano-HA	3D porous blocks	Freeze-drying	Osteoblast-like MC3T3-E1 (newborn mouse calvaria cell line)	Bone	[278]
Chitosan/ β-TCP	3D porous blocks	Freeze-drying	Osteoblast-like MG63 (human osteosarcoma cell line)	Bone	[279,280]
Chitosan/ coralline	3D porous cylinders	Freeze-drying	MSCs: CRL-12424 (mouse bone marrow cell line)	Bone	[281]
Chitosan/ gelatin/HA	3D porous disks	Freeze-drying	Osteoblasts (neonatal rats calvaria, primary culture)	Bone	[282]
Chitosan/ gelatin/ (HA)	3D porous disks	Freeze-drying	MSCs (human bone marrow, primary culture)	Bone	[283]
Chitosan	3D porous disks	Freeze-drying	Chondrocytes (pig knee and dog shoulder joint, primary culture)	Cartilage	[77,284,285]
Chitosan	3D porous cylinders	Particle aggregation	ADAS cells (human adipose tissue, primary culture)	Cartilage	Malafaya, Pedro *et al.*, 2006
Chitosan/ polyester	3D fiber meshes	Fiber extrusion	Chondrocytes (bovine knee, primary culture)	Cartilage	[202,274,275]
Chitosan/ gelatin	3D porous cylinders	Freeze-drying	Chondrocytes (rabbit knee and pig auricular cartilage, primary culture)	Cartilage	[78,287]

(*Continued*)

Table 16.2 Chitosan-Based Scaffolds for Different Tissue Engineering Applications—Cont'd

Material	Scaffold Structure	Processing Methodology	Cell type (Source)	TE Application	References
Chitosan/GP	Injectable gel	Gelation by polyol salts	Chondrocytes (calf knee, primary culture)	Cartilage	[288,289]
Chitosan/hyaluronan	3D fiber sheets	Wet spinning	Chondrocytes (rabbit knee, hip, and shoulder joints, primary culture)	Cartilage	[79]
Chitosan/alginate	3D porous cylinders	Freeze-drying	Chondrocyte-like HTB-94 (human bone chondrosarcoma cell line)	Cartilage	[43]
Chitosan/(HA)	Bilayered	Particle aggregation	ADAS cells (human adipose tissue, primary culture)	Osteochondral	[72]
Chitosan/hyaluronan	Bilayered with PLA	Freeze-drying	*In vivo* (rabbit femoral condyle)	Osteochondral	[154]
Chitosan	Porous membranes	Freeze-drying	-	Skin	[64]
Chitosan/gelatin	Bilayered porous membranes	Freeze-drying	Co-culture of fibroblasts and keratinocytes (human skin, primary culture)	Skin	[60,61]
Chitosan/collagen	Porous membranes	Freeze-drying	Co-culture of fibroblasts and keratinocytes (human foreskin, primary culture)	Skin	[290]
Chitosan	Tubular system	Wire-heating and freeze-drying	–	Neural	[291]
Chitosan	Porous hollow conduits	Thermal-induced phase separation	Neuro-2a cells (mouse neuroblastoma cell line)	Neural	[292–294]
Chitosan derivative	Tubular system	Crab tendon treatment	*In vivo* (rat sciatic nerve)	Neural	[295]
Chitosan/hyaluronan	3D fiber sheets	Wet spinning	Fibroblasts (rabbit patellar tendon, primary culture)	Ligament	[159,160]
Chitosan/collagen	3D porous blocks	Freeze-drying	Hepatocytes (rat liver, primary culture)	Liver	[94]
Chitosan/gelatin	3D hydrogel cylinders	Solvent casting	Respiratory epithelial cells (human tissue, primary culture)	Tracheal	[296]

(), with or without; HA, hydroxyapatite; b-TCP, b-tricalcium phosphate; GP, glycerophosphate disodium salt; PLA, polylactic acid; MSCs, mesenchymal stem cells; ROS, rat osteosarcoma; ADAS cells, adipose derived adult stem cells

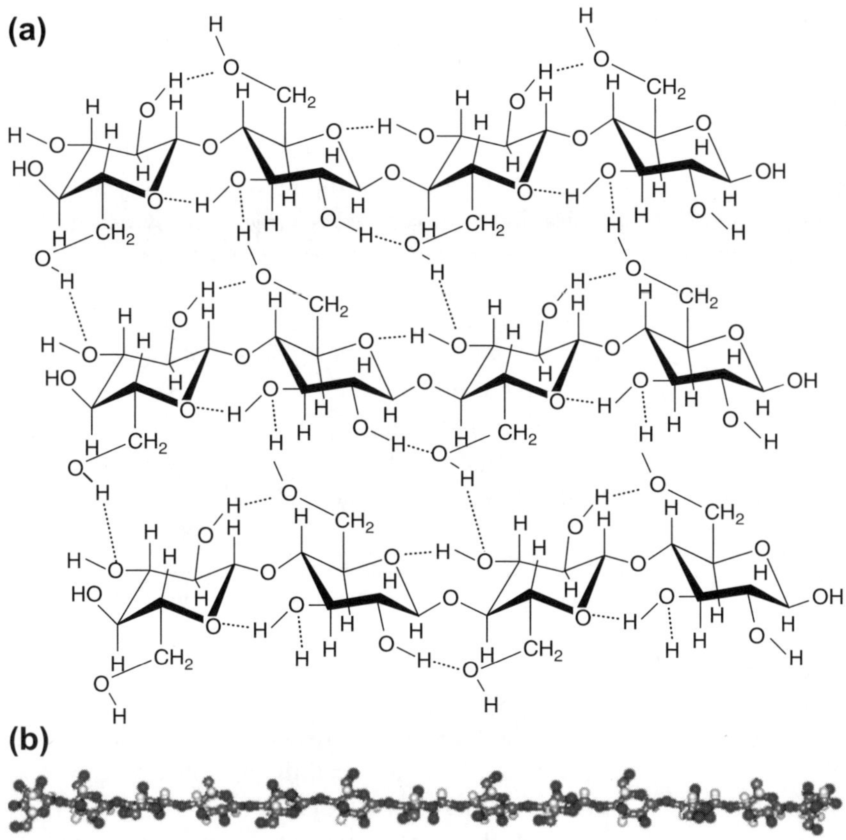

Figure 16.6 Structure of cellulose.

The biodegradability of cellulose is considered to be limited, if it occurs at all, because of the absence of hydrolases that attack the β(1→4) linkage [82]. This fact, together with the difficult processing, is the most limiting factor for the use of cellulose in tissue engineering applications. However, some partial degradation in processed cellulose sponges *in vivo* was reported [83]. The modification of the highly regular structural order of cellulose may also improve and tailor its degradation, as well as its tissue response [84]. For example, *in vitro* and *in vivo* studies on acetyl-cellulose and ethyl-cellulose sponges allowed to conclude that, for the first case, a gradual degradation over time could be detected, consistent with the observation on implanted sponges in Wistar rats [85].

Cellulose and its derivatives have been employed with success as biomaterials, and there are some indications that they could be an adequate source for tissue engineering applications. In orthopedic applications, it has been shown that cellulose sponges could support bone tissue ingrowths, suggesting that it could be used in bone tissue engineering [86]. Takata *et al.* [87] compared different membranes for guided bone regeneration and, among other materials, cellulose that also exhibited the ability to induce cell migration. Cellulose acetate and cellulose scaffolds also showed to be interesting for cardiac tissue regeneration, as they could promote cardiac cell growth, enhancing cell connectivity and electrical functionality [81].

Bacterial cellulose (BC) is a biotechnological method for producing pure nanofibrillar cellulose structures that have high mechanical strength, high water content, and high crystallinity [88]. BC is excreted extracellularly by *Acetobacter xylinum* bacteria, and the pellicle formed has been proposed to be used in tissue engineering-related applications. The biocompatibility of BC was confirmed *in vivo* where subcutaneous implantations in rats did not show substantial inflammatory response [89]. BC was shown to be able to support the proliferation of bovine-derived chondrocytes, and thus suggested the potential for use in tissue engineering of cartilage [90]. Moreover, the adequate mechanical properties of BC pellicles and the fact that smooth muscle cells adhere to and proliferate onto it suggested that BC could also be attractive for tissue engineering of blood vessels [88].

The properties of cellulose can be highly altered with chemical modification (e.g., through the substitution of the hydroxyl groups) allowing expansion and tailoring of the physical features and the response to tissues of this material. A few cellulose derivatives have been specifically proposed for tissue engineering purposes. For example, 2,3-dialdehydecellulose porous membranes were prepared from methylcellulose combining water-induced phase separation and salt leaching techniques; this material is biodegradable and has been used as a drug carrier. Human neonatal skin fibroblast cells attached and spread on these membranes [91]. Hydroxypropyl methylcellulose grafted with silanol groups was developed as an injectable and self-setting hydrogel, which could be used to deliver and fix cells into a site through a nonevasive procedure [92]. Chondrocytes from two different origins were found to maintain their viability and to proliferate when cultured into the hydrogel [92], indicating that cellulose derivatives could also be used as a carrier of chondrocytes in cartilage tissue engineering. Cellulose sulfate was found to be biocompatible and nonimmunogenic and also showed to be adequate to encapsulate cells, to be used, for example, in protecting pancreatic xenogeneic cells from the immune system as a potentially curative treatment option for diabetes [93].

Cellulose and its derivatives may be seen as a potential source of natural-based materials in tissue engineering applications. There is, however, more work to be done in order to enhance the degradation rate and to find more processing routes to produce scaffolds with controlled architectures.

16.3.4 Starch

Starch is the dominant carbohydrate reserve material of higher plants, being found in the leaf chloroplasts and in the amyloplasts of storage organs such as seeds and tubers [94]. Although there is a broad range of possible origins of native starch, most of the starch utilized worldwide comes from a relatively small number of crops, the most important being corn, potato, wheat, and tapioca, with smaller amounts from rice, sorghum, sweet potato, arrowroot, sago, and mung beans [94].

As a natural polymer, it has received great attention as a possible alternative to synthetic polymers in several applications, mainly for being one of the cheapest biopolymers, being totally biodegradable into carbon dioxide and water [95], and abundantly available [95–97].

Native starch is composed of granules of variable sizes and shapes depending on the source of the starch. Chemically, starch is a polysaccharide consisting only of homoglucan units [95,96]. Starch is constituted by α-D-glucose units, which can be organized to form two distinct molecules, amylose and amylopectin [94,95,98–100]. The typical structure of amylose consists of a linear, very sparsely branched, polymer basically linked by 1→4 bonds [95,96,98,100,101]. On the contrary, amylopectin is highly branched on multiple points of the backbone, and contains not only 1→4 bonds, but also 1→6 branching points, that tend to appear each 25–30 glucose units [95,98,100,101]. The correspondent molecular weights are around 10^5–10^6 for amylose and 10^7–10^9 g/mol for amylopectin [95,99]. The distinct molecular weights and degrees of branching of both molecules are responsible for the quite different properties of starch isolated from sources with diverse amylose/amylopectin relative ratios [95,96,98] (Fig. 16.7). Besides its two basic macromolecular constituents, traces of lipids, proteins, and minerals (mainly phosphates) may also be found on native starch [98].

One of the most important properties of native starch is its semicrystallinity. Depending on the source and the moisture content, the degree of crystallinity in native starch ranges between 15 and 50% [102]. The crystallinity of starch is due to amylose and amylopectin, but mostly depends on amylopectin. Even though amylopectin has a branched structure, the branches form double helices between branches. The starch granule has been found to have alternating crystalline and amorphous concentric layers. The amorphous may be due to areas where the α-1,6 branch points form the chains, while the crystalline regions arise when the joined α-1,4 joined branches intertwine with each other and form double helices [99], resulting in the formation of parallel crystalline lamellae [102].

Besides its use as a filler material, native starch must be modified by destructuring of its granular structure to find other applications. The destructuring agent is usually water. The disruption of the granule organization obtained with the combination of water and heat is termed "gelatinization" and is characterized by the swelling of starch, forming a viscous past with destruction of most intermolecular hydrogen links [96,98].

To be able to make a thermoplastic starch (TPS) that can be processed by conventional processing

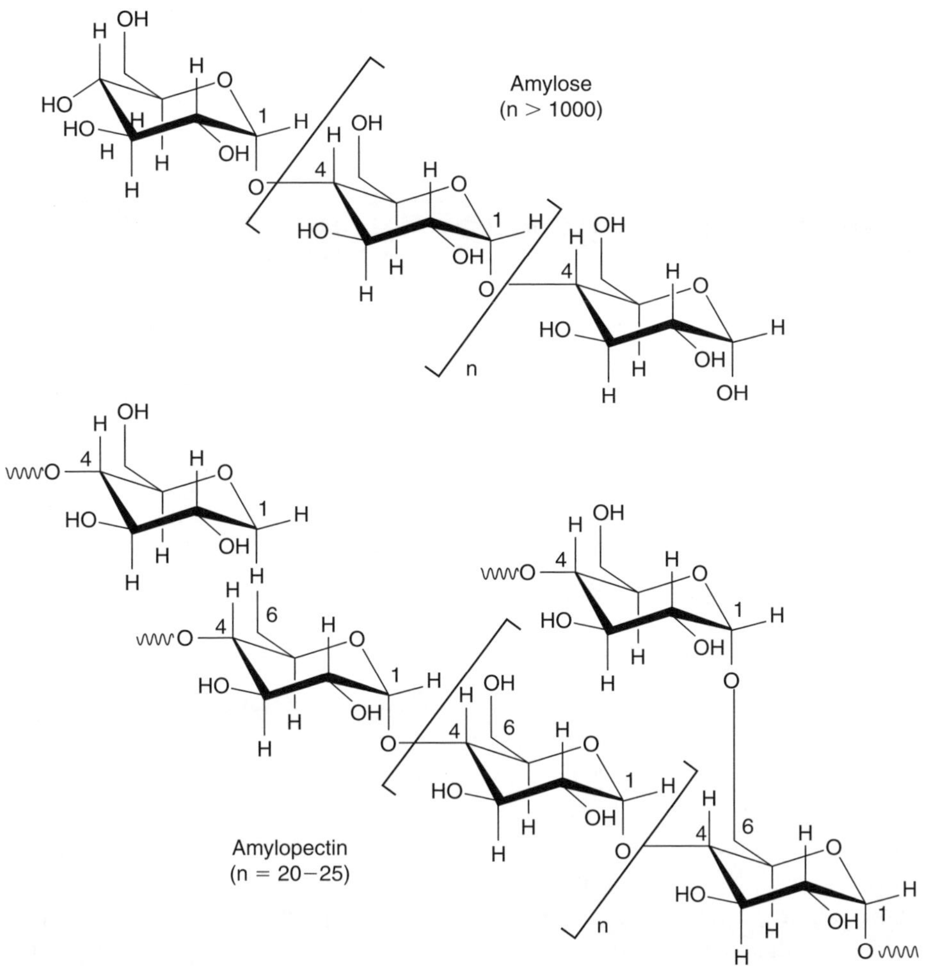

Starch
10–20% amylose (1–4-α-D-glucopyranoside) polymer
and
80–90% amylopectin (1–4, 1–6'-α-D-glucopyranoside)
>100 branched, interconnecting chains of 20–25 glucose units

Figure 16.7 Structure of the two molecules that constitute starch, amylose, and amylopectin. The amylose content of starch can vary between 10 and 20% and the amylopectin content from 80 to 90%. The different ratios of amylose/amylopectin found in starch isolated from different sources, determine its properties.

techniques such as extrusion or injection molding, it is necessary to disrupt the granule and melt the partially crystalline nature of starch in the granule [98,101,103]. For granular starch, the glass transition temperature (T_g) is above the T_d of the polymer chains due to the strong interactions by hydrogen bonding of the chains. Several authors have estimated the T_g of dry starch to be in the region of 230−250 °C [98]. Therefore, plasticizers have to be added to lower the T_g beneath the T_d. Very important factors that will determine the final properties of TPS products are, among others, the type and amount of used plasticizers, the amylose/amylopectin ratio, the molecular weight of the starch (both mainly depend on the plant of origin), and the final crystallinity of the products [96,98,101]. Important plasticizers are water and several polyols such as glycerol and glycol [96,98].

Nevertheless, the application of unblended TPS is limited because of the thermal sensitivity and degradation of starch due to water loss at elevated temperatures [95,96]. Generally, for temperatures exceeding 180−190 °C, rapid degradation occurs during processing of TPS. The behavior of TPS is glassy and materials can only be processed by the addition of water, other plasticizers, or melt flow accelerators.

To overcome difficulties associated with the limited applicability of unblended TPS, while the starch is being destructurized in the extruder, it is possible to add, together with the plasticizers and other additives, other polymers in order to create biodegradable blends that will confer a more thermoplastic nature to the TPS. Other aimed properties are a better resistance to thermomechanical

Table 16.3 Examples of the use of Starch Based Polymers in Tissue Engineering Research

Polymer	Scaffold	Processing Methodology	Cells/Animal Model	TE Application	References
Starch/ polycaprolactone	2D porous scaffold	Selective laser sintering	NIH-3T3 mouse fibroblasts	Not specified	[43]
Starch/dextran/ gelatine	3D porous cylindrical	Rapid prototyping technologies (3D printing)	Not shown	Not specified	[297]
Starch/ polycaprolactone	Fiber mesh	Fiber bonding	Rat marrow stromal cells	Bone, cartilage	[134,271,298]
			Micro- and macrovascular endothelial cells	Bone	[299]
	Nano-micro fiber mesh	Fiber bonding 1 electrospinning	Rat marrow stromal cells	Bone	[271]
Starch/ethylene-vinyl alcohol	3D porous	Extrusion/injection molding with blowing agents	Human osteoblast-like cells (SaOS-2)	Bone	[109,111, 132,133,300]

degradation, meaning that the blends are more readably processable, have a less brittle nature, and enhanced resistance to water and ageing, as compared to fully starch thermoplastics. This has led to the development of a large range of starch-based thermoplastic blends for several different applications, including in the biomedical field.

Reis *et al.* [103,101,108,109,110,176,179,186, 200,201,233,234,235,281] have worked extensively using blends of corn starch (in amounts varying from 30 to 50%wt) with several different synthetic polymers such as poly(ethylene vinyl alcohol) (SEVA-C), cellulose acetate (SCA), poly(ϵ-caprolactone) (SPCL), and polylactic acid (SPLA) [104–106]. These polymers can be designed into distinct structural forms and/or properties by tailoring the synthetic component of the starch-based blend, their processing methods, and the incorporation of additives and reinforcement materials. These polymeric blends are degraded by hydrolytic processes and several enzymes [107,108] can also be involved in the process, mainly α-amylase, β-amylase, α-glucosidase, and other debranching enzymes [108]. The biocompatibility and nonimmunogenicity of starch-based polymers have been well demonstrated by several *in vitro* [109–111] and *in vivo* studies [110,112]. For all these reasons, starch-based polymers have been suggested for a wide range of biomedical applications, such as partially degradable bone cements [113–117], as systems for controlled release of drugs [109,117–119], as bone substitutes in the orthopedic field [120–124], and as scaffolds for tissue engineering [111,125–132]. A wide range of starch-based scaffolds have been developed exhibiting different properties and porous architectures, using several different processing methodologies, from conventional melt based technologies, such as extrusion and injection molding using blowing agents [130,133] to innovative techniques, such as microwave baking [130]. Some of these scaffolds have been successfully used in bone tissue engineering studies using human osteoblasts [111,132] and rat bone marrow stromal cells [128,134,135]. These and some other examples of tissue engineering studies performed using starch-based polymers as scaffold materials are summarized in Table 16.3.

16.3.5 Hyaluronan

Hyaluronan, formerly known as hyaluronic acid (HA), is a natural and highly hydrophilic polysaccharide, which has been found to be a key constituent of native ECM and tissues. HA belongs to the family of GAGs and is synthesized as a large,

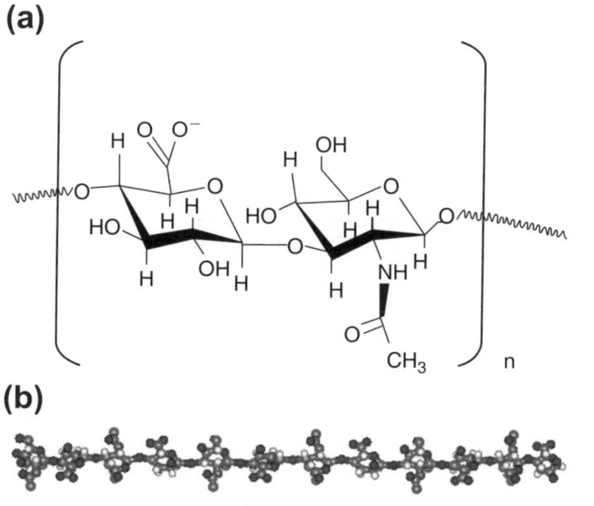

Figure 16.8 Structure of hyaluronan, which is composed of a repeating disaccharide of (1→3) and (1→4)-linked β-D-glucuronic acid and *N*-acetyl-β-D-glucosamine units.

negatively charged and linear polysaccharide of varying chain length (2225 μm) composed of repeating disaccharide units (Fig. 16.8) [136].

It is believed that interactions between HA with other ECM macromolecules and chondrocytes on the one hand, and its hydrodynamic characteristics, especially its viscosity and ability to retain water on the other, are critical for the maintenance of both cartilage homeostasis and biomechanical integrity. The characteristics of HA are, to a great extent, responsible for the regulation of the porosity and malleability of these matrices. Moreover, in the human body, HA is cleaved by enzymes called hyaluronidases [137], showing that the cells of the host or the ones present in the engineered tissue, may regulate the local clearance of the material, while the new tissue is being formed. Moreover, it has been demonstrated that HA is biocompatible [138] and has a greater bacteriostatic effect when compared with other matrices such as collagen type I, poly(lactide-*co*-glycolide) (PLGA), and hydroxyapatite [139].

Ehlers *et al.* [140,141] reported a double action of HA on chondrocytes, when added to the culture media. Results showed that chondrocytes presented a great tendency to differentiate and a higher rate of proliferation. These findings are corroborated by the works of other authors, which demonstrated that HA also stimulates bone marrow stromal cells proliferation and differentiation. These results are quite interesting since it is known that usually a differentiation-inducing stimuli leads to a lower cell proliferation. Therefore, HA possesses some of the features required when choosing a material suitable for tissue engineering scaffolding, although little is known about the mechanical properties of the HA molecules. The pioneering work of Fujii *et al.* [142] has demonstrated that the persistent length of single hyaluronan molecule is about 4.5 nm. This is important data for designing tissue engineering scaffolds, where the mechanical component is essential.

It was also found that HA interacts with cell surfaces in two ways, by binding to specific cell-surface receptors such as the hyaluronan receptor CD44 [143] and receptor for hyaluronan-mediated motility (RHAMM), and sustained transmembrane interactions with its synthetases [136]. The binding of chondrocytes to HA through the CD44 receptor greatly affects the functioning of these cells, thus cartilage homeostasis. In fact, the blocking of the CD44 receptors of chondrocytes results in the degradation of cartilage matrix. However, the physiological role of HA is not restricted to its participation in the synovial fluid of joints, umbilical cord, and vitreous body of the eye [137], but also has been described to be involved on the co-regulation of cell behavior during embryonic lung development, angiogenesis, wound healing processes, and inflammation [136]. In the last few years, HA and its derivatives have been showing interesting results when used in medicine, namely in the treatment of several soft and hard tissue defects such as skin [144,145], blood vessels [146,147], eye [148], ear [149], and bone [150] tissue. HA could certainly find other applications but the water solubility and rapid resorption preclude many clinical applications. To circumvent some of HA's limitations, several authors have been proposing the modification of the HA molecular structure, in an attempt to obtain more stable HA-based materials. Covalent cross-linking [151–153], partial or total esterification of its free carboxylic groups [149], and annealing [154] are allways to obtain a modified and stable form of HA. Most of hyaluronan-based polymers that can be obtained by cross-linking are water-insoluble gels or hydrogels (hylans), and much of which has still to be explored. In fact, HA and its derivatives offer a wide range of features that allow its prevalent use in tissue engineering as scaffolds since it can be used in the form of gels [155,156], sponges [157,158], films [151,153], fibers [159–161], and microparticles [162]. These materials met some of the

Table 16.4 Applications of Hyaluronan and Its Derivatives in Tissue Engineering

Polymer	TE Application	References
Hyaluronan/fibrin glue	Articular cartilage	[155]
Hyaluronan/alginate	Articular cartilage	[30,107,301]
Hyaluronan/chitosan	Articular cartilage and skin	[79,159,160,166]
Hyaluronan/collagen cross-linked via a poly(ethyleneglycol) diepoxide	Not specified	[164]
Hyaluronan/collagen cross-linked via pyridinoline	Articular cartilage	[302]
Hyaluronan modified with methacylic anhydride	Articular cartilage	[303]
Hyaluronan/elastin	Articular cartilage	[153,304]
Fibronectin-coated ACP™ (hyaluronan-based sponge)	Osteochondral	[158]
Hyaluronan/calcium phosphates	Bone and osteochondral	[150]
Hyaluronan/PLGA	Osteochondral and articular cartilage	[158,167]
Hyaluronan/laminin	Brain	[165]
Hyaff® (hyaluronan derivative obtained by esterifying the free carboxylic group)	Skin, cartilage, trachea, and other soft tissues	[149,157,161,305–309]
Laserskin® (hyaluronan 100% esterified with benzyl alcohol)	Skin	[144,145]
Disulfide-cross-linked hyaluronan	Soft tissues	[151]
Hyaluronan-graft-poloxamer	Eye	[148]
Hylans (hydrogels based on cross-linked hyaluronan)	Vascular and aortic heart valves	[146,147,310]
Nonwoven Hyaff (esterified Hyaluronan)	Vascular	[305,311]

criteria for their successful application not only in tissue engineering scaffolding [157], but also in drug delivery applications [148,162]. Despite this, a great deal of attention has been given to the development of alternative high-quality scaffolds. In this context, several authors have been proposing the blending with other polymers to utilize the benefits of each biomaterial. Hyaluronan has been combined with fibrin glue (useful cell delivery matrix) [155], alginate [163], collagen [164], gelatin [153], laminin [165], chitosan [79,166], polyesters [167], and calcium phosphates [150] to develop composite scaffolds for regeneration of several damaged tissues (Table 16.4).

It has been known for some time that HA also plays a key role in interactions with tumor cells [168]. In fact, there is an association of high levels of HA with malignancy of tumors [169]. These observations highlight the biological role of HA, demonstrating that this molecule can be a viable therapeutic target, which might be useful for developing more effective therapeutic strategies in the coming years.

16.4 Proteins

Proteins are the most abundant organic molecules within the cell extracellular and intracellular medium, where they ensure multiple biological functions, such as transport, regulation of pathways, protection against foreign molecules, structural support, protein storage, as well as being the catalyst for a great diversity of reactions, acting as biocatalysts (enzymes).

In a molecular perspective, proteins may be considered as polymer structures composed of 20 distinct amino acids linked by amide (or peptide) bonds. Amino acids are, therefore, the building blocks of polypeptides and proteins, which consist of a central carbon linked to an amine group, a carboxyl group, a hydrogen atom, and a side chain (R groups). R groups can be classified as nonpolar groups, uncharged polar groups or charged polar groups, in which their distribution along the protein backbone renders proteins with distinct characteristics.

The structure of a protein is not, however, as simple as a polysaccharide, or other polymer. Generally, the protein structure is described on four levels. The *primary structure* of a protein is its amino acid sequence, whereas the *secondary structure* refers to the local spatial arrangement of the polypeptide's backbone atoms without regard to the conformation of its side chains. The folding of the polypeptide chain is responsible for putting in close contact different parts of the chain to create binding sites to the substrate, etc. The *tertiary structure* is related to the 3D structure of the entire polypeptide. When proteins are composed of more than one polypeptide chain (referred as subunits), the resultant spatial arrangement of its subunits is known as the protein's *quaternary structure*.

The configuration assumed by a protein, and thus the one that determines its properties, is the one that minimizes the molecule's free energy. Protein conformation is determinant for protein bioactivity, being known that a certain 3D structure is essential for protein functionality. Most of the forces that stabilize the protein structure are weak (hydrogen bonding, ionic and hydrophobic interactions, van der Waals forces), giving some flexibility to the macromolecule. In general, nonpolar amino acid side chains (e.g., phenylalanine, leucine, tryptophan, valine, etc.) are located in the interior of the protein away from the aqueous solvent. The hydrophobic effects that promote this distribution are largely responsible for the 3D structure of native proteins. On the contrary, ionized side chains tend to be on the surface of the molecule to interact with the aqueous solvent. In addition, the polypeptide chains of larger proteins tend to exist in structural domains independently folded and connected by segments of peptide chains.

Taking into account the low stabilities of protein conformations, these molecules are easily susceptible to denaturation by changing the balance of the weak interactions that maintain the native conformation. Proteins can be denaturated by a variety of conditions and substances as heating, extreme pH, chaotropic agents, detergents, adsorption to certain surfaces, etc. [170].

We now focus on the most important proteins that have been studied for tissue engineering applications, such as collagen, elastin, soybean, and silk fibroin.

16.4.1 Collagen

Collagen is the most abundant protein in mammalian tissues (cornea, blood vessels, skin, cartilage, bone, tendon, and ligament) and is the main component of the ECM [171,172]. Its main function is to maintain the structural integrity of vertebrates and many other organisms. However, collagens also exert important functions in the cell microenvironment and are involved in the storage and release of cell mediators, like growth factors [171]. More than 20 genetically distinct collagens have been identified [171−174], but the basic structure of all collagens is composed of three polypeptide chains, which wrap around one another to form three-stranded rope structure (triple helix, Fig. 16.9b). Close-packing of the chains near the central axis imposes the requirement that glycine (Gly) occupies every third position, generating a (X-Y-Gly)*n* repeating sequence. Proline (Prol) and 4-hydroxyproline (Hyp), which in collagens constitute about 20% of all residues, are found almost exclusively in the X and Y positions, respectively. Therefore, the most common triplet in collagen is Prol-Hyp-Gly, which accounts for about 10% of the total sequence [175] (Fig. 16.9a). Peptides that contain Gly as every third residue and have large amounts of Prol and Hyp behave as triple helices in solution.

The individual triple helices are arranged to form fibrils which are of high tensile strength and can be further assembled and cross-linked (collagen fibrils are stabilized in the ECM by the enzyme lysyl oxidase).

In tissues that have to resist shear, tensile, or pressure forces, such as tendons, bone, cartilage, and skin, collagen is arranged in fibrils, with a characteristic 67 nm axial periodicity, which provides the tensile strength [176]. Only collagen types I, II, III, V, and XI self-assemble into fibrils [171]. The fibrils are composed of collagen molecules, which consist of a triple helix of approximately 300 nm in length and 1.5 nm in diameter [176]. Collagen fibril formation is an extracellular process, which occurs through the cleavage of terminal procollagen peptides by specific

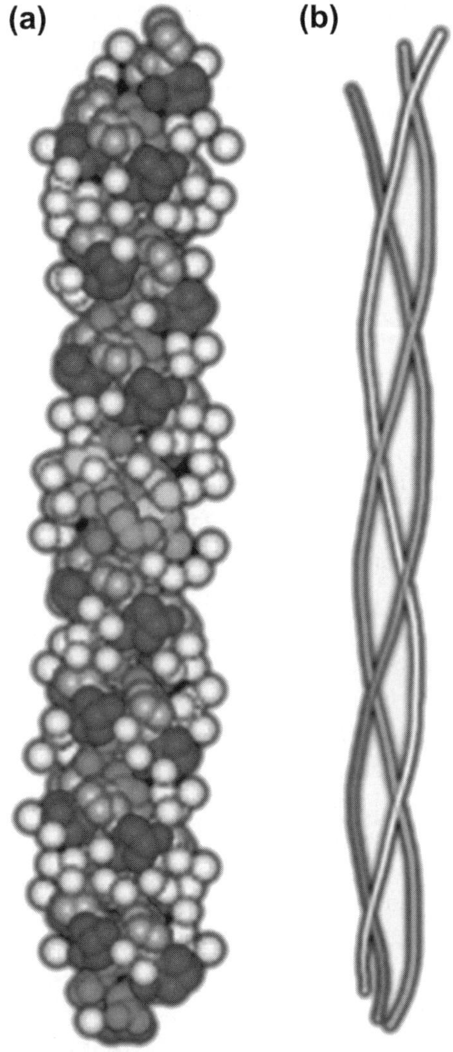

Figure 16.9 (a) CPK model of the structure of the triple-helical collagen-like peptide (1QSU, retrieved from Protein Data Bank at http://www.rcsb.org) showing Gly residues in gray, Prol resides in gray and Hyp (magenta). All residues are exposed to the solvent (water molecules are displayed in white); (b) Schematic representation of the collagen-like peptide showing the triple helix. The structures were generated using the WebLab ViewerLite 3.7 program (Molecular Simulations Inc, USA).

procollagen metalloproteinases. An article by Stevens and George [177] provides an interesting schematic diagram showing the natural assembly of collagen fibers. Further reading about collagen fibril formation and molecular packing in collagen fibrils may be found in previous publications [178–180].

During the 1970s and 1980s, academics and commercial researchers began to use collagen as a biomaterial in a variety of connective tissue applications because of its excellent biocompatibility [173,174], low antigenicity [173,174], high biodegradability [173,181,182], and good hemostatic and cell-binding properties [173,183].

The primary sources of industrial collagens are from animal tissues (porcine and calf skin, bovine tendon, rat tail, etc.). It may readily be purified from animal tissues with enzyme treatment and salt/acid extraction. However, the use of animal-derived collagen raises concerns over the possible transmission of infectious agents such as viruses and prions [172]. Transmissible bovine spongiform encephalopathy (BSE) is one of the most difficult contaminating agents to detect and remove from animal tissues. Therefore, different attempts have been made to find new and safer sources of collagen, namely from marine sources (e.g., jellyfish collagen) [184] or by producing recombinant human collagen (rhC) for clinical use [172] using different expression systems. The use of recombinant sources of human collagen provides a reliable, predictable, and chemically defined source of purified human collagens that is free of animal components (please see for instances the review of Yang *et al.* [172]; for more details about the application of rhC in tissue engineering). The triple-helical collagens made by recombinant technology have the same amino acid sequence as human tissue-derived collagen. In addition, collagen products can be purified from fibers, from molecules reconstituted as fibers, or from specific recombinant polypeptides with specific composition and conformation.

A feature common to all of these collagen materials is the need for stable chemical cross-linking to control the mechanical properties and the residence time in the body, and to some extent their potential immunogenicity. This can be achieved via chemical (glutaraldehyde, formaldehyde, carbodiimides, diphenylphosphoryl azide), physical (UV radiation, freeze-drying, heating, thermal dehydration), and enzymatic cross-linking. These cross-linking agents react with specific amino acid residues on the collagen molecule imparting individual biochemical, thermal, and mechanical stability to the biomaterial. Collagen may be an ideal scaffold material, as it is the major component of the ECM, and because it can be processed into a wide variety of structures and shapes (sponges, fibers, films, 3D gels, fleeces; Table 16.5). Furthermore, collagen substrates can modify the morphology, migration, and in certain cases the

Table 16.5 Examples of Application of Collagen Scaffolds in Tissue Engineering Research

Type of Scaffold	Processing Methodology	TE Application	References
Collagen sponge with 11 mm in diameter and 2 mm in thickness and pore volume fraction of 97.5%	Freeze-drying. Cross-linking by thermal dehydration	Tooth tissue engineering Guided tissue regeneration (GTR) in dentistry	[312]
Collagen membranes	Conversion of rhCI monomers into oligomers and reconstitution into collagen fibrils. The resulting fibrillar networks were subsequently cross-linked with ethyl-3-(3-dimethylaminopropyl) carbodiimide (EDC)		[172]
Scaffolds with predefined and reproducible internal channels with widths of 135 μm	Rapid prototyping (solid freeform fabrication technology and critical point drying technique)	Cardiovascular (aortic valve, blood vessel) tissue engineering	[313,314]
Scaffolds of 6 mm in diameter and 0.75 mm in thickness	Freeze-drying, cross-linked with hexamethylene diisocyanate		[182]
Flat sheets of collagen type I	Solvent evaporation		[183]
Porous tubular scaffolds with an inner diameter of 3 mm, an outer diameter of 6 mm and a length of 4 cm	Freeze-drying of a suspension of type I insoluble collagen and insoluble elastin. Cross-linked with a carbodiimide		[315,316]
Collagen-gel tubular constructs	Polymerization into glass test tubes		[317]
Type I collagen sponge with interconnected pores	Discs cored from sheets of Ultrafoam® collagen hemostat (Davol Inc., Cranston RI)	Bone tissue engineering Bone graft substitutes	[318]
Recombinant collagen sponges (porous micromatrice structures interconnected by homogenous thin sheets of recombinant human collagen I fibrils)	In-mold fibrillogenesis/cross-linking process followed by lyophilization. Cross-linking with EDC		[172]
Collagen gel (Atelocollagen gel from Koken Co., Tokyo, Japan). Dome shape of 0.8 cm diameter and 0.2 cm top height	Gelation at 37 °C for 60 min	Cartilage tissue engineering	[319]
Matriderm®. 3D structure made of purified collagen I of bovine epidermis and small amounts of elastine	Freeze drying. Cross-linking with a carbodiimide		[320]
Collagen-based wound dressings (membranes, fibers, sponges)	Several	Dermal tissue engineering (artificial skin, skin substitutes)	[180]

(*Continued*)

Table 16.5 Examples of Application of Collagen Scaffolds in Tissue Engineering Research—Cont'd

Type of Scaffold	Processing Methodology	TE Application	References
Type I collagen contracted gels (discs)	Gelation at 37 °C for 60 min		[321]
Fibrous scaffolds	Electrospinning. Cross-linking with 1,6-diisocyanatohexane	TE applications in general	[187]

differentiation of cells due to the presence of cell adhesion sequences present on its structure (e.g., RGD). Collagen is naturally degraded by matrix metalloproteinases, specifically collagenase, and serine proteases [180]. These enzymes are secreted by neutrophils during the foreign body reaction, allowing the collagen degradation to be controlled by the cells present at the implantation site [182]. However, its low thermal stability, due to its protein nature, does not allow collagen to be processed by melt-based techniques, limiting its processing to solvent-based methods and consequently the final properties of the scaffolds, normally characterized by poor mechanical strength. Another drawback related with the use of collagen materials is the requirement of additional chemical or physical cross-linking to confer mechanical strength and enzymatic resistance. Intermolecular cross-linking reduces the degradation rate by making collagen less susceptible to enzymatic attack. Collagen has long been known to elicit minimal inflammatory and antigenic responses and has been approved by the United States Food and Drug Administration (FDA) for many types of medical applications, including wound dressings and artificial skin [180,185]. These properties of collagen emphasize its significance in tissue regeneration and its value as a scaffold material, being currently used in a great number of tissue engineering applications. Table 16.5 presents some examples of TE applications using collagen scaffolds. In addition, composites of collagens with GAGs, as well as with synthetic biodegradable polymers and ceramics, have also been extensively studied for their potential application as scaffolds for tissue engineering. A vast number of publications can be found in the literature, covering a diversity of clinical applications, such as general surgery, orthopedics, cardiovascular, dermatology, otorhinolaryngology, urology, dentistry, ophthalmology, and plastic and reconstructive surgery [180].

Although these examples offer encourage applications of collagen in tissue engineering, its low mechanical properties, the risk of viral infection, its antigenicity potential, and fast biodegradation when implanted in the human body are, to some extent, limiting the clinical applications of this natural biomaterial.

16.4.2 Elastin

Elastin is another key structural protein found in the ECMs of connective tissues (e.g., blood vessels, esophagus, skin) that need to stretch and retract following mechanical loading and release [186,187]. It is found predominantly in the walls of arteries, lungs, intestines, and skin, as well as other elastic tissues. However, unlike type I collagen, elastin has found little use as a biomaterial, for two main reasons [188,189]: (i) elastin preparations have a strong tendency to calcify upon implantation, probably because of the microfibrillar components (mainly fibrillin) within the elastic fiber that are difficult to remove and (ii) the purification of elastin is complex [188]. The insoluble nature of elastin has also limited in its use in traditional reconstituted matrix fabrication techniques [190] and when applied only poorly defined elastin preparations have been used [189].

Elastin consists of several repetitive amino acid sequences, including VPGVG, APGVGV, VPGF GVGAG, and VPGG [191]. Highly insoluble and extensively cross-linked, mature elastin is formed from tropoelastin, its soluble precursor [192]. Tropoelastin is secreted from elastogenic cells as a 60-kDA monomer that is subjected to oxidation by lysyl oxidase. Subsequent protein–protein associations give rise to massive macroarrays of elastin [187]. The structure of tropoelastin consists of an alteration of hydrophobic regions, responsible for elasticity, and cross-linking domains. Additionally, it ends with a hydrophilic carboxy-terminal sequence containing its only two cysteine residues [192]. As a consequence, elastin is a substantially insoluble protein network [187,192]. Soluble material is typically derived either as a fragmented elastin in the form of alpha- and kappa-elastin or preferably through expression of the natural monomer tropoelastin [187]. In the production of α-elastin, bovine ligament elastin is treated [193] with a mild acid hydrolysis to yield a high-molecular-weight digest that retains the amino acid composition of native elastin. Despite structural heterogeneities resulting from the hydrolysis, α-elastin retains several key physicochemical properties of the nascent elastin. Nevertheless, the development of α-elastin based biomaterials is still a quite unexplored area (Fig. 16.10).

Recombinant protein technologies have allowed the synthesis of well-defined elastin-derived polypeptides, which have driven insightful structure-function studies of tropoelastin, as well as several discrete elastin domains. Elastin-like polypeptides (ELPs) are artificial polypeptides with unique properties that make them attractive as biomaterial for tissue engineering, as it has been demonstrated by the work of Urry *et al.* [63,194,195]. ELPs consist of oligomeric repeats of the pentapeptide sequence Val-Pro-Gly-Xaa-Gly (Xaa is any amino acid except proline), a naturally occurring sequence in the protein elastin. ELPs are soluble in aqueous solution below their transition temperature (T_t) but when the solution temperature is raised above their T_t, the polymers start a complex self-assembly process that leads to an aggregation of the polymer chains, initially forming nano- and microparticles, which segregate from the solution [77,196]. This "smart" nature may not be of particular interest for the final application of ELPs as ECM, but it is extremely important to simplify several steps in the production of ELPs and preparation of the ECM [197]. ELPs have also demonstrated an outstanding biocompatible behavior. Apparently, the immune response system of the human body does not differentiate the ELPs from endogenous elastin. Moreover, because of their protein nature, their bioabsorption is carried out by conventional metabolic routes, yielding just natural amino acids [197]. In addition, the matrices resulting from cross-linking of ELPs show a mechanical response quite similar to the natural elastin [198]. This characteristic is very important for their application in tissue engineering, as the scaffold (artificial ECM) has to properly transmit the forces from the surrounding environment to the attached cells so that they can build new tissue that can eventually replace the artificial ECM [197].

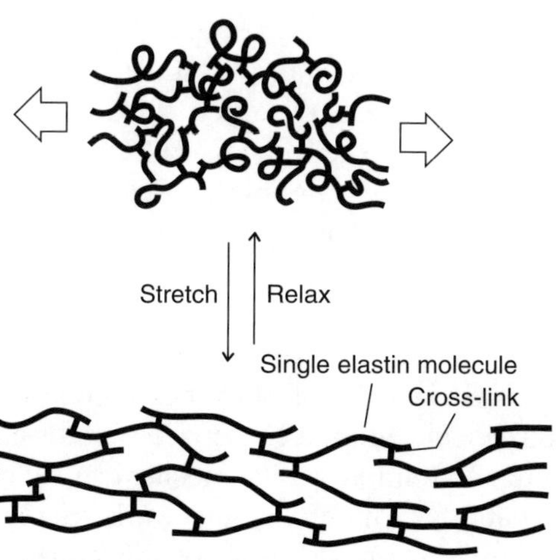

Figure 16.10 Structure of elastin.

However, the broad application of these materials is limited by the inherent challenges of synthesizing recombinant proteins (e.g., residual endotoxin, capital cost and expertise, scale-up).

Table 16.6 gives several examples of tissue engineering studies in which elastin-based scaffolds were used.

16.4.3 Soybean

Soybeans belong to the legume family and can be processed into three kinds of protein-rich products: soy flour [199,200], soy concentrate [200,201], and soy isolate [200,203]. Soy protein, the major component of the soybean (30−45%) is readily available from renewable resources, is economically competitive, and presents good water resistance as well as storage stability [204]. About 90−95% of the soy is storage protein, with two subunits, namely 35% conglycinin (7S) and 52% glycinin (11S) [205]. Due to its low cost and surface active properties, soy protein is of great importance to the food industry, especially as it provides stability against phase separation in food systems [206]. Nevertheless, the combination of its properties with a similarity to tissue constituents and a reduced susceptibility to thermal degradation makes soy an ideal template for use in biodegradable polymer for biomedical applications [207]. Membranes, microparticles, and thermoplastics-based soy materials have been developed for tissue regeneration [208]. Biodegradable soy plastics have been developed by melt-based methods

Table 16.6 Examples of Application of Elastin Based Scaffolds in Tissue Engineering Research

Polymer	Scaffold	Processing Methodology	Cells/Animal Model	TE Application	References
α-Elastin	Films	Cross-linking	Bovine aortic smooth muscle cells	Vascular tissue	[193]
Elastin and tropoelastin	Fibers	Electrospinning followed by cross-linking	HEPM	Not specified	[187]
Aortic elastin	3D porous structure	Cyanogen bromide treatment for decellularization and removal of collagen and other ECM components	3T3 mouse fibroblast cell line (ATCC)	Not specified	[322]
Collagen/ elastin/PLGA	Electrospun fiber meshes	Electrospinning	Bovine endothelial and smooth cells	Vascular tissue	[323]
Elastin/ collagen	3D structure composed of thin sheets and fibrils (collagen) and thick fibers (elastin)	Lyophilization followed by cross-linking	Sprague-Dawley rats (subcutaneous pockets	Not specified	[188]
Aortic elastin	3D porous structure	Cyanogen bromide treatment (CNBr)	Sprague-Dawley rats (subdermal implantation)	Vascular tissue	[73]
Collagen and elastin (1:1)	Tubular porous structures	Freeze-drying followed by cross-linking	Human smooth muscle cells	Vascular tissue	[315,316,324]
Elastin-like polypeptides (ELPs)	Injectable scaffolds	Gene design and synthesis	Pig chondrocytes	Cartilage	[77,196]

PLGA, poly(D,L-lactide-co-glycolide); HEPM, human embryonic palatal mesenchyme

such as extrusion and injection molding [207]. Mano *et al.* [207] reported that soy protein-based thermoplastics presented a suitable range of mechanical and dynamical properties that might allow their use as biomaterials, namely in controlled release applications.

Soy protein has many reactive groups, such as $2NH_2$, 2OH, and 2SH, that are susceptible to chemical and physical modifications [209]. Some studies reported that the combination of soy protein with other proteins (e.g., wheat gluten [210], casein [207], and polysaccharides) such as cellulose [97], dialdehyde starch [211], and chitosan [212,213] in film form may promote physical and chemical interactions, which improve some properties. Silva *et al.* [214] have reported that, by means of combining a sol–gel process with the freeze-drying technique, it was possible to develop cross-linked porous structures based on chitosan and soy protein. It was demonstrated that the developed porous structures possess a suitable porosity and adequate interconnectivity. Furthermore, tetraethylorthosilicate (TEOS) can be used to introduce specific interactions in the interfaces between chitosan and soy protein, and improve its mechanical stability and degradability. Therefore, this work has shown that these structures have great potential for tissue engineering of cartilage.

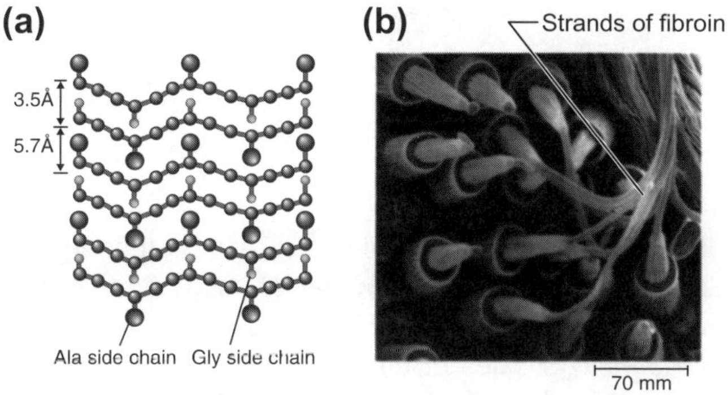

Figure 16.11 Structure of silk. The fibers used to make silk cloth or spiderweb are made up of the protein fibroin. (a) Fibroin consists of layers of antiparallel β sheets rich in Ala and Gly residues. The small side chains interdigitate and allow close packing of each layered sheet, as shown in this side view. (b) Strands of fibroin emerge from the spinnerets of a spider in this colorized electron micrograph. *Source: Nelson, D.L. and Cox, M.M. (2003). Lehninger Principles of Biochemistry, 3rd edn, Worth Publishers, New York, NY, p. 174.*

16.4.4 Silk Fibroin

Silk fibroin is a highly insoluble fibrous protein produced by domestic silk worms (*Bombyx mori*) containing up to 90% of the amino acids glycine, alanine, and serine leading to antiparallel β-pleated sheet formation in the fibers [215]. Fibroin is a structural protein of silk fibers and sericins are the water-soluble glue-like proteins that bind the fibroin fibers together [216] (Fig. 16.11). High purity silk fibroin fiber can be obtained easily from degummed silk (boiling-off), which refers to partial or complete removal of the sericin. Removal of the sericin coating before use removes the thrombogenic and inflammatory response of silk fibroin [217]. *Bombyx mori* silk fibroin can be dissolved with neutral salt solutions such as lithium bromide (LiBr), lithium thiocyanate (LiSCN), hexafluoroisopropyl alcohol (HFIP), and calcium nitrate-methanol [$Ca(NO_3)_2$-MeOH] [218]. Their mixtures are dialyzed to get pure fibroin solution, which can be used to prepare silk fibroin membranes, fiber, hydrogel, scaffolds, and others types of matcrials [216]. Traditionally, silk fibroin has been used for decades as suture material [216]. Nowadays, several studies demonstrate the utility of silk matrices in films [216,217,219], nanofibers [215,220,221], hydrogels [222], and porous matrices [223,224] for biomaterials and tissue engineering with stem cells for cartilage and bone applications. These applications of silk fibroin are related to its permeability to oxygen and water, cell adhesion and growth characteristics, slow degradability, low inflammatory response, and high-tensile strength with flexibility [216]. Porous 3D scaffolds with silk fibroin have been obtained using various processing techniques (Table 16.7); these include salt leaching [223–225], electrospinning [220,221,226], freeze-drying [224,227–230], and gas-foaming [224]. Li *et al.* [228–230] reported a series of studies on preparation conditions of porous silk fibroin materials and its relationship between the structure and properties. These materials were prepared by means of freeze-drying. A new process to form a silk fibroin spongy porous 3D structure with both good porous structures and mechanical properties has also been reported [231]. This process involves freezing and a thawing fibroin aqueous solution in the presence of a small amount of water-miscible organic solvent. It requires no freeze-drying, no cross-linking chemicals, or the aid of other materials. In general, the silk scaffolds produced by different methods

Table 16.7 Examples of Application of Soy and Silk-based Materials in Tissue Engineering Studies

Polymer	Scaffold	Processing Methodology	TE Application	References
Soy protein/ chitosan blend	Not specified	Freeze-drying and sol–gel process	Cartilage	[214]
Silk fibroin	Not specified	Freeze-drying	Not specified	[227–230]
Silk fibroin	Not specified	Salt leaching	Cartilage, bone	[225,235,236]
Silk fibroin	Not specified	Gas-foaming	Not specified	[224]
Silk fibroin	Nanofiber	Electrospinning	Bone	[226]

Source: Invited chapter, in: Tissue Engineering, Clemens van Blitterswijk, Anders Lindahl, Peter Thomsen, David Williams, Jeffrey Hubbell, Ranieri Cancedda (Editors), Elsevier.

described here presented good porosity and mechanical properties that can be controlled by silk fibroin concentration, freezing temperature, and particle size of salt used in the process. Other approaches to form silk scaffolds involved the blending of polymers, such as poly(ethylene oxide) [215], chitosan [227], or the surface modification of synthetic polymers such as poly(ε-caprolactone) [232] and polyurethane [233] with silk fibroin coating in order to improve their collective properties, especially processability, mechanical properties, and biocompatibility, respectively.

With respect to using silk fibroin for cell culture, many researchers have investigated the effects of the silk matrices in nanofiber and porous matrix obtained by methodologies described previously on the culture of osteoblasts-like cells [234], human mesenchymal stem cells [221,225,226,235,236] have shown very promising results regarding their application in cartilage and bone tissue engineering. Jin *et al.* [221] concluded that electrospun silk matrices support bone marrow mesenchymal stem cells attachment, spreading, and growth *in vitro*. Meinel *et al.* [235] reported the feasibility of silk-based implants with engineered bone for the (re-)generation of bone tissues. Recently, the potential of electrospun silk fibrous scaffold for bone formation from human bone marrow-derived mesenchymal stem cells (hMSCs) was explored by combining the unique structural features generated by electrospinning with functional factors, such as bone morphogenic protein-2 (BMP-2) and nanohydroxyapatite particles [226].

16.5 Polyhydroxyalkanoates

Polyhydroxyalkanoates (PHAs) are naturally occurring biodegradable polymers. PHAs are synthesized and stored as water-insoluble inclusions in the cytoplasm of several bacteria and used as carbon and energy reserve materials [237,238]. The first PHA to be identified was poly(3-hydroxybutyric acid) (P[HB]). This homopolymer is the most abundant bacteria synthesized polyester and its 3-hydroxybutyrate (HB) monomer was thought to be the unique PHA constituent in bacteria [238,239]. Further research [240] reported heteropolymers in chloroform extracts of activated sewage sludge, like 3-hydroxyvalerate (HV) among others. The introduction of other units in the PHA chain (besides 3HB) has a significant effect on the mechanical

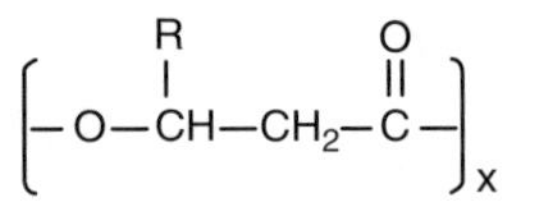

Figure 16.12 Structure of polyhydroxyalkanoates.

behavior of polyester [241,242]. The homopolymer of PHB is a brittle material, while the increase in HV content turns the HB-*co*-HV copolymer more ductile [243–245]. The mechanical behavior of PHAs depends on both the length of the pendant groups and the distance between ester linkages. PHAs with short pendant groups are prone to crystallization but exhibit stiff and brittle behavior, while PHAs with longer pendant groups are ductile [246]. The wide performance range of PHA copolymers justified additional scientific and industrial interest, which led to the discovery of further bacterial PHAs. PHAs can be synthesized in molecular weights that depend on the growth conditions and on the microorganism species—between, 200,000 and 3,000,000 Da [238] (Fig. 16.12).

The wide range of mechanical properties [243,245,247–254] coupled with the biodegradable [248,253–257] and the biocompatible behaviors [256,258–261] of PHAs makes them potential biomedical candidates including drug delivery and tissue engineering applications [246,257]. The biocompatibility assessment of PHAs has indicated that cell response also depends on the type of polyester. In a research study, the viable cell number of mouse fibroblasts (cell line L929) on polyhydroxybutyrate (PHB) films have increased more than two orders of magnitude upon blending with poly(hydroxybutyrate-*co*-hydroxyhexanoate) (PHBHH) [262]. The influence of PHB content on mechanical behavior is also evident from the strong ductility increase which occurs with the introduction of PHBHH in PHBHH/PHB blends [263]. Several studies [264–270] reported the investigation of PHAs as potential scaffold materials in diverse range of tissue engineering applications. Sodian *et al.* developed a trileaflet heart valve from a porous PHA scaffold produced by salt leaching. Constructs were produced using vascular cells harvested from an ovine carotid artery and placed into a pulsatile flow bioreactor [268]. Results indicated that cells were mostly viable and grew into scaffolds pores. The formation of connective tissue between the inside and the outside of the porous heart valve scaffold was also observed. Other studies have focused the assessment of PHA scaffolds for bone and cartilage tissue engineering. A study by Rivard *et al.* [265] investigated the proliferation of ovine chondrocytes and osteoblasts in poly(β-hydroxybutyrate-β-hydroxyvalerate) scaffolds. Another study assessed the performance of porous PHBHH/PHB scaffolds, produced by salt leaching method, as matrices for 3D growth of chondrocytes. Cell densities were higher for PHBHH/PHB scaffolds as compared to PHB scaffolds alone. The authors explained this discrepancy based on eventual differences in crystalline and amorphous arrangements between PHB and PHBHH/PHB scaffolds, as the presence of PHB crystalline domains may reduce oxygen permeability [264]. Another study [270] has shown that PHB/PHBHH blends with 1:1 ratio have higher surface free energy as compared to PHB alone, which maximizes chondrocytes adhesion. Furthermore, polarity of the PHA on the scaffold seems to play an important role on what concerns cell morphology. In the case of PHBHH/PHB substrates, PHB content affects blend polarity, which has an important effect on cell shape. Polarity increases with decreasing blend crystallinity, which affects chondrocyte shape, by altering it from spherical to flat.

16.6 Future Developments

Several scaffolds based on natural origin polymers have been widely studied for tissue engineering application. Many of them exhibit unique advantageous features concerning intrinsic cellular interaction and degradability. However, these materials do also exhibit some disadvantages that limit their widespread use. Therefore, it is necessary to increase the knowledge about these natural polymers in order to enable the development of new approaches, including methods for production, purification, controlling material properties (molecular weight, mechanical, degradation rate), and for enhancing material biocompatibility (for instance by using nonanimal derived production), in order to design better and more versatile scaffolding materials.

Tissue engineering scaffolding will also benefit from advances in recombinant protein technologies, which have proven to be a very powerful tool for the design and production of complex protein polymers with well-defined molecular weights, monomer compositions, sequences, and stereochemistry. Very little has been explored within this new class of polymers and therefore much remains to be investigated about their versatility and possibilities of obtaining tailored properties for target applications. Accordingly, special interest has emerged for the use of these protein-based polymers for tissue engineering and other biomedical applications.

Further studies are expected to widen the range of natural origin materials (and combination of these with synthetic polymers) and the tailoring of their properties in order to make them even more suitable for applications within tissue engineering.

16.7 Summary

1. A wide range of natural origin polymers have frequently been used and might in future be potentially useful in tissue engineering.
2. Tissue engineering scaffolds comprised of naturally derived macromolecules have potential advantages of biocompatibility, cell-controlled degradability, and intrinsic cellular interaction.
3. However, they may exhibit batch variations and, in many cases, exhibit a narrow and limited range of mechanical properties. In many cases, they can also be difficult to process by conventional methods.
4. In contrast, synthetic polymers can be prepared with precisely controlled structures and functions. However, many synthetic polymers do not degrade as desired in physiological conditions, and the use of toxic chemicals in their synthesis or processing may require extensive purification steps. Many of them are also not suitable for cell adhesion and proliferation.
5. The combination of natural origin polymer with synthetic polymers and the further development in emerging methodologies such as recombinant protein technologies is expected to lead to outstanding developments in improved materials to be used in tissue engineering applications.
6. No one material alone will satisfy all design parameters in all applications within the tissue engineering field, but a wide range of materials can be tailored for discrete applications.

References

[1] K. Kurita, Controlled functionalization of the polysaccharide chitin, Prog. Polym. Sci. 26 (9) (2001) 1921–1971.

[2] S.X. Xie, Q. Liu, et al., Starch modification and applications, in: S.W. Cui (Ed.), Food Carbohydrates: Chemistry, Physical Properties, and Applications, CRC Press Taylor & Francis Group, Boca Raton, 2005, pp. 357–405.
[3] E.P. Broderick, D.M. O'Halloran, et al., Enzymatic stabilization of gelatin-based scaffolds, J. Biomed. Mater. Res. B Appl. Biomater. 72B (1) (2005) 37–42.
[4] T.H. Chen, H.D. Embree, et al., Enzyme-catalyzed gel formation of gelatin and chitosan: potential for *in situ* applications, Biomaterials 24 (17) (2003) 2831–2841.
[5] A. Steinbüchel, S.K. Rhee, Polysaccharides and Polyamides in the Food Industry. Properties, Production and Patents, WILEY-VCH Verlag GmbH & Co, Weinheim, 2005. KGaA.
[6] G. Franz, W. Blaschek, in: P.M. Dey (Ed.), Cellulose. Methods in plant biochemistry. Carbohydrates, vol. 2, Academic Press Limited, London, 1990, pp. 291–322.
[7] W.R. Morrison, J. Karkalas, Starch, in: P.M. Dey (Ed.), Methods in Plant Biochemistry, Carbohydrates, vol. 2, Academic Press Limited, London, 1990, pp. 323–352.
[8] A.M. Stephen, S.C. Churms, et al., Exudate gums, in: P.M. Dey (Ed.), Methods in Plant Biochemistry. Carbohydrates, vol. 2, Academic Press Limited, London, 1990, pp. 483–522.
[9] M. Izydorczyk, S.W. Cui, et al., Polysaccharide gums: structures, functional properties, and applications, in: S.W. Cui (Ed.), Food Carbohydrates: Chemistry, Physical Properties, and Applications, CRC Press Taylor & Francis Group, Boca Raton, FL, 2005, pp. 263–307.
[10] R.P. Lezica, L. Quesada-Allué, Chitin, in: P.M. Dey (Ed.), Methods in plant biochemistry, Carbohydrates, vol. 2, Academic Press Limited, London, 1990, pp. 443–481.
[11] E. Percival, R.H. McDowell, Algal polysaccharides, in: P.M. Dey (Ed.), Methods in Plant Biochemistry. Carbohydrates, vol. 2, Academic Press Limited, London, 1990, pp. 523–547.
[12] M. Naessens, A. Cerdobbel, et al., Leuconostoc dextransucrase and dextran: production, properties and applications, J. Chem. Technol. Biotechnol. 80 (8) (2005) 845–860.
[13] B. Widner, R. Behr, et al., Hyaluronic acid production in *Bacillus subtilis*, Appl. Environ. Microbiol. 71 (7) (2005) 3747–3752.
[14] S. Kobayashi, S. Fujikawa, et al., Enzymatic synthesis of chondroitin and its derivatives catalyzed by hyaluronidase, J. Am. Chem. Soc. 125 (47) (2003) 14357–14369.
[15] I.V. Yannas, J.F. Burke, et al., Wound tissue can utilize a polymeric template to synthesize a functional extension of skin, Science 215 (4529) (1982) 174–176.
[16] N. Dagalakis, J. Flink, et al., Design of an artificial skin. Part III. Control of pore structure, J. Biomed. Mater. Res. 14 (4) (1980) 511–528.
[17] I.V. Yannas, J.F. Burke, Design of an artificial skin. I. Basic design principles, J. Biomed. Mater. Res. 14 (1) (1980) 65–81.
[18] I.V. Yannas, J.F. Burke, et al., Design of an artificial skin. II. Control of chemical composition, J. Biomed. Mater. Res. 14 (2) (1980) 107–132.
[19] J.M. Dang, K.W. Leong, Natural polymers for gene delivery and tissue engineering, Adv. Drug Deliv. Rev. 58 (4) (2006) 487–499.
[20] S.H. Lim, I.C. Liao, et al., Nonviral gene delivery from nonwoven fibrous scaffolds fabricated by interfacial complexation of polyelectrolytes, Mol. Ther. 13 (6) (2006) 1163–1172.
[21] M. Izydorczyk, Understand the chemistry of food carbohydrates, in: S.W. Cui (Ed.), Food Carbohydrates: Chemistry, Physical Properties, and Applications, CRC Press Taylor & Francis Group, Boca Raton, FL, 2005, pp. 1–65.
[22] G.T. Grant, E.R. Morris, et al., Biological interactions between polysaccharides and divalent cations: the egg-box model, FEBS Lett. 32 (1973) 195–198.
[23] M.A. LeRoux, F. Guilak, et al., Compressive and shear properties of alginate gel: effects of sodium ions and alginate concentration, J. Biomed. Mater. Res. 47 (1) (1999) 46–53.
[24] W.R. Gombotz, S.F. Wee, Protein release from alginate matrices, Adv. Drug Deliv. Rev. 31 (3) (1998) 267–285.
[25] D. Lemoine, F. Wauters, et al., Preparation and characterization of alginate microspheres containing a model antigen, Int. J. Pharm. 176 (1) (1998) 9–19.
[26] J.A. Rowley, G. Madlambayan, et al., Alginate hydrogels as synthetic extracellular matrix materials, Biomaterials 20 (1) (1999) 45–53.

[27] C.K. Kuo, P.X. Ma, Ionically cross-linked alginate hydrogels as scaffolds for tissue engineering: part 1. Structure, gelation rate and mechanical properties, Biomaterials 22 (6) (2001) 511–521.
[28] S.R. Van Tomme, M.J. van Steenbergen, et al., Self-gelling hydrogels based on oppositely charged dextran microspheres, Biomaterials 26 (14) (2005) 2129–2135.
[29] G.M. Williams, T.J. Klein, et al., Cell density alters matrix accumulation in two distinct fractions and the mechanical integrity of alginate-chondrocyte constructs, Acta Biomater. 1 (6) (2005) 625–633.
[30] G. Fundueanu, C. Nastruzzi, et al., Physico-chemical characterization of Ca-alginate microparticles produced with different methods, Biomaterials 20 (15) (1999) 1427–1435.
[31] S.G. Levesque, R.M. Lim, et al., Macroporous interconnected dextran scaffolds of controlled porosity for tissue-engineering applications, Biomaterials 26 (35) (2005) 7436–7446.
[32] S. Massia, J. Stark, Immobilized RGD peptides on surfaces-grafted dextran promote biospecific cell attachment, J. Biomed. Mater. Res. 56 (2001) 390–399.
[33] J. Guo, G.W. Jourdian, et al., Culture and growth characteristics of chondrocytes encapsulated in alginate beads, Connect. Tissue Res. 19 (1989) 277–297.
[34] O. Smidsrød, G. Skjak-Bræk, Alginate as immobilization matrix for cells, Trends Biotechnol. 8 (1990) 71–78.
[35] S.C. Chang, J.A. Rowley, et al., Injection molding of chondrocyte/alginate constructs in the shape of facial implants, J. Biomed. Mater. Res. 55 (4) (2001) 503–511.
[36] E. Fragonas, M. Valente, et al., Articular cartilage repair in rabbits by using suspensions of allogenic chondrocytes in alginate, Biomaterials 21 (8) (2000) 795.
[37] K. Kataoka, Y. Suzuki, et al., Alginate, a bioresorbable material derived from brown seaweed, enhances elongation of amputated axons of spinal cord in infant rats, J. Biomed. Mater. Res. 54 (3) (2001) 373–384.
[38] G. Miralles, R. Baudoin, et al., Sodium alginate sponges with or without sodium hyaluronate: *in vitro* engineering of cartilage, J. Biomed. Mater. Res. 57 (2) (2001) 268–278.
[39] K.T. Paige, L.G. Cima, et al., De novo cartilage generation using calcium alginate2chondrocyte constructs, Plast. Reconstr. Surg. 97 (1996) 168–180.
[40] C.S. Cho, S.J. Seo, et al., Galactose-carrying polymers as extracellular matrices for liver tissue engineering, Biomaterials 27 (4) (2006) 576.
[41] M.D. Buschmann, Y.A. Gluzband, et al., Mechanical compression modulates matrix biosynthesis in chondrocyte/agarose culture, J. Cell Sci. 108 (1995) 1497–1508.
[42] S.H. Cho, S.H. Oh, et al., Fabrication and characterization of porous alginate/polyvinyl alcohol hybrid scaffolds for 3D cell culture, J. Biomater. Sci. Polym. Ed. 16 (8) (2005) 933–947.
[43] G. Ciardelli, V. Chiono, et al., Blends of poly-(epsilon-caprolactone) and polysaccharides in tissue engineering applications, Biomacromolecules 6 (4) (2005) 1961–1976.
[44] Z. Li, H.R. Ramay, et al., Chitosan-alginate hybrid scaffolds for bone tissue engineering, Biomaterials 26 (18) (2005) 3919.
[45] M. Sivakumar, K. Panduranga Rao, Preparation, characterization, and *in vitro* release of gentamicin from coralline hydroxyapatite-alginate composite microspheres, J. Biomed. Mater. Res. 65 (2003) 222–228.
[46] S. Sotome, T. Uemura, et al., Synthesis and *in vivo* evaluation of a novel hydroxyapatite/collagen-alginate as a bone filler and a drug delivery carrier of bone morphogenetic protein, Mater. Sci. Eng. C Biomim. Supramol. Syst. 24 (3) (2004) 341–347.
[47] M. Dvir-Ginzberg, I. Gamlieli-Bonshtein, et al., Liver tissue engineering within alginate scaffolds: effects of cell-seeding density on hepatocyte viability, morphology, and function, Tissue Eng. 9 (4) (2003) 757–766.
[48] T. Elkayam, S. Amitay-Shaprut, et al., Enhancing the drug metabolism activities of C3A – A human hepatocyte cell line – By tissue engineering within alginate scaffolds, Tissue Eng. 12 (5) (2006) 1357–1368.
[49] R. Glicklis, L. Shapiro, et al., Hepatocyte behavior within three-dimensional porous alginate scaffolds, Biotechnol. Bioeng. 67 (3) (2000) 344–353.
[50] H.A. Awad, M.Q. Wickham, et al., Chondrogenic differentiation of adipose-derived adult

stem cells in agarose, alginate, and gelatin scaffolds, Biomaterials 25 (16) (2004) 3211–3222.
[51] E. Alsberg, K.W. Anderson, et al., Cell-interactive alginate hydrogels for bone tissue engineering, J. Den. Res. 80 (11) (2001) 2025–2029.
[52] V. Dodane, V.D. Vilivalam, Pharmaceutical applications of chitosan, Pharm. Sci. Technol. Today 1 (6) (1998) 246–253.
[53] J.L. Drury, D.J. Mooney, Hydrogels for tissue engineering: scaffold design variables and applications, Biomaterials 24 (24) (2003) 4337.
[54] A. Di Martino, M. Sittinger, et al., Chitosan: a versatile biopolymer for orthopaedic tissue-engineering, Biomaterials 26 (30) (2005) 5983.
[55] F. Chellat, M. Tabrizian, et al., *In vitro* and *in vivo* biocompatibility of chitosan-xanthan polyionic complex, J. Biomed. Mater. Res. 51 (1) (2000) 107–116.
[56] A. Chenite, C. Chaput, et al., Novel injectable neutral solutions of chitosan form biodegradable gels in situ, Biomaterials 21 (21) (2000) 2155–2161.
[57] A. Vila, A. Sanchez, et al., Design of biodegradable particles for protein delivery, J. Control Release 78 (1-3) (2002) 15–24.
[58] U. Bertram, R. Bodmeier, *In situ* gelling, bioadhesive nasal inserts for extended drug delivery: *in vitro* characterization of a new nasal dosage form, Eur. J. Pharm. Sci. 27 (1) (2006) 62.
[59] S. Govender, V. Pillay, et al., Optimization and characterization of bioadhesive controlled release tetracycline microspheres, Int. J. Pharm. 306 (1–2) (2005) 24.
[60] J. Mao, L. Zhao, et al., Study of novel chitosan-gelatin artificial skin in vitro, J. Biomed. Mater. Res. 64A (2) (2003a) 301–308.
[61] J.S. Mao, L.G. Zhao, et al., Structure and properties of bilayer chitosan-gelatin scaffolds, Biomaterials 24 (6) (2003b) 1067–1074.
[62] Y.J. Park, Y.M. Lee, et al., Controlled release of platelet-derived growth factor-BB from chondroitin sulfate-chitosan sponge for guided bone regeneration, J. Control Release 67 (2-3) (2000) 385–394.
[63] D.W. Urry, A. Pattanaik, et al., Elastic protein-based polymers in soft tissue augmentation and generation, J. Biomater. Sci. Polym. Ed. 9 (10) (1998) 1015–1048.
[64] I. Adekogbe, A. Ghanem, Fabrication and characterization of DTBP-cross-linked chitosan scaffolds for skin tissue engineering, Biomaterials 26 (35) (2005) 7241.
[65] T. Kiyozumi, Y. Kanatani, et al., Medium (DMEM/F12)-containing chitosan hydrogel as adhesive and dressing in autologous skin grafts and accelerator in the healing process, J. Biomed. Mater. Res. B Appl. Biomater. 79B (1) (2006) 129–136.
[66] E.T. Baran, R.L. Reis. Development and in vitro evaluation of chitosan and soluble starch-chitosan nano-microparticles to be used as drug delivery vectors. Society for Biomaterials 29th Annual Meeting Transactions, Reno, USA, 2003
[67] K.C. Gupta, M.N.V.R. Kumar, Trends in controlled drug release formulations using chitin and chitosan, J. Sci. Ind. Res. 59 (3) (2000) 201–213.
[68] S. Patashnik, L. Rabinovich, et al., Preparation and evaluation of chitosan microspheres containing bisphosphonates, J. Drug Target. 4 (6) (1997) 371–380.
[69] C. Peniche, M. Fernandez, et al., Drug delivery systems based on porous chitosan/polyacrylic acid microspheres, Macromol. Biosci. 3 (10) (2003) 540–545.
[70] J.K. Francis Suh, H.W.T. Matthew, Application of chitosan-based polysaccharide biomaterials in cartilage tissue engineering: a review, Biomaterials 21 (24) (2000) 2589.
[71] M.I. Haque, A.C. Beekley, et al., Bioabsorption qualities of chitosan-absorbable vascular templates(1), Curr. Surg. 58 (1) (2001) 77–80.
[72] P.B. Malafaya, A. Pedro, et al., Chitosan particles agglomerated scaffolds for cartilage and osteochondral tissue engineering approaches with adipose tissue derived stem cells, J. Mater. Sci. Mater. Med. 16 (12) (2005) 1077.
[73] D.T. Simionescu, Q. Lu, et al., Biocompatibility and remodeling potential of pure arterial elastin and collagen scaffolds, Biomaterials 27 (5) (2006) 702–713.
[74] F. Mwale, M. Iordanova, et al., Biological evaluation of chitosan salts cross-linked to genipin as a cell scaffold for disk tissue engineering, Tissue Eng. 11 (1-2) (2005) 130–140.
[75] P.J. VandeVord, H.W.T. Matthew, et al., Evaluation of the biocompatibility of a chitosan scaffold in mice, J. Biomed. Mater. Res. 59 (3) (2002) 585–590.

[76] J.K.F. Suh, H.W.T. Matthew, Application of chitosan-based polysaccharide biomaterials in cartilage tissue engineering: a review, Biomaterials 21 (24) (2000) 2589.

[77] S.R. Ong, K.A. Trabbic-Carlson, et al., Epitope tagging for tracking elastin-like polypeptides, Biomaterials 27 (9) (2006) 1930−1935.

[78] W. Xia, W. Liu, et al., Tissue engineering of cartilage with the use of chitosan-gelatin complex scaffolds, J. Biomed. Mater. Res. B Appl. Biomater. 71B (2) (2004) 373−380.

[79] S. Yamane, N. Iwasaki, et al., Feasibility of chitosan-based hyaluronic acid hybrid biomaterial for a novel scaffold in cartilage tissue engineering, Biomaterials 26 (6) (2005) 611.

[80] D. Voet, Biochemistry, John Wiley & Sons, Inc, New York, 1995.

[81] E. Entcheva, H. Bien, et al., Functional cardiac cell constructs on cellulose-based scaffolding, Biomaterials 25 (26) (2004) 5753−5762.

[82] P. Beguin, J.P. Aubert, The biological degradation of cellulose, FEMS Microbiol. Rev. 13 (1) (1994) 25−58.

[83] M. Martson, J. Viljanto, et al., Is cellulose sponge degradable or stable as implantation materialb An *in vivo* subcutaneous study in the rat, Biomaterials 20 (21) (1999) 1989−1995.

[84] T. Miyamoto, S. Takahashi, et al., Tissue biocompatibility of cellulose and its derivatives, J. Biomed. Mater. Res. 23 (1) (1989) 125−133.

[85] A.E. Elcin, *In vitro* and *in vivo* degradation of oxidized acetyl- and ethyl-cellulose sponges, Artif. Cells Blood Substit. Immobil. Biotechnol. 34 (4) (2006) 407−418.

[86] M. Martson, J. Viljanto, et al., Biocompatibility of cellulose sponge with bone, Eur. Surg. Res. 30 (6) (1998) 426−432.

[87] T. Takata, H.L. Wang, et al., Migration of osteoblastic cells on various guided bone regeneration membranes, Clin. Oral Implants Res. 12 (4) (2001) 332−338.

[88] H. Backdahl, G. Helenius, et al., Mechanical properties of bacterial cellulose and interactions with smooth muscle cells, Biomaterials 27 (9) (2006) 2141−2149.

[89] G. Helenius, H. Backdahl, et al., *In vivo* biocompatibility of bacterial cellulose, J. Biomed. Mater. Res. A 76A (2) (2006) 431−438.

[90] A. Svensson, E. Nicklasson, et al., Bacterial cellulose as a potential scaffold for tissue engineering of cartilage, Biomaterials 26 (4) (2005) 419−431.

[91] P. RoyChowdhury, V. Kumar, Fabrication and evaluation of porous 2,3-dialdehydecellulose membrane as a potential biodegradable tissue-engineering scaffold, J. Biomed. Mater. Res. A 76A (2) (2006) 300−309.

[92] C. Vinatier, D. Magne, et al., A silanized hydroxypropyl methylcellulose hydrogel for the three-dimensional culture of chondrocytes, Biomaterials 26 (33) (2005) 6643−6651.

[93] S. Schaffellner, V. Stadlbauer, et al., Porcine islet cells microencapsulated in sodium cellulose sulfate, Transplantat. Proc. 37 (1) (2005) 248−252.

[94] T.L. Wang, T.Y. Bogracheva, et al., Starch: as simple as A, B, C? J. Exp. Bot. 49 (320) (1998) 481−502.

[95] T. Galliard, Starch: Properties and Potential. John Wiley, New York, 1987.

[96] R. Reis, A. Cunha, et al., Starch and starch based thermoplastics, in: K.H. Jurgen, et al. (Eds.), Encyclopedia of Materials Science and Technology, Elsevier Science, Amsterdam, 2001, p. 8810.

[97] F. Xie, L. Yu, et al., Starch modification using reactive extrusion, Starch/Starke 58 (2006) 131−139.

[98] L. Avérous, Biodegradable multiphase systems based on lasticized starch: a review, J. Macromol. Sci. 44 (3) (2004) 231−274.

[99] K.E. Beery, M.R. Ladisch, Chemistry and properties of starch based desiccants, Enzyme Microb. Technol. 28 (7-8) (2001) 573−581.

[100] D. Eagles, D. Lesnoy, et al., Starch fibers: processing and characteristics, Text. Res. J. 66 (4) (1996) 277−282.

[101] J.L. Willett, B.K. Jasberg, et al., Rheology of thermoplastic starch: effects of temperature, moisture content and additives on melt viscosity, Polym. Eng. Sci. 35 (2) (1995) 202−210.

[102] G. Yilmaz, R.O.J. Jongboom, et al., Thermoplastic starch as a biodegradable matrix for encapsulation and controlled release, in: S. Mallapragada, B. Narasimhan (Eds.), Handbook of Biodegradable Polymeric Materials and their Applications, American Scientific Publishers, California, 2006, p. 2.

[103] A.L. Da Róz, A.J.F. Carvalho, et al., The effect of plasticizers on thermoplastic starch compositions obtained by melt spinning, Carbohydr. Polym. 63 (2006) 417–424.
[104] C. Bastiolli, Starch-polymer composites, in: G. Scott, D. Gilead (Eds.), Degradable Polymers, Chapman and Hall, London, 1995, pp. 112–137.
[105] C. Bastiolli, C. Belloti, et al., Mater-Bi: properties and biodegradability, J. Env. Polym. Deg. 1 (1993) 181–191.
[106] R.L. Reis, S.C. Mendes, et al., Processing and *in vitro* degradation of starch/EVOH thermoplastic blends, Polym. Int. 43 (4) (1997) 347–352.
[107] M. Araújo, C. Vaz, et al., *In-vitro* degradation behaviour of starch/EVOH biomaterials, Polym. Degrad. Stab. 73 (2001) 237–244.
[108] H.S. Azevedo, F.M. Gama, et al., *In vitro* assessment of the enzymatic degradation of several starch based biomaterials, Biomacromolecules 4 (6) (2003) 1703–1712.
[109] M.E. Gomes, R.L. Reis, et al., Cytocompatibility and response of osteoblastic-like cells to starch-based polymers: effect of several additives and processing conditions, Biomaterials 22 (13) (2001) 1911–1917.
[110] A.P. Marques, R.L. Reis, et al., The biocompatibility of novel starch-based polymers and composites: *In vitro* studies, Biomaterials 23 (6) (2002) 1471–1478.
[111] A.J. Salgado, O.P. Coutinho, et al., Novel starch-based scaffolds for bone tissue engineering: cytotoxicity, cell culture, and protein expression, Tissue Eng. 10 (3-4) (2004) 465–474.
[112] S.C. Mendes, R.L. Reis, et al., Biocompatibility testing of novel starch-based materials with potential application in orthopaedic surgery: a preliminary study, Biomaterials 22 (14) (2001) 2057–2064.
[113] L.F. Boesel, M.H. Fernandes, et al., The behavior of novel hydrophilic composite bone cements in simulated body fluids, J. Biomed. Mater. Res. 70B (2) (2004a) 368–377.
[114] L.F. Boesel, J.F. Mano, et al., Optimization of the formulation and mechanical properties of starch based partially degradable bone cements, J. Mater. Sci. Mater. Med. 15 (1) (2004b) 73–83.
[115] L.F. Boesel, R.L. Reis, Hydrophilic matrices to be used as bioactive and degradable bone cements, J. Mater. Sci. Mater. Med. 15 (4) (2004) 503–506.
[116] I. Espigares, C. Elvira, et al., New partially degradable and bioactive acrylic bone cements based on starch blends and ceramic fillers, Biomaterials 23 (8) (2002) 1883–1895.
[117] C. Pereira, A. Cunha, et al., New starch-based thermoplastic hydrogels for use as bone cements or drug-delivery carriers, J. Mater. Sci. Mater. Med. 9 (1998) 825–833.
[118] C. Elvira, J.F. Mano, et al., Starch-based biodegradable hydrogels with potential biomedical applications as drug delivery systems, Biomaterials 23 (9) (2002) 1955–1966.
[119] A.L. Oliveira, P.B. Malafaya, et al., Sodium silicate gel as a precursor for the *in vitro* nucleation and growth of a bone-like apatite coating in compact and porous polymeric structures, Biomaterials 24 (15) (2003) 2575–2584.
[120] R.L. Reis, A.M. Cunha, et al., Mechanical behavior of injection-molded starch-based polymers, Polym. Adv. Technol. 7 (10) (1996) 784–790.
[121] R.L. Reis, A.M. Cunha, et al., Structure development and control of injection-molded hydroxyl-apatite-reinforced starch/EVOH composites, Adv. Polym. Technol. 16 (4) (1997) 263–277.
[122] R. Sousa, R. Reis, et al., Processing and properties of bone-analogue biodegradable and bioinert polymeric composites, Compos. Sci. Tech. 63 (2003) 389–402.
[123] R.A. Sousa, G. Kalay, et al., Injection molding of a starch/EVOH blend aimed as an alternative biomaterial for temporary applications, J. Appl. Polym. Sci. 77 (6) (2000) 1303–1315.
[124] R.A. Sousa, J.F. Mano, et al., Mechanical performance of starch based bioactive composite biomaterials molded with preferred orientation, Polym. Eng. Sci. 42 (5) (2002) 1032–1045.
[125] M. Gomes, A. Salgado, et al., Bone tissue engineering using starch based scaffolds obtained by different methods, in: R. Reis, D. Cohn (Eds.), Polymer Based Systems on Tissue Engineering, Replacement and Regeneration, Kluwer Academic Publishers, Amsterdam, 2002, pp. 221–249.

[126] M.E. Gomes, J.S. Godinho, et al., Alternative tissue engineering scaffolds based on starch: processing methodologies, morphology, degradation and mechanical properties, Mater. Sci. Eng. C-Biomim. Supramol. Syst 20 (1-2) (2002a) 19–26.

[127] M.E. Gomes, J.S. Godinho, et al., Alternative tissue engineering scaffolds based on starch: processing methodologies, morphology, degradation and mechanical properties, Mater. Sci. Eng. C 20 (1–2) (2002b) 19–26.

[128] M.E. Gomes, H.L. Holtorf, et al., Influence of the porosity of starch-based fiber mesh scaffolds on the proliferation and osteogenic differentiation of bone marrow stromal cells cultured in a flow perfusion bioreactor, Tissue Eng. 12 (4) (2006) 801–809.

[129] M.E. Gomes, P.B. Malafaya, et al., Methodologies for processing biodegradable and natural origin scaffolds for bone and cartilage tissue-engineering applications, in: A. Hollander, P. Hatton (Eds.), Methods in Molecular Biology Series, vol. 238, The Humana Press Inc, Totowa, USA, 2004, pp. 65–76.

[130] M.E. Gomes, A.S. Ribeiro, et al., A new approach based on injection moulding to produce biodegradable starch-based polymeric scaffolds: morphology, mechanical and degradation behaviour, Biomaterials 22 (9) (2001) 883–889.

[131] C. Pereira, M.E. Gomes, et al., Hard cellular materials in the human body: properties and production of foamed polymers for bone replacement, in: N. Rivier, J. Sadoc (Eds.), NATO/ASI Series, Kluwer Press, Dordrecht, Kluwer Press, Dordrecht, 1998, pp. 193–204.

[132] A.J. Salgado, M.E. Gomes, et al., Preliminary study on the adhesion and proliferation of human osteo-blasts on starch-based scaffolds, Mater. Sci. Eng. C 20 (1-2) (2002) 27–33.

[133] M. Gomes, J. Godinho, et al., Design and processing of starch based scaffolds for hard tissue engineering, J. Appl. Med. Polym. 6 (2) (2002c) 75–80.

[134] M.E. Gomes, V.I. Sikavitsas, et al., Effect of flow perfusion on the osteogenic differentiation of bone marrow stromal cells cultured on starch-based three-dimensional scaffolds, J. Biomed. Mater. Res. A 67 (1) (2003) 87–95.

[135] S.C. Mendes, J. Bezemer, et al., Evaluation of two biodegradable polymeric systems as substrates for bone tissue engineering, Tissue Eng. 9 (Suppl. 1) (2003) S91–S101.

[136] B.P. Toole, Hyaluronan: from extracellular glue to pericellular cue, Nature Rev. Cancer 4 (2004) 528–539.

[137] E.J. Menzel, C. Farr, Hyaluronidase and its substrate hyaluronan: biochemistry, biological activities and therapeutic uses, Cancer Lett. 131 (1998) 3–11.

[138] E.M. Ehlers, P. Behrens, et al., Effects of hyaluronic acid on the morphology and proliferation of human chondrocytes in primary cell culture, Ann. Anat. – Anat. Anz. 183 (1) (2001) 13.

[139] G.A. Carlson, J.L. Dragoo, et al., Bacteriostatic properties of biomatrices against common orthopaedic pathogens, Biochem. Biophys. Res. Commun. 321 (2) (2004) 472.

[140] A.A. Hegewald, J. Ringe, et al., Hyaluronic acid and autologous synovial fluid induce chondrogenic differentiation of equine mesenchymal stem cells: a preliminary study, Tissue Cell (2004) 431–438.

[141] X. Zou, H. Li, et al., Stimulation of porcine bone marrow stromal cells by hyaluronan, dexamethasone and rhBMP-2, Biomaterials 25 (23) (2004) 5375.

[142] T. Fujii, Y.-L. Sun, et al., Mechanical properties of single hyaluronan molecules, J. Biomech. 35 (4) (2002) 527.

[143] P.M. van der Kraan, P. Buma, et al., Interaction of chondrocytes, extracellular matrix and growth factors: relevance for articular cartilage tissue engineering, Osteoarthr. Cartil. 10 (8) (2002) 631.

[144] R.E. Horch, J. Kopp, et al., Tissue engineering of cultured skin substitutes, J. Cell. Mol. Med 9 (3) (2005) 592–608.

[145] E. Pianigiani, A. Andreassi, et al., A new model for studying differentiation and growth of epidermal cultures on hyaluronan-based carrier, Biomaterials 20 (18) (1999) 1689.

[146] L.P. Amarnath, A. Srinivas, et al., *In vitro* hemocompatibility testing of UV-modified hyaluronan hydrogels, Biomaterials 27 (8) (2006) 1416.

[147] R.A. Peattie, A.P. Nayate, et al., Stimulation of *in vivo* angiogenesis by cytokine-loaded

hyaluronic acid hydrogel implants, Biomaterials 25 (2004) 2789–2798.
[148] K.Y. Cho, T.W. Chung, et al., Release of ciprofloxacin from poloxamer-graft-hyaluronic acid hydrogels in vitro, Int. J. Pharm. 260 (1) (2003) 83.
[149] D. Campoccia, P. Doherty, et al., Semisynthetic resorbable materials from hyaluronan esterification, Biomaterials 19 (23) (1998) 2101.
[150] J. Gao, J.E. Dennis, et al., Repair of osteochondral defect with tissue-engineered two-phase composite material of injectable calcium phosphate and hyaluronan sponge, Tissue Eng. 8 (2002) 827–837.
[151] Y. Liu, X.Z. Shu, et al., Biocompatibility and stability of dissulfide-cross-linked hyaluronan films, Biomaterials 26 (2005) 4737–4746.
[152] G.D. Prestwich, D.M. Marecak, et al., Controlled chemical modification of hyaluronic acid: synthesis, applications, and biodegradation of hydrazide derivatives, J. Control Release 53 (1-3) (1998) 93.
[153] X.Z. Shu, Y. Liu, et al., Disulfide-cross-linked hyaluronan-gelatin hydrogel films: a covalent mimic of the extracellular matrix for *in vitro* cell growth, Biomaterials 24 (2003) 3825–3834.
[154] S.R. Frenkel, G. Bradica, et al., Regeneration of articular cartilage – Evaluation of osteochondral defect repair in the rabbit using multiphasic implants, Osteoarthr. Cartil. 13 (9) (2005) 798.
[155] S.-H. Park, S. Park, et al., Tissue-engineered cartilage using fibrin/hyaluronan composite gel and its *in vivo* implantation, Artif. Organs 29 (10) (2005) 838–845.
[156] X.Z. Shu, Y. Liu, et al., *In situ* cross-linkable hyaluronan hydrogels for tissue engineering, Biomaterials 25 (7–8) (2004) 1339.
[157] M. Halbleib, T. Skurk, et al., Tissue engineering of white adipose tissue using hyaluronic acid-based scaffolds. I: in vitro differentiation of human adipocyte precursor cells on scaffolds, Biomaterials 24 (18) (2003) 3125.
[158] L.A. Solchaga, J.S. Temenoff, et al., Repair of osteochondral defects with hyaluronan- and polyester-based scaffolds, Osteoarthr. Cartil. 13 (4) (2005) 297.
[159] T. Funakoshi, T. Majima, et al., Novel chitosan-based hyaluronan hybrid polymer fibers as a scaffold in ligament tissue engineering, J. Biomed. Mater. Res. A 74A (3) (2005a) 338–346.
[160] T. Funakoshi, T. Majima, et al., Novel chitosan-based hyaluronan hybrid polymer fibers as a scaffold in ligament tissue engineering, J. Biomed. Mater. Res. 74A (2005b) 338–346.
[161] E. Milella, E. Brescia, et al., Physico-chemical properties and degradability of non-woven hyaluronan benzylic esters as tissue engineering scaffolds, Biomaterials 23 (4) (2002) 1053.
[162] J.-E. Lee, J.-G. Park, et al., Preparation of collagen modified hyaluronan microparticles as antibiotic carrier, Yonsei Med. J. 42 (3) (2001) 291–298.
[163] S. Oerther, E. Payan, et al., Hyaluronate-alginate combination for the preparation of new biomaterials: investigation of the behaviour in aqueous solutions, Biochim. Biophys. Acta 1426 (1999) 185–194.
[164] T. Segura, B.C. Anderson, et al., Cross-linked hyaluronic acid hydrogels: a strategy to functionalize and pattern, Biomaterials 26 (4) (2005) 359.
[165] S. Hou, Q. Xu, et al., The repair of brain lesion by implantation of hyaluronic acid hydrogels modified with laminin, J. Neurosci. Methods 148 (1) (2005) 60.
[166] A. Denuziere, D. Ferrier, et al., Chitosan-chondroitin sulfate and chitosan-hyaluronate polyelectrolyte complexes: biological properties, Biomaterials 19 (14) (1998) 1275.
[167] H.S. Yoo, E.A. Lee, et al., Hyaluronic acid modified biodegradable scaffolds for cartilage tissue engineering, Biomaterials 26 (14) (2005) 1925.
[168] K. Knudson, C. Biswas, et al., The role and regulation of tumour-associated hyaluronan. The Biology of Hyaluronan, Ciba Found. Symp. 143 (1989) 150–159.
[169] B. Toole, T. Wight, et al., Hyaluronan-cell interactions in cancer and vascular disease, J. Biol. Chem. 277 (2002) 4593–4596.
[170] J.E. Bailey, D.F. Ollis, Biochemical Engineering Fundamentals, McGraw-Hill, Singapore, 1986.
[171] K. Gelse, E. Poschl, et al., Collagens – structure, function, and biosynthesis, Adv. Drug Deliv. Rev. 55 (12) (2003) 1531–1546.
[172] C.L. Yang, P.J. Hillas, et al., The application of recombinant human collagen in tissue engineering, Biodrugs 18 (2) (2004) 103–119.

[173] W. Friess, Collagen – biomaterial for drug delivery, Eur. J. Pharm. Biopharm. 45 (2) (1998) 113–136.
[174] C.H. Lee, A. Singla, et al., Biomedical applications of collagen, Int. J. Pharm. 221 (1–2) (2001) 1–22.
[175] J. Bella, M. Eaton, et al., Crystal-structure and molecular-structure of a collagen-like peptide at 1.9-Angstrom resolution, Science 266 (5182) (1994) 75–81.
[176] F. Rosso, G. Marino, et al., Smart materials as scaffolds for tissue engineering, J. Cell. Physiol. 203 (3) (2005) 465–470.
[177] M.M. Stevens, J.H. George, Exploring and engineering the cell surface interface, Science 310 (5751) (2005) 1135–1138.
[178] P. Fratzl, Cellulose and collagen: from fibres to tissues, Curr. Opin. Colloid Interface Sci. 8 (1) (2003) 32–39.
[179] K.E. Kadler, D.F. Holmes, et al., Collagen fibril formation, Biochem. J. 316 (1996) 1–11.
[180] J.M. Pachence, Collagen-based devices for soft tissue repair, J. Biomed. Mater. Res. 33 (1) (1996) 35–40.
[181] E.D. O'Cearbhaill, V. Barron, et al., Characterization of a collagen membrane for its potential use in cardiovascular tissue engineering applications, J. Mater. Sci. Mater. Med. 17 (3) (2006) 195–201.
[182] M.J. van Amerongen, M.C. Harmsen, et al., The enzymatic degradation of scaffolds and their replacement by vascularized extracellular matrix in the murine myocardium, Biomaterials 27 (10) (2006) 2247–2257.
[183] F. Boccafoschi, J. Habermehl, et al., Biological performances of collagen-based scaffolds tor vascular tissue engineering, Biomaterials 26 (35) (2005) 7410–7417.
[184] E. Song, S. Yeon Kim, et al., Collagen scaffolds derived from a marine source and their biocompatibility, Biomaterials 27 (15) (2006) 2951–2961.
[185] B.S. Kim, C.E. Baez, et al., Biomaterials for tissue engineering, World J. Urol. 18 (1) (2000) 2–9.
[186] B. Joddar, A. Ramamurthi, Fragment size- and dose-specific effects of hyaluronan on matrix synthesis by vascular smooth muscle cells, Biomaterials 27 (15) (2006) 2994–3004.
[187] M. Li, M.J. Mondrinos, et al., Electrospun protein fibers as matrices for tissue engineering, Biomaterials 26 (30) (2005) 5999–6008.
[188] W.F. Daamen, S.T. Nillesen, et al., Tissue response of defined collagen-elastin scaffolds in young and adult rats with special attention to calcification, Biomaterials 26 (1) (2005) 81–92.
[189] W.F. Daamen, H.T. van Moerkerk, et al., Preparation and evaluation of molecularly-defined collagen-elastin-glycosaminoglycan scaffolds for tissue engineering, Biomaterials 24 (22) (2003) 4001–4009.
[190] J.D. Berglund, R.M. Nerem, et al., Incorporation of intact elastin scaffolds in tissue-engineered collagen-based vascular grafts, Tissue Eng. 10 (9-10) (2004) 1526–1535.
[191] M. Haider, Z. Megeed, et al., Genetically engineered polymers: status and prospects for controlled release, J. Control Release 95 (1) (2004) 1–26.
[192] L. Debelle, A.J. Alix, et al., The secondary structure and architecture of human elastin, Eur. J. Biochem. 258 (2) (1998) 533–539.
[193] J.B. Leach, J.B. Wolinsky, et al., Cross-linked [alpha]-elastin biomaterials: towards a processable elastin mimetic scaffold, Acta Biomater. 1 (2) (2005) 155–164.
[194] D.W. Urry, T. Hugel, et al., Elastin: a representative ideal protein elastomer, Philos. Trans. R. Soc. Lond. B Biol. Sci. 357 (1418) (2002) 169–184.
[195] D.W. Urry, T.M. Parker, Mechanics of elastin: molecular mechanism of biological elasticity and its relationship to contraction, J. Muscle Res. Cell Motil. 23 (5-6) (2002) 543–559.
[196] H. Betre, L.A. Setton, et al., Characterization of a genetically engineered elastin-like polypeptide for cartilaginous tissue repair, Biomacromolecules 3 (5) (2002) 910–916.
[197] A. Girotti, J. Reguera, et al., Design and bioproduction of a recombinant multi(bio)functional elastin-like protein polymer containing cell adhesion sequences for tissue engineering purposes, J. Mater. Sci. Mater. Med. 15 (4) (2004) 479–484.
[198] J. Lee, C.W. Macosko, et al., Mechanical properties of cross-linked synthetic elastomeric polypentapeptides, Macromolecules 34 (17) (2001) 5968–5974.

[199] S. Chabba, G.F. Mattews, et al., Green composites using cross-linked soy flour and flax yarns, Green Chem. 7 (2005) 576−581.

[200] S.N. Swain, S.M. Biswal, et al., Biodegradable soy-based plastics: opportunities and challenges, J. Polym. Environ. 12 (1) (2004) 35−42.

[201] Z. Alibhai, M. Mondor, et al., Production of soy protein concentrates/isolates: traditional and membrane technologies, Desalination 191 (2006) 351−358.

[202] Oliveira, J.T., Correlo, V.M., et al. (2005). Chitosan-Polyester Scaffolds Seeded with Bovine Articular Chondrocytes for Cartilage Tissue Engineering Applications. Bologna: Artificial Organs.

[203] V. Schmidt, C. Giacomelli, et al., Soy protein isolate based films: influence of sodium dodecyl sulfate and polycaprolactone-triol on their properties, Macromol. Symp. 229 (2005) 127−137.

[204] M.T. Hinds, R.C. Rowe, et al., Development of a reinforced porcine elastin composite vascular scaffold, J. Biomed. Mater. Res. A 77 (2006) 458−469.

[205] B.-S. Kim, J. Nikolovski, et al., Engineered smooth muscle tissues: regulating cell phenotype with the Scaffold, Exp. Cell Res. 251 (2) (1999) 318−328.

[206] B.E. Elizalde, G.B. Bartholomai, et al., The effect of pH on the relationship between hydrophilic/lipophilic characteristics and emulsification properties of soy proteins, Food Sci. Technol.-Lebensmittel-Wissenschaft & Technologie 29 (4) (1996) 334−339.

[207] J.F. Mano, C.M. Vaz, et al., Dynamic mechanical properties of hydroxyapatite-reinforced and porous starch-based degradable biomaterials, J. Mater. Sci. Mater. Med. 10 (12) (1999) 857−862.

[208] C.M. Vaz, L.A. Graaf, et al., Soy protein-based systems for different tissue regeneration applications, in: R. Reis, D. Cohn (Eds.), Polymer Based Systems on Tissue Engineering, Replacement and Regeneration, Kluwer Academic Publishers, Amsterdam, 2002, pp. 93−110.

[209] S.N. Swain, K.K. Rao, et al., Biodegradable polymers. III. Spectral, thermal, mechanical, and morphological properties of cross-linked furfural2Soy protein concentrate, J. Appl. Polym. Sci. 93 (2004) 2590−2596.

[210] L. Were, N.S. Hettiarachchy, et al., Properties of cysteine-added soy protein-wheat gluten films, J. Food Sci. 64 (3) (1999) 514−518.

[211] J.W. Rhim, A. Gennadios, et al., Soy protein isolate dialdehyde starch films, Ind. Crops and Prod. 8 (3) (1998) 195−203.

[212] R.M. Silva, C. Elvira, et al., Influence of beta-radiation sterilization in properties of new chitosan/soybean protein isolate membranes for guided bone regeneration. J. Mater. Sci. Mater. Med 15 (2004) 523−528.

[213] S.S. Silva, M.I. Santos, et al., Physical properties and biocompatibility of chitosan/soy blended membranes. J. Mater. Sci. Mater. Med 16 (6) (2005) 575−579.

[214] S.S. Silva, J.M. Oliveira, et al., (2006). Physicochemical Characterization of Novel Chitosan-Soy Protein/TEOS Porous Hybrids for Tissue Engineering Applications.

[215] H.J. Jin, S.V. Fridrikh, et al., Electrospinning Bombyx mori silk with poly(ethylene oxide), Biomacromolecules 3 (2005) 1233−1239.

[216] G.H. Altman, F. Diaz, et al., Silk-based biomaterials, Biomaterials 24 (3) (2003) 401−416.

[217] M. Santin, A. Motta, et al., *In vitro* evaluation of the inflammatory potential of the silk fibroin, J. Biomed. Mater. Res. 46 (3) (1999) 382−389.

[218] S.W. Ha, Y.H. Park, et al., Dissolution of Bombyx mori silk fibroin in the calcium nitrate tetrahydrate-methanol system and aspects of wet spinning of fibroin solution, Biomacromolecules 4 (3) (2003) 488−496.

[219] E. Servoli, D. Maniglio, et al., Surface properties of silk fibroin films and their interaction with fibroblasts, Macromol. Biosci. 5 (12) (2005) 1175−1183.

[220] J. Ayutsede, M. Gandhi, et al., Regeneration of *Bombyx mori* silk by electrospinning. Part 3: characterization of electrospun nonwoven mat, Polymer 46 (5) (2005) 1625−1634.

[221] H.J. Jin, J.S. Chen, et al., Human bone marrow stromal cell responses on electrospun silk fibroin mats, Biomaterials 25 (6) (2004) 1039−1047.

[222] A. Motta, C. Migliaresi, et al., Fibroin hydrogels for biomedical applications: preparation, characterization and *in vitro* cell culture

studies, J. Biomater. Sci. Polym. Ed. 15 (7) (2004) 851–864.
[223] U.J. Kim, J. Park, et al., Three-dimensional aqueous-derived biomaterial scaffolds from silk fibroin, Biomaterials 26 (15) (2005) 2775–2785.
[224] R. Nazarov, H.J. Jin, et al., Porous 3D scaffolds from regenerated silk fibroin, Biomacromolecules 5 (3) (2004) 718–726.
[225] H.J. Kim, U.J. Kim, et al., Influence of macroporous protein scaffolds on bone tissue engineering from bone marrow stem cells, Biomaterials 26 (21) (2005) 4442–4452.
[226] C. Li, C. Vepari, et al., Electrospun silk-BMP-2 scaffolds for bone tissue engineering, Biomaterials 27 (2006) 3115–3124.
[227] A.S. Gobin, V.E. Froude, et al., Structural and mechanical characteristics of silk fibroin and chitosan blend scaffolds for tissue regeneration, J. Biomed. Mater. Res. A 74A (3) (2005) 465–473.
[228] M.Z. Li, S.Z. Lu, et al., Study on porous silk fibroin materials. I. Fine structure of freeze dried silk fibroin, J. Appl. Polym. Sci. 79 (12) (2001) 2185–2191.
[229] M.Z. Li, Z.Y. Wu, et al., Study on porous silk fibroin materials. II. Preparation and characteristics of spongy silk fibroin materials, J. Appl. Polym. Sci. 79 (12) (2001) 2192–2199.
[230] M.Z. Li, C.S. Zhang, et al., Study on porous silk fibroin materials: 3. Influence of repeated freeze-thawing on the structure and properties of porous silk fibroin materials, Polym. Adv. Technol. 13 (8) (2002) 605–610.
[231] Y. Tamada, New process to form a silk fibroin porous 3-D structure, Biomacromolecules 6 (6) (2005) 3100–3106.
[232] G. Chen, P. Zhou, et al., Silk fibroin modified porous poly(E-caprolactone) scaffold for human fibro-blast culture in vitro, J. Mater. Sci. Mater. Med. 15 (6) (2004) 671–677.
[233] P. Petrini, C. Parolari, et al., Silk fibroin-polyurethane scaffolds for tissue engineering. J. Mater. Sci. Mater. Med 12 (10-12) (2001) 849–853.
[234] R.E. Unger, M. Wolf, et al., Growth of human cells on a non-woven silk fibroin net: a potential for use in tissue engineering, Biomaterials 25 (6) (2004) 1069–1075.
[235] L. Meinel, R. Fajardo, et al., Silk implants for the healing of critical size bone defects, Bone 37 (5) (2005) 688–698.
[236] L. Meinel, S. Hofmann, et al., Engineering cartilage-like tissue using human mesenchymal stem cells and silk protein scaffolds, Biotechnol. Bioeng. 88 (3) (2004) 379–391.
[237] A.J. Anderson, E.A. Dawes, Occurrence, metabolism, metabolic role, and industrial uses of bacterial polyhydroxyalkanoates, Microbiol. Rev. 54 (4) (1990) 450–472.
[238] K. Sudesh, H. Abe, et al., Synthesis, structure and properties of polyhydroxyalkanoates: biological polyesters, Prog. Polym. Sci. (Oxford) 25 (10) (2000) 1503–1555.
[239] E.A. Dawes, P.J. Senior, The role and regulation of energy reserve polymers in microorganisms, Adv. Microb. Physiol. 10 (1973) 135–266.
[240] L.L. Wallen, W.K. Rohwedder, Poly B hydroxyalkanoate from activated sludge, Env. Sci. Technol. 8 (6) (1974) 576–579.
[241] P.J.Barham. Physical properties of poly (hydroxybutyrate) and poly(hydroxybutyrate-co-hydroxyvalerate). Novel Biodegradable Microbial Polymers, 1990, pp. 81–96.
[242] P.J. Barham, P. Barker, et al., Physical properties of poly(hydroxybutyrate) and copolymers of hydroxybutyrate and hydroxyvalerate, FEMS Microbiol. Rev. 103 (2–4) (1992) 289–298.
[243] H. Bauer, A.J. Owen, Some structural and mechanical properties of bacterially produced poly-beta-hydroxybutyrate-co-beta-hydroxyvalerate, Colloid Polym. Sci. 266 (3) (1988) 241–247.
[244] H. Mitomo, P.J. Barham, et al., Temperature dependence of mechanical properties of poly (b-hydroxybutyrate-b-hydroxyvalerate), Polym. Commun. Guildford 29 (4) (1988) 112–115.
[245] A.J. Owen, Some dynamic mechanical properties of microbially produced poly-beta-hydroxybutyrate/ beta-hydroxyvalerate copolymers, Colloid Polym. Sci. 263 (10) (1985) 799–803.
[246] S.F. Williams, D.P. Martin, et al., PHA applications: addressing the price performance issue I, Tissue engineering. Int. J. Biol. Macromol. 25 (1–3) (1999) 111–121.
[247] P.J. Barham, A. Keller, Relationship between microstructure and mode of fracture in

polyhydroxybutyrate, J. Polym. Sci. A 24 (1) (1986) 69–77.
[248] F. Gassner, A.J. Owen, Some properties of poly(3-hydroxybutyrate) – Poly(3-hydroxyvalerate) blends, Polym. Int. 39 (3) (1996) 215–219.
[249] J.K. Hobbs, The fracture of poly (hydroxybutyrate): part I fracture mechanics study during ageing, J. Mater. Sci. 33 (10) (1998) 2509–2514.
[250] J.K. Hobbs, P.J. Barham, The fracture of poly(hydroxybutyrate): part II fracture mechanics study after annealing, J. Mater. Sci. 33 (10) (1998) 2515–2518.
[251] J.K. Hobbs, P.J. Barham, The fracture of poly(hydroxybutyrate): part III fracture morphology in thin films and bulk systems, J. Mater. Sci. 34 (19) (1999) 4831–4844.
[252] K. Ishikawa, Y. Kawaguchi, et al., Plasticization of bacterial polyester by the addition of acylglycerols and its enzymatic degradability, Kobunshi Ronbunshu 48 (4) (1991) 221–226.
[253] J.C. Knowles, Development of a natural degradable polymer for orthopaedic use, J. Med. Eng. Technol. 17 (4) (1993) 129–137.
[254] M. Scandola, M.L. Focarete, et al., Polymer blends of natural poly(3-hydroxybutyrate-co-3-hydroxyvalerate) and a synthetic atactic poly(3-hydroxybutyrate). Characterization and biodegradation studies, Macromolecules 30 (9) (1997) 2568–2574.
[255] C. Chaput, L.H. Yahia, et al., Natural poly (hydroxybutyrate-hydroxyvalerate) polymers as degradable biomaterials, in: Materials Research Society Symposium -Proceedings, San Francisco, CA, Materials Research Society, 1995.
[256] T. Freier, C. Kunze, et al., *In vitro* and *in vivo* degradation studies for development of a biodegradable patch based on poly (3-hydroxybutyrate), Biomaterials 23 (13) (2002) 2649–2657.
[257] C.W. Pouton, S. Akhtar, Biosynthetic polyhydroxyalkanoates and their potential in drug delivery, Adv. Drug Deliv. Rev. 18 (2) (1996) 133–162.
[258] S. Gogolewski, M. Jovanovic, et al., Tissue response and *in vivo* degradation of selected polyhydroxyacids: polylactides (PLA), poly(3-hydroxybutyrate) (PHB), and poly (3- hydroxybutyrate-co-3-hydroxyvalerate) (PHB/VA), J. Biomed. Mater. Res. 27 (9) (1993) 1135–1148.
[259] Y.Y. Shangguan, Y.W. Wang, et al., The mechanical properties and *in vitro* biodegradation and biocompatibility of UV-treated poly(3-hydroxybutyrate-co-3-hydroxyhexanoate), Biomaterials 27 (11) (2006) 2349–2357.
[260] T. Volova, E. Shishatskaya, et al., Results of biomedical investigations of PHB and PHB/PHV fibers, Biochem. Eng. J. 16 (2) (2003) 125–133.
[261] Y.W. Wang, F. Yang, et al., Effect of composition of poly(3-hydroxybutyrate-co-3-hydroxyhexanoate) on growth of fibroblast and osteoblast, Biomaterials 26 (7) (2005) 755–761.
[262] X. Yang, K. Zhao, et al., Effect of surface treatment on the biocompatibility of microbial polyhydroxyalkanoates, Biomaterials 23 (5) (2002) 1391–1397.
[263] K. Zhao, Y. Deng, et al., Polyhydroxyalkanoate (PHA) scaffolds with good mechanical properties and biocompatibility, Biomaterials 24 (6) (2003) 1041–1045.
[264] Y. Deng, K. Zhao, et al., Study on the three-dimensional proliferation of rabbit articular cartilage-derived chondrocytes on polyhydroxyalkanoate scaffolds, Biomaterials 23 (20) (2002) 4049–4056.
[265] C.H. Rivard, C. Chaput, et al., Bioabsorbable synthetic polyesters and tissue regeneration: a study on the three-dimensional proliferation of ovine chondrocytes and osteoblasts [Polyesters biosynthetiques absorbables et regeneration tissulaire. Etude de la proliferation tridimensionnelle de chondrocytes et osteoblastes ovins], Ann. Chir. 50 (8) (1996) 651–658.
[266] R. Sodian, S.P. Hoerstrup, et al., Early *in vivo* experience with tissue-engineered trileaflet heart valves, Circulation 102 (19) (2000a).
[267] R. Sodian, S.P. Hoerstrup, et al., Tissue engineering of heart valves: *In vitro* experiences, Ann. Thorac. Surg. 70 (1) (2000b) 140–144.
[268] R. Sodian, J.S. Sperling, et al., Fabrication of a trileaflet heart valve scaffold from a polyhydroxyalkanoate biopolyester for use in tissue engineering, Tissue Eng. 6 (2) (2000) 183–188.

[269] Y.W. Wang, Q. Wu, et al., Attachment, proliferation and differentiation of osteoblasts on random biopolyester poly(3-hydroxybutyrate-co-3-hydroxyhexanoate) scaffolds, Biomaterials 25 (4) (2004) 669–675.
[270] Z. Zheng, F.F. Bei, et al., Effects of crystallization of polyhydroxyalkanoate blend on surface physicochemical properties and interactions with rabbit articular cartilage chondrocytes, Biomaterials 26 (17) (2005) 3537–3548.
[271] K. Tuzlakoglu, N. Bolgen, et al., Nano- and micro-fiber combined scaffolds: a new architecture for bone tissue engineering, J. Mater. Sci. Mater. Med. 16 (12) (2005) 1099–1104.
[272] M.-H. Ho, P.-Y. Kuo, et al., Preparation of porous scaffolds by using freeze-extraction and freeze-gelation methods, Biomaterials 25 (1) (2004) 129.
[273] M.-H. Ho, D.-M. Wang, et al., Preparation and characterization of RGD-immobilized chitosan scaffolds, Biomaterials 26 (16) (2005) 3197.
[274] V.M. Correlo, L. Boesel, et al., Hydroxyapatite reinforced chitosan and polyester blends for biomedical applications, Macromol. Mater. Eng. 290 (12) (2005a) 1157–1165.
[275] V.M. Correlo, L. Boesel, et al., Properties of melt processed chitosan and aliphatic polyester blends, Mater. Sci. Eng. A 403 (1–2) (2005b) 57–68.
[276] A.R. Pinto, V.M. Correlo, et al., Behaviour of human bone marrow mesenchymal stem cells seeded on fiber bonding chitosan polyester based for bone tissue engineering Scaffolds. in: 8th TESI Annual Meeting, Shanghai, 2005.
[277] D.-J. Park, B.-H. Choi, et al., Injectable bone using chitosan-alginate gel/mesenchymal stem cells/BMP-2 composites, J. Craniomaxillofac. Surg. 33 (1) (2005) 50.
[278] L. Kong, Y. Gao, et al., Preparation and characterization of nano-hydroxyapatite/chitosan composite scaffolds, J. Biomed. Mater. Res. A 75A (2) (2005) 275–282.
[279] Y. Zhang, M. Ni, et al., Calcium phosphate-chitosan composite scaffolds for bone tissue engineering, Tissue Eng. 9 (2) (2003) 337–345.
[280] Y. Zhang, M. Zhang, Synthesis and characterization of macroporous chitosan/calcium phosphate composite scaffolds for tissue engineering, J. Biomed. Mater. Res. 55 (3) (2001) 304–312.
[281] M. Gravel, T. Gross, et al., Responses of mesenchymal stem cell to chitosan-coralline composites microstructured using coralline as gas forming agent, Biomaterials 27 (9) (2006) 1899.
[282] F. Zhao, Y. Yin, et al., Preparation and histological evaluation of biomimetic three-dimensional hydroxyapatite/chitosan-gelatin network composite scaffolds, Biomaterials 23 (15) (2002) 3227.
[283] F. Zhao, W.L. Grayson, et al., Effects of hydroxyapatite in 3-D chitosan-gelatin polymer network on human mesenchymal stem cell construct development, Biomaterials 27 (9) (2006) 1859.
[284] S.E. Kim, J.H. Park, et al., Porous chitosan scaffold containing microspheres loaded with transforming growth factor-[beta]1: implications for cartilage tissue engineering, J. Control Release 91 (3) (2003) 365.
[285] A. Subramanian, H.-Y. Lin, Cross-linked chitosan: its physical properties and the effects of matrix stiffness on chondrocyte cell morphology and proliferation, J. Biomed. Mater. Res. A 75A (3) (2005) 742–753.
[286] Oliveira, J.T., Correlo, V.M., et al. (2005). Chitosan-Polyester Scaffolds Seeded with Bovine Articular Chondrocytes for Cartilage Tissue Engineering Applications. Bologna: Artificial Organs.
[287] T. Guo, J. Zhao, et al., Porous chitosan-gelatin scaffold containing plasmid DNA encoding transforming growth factor-[beta]1 for chondrocytes proliferation, Biomaterials 27 (7) (2006) 1095.
[288] A. Chenite, C. Chaput, et al., Novel injectable neutral solutions of chitosan form biodegradable gels in situ, Biomaterials 21 (21) (2000) 2155.
[289] C.D. Hoemann, J. Sun, et al., Tissue engineering of cartilage using an injectable and adhesive chitosan-based cell-delivery vehicle, Osteoarthr. Cartil. 13 (4) (2005) 318.
[290] A.F. Black, C. Bouez, et al., Optimization and characterization of an engineered human skin equivalent, Tissue Eng. 11 (5–6) (2005) 723–733.
[291] Y.-C. Huang, Y.-Y. Huang, et al., Manufacture of porous polymer nerve conduits through a lyophilizing and wire-heating process,

J. Biomed. Mater. Res. B Appl. Biomater. 74B (1) (2005) 659–664.

[292] Q. Ao, A. Wang, et al., Manufacture of multimicrotubule chitosan nerve conduits with novel molds and characterization in vitro, J. Biomed. Mater. Res. A 77A (1) (2006) 11–18.

[293] T. Chandy, G.H. Rao, et al., The development of porous alginate/elastin/PEG composite matrix for cardio-vascular engineering, J. Biomater. Appl. 17 (4) (2003) 287–301.

[294] Y.W. Wang, Q. Wu, et al., Evaluation of three-dimensional scaffolds made of blends of hydroxyapatite and poly(3-hydroxybutyrate-co-3-hydroxyhexanoate) for bone reconstruction, Biomaterials 26 (8) (2005) 899–904.

[295] S. Itoh, A. Matsuda, et al., Effects of a laminin peptide (YIGSR) immobilized on crab-tendon chitosan tubes on nerve regeneration, J. Biomed. Mater. Res. B Appl. Biomater. 73B (2) (2005) 375–382.

[296] M. Risbud, M. Endres, et al., Biocompatible hydrogel supports the growth of respiratory epithelial cells: possibilities in tracheal tissue engineering, J. Biomed. Mater. Res. 56 (1) (2001) 120–127.

[297] C.X.F. Lam, X.M. Mo, et al., Scaffold development using 3D printing with a starch-based polymer, Mater. Sci. Eng. C 20 (1-2) (2002) 49–56.

[298] M.E. Gomes, C.M. Bossano, et al., *In vitro* localization of bone growth factors in constructs of biodegradable scaffolds seeded with marrow stromal cells and cultured in a flow perfusion bioreactor, Tissue Eng. 12 (1) (2006) 177–188.

[299] M.I. Santos, S. Fuchs, et al., Response of micro- and macrovascular endothelial cells to starch-based fiber meshes for bone tissue engineering, Biomaterials 28 (2) (2007) 240–248.

[300] A.J. Salgado, J.E. Figueiredo, et al., Biological response to pre-mineralized starch based scaffolds for bone tissue engineering, J. Mater. Sci. Mater. Med. 16 (3) (2005) 267–275.

[301] Y. Dausse, L. Grossin, et al., Cartilage repair using new polysaccharidic biomaterials: macroscopic, histological and biochemical approaches in a rat model of cartilage defect, Osteoarthr. Cartil. 11 (1) (2003) 16.

[302] M. Abe, M. Takahashi, et al., The effect of hyaluronic acid with different molecular weights on collage cross-link synthesis in cultured chondrocytes embedded in collagen gels, J. Biomed. Mater. Res. 75A (2005) 494–499.

[303] J.A. Burdick, C. Chung, et al., Controlled degradation and mechanical behavior of photopolymerized hyaluronic acid networks, Biomacromolecules 6 (2005) 386–391.

[304] N.J. Goodstone, A. Cartwright, et al., Effects of high molecular weight hyaluronan on chondrocytes cultured within a resorbable gelatin sponge, Tissue Eng. 10 (2004) 621–631.

[305] C. Arrigoni, D. Camozzi, et al., The effect of sodium ascorbate on the mechanical properties of hyaluronan-based vascular constructs, Biomaterials 27 (4) (2006) 623.

[306] K. Hemmrich, D. von Heimburg, et al., Implantation of preadipocyte-loaded hyaluronic acid-based scaffolds into nude mice to evaluate potential for soft tissue engineering, Biomaterials 26 (34) (2005) 7025.

[307] C. Tonello, B. Zavan, et al., *In vitro* reconstruction of human dermal equivalent enriched with endothelial cells, Biomaterials 24 (7) (2003) 1205.

[308] B. Zavan, P. Brun, et al., Extracellular matrix-enriched polymeric scaffolds as a substrate for hepatocyte cultures: *in vitro* and *in vivo* studies, Biomaterials 26 (34) (2005) 7038.

[309] B.W. Ziegelaar, J. Aigner, et al., The characterization of human respiratory epithelial cells cultured on resorbable scaffolds: first steps towards a tissue engineered tracheal replacement, Biomaterials 23 (6) (2002) 1425.

[310] A. Ramamurthi, I. Vesely, Evaluation of the matrix-synthesis potential of cross-linked hyaluronan gels for tissue engineering of aortic heart valves, Biomaterials 26 (9) (2005) 999.

[311] N.J. Turner, C.M. Kielty, et al., A novel hyaluronan-based biomaterial (Hyaff-11(R)) as a scaffold for endothelial cells in tissue engineered vascular grafts, Biomaterials 25 (28) (2004) 5955.

[312] Y. Sumita, M.J. Honda, et al., Performance of collagen sponge as a 3-D scaffold for tooth-tissue engineering, Biomaterials 27 (17) (2006) 3238–3248.

[313] E. Sachlos, N. Reis, et al., Novel collagen scaffolds with predefined internal morphology

made by solid freeform fabrication, Biomaterials 24 (8) (2003) 1487–1497.
[314] P.M. Taylor, E. Sachlos, et al., Interaction of human valve interstitial cells with collagen matrices manufactured using rapid prototyping, Biomaterials 27 (13) (2006) 2733–2737.
[315] P. Engbers-Buijtenhuijs, L. Buttafoco, et al., Biological characterization of vascular grafts cultured in a bioreactor, Biomaterials 27 (11) (2006a) 2390–2397.
[316] P. Engbers-Buijtenhuijs, L. Buttafoco, et al., Biological characterization of vascular grafts cultured in a bioreactor, Biomaterials 27 (11) (2006b) 2390–2397.
[317] D. Seliktar, R.A. Black, et al., Dynamic mechanical conditioning of collagen-gel blood vessel constructs induces remodeling, in vitro. Ann. Biomed. Eng. 28 (4) (2000) 351–362.
[318] L. Meinel, V. Karageorgiou, et al., Bone tissue engineering using human mesenchymal stem cells: effects of scaffold material and medium flow, Ann. Biomed. Eng. 32 (1) (2004) 112–122.
[319] M. Kino-Oka, Y. Maeda, et al., A kinetic modeling of chondrocyte culture for manufacture of tissue-engineered cartilage, J. Biosci. Bioeng. 99 (3) (2005) 197–207.
[320] Y. Stark, K. Suck, et al., Application of collagen matrices for cartilage tissue engineering, Exp. Toxicol. Pathol. 57 (4) (2006) 305–311.
[321] Z. Feng, M. Yamato, et al., Investigation on the mechanical properties of contracted collagen gels as a scaffold for tissue engineering, Artif. Organs 27 (1) (2003) 84–91.
[322] Q. Lu, K. Ganesan, et al., Novel porous aortic elastin and collagen scaffolds for tissue engineering, Biomaterials 25 (22) (2004) 5227–5237.
[323] J. Stitzel, J. Liu, et al., Controlled fabrication of a biological vascular substitute, Biomaterials 27 (7) (2006) 1088–1094.
[324] L. Buttafoco, P. Engbers-Buijtenhuijs, et al., Physical characterization of vascular grafts cultured in a bioreactor, Biomaterials 27 (11) (2006) 2380–2389.

17 Fabrication of Tissue Engineering Scaffolds

Adam Kramschuster and Lih-Sheng Turng

OUTLINE

17.1 Introduction

17.1.1 Tissue Engineering

Tissue engineering is an interdisciplinary field aimed at the development of biological substitutes that restore, maintain, or improve tissue function [1]. A highly porous biodegradable scaffold is essential to accommodate mammalian cells and guide their growth in three dimensions [2]. In the past, natural and synthetic polymers have routinely been used as substrates to provide this temporary scaffolding for transplanted cells as they excrete their extracellular matrix (ECM) and form new tissues or organs [1–4]. Although extensive research has been performed with both types of polymers, synthetic polymers offer several advantages over natural polymers such as collagen and fibrin. They can be prepared in a reproducible manner in almost unlimited quantities, and their physical, chemical, and mechanical properties may be easily altered by chemical modifications. In addition, they can be easily processed with conventional polymer processing equipment [5]. Some common synthetic biodegradable polymers currently used as scaffolding materials include polylactide (PLA), polyglycolide (PGA), and their

Ebnesajjad: Handbook of Biopolymers and Biodegradable Plastics. http://dx.doi.org/10.1016/B978-1-4557-2834-3.00017-3
Copyright © 2013 Elsevier Inc. All rights reserved.

copolymers, and polycaprolactone (PCL) [2–4]. Myriads of new materials including tyrosine-derived polycarbonates and trimethylene carbonate-based materials are also being explored as alternative synthetic polymers for tissue engineering scaffolds [5,6]. Many tissue engineering scaffold fabrication processes have been developed in order to meet the demands of tissue engineers. However, the majority of current scaffold fabrication techniques can be described as batch processes and/or use organic solvents, which can be detrimental to cell survival and tissue growth [7]. While these techniques may be adequate and essential for studying the effects of the substrate material, porosity, pore size, interconnectivity of the pores, mechanical and chemical properties, growth factors, and nutrient transport on the effects of tissue regeneration both *in vitro* and *in vivo*, they do not address the need for cost-effective manufacturing processes to meet patient needs. The ability to mass produce highly porous, highly interconnected scaffolds with complex geometries is essential to provide off-the-shelf availability [8]. Some of the most important and widely used fabrication methods using synthetic biodegradable polymers are explored here.

17.1.2 Tissue Engineering Overview

Conventional clinical therapies are commonly practiced to replace, repair, or facilitate the regeneration of damaged tissue, and can be classified as medical (primarily pharmacological) or surgical. Among surgical therapies, treatments provide either temporary or permanent patient support and can typically be categorized as one of the following: *repair*, *replacement*, *reconstruction*, or *removal* [9]. The majority of this overview will cover the replacement of damaged or diseased tissue.

Some common examples of repair involve stitches in skin and pins or screws in bone to facilitate the natural wound-healing response. These often require a second surgery to remove the fixation devices. Another more advanced example of repair can be suturing of peripheral nerve injuries.

The reconstruction of tissue generally involves the replacement of the damaged tissue with an autologous tissue of a different type. Two examples of this include coronary bypass surgery, where the saphenous vein is harvested from the leg of the patient, and the use of the small intestine for reconstruction of bladder defects and the esophagus [9]. However, especially in the case of coronary bypass surgery, donor site morbidity is an issue and subsequent loss of the leg supplying the donor vein can result. Tissue dysfunction can sometimes be remedied by surgical removal of the tissue or organ. Examples of this include removal of the appendix and a herniated vertebral disk.

While these surgical procedures have produced exceptional and sometimes life-saving results, there are many drawbacks to these conventional techniques. In the case of surgical repair, a second surgery is often needed to remove the fixation device, increasing costs and lengthening the recovery time. When discussing the replacement of tissue or organs, donor site morbidity, immune rejection, organ donor shortages, and a myriad of other challenges need to be overcome in order to successfully treat all patients in need. The shortcomings of these techniques led researchers to investigate new methods of treating disease and providing tissue and organs derived from the patient's own cells (or embryonic stem cells). If new tissue could be created from the patient's own cells, it would alleviate the organ donor shortage problem, as well as reduce the chance of immune rejection and the introduction of foreign pathogens. Additionally, if the tissue or organ could be created from only a handful of cells, issues associated with donor site morbidity could also be addressed. Therefore, tissue engineering was born in an attempt to overcome the shortcomings previously mentioned with regard to conventional clinical therapies while still working to restore, maintain, or improve tissue function.

Although the term "tissue engineering" was not coined until 1987, research into organ and tissue transplantation has been occurring since 3000 B.C. with attempts at the transplantation of noses. Since then, much progress has been made, with the 1950s marking successful transplants of kidneys, livers, and skin. The concept of bioresorbable vascular grafts and bioresorbable sutures was introduced in the 1960s, and by the 1980s research had progressed to seeding cells on 3D polymeric scaffolds [1]. Figure 17.1 displays several areas of tissue engineering currently being explored [10]. The interdisciplinary nature of tissue engineering makes it impossible to cover all aspects without providing a full textbook on the subject. Therefore, the following information is meant to provide a brief overview of tissue engineering strategies. For more detailed information regarding tissue engineering,

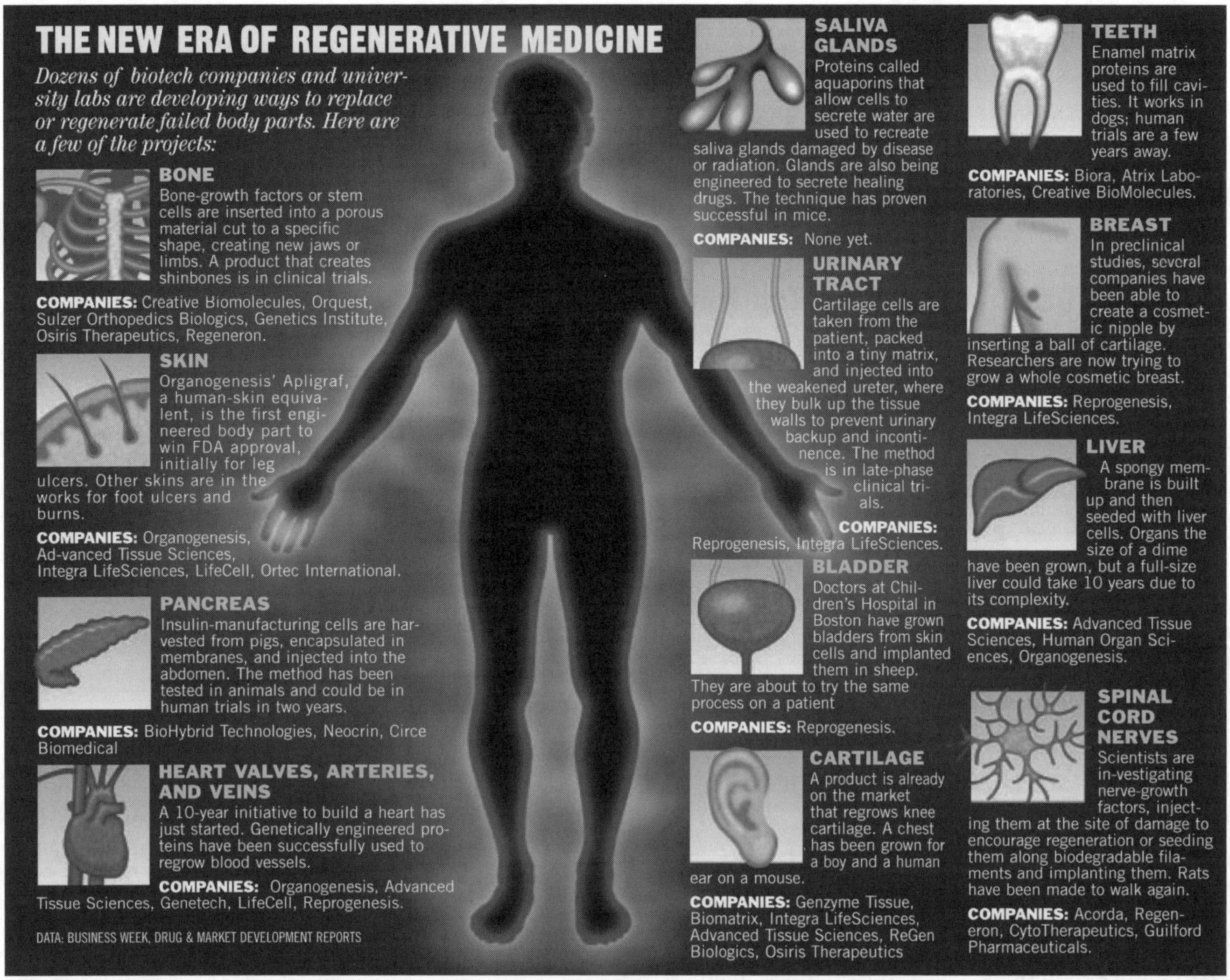

Figure 17.1 Several researchers in academia and industry have applied tissue engineering strategies in the hopes of producing a variety of tissues and/or organs [10].

the reader is referred to full texts on this subject [11,12].

There are three basic tissue engineering strategies that are used to restore, maintain, or improve tissue function, and they can be summarized as *cell transplantation*, *scaffold-guided regeneration*, and *cell-loaded scaffold implantation*.

1. Cell transplantation involves the removal of healthy cells from a biopsy or donor tissue and then injecting the healthy cells directly into the diseased or damaged tissue. However, this technique does not guarantee tissue formation and generally has less than 10% efficiency.
2. Scaffold-guided regeneration involves the use of a biodegradable scaffold implanted directly into the damaged area to promote tissue growth. An example of this could be a severed peripheral nerve that utilizes a conduit-shaped scaffold to guide the nerve ends toward each other to restore function. Another example could involve a defect in bone where a scaffold in the shape of the defect is implanted.
3. Cell-loaded scaffold implantation involves the isolation of cells from a patient and a biodegradable scaffold that is seeded with cells and then implanted into the defect location. Prior to implantation, the cells can be subjected to an *in vitro* environment that mimics the *in vivo* environment in which the cell–polymer constructs can develop into functional tissue. This *in vitro* environment is generally the result

of a bioreactor, which provides growth factors and other nutrients while also providing mechanical stimuli to facilitate tissue growth. Bioreactors will not be discussed in further detail in this chapter; for more information, the reader is referred to Ref. [13] and the references cited therein.

17.2 Tissue Engineering Scaffolds

17.2.1 Function of Scaffolds

The function of tissue engineering scaffolds is to direct the growth of cells migrating from surrounding tissue (in the case of scaffold-guided regeneration) or the growth of cells seeded within the porous structure of the scaffold [3]. The scaffold must provide temporary structural support for cell adhesion, proliferation and differentiation, nutrient transport, and the excretion of waste while the cells secrete their own ECM.

17.2.2 Types of Scaffolds

Polymeric tissue engineering scaffolds can be divided into three categories: *hydrogels*, *meshes*, and *sponges*. Hydrogels contain over 99% water and constitute a completely different array of properties and materials when compared to meshes and sponges. They do not employ traditional polymer processing techniques. Therefore, they will not be discussed further, though the reader is referred to Ref. [14] and the references cited within for more information regarding hydrogels. Meshes and sponges have been fabricated from a variety of synthetic biodegradable polymers and polymer processing techniques [15]. The types of polymers and techniques used to fabricate these types of scaffolds will be discussed in the following sections.

17.2.3 Synthetic Biodegradable Polymer Scaffold Materials

Dozens of synthetic biodegradable polymers have been used and developed for use as tissue engineering scaffold materials. The term "biodegradable" encompasses "bioresorbable," "bioerodible," and "bioabsorbable" in this text, though technically speaking, they all have distinct definitions that are found in Ref. [16]. Materials that undergo bulk hydrolysis (or bulk erosion) absorb water and then break down rather abruptly, causing a rapid decrease in mechanical properties. However, materials that undergo surface hydrolysis (or surface erosion) tend to maintain their mechanical properties for an extended period of time as they degrade. The majority of synthetic polymers that have traditionally been used for tissue engineering scaffolds (PLA, PGA, and their copolymers, and PCL) all undergo bulk erosion, while materials like polyanhydrides (PAs) undergo surface erosion and have gained increased interest. In general, the synthetic polymer selected for a specific tissue engineering application should degrade at the same rate as the tissue forms, be biodegradable, and be easily fabricated. Some common polymers used for fabricating tissue engineering scaffolds are discussed in the following section.

17.2.3.1 Polylactide

PLA is an aliphatic polyester prepared by either direct condensation of lactic acid or by ring-opening polymerization of the cyclic lactide dimer [17]. It is derived from corn starch and consists of two stereoisomeric forms: L-lactide and D-lactide. Poly(L-lactide) (PLLA) and poly(D-lactide) are the two homopolymers, and poly(D,L-lactide) is a copolymer obtained from a mixture of D- and L-lactic acid. PLLA and poly(D-lactide) are semicrystalline materials, though incorporation of more than 15% D-lactide into PLLA makes the polymer amorphous [16]. Poly(D,L-lactide) also exhibits a lower melting point, lower tensile strength, higher elongation at break, and a faster degradation rate than either PLLA or poly(D-lactide) [5,17,18]. When broken down via hydrolytic mechanisms, lactic acid is absorbed into the body, sometimes causing an inflammatory response. PLA can have up to 40% crystallinity with a melting point of 170 °C and a glass transition temperature of 56 °C [19]. It can be easily processed with conventional polymer processing equipment and is soluble in most organic solvents, namely methylene chloride and chloroform. These properties make PLA ideal for injection molding and extrusion, as well as most solvent-based scaffold manufacturing techniques.

17.2.3.2 Polyglycolide

PGA is the simplest linear aliphatic polyester and is prepared by either direct condensation of glycolic acid or by ring-opening polymerization of the cyclic glycolide dimer. PGA is semicrystalline (45–55%), with a melting point of 224–226 °C and a glass transition temperature of 36–40 °C [18]. When compared to PLA, PGA degrades at a faster rate (4–12 months for PGA and up to 2 years for PLA) and also has a higher tensile strength and modulus [4]. When degraded via hydrolytic mechanisms, PGA breaks down into glycolic acid, which can be excreted by urine [19]. PGA was used to develop the first totally synthetic absorbable suture, marketed as Dexon in the 1960s by Davis and Geck, Inc. (Danbury, CT) [18]. Like PLA, PGA can be easily processed with extrusion and injection molding equipment, though it is not soluble in most organic solvents (hexafluoroisopropanol is an exception).

17.2.3.3 Poly(lactide-co-glycolide)

Poly(lactide-co-glycolide) (PLGA) is a copolymer of PLA and PGA used to modify the mechanical and degradation properties of the two homopolymers. However, it is important to note that there is no linear relationship between the copolymer composition and the mechanical and degradation properties of the individual materials [20]. Morphological changes lead to an increase in the rate of hydrolysis and 50:50 copolymers degrade more rapidly than either PLA or PGA alone [21].

17.2.3.4 Polycaprolactone

PCL is a semicrystalline polymer that is prepared by the ring-opening polymerization of the cyclic monomer ε-caprolactone. It has a glass transition temperature of −60 °C and a melting temperature of approximately 59–64 °C [20]. PCL has a slow degradation rate (up to 2 years) and a high permeability; therefore, it is well-suited for long-term implants and drug delivery systems [4,5,18,22]. Due to its slow degradation rate, copolymers of ε-caprolactone and D,L-lactide have been synthesized to yield materials with more rapid degradation rates [20].

17.2.3.5 Polyanhydride

PA is synthesized by the dehydration of diacid molecules by melt polycondensation. Degradation times can be adjusted from days to years by the degree of hydrophobicity of the monomer selection [20]. The main application of PA is drug delivery due to the inherent surface erosion properties of the PA that allows for the drug to be released at a known rate [19,22]. A contributing factor to the use of PA for drug delivery devices is its low mechanical properties, which hinder its use as load-bearing scaffolds for orthopedic applications [5,19].

17.2.3.6 Other Biodegradable Polymers

Many other biodegradable polymers, such as polypropylene fumarate (PPF), polyorthoesters (POE), polyurethanes (PU), polypyrroles (PPy), polydioxanones (PDS), tyrosine-derived polycarbonates, and polyphosphazenes, have been developed and are being tested for use as scaffolding materials. These synthetic, biodegradable polymers all exhibit unique mechanical, chemical, and degradation properties that make them suitable materials for tissue engineering scaffolds. Extensive reviews on these polymers as well as many others can be found in Refs. [6,19,22–24] and the references cited therein.

17.2.4 Scaffold Requirements

Ideally, a scaffold should have the following characteristics: (1) 3D and highly porous structure with an interconnected pore network for cell growth and flow transport of nutrients and metabolic waste; (2) biocompatible and bioresorbable with a controllable degradation and resorption rate to match cell/tissue growth *in vitro* and/or *in vivo*; (3) suitable surface chemistry for cell attachment, proliferation, and differentiation; (4) mechanical properties to match those of the tissues at the site of implantation; and (5) able to be mass produced with consistent part-to-part quality and without the use of organic solvents [25].

17.2.4.1 Porosity, Pore Size, and Interconnectivity

Highly porous scaffolds are required to allow for cells to infiltrate and attach to the scaffold, to provide a high surface area-to-volume ratio for polymer–cell interactions, and to obtain minimal diffusion constraints during cell culture. Past research has stated that a scaffold porosity of greater than 90% is important for tissue engineering applications [26]. However, much of the basis for this is due to the early methods (namely solvent casting (SC)/particulate leaching (PL)) being unable to achieve high interconnectivity at porosities less than 90%. In fact, researchers have used scaffolds with porosities ranging from 55% to 74% for bone growth due to increased mechanical properties at lower porosities [27–29].

The optimal pore size for tissue regeneration is dependent on the type of tissue [2]. However, even for bone regeneration, no consensus regarding the optimal pore size in scaffolds has been determined. Pore sizes ranging from 50 to 710 μm have been suggested for bone regeneration, with many studies stating macropores in the range of 150–350 μm are ideal [2,24,30]. On the other hand, the recommended pore sizes range from 200 to 300 μm for the growth of fibrocartilaginous tissue [31] and pores on the order of 5 μm are necessary for neovascularization [2]. Therefore, macropores on the order of hundreds of micrometers with an interconnecting network on the order of tens of micrometers have been deemed suitable for tissue regeneration. At the same time, the effects of surface chemistry, culture conditions, mechanical properties, and degradation rate play a large role in tissue formation, and will be discussed further.

17.2.4.2 Biocompatibility and Degradation

The material used for the polymeric scaffold must exhibit good biocompatibility, meaning that it must not elicit an unresolved inflammatory response nor demonstrate extreme immunogenicity or cytotoxicity [7]. For example, PLA hydrolyzes to lactic acid, which is a normal by-product of muscle contractions in animals. The lactic acid is then further metabolized through the tricarboxylic acid cycle and excreted as carbon dioxide (CO_2) and water [20]. The issue of biocompatibility is directly linked to the degradation of the material as well. For instance, if the degradation rate of the polymer is too fast, not only will it cease to provide the necessary mechanical support for the tissue, but also the surrounding tissue cannot eliminate the acid by-products, resulting in an inflammatory or toxic response [5,20]. If it is too slow, it may impede the growth of new tissue. To remedy this, synthetic biodegradable polymers may be tailored to elicit specific degradation rates based on their composition (e.g., PLGA).

17.2.4.3 Surface Chemistry

The nature of the polymer surface can affect the ability of cells and proteins to attach to its surface, proliferate, and differentiate. Polymers like PLA and PGA are relatively hydrophobic, and it is difficult to efficiently and evenly seed cells into porous matrices fabricated from these polymers [32,33]. The use of hydrophilic polyvinyl alcohol (PVOH), prewetting with ethanol, hydrolysis with NaOH, oxidation with a perchloric acid mixture solution, oxygen or ammonia plasma discharge treatment, or adding cell-adhesive proteins [34–39] can be used to improve cell adhesion in hydrophilized scaffolds.

17.2.4.4 Mechanical Properties

The mechanical properties of the scaffold fabricated should closely match the mechanical properties of the neotissues to be generated in order to provide support during the initial stages of tissue growth. If the mechanical properties such as compressive strength are too low, the scaffold could be deformed or crushed, leading to deformed tissue growth, or no tissue growth at all. At the same time, if the compressive strength is too high, the cells may not be subjected to the proper *in vivo* conditions needed to support cell growth. Additionally, especially in load-bearing instances, scaffold mechanical properties that exceed the surrounding bone may lead to "stress-shielding," a condition where the surrounding bone experiences a decrease in density. However, this is more prevalent with permanent implants. Also important to consider is the elastic modulus of the scaffolds. Differences in elastic moduli between the scaffold and natural tissue can result in differing amounts of strain at the same stress levels. This can cause delamination of the scaffold from the surrounding tissue.

17.2.4.5 Cost-Effective Mass Production with Controlled Properties

In order for affordable, biodegradable tissue engineering scaffolds to be routinely used in the medical field, a manufacturing process must be developed to mass produce geometrically complex scaffolds without the use of organic solvents. It is vital that the biodegradable tissue engineering scaffolds be fabricated in a consistent and cost-effective fashion so that researchers, surgeons, and patients alike will receive ample supply of scaffolds with known properties and proven performance to meet the growing demands and accelerate the research progress.

> *The patient need that exists cannot be met by making one construct at a time on a bench top in some research laboratory. Accepting the*

challenge of imitating nature must include the development of cost-effective manufacturing processes [11].

17.3 Scaffold Manufacturing Methods

17.3.1 Introduction

Since the mid-1980s, many techniques have been developed and modified in order to fabricate polymeric scaffolds to meet the needs for tissue engineering. The properties and characteristics of these scaffolds are dependent not only on the type of materials used but also on the fabrication technique. Each technique offers advantages and disadvantages in the form of control over pore morphology, ability to incorporate bioactive molecules, mechanical properties, and cost. Techniques that involve heat and pressure and/or organic solvents inhibit the use of incorporating cells and bioactive molecules directly into the polymer. Other techniques that do not utilize heat or organic solvents may not be ideal for mass production of complex geometries. Each specific tissue engineering application needs to be evaluated and the proper technique selected to provide the optimal solution. The common tissue engineering scaffold fabrication techniques are discussed next.

17.3.2 Solvent Casting/Particulate Leaching

This process begins with a biodegradable polymer (generally PLA or PLGA) that is dissolved in an organic solvent (generally chloroform or methylene chloride). A water-soluble porogen such as sodium chloride (NaCl), which has been sieved to a desired size range, is added to the polymer solution which is then cast into a mold of desired shape. After the solvent has evaporated, the material is vacuum dried to remove any residual solvent. The polymer/salt composite is then leached in water to extract the salt and dried again. This method was first described by Mikos *et al.* [40].

Advantages of the SC/PL process include control over the porosity (up to 97%) and pore size of the scaffold based on the percentage and size range of the porogen selected, respectively. Additionally, porogen selection can also be used to impart a desired pore shape. Scaffolds with spherical pores using paraffin spheres as a porogen and hexane as a solvent have also been investigated [41]. Methods devised to overcome the lack of pore interconnectivity of scaffolds fabricated using SC/PL include membrane lamination [42] and a salt fusion technique [43]. Widmer *et al.* [44] used the technique developed by Mikos *et al.* to extrude porous biodegradable polymer conduits for guided tissue regeneration, though the plunger style extruder still classifies this as a batch process. Concerns have also been raised regarding the issue of residual organic solvents affecting the viability of seeded cells [7], though chondrocytes for cartilage formation were successfully seeded onto scaffolds fabricated by this technique [45].

17.3.3 Gas Foaming

This process eliminates the need for organic solvents by using a gas (usually CO_2) to foam the polymeric material. Solvent cast polymer disks or polymer disks formed by compression molding polymer pellets are saturated with CO_2 in a high-pressure chamber. Once the polymer has become saturated with CO_2, a rapid pressure drop triggers a thermodynamic instability in the polymer/gas solution, leading to the nucleation and growth of cells/pores. This method was first reported by Mooney *et al.* [46].

Utilizing the gas foaming (GF) technique allows for porosities up to 97% with pores up to 100 μm in diameter without the use of organic solvents. The absence of high processing temperatures also allows for the possibility of incorporating growth factors into the scaffold. However, a solid skin layer at the surface was observed, which makes it difficult to uniformly seed cells throughout the matrix and deliver nutrients. Also, the majority of the pores are not interconnected. Following a method proposed by Mooney *et al.*, and Wang and co-workers [47] used ultrasonic cavitation in an attempt to rupture the cell walls and increase interconnectivity. Finally, the free boundary needed to create gas-foamed structures with high porosity negates the possibility of creating complex 3D shapes without a secondary forming process.

17.3.4 Gas Foaming/Particulate Leaching

Due to the aforementioned drawbacks of the SC/PL and GF processes, namely the use of organic solvents, batch processes, and poor interconnectivity, processes consisting of GF/PL have been developed. Tsuji *et al.* [48] solution-cast PLA with water-soluble polyethylene oxide (PEO) into thin films (10 and 50 μm thick), then subsequently leached the PEO in water. Porosities of 90% were achieved and an increase in pore size ranged from 1 to 1000 μm. Harris *et al.* [49] expanded upon the GF technique developed by Mooney *et al.* by incorporating salt as a particulate. In the study, ground pellets of PLGA and NaCl were compression molded into disks with a thickness of 3.4 mm. After undergoing the GF process, porosities up to 95% were achieved, with the pore size dependent on the size of NaCl used. The majority of the pores were interconnected, though some closed pores remained (up to 6% by volume). Note that Riddle and Mooney [50] found a smaller initial PLGA particle size (<75 μm) that led to higher compressive moduli and increased interconnectivity of the pores when compared to a larger PLGA particle size (250–425 μm). This approach was also adapted by Sheridan *et al.* [51] to incorporate growth factors into the scaffolds, which are capable of sustained delivery.

Nam *et al.* [52] utilized solvents and GF salts to further study the GF/SC/PL methods described previously. PLLA was dissolved in methylene chloride or chloroform and ammonium bicarbonate was added to the solutions as the salt. The mixtures were cast into a disk-shaped mold and dried. The GF process was accomplished either by drying under vacuum or submersing in 90 °C water until no more gas was generated. Foams of up to 95% porosity were created, with the pore sizes highly dependent on the size range of the ammonium bicarbonate particulates. Based on this work, Yoon and Park [53] incorporated citric acids to the process and they were able to control the porosity and mechanical properties of the scaffolds.

17.3.5 Emulsion Freeze-Drying

The emulsion freeze-drying (EFD) method is also used to produce both tissue engineering scaffolds and drug delivery matrices. The method involves combining a polymer solution (e.g., PLGA dissolved in methylene chloride) and ultrapure distilled water, forming two immiscible layers. The immiscible phases are then homogenized to form an emulsion where water acts as the dispersed phase and the polymer solution acts as the continuous phase. After casting into a mold, the mixture is rapidly quenched with liquid nitrogen to solidify. Upon vacuuming at −55 °C and under 30 mTorr, the water and methylene chloride are sublimed. A vacuum desiccator is also used to remove the remaining solvents. The EFD method was first used for fabricating porous biodegradable PLGA scaffolds by Whang *et al.* [54].

This method is able to produce scaffolds with a high volume of an interconnected porous structure. The porosity ranges from 91% to 95%, and the median pore size is around 13–35 μm. However, one disadvantage of this process is its potential for closed-pore morphology [55]. Many parameters can be varied in the process such as the types of solvent, polymer, and their ratio. Biocompatible polymers that have been utilized with this method include PGA, PLLA, PLGA, and PPF blends [3]. Natural polymers, such as collagen, chitin, and alginate can also be used in emulsion freeze-drying.

17.3.6 Thermally Induced Phase Separation (TIPS)

The thermally induced phase separation (TIPS) method uses a high-temperature polymer solution, consisting of a biocompatible polymer in a solvent (e.g., phenol, dioxane, or naphthalene) [3]. When reducing the solution temperature to below the melting point of the solvent, phase separation occurs, forming a polymer-rich phase and a solvent-rich phase. The solvent of the solidified solvent-rich phase is sublimed, changing from the solid phase to the gaseous state directly (Fig. 17.2). TIPS was first reported for fabricating tissue engineering scaffolds by Lo *et al.* [57].

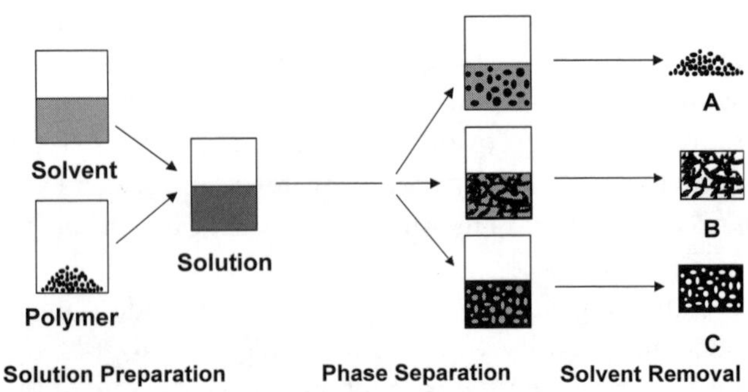

Figure 17.2 Schematic of thermally induced phase separation technique [56].

With this technique, a variety of foams with pore sizes ranging from 20 to 500 μm have been fabricated, and many of them use PLLA, poly(bisphenol A-phenylphosphonate) (BPA/PP), and its copolymer with poly[bis(2-ethoxy)-hydrophosphonic terephthalate] (PP/PPET) as the polymer base [56]. Lo *et al.* [57] were also able to incorporate proteins into the foam during processing with naphthalene, suffering only a 25% loss in activity. The type of polymer, type of solvent, polymer-to-solvent ratio, and the cooling rate all play a large role in the pore morphology [58]. Ma and Zhang [59] developed a novel liquid–liquid phase separation technique to create a 3D interconnected fibrous network with porous channels on the order of 50–500 nm.

In order to incorporate a macroporous structure with a nanofibrous structure, TIPS has been combined with PL techniques utilizing paraffin microspheres, salt or sugar particles, or sugar fibers [41,60]. As discussed previously, it is advantageous to provide macropores or channels for delivery of cells and growth factors, while smaller pore sizes are better for inducing neovascularization. Further research utilizing solid freeform fabrication (SFF) techniques for creating wax molds with a macroporous structure along with TIPS to create a nanofibrous morphology has been evaluated [61].

17.3.7 Solid Freeform Fabrication Techniques

SFF techniques were developed in the late 1980s and early 1990s as a way to rapid prototype part designs in a quick, cost-effective manner. Since then, advancements in computer aided design (CAD) and the types of materials processed have led to a plethora of research into the use of SFF techniques for manufacturing tissue engineering scaffolds. SFF techniques require a solid model of the scaffold to be developed prior to processing. Software programs then slice the model into several subunits that are then built up consecutively using an additive manufacturing technique to create a complex 3D structure. The main advantage of SFF techniques is the ability to precisely control porosity, pore size, pore shape, and interconnectivity. This aids in the viability of cells as well as being able to control the mechanical properties to a certain extent, as the bulk properties of the scaffold are highly dependent on porosity, material used, and the strength of the bond holding the polymer particles together.

17.3.7.1 Stereolithography

The stereolithography apparatus, or SLA, was the first commercial SFF technique available, supplied by 3D Systems, Inc. in 1988 [62]. It utilizes a photo-curable resin and an ultraviolet (UV) laser to polymerize the polymer layer by layer to create a 3D construct. The process (shown in Fig. 17.3) begins as a moveable platform and is lowered into a vat of photo-curable resin far enough so that the resin completely covers the platform. The platform is then raised and a leveling wiper or vacuum blade moves across the surface to provide an evenly coated layer of resin. To build each layer, the laser is guided across the surface, drawing a cross-sectional pattern in the *x*–*y* plane to form a solid section (through curing of the resin) [64]. This step can also be performed through photomasking, where each cross-section is imaged onto an erasable mask before it is placed over the liquid resin and subjected to UV light to selectively cure the material [64]. The platform is then lowered along with the cured layer, which is attached to the platform, and the process is repeated with the next slice of the model until the 3D geometry is completed. Once completed, the uncured resin is washed away and the 3D construct is then subjected to a postcuring process, yielding a fully cured part.

One of the main challenges to overcome with SLA for tissue engineering applications is the need for a photo-curable resin to be used as the scaffold material. Chu *et al.* [65] were the first group to use the SLA technique to fabricate tissue engineering scaffolds. In their process, they incorporated hydroxyapatite (HA) powder, a mineral found in human teeth and bone, into a low viscosity acrylate mixed with photoinitiators to form a UV-curable

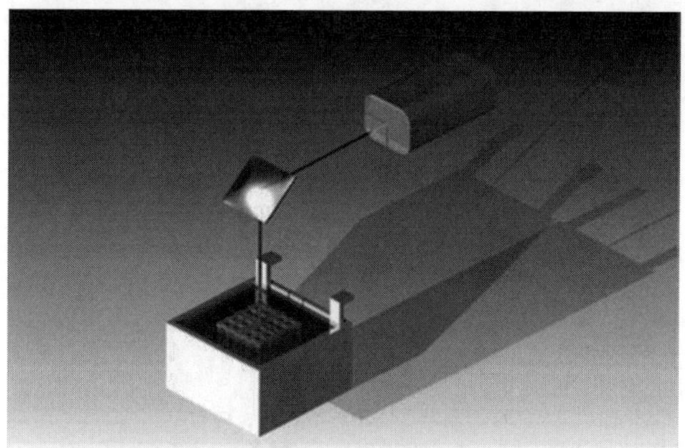

Figure 17.3 Stereolithography system [63].

slurry [66]. After exposing to the UV light and being cured, the resin was sintered to create a fully ceramic scaffold with 50% porosity. Similarly, Cooke *et al.* [67] used a solution of PPF, dimethyl fumarate, and bisacylphosphine oxide (photoinitiator), and fabricated tissue engineering scaffolds for bony substrates.

Limitations generally arise in the form of attainable resolution and the requirement for the use of photo-curable materials. While the thickness of the cross-sections can be selected through computer software, the realistic cross-sectional thickness is physically limited by the precision of the mechanical stepping system that moves the platform, which is typically no less than 100 μm [68]. The spot size of the laser can also be a concern, though a spot size of less than 5 μm has been reported [64]. Regardless, as the resolution increases, the time to manufacture increases dramatically [68].

17.3.7.2 Fused Deposition Modeling

The fused deposition modeling (FDM) process was first developed and commercialized by Stratasys, Inc. in 1992 [64]. Polymer filament is supplied on a spool and fed to an extrusion head where it is heated above the glass transition temperature (T_g) for amorphous polymers and just above the melt temperature (T_m) for semicrystalline polymers. The molten polymer is extruded as the extrusion head moves in the x–y plane to form a thin slice of the 3D model created by CAD software (Fig. 17.4). The polymer melt solidifies quickly and upon completion of the first slice, the base is lowered and the process is repeated until the 3D construct is created. The direction of material deposition, or laydown pattern, can be changed for each layer to provide variations in mechanical properties and pore morphology in the case of porous structures. Support structures can also be created from the same material for overhanging features and slight perforations allow for easy removal, though generally a separate material is used for the support structures [64,69].

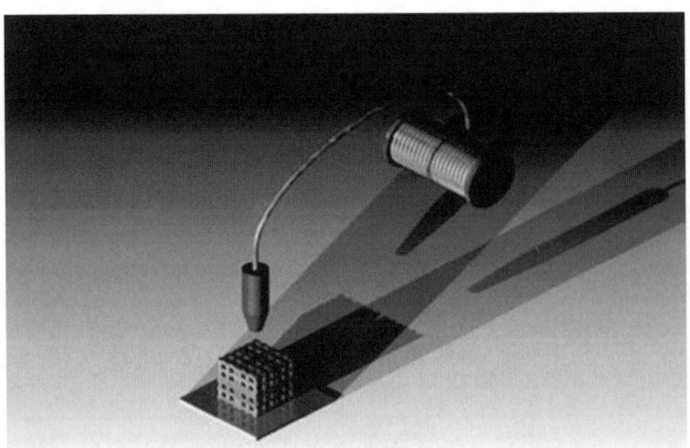

Figure 17.4 Fused deposition modeling process [63].

PCL was the first biodegradable polymer that was used to create 3D constructs with the FDM technique for use as tissue engineering scaffolds by Hutmacher *et al.* [70]. This work resulted in scaffolds with a porosity of 61% and complete pore interconnectivity. Further research by Zein *et al.* [71] resulted in scaffolds with a porosity ranging from 48% to 77% and pore sizes ranging from 160 to 700 μm. The porosity at small pore sizes was limited by the relatively large filaments required for the FDM process, which were 260 and 370 μm. The compressive moduli of the PCL scaffolds fabricated by FDM ranged from 4 MPa to 77 MPa.

Advantages of the FDM process include the ability to form a fully interconnected pore network in complex 3D structures. However, though a high degree of precision can be achieved in the x–y plane, control of the z-direction is limited and governed by the diameter of the material extruded through the extrusion head [68]. A wide variety of biodegradable polymers, including PCL and PLGA, can be used with FDM [3]. Additionally, no organic solvents are used in the FDM process. However, the extrusion process precludes the use of natural polymers and the introduction of bioactive molecules, and the large-diameter filaments used limit the incorporation of micron-sized pores or channels. Lastly, the use of secondary support structures may carry a risk of material contamination [72]. Nevertheless, the wide range of porosities and achievable mechanical properties make FDM a highly researched process for fabricating tissue engineering scaffolds.

17.3.7.3 Selective Laser Sintering

The selective laser sintering (SLS) process was first developed at the University of Texas at Austin in 1986 and commercialized by DTM Corporation in 1992 [64]. While this process can be used with a variety of materials including polymers, metals, and ceramics, the focus of this discussion will be on the use of polymers. This process (Fig. 17.5) begins with a layer of thermoplastic powder deposited and

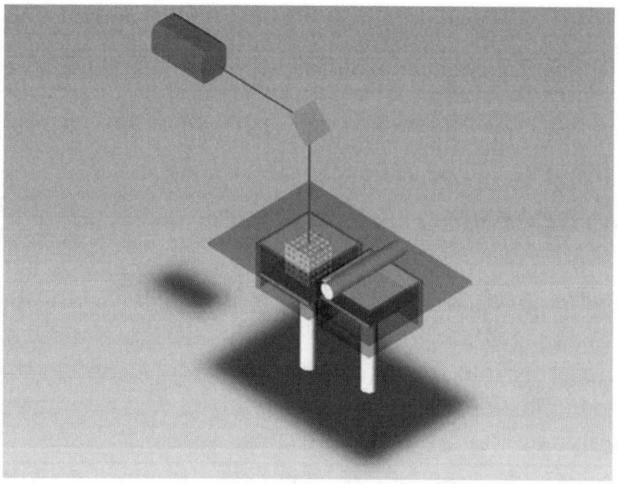

Figure 17.5 Selective laser sintering process [63].

leveled on a build platform. A heat-generating CO_2 laser is guided along the cross-section of the 3D model in the $x-y$ plane, selectively fusing the polymer particles together [73]. The powder bed may be heated to reduce the time and laser energy required to fuse the particles. The unfused particles remain in the plane and act as a support structure for subsequent layers. After the first layer is fused, the platform is lowered, another layer of material is deposited, and the process is repeated. The heat generated from the laser fuses each subsequent layer to the layer beneath it until the 3D construct is completed. The unfused polymer particles are then removed with a post-processing treatment, leaving behind the 3D model created by the CAD software.

The use of SLS was first reported for implantable devices using calcium phosphate in 1996 by Lee *et al.* [74]. The biocompatible materials used for SLS were expanded to produce a porous polymer– ceramic composite of polyetheretherketone (PEEK) and HA by Tan *et al.* in 2003 [75]. In 2004, reports on the use of SLS for fabricating biodegradable polymer scaffolds were published, utilizing biodegradable polymers such as PCL [76] and PLA [77]. Popov *et al.* [77] modified the SLS process by incorporating a small amount of carbon black on the polymer surface to prevent significant overheating. They coined this process "surface selective laser sintering".

SLS can be used to create porous structures that have excellent interconnectivity, though the achievable pore size is generally on the order of 50 μm or less [72]. The pore size and porosity are dependent on the size of the polymer particles, the amount of force used to compact the particles when they are deposited, and the spot size of the laser [72]. SLS has also been used to manufacture scaffolds with a compressive modulus and yield strength ranging from 52 to 67 MPa and 2.0 to 3.2 MPa, respectively, lying within the lower range of properties reported for human trabecular bone [78]. Although no organic solvents are used in SLS, the incorporation of bioactive molecules is limited due to the high temperatures used for fusing the polymer particles together. Additionally, the use of SLS makes it possible to accurately produce porous, complex 3D geometries that can act as tissue engineering scaffolds [79].

17.3.7.4 Three-Dimensional Printing

The three-dimensional printing (3DP) process was first developed at the Massachusetts Institute of Technology and commercialized by Soligen Corporation in 1993 [64]. Though originally developed for ceramics, 3DP has been adapted to use thermoplastic powder, which will be the focus of this discussion. In this process (Fig. 17.6), a layer of powder is deposited onto a build piston and leveled by a roller. An ink-jet printing head scans the surface in the $x-y$ plane, selectively depositing a binder to create the first layer of the 3D construct. The particles not subjected to the binder remain in the layer for structural support after the ink-jet head has completed scanning the surface. Once the solvent has properly bonded the particles together, the build piston is lowered, another layer of powder is deposited, and the process is repeated for each subsequent layer until the 3D model created by the CAD software has been completed. The unfused particles are then removed via vacuum, leaving behind the

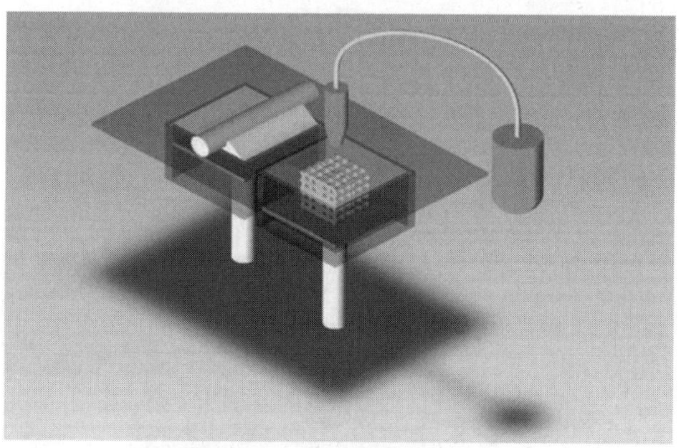

Figure 17.6 3DP process [63].

particles selectively fused by the binder in the form of the 3D model.

In order to use the 3DP process with biodegradable polymers, chloroform was explored as a binder [80]. Chloroform acts as a solvent for a wide variety of polymers, including PLA, PGA, and PLGA. Kim *et al.* [81] were the first to publish results of fabricating porous 3D scaffolds with PLGA and the 3DP process, combining the process with PL and achieving interconnected channels of 800 μm and 60% porosity. Sherwood *et al.* [28] constructed a scaffold for encouraging bone and cartilage growth by fabricating three distinct regions into one structure using the 3DP process. Because of the precise level of control over pore architecture, Sherwood *et al.* were able to fabricate fully interconnected PLGA scaffolds reinforced with tricalcium phosphate (TCP) with a porosity of 55% that had mechanical properties comparable to cancellous bone.

The 3DP process can incorporate different materials in subsequent layers and allows for excellent interconnectivity, as long as the unbound particles can be successfully removed. However, complete removal of the unbound particles, along with the organic solvents used as binders, can sometimes be difficult [82]. However, the use of water-based binders such as PVOH eliminates the need for removal, and also allows for the incorporation of bioactive molecules during the fabrication stage [72]. Though the horizontal and vertical resolutions are generally on the order of 200 and 100 μm, respectively [83], the incorporation of salt particulates allows for a microporous network to be formed in the part along with the macroporous structures formed by the 3DP process. The flexibility of materials used to fabricate complex 3D structures, the achievable mechanical properties, and the high interconnectivity make 3DP one of the most highly researched SSF techniques for fabricating tissue engineering scaffolds.

17.3.8 Traditional Polymer Processing Techniques

Plastic processing encompasses the methods and techniques used to convert plastics materials in the form of pellets, granules, powders, sheets, fluids, or preforms into formed shapes or parts. Although the term plastics has been used loosely as a synonym for polymers and resins, plastics generally refer to polymeric compounds that are formulated with plasticizers, stabilizers, fillers, and other additives for purposes of processability, costs, and performance. After forming, the part may be subjected to a variety of secondary operations such as welding, adhesive bonding, machining, or surface decorating (painting, metalizing). The choice of process is influenced by economic considerations, product specifications, number and size of finished parts, and complexity of postfinishing operations, as well as the adaptability of the plastics to the process. A variety of processes have been employed to produce tissue engineering scaffolds, as discussed in the following sections.

17.3.8.1 Extrusion

In this typically continuous process (Fig. 17.7), plastic pellets or granules are plasticized and homogenized through the rotation action of a screw (or screws in cases of twin-screw extruders) inside of a barrel. The melt is continuously pushed under pressure through a shaping die to form the final product. As material is passing through the die, the extrudate initially acquires the shape of the die

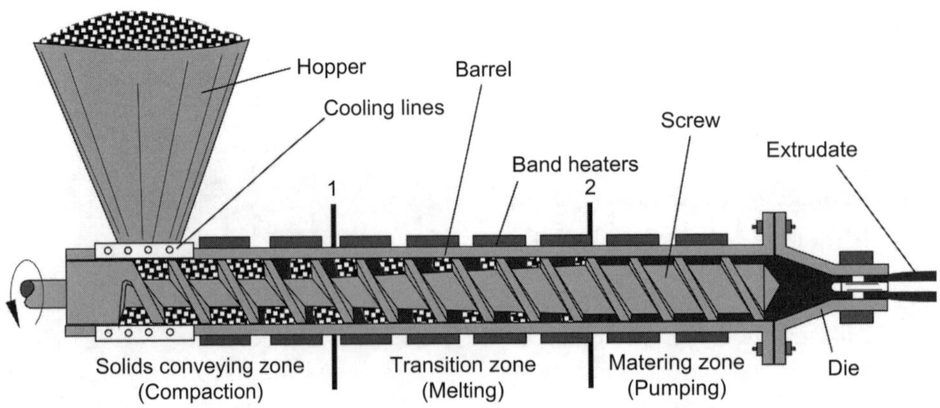

Figure 17.7 Schematic of a plasticating single-screw extruder [84].

opening, but changes its shape due to the structural recovery. Depending on the types of die, products of various shapes can be made such as tubing, pipe, film, sheet, wire, substrate coatings, and other profiles.

Two different groups have used extrusion processes producing porous scaffolds from PCL. Washburn *et al.* utilized a research-scale mini-compounder to create a co-continuous blend of PCL and water-soluble PEO [85,86]. Similar to NaCl leaching, the co-continuous blends were placed in water to selectively extract the PEO and leave a porous PCL scaffold. Blends of 50/50 and 30/70 (PCL/PEO) were extruded and scaffolds with porosities of 50% and 70% were obtained, respectively. The channel diameter generated by the extraction of the PEO was approximately 1 micron. By annealing the PCL/PEO extrudates at 80 °C prior to leaching, coalescence of the PEO resulted in channels with a diameter of approximately 100 μm. However, this annealing process resulted in a loss of the extruded geometry, as the annealing temperature was well above the T_m of either PCL or PEO (T_m~ 60 °C). Therefore, the blends needed to be compression molded and subsequently leached in order to possess pore sizes in the order of 100 μm.

Reignier and Huneault [87] expanded on the work originally performed by Washburn *et al.* A ZSK-30 mm twin-screw extruder operating at 100 °C was used to extrude PCL and PEO blended with NaCl to create PCL matrices with porosities up to 88%. The PCL/PEO blend ratio was kept between 50/50 and 40/60 to maintain a co-continuous morphology, and the NaCl content was varied from 50 to 70 wt%. This work more realistically displayed the feasibility of using co-continuous blends for manufacturing tissue engineering scaffolds by using an industrial-sized twin-screw extruder. Additionally, no annealing step was used, allowing for the extrudates to exhibit a complex geometry in two dimensions.

However, due to the intensive mixing in the twin-screw extruder, the NaCl particulates underwent erosion and breakdown. This could result in a pore size that did not accurately match the sieved size range and left up to 17 ppm of residual salt entrapped in the PCL matrix.

17.3.8.2 Melt Spinning

Melt spinning is an effective method for manufacturing polymer fibers. The polymer is melted and extruded from a spinneret, forming long strands of polymer. The spinneret may have any number of holes, depending on the application. Often the material is extruded into a monofilament or multi-filament yarn, which is drawn and solidified by cooling (air is commonly used), and then wound onto spools, or processed further, such as by texturizing, weaving, or braiding, to make a more complex application [88].

The most common medical application for melt-spun fibers is as sutures. However, they are increasingly being explored for usage as tissue engineering scaffolds [89]. Multifilament yarns created through melt spinning are often used as vascular grafts [88]. Applications for melt spinning are somewhat limited due to the large fiber size (10−12 μm), which is a frequent drawback of typical extrusion methods [90].

17.3.8.3 Electrospinning

Electrospinning, or electrostatic fiber spinning, is an old polymer processing technology dating back to 1934 [91], which has many applications in tissue engineering scaffolds, drug delivery vehicles, textiles, and filters. This method involves applying a high voltage electrical charge (5−30 kV) to a liquid polymer solution or polymer melt. This charge overcomes the surface tension of the polymer, forming a Taylor cone, which ejects a charged liquid jet of polymer which is elongated by a series of movements called electrostatic repulsion, forming an interconnected web of fibers [92,93]. Due to the high speeds of the ejected liquid and the dynamic bending motion, it is difficult to control the deposition of the fibers [93].

Electrospun scaffolds may be formed from biocompatible synthetic polymers such as PLLA, PLGA, PVOH, poly(ethylene-*co*-vinyl acetate), PEO, PU, and polycarbonates [88], natural polymers including collagen/gelatin, chitosan, hyaluronic acid (HA), elastin, or silk fibroin, or a blend of natural and synthetic polymers [93]. The polymers are dissolved in a solvent and the solution or melt is expelled from a syringe at a constant rate. The solvent evaporates and the polymer dries as the fibers land on a grounded collecting drum, plate, or other specially shaped collecting device, forming the scaffold. These scaffolds can be highly porous (greater than 90%) with nonwoven fibers on the micro- or nanoscale [3]. Typical fiber diameters range from 200 nm to 5 μm [93].

Many factors affect the properties of the scaffold such as scaffold size, pore size, and diameter and orientation of the fibers. The properties can be changed by varying the type of solvent, rate of ejection, melt viscosity, concentration of polymers, voltage and uniformity of the electric field, capillary diameter, type of collection device, or distance from the capillary or nozzle to the collection device [3,93].

They may also utilize other materials within the fibers or as coating to promote cell differentiation, growth, and adhesion, such as using different types of collagen, laminin, or glycosaminoglycans. Additionally, living cells can be incorporated into the scaffold by concurrently electrospinning with the polymer [93]. Electrospun tissue engineering scaffolds have many applications, including use as bone, cartilage, vascular, and neural scaffolds. Electrospun bone scaffolds are currently being explored that include bone grafts and scaffold membranes which will assist with guided bone regeneration. One promising use for electrospun scaffolds involves the production of vascular scaffolds using natural polymers collagen and elastin or collagen and synthetic polymers such as PLGA, which may be utilized as vascular grafts, heart tissue scaffolds, or new blood vessels with the proper mechanical properties and biocompatibility [93].

17.3.8.4 Injection Molding

Injection molding (Fig. 17.8) is a "continuous" cyclic process of forming plastic into a desired shape by forcing the material under pressure into a cavity that has the shape of the final part [95]. The shaping is achieved by cooling (thermoplastics) or by a chemical reaction (thermosets). It is one of the most common and versatile operations for mass production of complex plastics parts with excellent dimensional tolerance and net-shape. It requires minimal or no finishing or assembly operations. In addition to thermoplastics and thermosets, the process is being extended to such materials as fibers, ceramics, and powdered metals, with polymers as binders. Moreover, numerous attempts have been made to develop various special injection molding processes to produce parts with special design features and properties [96]. Some of these alternative processes derived from conventional injection molding, such as microcellular injection molding [96], have created a new era for tissue engineering scaffold fabrication.

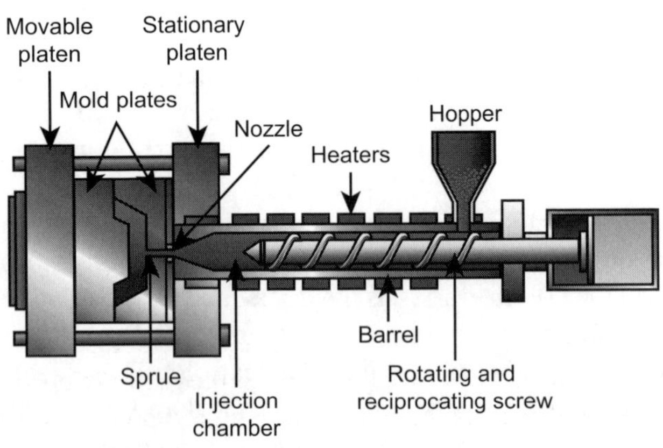

Figure 17.8 The reciprocating screw injection molding machine [94].

It has been mentioned in the literature that while synthetic biodegradable polymers can easily be fabricated into 3D shapes using injection molding, it is difficult to fabricate scaffolds because of the need to create high porosity [24]. The first published attempt at injection molding porous scaffolds was released in 2001 by Gomes *et al.* [97]. The materials used consisted of corn starch blended with (1) ethylene vinyl acetate (SEVA-C) and (2) cellulose acetate (SCA). A blowing agent with the trade name Hostalon P 9947 was used to produce the foamed structure. The porosity and interconnectivity were not reported, but pores sizes ranging from 10 to 1000 μm encapsulated by a dense surface layer were observed.

This same group further investigated the use of starch–polymer blends using a different blowing agent with injection molding in a paper published in 2005 by Neves *et al.* [98]. The materials used consisted of 50 wt% starch blended with SEVA-C and 30 wt% starch blended with polylactide (SPLA). The blowing agent chosen for this study was based on azo-dicarbonamide. Similarly, the pore sizes obtained ranged from 100 to 500 μm. The interconnectivity was claimed to be "rather poor" and the porosity was not reported, though visual examination of the morphology suggests that a maximum porosity of 50% or less was obtained. While these two research papers involving injection molding were the first of their kind, the porosity and interconnectivity observed leave much to be desired.

Wintermantel's group published two papers discussing injection molding polyether–urethane (PEU) using benign blowing agents. Leicher *et al.* [99] released the results of using microcellular injection molding with CO_2 to create porous structures in 2005. During screw recovery, supercritical CO_2 was injected into the polymer melt. Under high pressure, the CO_2 dissolved into the PEU and upon injection, the rapid pressure drop triggered cell nucleation and growth in the molded parts. This research claimed to produce porosities of over 70% using a technique to measure porosity called mercury intrusion porosimetry. Pore sizes ranging from 40 to 1102 μm could be achieved, with the actual range dependent upon injection molding parameters such as injection speed, porosity, and the amount of CO_2. Though not shown, the presence of a dense surface layer was reported.

Haugen *et al.* [100] released the group's second paper in 2006 detailing the use of NaCl PL with water as a blowing agent. The PEU pellets were compounded with NaCl in a twin-screw extruder and allowed to absorb moisture in a controlled environment (50% relative humidity). Under high pressure and heat in the injection molding barrel, the water dissolved into the polymer melt in a supercritical state. Upon injection molding, the thermodynamic instability resulted in the foaming of the material, and a subsequent leaching step removed the NaCl from the PEU−NaCl composite. This research resulted in a pore size distribution from 30 to 450 μm and porosity up to 64%. Additionally, the pores were observed to be relatively well interconnected with interconnected channels in the range of 5−58 μm in diameter. The use of the NaCl particles apparently eliminated the presence of the dense surface layer, which had been an issue observed in the aforementioned studies involving injection molding. However, the researchers noted that an improvement in the obtainable porosity is desired. Additionally, it has been shown that the processing of biodegradable polymers can lead to a decrease in the molecular weight, ultimately leading to decreased mechanical properties [101].

Wu *et al.* [102] adapted the SC/PL technique to injection molding in 2006. PLGA was dissolved in chloroform and sieved NaCl particles were added, identical to the SC/PL process described in Section 17.3.2. However, in order to injection mold the composite at room temperature, the solvent was not fully evaporated, leaving a PLGA/NaCl/chloroform slurry that could be injection molded into the desired shape. Upon ejection from the mold, the solvent was then evaporated completely and the salt was leached in water, resulting in porous scaffolds with porosity as high as 94%. The pore size could be controlled by the size range of salt incorporated into the polymer/solvent solution, and the scaffolds were deemed to be completely interconnected. Even though this research requires the use of organic solvents, it describes a method to mass produce highly porous, interconnected 3D scaffolds. Further research is needed to validate this method using an industrial-scale injection molding machine, as it has been shown that the intensive mixing observed during twin-screw compounding and injection molding can drastically reduce NaCl particle size and affect the ability to completely leach the scaffolds.

Kramschuster and Turng [103] applied PL and water-soluble polymers in combination with microcellular injection molding to create highly porous and interconnected PLA foams. PLA, water-soluble PVOH, and NaCl particulates were compounded in a twin-screw extruder and pelletized. The composite blend contained approximately 60 vol% NaCl and 20 vol% PVOH, resulting in a composite blend with approximately 80 vol% water-soluble content. Using CO_2 as the blowing agent, the composite blend was microcellular injection molded and the parts were leached in deionized water and subsequently dried. Porosities of up to 76% were obtained with this method. The discrepancy between water-soluble content and achievable porosity was likely due to variations in the gravimetric feeders during twin-screw extrusion and a loss of NaCl particulates during the pelletizing step. Control over pore size and channel size is dictated partially by the size of the NaCl particulates incorporated into the composite blend, though aggressive mixing has resulted in a size reduction of the NaCl particles. However, the ability to mass produce highly porous and interconnected PLA matrices without the use of organic solvents makes this a promising method for fabricating tissue engineering scaffolds. Unpublished research using this technique has resulted in PLA foams with up to 84% porosity.

Ghosh *et al.* [104] utilized conventional injection molding to create PLLA foams with porosities ranging from 57 to 74%. PLLA was blended with water-soluble PEO at a 1:1 weight ratio and injection molded. A subsequent leaching step in water resulted in a porous lamellar structure. The variation in porosity was attributed to swelling during the leaching step. Both macropores (from leaching a PEO-rich phase) between 50 and 100 μm and micropores (from leaching a PLLA/PEO phase) were obtained for the structure. Mechanical testing revealed tensile moduli between 580 and 800 MPa were achieved. The ability to mold structures with 57−74% porosity without organic solvents on conventional injection molding equipment makes this an interesting process for the evaluation of high-volume manufacturing tissue engineering scaffolds.

References

[1] R. Langer, J.P. Vacanti, Tissue engineering, Science 260 (5110) (1993) 920−926.

[2] S. Yang, K. Leong, Z. Du, C. Chua, The design of scaffolds for use in tissue engineering. Part I. Traditional factors, Tissue Eng. 7 (6) (2001) 679−689.

[3] M.B. Murphy, A.G. Mikos, Chapter 22 Polymer scaffold processing, in: R.P. Lanza, R. Langer, J.P. Vacanti (Eds.), Principles of Tissue Engineering, third ed., Academic Press, San Diego, 2007, pp. 309−321.

[4] H. Cheung, K. Lau, T. Lu, D. Hui, A critical review on polymer-based bio-engineered materials for scaffold development, Compos. Part B 38 (3) (2007) 291−300.

[5] D.W. Hutmacher, in: Muellhaupt, et al. (Eds.), Polymers for medical applications, in Encyclopedia of Materials: Science and Technology, Elsevier Science, 2001.

[6] M. Martina, D.W. Hutmacher, Biodegradable polymers applied in tissue engineering research: a review, Polym. Int. 56 (2) (2007) 145−157.

[7] S. Forman, J. Kas, F. Fini, M. Steinberg, T. Ruml, The effect of different solvents on the ATP/ADP content and growth properties of HeLa Cells, J. Biochem. Mol. Toxicol. 13 (1) (1999) 11−15.

[8] R.M. Nerem, Chapter 2 The challenge of imitating nature, in: R.P. Lanza, R. Langer, J.P. Vacanti (Eds.), Principles of Tissue Engineering, third ed., Academic Press, San Diego, 2007, pp. 7−14.

[9] B.O. Palsson, S.N. Bhatia, Chapter 17 Conventional clinical approaches to tissue dysfunction, in: Tissue Engineering, first ed., Pearson Prentice Hall, Upper Saddle River, NJ, 2004, pp. 290−302.

[10] "The New Era of Regenerative Medicine", Figure found at http://www.businessweek.com/pdfs/biobodies.pdf.

[11] R.P. Lanza, R. Langer, J.P. Vacanti, Principles of Tissue Engineering, third ed., Academic Press, San Diego, CA, 2007.

[12] B.O. Palsson, S.N. Bhatia, Tissue Engineering, first ed., Pearson Prentice Hall, Upper Saddle River, NJ, 2004.

[13] R.I. Freshney, B. Obradovic, W. Grayson, C. Cannizzaro, G. Vunjak−Novakovic, Chapter 12 Principles of tissue culture and bioreactor design, in: R.P. Lanza, R. Langer, J.P. Vacanti (Eds.), Principles of Tissue Engineering, third ed., Academic Press, San Diego, 2007, pp. 155−183.

[14] S.J. Bryant, K.S. Anseth, Chapter 6 Photopolymerization of hydrogel scaffolds, in: P.X. Ma, J. Elisseeff (Eds.), Scaffolding in Tissue Engineering, first ed., CRC Press, Boca Raton, FL, 2006, pp. 71−90.

[15] S.J. Hollister, Porous scaffold design for tissue engineering, Nature Mater. 4 (7) (2005) 518−524.

[16] M. Vert, S.M. Li, G. Spenlehauer, P. Guerin, Bioresorbability and biocompatibility of aliphatic polyesters, J. Mater. Sci. Mater. Med. 3 (6) (1992) 432−446.

[17] R.E. Drumright, P.R. Gruber, D.E. Henton, Polylactic acid technology, Adv. Mater. 12 (23) (2000) 1841−1846.

[18] D.W. Hutmacher, Biodegradable polymeric materials, in: Muellhaupt, et al. (Eds.), Encyclopedia of Materials: Science and Technology, Elsevier Science, 2001.

[19] P.A. Gunatillake, R. Adhikari, Biodegradable synthetic polymers for tissue engineering, Eur. Cell Mater. 5 (1) (2003) 1−16.

[20] J.C. Middleton, A.J. Tipton, Synthetic biodegradable polymers as orthopedic devices, Biomat. 21 (23) (2000) 2335−2346.

[21] J. Kohn, S. Abramson, R. Langer, Chapter 2.7 Bioresorbable and bioerodible materials, in: B.D. Ratner, A.S. Hoffman, F.J. Schoen, J.E. Lemons (Eds.), Biomaterials Science, second ed., Academic Press, San Diego, 2004, pp. 115−127.

[22] J.M. Pachence, M.P. Bohrer, J. Kohn, Chapter 23 Biodegradable polymers, in: R.P. Lanza, R. Langer, J.P. Vacanti (Eds.), Principles of Tissue Engineering, third ed., Academic Press, San Diego, 2007, pp. 323−339.

[23] L.G. Griffith, Polymeric biomaterials, Acta Mater. 48 (1) (2000) 263−277.

[24] C.M. Agrawal, R.B. Ray, Biodegradable polymeric scaffolds for muskuloskeletal tissue engineering, J. Biomed. Mater. 55 (2) (2001) 141–150.
[25] D.W. Hutmacher, Scaffolds in tissue engineering bone and cartilage, Biomat. 21 (24) (2000) 2529–2543.
[26] L.E. Freed, G. Vunjaknovakovic, R.J. Biron, D.B. Eagles, D.C. Lesnoy, S.K. Barlow, R. Langer, Biodegradable polymer scaffolds for tissue engineering, Biotech. 12 (7) (1994) 689–693.
[27] K. Whang, K.E. Healy, D.R. Elenz, E.K. Nam, D.C. Tsai, C.H. Thomas, G.W. Nuber, F.H. Glorieux, R. Travers, S.M. Sprague, Engineering bone regeneration with bioabsorbable scaffolds with novel microarchitecture, Tissue Eng. 5 (1) (1999) 35–51.
[28] J.K. Sherwood, S.L. Riley, R. Palazzolo, S.C. Brown, D.C. Monkhouse, M. Coates, L.G. Griffith, L.K. Landeen, A. Ratcliffe, A three-dimensional osteochondral composite scaffold for articulate cartilage repair, Biomat. 23 (24) (2002) 4739–4751.
[29] X.X. Shao, D.W. Hutmacher, S.T. Ho, J. Goh, E.H. Lee, Evaluation of a hybrid scaffold/cell construct in repair of high-load-bearing osteochondral defects in rabbits, Biomat. 27 (7) (2006) 1071–1080.
[30] T.S. Karande, J.L. Ong, C.M. Agrawal, Diffusion in musculoskeletal tissue engineering scaffolds: design issues related to porosity, permeability, architecture, and nutrient mixing, Ann. Biomed. Eng. 32 (12) (2004) 1728–1743.
[31] H. Elema, J.H. de Groot, A.J. Hijenhuis, A.J. Pennings, R.P.H. Veth, J. Klompmaker, H.W.B. Jansen, Use of porous biodegradable polymer implants in meniscus reconstruction. 2) Biological evaluation of porous biodegradable implants in menisci, Colloids Polym. Sci. 268 (12) (1990) 1082–1088.
[32] D.J. Mooney, Tissue Engineering with Biodegradable Polymer Matrices, Southern Biomedical Engineering Conference – Proceedings (1996) 537–540.
[33] B.D. Boyan, T.W. Hummert, D.D. Dean, Z. Schwartz, Role of material surfaces in regulating bone and cartilage cell response, Biomat. 17 (2) (1996) 137–146.
[34] S.H. Oh, S.G. Kang, E.S. Kim, S.H. Cho, J.H. Lee, Fabrication and characterization of hydrophilic poly(lactic-co-glycolic acid)/poly(vinyl alcohol) blend cell scaffolds by melt-molding particulate leaching method, Biomat. 24 (22) (2003) 4011–4021.
[35] W.M. Saltzman, T.R. Kyriakides, Chapter 20 Cell interactions with polymers, in: R.P. Lanza, R. Langer, J.P. Vacanti (Eds.), Principles of Tissue Engineering, third ed., Academic Press, San Diego, 2007, pp. 279–296.
[36] X. Liu, P.X. Ma, Polymeric scaffolds for bone tissue engineering, Ann. Biomed. Eng. 32 (3) (2004) 477–486.
[37] Y. Cao, T.I. Croll, J.J. Cooper–White, A.J. O'Connor, G.W. Stevens, Production and surface modification of polylactide-based polymeric scaffolds for soft tissue engineering, Meth. Mol. Biol. 238 (2004) 87–112.
[38] C.M. Agrawal, J. Carter, J.L. Ong, Basics of polymeric scaffolds for tissue engineering, J. ASTM Int. 3 (9) (2006) 1–10.
[39] C.Z. Liu, J.T. Czernuszka, Development of biodegradable scaffolds for tissue engineering: a perspective on emerging technology, Mater. Sci. Technol. 23 (4) (2007) 379–391.
[40] A.G. Mikos, A.J. Thorsen, L.A. Czerwonka, Y. Bao, R. Langer, Preparation and characterization of poly(L-lactic acid) foams, Polymer 35 (5) (1994) 1068–1077.
[41] P.X. Ma, J.–W. Choi, Biodegradable polymer scaffolds with well-defined interconnected spherical pore network, Tissue Eng. 7 (1) (2001) 23–33.
[42] A.G. Mikos, S. Georgios, S.M. Leite, J.P. Vacanti, R. Langer, Laminated three-dimensional biodegradable foams for use in tissue engineering, Biomat. 14 (5) (1993) 323–330.
[43] W.L. Murphy, R.G. Dennis, J.L. Kileny, D.J. Mooney, Salt fusion: an approach to improve pore interconnectivity within tissue engineering scaffolds, Tissue Eng. 8 (1) (2002) 43–52.
[44] M.S. Widmer, P.K. Gupta, L. Lu, R.K. Meszlenyi, G. Evans, K. Brandt, T. Savel, A. Gurlek, C.W. Patrick Jr., A.G. Mikos, Manufacture of porous biodegradable polymer conduits by an extrusion process for guided

tissue regeneration, Biomat. 19 (21) (1998) 1945–1955.

[45] L.E. Freed, J.C. Marquis, A. Nohria, J. Emmanual, A.G. Mikos, R. Langer, Neocartilage formation in vitro and in vivo using cells cultured on synthetic biodegradable polymers, J. Biomed. Mater. Res. 27 (1) (1993) 11–23.

[46] D.J. Mooney, D.F. Baldwin, N.P. Suh, J.P. Vacanti, R. Langer, Novel approach to fabricate porous sponges of poly(D, L-lactic-co-glycolic acid) without the use of organic solvents, Biomat. 17 (14) (1996) 1417–1422.

[47] X. Wang, L. Wei, V. Kumar, Solvent free fabrication of biodegradable porous polymers, ASME Manuf. Eng. Div. 15 (2004) 595–602.

[48] H. Tsuji, R. Smith, W. Bonfield, Y. Ikada, Porous biodegradable polyesters. I. Preparation of porous poly(L-lactide) films by extraction of poly(ethylene oxide) from their blends, J. App. Polym. Sci. 75 (5) (2000) 629–637.

[49] L.D. Harris, B.–S. Kim, D.J. Mooney, Open pore biodegradable matrices formed with gas foaming, J. Biomed. Mater. Res. 42 (3) (1998) 396–402.

[50] K.W. Riddle, D.J. Mooney, Role of poly(lactide-co-glycolide) particle size on gas-foamed scaffolds, J. Biomed. Mater. Res. Polym. Ed. 15 (12) (2004) 1561–1570.

[51] M.H. Sheridan, L.D. Shea, M.C. Peters, D.J. Mooney, Bioabsorbable polymer scaffolds for tissue engineering capable of sustained growth factor delivery, J. Control. Release 64 (1) (2000) 91–102.

[52] Y.S. Nam, J.J. Yoon, T.G. Park, A novel fabrication method of macroporous biodegradable polymer scaffolds using gas foaming salt as a porogen additive, J. Biomed. Mater. Res. B 53 (1) (2000) 1–7.

[53] J.J. Yoon, T.G. Park, Degradation behaviors of biodegradable macroporous scaffolds prepared by gas foaming of effervescent salts, J. Biomed. Mater. Res. 55 (3) (2001) 401–408.

[54] K. Whang, C.H. Thomas, K.E. Healy, G. Nuber, A novel method to fabricate bioabsorbable scaffolds, Polymer 36 (4) (1995) 837–842.

[55] Y.S. Nam, T.G. Park, Porous biodegradable polymeric scaffolds prepared by thermally induced phase separation, J. Biomed. Mater. Res. 47 (1) (1999) 8–17.

[56] R. Zhang, P.X. Ma, Chapter 62 Processing of polymer scaffolds: phase separation, in: A. Atala, R.P. Lanza (Eds.), Methods of Tissue Engineering, Academic Press, San Diego, 2002, pp. 715–724.

[57] H. Lo, M.S. Ponticello, K.W. Leong, Fabrication of controlled release biodegradable foams by phase separation, Tissue Eng. 1 (1) (1995) 15–28.

[58] V.J. Chen, P.X. Ma, Chapter 9 Polymer phase separation, in: P.X. Ma, J. Elisseeff (Eds.), Scaffolding in Tissue Engineering, CRC Press, Boca Raton, FL, 2006, pp. 125–137.

[59] P.X. Ma, R.Y. Zhang, Synthetic nano-scale fibrous extracellular matrix, J. Biomed. Mater. Res. 46 (1) (1999) 60–72.

[60] R.Y. Zhang, P.X. Ma, Synthetic nano-fibrillar extracellular matrices with predesigned macroporous architectures, J. Biomed. Mater. Res. 52 (2) (2000) 430–438.

[61] V.J. Chen, L.A. Smith, P.X. Ma, Collagen-inspired nano-fibrous poly(L-lactic acid) scaffolds for bone tissue engineering created from reverse solid freeform fabrication, Mater. Res. Soc. Symp. Proc. 823 (2004) 213–218.

[62] C.K. Chua, K.F. Leong, C.S. Lim, Chapter 3 Liquid-based rapid protoyping systems" in Rapid Prototyping, second ed., World Scientific Publishing Co., Singapore, 2003, pp. 35–109.

[63] P.J. Bartolo, H.A. Almeida, R.A. Rezende, T. Laoui, B. Bidanda, Chapter 8 Advanced processes to fabricate scaffolds for tissue engineering, in: B. Bidanda, P. Bartolo (Eds.), Virtual Prototyping & Bio Manufacturing in Medical Applications, Springer, 2008, pp. 149–170.

[64] J.J. Beaman, J.W. Barlow, D.L. Bourell, R.H. Crawford, H.L. Marcus, K.P. McAlea, Chapter 2 Process methods, in: first ed., Solid Freeform Manufacturing: A New Direction in Manufacturing Kluwer Academic Publishers, Boston, 1997, pp. 23–49.

[65] T.M. Chu, J.W. Halloran, W.C. Wagner, Ultraviolet curing of highly loaded hydroxyapatite suspension, in: R.P. Rusin, G.S. Fischman (Eds.), Bioceramics: Materials and Applications II, American Ceramic Society, Westerville, OH, 1996, pp. 57–66.

[66] T.M. Chu, Chapter 10 Solid freeform fabrication of tissue engineering scaffolds, in: P.X. Ma,

J. Elisseeff (Eds.), Scaffolding in Tissue Engineering, CRC Press, Boca Raton, FL, 2006, pp. 139–153.
[67] M.N. Cooke, J.P. Fisher, D. Dean, C. Rimnac, A.G. Mikos, Use of stereolithography to manufacture critical-sized 3D biodegradable scaffolds for bone ingrowth, J. Biomed. Mater. Res. part B App. Biomat. 64 (2) (2003) 65–69.
[68] C.J. Bettinger, J.T. Borenstein, R.S. Langer, Chapter 24 Micro- and nanofabricated scaffolds, in: R.P. Lanza, R. Langer, J.P. Vacanti (Eds.), Principles of Tissue Engineering, third ed., Academic Press, San Diego, 2007, pp. 341–358.
[69] C.K. Chua, K.F. Leong, C.S. Lim, Chapter 4 Solid-based rapid prototyping systems, in: Rapid Prototyping, second ed., World Scientific Publishing Co., Singapore, 2003, pp. 111–171.
[70] D.W. Hutmacher, T. Schantz, I. Zein, K.W. Ng, S.H. Teoh, K.C. Tan, Mechanical properties and cell cultural response of polycaprolactone scaffolds designed and fabricated via fused deposition modeling, J. Biomed. Mater. Res. 55 (2) (2001) 203–216.
[71] I. Zein, D.W. Hutmacher, K.C. Tan, S.H. Teoh, Fused deposition modeling of novel scaffold architectures for tissue engineering applications, Biomat. 23 (4) (2002) 1169–1185.
[72] K.F. Leong, C.M. Cheah, C.K. Chua, Solid freeform fabrication of three-dimensional scaffolds for engineering replacement tissues and organs, Biomat. 24 (13) (2003) 2363–2378.
[73] C.K. Chua, K.F. Leong, C.S. Lim, Chapter 5 Powder-based rapid prototyping systems, in: Rapid Prototyping, second ed., World Scientific Publishing Co., Singapore, 2003, pp. 173–235.
[74] G. Lee, J.W. Barlow, Selective laser sintering of bioceramic materials for implants, in: Proceedings of Solid Freeform Fabrication Symposium, Austin, TX, August 9–11, 1996, pp. 376–380.
[75] K.H. Tan, C.K. Chua, K.F. Leong, C.M. Cheah, P. Cheang, M.S. Abu Bakar, S.W. Cha, Scaffold development using selective laser sintering of polyetheretherketone-hydroxyapatite biocomposite blends, Biomat. 24 (18) (2003) 3115–3123.
[76] B. Partee, S.J. Hollister, S. Das, Fabrication of polycaprolactone bone tissue engineering scaffolds using selective laser sintering, ASME — Manuf. Eng. Div. 15 (2004) 525–536.
[77] V.K. Popov, E.N. Antonov, B.N. Bagratashvili, S.M. Howdle, Selective laser sintering of 3-D biodegradable scaffolds for tissue engineering, Mater. Res. Soc. Symp. Proc. EXS (1) (2004) 51–53.
[78] J.M. Williams, A. Adewunmi, R.M. Schek, C.L. Flanagan, P.H. Krebsbach, S.E. Feinberg, S.J. Hollister, S. Das, Bone tissue engineering using polycaprolactone scaffolds fabricated via selective laser sintering, Biomat. 26 (23) (2005) 4817–4827.
[79] K.H. Tan, C.K. Chua, K.F. Leong, C.M. Cheah, W.S. Gui, W.S. Tan, F.E. Wiria, Selective laser sintering of biocompatible polymers for applications in tissue engineering, Biomed. Mater. Eng. 15 (1–2) (2005) 113–124.
[80] R.A. Giordano, B.M. Wu, S.W. Borland, L.G. Cima, E.M. Sachs, M.J. Cima, Mechanical properties of dense polylactic acid structures fabricated by three dimensional printing, J. Biomater. Sci. Polym. Ed. 8 (1) (1996) 63–75.
[81] S.S. Kim, H. Utsunomiya, J.A. Koski, B.M. Wu, M.J. Cima, J. Sohn, K. Mukai, L.G. Griffith, J.P. Vacanti, Survival and function of hepatocytes on a novel three-dimensional synthetic biodegradable polymer scaffold with an intrinsic network of channels, Ann. Surg. 228 (1) (1998) 8–13.
[82] G. Vozzi, A. Ahluwalia, in: P.K. Chu, X. Liu (Eds.), Chapter 4 Rapid Prototyping Methods for Tissue Engineering Applications, first ed., CRC Press, Boca Raton, 2008, pp. 95–114.
[83] C.Z. Liu, E. Sachlos, D.A. Wahl, Z.W. Han, J.T. Czernuszka, On the manufacturability of scaffold mould using a 3D printing technology, Rapid Proto. J. 13 (3) (2007) 163–174.
[84] T.A. Osswald, G. Menges, in: T.A. Osswald, G. Menges (Eds.), Chapter 6 Introduction to processing in Materials Science of Polymers for Engineers, second ed., Hanser, 2003, pp. 185–281.
[85] N.R. Washburn, C.G. Simon Jr., A. Karim, E.J. Amis, Development of biodegradable polymer scaffolds using co-extrusion, Mater. Res. Soc. Symp. Proc. 662 (2001) LL1.6.1–LL1.6.5.

[86] N.R. Washburn, C.G. Simon Jr., A. Tona, H.M. Elgendy, A. Karim, E.J. Amis, Co-extrusion of biocompatible polymers for scaffolds with co-continuous morphology, J. Biomed. Mater. Res. 60 (1) (2002) 20–29.
[87] J. Reignier, M.A. Huneault, Preparation of interconnected poly(ε-Caprolactone) porous scaffolds by a combination of polymer and salt particulate leaching, Polymer 47 (13) (2006) 4703–4717.
[88] S. Weinberg, M.W. King, Chapter 2.4 Medical fibers and biotextiles, in: B.D. Ratner, A.S. Hoffman, F.J. Schoen, J.E. Lemons (Eds.), Biomaterials Science, second ed., Elsevier Academic Press, San Diego, 2004, pp. 86–100.
[89] E.M. Pirhonen, V. Ella, Meltspinning in Encyclopedia of Biomaterials and Biomedical Engineering, in: G.L. Bowlin, G. Wnek (Eds.), Informa Healthcare, 1816, 2004.
[90] D.J. Mooney, R.S. Langer, Engineering biomaterials for tissue engineering to 10–100 micron scale, in: J. Bronzino (Ed.), The Biomedical Engineering Handbook, CRC Press, Boca Raton, FL, 1997, pp. 1609–1618.
[91] A. Formhals, Process and Apparatus for Preparing Artificial Threads, US Patent 1,975,504, (1934).
[92] D.H. Reneker, A.L. Yarin, H. Fong, S. Koombhongse, Bending instability of electrically charged liquid jets of polymer solutions in electrospinning, J. Appl. Phys. Part 1 87 (9) (2000) 4531–4547.
[93] D.W. Hutmacher, A.K. Ekaputra, Chapter 5 Design and fabrication principles of electrospinning of scaffolds, in: P.K. Chu, X. Liu (Eds.), Biomaterials Fabrication and Processing Handbook, first ed., CRC Press, Boca Raton, 2008, pp. 115–139.
[94] J. Shoemaker, Appendix B injection-molding machine: system and operations, in: J. Shoemaker (Ed.), Moldflow Design Guide, Hanser, 2006, pp. 247–257.
[95] T.A. Osswald, L.S. Turng, P. Gramann, Injection Molding Handbook, second ed., Hanser Publishers, 2008.
[96] L.S. Turng, Special emerging injection molding processes, J. Inj. Mold. Technol. 5 (3) (2001) 160–179.
[97] M.E. Gomes, A.S. Ribeiro, P.B. Malafaya, R.L. Reis, A.M. Cunha, A new approach based on injection moulding to produce biodegradable starch-based polymeric scaffolds: morphology, mechanical and degradation behaviour, Biomat. 22 (9) (2001) 883–889.
[98] N.M. Neves, A. Kouyumdzhiev, R.L. Reis, The morphology, mechanical properties and ageing behavior of porous injection molded starch-based blends for tissue engineering scaffolding, Mater. Sci. Eng. C 25 (2) (2005) 195–200.
[99] S. Leicher, J. Will, H. Haugen, E. Wintermantel, MuCell® technology for injection molding: a processing method for polyether–urethane scaffolds, J. Mater. Sci. 40 (17) (2005) 4613–4618.
[100] H. Haugen, J. Will, W. Fuchs, E. Wintermantel, A novel processing method for injection-molded polyether–urethane scaffolds. Part 1: processing, J. Biomed. Mater. Res. Part B 77 (1) (2006) 65–72.
[101] R. von Oepen, W. Michaeli, Injection moulding of biodegradable implants, Clin. Mater. 10 (1) (1992) 21–28.
[102] L. Wu, D. Jing, J. Ding, A room-temperature injection molding/particulate leaching approach for fabrication of biodegradable three-dimensional porous scaffolds, Biomat. 27 (2) (2006) 185–191.
[103] A. Kramschuster, L.S. Turng, Highly porous injection-molded biodegradable polymer foams for tissue engineering scaffolds, in: Biofoams, Capri, Italy, 26–28 September, 2007.
[104] S. Ghosh, J.C. Viana, R.L. Reis, J.F. Mano, Development of porous lamellar poly(L-lactic acid) scaffolds by conventional injection molding process, Acta Biomat. 4 (2008) 887–896.

Index

Note: Page numbers followed by "f" indicate figures and "t" indicate tables.

C

D

E

F

Q

R

S

T

U

V

W

X

Y

Z